Decellularized Materials

Xiaoming Li • Huiqi Xie
Editors

Decellularized Materials

Preparations and Biomedical Applications

Springer

Editors
Xiaoming Li
Key Laboratory for Biomechanics and Mechanobiology of Ministry of Education, Beijing Advanced Innovation Center for Biomedical Engineering, School of Biological Science and Medical Engineering
Beihang University
Beijing, China

Huiqi Xie
Laboratory of Stem Cell and Tissue Engineering
Orthopedic Research Institute
State Key Laboratory of Biotherapy and Cancer Center, West China Hospital
Sichuan University and Collaborative Innovation Center of Biotherapy
Chengdu, China

ISBN 978-981-33-6961-0 ISBN 978-981-33-6962-7 (eBook)
https://doi.org/10.1007/978-981-33-6962-7

This Springer imprint is published by the registered company Springer Nature Singapore Pte Ltd.
The registered company address is: 152 Beach Road, #21-01/04 Gateway East, Singapore 189721, Singapore

Acknowledgements

The authors acknowledge the financial supports from the National Key R&D Program of China *(No. 2017YFC1104700)*, the National Natural Science Foundation of China (*No. 31771042*), the Sichuan Science and Technology Program (*No. 2019JDRC0020*), the 1.3.5 project for disciplines of excellence, West China Hospital, Sichuan University (*No. ZYJC18002*), the International Joint Research Center of Aerospace Biotechnology and Medical Engineering, Ministry of Science and Technology of China, and the 111 Project (*No. B13003*).

Contents

About the Editors

Xiaoming Li is a professor at the School of Biological Science and Medical Engineering in Beihang University. He got his PhD from Tsinghua University in 2006 and then went to University of Twente as a researcher. From 2007 to 2009, he worked in Hokkaido University as a JSPS postdoctoral fellow. Since 2009, he has been a faculty member in Beihang University. He has been an invited lead guest editor for nine special issues published in SCI journals. Further, as the first author or corresponding author, he has published more than 80 SCI articles in the fields of biomaterials, tissue engineering, and regenerative medicine, which have been cited more than 3000 times. He was awarded by the Beijing Nova Program, the Program for New Century Excellent Talents in University of China, and Fok Ying Tung Education Foundation.

Huiqi Xie is a professor of State Key Laboratory of Biotherapy in West China Hospital of Sichuan University. She received her PhD in clinical medicine in 2001 from Sichuan University. Her main research focus is on decellularized materials, stem cells, and tissue regeneration. She has published more than 60 peer-reviewed papers and obtained more than 30 international or national patents for invention, four of which have been issued national registration certificate of medical devices. She is currently the chief scientist of the National Key R&D Program of China and was awarded by the Program for New Century Excellent Talents in University of China and Fok Ying Tung Education Foundation.

Chapter 1
Overview of Decellularized Materials for Tissue Repair and Organ Replacement

Jie Liao, Qi Guo, Bo Xu, and Xiaoming Li

Abstract Decellularized materials (DMs) including decellularized tissues and organs are derived from natural tissues and organs through decellularization process removing cells and other antigenic components. They keep main components of natural ECM, exhibiting satisfactory bioactivities and being widely applied in tissue repair. Especially, the remain of macro and microstructures of native organs to some extent provides them superiorities on organ replacement. Hence, they have attracted wide attention, and further studies about them would be conducted. In this chapter, the background, preparation, and composition and structure of DMs were introduced. Then, the degradation of DMs and their mechanisms of promoting tissue regeneration/organ replacement was discussed in detail through two aspects including: (1) initiating relatively low host tissue response to provide appropriate regeneration microenvironment; (2) containing bioactive factors to recruit endogenous stems/progenitor cells and promote matrix production and angiogenesis. In addition, decellularized tissues and their applications, as well as decellularized organs and their recellularization were briefly introduced. Finally, the structure and main content of this book were expatiated.

Keywords Decellularized materials · Tissue repair · Organ replacement · Decellularization · Recellularization

1.1 Background of Decellularized Materials

Because of the development of economic, serious aging population, and the ever-increasing degenerative diseases, congenital abnormalities, and trauma, the requirements of tissue repair and organ reconstruction are increasing. Regeneration could be

J. Liao · Q. Guo · B. Xu · X. Li (✉)
Key Laboratory for Biomechanics and Mechanobiology of Ministry of Education, Beijing Advanced Innovation Center for Biomedical Engineering, School of Biological Science and Medical Engineering, Beihang University, Beijing, China
e-mail: lixm@buaa.edu.cn; x.m.li@hotmail.com

X. Li, H. Xie (eds.), *Decellularized Materials*,
https://doi.org/10.1007/978-981-33-6962-7_1

regarded as the repair of tissues to similar structure and function of uninjured tissues by the activation and proliferation of progenitor or endogenous stem cells and the generation of new matrix. However, if the original tissue integrity was seriously damaged because of devastating deficits or tissue dysfunction, human body could not be capable of correctly auto-regenerating most of its main tissues and organs [1]. Simultaneously, the potential of natural living tissues to auto-regenerate might be impaired by many factors including the natural low regenerative abilities of certain tissues, the negative inflammation, and the reduced quality of host stem and progenitor cells populations because of the increasing age [2]. Hence, tissue engineering and regenerative medicine creating replacement biological tissues even whole organs to displace damaged, deteriorated or even lost body parts, or accelerating the regeneration by stimulating the patient's own inherent healing potential is hopeful to develop new biological therapeutics to treat currently intractable diseases and has been an emerging trend in medical science [3, 4].

As one of three key factors of tissue engineering, biomaterials act as temporary extracellular matrix (ECM), providing mass transport and provisional mechanical support to promote cell adhesion, proliferation and differentiation, and presenting chemical and physical signals with spatiotemporal accuracy to modulate cell performance and function and to guide correct tissue regeneration [5, 6]. Specifically, ideal biomaterials should provide a transient structural support over a longer period and be controllably degraded or reabsorbed with the regrowth rate similar to that of new tissue [7]. Simultaneously, biomaterials could stimulate human host responses to exogenously implanted cells in the host environment, which could positively regulate cell behavior and function and finally promote the desired tissue formation [8]. In addition, through optimized designation, biomaterials containing active molecules such as cell homing factors, cell adhesion peptides, and growth/differentiation and mechanical signals could promote stem cells recruitment and their subsequent differentiation into a great deal of daughter cells, and finally guiding the formation and integration of new tissue [9–12]. Biomaterials mainly play the role of homing the endogenous cells for in situ tissue regeneration in damaged sites that could provide enough repair cells, whereas they act as material templates to deliver exogenously expanded cell populations to regulate the body's cell niche in some other conditions [12, 13]. Crucial factors (mechanical, structural, biological, and biochemical) involved in the design of biomaterials for tissue engineering were summarized in Fig. 1.1, which could guide cells to behave in similar or the same behaviors as their natural in vivo counterparts. In a word, biomaterials as tissue engineering scaffolds need at least partially to mimic the natural ECM (composition, the structure, and the mechanical and biochemical properties), and to provide an appropriate microenvironment to direct tissue formation.

The ECM, which is generated and assembled by cells, mainly contains ubiquitous structural biomacromolecules like collagens, proteoglycans, glycoproteins, and elastic fibers [15]. As a complex, fibrillar 3D extracellular macromolecule network with tightly regulated fiber diameter, composition, and organization between and around niche cells, it has a highly adjustable, tissue-specific composition, architecture, and corresponding physical properties in most tissues and organs in animal and human beings [16]. The ECM of living organisms could be organized as interacting fibrous

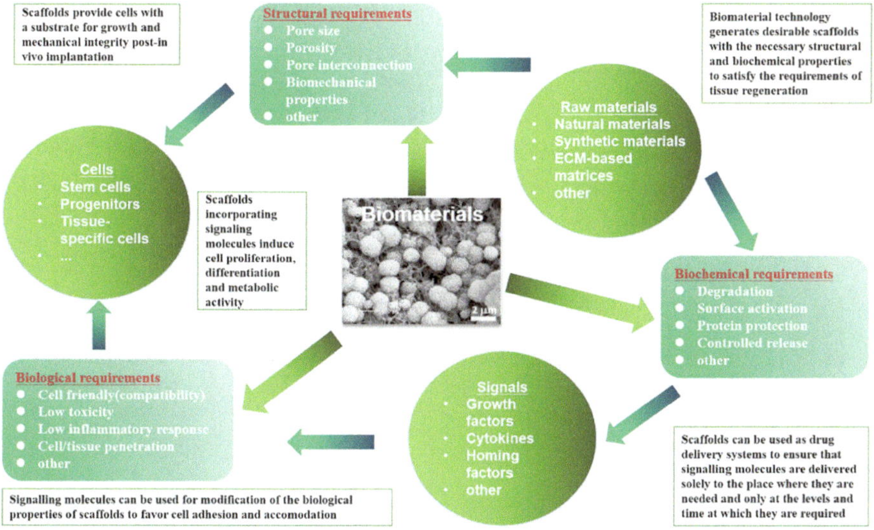

Fig. 1.1 The schematic diagram to illustrate that pivotal factors (mechanical, structural, biological, and biochemical) participated in the design of biomaterials that induce cells to behave in a similar or the same behavior as their natural in vivo counterparts [14]

structures arranged in a tissue-specific manner and/or dispersed as an amorphous "ground substance." It imparts physical cohesiveness on a tissue, providing stability and support to the surrounding cells and participating in the formation, development, and regeneration of various tissues and organs. Simultaneously, it also acts as a reservoir for various cytokines (such as chemokines and growth factors), which increases the local concentrations of agonists [17]. Moreover, it plays fundamental functional roles of influencing the cell fate (including cell survival, morphology, growth, proliferation, migration, differentiation, autophagy, and plasticity) by its multifactorial and dynamic nature and producing a large number of dynamic signals through its structural and biochemical variability [18]. The role of ECM in cell proliferation, development, and differentiation is showed in Fig. 1.2. The effects that ECM exerts on cells are mainly mediated through three different ways including directly binding receptors or co-receptors on cell surface to mediate cell anchorage and regulate several pathways (intracellular signaling and mechanotransduction), acting by non-canonical growth factor expression, and being remodeled by enzymes to release functional fragments [19]. Among them, providing an adhesive and structural substrate to which adhesive cell receptors could bind is one of the most representative functions of ECM. These interactions are participated in the activation and modulation of pro-survival signaling cascades. Furthermore, other bio-responsive molecules could provide signals by cell–ECM and cell–cell communications, which modulate cell adhesion, differentiation, and development [20]. It has been confirmed that these communications are mediated by the interactions between both soluble and cell surface receptors and ECM components

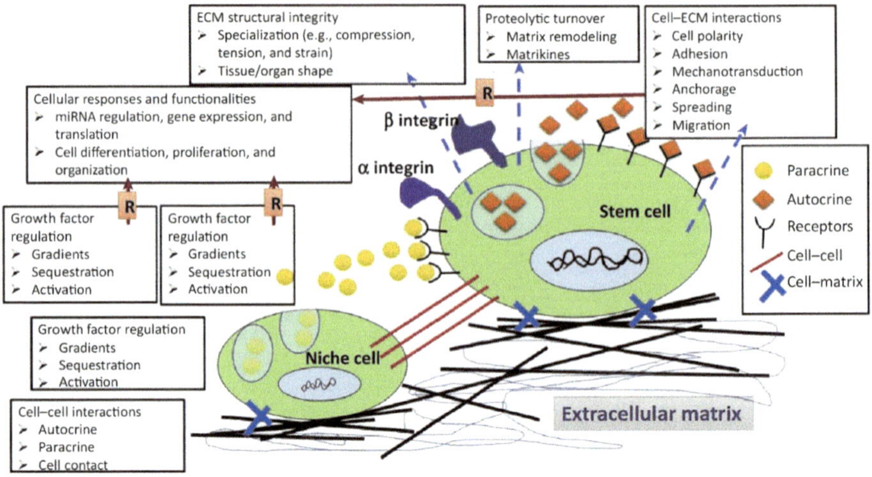

Fig. 1.2 The schematic diagram to illustrate the role of ECM in cell proliferation, development, and differentiation. ECM growth factors adjust the interactions between stem and niche cells through autocrine/paracrine modes. The stem cells depict accurate biological responses including gene expression and cell differentiation when they are stimulated through the receptors on cell membrane. And, the ECM regulates cell manner via its biological and physical properties. Soluble matrix-binding factors together with other components decide the destiny of key properties of stem cells mainly including cell death, adhesion, anchorage, migration, proliferation, and differentiation. Cells close the feedback loop via ensuring ECM structural integrity and acting through proteolytic turnover [22]

[21]. Representative receptors and corresponding interactors and cellular functions involved in cell–ECM interactions are summarized in Table 1.1.

As a result, taking efforts to construct such "cell-instructive" materials by incorporating well-defined chemical and physical properties, biological signals designed with high degrees of functional and compositional definition and multi-component frameworks to affect surrounding cells in a specific manner is one developing direct of bioactive materials [24]. However, the balance between strength, toughness, and complex microscopic three-dimensional structures of ECM in specific tissue is tightly regulated, so each organization could perform its special function as a whole [23]. For example, as is shown in Fig. 1.3, different tissues or organs exhibited different stiffness. In addition, ECM is not static even in mature tissues in adult mammals, and is produced, degraded and remodeled by cells. Hence, it must be taken into consideration that the high complexity of the ultrastructure and composition of ECM are still difficult to be fully understood. Even though the tools provided by protein engineering and synthetic biology, and electrospinning and 3D printing have offered an unprecedented level of biomimicry, the design of materials that highly and closely simulate the hierarchical structure and the biophysical and biochemical properties of natural ECM of human tissues reserve a great challenge. Artificial biomaterials are usually far from the natural and ideal microenvironments. Thus, it is not suitable for new tissue formation and reconstruction.

Table 1.1 Representative receptor involved in cell-ECM communication [23]

ECM receptors	ECM interactors	Cellular functions
Integrins	Collagen, LM, FN, soluble galectins, and several matrix glycoproteins	Cell adhesion, regulation of stress transmission and bidirectional signaling, and angiogenesis
Discoidin domain receptors (DD1 and DD2)	Different fibrillar collagen types	Embryo development, cell survival, cell migration, proliferation and differentiation, and remodeling of extracellular matrices
Syndecans	FN and TSP, collagens, βFGF, VEGF (vascular endothelial growth factor), βTGF, and PDGF (platelet-derived growth factor)	Growth factor receptor, activation, cell–cell communication, cell adhesion, cell proliferation, differentiation, and migration
Dystroglycan	LM, agrin, and perlecan in basement membranes and neurexins transmembrane	Cell development, basement membrane formation, epithelial morphogenesis, membrane stability, cell adhesion, polarization, and migration
Lectins	LM, FN, integrins, VN and TSP, GAGs, and other glycoproteins	Cell adhesion, migration, growth and differentiation, apoptosis
CD44	GAGs	Cellular motility, cell–cell and cell–ECM adhesions
CD36	collagen	Cell adhesion, fatty acid uptake, and angiogenesis

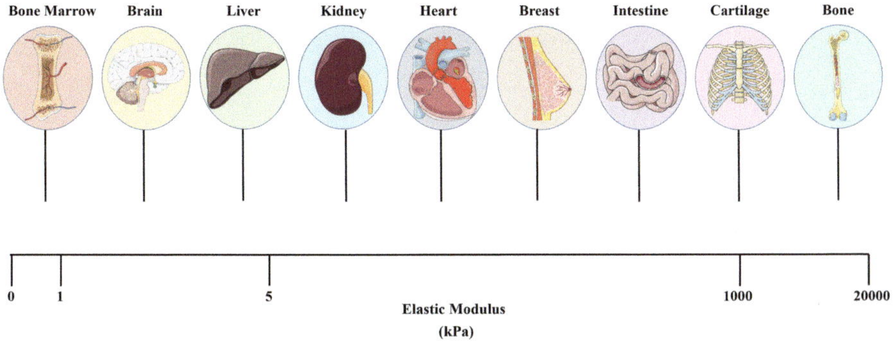

Fig. 1.3 Organ-specific stiffness values. Different tissues or organs exhibited different stiffness

Consequently, decellularized material (DM), or acellular extracellular matrix (AEM), which derives from natural tissues or organs and aims to remove antigen components in ECM and retain the components, such as polysaccharides, collagen, glycoproteins, etc., sticks out in various biomaterials and has been widely studied because it maintains the main compositions and structure of ECM.

1.2 Preparation of Decellularized Materials

Decellularized materials are prepared from natural tissues/organs via various decellularization methods mainly including chemical, physical, biological, and combinational methods, as is shown in Fig. 1.4 [25, 26]. After decellularization, cells and antigenic components would be removed as clean as possible, and the main components and structures of natural ECM would be remained, which are almost impossible to be imitated by artificial materials through traditional technologies, exhibiting their unique superiorities over all kinds of biomaterials.

In 1964, Grillo and Mckhann et al. firstly destroyed and removed cells by freezing and thawing method, developing the physical methods for decellularization [28]. Through physical methods, some physical means including force, electric current, and temperature would be applied to destruct cell membrane of tissues or organs. Then, cells would be lysis and cell debris would be finally removed. Up to mow, commonly used physical methods mainly involve mechanical oscillations, pressure gradient method, electroporation, freeze-thaw method, and supercritical fluid dissolution method. And, they are widely used in the decellularization of cornea, blood vessels [29, 30], bladder [31], aorta [32], etc., because their

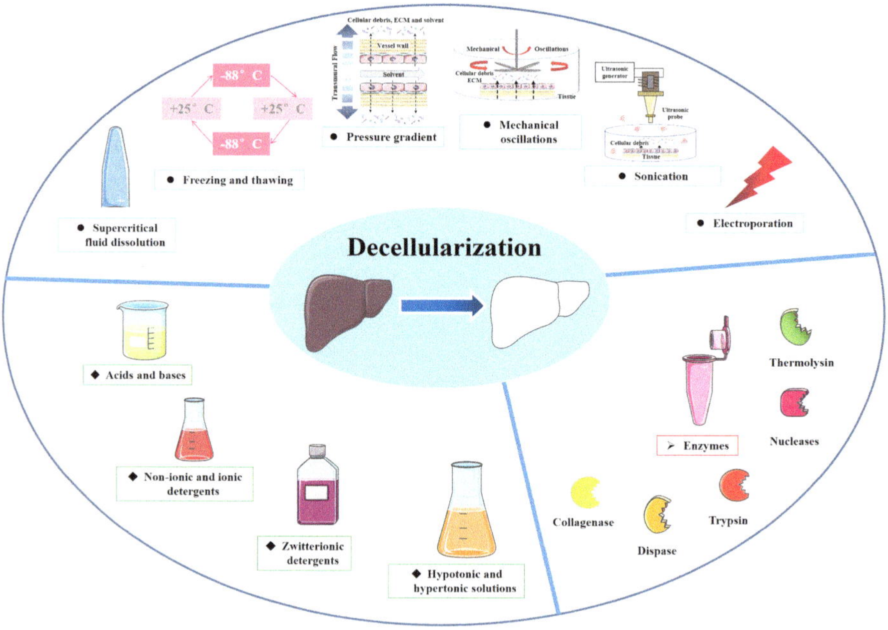

Fig. 1.4 Decellularization methods could be divided into four kinds including physical methods (multiple water or buffer washes, sonication, electroporation, freeze/thaw cycles, supercritical fluid dissolution, pressure gradient, mechanical oscillations), chemical (using acids and bases, hypotonic or hypertonic solution, zwitterionic detergents, and ionic or nonionic detergents), biological (enzymatic digestion), and combinational methods [27]

characteristics of easy operation and hardly no application of chemical reagents, which would not trigger adverse reactions [33]. And, it has been demonstrated that using supercritical fluid (CO_2) is the method having minimal harm on the mechanical properties of ECM among all kinds of physical methods [32]. However, the possibility of damaging ECM microstructures and reducing the mechanical properties in a certain extent, difficulty in decellularizing thicker tissues, and the limitations of some technologies limit the applications of physical methods. Specifically, electroporation is only fit for thin and small tissues.

In 1975, Meezan et al. used 4% sodium deoxycholate to successfully decellularize acellular basement membrane, which is one of the earliest attempts of chemical decellularization [34]. Different from physical methods, chemical methods removing cells mainly through the dissolution of chemical agent. After chemical decellularization, soluble proteins including cells, some proteins, antigens, lipids, and others would be removed, and the insoluble components in tissues such as elastin, collagen, glycosaminoglycans, proteoglycans, non-collagenous glycoproteins, etc., would be remained to maintain the appearance and ultrastructure of natural tissues. Up to date, most used chemical reagents include acid (e.g. sulfuric, acetic, hydrochloric, peracetic) or alkali (e.g. NaOH, calcium hydroxide, ammonium hydroxide, sodium sulfide), hypotonic, hypertonic solutions, eluent, ethanol, etc. [35] It could be found that the cleanness of cell removal of chemical methods is significantly larger than physical methods because the powerful chemical reactions could go deeply into tissues, which might overcome some limitations of physical methods. Specifically, Triton X-100 has been demonstrated to effectively remove cells and residues in thicker valves with less protein loss and few adverse reactions [36]. Furthermore, they have relatively small effect on the ingredient and structure of native ECM [37]. Among various chemical methods, zwitterionic and nonionic eluents are outstanding in maintaining ECM ultrastructure, and ion eluents stands out in removing cells and residues [38]. As a result, chemical methods have been widely applied to decellularize organs or tissues such as corneal [39], sciatic nerve [40] , rat brain tissue [41], blood vessels [42], heart valves [43], adipose tissue [44], etc. However, it is worthy of noting that the reagent remnants would increase the appearance probability of unwanted immune response, and excess chemical reagents would destroy collagen fibers, affect mechanical properties, and might remove some bioactive substances, reducing the bioactivity of decellularized materials [45]. Furthermore, it has been demonstrated that using the mixture of various eluents would aggravate the destruction of protein [46].

Biological method is the earliest decellularization method. Under the action of specific enzyme reagent, cells and some unwanted tissue components would be specifically removed, causing less damage to other components. Commonly used enzymes involve lipase, collagenase, trypsin, nuclease, thermolysin, alpha-galactosidase, dispersible enzyme, etc. [47–49]. However, it worth to note the possible bad effect on microstructure of native ECM bringing by enzymes and the negative immune response produced by residual enzymes. In addition, the required long time and difficulty in modulating appropriate enzyme treatment time limit their applications. For instance, nucleases were used to deprotect nucleotides after cell lysis, but it had strong damage to cells, so the treatment time should be short [50]. However, the nuclei could be successfully removed from dense tissues only when cells have sufficient trypsin contact time [48].

Combining at least two kinds of above methods to decellularize tissues or organs is the combination method, which could make the best of the both worlds, overcoming the drawbacks of one single method to achieve best decellularization effect. Numerous tissues or organs (e.g. spleen, liver, aorta) have applied this method to successfully decellularized with good properties [51]. For example, it has been demonstrated the combination of hypotonic buffer and freeze/thaw cycling could efficiently decellularized tissues with 97% reduction of DNA and complete remaining of major components [52]. Wei combined low-temperature and chemical method, and successfully decellularized liver with appropriate mechanical properties and good biocompatibility [53]. Apart from that, Wainright et al. combined chemical and physical methods to decellularize the epidermis and dermis, during which all cell components were successfully removed and the microstructure and the basement membrane of ECM was completely retained [54]. Besides, intact biliary and vascular structures can be obtained by non-thermal irreversible electroporation (N-TIRE) during the decellularization of liver [55]. However, finding an appropriate combination scheme for specific tissues or organs is the key factor of combination methods, which is factually difficult because differences in physiological structures between individuals and different tissues or organs. In order to find the suitable combination, massive efforts have been made in comparing different decellularization combination scheme. For example, Sarig et al. compared the effect of decellularization methods for thick cardiac tissue including using perfusion replacing Trypsin with 1% SDS (1), using a Trypsin-Triton sequence with perfusion (2), sonication or stirring through the built-in coronary arteries (3), and using agitation and stirring (4), and found that the third caused approximate acellularity compared to the fourth method and worse results compared to the first method because of their less spacious fiber morphology [56]. In the process of decellularizing the common carotid artery, the enzyme and eluent could be used together to reduce the probability of neointimal hyperplasia or the degeneration of aneurysms [42]. Therefore, further research is needed to render all methods more effective and to correlate better with the type of tissue being treated.

Even though many accepted procedures for decellularization as above, no method has been recognized as optimal because of variability in tissue composition. Various factors have effect on the effectiveness of decellularization, which include the composition, density, thickness of matrix, and the decellularization methods. With the development of decellularization methods and technologies, the removal of antigenic components is clearer and clearer, reducing antigenicity of acellular matrix to a large extent. Simultaneously, the damage on ECM is smaller and smaller, improving the mechanical properties. However, it is still a challenging problem that achieving a coordination between undamaged retention of non-antigen components as well as active ingredients and precise and thorough decellularization, which directly determines the host response after their implantation in vivo and limits the types of DMs. In addition, some DMs with poor mechanical properties might be not capable of providing good attachment sites for cells, thereby, affecting their subsequently proliferation and differentiation. Further studies are needed to be conducted to optimize the decellularization.

It is no doubt that a unified criterion of decellularization to judge the effectiveness is important for the preparation, application, and the mass production of decellularized materials, which include the quantitative requirements of the main components and the mechanical and biological properties. The existing index requirement for decellularization is a range interval. The current provisions including: (1) Less than 50 ng dsDNA/mg ECM dry weight; (2) Less than 200 bps DNA fragment length; (3) Lack of visible nuclear materials in tissue sections stained with 4′,6-diamino-2-phenylindole (DAPI) or H&E [31]. However, DMs could be put into clinical treatment only after the examination criterion is established, so that it could be examined by the indicators of safety supervision and management. To evaluate the effect of decellularization, it is necessary to develop a procedure to inspect the quantity of residual nuclear content and the characteristics of the structural properties. For example, Ozlu et al. proved that the alignment and diameter of fibers could regard as the additional parameters, which evaluated the characterization of biological heart scaffolds, whereas provided valuable input parameter to evaluate the preservation about structure for decellularized heart [57].

In addition, the disinfection and sterilization of DMs should be taken into consideration. The existing methods for the final sterilization of DMs could be divided into two categories, each of which has problems that need to be improved. The first method is sterilization by peracetic acid [58]. The advantage of this method is that the biological activity of DMs could be guaranteed effectively, but the problem is that it cannot realize the effective and thorough sterilization. The second method is sterilization by more drastic ways, such as by exposing to ethylene oxide, irradiating by gamma radiation, electron beam irradiation, etc. [59, 60]. The advantage of this method lies in thorough sterilization. However, it not only damages the ultrastructure and mechanical properties of DMs to a certain extent but also affects the activity of bioactive factors contained. Therefore, how to find an effective disinfection and sterilization method is also the basic problem in wider applications of DMs.

The traditional and emerging decellularization methods, the effective decellularization index, the subsequent cleaning and sterilization, and the effect of decellularization process on the performance of decellularized tissues will be introduced in detail in Chap. 2. Specially, the decellularization of organs will be introduced in Chap. 5.

1.3 Composition and Structure of Decellularized Materials

Decellularized materials with tissue-specificity are derived natural tissues or organs through decellularization process. Ideal decellularization process should thoroughly and accurately remove the cells and other antigenic components, remaining other components and the structure constituted by the remained components of native ECM. Even though, there are some damages on ECM during the degradation progress to some extent, the obtained DMs possess the compositions and structures highly similar to those of native ECM.

1.3.1 Composition of Decellularized Materials

As mentioned above, the ECM is composed of different kinds of tissue-specific macromolecules including collagen, laminin, elastin, soluble growth factors, glycosaminoglycans, etc. DMs have been demonstrated to have the same components with their deriving ECMs. For example, the DNA of decellularized porcine brain (DPB) was significantly reduced. And, it was reported that DPB contain GAG, collagen, and various molecules, such as Endostatin, CD26, tissue factor, fibroblast growth factor-1 (FGF-1), FGF-2, IGFBP-9, and osteopontin, as is shown in Fig. 1.5a–d [61]. Simultaneously, the decellularized liver was demonstrated to remain the type I collagen, type IV collagen, fibronectin, and laminin (Fig. 1.5e) [53]. Decellularized materials derived from different tissues have their tissue-specific components. And, the main components contained in most kinds of tissues include collagen, elastin, glycosaminoglycan, proteoglycan, fibronectin, and growth factor. The characteristics and functions of each component are as follows.

1.3.1.1 Collagen

Collagen, as the most prominent protein (about 30% of the whole protein content) and the main structural protein component of the ECM, is the main protein in tendons, skin, ligaments, cornea, cartilage, and other tissues, and is mainly secreted by fibroblasts [62]. It specifically consists of a right-handed bundle of three parallel, left-handed polyproline II-type helices (Fig. 1.6). And, collagen fibrils bundle combines to form fibers with uniaxial mechanical properties, providing almost all the mechanical strength in the tissue [63]. Fibrous extracellular networks constituted by collagen, keratin, and elastin provide tissues and organs ability to support tensile stress and repetitive stresses [64]. Collagen fibers could produce intracellular bonds, which supply cementing mechanisms that are required for cell connection, protection, and strengthening, and even provide nutrients and oxygen. In addition, sufficient collage play an important role in maintaining healthy cellular microenvironment, their insufficiency could do harm to body health. Based on its bioactive and functional properties, collagen has been regarded as an outstanding biomaterial that could be produced to highly organized scaffolds (e.g., dressings, skin grafts, sponges, and films) with good mechanical properties, biocompatibility, and biodegradability, which provide collagen superiority in the fields of drug delivery, tissue regeneration, etc. For example, studies demonstrated that collagen-based matrix could obviously induce endothelial cell adhesion and capillary-like

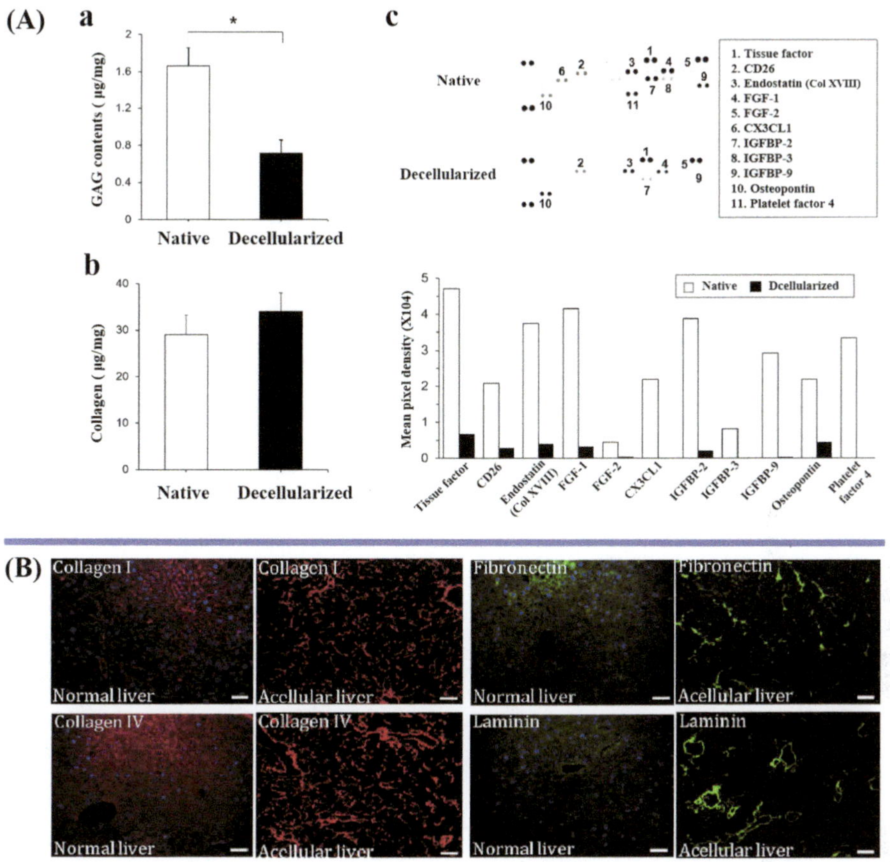

Fig. 1.5 The compositions of decellularized materials. Two examples of decellularized porcine brain (DPb) (**a–d**) and decellularized mice liver (**e**). Quantification of DNA (**a**), GAG (**b**), and collagen (**c**). (**d**) Cytokine array of molecules in DPb and the quantification of representative ones. (**e**) H&E staining and immunofluorescent staining of normal and acellular liver tissues. Data shows that tissue-specific decellularized materials keep the similar components to native ECM including GAG, collagen, fibronectin, laminin, cytokines, etc. [53, 61]

network formation, promote periodontal tissue cells proliferation and differentiation, promote fibroblasts, and liver cells proliferation [65–67]. However, there are some controversy persists about the collagen source. It must be taken into consideration that possible immunogenicity related to allogenic or animal sources and disease transmission might initiate unexpected immune response. And, reduced bioactivity caused by production of recombinant techniques could get a decreased biofunction [68]. Up to now, most kinds of collagens have been obtained from tissues of human or animal cadaveric (e.g., the skin or tendons; or the placenta or extracted adipose tissue from discarded human tissues). Compared with animal-derived collagen, collagen extracted from human tissues for scaffold production reduces the risk of

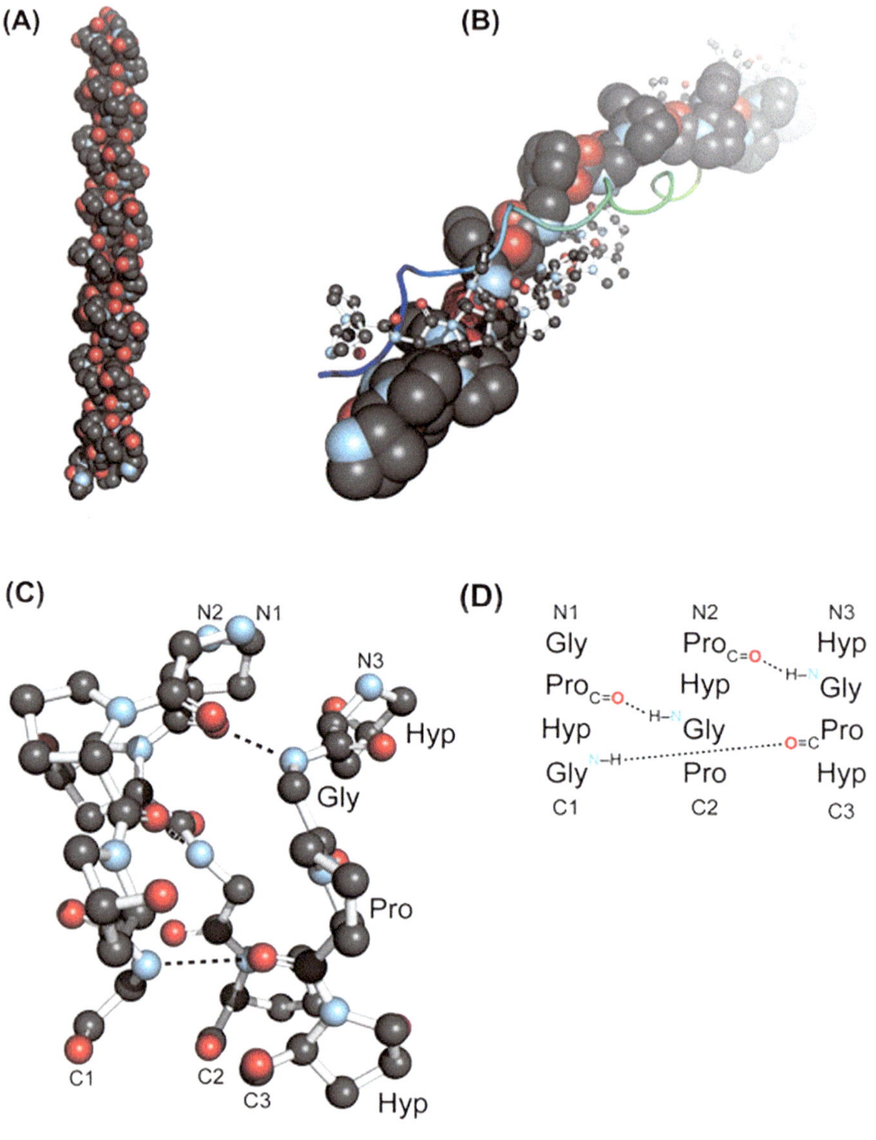

Fig. 1.6 Schematic diagram of collagen triple helix. (**a**) The crystal structure of collagen triple helix, which is consisted of (ProHypGly)4–(ProHypAla)–$(ProHypGly)_5$. (**b**) Planform of collagen tripe helix (view down the axis of the $(ProProGly)_{10}$ triple helix). Ribbon, ball-and-stick, and space-filling could be found (**c**). The ball-and-stick image in the collagen triple helix, striking the ladder of inter-strand hydrogen bonds (**d**). Staggering of the three strands [14]

immunogenicity and hypersensitivity, however, there is still potential to cause disease transmission and/or pathogenic contamination [69]. In this aspect, plant materials consisting of recombinant human collagen supplies an alternate native collagen source without concerns regarding variability or the risk of disease transmission [70].

There are at least 29 types of collagens have been uncovered in the decades of research. And, they could be roughly divided into eight types with different chemical structures and immunological characteristics, namely the type of I, II, III, V, VII, IX, XI, and XII collagens. Among them, the fibrils of type I collagen integrate as parallel bundles, and can resist the strong traction from skeletal muscles, such as tendons, ligaments, etc., is the most abundant in the interstitial matrix (approximately 90%) [71]. This protein adheres to a variety of cell types and could provide a satisfactory substrate which help cells thrive via cell chemotaxis and capillary formation [72]. Due to their unique properties and relative abundance, collagen I have been widely applied in tissue engineering.

Type II collagen is mainly present in cartilage and the nucleus pulposus (NP). It could be used for limited applications because it is a potential autoantigen in inflammatory synovial disease. An example of its application is that small doses of collagen II could modulate joint health in both osteoarthritis and rheumatoid arthritis [73]. The ingestion of lower-dose non-denatured type II collagen in the animal model of collagen-induced arthritis triggered an increase in anti-inflammatory molecule such as IL-4 and transforming growth factor-β (TGF-β), and an obviously decrease of pro-inflammatory mediators such as IL-2 and IL-17 production in the circulating levels, thus reducing both the incidence and the severity of arthritis [74].

In addition, type III collagen is sheet-like and has an outstanding multi-directional tensile resistance of the tissue such as skin. Type IV collagen with a structure of two-dimensional network is mainly involved in the assembly of cell basement membranes. Specifically, the presences of laminin and collagen IV have been observed in developing and mature articular cartilage, in vitro tissue-engineered cartilaginous constructs, as well as in vivo cartilage repair implants. In addition, collagen IV and laminin could connect a cell to and procuring a cell from its microenvironment, both macromolecules play a pivotal role in endothelial cell function and tissue regeneration [75]. Type VII collagen is particularly rich in skin and can connect the basement membrane of stratified epithelium with the connective tissue below. Moreover, collagen VII has been considered as a pivotal player in the physiological wound healing cascade in human [76]. Type IX and XII collagen are called as fibril-associated collagen, which can connect fibrils with other components in ECM.

1.3.1.2 Elastin

Elastin, as the main component of elastic fibers, is a highly hydrophobic non-glycosylated protein synthesized and secreted by fibroblasts and smooth muscle cells. It is rich in glycine, proline, and hydrophobic amino acids [77]. The elastin fiber network can automatically return to its original state after being stretched. Its stretch ability is at least five times larger than a rubber strip with the same cross-sectional area, allowing those tissues and organs requiring greater elasticity such as skin, elastic cartilage, aorta, dermis, blood vessels, tendons, lung parenchyma, ligaments, etc. [78].

Elastin is the most abundant ECM protein in arterial wall, which comprises up to 50% of the non-hydrated mass in vessel [79]. The elastic matrix consisting elastin provides vessel integrity, arterial elasticity, and extensibility, and their ability to recoil after stretch of pulsatile flow. In addition, elastin plays a quite pivotal role in modulating cell signaling pathways, such as smooth muscle cells in the arterial wall and luminal endothelial cells, which involved in morphogenesis, inflammation, and injury response via biomechanical transduction [80, 81]. Moreover, it is also an important autocrine factor to ensure vascular homeostasis via a combination of biomechanical support and biologic signaling. The elastic lamina alternates with smooth muscle rings because of the interaction of elastin-smooth muscle cell to improve the assembly of elastic matrix superstructures and cellular synthesis, forming a strong and flexible part of arterial wall [79, 82]. As a result, it is important to generate the unique elastic matrix superstructures of its target tissue for biomaterials to maintain vascular homeostasis. However, because adult vascular cells are inherently difficult to synthesize elastin precursors, there is limited possibility to produce the vascular elastic matrix by tissue engineering based on strategies such as contact guidance and dynamic stretch. Furthermore, replicating the process of elastin precursors organizing into the mature elastic matrix is an even more challenging task [83]. Hence, although a small amount yield of natural elastin could be extracted from tissues through harsh alkaline treatments, incorporating elastin into biomaterials is not a good choice due to the difficulty to obtain pure and homogeneous human elastin [83]. And, using short elastin-mimetic peptide sequences to simulate the active motifs of human elastin might engineer the crosslinking of tropoelastin in the developing elastin matrix, which could be a new concept for elastin tissue engineering [84].

1.3.1.3 Proteoglycan and Glycosaminoglycan

Proteoglycan (PG) is found in all connective tissues and extracellular matrices and many cell surfaces. It is a complex of sugars and various proteins, including transmembrane glycoprotein, and transmembrane proteoglycan and adsorbed glycoprotein. It can be combined with cytokines and growth factors to enhance or inhibit their activity. In addition, as one part of fibrous network structure, PG could serve as a permanent or temporary site for cell adhesion, which might improve the coherence of ECM and affect the activity of individual cells and the whole cell layers. Synthetic proteoglycan can be used for wound repair and suture. The proteoglycan complex prepared by collagen and mucopolysaccharide has good mechanical properties and can be used as artificial dermis for transplantation.

Glycosaminoglycans (GAGs) are anionic, linear, and highly heterogeneous carbohydrate polymers with repeating disaccharides, which are commonly a hexosamine (galactosamine or glucosamine) and an uronic acid component. They are proverbially present in the ECM and at the cell surface, and could regulate interactions of cell–matrix and cell–cell matrix and assembly and remodeling via interacting with various signaling molecules (e.g., collagen and fibronectin) and signaling molecules (e.g., chemokines and growth factors), structural proteins (e.g., collagen and fibronectin) [85]. The supramolecular presentation of GAG chains, together with other molecules on cell surface or in ECM, is possible to be functionally important for cell adhesion, which is widely applied in creating biomimetic multifunctional surfaces in biomaterials [86]. In addition, they are highly hydrophilic, and can form porous hydrocolloids with their volume increasing several times to fill the most space of ECM, providing mechanical support for the tissue. It hardly exists in dense connective tissues such as tendons, while it is the main component of the gelatinous material in eyeballs. Their ionic properties make them the property of osmotic swelling, which provides compressive strength on an ECM product, within which the content of GAGs is significantly decided by the decellularization methods. Additionally, sulfated GAGs are hopeful components for functional scaffolds because of the determination role of sulfate groups on growth factor binding and subsequently effect on wound repair [87]. By producing the stable haptotactic gradients needed for directional cell migration under shear flow or through immobilizing chemokines either on cell surfaces or in the ECM, GAGs also regulate the in vivo bioactivity of chemokines [88]. Utilizing the interactions between chemokines and GAGs, GAG-based biomaterials could be prepared to induce expected cell activities, of which the degree of GAG sulfation could be regulated via bioinspired changes in the GAG content (decreased sulfation from heparin to CS to HA), which help to adjust the sequestration of growth factor signals and the corresponding biofunction [89]. The strong interactions between chemokines and GAGs can protect chemokines from proteolysis and format stable various chemokine oligomers and structures that would not form in solution, which play a significantly important role in the overall chemotactic function of specific chemokines [90]. Moreover, interactions between stromal cell-derived factor-1α (CXCL12 or SDF-1α,) isoforms and GAGs not only contribute to their retention of hematopoietic stem cells (HSCs) in the bone marrow in homeostatic conditions, but also play a distinct role in stem cell recruitment and satisfactory tissue revascularization after acute ischemia [91].

GAGs can be divided into five types: hyaluronic acid, dermatan sulfate, sulfuric acid chondroitin sulfate, keratan sulfate, and heparan sulfate (HS). Among them, hyaluronic acid (HA), also known as hyaluronan, is the only non-sulfated amino glycan presenting as high-molecular-weight chains and is found in all tissues and body fluids of animals. It is synthesized by membrane-bound hyaluronan synthases and is comprised of disaccharide units including glucuronic acid and N-acetyl-d-glucosamine [92]. It is very abundant in early embryonic tissues as void filler, keeping the structure of the tissue in a certain shape, and is also one of the main components of skin tissue. Hyaluronan has also been suggested to attribute to the cells' final anchorage in specific niches in the bone matrix and activate human $CD34^+$ stem/progenitor cells migrate to sites with low CXCL12 concentrations (SDF-1-dependent transendothelial migration) [93]. It is now diffusely confirmed that natural HA exerts a pro-survival effect on interacting cells by protecting the cells against toxic insults [94] or by activating cell anti-apoptotic Akt pathways [95]. Additionally, hyaluronic acid can change the interactions between cell and cell, as well as cell and matrix by binding extracellular matrix molecules, and can activate intracellular signaling pathways, which directly leads to rearrangement of cytoskeleton and increased cell migration. These advantages make it widely applied in bone and skin tissue engineering. However, it needs to be taken into consideration that low-molecular-weight HA and high-molecular-weight HA present negative effects on orchestrating cell function at the cellular level. For example, high-molecular-weight HA acts as a barrier to block the migration of endothelial cell and subsequent angiogenesis in the fetal development of rat follicles. However, low-molecular-weight HA prepared by cleaving the polymer with hyaluronidase promotes endothelial cells migration and activate angiogenesis [96].

Other chains are bound to a central protein including perlecan, glypican or syndecan, to form proteoglycans. Especially, HS, a carbohydrate–protein complex, is a highly sulfated proteoglycan containing iduronic/glucuronic acid repeating disaccharide units and heparin chains or glucosamine. As a component of ECMs and endothelial cell membranes, HS proteoglycans are participated in virous critical functions of antigen-presenting cells and endothelium cells, and are required for the self-assembly, insolubility, and barrier characteristics of BMs. Hence, heparin is widely utilized as barrier compound [97–99]. Additionally, the polysaccharide side chains of HS proteoglycans in various tissues are different in the composition and structure of their sulfated domains, which would cause selective protein binding. And, through selective accumulation on GAG coating, HS may also promote the specificity of chemokine function. HS plays a crucial role in lots of biological processes through its interactions with various proteins. It is well recognized that SDF-1, a potent cell-recruiting and cell-mobilizing molecule, could be bound to cell surfaces by HS proteoglycans, which could significantly influence the chemokine's biological properties. However, there are obvious differences in the SDF-1 binding and glycocalyx GAG pattern between the two cell types (bone marrow endothelial cells and human umbilical vein endothelial cells) [100]. In addition, soluble HS was reported to promote SDF-1-driven migration in a dose-dependent manner, which indicated that SDF-1 presentation could be optimized by SDF-1–HS complexes rather than SDF-1 alone. This strategy for SDF-1 immobilization could be also used to recruit progenitor cells toward a targeted place [101]. Except for SDF-1, it also has been revealed that HS proteoglycans could bind other cytokines, such as platelet-derived growth factors (PDGFs), fibroblast growth factors (FGFs), and VEGFs, to adjust progenitor cell migration, recruitment, and angiogenesis, which could help to achieve HS-induced control of cell fate commitments [102]. Studies demonstrated HS is necessary for the growth factor-stimulated differentiation of osteoprogenitor cells and for endogenous FGF-2 signaling (as a coreceptor), which suggested that purified GAGs might be hopeful alternatives of certain growth factors to enhance the in vitro proliferation and differentiation of MSCs [103]. Hydrogels comprised of heparin exhibit attractive properties, such as growth factor binding, anticoagulant activity, antiangiogenic and apoptotic effects, making them appropriate candidates for emerging applications. Simultaneously, the anti-fibrotic effect of heparin on kidney and hepatitis fibrosis has been experimentally demonstrated in mouse models [104, 105].

1.3.1.4 Fibronectin and Laminin

Fibronectin, an elongated 45 nm protein, consists of two nearly the same outer globular domains (subunits of approximately 250 KDa). It is composed of a sequence of specific structural unit with multiple binding sites and domains, of which α-helical coiled-coil segment of which is covalently contacted to each end of the central domain (C-terminus) by a pair of disulfide bonds. Except for being quite crucial for vertebrate development, fibronectin could also regulate multiply cellular behaviors such as cell adhesion, growth, migration, and differentiation by recognizing and binding receptors on cell surface and molecules in ECM. Because fibronectin can elicit numerous responses in diverse cell types, many studies have tried to procure and apply specific bioactive domains in the fibronectin molecule. For example, the arginine–glycine–aspartic acid (RGD) tripeptide or arginine–glycine–aspartic acid, which is in the tenth Fn3 module, has effect on multiple cell fates, and has been widely applied in tissue engineering. Specifically, the modification of RGD sequences on biomaterials could increase the degree of MSCs migrating to biomaterials and the rate of MSCs populating the constructs [106, 107]. There were little cell elongation and no network formation when endothelial cells were encapsulated into 3D hydrogel scaffolds without binding sites. However, robust microvascular network formation was induced in hydrogels contained RGD binding sites, the extent of which was inversely proportional to the matrix stiffness. And, the absence of RGD caused round morphology at all stiffness [108]. Except for binding to cell surface integrins, fibronectin has a remarkably various of functional bioactivities by binding to numerous biologically significant biomolecules, including heparin, collagen/gelatin, and fibrin. These interactions are modulated by several distinct structural and functional domains [109]. It is widely used in wound repair such as burn wounds, chronic skin ulcer wounds, etc.

Laminins are a large glycoprotein family containing at least 16 isoforms, which associate with different heterotrimers composed of α-helical domains and laminin-type, globular epidermal growth factor (EGF)-like repeats. Specifically, laminins are composed of α, β, and γ polypeptide chains, and the triple-helical coiled-coil part in the center of each chain could connect to form different assembled structures taking part in different cellular events, such as cell survival, migration, adhesion, and differentiation [110]. It has been demonstrated that various laminin subunits of BM laminins distinctly influence tissue morphogenesis through inducing and remaining cell polarity, setting up tissue compartment barriers, organizing cells into tissues, and protecting adherent cells from detachment-induced cell anoikis and death [111]. And, laminin-1 was reported to induce endothelial differentiation *ex vivo* and promote the formation of new blood vessels in vivo, being capable of stimulating angiogenesis [112] and supporting vasculogenesis through guiding smooth muscle cell proliferation [113]. In addition, it has been widely reorganized that laminin-rich microenvironment plays a quite critical role in muscle regeneration. Specifically, using laminin-111 isoform to treat dystrophic muscles in the mdx mouse model of Duchenne muscular dystrophy (DMD) significantly decreased the amount of muscle damage [114], which is because that laminin-111 acts as not only a powerful stimulator of myoblast migration and proliferation ex vivo, but also as the factor increasing strength and resistance in treated muscles. As a result, laminin has been demonstrated to be a powerful signaling molecule in endogenous regenerative therapies to guide skeletal myoblasts to damaged areas, improving the in situ regeneration of skeletal muscles [72].

The cell-laminin interaction has been applied in biomaterial design, laminin based biomaterial has been demonstrated to increase axonal outgrowth and retinal explant attachment, as well as neurite survival, expansion, and outgrowth for neurons [115, 116]. For example, laminin-modified linear ordered collagen strates loading with laminin-binding ciliary neurotrophic factor (CNTF) could promote the regeneration and functional recovery of rat sciatic nerve [117]. In addition, a methylcellulose material functionalized with laminin-1 was demonstrated to modulate neural stem cell survival, apoptosis, migration, differentiation, and matrix production [118].

1.3.1.5 Growth Factor

Growth factor is a kind of polypeptide substance that regulates multiple effects such as cell growth and other cell functions by binding to specific and highly compatible cell membrane receptors. There are numerous types of growth factors in human body, such as platelet-like growth factors (platelet-derived growth factors, PDGF; osteosarcoma-derived growth factor, ODGF), epidermal growth factor (EGF), transforming growth factor (TGFα and TGFβ), fibroblast growth factor (αFGF, βFGF), nerve growth factor (NGF), erythropoietin (EPO). Extracellular matrix can bind and release specific growth factors, thus directly controlling their activity. ECM proteins such as collagens, proteoglycans (PGs), fibronectin, and vitronectin could bind a number of growth factors, such as HGF, FGFs, and VEGFs by themselves or in combination with heparin and heparin sulfate [119]. In this way, extracellular matrix acts as isolation and storage of growth factors, concentrating the activity of them in the vicinity of cells to prevent their degradation [120]. On the one hand, extracellular matrix is the repository of growth factors, directing their local availability. On the other hand, various proteins and proteoglycans of extracellular matrix can function as distributors of growth factors and release factors which are originally in an insoluble state under specific conditions [121].

The existence of growth factors is an important component of the spatiotemporal coordination of cell activities to ensure the correct formation of tissues/organs during wound healing [122]. For example, during wound healing, fibroblast growth factor (FGF) can induce the growth, mitogenesis, and angiogenesis of various cells, such as fibroblasts and endothelial cells. Heparan sulfate is a member of the highly sulfated glycosaminoglycan family, which is widely distributed on mammalian cell surface and extracellular matrix. Early studies have confirmed that heparan sulfate is required for cell reaction to FGF. It was confirmed that inhibiting the binding of FGF with heparan sulfate not only prevents the supporting effect of FGF on fibroblast growth, but also reduces the binding of FGF with cell surface receptors [123]. Later, Schlessinger et al. have explored its detailed mechanism and considered that in the presence of heparan sulfate or proteoglycan heparin, FGF molecules combine with receptors to form tetramer complexes, which stabilize two FGF receptors (FGFR) and promote the interaction between FGF and FGFR [124].

VEGF is a cytomitogen that specifically acts on glycosylation of endothelial cells, induces angiogenesis, and promotes the growth and formation of blood vessel. VEGF has various subtypes, among which V145 and V189 are combined with heparin sulfate. Plasmin releases V189 and V145 from their bound state, thus enhancing the mitogenic activity of endothelial cells and the vascular permeability. Heparan sulfate has a similar effect in its interaction with epidermal growth factor (EGF) and hepatocyte growth factor (HGF) [120].

Interstitial ECM proteins can also combine with growth factors and regulate their activities. Fibronectin, which is the survival factor for fibroblasts, is required for wound healing. Macri et al. revealed that specific fibronectin functional domains (FNfds) are combined with TGF-β, PDGF-BB, and bFGF, which maintain their activity and ensure them remains in the wound surface, providing local signals for the growth of epidermal cells, fibroblasts, and other tissue cells [122]. HGF is one of the most widely known growth factors with biological activity, which can stimulate mitosis and movement of epithelial and endothelial cells. Rahman et al. certified that the migration of endothelial cells was enhanced under the synergistic effect of FN and HGF [125]. The possible reason for this phenomenon is that fibronectin can be used as a reservoir to immobilize HGF in the ECM and generate a local concentration gradient that can modulate cell reactions.

1.3.1.6 Matrix-Bound Nanovesicles

Except for the above components, matrix-bound nanovesicles (MBVs) have been recently reported as an essential and functional component of ECM bio-scaffolds, which was firstly found by Badylak et al. in 2016 [126, 127]. MBVs could only be confined from ECM scaffolds after the enzymatic digestion of the matrix, which might benefit from their nanometer size, the close relationship of MBVs with collagen fibers, and the native stability of EVs (including capacity to withstand lyophilization and extreme changes in temperature and pH, and resistance to RNase degradation) that could protect MBVs from being destroyed by the decellularization agents [128, 129].

MBVs could be defined as transmission electron microscopy (TEM) on an osmium tetroxide–post fixed ECM bio-scaffolds. As is shown in Fig. 1.7a–c, MBVs are in rounded structures with a diameter of 10–1000 nm. Obviously, the signature of MBVs varies between the diverse ECM source bio-scaffolds derived from different tissues, which means that a more effective and defined use of ECM scaffold materials for clinical applications could be developed. Specifically, the

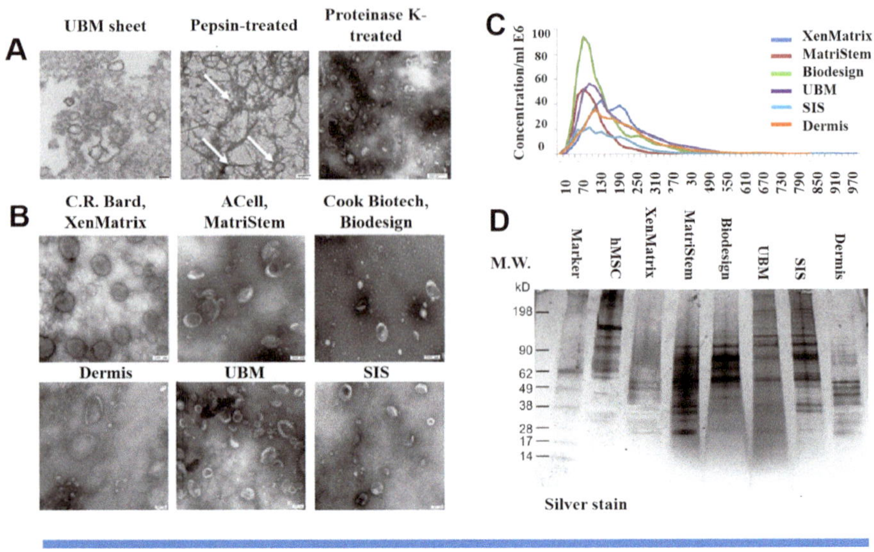

E

Product / description	C.R. Bard, XenMatrix	Dermis	ACell, MatriStem	UBM	Cook Biotech, Biodesign	SIS
miRNA count	46	33	223	240	62	53
Mutual miRNAs	25		185		35	

Fig. 1.7 The characteristics of matrix-bound nanovesicles (MBVs). (**a**) TEM imaging of MBVs existed in a UBM sheet stained, pepsintreated UBM, or proteinase K-treated UBM. (**b**) TEM imaging of MBVs existed in proteinase K–treated ECM from three commercial and three laboratory-produced scaffolds. (**c**) Validation of MBV size, which was assessed with NanoSight. (**d**) MBV protein cargo signature was different between MBVs and hMSCs as analyzed using SDS-PAGE and silver stain imaging. (**e**) Numbers of different miRNAs within each sample. Data showed the differences between different source-derived MBVs [127]

structures and sizes of different source-derived MBVs are distinctly different (Fig. 1.8b), the size of various of EVs 10–1000 nm (Fig. 1.7c) and the protein cargo evaluated by SDS–polyacrylamide gel electrophoresis (SDS-PAGE) present to be extremely diverse from exosomes isolated from hMSCs (Fig. 1.7d). Compared to dermis MBVs, UBM and SIS MBVs have an analogous protein cargo signature. However, even though the numbers of miRNAs identified in per sample are different (between 33 and 240), more than 50% were mutual between the commercial, and 22 miRNAs expressed in all samples were identified regardless of their tissue origin (Fig. 1.7e).

MBVs have features similar to microvesicles and exosomes, including the presence of miRNA cargo and nanometer size. However, they existed in the interstitial matrix of soft tissue and lacked identifiable markers, such as CD9, CD63, CD81, and Hsp70, which indicated that MBVs represent a different population of signaling vesicles. EVs, secreted by a variety of cell types [130], are membrane-bound, nanosized vesicles with diameters ranging from 30 to 1000 nm and could be

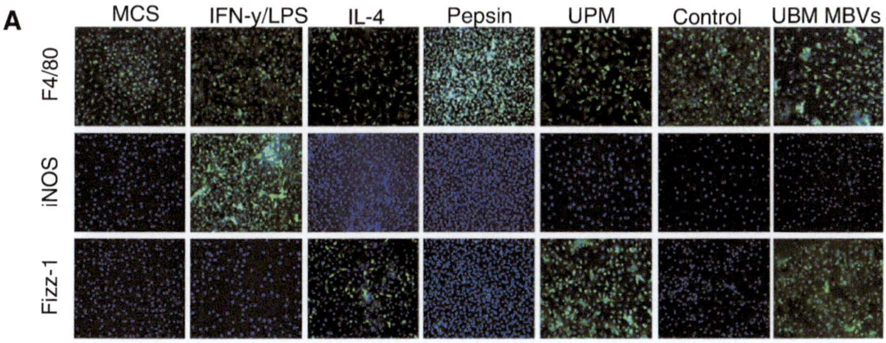

B

Product/molecular and cellular functions	C.R. Bard XenMatrix	Dermis	ACell, MatriStem	UBM	Cook Biotech, Biodesign	SIS
Cellular development	26	18	50	74	30	30
Cellular growth and proliferation	27	19	48	73	30	29
Cell death and survival	21	14	36	64	22	23
Cellular movement	17	12	32	58	22	20
Cell cycle	10	7	19	37	11	10

Product/physiological system development and function	C.R. Bard XenMatrix	Dermis	ACell, MatriStem	UBM	Cook Biotech, Biodesign	SIS
Connective tissue development and function	7	6	12	0	7	4
Organismal development	12	0	17	33	0	14
Organ development	0	9	13	32	11	11

Fig. 1.8 The biological activity of MBVs. (**a**) MBV-treated macrophages are mainly F4/80$^+$ Fizz-1$^+$ macrophages, which indicates an M2-like phenotype. (**b**) Molecular and cellular functions, physiological system development, as well as function pathways involved in identified miRNAs were generated using IPA. Every box means the numbers of different miRNAs related to each pathway. Data showed the differences in the biological activity of MBVs with diverse resources [127]

categorized into three main groups of microvesicles, exosomes, and apoptotic bodies according to their origin, size, mode of release, and markers [131]. They have attracted great attention over the past decade as potent vehicles of intercellular communication because of their ability to transfer proteins, lipids, RNA, and enzymes, thereby influencing various pathologic and physiologic processes. Some reports have demonstrated that EVs could anchor to ECM constituents via adhesion molecules, such as members of the integrin family (α_M and β_2 integrins) and intercellular adhesion molecule–1(ICAM-1) [132]. The biologic effects of EV signaling mainly include inhibition of cell senescence and apoptosis [133], transfer of proangiogenic proteins and miRNAs [134], modulation of the M1/M2 macrophage phenotype, promotion of anti-inflammatory cytokine secretion [135], and ECM production and remodeling [136], all of which are involved in the processes of wound healing and regeneration [137].

Similarly, the biologic effects of the MBV miRNA mainly include cell survival [138], growth and proliferation [139], migration and differentiation [31], which mainly attributed to that miRNAs have been demonstrated to be highly conserved in both invertebrate and vertebrate animal species. MBV miRNA isolated from porcine DMs could exert positive biologic effects on human and murine-derived macrophages cells in clinical applications [140, 141]. In addition, MBVs might influence stem cell differentiation and promote macrophage polarization toward anti-inflammatory M2-like phenotype. These results also indicated possible physiologic roles of MBVs in wound healing, and tissue development, homeostasis. And, MBVs with diverse resources exhibited different biologic effects, as is shown in Fig. 1.8.

1.3.2 Structure of Decellularized Materials

The removal of cells and other antigen components in natural tissues would cause porous structures of most DMs. It has been demonstrated that the decellularized porcine cardiac performed similar myocardial fiber orientation in natural cardiac with loose porous structure [56]. These internally connected holes possessing suitable diameter are quite important for tissue engineering scaffolds to provide structural support for cell metabolism and better environment for the exchange of nutrients. Simultaneously, evaluating mechanical properties of decellularized porcine cardiac on an Instron 5567 machine using stress-relaxation, dynamic cyclic loading, and strain-to-break, it could be found that the perfusion decellularized porcine cardiac exhibited a analogous cyclic response profile up to 15% strain, and, energy dissipation of the decellularized porcine cardiac during the "working cycle" was extremely different from that of the original tissue throughout the whole dataset, which suggested the sufficient mechanical stability of decellularized porcine cardiac [56].

In addition, an obviously crucial advantage of the DMs macro and microstructures is that they are formed from components in extracellular matrix via a self-assembly way, which could provide the cells with most comfortable microenvironment resulting from the long evolution of life. The natural complex 3D structures are hardly impossible for artificial material to imitate using traditional science and technology, which provides DMs superiority in the field of tissue repair and organ reconstruction, especially those complex tissues and organs.

It must be taken into consideration that some natural microstructures such as vascular structure remained in DMs provide them priority in tissue engineering, especially the repair of those vascular tissues. It must be taken into consideration that reconstructing tissues or organs with complete vascular system is still a big challenge, which directly decided whether or not the eventual tissue reconstruction is successful. For example, a complete vascular tree preserved in decellularized porcine cardiac could be found in the corrosion casting evaluation, which was highly similar to that of natural cardiac. And, the functionality and patency of the acellular vasculature ECM was confirmed by the perfusion of fluorescently labeled Rhodamine Dextran at successive time points. In addition, it has been demonstrated that decellularized bovine saphenous vein (SV) prepared through antigen removal process remained natural SV structure-function relationships and correlated venous valves function and had potential to perform as off-the-shelf venous valve replacements for chronic venous insufficiency [142].

In a word, the retention of the structural integrity at the micro and macroscale of DMs results a proximate tensile strength to original tissues of scaffolds [143] and retains most of the native vasculature, to about the third or even fourth level of branching [56, 144], which provides DMs great advantages and potentials in tissue engineering. Under the synergistic effect of retained 3D structure of the original tissue and necessary signaling for regulating cell functions, DMs have been demonstrated to be superior in remaining [145] and/or directing stem cell differentiation [146] in comparable of native biomaterials such as collagen or tissue culture plastics. Even powdered or dECM hydrogels supports cell differentiation and survival [147, 148] better than collagen matrices.

1.4 Degradation of Decellularized Materials

As is discussed in part 1.31, DMs are mainly composed of natural degradable collagens, glycoproteins, proteoglycans, polysaccharides, growth factors, etc. Thus, they could be degraded completely as non-toxic degradation products through cellular or enzymatic pathways. Specifically, the inflammatory cells could mediate the process by producing oxidants and proteinases [149]. And, the main families of proteinases related to degradation of the ECM are metalloproteinase with thrombospondin motif families (ADAMTS) (19 identified family members) and the matrix metalloproteinases (MMP) (23 identified family members) [150]. The good biodegradability of DMs has been suggested in vast researches. For example, Rae et al. demonstrated that the ^{14}C-labeled SIS matrix could be fleetly degraded more than 90% in 3-month implantation. And then, their degradation products could be completely excreted outside the body via a hematogenous way following with urinary excretion [151]. In addition, Brinker et al. found that the 10 layer ^{14}C-labeled SIS matrix could be 60% degraded after 30-days post implantation and be completely degraded after 90-days post-surgery [152]. Likewise, it has also been demonstrated that the unprocessed decellularized tendon matrix could be 100% degraded after 9-weeks post implantation [153]. Their relatively good biodegradability enables them prior in tissue engineering as one kind of fully absorbable and bioactive biomaterials.

However, it is a pity that the degradation of DMs at the implantation site is still uncontrollable. Usually, in vivo degradation exhibits quickly, which is mainly because of three reasons including their poor mechanical properties and the cell-mediated decrease of strength, avoiding the foreign body reaction that appears in response to non-degradable biomaterials, and the bioactive degradation products that accelerate the cell invasion [154]. The too fast degradation will have effect on tissue regeneration. For example, Zheng et al. proved that there were few vessels or parenchymal cells found after a period of observation, but only some inflammatory cells and fibroblast because of the fast degradation of the decellularized liver scaffold [51]. Cao et al. also demonstrated that the too fast degradation of decellularized scaffolds would affect the cartilage regeneration [155]. It needs to decrease the degradation rate of DMs for a better tissue repair. Inhibition of degradation will occur when threated DMs with chemical crosslinking, yet, it will prevent the release of these bioactive molecules during remodeling process [156].

Under the actions of proteolytic degradation, ECM molecules will release cryptic segments, such as matricryptic peptides provided by cleavage of parent molecules such as collagen [45], which exhibit antimicrobial activity or activities to promote chemotaxis, angiogenesis, mitogenesis, and differentiation [157], and influence ECM assembly, polymerization, as well as the production of ECM growth factor complexes, playing an important role in tissue repair and organ replacement [158].

And, with the degradation of DMs, tissue regeneration could happen gradually. As is shown in Fig. 1.9, the DM-based hydrogels with different concentrations of protein were implanted into the stroke cavity to attract endogenous cells and repair brain, the repair effects of which were dependent on their degradation rates. Specifically, only 32% of the hydrogel with 8 mg/mL of proteins was resorbed by 90 days, whereas less concentrated hydrogels (3 and 4 mg/mL) were precisely degraded with a 95% reduction in volume. And, the macrophage infiltration and density, as well as the invasion of neural cells and endothelial cells with neovascularization within the bio-scaffold progressively increased in the lesion-tissue boundary along with hydrogels degradation.

1.5 Mechanisms of Decellularized Materials on Promoting Tissue Repair/Organ Replacement

Tissue repair or organ replacement occurs along the degradation of scaffolds. For functional tissue restoration promoted by DMs, this process is also identified as "constructive remodeling," which is remarkably similar to the process occurring naturally in damaged muscle that is capable of full recovery [160]. Specifically, it is a process with a temporal sequence of myogenesis, remodeling, maturation/ functional repair, and robust cellular infiltrate, including the participation of macrophages and specific tissue progenitor cells, restoration of innervation and vascularization, and site suitable spatial reorganization of stroma and cells in response to mechanical loading [161].

For example, the natural repair process of skeletal muscle following recoverable injury mainly involves three phases [162]: (1) the degeneration phase, which includes the disruption of myocyte and ECM and the influx of inflammatory cells [163, 164]. (2) the repair phase, during which quiescent myogenic stem cells will be inspired to enter the cell cycle and migrate toward the site of damage [164, 165], differentiate, and fuse to form multinucleated myofibers [166, 167]. (3) The remodeling phase, during which regenerated myofibers spatially organize, mature,

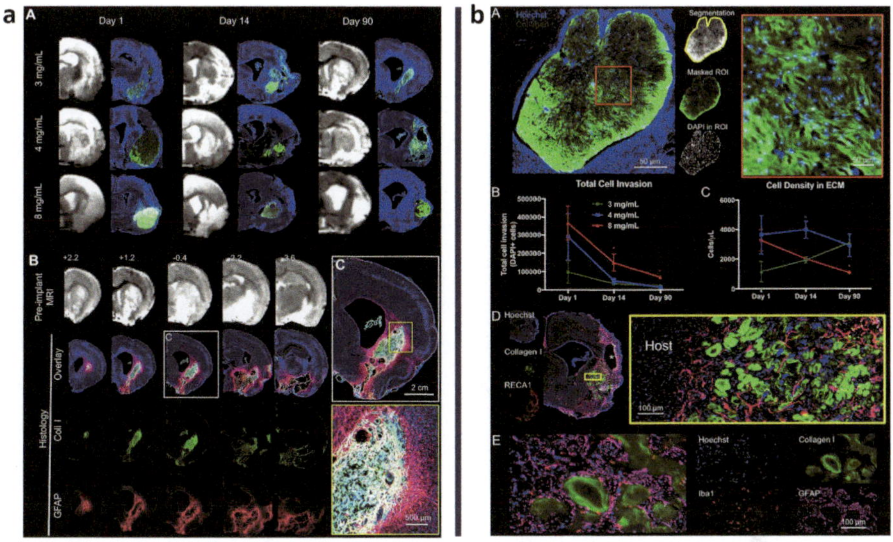

Fig. 1.9 The cell infiltration and tissue regeneration occurred with the degradation of DM-based scaffold. (**a**) Macroscopic distribution of DM hydrogel in the stroke cavity. (A) Using the pre-implantation T2-weighted magnetic resonance images to define calculate volumes and stereotactic coordinates of DM hydrogel precursor for injection. Data showed that degradation rates and structural remodeling were decreased with the concentration ECM. (B) The location and volume of the lesion cavity (in Anterior-Posterior pre-implantation MRI scans) and 4 mg/mL DM hydrogel after 90-days post-implantation. Data showed significant active tissue remodeling around the lesion cavity and in DM hydrogel scaffold. (C) At the lesion-tissue boundary, astrocytes (GFAP+ cells) cross the glial scar and invade the scaffold that is replacing the stroke cavity. (DAPI in Blue, Collagen I in green, GFAP in red) (**b**) The cells in ECM hydrogel. (A) The quantification of the cells number inside hydrogel (8 mg/mL). (B) Total cell infiltration, which indicated a gradual decrease in total number of cells over the biodegradation time. (C) Cell density were decreased with the increased hydrogels concentrations, which indicated the changes of DM hydrogel (4 mg/mL) volume because of biodegradation. (approximately 4000 cells/L). (D) Cell infiltration and density focused on the DM hydrogel scaffold (4 mg/mL). However, numbers of cells are obvious inside the previous cavity between DM hydrogel fragments. (E) GFAP^{+} astrocytes and Iba^{-1+} macrophages were common phenotypes, showing few foreign body response or scar [159]

and promote the ability to contract [163]. Besides, it must be taken into consideration that the microenvironmental cues before these processes are provided in part by the host innate immune response, especially macrophages [168, 169]. When skeletal muscle is injured, cytokines released by acute responder cells such as neutrophils and damaged myocytes will activate the host macrophages toward a pro-inflammatory (M1-like) phenotype, which will subsequently migrate to the injured area. These pro-inflammatory macrophages will excite the proliferation and mobilization of resident satellite stem cells through paracrine effects [169]. The pro-inflammatory (M1-like) phenotype will then be activated and be transited to a major anti-inflammatory pro-remodeling (M2-like) phenotype, which promote cell cycle exit and differentiation of the expanded myogenic satellite cells

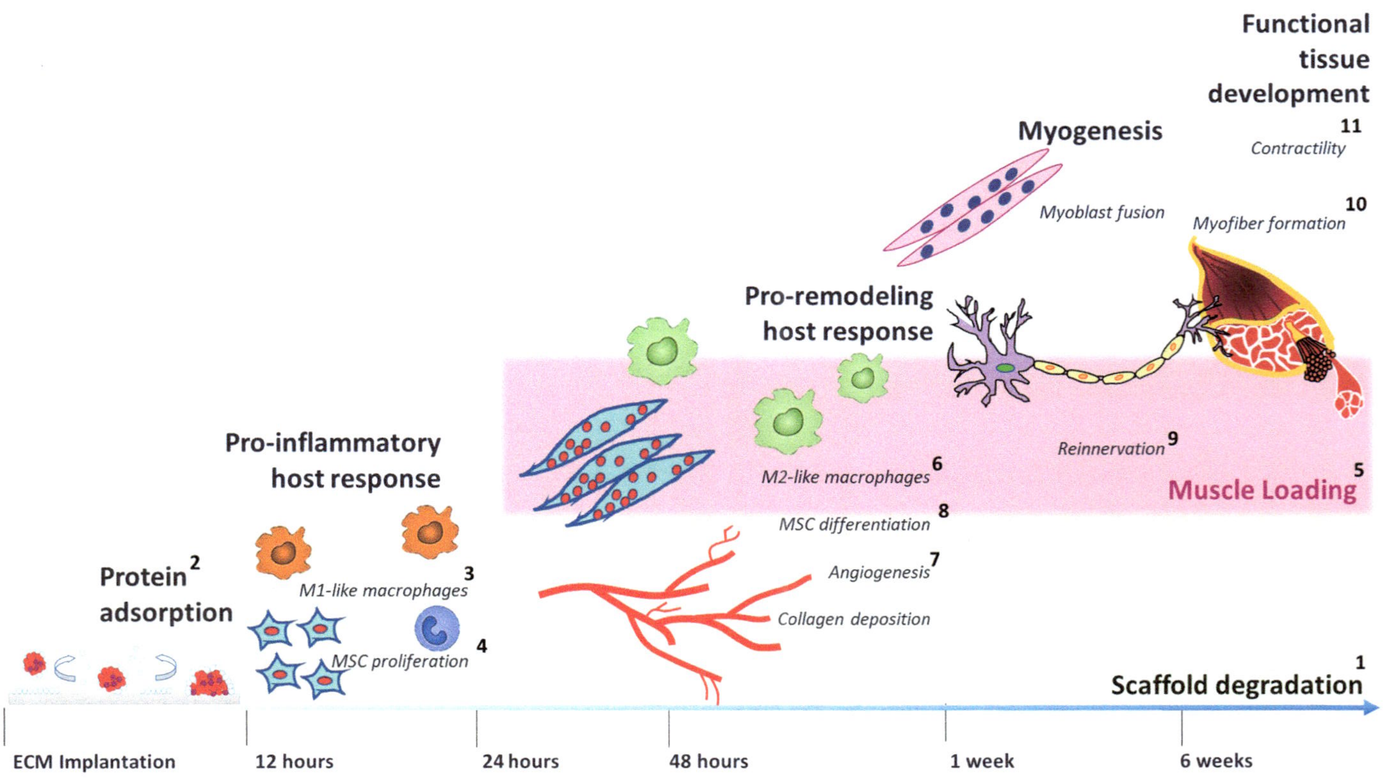

Fig. 1.10 The schematic diagram to illustrate the mechanisms of DMs-mediated skeletal muscle repair. After implantation, DMs will be degraded, releasing bioactive components that induce the infiltration and activation of macrophage as well as the recruitment, proliferation, and differentiation of myogenic stem/progenitor cell. Under concomitant mechanical load, clinical results have demonstrated that the formatted skeletal muscle tissue is innervated and functionally contractile [171].

into functional contractile skeletal muscle under the paracrine effects [169, 170]. Thus, the secreted products of macrophages are constituted of some of the mediators for skeletal muscle regeneration. As is shown in Fig. 1.10, it is the mechanism of DMs-mediated skeletal muscle repair. DMs degrade after implantation into the defect site, releasing bioactive substances, which induce macrophage infiltration and activation, as well as the recruitment, proliferation, and differentiation of myogenic progenitor/stem cells. Clinical results have demonstrated that the formation of skeletal muscle tissue is functionally contractile and innervated under concomitant mechanical load.

In a word, the mechanisms of decellularized materials to promote tissue repair/ organ replacement could be summarized as tow aspects: (1) Initiating relatively low host tissue response to provide regeneration microenvironment; (2) Containing bioactive factors to recruit endogenous stems/progenitor cells and promote matrix production and angiogenesis. Both of them will be comprehensively discussed in the following parts.

1.5.1 Initiating Relatively Low Host Tissue Response to Provide Regeneration Microenvironment

It has been demonstrated that immune response regulated by biomaterials might create a more suitable environment for tissue regeneration, which indicates the necessarily of appropriate immune response [172]. Studies have proved that ECM can directly adjust macrophage phenotype, which in turn induces that factors within DMs are capable of fostering this phenotypic switch. It has been demonstrated that the components of DMs can alter the plasticity of adherent macrophages, thereby promoting inflammation and immune regulation. Some studies have confirmed that circulating macrophages may cause necessary reactions after activated by heterogeneous ECM, thus leading to anti-inflammatory or constructive remodeling reactions to scaffolds or xenografts [173].

Macrophages are the key components of immune response and can be divided into M1 and M2 macrophages according to the expression of their receptors, the difference of secreted cytokines, effector molecules and the distinction of their functions [174]. Specifically, M1 macrophages are pro-inflammatory cells, have cytotoxicity, can promote the killing of antigens, and are associated with inflammatory signals of classical inflammatory reactions, especially chronic inflammatory reactions. Conversely, M2 type macrophages are anti-inflammatory cells, which have the characteristics of promoting immune regulation, tissue repair, and structural tissue remodeling [175]. Relevant identification research on M2 macrophages suggests that it may be beneficial in inflammatory response of tissue regeneration, which is an area in which a large number of researches are being conducted. Macrophages are called up in response to tissue injury, infection or the presence of external antigens, and then release various cytokines and chemokines [176].

Many studies have shown that the degradation products of DMs could directly and indirectly (i.e. paracrine/cell-mediated) modulate the immune response by regulating the macrophage/T cell phenotype [177, 178]. For example, it was demonstrated that porcine-derived decellularized scaffolds could promote the change in the host innate immune response into Th2-restricted response [179]. Besides, Deng et al. reported that 3D-printed poly (lactide-co-glycolide) (PLGA) matrix modified with hUCMSC-derived ECM could effectively reduce the external body response, further, it could significantly promote bone regeneration via increasing the M2 macrophages migration and the repolarization of M1 phenotype toward M2 macrophages [180]. Badylak et al. also found that DMs could promote the transition of macrophage phenotype from M1-like macrophage (pro-inflammatory, cytotoxic) population to M2-like macrophages (anti-inflammatory and pro-healing) in 1–2 weeks post-implantation [156]. And, Hong et al. introduced decellularized porcine brain to direct the macrophages in the damaged spinal cord. They found that dECM hydrogels could enhance the neurite outgrowth of cortical and hippocampal neurons, as well as the polarization of macrophages toward M2 phase, as is shown in Fig. 1.11a. And, the proportion of M1 and M2 macrophages in the damaged spinal cord was substantially altered in rat SCI models (Fig. 1.11b), especially, the expression of molecules related to M2 (arginase1, CD206, and IL-10) was remarkably increased with the received dECM concentration of 5 mg/ml. In addition, some peptides released during DM degradation exhibited prominent antimicrobial activity to keep wounds sterile during regeneration process, which also help to reduce inflammation responses [181].

Besides the immunomodulatory abilities of DMs, the low immunogenicity plays an important role in the relatively weak immune responses. In most cases, cells and antigen components could be removed as clean as possible after the process of decellularization, resulting relatively low immunogenicity of DMs. It has been demonstrated that DMs would trigger few inflammatory responses when the DNA was less than 200 bp in length [182]. Furthermore, macrophages revealed to degradation products of DMs have been reported to show the phenotype with high antigen-presenting and anti-inflammatory abilities [183], even in the microenvironments with severe pro-inflammatory (e.g. ulcerative colitis [184] and volumetric muscle loss [185]). In addition, the reported negative inflammatory responses to DMs were triggered by the remained antigen components in most cases. As a result, improving the processes of decellularization and post-process is hopeful to further decrease DMs immunogenicity. For example, Conklin et al. found that the immune reactions of the decellularized porcine common carotid arteries could be reduced obviously by dealing with enzyme and detergent [42]. In addition, Wang et al. found that through crosslinking with glutaraldehyde could decrease the immune reactions of decellularized porcine whole-liver matrix. Furthermore, they concluded that the natural genipin crosslinking performed better in reducing the immunogenic problem of xenogeneic decellularized whole-liver scaffolds [186]. Wu et al. transformed decellularized bone particles into gel scaffolds to optimize the immunomodulatory properties, and they found that the decellularized bone gels exhibited a relatively low immunogenicity and achieved accelerated tissue regeneration progress in a rat

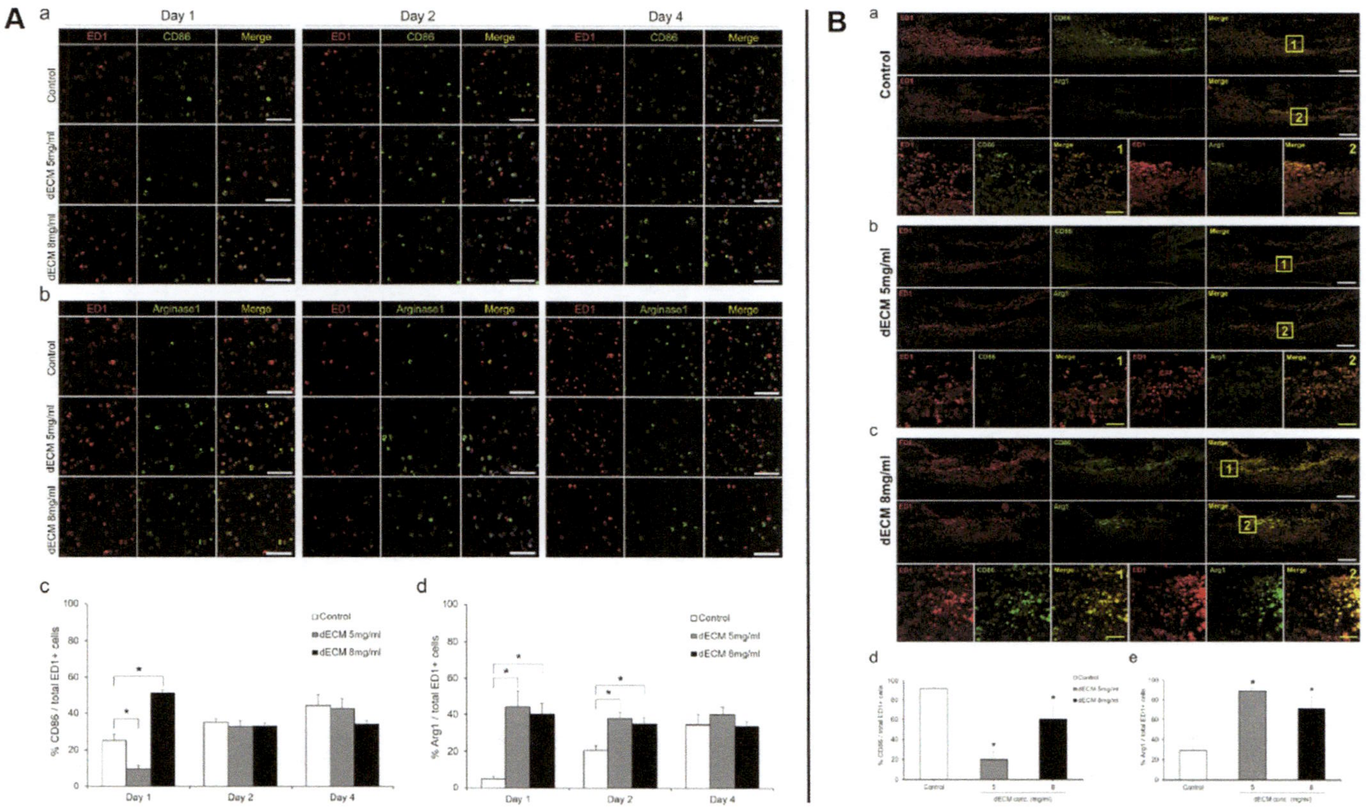

Fig. 1.11 The DMs (decellularized porcine brain) hydrogels were demonstrated to promote the neurite outgrowth of cortical and hippocampal neurons and the polarization of macrophages toward M2 phase. (**a**) Changes in the population of primary cultured M1 and M2 macrophages in DMs hydrogels for days 1, 2, and 4. Representative immunocytochemical images of (a) M1 macrophages (CD86 positive, double-stained with ED1) and (b) M2 macrophages (Arginase1 positive, double-stained with ED1) cultured in the hydrogels for 3 days, and the quantified cell population of (c) ED1-positive/CD86-positive cells and

periodontal defect model because they could induce macrophages toward anti-inflammatory (M2) polarization, which might mainly benefit from the effect of some unique biomolecules, including microRNA, proteins, and lipids, released during preparation of the decellularized bone gel, thereby regulating macrophage responses [187].

Overall, DMs with relatively low immunogenicity and their immunomodulatory abilities help to initiate relatively low host tissue response to provide an appropriate regeneration microenvironment for tissue repair and organ replacement.

1.5.2 Containing Bioactive Factors to Recruit Endogenous Stems/Progenitor Cells and Promote Matrix Production and Angiogenesis

A stem cell bank plays a key role in maintaining the integrity of the tissue throughout the life of the animal. For example, skin stem cells are involved in skin metabolism, hair growth, and wound healing. And endogenous neural precursor cells in brain tissue can regenerate nerve cells. The microenvironment, also named as stem cell niche, contains extracellular matrix components, growth factors, cytokines, and adhesion molecules providing various complex signals to maintain the self-renewal and differentiation functions of stem cells and determining their fate. In addition, the microenvironment also controls the flow of stem cells, and decides how many stem cells continue to sleep and how many stem cells should be mobilized into the cell cycle to maintain the dynamic balance of the number of tissue cells. For example, angiopoietin-1 (Ang-1) produced by osteoblasts can induce hematopoietic stem cells (HSCs) to enter a dormant state. Changes in the concentration of cytokines regulate the retention of stem cells in the microenvironment. Specifically, cathepsins and neutrophil elastase can regulate the reservation of stem cells in bone marrow by cleaving receptors. And, the dormancy and activity of stem cells in the bone marrow are determined by various cytokines such as colony stimulating factors and angiogenic factors [188–190].

In natural tissues, matrikines are produced by the proteolytic action of ADAMs (a disintegrin and metalloproteinase) and matrix metalloproteinases (MMPs) family members [191], and are well known to recruit stem/progenitor cells and direct cell adhesion, migration, and differentiation [19, 158, 192]. One example of these

---

Fig. 1.11 (continued) (d) Arginase1-positive/ED1-positive cells. White scale bar = 50μm. (**b**) Immunohistochemical analysis of macrophages for M1/M2 polarization. After 8 weeks of treatment with hydrogels, tissue samples were analyzed by immunohistochemistry. (a–c) M1 macrophages positive for CD86 and M2 macrophages positive for Arginase1. Cell was double-stained with ED1 pan-macrophage marker. (d and e) Quantification of fraction of cells positive for CD86 (d) and Arginase1 (e), normalized to ED1-positive cells. Statistically significant difference noted at $*p < 0.05$ vs. control. White scale bar = 500μm, yellow scale bar = 100μm [61]

Table 1.2 Some sequences and their receptors and stimuli [23]

Sequences/ motifs	Receptor	Stimuli
RGD and cyclic peptides	Integrins α5β1, αvβ3	Cell adhesion
IKVAV	Integrin αvβ3	Neuronal differentiation
YIGSR	Integrins α1β1, α3β1	Angiogenesis, epidermal development of skin, and inhibition of tumor growth and metastasis
FRHRNRKGY	Heparin	Human osteoblast cell adhesion
CGG-QPPRARITGYII	Integrin α4β1	Syndecan

"cryptic peptides" is the Arg-Gly-Asp (RGD) peptide, which is mainly prepared from collagen and fibronectin, is the most crucial peptide relating to cell adhesion. And, its conformation variations from linear to cyclic could remarkably transform the integrin recognition specificity, initiating diverse cellular responses (e.g. phenotype maintenance and differentiation) [191]. In addition, Lamin (LM) derived peptide are also involved for their unique biofunctions, which mainly include IKVAV (improving neurite outgrowth in cortical neurons and inducing neuronal differentiation) [193] and YIGSR that is capable of inducing angiogenesis, restraint of tumor growth and metastasis, and promotion of skin development [194]. In addition, FRHRNRKGY, as a VN-derived peptide, communicates with heparin and is specifically related to human osteoblast adhesion [195]. Simultaneously, the hydrophobic sequences VGVPG and VGVAPG derived from elastin could stimulate smooth muscle cell proliferation, and fibroblast, monocyte, and macrophage chemotaxis, respectively [196]. Some sequences and their receptors and stimuli are summarized in Table 1.2. It has also been demonstrated that matricryptic peptides produced by some large molecules such as collagen showed antimicrobial activity [197], and could act as chemokines of stem/progenitor cells [198]. In addition, fragments of other ECM products, such as hyaluronic acid produced during tissue injury and inflammation, have been confirmed to inspire the production of angiogenesis and MMP [191] and to modulate macrophage phenotype [199]. Moreover, remained glycosaminoglycans (GAGs) in the decellularized tissues are able to bind and sequester soluble growth factors, directing their activity and availability [200].

These abundant bioactive substances contained in DMs have been demonstrated to stimulate the migration of stem cells and progenitor cells to the regenerative sites, providing a suitable and stable living environment for cell differentiation in the desired direction [201, 202]. For example, it was reported that multipotent progenitor cells migrated toward the injury sites in the tendon repair model under the action of decellularized urinary bladder [157]. In addition, researches have demonstrated the abilities of DMs recruiting Sox2þ cells in vivo [203], multipotent perivascular stem cells (PVSC) in vitro and in vivo [204, 205], skeletal muscle myoblasts in vitro and in vivo [206], and myogenic progenitor and stem cells in vitro [139]. Hence,

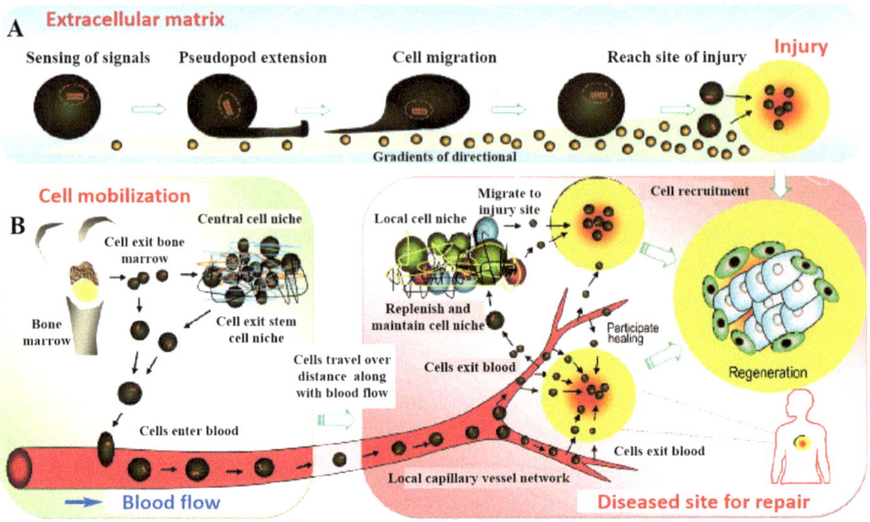

Fig. 1.12 Schematic diagram of cell migration mechanisms within the body that may be regulated to promote the recruitment or homing of endogenous stem cells for wound healing and regeneration. (**a**) Stem cells in cell niches neighboring the injury could identify and submit to the signals gradients inside ECM (such as directional cues) and could arrive to injury sites independent of blood flow through chemotaxis-guided interstitial migration or active amoeboid movement in most cases. (**b**) Stem cells located in central cell niches could be stimulated into blood and disseminated throughout the whole blood circulation system until they migrate to the local capillary vessel network around the injury sites. Then, the cells recognize and communicate with microvascular endothelial cells. Subsequently, they get out of the blood to supplement and keep the cell niche neighboring the damage, promoting the regenerative potential of the damaged tissue. In another way, the cells could directly arrive to the damage sites to take part in the process of wound healing and tissue remodeling if they could successfully escape from the circulation [14]

DMs are hopeful to achieve tissue remodeling without foreign stem cell transplantation. Namely, collecting endogenous stem and progenitor cells from patient's own body to maintain an effectively stable stem cell bank is quite crucial for tissue repair [198, 207]. Figure 1.12 showed the schematic diagram of cell migration mechanisms within the body that may be regulated to promote the recruitment or homing of endogenous stem cells for wound healing and regeneration. And, the stem/progenitor cell-instructive effects mediated by DMs are showed in Fig. 1.13.

In addition, DMs are also known as repository for a variety of growth factors, such as fibroblast growth factor (FGF), insulin-like growth factor (IGF), stromal derived growth factor (SDF-1), keratinocyte growth factor (KGF), transforming growth factor (TGF), hepatocyte growth factor (HGF), vascular endothelial growth factor (VEGF), bone morphogenetic protein (BMP-2), etc., which directly regular cell behavior by binding to receptors on cell surface or interacting with other components of the ECM, showing biologic activities including angiogenesis, antimicrobial effects, chemotactic effects, etc.

Growth factors are crucial for proper regulation of cell division, differentiation, and regeneration of damaged tissues. By combining with proteoglycan and ECM

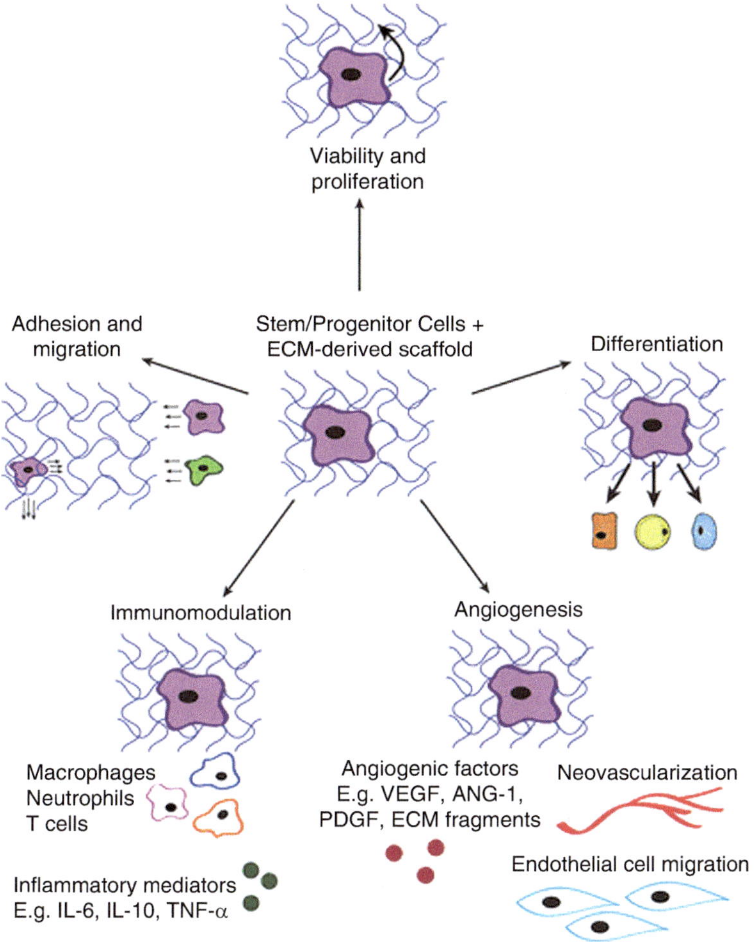

Fig. 1.13 Stem/progenitor cell-instructive effects mediated by DMs. DMs can induce stem/progenitor cell adhesion, migration, viability, proliferation, and differentiation along multiple lineages. Furthermore, these scaffolds can affect stem/progenitor cell secretion of paracrine factors to enhance angiogenesis and regulate the immune response. *IL-6* interleukin 6, *IL-10* interleukin 10, *TNF-α* tumor necrosis factor-alpha [208]

protein in decellularized materials, GF activity can be regulated in many different ways, thus affecting cell behavior. Therefore, after the acellular materials are implanted into the damage sites, the dECM proteins regulate cell behavior through typical interaction with its standard adhesion receptors or through atypical growth factors presentation [119]. Both approaches are absolutely important in promoting wound healing and tissue regeneration. The synergistic effect of dECM and growth factor has the ability to induce the differentiation of stem cells. He et al. [209] cultured adipose-derived mesenchymal stem cells (ADSCs) in plastic bottles, CECM derived from ADSCs, and CECM derived from synovial mesenchymal stem cells

(SDSCs), respectively, and used additive transformation for growth. Cells were cultured in serum-free induction medium containing factor-β3 (TGF-β3) or TGF-β3 and bone morphogenetic protein-6 (BMP-6). After culturing for a period of time, it was found that the proliferation number of cells cultured using ADSCs and SDSCs-derived CECM was 6–8 times that of cells cultured in ordinary plastic bottles and showed similar cartilage formation index. These experimental results showed that the synergistic effect of CECM and growth factors can enhance the effect of factors in inducing stem cells to differentiate into chondrocytes and promote the potential of ADSCs to differentiate into chondrocytes.

The ability to induce cell differentiation of DMs have been demonstrated by numerous studies. Specifically, acellular heart could guide the differentiation of stem cells towards cardiac lineage [210]. The decellularized lung was reported to be capable of enhancing the lung-specific lineage differentiation of embryonic stem cell (ESC) [211]. Furthermore, many decellularized tissues or organs such as decellularized tendon, central nervous system, liver, muscle, adipose, cartilage, etc., have also exhibited positive effect on corresponding cell differentiation to different degrees [212, 213]. For instance, it has been demonstrated that acellular hyaline cartilage could enhance the differentiation of stem cells and human BMSCs towards chondrogenic lineage, showing the good chondrogenic inductivity of decellularized cartilage [214].

In a world, these bioactive substances within DMs provide them the capability of recruiting stem/progenitor cells, induce desired differentiation of cells, and promote matrix production and angiogenesis, promoting tissue repair and organ replacement.

1.6 Decellularized Tissues and Their Applications

As discussed above, the bioactive unique composition and structure provide decellularized materials priority in tissue engineering. Until now, hardly all kinds of tissues and organs have been successfully decellularized and widely applied in tissue reconstruction and organ replacement. Among various kinds of decellularized tissues, acellular small intestinal submucosa (SIS) and dermal might be two of the most researched acellular tissues. First of all, decellularized SIS was reported to be constituted of type I collagen, type II collagen, type IV collagen, and silk-fibroin [215]. It has been researched in the fields of esophageal reconstruction [215–218], urethral reconstruction [219, 220], bladder repair [49, 221–224], abdominal defect repair [225–227], dura defect repair [228, 229], gastric mucosa repair [230], tendon and ligament repair [231, 232], cardiovascular system repair [215, 233–235], etc. Its products approved by FDA include Oasis®, Surgisis®, CorMatrix® ECM, CuffPatch™, Meso BioMatrix Surgical Mesh, Restore, Stratasis, Surgisis, Durasis, etc. [191, 236]. For acellular dermal, which was mainly composed of elastic and collagen fibers, its applications involve burn treatment [54, 237, 238], breast surgery [42, 239–242], ENT surgery [243, 244], oral and maxillofacial surgery [245, 246], tendon repair [236, 247], abdominal defect repair [54, 248], calvarial defect repair

[249], etc. And, their products mainly include DermaMatrix®, GraftJacket, Epiflex®, Matriderm®, Medeor Matrix, Zimmer Collagen™, Gelfoam, Permacol™, InteXen™, Strattice™, XenMatriX™, Xenform™, Repliform®, SurgiMend™--SurgiMend PRS, FlexHD®, Collamend™, Axis dermis, Axis™, AlloDerm®, AlloPatch, AlloMax™, etc. [191, 250] The clinical applications of decellularized dermal are mainly involved soft tissue defects, breast reconstruction, larynx repair, vaginas repair, ventral hernia repair, covered prostheses, and pelvis and abdominal wall reconstruction, etc. [191, 250].

Because of different difficulties in decellularization and post-process, different tissues are developing at different phases including small animal experiments, big animal experiments, clinical trials, or products. Except for acellular dermal and SIS, some acellular tissues such as acellular mesothelium [251–254], acellular bone matrix (ABM) [249, 255–259], acellular blood vessel [252, 253, 260–262], acellular pericardium [263, 264], acellular valve [265, 266], acellular corneal [29, 267–270], acellular meniscus [271–274], and acellular nerve [40, 275] have been used in clinical applications as products. Besides, the latest researches of acellular trachea [276, 277] and acellular adipose tissue [273, 274] were mainly in the phases of clinical trials. And, decellularized complex tissues such as decellularized esophageal epithelium were still on the phase of large animal experiments [218, 262]. The studies of acellular muscle [278] and acellular cartilage [218] were mainly focused on small animal experiments.

Besides the DMs mentioned above, there are still a number of DMs under study. For example, acellular cochlea and brain matrix have been used in the repair of inner ear and nerve and spinal cord, respectively [61, 279, 280]. In recent years, decellularized urethra and bladder were researched more and more for bladder wall and urethral reconstruction [49, 219, 222]. Furthermore, the researches on decellularized ovary [281–283], uterine [284, 285], fetal membranes [224, 286, 287], placental [288–290], and amniotic membrane [291] were gradually deepen, and have been applied in clinical utility.

Although the studies of these DMs are not sufficient at present, they still have great potential and will play a more and more important role in the area of tissue engineering with the continuous development of decellularized methods and technology in future. The characteristics and the applications in repairing all kinds of soft and hard tissue of different decellularized tissues will be elaborated in character 4.

DMs derived from different tissues could be processed into different forms including sheet, tubular, powder, hydrogels, and 3D bio printing oink. Exactly, there are mainly six methods of their applications, including using DMs directly as scaffold, DMs powder, DMs gels, DMs solutions for coating, DMs 3D printing bio ink, DMs-based electrospinning, as shown in Fig. 1.14.

Firstly, DMs in sheet and tubular were widely used as biological patches or tubular scaffolds to repair the defects in sheet or tubular, including cornea [292], skin wounds [293], vessels [294], intestinal wall [229, 295], and other tissues. However, because the performances of DMs were damaged during preparation to some extent, it usually needs a further processing for these DMs to improve their mechanical properties, biocompatibility, etc. Specially, the sheet-like DMs could

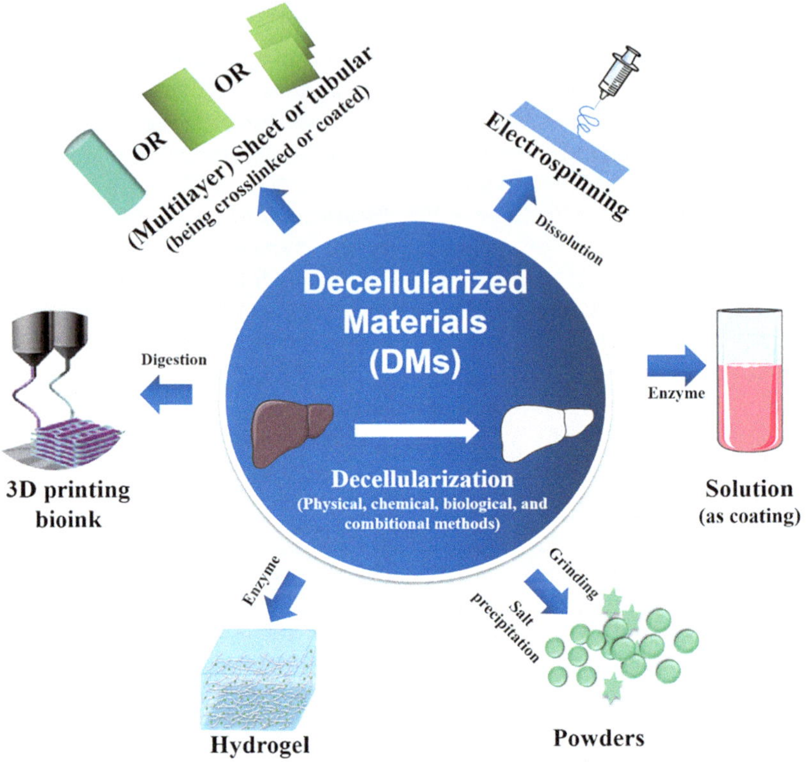

Fig. 1.14 The main six methods of DMs applications, including being used directly as (multilayer) sheet or tubular scaffold, powder, gels, solutions for coating, 3D printing bio ink, and DMs-based electrospinning [27]

also enhance mechanical properties by stacking multiple layers to meet the requirements. Usually, constructing coating such as CNT coating [296], chitosan/nanohydroxyapatite coating [297], HG-VEGF coating [298], fucoidan/VEGF PEM coating [299] was the common method. Specifically, Marinval et al. coated fucoidan/VEGF on decellularized pulmonary heart valve and significantly improved the antithrombotic and re-endothelialization potential of bioprostheses (Fig. 1.15). In addition, Yang et al. prepared hybrid small-diameter vascular grafts (HTEV) and RM-loaded hybrid tissue-engineered vascular graft (RM-HTEV) by constructing electrospinning polycaprolactone (PCL) and PCL nanofibers blended with rapamycin (RM) coating on decellularized rat aorta (DRA) (Fig. 1.16a), respectively, and they found that the mechanical properties of HTEV and RM-HTEV were distinctly stronger than DRA (Fig. 1.16b) [300].

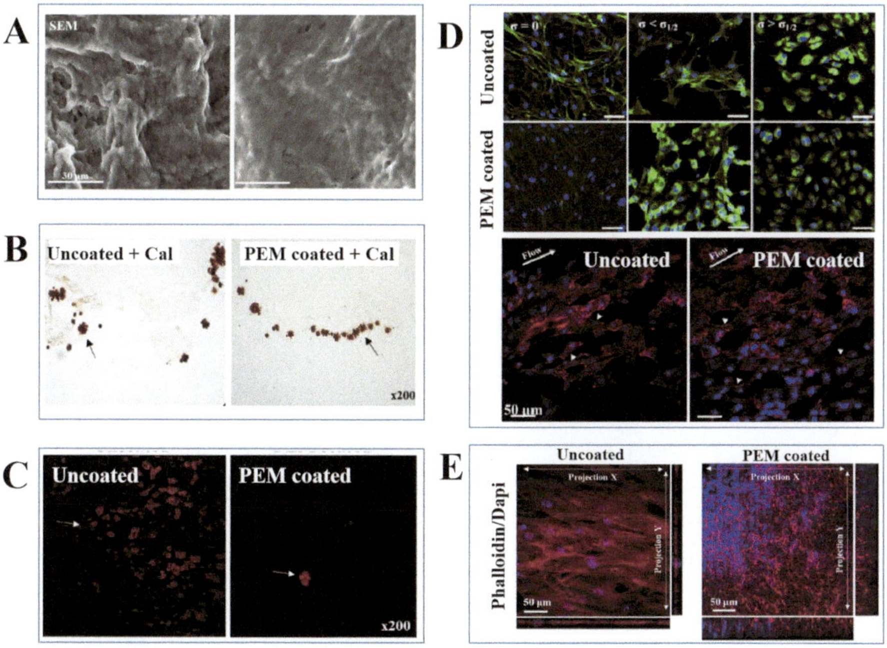

Fig. 1.15 One study example of constructing fucoidan/VEGF PEM coating to improve the antithrombotic and re-endothelialization potential of bioprostheses decellularized pulmonary heart valve. (**a**) SEM images of uncoated and PEM coated valve scaffold under; (**b**) Calcification assay of decellularized valve scaffolds coated or not with the PEM, the arrows showed the calcification. (**c**) Luminal-activated platelet adhesion and aggregation to the valve surface after flow observed in fluorescent microscopy. The arrows showed the adherent activated platelets (red). (**d**) Cell morphology and density before ($\sigma = 0$) and after flow at different values of shear stress: above $\sigma^{1/2}$ value ($\sigma > \sigma^{1/2}$); and flow below $\sigma^{1/2}$ ($\sigma < \sigma^{1/2}$). And, HUVECs morphology and their alignment toward the flow direction (arrows when whole valve scaffold was seeded with HUVECs in bioreactor under physiological flow and cultured for 6 h). (**e**) Cell morphology was observed with phalloidin/dapi staining by confocal microscopy. Data showed the fucoidan/VEGF coating on decellularized pulmonary heart valve significantly improved the antithrombotic and re-endothelialization potential of bioprostheses [299]

Secondly, DMs powders prepared through salt precipitation or grinding methods were applied as bioactive substances to combine with other biomaterials constructing composites [301]. For instance, we added the SIS powders into PU matrix to prepare PU/SIS composites (Fig. 1.16a), and found that the angiogenesis, cell infiltration, and tissue regeneration of PU/SIS composites were significantly prior than the pure PU scaffolds [302]. Likewise, Zhang et al. added decellularized human amniotic powders into GelMA hydrogels to fabricate composite hydrogel (Fig. 1.17b), in other words, the GelMA hydrogels played the role of loading DMs powders and delaying their release, and they demonstrated the ameliorative repair effect of oral mucosal defects [303]. Besides, there were some researches about suspending DMs powders in liquid to form turbid liquid, which could be injected

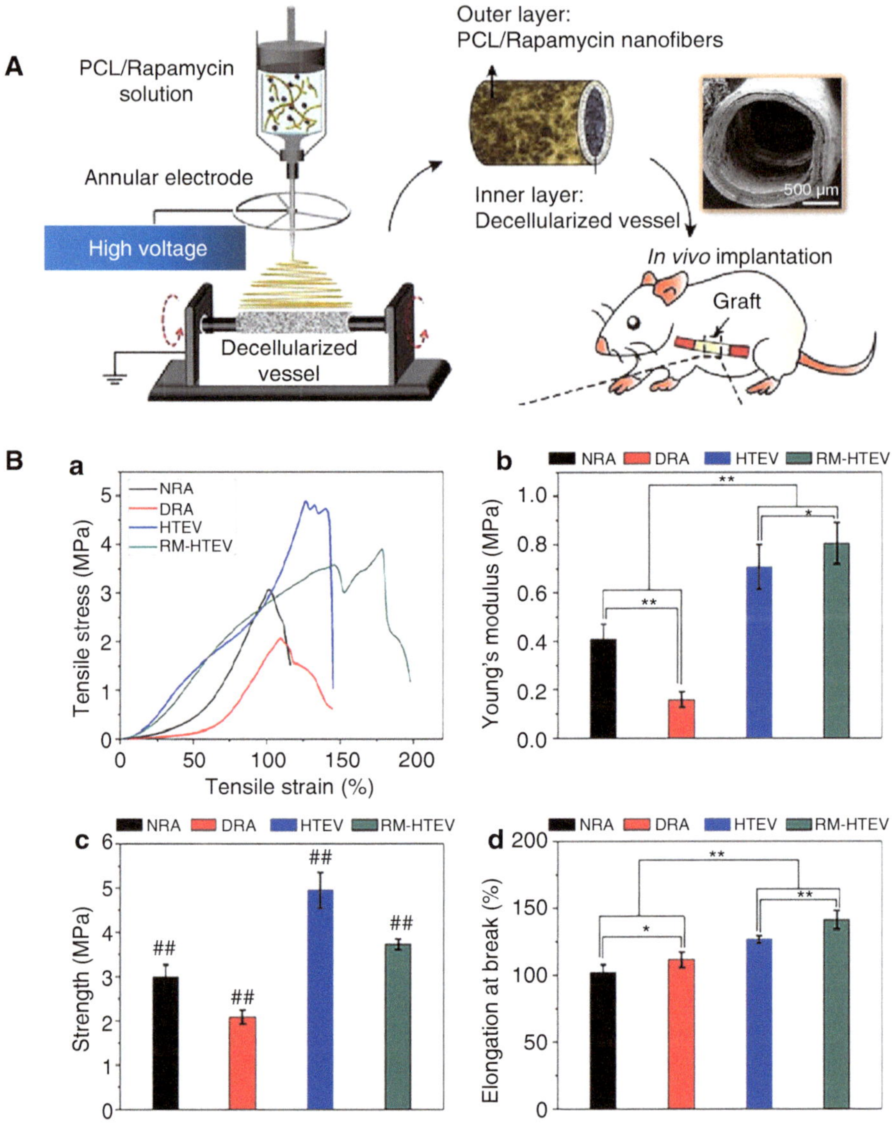

Fig. 1.16 One study example of constructing electrospinning fiber coating to improve the mechanical properties of DMs, showing that hybrid small-diameter vascular grafts (HTEV) and RM-loaded hybrid tissue-engineered vascular graft (RM-HTEV) fabricated by coating electrospinning polycaprolactone (PCL) and electrospinning PCL nanofibers blended with rapamycin (RM) on decellularized rat aorta (DRA), respectively (c1), exhibited stronger mechanical properties including elongation at breakage, tensile strength and Young's moduli than those of DRA [300]

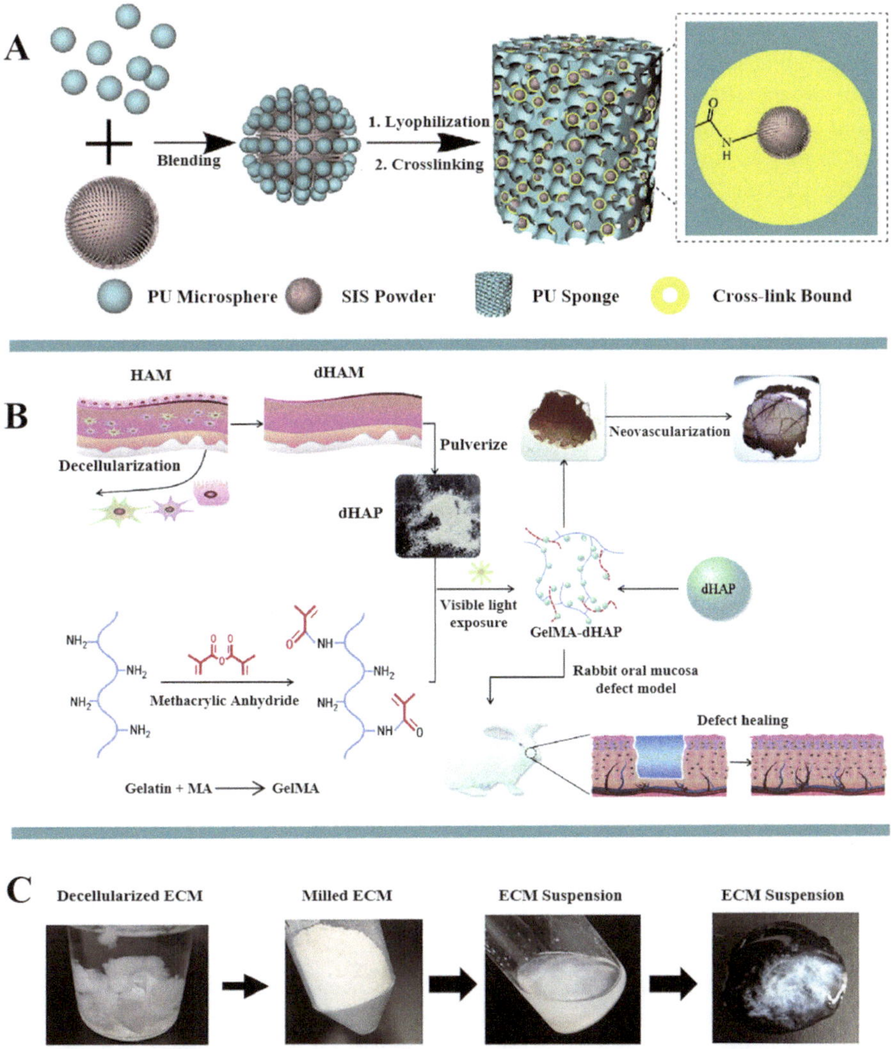

Fig. 1.17 Application methods of DMs including being used as bioactive components mixing with other biomaterials (**a**) [302], being loaded with hydrogel (**b**) [303], and being processed into suspensions (**c**) [311]

into in injury sites to help wound healing and tissue remodeling (Fig. 1.17c). For example, the suspensions composed of human DMs powders and human adipose-derived stem cells (hASCs) had been demonstrated to accelerate the adipogenesis and tissue reconstruction when they were subcutaneously injected into nude mice [304]. Currently, many kinds of DMs derived from diverse tissues, such as human lipoaspirate [305], liver tissues [306], cardiac tissue [307], etc., have been successfully applied in the preclinical models. Furthermore, there are some products made by DMs powders, which mainly include demineralized bone matrix [308], bladder matrix (MatriStem) [309], micronized acellular dermis (micronized AlloDerm) [310], etc. The advantage of DMs powders compared to DMs hydrogels might be their relatively complete remaining of natural ECM ingredients. However, it must be taken into consideration that their properties would be greatly influenced by the particle conformation (e.g., porous, fiber-like or sheet-like particles) and powder concentration, which must be well controlled during application.

Thirdly, DMs gels were prepared through two steps: firstly, DMs powders were solubilized into protein monomeric components by "Freytes method" [197] or "Voytik-Harbin method" [312]; then, DMs solution could be crosslinked into hydrogel through regulating temperature and/or pH neutralization to accelerate spontaneous bonding of intramolecular bonds in monomeric components. Up to now, many animals, including cow, goat, rat, pig, and human cadavers have been used as tissue sources for DMs gels [312, 313]. Because of their complex nanoscale mesh of fiber, DMs gels could overcome the decrease of physical strength during decellularization and shortcomings of microstructure disintegration to some extents [314]. DMs gels have been applied in the repair of various tissues and showed a prior biofunction compared with other hydrogels [315–317]. In addition, compared with suspensions of DMs powders, DMs hydrogels possessed greater ease and performed more homogenous concentration, which might help to realize a better tissue regeneration effect. Moreover, by modulating gelation parameters including concentration, pH, temperature, and ionic strength, researchers could control the gelation kinetics, nanoscale architecture, and mechanical properties [318]. Except for the hydrogels only containing DMs, DMs could be also used to construct composite hydrogel scaffolds acting as cell-supportive matrix to promote cell viability [61, 212]. Above all, because the minimally invasive treatment using injectable hydrogels could decrease patients' secondary injury, it has been one of the most famous clinical repair methods, promoting the development of DMs injectable hydrogels together with their characteristics of variable shapes for unique applications [319].

The applications of DMs injectable hydrogels were showed in Fig. 1.18.

Fourthly, DMs solutions could be applied as coating for tissue engineering scaffolds, and have exhibited prominent improvement on cell survival, adhesion, proliferation, migration, and differentiation, as well as alleviation of host foreign body and chronic inflammatory response [320–322]. Currently, the applications of DMs as a scaffold coating in cell 2D culture have been evaluated in many studies [323]. For instance, the caprine liver extracellular matrix (CLECM) coating constructed by Agarwal et al. was used as hepatocyte 2D culture environment and significantly promoted the synthesis of urea, GAGs, albumin, and glycogen of HepG2 cells compared to collagen-coated surface [316] (Fig. 1.19). In addition, liver DMs coating has also been demonstrated to obviously improved the differentiation of adipose-derived mesenchymal stem cells (ADSCs) to hepatic cells compared to pure collagen coating, fibronectin coating, and matrigel coating [213]. Furthermore, Kim et al. fabricated polycaprolactone (PCL)/beta-tricalcium

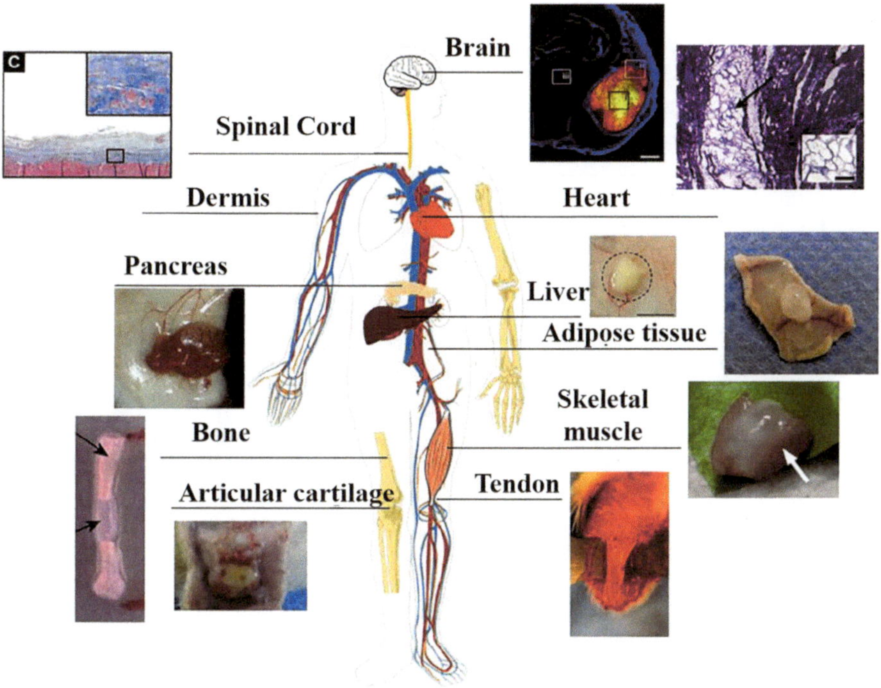

Fig. 1.18 DMs hydrogels injected for therapeutic applications throughout the body [311]

phosphate (β-TCP) scaffold using 3D printing technology /bone with decellularized extracellular matrix (dECM) coating, results of which showed that PCL/β-TCP/bone dECM scaffolds presented outstanding in vitro cell seeding efficiency, proliferation, and early and late osteogenic differentiation capacity, as well as excellent results of in vivo bone regeneration in the rabbit calvarial defect model. In a word, these studies showed huge potential of DM as a substrate coating material [324].

Fifthly, DMs have been applied in 3D bioprinting as bio ink. Up to now, diverse tissues and organs, such as vascular tissue, skin, adipose, skeletal muscle, cartilage, heart, and liver, have been fabricated as DMs bio inks, which exhibited outstanding abilities to improve and regulate cellular functions of induced pluripotent stem cell (iPSC)-derived differentiated cells, specific cell lines, or primary cells [325, 326]. However, the species-specific differences in the composition of the DMs could affect the mechanical, biological, and biochemical properties of bio inks. Choudhury et al. summarized a large number of unique bioprinting methods for printing tissue constructs by using decellularized extracellular matrix (dECM) bioinks [22].

Specially, it must be taken into consideration that the recellularization in vitro before implantation might be effective. Especially, for the whole organ replacement, decellularized organ must be recellularized to form the composition, structure, and

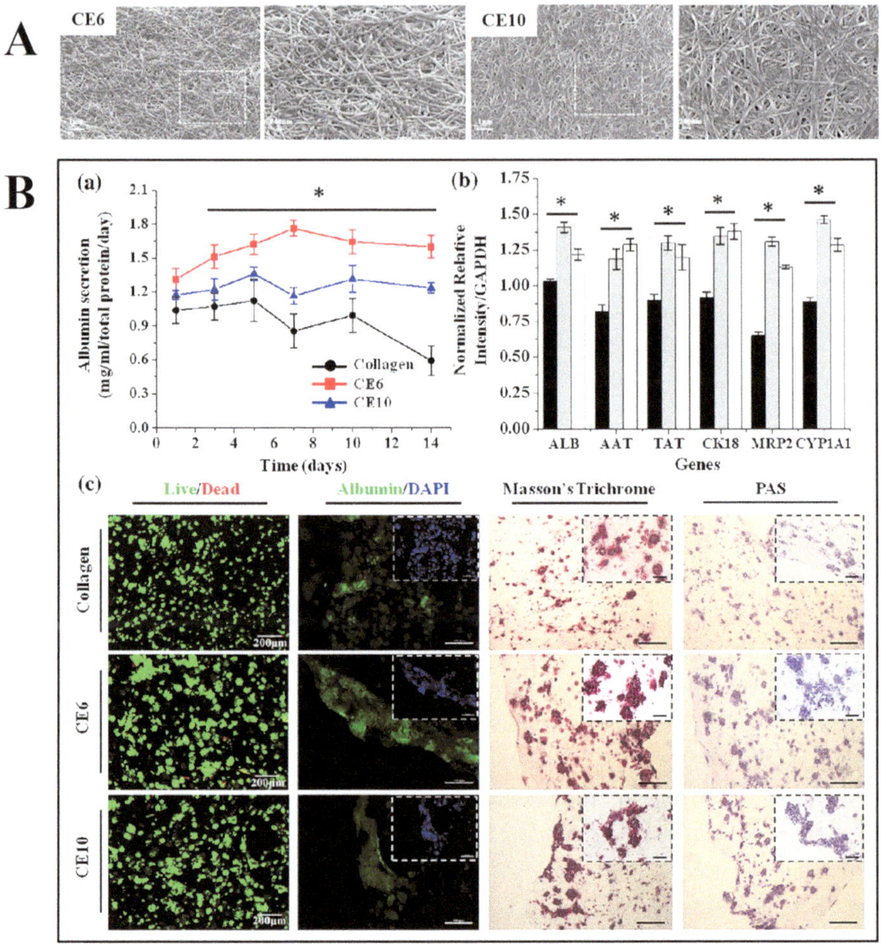

Fig. 1.19 One study example of the application of DMs coating. (**a**) Morphological characterization of caprine liver extracellular matrix (CLECM) hydrogels. A dense networking of randomly arranged collagen bundles could be easily seen. (**b**) Caprine liver extracellular matrix 3D hydrogels for hepatocyte culture. (a) Albumin secretion by hepatocytes, (b) Semi-qRT-PCR-based quantification of gene expression profiles of the hepatocytes, and (c) Calcein-AM/ethidium-based liver dead staining; immunohistological staining of albumin and nucleus; histological evaluation using Masson's trichrome and PAS staining of the hepatocytes cultured in collagen, CE6, and CE10 hydrogels. Data showed that HepG2 cells cultured on CLECM-coated surface presented higher glycogen, albumin, GAGs, and urea synthesis compared to cells on the collagen-coated surface, indicating the outstanding bioactivity of DMs coating [316]

function similar to nature organ before implantation [327]. Although decellularized material is not as easy as artificial scaffold to implant cells in to some extent, there have been quite extensive researches in this area [327–330]. Meanwhile, the related clinical applications have been constantly explored [236]. DM experiencing a process of recellularization performed a better effect of repairing defect compared with the pure DM. Mohsina et al. compared the efficacy in repairing abdominal wall defects of rats between the scaffolds made by decellularized rabbit skin with and without primary mouse embryonic fibroblasts (p-MEF). Results showed that the group repaired with p-MEF seeded ADM had less adhesion and weaker immune response of the skin and abdominal organs than that of the group repaired only with ADM [248]. Parallel to that, researchers also attempted to re-cell for some special tissues, such as lymphomas [331], however, the function of corresponding tissue is still needed to be studied. In a word, there are still some problems to be solved for the recellularization of DM, and of which the seeded cell is the most fundamental and critical factor. Its type, source, number, arrangement, and distribution have a significant impact on recellularization, including function generation, structural maintenance, revascularization, and so on.

The preparation methods and the characteristics of decellularized tissues in different forms, the applications of decellularized tissues in different forms in tissue repair, as well as the efforts in performance optimization will be comprehensively introduced in Chap. 3.

1.7 Decellularized Organs and Their Recellularization

In the past few decades, with the development of serious phenomenon of global population aging, which means that the life span of healthy tissues and organs would gradually become longer than they could carry, the applications of medical implants and the cost of treating diseases are increasing. Statistically, over 123,000 patients needed organ transplant in May 2015 in the USA alone, and 6885 patients died in 2014 because they had not waiting for suitable organ transplants [332]. And, the expenses for treating diseases and other problems derived from loss of tissue function at present surpassed US $39 billion in North America alone [333]. Hence, the transplantation of organs has become a serious problem to be solved.

Allograft transplantation is the traditional method for organ replacement, and has been successful with satisfactory peri-operative survival rates [334]. However, there are still some serious problems. First of all, the matching process is a demanding and unsatisfying matter, which often requires a long time to wait for the right donor organ and might cause the patient's illness not be treated in time, reducing the patient's recovery rate or even leading to the patient's death. Secondly, even though there is an organ matching before operation, the possible serious immune rejections caused by the transplanted organ might lead to failure. Thirdly, the limited organ source would make organ transplantation expensive and even cause a series of social problems like selling organs in black markets. Therefore, developing artificial organs to obtain more organ transplants and reducing the immunogenicity by using suitable materials and advanced technologies has been a challenging but promising direction.

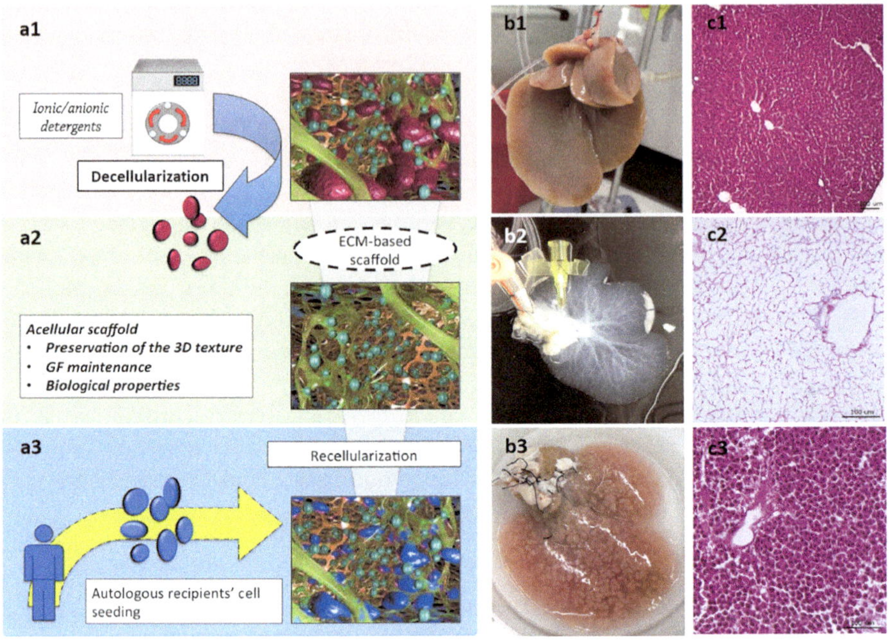

Fig. 1.20 Concept of decellularization and recellularization (an example is shown for the liver). (a1–a3) Schematic representation of decellularization and recellularization. (b1) Native, (b2) decellularized, and (b3) recellularized rat liver. H&E staining of (c1) native rat liver, (c2) decellularized rat liver, (c3) recellularized rat liver [338, 339]

Some methods have been applied for constructing artificial organs including directed self-assembly [335] that could be achieved through chemical or genetic engineering and rapid prototyping [336], which could achieve precise spatial control of materials, growth factors, and cells according to specific requirements. However, because the complex composition and spatial structure of organs, it is difficult to successfully develop a functional artificial organ. As a result, acellular organs exhibited big priority in organ replacement because they reserved the highly bionic macro and microstructures and abundant bioactive substances, which might be hopeful to reconstruct functional organs. Especially, the native organ vascular system including the capillary have been demonstrated to be preserved after perfusion-based decellularization [187, 337], which would help to achieve the function of recellularized organs.

Acellular organs keeping the similarly native organ-specific structure were prepared through decellularizing natural organs, during which all the native cells were removed from the ECM framework by using chemical regents (ionic and anionic detergents) with different concentrations and timings, as is shown in Fig. 1.20a1. They needed to be completely analyzed to check the effective reservation of the original texture and the growth factors, and to study their biological properties, as is shown in Fig. 1.20a2. After that, organ-specific cells would be seeded on acellular

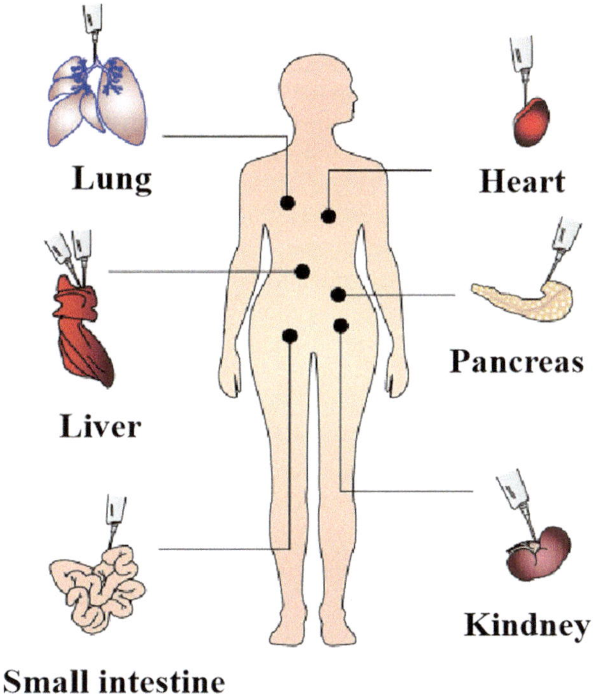

Fig. 1.21 The organs that have been decellularized and recellularized, mainly including lung, heart, liver, pancreas, small intestine, and kidney

organs to recellularize (Fig. 1.20a3), and these cells might be autologous cells to avoid immunological responses. An example of the process from natural liver to recellularized liver is showed in Fig. 1.20b1–b3, and the histological differences could be seen from Fig. 1.20c1–c3.

Since Ott et al. [340] successfully decellularized rat whole heart by using coronary perfusion with detergents, and recellularized with neonatal rodent cardiac cells to obtain artificial heart with functions of macroscopic contractions and rudimental pump in 2008, which means the feasibility of acellular organs for organ replacement, numerous studies have been conducted to develop more types of organs. Up to date, many types of organs including lung [341–343], heart [144, 340, 344, 345], liver [337, 346–348], pancreas [287], small intestine [349], kidney [350–352], etc., have been successfully decellularized and tried to recellularize to reconstruct functional organs (Fig. 1.21). For example, the decellularized kidney prepared by Nakayama et al. exhibited semblable macro and microstructures, mechanical and physiological properties to those of natural kidney [351]. Besides, other decellularized organs such as decellularized heart, intestinal, kidney, lung, and liver, have also been tested in some small rodent models, and performed semblable characteristics and functions to those of native organs [297]. For example, Mirmalek et al. created a pancreas acellular scaffold via using a detergent-based infusion technique, and subsequently seeded pancreatic islets on the scaffold. The results showed that the reconstructed pancreas could obviously secret insulin, exhibiting excellent pancreas function [353]. At the same time, Baptista et al. used human progenitor cells to recellularize the acellular liver scaffolds and eventually constructed a functional humanized rat liver with hepatic characteristics including clusters of hepatocytes, biliary duct structures, and hepatic functions, such as secreting urea and albumin, expressing endothelial cell nitric oxide synthase, and metabolizing drugs [354]. Recently, a study conducted by Tanya et al. [355] indicated that the acellular rat liver could be effectively recellularized for whole organ regeneration. They used 0.1% SDS (Sodium Dodecyl Sulfate) for decellularization and used primary hepatocytes for recellularization. The recellularized liver scaffold could product albumin and urea, express Cytokeratin-19 (CK-19), Glucose 6-Ph osphatase (G6P), Gamma Glutamyl Transferase (GGT) genes.

It is worth of noting that the vascular structure of native organ reserved during decellularization and the generation of complete vascular system with endothelial cells during recellularization are quite important for preventing the thrombogenicity induced because of the exposed vascular ECM and possible loss of the neo-organ graft. Some studies have reported breakthrough in this aspect. For example, Song et al. transplanted the appropriate epithelial and endothelial cells into rat acellular renal matrix via ureter and renal artery and found the formation of vascular endothelial and tubular epithelial-like structure after 3-days culture in bioreactor [356]. Recently, Aram et al. decellularized and assessed ovine hearts through coronary perfusion and transplanted a decellular graft heterotopically into the omental wrap to evaluate in situ recellularization. Angiography and blood circulation indicated an intact vascular network, and the implantation led to proper vascularization, showing the potency of recellularization for the functional organ in the future [357]. However, it is quite regretful that none of the re-endothelialization studies currently published has presented the full coverage of the vascular framework from the major branches to the capillaries. As a result, other measures such as using different covering agents and antithrombotic chemical crosslinking have been described to cover residual exposed ECM areas [358, 359].

Even though substantial achievements in the field of organ decellularization and the subsequent recellularization have been reported in the last decades, the change of this technique from an experimental method to a clinically relevant treatment is still a major obstacle to overcome. Firstly, the cell source remains the greatest challenge. On the one hand, few of the previously published researches on recellularization produced a relevant cell mass in regard to the model because of the large amounts of cells needed to occupy the whole volume of the acellular scaffold and the difficulty to remain specific cell type proportions to generate a physiologically functional organ. On the other hand, some other organs, such as heart and pancreas, autologous cells are often insufficient and difficult to collect, which limits their clinical applications. Therefore, some researchers turn attentions to allogeneic cells, which have the characteristics of wide source, large quantity, and easy access, however, have a risk of spreading the disease, and may lead to an immune response. Secondly, the method of recellularization also needs improvement. The most common method of recellularization currently used is to introduce dense cell suspensions into the scaffold by magnetic, dynamic, electrostatic, vacuum, and centrifugal sowing, which clearly lacks precise control of the process of recellularization. Thirdly, a more suitable culture system that allows the implanted cells to survive for at least a few weeks to provide enough time for them to proliferate and differentiate is needed for the recellularization process.

1.8 The Structure and Main Content of this Book

All in all, with the development of medical technology, higher and newer requirements have been put forward for biomaterials used in tissue engineering that they should not only have good biocompatibility and bioactivity, but also have effects similar to that of extracellular matrix (ECM). Decellularized material (DM) sticks out in biomaterials because it maintains highly similar composition and structure to ECM. DMs mainly derived from natural tissues or organs are attracting more and more attentions because of their native structure, relatively high bioactivity, low immunogenicity, and good biodegradability, which are difficult or even impossible to be imitated by synthetic materials. They overcome some shortcomings of self-transplantation and other artificial tissue engineering scaffold materials to a certain extent. Hence, they are hopeful to play an increasingly crucial role in tissue repair and organ replacement. Simultaneously, DMs themselves have some inevitable deficiencies, such as uncontrollable degradation, insufficient mechanical properties, etc. Further studies are needed for the deep development of DMs. In order to provide a relatively comprehensive understanding on the characteristics of DMs and their research status, we wrote this book based on our decades of studies and extensive inquiries and research of the related references. Specifically, this book includes other 3 parts or 7 chapters besides the first one as follows:

Part I: The content of decellularized tissues for tissue repair will be introduced.

Chapter 2: The traditional and emerging decellularization methods, the effective decellularization index, and the subsequent cleaning and sterilization will be introduced; the effect of inefficient/ineffective decellularization, and the effect of decellularization process on the performance of decellularized tissues will be discussed in this chapter.

Chapter 3: The preparation methods and the characteristics of decellularized tissues in different forms will be elaborated; the applications of decellularized tissues in different forms in tissue repair, as well as the efforts in performance optimization will be introduced.

Chapter 4: The characteristics of decellularized tissues from different tissues will be elaborated; the applications of different decellularized tissues in repairing all kinds of soft and hard tissue will be expatiated.

Part II: The content of decellularized organs for organ replacement will be introduced.

Chapter 5: The decellularization methods of different organs, the effective decellularization index, and the decellularization equipment will be introduced.

Chapter 6: The recellularization methods of all kinds of organs and the bio-reactors for whole organ recellularization will be set forth; the cell types for parenchymal recellularization and the vasculature of decellularized organs will be discussed.

Chapter 7: The applications of decellularized organs in organ replacement will be stated; the effect of organ reconstruction compared to other types of organ transplant will be discussed.

Part III

Chapter 8: The challenges faced by decellularized materials will be stated; The potential directions will be discussed.

All in all, this book provides a review of the present understanding of the decellularized materials, mainly including the physical and chemical properties of DMs as well as the degradation and the effects of decellularized materials on promoting tissue repair/organ replacement, the preparation and corresponding decellularization criterion as well as the subsequent cleaning and sterilization of decellularized materials, the preparation methods and the characteristics of decellularized tissues in different forms as well as their applications, the characteristics of decellularized tissues from different tissues and their applications, the decellularization of different organs, the recellularization of acellular organs and their applications, and the challenges and the potential directions of decellularized materials. It provides a comprehensive and accurate summary of the recent related study progress besides primary knowledge and provides a valuable reference for the future development of this field. Moreover, it may provide a general guide or the design and fabrication of biomaterials and regenerative medicine.

References

1. Gurtner GC, Callaghan MJ, Longaker MT. Progress and potential for regenerative medicine. Annu Rev Med. 2007;58:299–312.
2. Place ES, Evans ND, Stevens MM. Complexity in biomaterials for tissue engineering. Nat Mater. 2009;8:457–70.
3. Langer R, Vacanti JP. Tissue engineering. Science. 1993;260:920–6.
4. Feinberg AW. Engineered tissue grafts: opportunities and challenges in regenerative medicine. Wiley Interdiscip Rev. 2012;4:207–20.
5. Hollister SJ. Porous scaffold design for tissue engineering. Nat Mater. 2005;4:518–24.
6. Dutta RC, Dutta AK. Cell-interactive 3D-scaffold; advances and applications. Biotechnol Adv. 2009;27:334–9.
7. Kretlow JD, Young S, Klouda L, Wong M, Mikos AG. Injectable biomaterials for regenerating complex craniofacial tissues. Adv Mater. 2009;21:3368–93.
8. Franz S, Rammelt S, Scharnweber D, Simon JC. Immune responses to implants - a review of the implications for the design of immunomodulatory biomaterials. Biomaterials. 2011;32:6692–709.

9. Tabata Y. Biomaterial technology for tissue engineering applications. J R Soc Interface. 2009;6:S311–24.
10. Chen FM, Zhang J, Zhang M, An Y, Chen F, Wu ZF. A review on endogenous regenerative technology in periodontal regenerative medicine. Biomaterials. 2010;31:7892–927.
11. Chen FM, Zhang M, Wu ZF. Toward delivery of multiple growth factors in tissue engineering. Biomaterials. 2010;31:6279–308.
12. Chen FM, An Y, Zhang R, Zhang M. New insights into and novel applications of release technology for periodontal reconstructive therapies. J Control Release. 2011;149:92–110.
13. Chen FM, Sun HH, Lu H, Yu Q. Stem cell-delivery therapeutics for periodontal tissue regeneration. Biomaterials. 2012;33:6320–44.
14. Chen F-M, Liu X. Advancing biomaterials of human origin for tissue engineering. Prog Polym Sci. 2016;53:86–168.
15. Halper J, Kjaer M. Basic components of connective tissues and extracellular matrix: elastin, fibrillin, fibulins, fibrinogen, fibronectin, laminin, tenascins and thrombospondins. In: Halper J, editor. Progress in heritable soft connective tissue diseases. Dordrecht: Springer; 2014. p. 31–47.
16. Hubmacher D, Apte SS. The biology of the extracellular matrix: novel insights. Curr Opin Rheumatol. 2013;25:65–70.
17. Lane SW, Williams DA, Watt FM. Modulating the stem cell niche for tissue regeneration. Nat Biotechnol. 2014;32:795–803.
18. Multhaupt HAB, Leitinger B, Gullberg D, Couchman JR. Extracellular matrix component signaling in cancer. Adv Drug Deliv Rev. 2016;97:28–40.
19. Gattazzo F, Urciuolo A, Bonaldo P. Extracellular matrix: a dynamic microenvironment for stem cell niche. Biochim Biophys Acta. 2014;1840:2506–19.
20. Iozzo RV, Schaefer L. Proteoglycan form and function: a comprehensive nomenclature of proteoglycans. Matrix Biol. 2015;42:11–55.
21. Roskelley CD, Srebrow A, Bissell MJ. A hierarchy of ECM-mediated signalling regulates tissue-specific gene expression. Curr Opin Cell Biol. 1995;7:736–47.
22. Choudhury D, Tun HW, Wang T, Naing MW. Organ-derived decellularized extracellular matrix: a game changer for bioink manufacturing? Trends Biotechnol. 2018;36:787–805.
23. Nicolas J, Magli S, Rabbachin L, Sampaolesi S, Nicotra F, Russo L. 3D extracellular matrix mimics: fundamental concepts and role of materials chemistry to influence stem cell fate. Biomacromolecules. 2020;21:1968–94.
24. Viswanathan P, Chirasatitsin S, Ngamkham K, Engler AJ, Battaglia G. Cell instructive microporous scaffolds through interface engineering. J Am Chem Soc. 2012;134:20103–9.
25. Jacob S, Dunphy JE. Transplantation of tissues. Philadelphia: Williams & Wilkins; 1955.
26. Gaffney L, Wrona EA, Freytes DO. Potential synergistic effects of stem cells and extracellular matrix scaffolds. ACS Biomater Sci Eng. 2018;4:1208–22.
27. Liao JXB, Zhang RH, Fan YB, Xie HQ, Li XM. Applications of decellularized materials in tissue engineering: advantages, drawbacks and current improvements, and future perspectives. J Mater Chem B. 2020; https://doi.org/10.1039/D0TB01534B.
28. Grillo HC, McKhann CF. The acceptance and evolution of dermal homografts freed of viable cells. Transplantation. 1964;2:48–59.
29. Sasaki S, Funamoto S, Hashimoto Y, Kimura T, Honda T, Hattori S, Kobayashi H, Kishida A, Mochizuki M. In vivo evaluation of a novel scaffold for artificial corneas prepared by using ultrahigh hydrostatic pressure to decellularize porcine corneas. Mol Vis. 2009;15:2022–8.
30. Funamoto S, Nam K, Kimura T, Murakoshi A, Hashimoto Y, Niwaya K, Kitamura S, Fujisato T, Kishida A. The use of high-hydrostatic pressure treatment to decellularize blood vessels. Biomaterials. 2010;31:3590–5.
31. Crapo PM, Gilbert TW, Badylak SF. An overview of tissue and whole organ decellularization processes. Biomaterials. 2011;32:3233–43.
32. Sawada K, Terada D, Yamaoka T, Kitamura S, Fujisato T. Cell removal with supercritical carbon dioxide for acellular artificial tissue. J Chem Technol Biotechnol. 2008;83:943–9.

33. Patel N, Solanki E, Picciani R, Cavett V, Caldwell-Busby JA, Bhattacharya SK. Strategies to recover proteins from ocular tissues for proteomics. Proteomics. 2008;8:1055–70.
34. Meezan E, Hjelle JT, Brendel K, Carlson EC. A simple, versatile, nondisruptive method for the isolation of morphologically and chemically pure basement membranes from several tissues. Life Sci. 1975;17:1721–32.
35. Yi S, Ding F, Gong L, Gu X. Extracellular matrix scaffolds for tissue engineering and regenerative medicine. Curr Stem Cell Res Ther. 2017;12:233–46.
36. Meyer SR, Chiu B, Churchill TA, Zhu L, Lakey JR, Ross DB. Comparison of aortic valve allograft decellularization techniques in the rat. J Biomed Mater Res A. 2006;79:254–62.
37. Hoshiba T, Lu H, Yamada T, Kawazoe N, Tateishi T, Chen G. Effects of extracellular matrices derived from different cell sources on chondrocyte functions. Biotechnol Prog. 2011;27:788–95.
38. Hudson TW, Liu SY, Schmidt CE. Engineering an improved acellular nerve graft via optimized chemical processing. Tissue Eng. 2004;10:1346–58.
39. Singh K, Gopinathan A, Sangeetha P, Kumar N, Singh KP, Raina OK. Development and clinical application of decellularized porcine SIS and cornea for the repair of corneal defects in animals. Indian J Anim Sci. 2016;86:1391–5.
40. Sondell M, Lundborg G, Kanje M. Regeneration of the rat sciatic nerve into allografts made acellular through chemical extraction. Brain Res. 1998;795:44–54.
41. Ribatti D, Conconi MT, Nico B, Baiguera S, Corsi P, Parnigotto PP, Nussdorfer GG. Angiogenic response induced by acellular brain scaffolds grafted onto the chick embryo chorioallantoic membrane. Brain Res. 2003;989:9–15.
42. Conklin BS, Richter ER, Kreutziger KL, Zhong DS, Chen C. Development and evaluation of a novel decellularized vascular xenograft. Med Eng Phys. 2002;24:173–83.
43. Nakamura N, Kimura T, Kishida A. Overview of the development, applications, and future perspectives of decellularized tissues and organs. ACS Biomater Sci Eng. 2016;3:1236–44.
44. Lumpkins SB, Pierre N, McFetridge PS. A mechanical evaluation of three decellularization methods in the design of a xenogeneic scaffold for tissue engineering the temporomandibular joint disc. Acta Biomater. 2008;4:808–16.
45. Reing JE, Brown BN, Daly KA, Freund JM, Gilbert TW, Hsiong SX, Huber A, Kullas KE, Tottey S, Wolf MT, Badylak SF. The effects of processing methods upon mechanical and biologic properties of porcine dermal extracellular matrix scaffolds. Biomaterials. 2010;31:8626–33.
46. Alhamdani MSS, Schroder C, Werner J, Giese N, Bauer A, Hoheisel JD. Single-step procedure for the isolation of proteins at near-native conditions from mammalian tissue for proteomic analysis on antibody microarrays. J Proteome Res. 2010;9:963–71.
47. Brooker JE, Camison LB, Bykowski MR, Hurley ET, Yerneni SS, Campbell PG, Weiss LE, Mooney MP, Cray J, Gilbert JR, Cooper GM, Losee JE. Reconstruction of a calvarial wound complicated by infection: comparing the effects of biopatterned bone morphogenetic protein 2 and vascular endothelial growth factor. J Craniofac Surg. 2019;30:260–4.
48. Yang B, Zhang Y, Zhou L, Sun Z, Zheng J, Chen Y, Dai Y. Development of a porcine bladder acellular matrix with well-preserved extracellular bioactive factors for tissue engineering. Tissue Eng Part C Methods. 2010;16:1201–11.
49. Yoo JJ, Meng J, Oberpenning F, Atala A. Bladder augmentation using allogenic bladder submucosa seeded with cells. Urology. 1998;51:221–5.
50. Jiang T. Development of acellular matrix stromal scaffolds. Chin J Spine Spinal Cord. 2010;2010:782–5.
51. Xiang JX, Zheng XL, Gao R, Wu WQ, Zhu XL, Li JH, Lv Y. Liver regeneration using decellularized splenic scaffold: a novel approach in tissue engineering. Hepatobiliary Pancreat Dis Int. 2015;14:502–8.
52. Pati F, Song TH, Rijal G, Jang J, Kim SW, Cho DW. Ornamenting 3D printed scaffolds with cell-laid extracellular matrix for bone tissue regeneration. Biomaterials. 2015;37:230–41.

53. Jiang WC, Cheng YH, Yen MH, Chang Y, Yang VW, Lee OK. Cryo-chemical decellularization of the whole liver for mesenchymal stem cells-based functional hepatic tissue engineering. Biomaterials. 2014;35:3607–17.
54. Wainwright DJ. Use of an acellular allograft dermal matrix (AlloDerm) in the management of full-thickness burns. Burns. 1995;21:243–8.
55. Sano MB, Neal RE, Garcia PA, Gerber D, Robertson J, Davalos RV. Towards the creation of decellularized organ constructs using irreversible electroporation and active mechanical perfusion. Biomed Eng Online. 2010;9:83.
56. Sarig U, Au-Yeung GCT, Wang Y, Bronshtein T, Dahan N, Boey FYC, Venkatraman SS, Machluf M. Thick acellular heart extracellular matrix with inherent vasculature: a potential platform for myocardial tissue regeneration. Tissue Eng A. 2012;18:2125–37.
57. Ozlu B, Ergin M, Budak S, Tunali S, Yildirim N, Erisken C. A bioartificial rat heart tissue: perfusion decellularization and characterization. Int J Artif Organs. 2019;42:757–64.
58. Badylak SF. Decellularized allogeneic and xenogeneic tissue as a bioscaffold for regenerative medicine: factors that influence the host response. Ann Biomed Eng. 2014;42:1517–27.
59. Caralt M, Uzarski JS, Iacob S, Obergfell KP, Berg N, Bijonowski BM, Kiefer KM, Ward HH, Wandinger-Ness A, Miller WM, Zhang ZJ, Abecassis MM, Wertheim JA. Optimization and critical evaluation of decellularization strategies to develop renal extracellular matrix scaffolds as biological templates for organ engineering and transplantation. Am J Transplant. 2015;15:64–75.
60. Yoganarasimha S, Trahan WR, Best A, Bowlin GL, Kitten TO, Moon PC, Madurantakam PA. Peracetic acid: a practical agent for sterilizing heat-labile polymeric tissue-engineering scaffolds. Tissue Eng Pt C Methods. 2014;20:714–23.
61. Hong JY, Seo Y, Davaa G, Kim H-W, Kim SH, Hyun JK. Decellularized brain matrix enhances macrophage polarization and functional improvements in rat spinal cord injury. Acta Biomater. 2020;101:357–71.
62. Frantz C, Stewart KM, Weaver VM. The extracellular matrix at a glance. J Cell Sci. 2010;123:4195–200.
63. Gilbert TW, Wognum S, Joyce EM, Freytes DO, Sacks MS, Badylak SF. Collagen fiber alignment and biaxial mechanical behavior of porcine urinary bladder derived extracellular matrix. Biomaterials. 2008;29:4775–82.
64. Muiznieks LD, Keeley FW. Molecular assembly and mechanical properties of the extracellular matrix: a fibrous protein perspective. Biochim Biophys Acta. 2013;1832:866–75.
65. Piazuelo E, Jimenez P, Lanas A, Garcia A, Esteva F, Sainz R. Platelet-derived growth factor and epidermal growth factor play a major role in human colonic fibroblast repair activities. Eur Surg Res. 2000;32:191–6.
66. Wang T, Feng ZQ, Leach MK, Wu J, Jiang Q. Nanoporous fibers of type-I collagen coated poly(l-lactic acid) for enhancing primary hepatocyte growth and function. J Mater Chem B. 2013;1:339–46.
67. Zhu J, He P, Lin L, Jones DR, Marchant RE. Biomimetic poly(ethylene glycol)-based hydrogels as scaffolds for inducing endothelial adhesion and capillary-like network formation. Biomacromolecules. 2012;13:706–13.
68. Lynn AK, Yannas IV, Bonfield W. Antigenicity and immunogenicity of collagen. J Biomed Mater Res Pt B. 2004;71B:343–54.
69. Bayrak A, Pruger P, Stock UA, Seifert M. Absence of immune responses with xenogeneic collagen and elastin. Tissue Eng A. 2013;19:1592–600.
70. Willard JJ, Drexler JW, Das A, Roy S, Shilo S, Shoseyov O, Powell HM. Plant-derived human collagen scaffolds for skin tissue engineering. Tissue Eng A. 2013;19:1507–18.
71. Kim BS, Choi JS, Kim JD, Yoon HI, Choi YC, Cho YW. Human collagen isolated from adipose tissue. Biotechnol Prog. 2012;28:973–80.
72. Kuraitis D, Giordano C, Ruel M, Musaro A, Suuronen EJ. Exploiting extracellular matrix-stem cell interactions: a review of natural materials for therapeutic muscle regeneration. Biomaterials. 2012;33:428–43.

73. Crowley DC, Lau FC, Sharma P, Evans M, Guthrie N, Bagchi M, Bagchi D, Dey DK, Raychaudhuri SP. Safety and efficacy of undenatured type II collagen in the treatment of osteoarthritis of the knee: a clinical trial. Int J Med Sci. 2009;6:312–21.
74. Tong T, Zhao W, Wu YQ, Chang Y, Wang QT, Zhang LL, Wei W. Chicken type II collagen induced immune balance of main subtype of helper T cells in mesenteric lymph node lymphocytes in rats with collagen-induced arthritis. Inflamm Res. 2010;59:369–77.
75. Anderson DEJ, Hinds MT. Extracellular matrix production and regulation in micropatterned endothelial cells. Biochem Biophys Res Commun. 2012;427:159–64.
76. Nystrom A, Velati D, Mittapalli VR, Fritsch A, Kern JS, Bruckner-Tuderman L. Collagen VII plays a dual role in wound healing. J Clin Investig. 2013;123:3498–509.
77. Lv S, Dudek DM, Cao Y, Balamurali MM, Gosline J, Li HB. Designed biomaterials to mimic the mechanical properties of muscles. Nature. 2010;465:69–73.
78. Gray WR, Sandberg LB, Foster JA. Molecular model for elastin structure and function. Nature. 1973;246:461–6.
79. Brooke BS, Bayes-Genis A, Li DY. New insights into elastin and vascular disease. Trends Cardiovasc Med. 2003;13:176–81.
80. Bashur CA, Venkataraman L, Ramamurthi A. Tissue engineering and regenerative strategies to replicate biocomplexity of vascular elastic matrix assembly. Tissue Eng Pt B Rev. 2012;18:203–17.
81. Waterhouse A, Wise SG, Ng MKC, Weiss AS. Elastin as a nonthrombogenic biomaterial. Tissue Eng Pt B Rev. 2011;17:93–9.
82. Simionescua DT, Lua QJ, Song Y, Lee JS, Rosenbalm TN, Kelley C, Vyavahare NR. Biocompatibility and remodeling potential of pure arterial elastin and collagen scaffolds. Biomaterials. 2006;27:702–13.
83. Daamen WF, Hafmans T, Veerkamp JH, van Kuppevelt TH. Isolation of intact elastin fibers devoid of microfibrils. Tissue Eng. 2005;11:1168–76.
84. Patel D, Menon R, Taite LJ. Self-assembly of elastin-based peptides into the ECM: the importance of integrins and the elastin binding protein in elastic fiber assembly. Biomacromolecules. 2011;12:432–40.
85. Tumova S, Woods A, Couchman JR. Heparan sulfate proteoglycans on the cell surface: versatile coordinators of cellular functions. Int J Biochem Cell Biol. 2000;32:269–88.
86. Migliorini E, Thakar D, Sadir R, Pleiner T, Baleux F, Lortat-Jacob H, Coche-Guerente L, Richter RP. Well-defined biomimetic surfaces to characterize glycosaminoglycan-mediated interactions on the molecular, supramolecular and cellular levels. Biomaterials. 2014;35:8903–15.
87. van der Smissen A, Samsonov S, Hintze V, Scharnweber D, Moeller S, Schnabelrauch M, Pisabarro MT, Anderegg U. Artificial extracellular matrix composed of collagen I and highly sulfated hyaluronan interferes with TGFβ1 signaling and prevents TGFβ1-induced myofibroblast differentiation. Acta Biomater. 2013;9:7775–86.
88. Andreas K, Sittinger M, Ringe J. Toward in situ tissue engineering: chemokine-guided stem cell recruitment. Trends Biotechnol. 2014;32:483–92.
89. Hortensius RA, Harley BAC. The use of bioinspired alterations in the glycosaminoglycan content of collagen-GAG scaffolds to regulate cell activity. Biomaterials. 2013;34:7645–52.
90. Salanga CL, Handel TM. Chemokine oligomerization and interactions with receptors and glycosaminoglycans: the role of structural dynamics in function. Exp Cell Res. 2011;317:590–601.
91. Rueda P, Richart A, Recalde A, Gasse P, Vilar J, Guerin C, Lortat-Jacob H, Vieira P, Baleux F, Chretien F, Arenzana-Seisdedos F, Silvestre JS. Homeostatic and tissue reparation defaults in mice carrying selective genetic invalidation of CXCL12/proteoglycan interactions. Circulation. 2012;126:1882–U262.
92. Kogan G, Soltes L, Stern R, Gemeiner P. Hyaluronic acid: a natural biopolymer with a broad range of biomedical and industrial applications. Biotechnol Lett. 2007;29:17–25.

93. Avigdor A, Goichberg P, Shivtiel S, Dar A, Peled A, Samira S, Kollet O, Hershkoviz R, Alon R, Hardan I, Ben-Hur H, Naor D, Nagler A, Lapidot T. CD44 and hyaluronic acid cooperate with SDF-1 in the trafficking of human CD34(+) stem/progenitor cells to bone marrow. Blood. 2004;103:2981–9.
94. Ratliff BB, Ghaly T, Brudnicki P, Yasuda K, Rajdev M, Bank M, Mares J, Hatzopoulos AK, Goligorsky MS. Endothelial progenitors encapsulated in bioartificial niches are insulated from systemic cytotoxicity and are angiogenesis competent. Am J Physiol. 2010;299:F178–86.
95. Toole BP. Hyaluronan: from extracellular glue to pericellular cue. Nat Rev Cancer. 2004;4:528–39.
96. Tempel C, Gilead A, Neeman M. Hyaluronic acid as an anti-angiogenic shield in the preovulatory rat follicle. Biol Reprod. 2000;63:134–40.
97. Glynn JJ, Hinds MT. Bioactive anti-thrombotic modification of decellularized matrix for vascular applications. Adv Healthc Mater. 2016;5:1439–46.
98. Koobatian MT, Row S, Smith RJ Jr, Koenigsknecht C, Andreadis ST, Swartz DD. Successful endothelialization and remodeling of a cell-free small-diameter arterial graft in a large animal model. Biomaterials. 2016;76:344–58.
99. Dimitrievska S, Cai C, Weyers A, Balestrini JL, Lin T, Sundaram S, Hatachi G, Spiegel DA, Kyriakides TR, Miao J, Li G, Niklason LE, Linhardt RJ. Click-coated, heparinized, decellularized vascular grafts. Acta Biomater. 2015;13:177–87.
100. Netelenbos T, Drager AM, van het Hof B, Kessler FL, Delouis C, Huijgens PC, van den Born J, van Dijk W. Differences in sulfation patterns of heparan sulfate derived from human bone marrow and umbilical vein endothelial cells. Exp Hematol. 2001;29:884–93.
101. Netelenbos T, van den Born J, Kessler FL, Zweegman S, Merle PA, van Oostveen JW, Zwaginga JJ, Huijgens PC, Drager AM. Proteoglycans on bone marrow endothelial cells bind and present SDF-1 towards hematopoietic progenitor cells. Leukemia. 2003;17:175–84.
102. Cool SM, Nurcombe V. Heparan sulfate regulation of progenitor cell fate. J Cell Biochem. 2006;99:1040–51.
103. Dombrowski C, Song SJ, Chuan PY, Lim XH, Susanto E, Sawyer AA, Woodruff MA, Hutmacher DW, Nurcombe V, Cool SM. Heparan sulfate mediates the proliferation and differentiation of rat mesenchymal stem cells. Stem Cells Dev. 2009;18:661–70.
104. Pecly IM, Goncalves RG, Rangel EP, Takiya CM, Taboada FS, Martinusso CA, Pavao MS, Leite M Jr. Effects of low molecular weight heparin in obstructed kidneys: decrease of collagen, fibronectin and TGF-beta, and increase of chondroitin/dermatan sulfate proteoglycans and macrophage infiltration. Nephrol Dial Transplant. 2006;21:1212–22.
105. Abe W, Ikejima K, Lang T, Okumura K, Enomoto N, Kitamura T, Takei Y, Sato N. Low molecular weight heparin prevents hepatic fibrogenesis caused by carbon tetrachloride in the rat. J Hepatol. 2007;46:286–94.
106. Dong XC, Wei XF, Yi W, Gu CH, Kang XJ, Liu Y, Li Q, Yi DH. RGD-modified acellular bovine pericardium as a bioprosthetic scaffold for tissue engineering. J Mater Sci. 2009;20:2327–36.
107. Liao J, Wu S, Li K, Fan YB, Dunne N, Li XM. Peptide-modified bone repair materials: factors influencing osteogenic activity. J Biomed Mater Res A. 2019;107:1491–512.
108. Stevenson MD, Piristine H, Hogrebe NJ, Nocera TM, Boehm MW, Reen RK, Koelling KW, Agarwal G, Sarang-Sieminski AL, Gooch KJ. A self-assembling peptide matrix used to control stiffness and binding site density supports the formation of microvascular networks in three dimensions. Acta Biomater. 2013;9:7651–61.
109. Pankov R, Yamada KM. Fibronectin at a glance. J Cell Sci. 2002;115:3861–3.
110. Kruegel J, Miosge N. Basement membrane components are key players in specialized extracellular matrices. Cell Mol Life Sci. 2010;67:2879–95.
111. Yurchenco PD, Amenta PS, Patton BL. Basement membrane assembly, stability and activities observed through a developmental lens. Matrix Biol. 2004;22:521–38.

112. Lozito TP, Kuo CK, Taboas JM, Tuan RS. Human mesenchymal stem cells express vascular cell phenotypes upon interaction with endothelial cell matrix. J Cell Biochem. 2009;107:714–22.
113. Suzuki S, Narita Y, Yamawaki A, Murase Y, Satake M, Mutsuga M, Okamoto H, Kagami H, Ueda M, Ueda Y. Effects of extracellular matrix on differentiation of human bone marrow-derived mesenchymal stem cells into smooth muscle cell lineage: utility for cardiovascular tissue engineering. Cells Tissues Organs. 2010;191:269–80.
114. Rooney JE, Gurpur PB, Burkin DJ. Laminin-111 protein therapy prevents muscle disease in the mdx mouse model for Duchenne muscular dystrophy. Proc Natl Acad Sci U S A. 2009;106:7991–6.
115. Smalheiser NR, Crain SM, Reid LM. Laminin as a substrate for retinal axons in vitro. Dev Brain Res. 1984;12:136–40.
116. Ma W, Tavakoli T, Derby E, Serebryakova Y, Rao MS, Mattson MP. Cell-extracellular matrix interactions regulate neural differentiation of human embryonic stem cells. BMC Dev Biol. 2008;8:90.
117. Cao JI, Sun CK, Zhao H, Xiao ZF, Chen B, Gao J, Zheng TZ, Wu W, Wu S, Wang JY, Dai JW. The use of laminin modified linear ordered collagen scaffolds loaded with laminin-binding ciliary neurotrophic factor for sciatic nerve regeneration in rats. Biomaterials. 2011;32:3939–48.
118. Stabenfeldt SE, Munglani G, Garcia AJ, LaPlaca MC. Biomimetic microenvironment modulates neural stem cell survival, migration, and differentiation. Tissue Eng A. 2010;16:3747–58.
119. Brizzi MF, Tarone G, Defilippi P. Extracellular matrix, integrins, and growth factors as tailors of the stem cell niche. Curr Opin Cell Biol. 2012;24:645–51.
120. Schultz GS, Wysocki A. Interactions between extracellular matrix and growth factors in wound healing. Wound Repair Regen. 2009;17:153–62.
121. Martino MM, Briquez PS, Guc E, Tortelli F, Kilarski WW, Metzger S, Rice JJ, Kuhn GA, Muller R, Swartz MA, Hubbell JA. Growth factors engineered for super-affinity to the extracellular matrix enhance tissue healing. Science. 2014;343:885–8.
122. Macri L, Silverstein D, Clark RAF. Growth factor binding to the pericellular matrix and its importance in tissue engineering. Adv Drug Deliv Rev. 2007;59:1366–81.
123. Laslett AL, McFarlane JR, Hearn MTW, Risbridger GP. Requirement for heparan sulphate proteoglycans to mediate basic fibroblast growth factor (FGF-2)-induced stimulation of Leydig cell steroidogenesis. J Steroid Biochem Mol Biol. 1995;54:245–50.
124. Schlessinger J, Plotnikov AN, Ibrahimi OA, Eliseenkova AV, Yeh BK, Yayon A, Linhardt RJ, Mohammadi M. Crystal structure of a ternary FGF-FGFR-heparin complex reveals a dual role for heparin in FGFR binding and dimerization. Mol Cell. 2000;6:743–50.
125. Rahman S, Patel Y, Murray J, Patel KV, Sumathipala R, Sobel M, Wijelath ES. Novel hepatocyte growth factor (HGF) binding domains on fibronectin and vitronectin coordinate a distinct and amplified Met-integrin induced signalling pathway in endothelial cells. BMC Cell Biol. 2005;6:8.
126. Hussey GS, Pineda Molina C, Cramer MC, Tyurina YY, Tyurin VA, Lee YC, El-Mossier SO, Murdock MH, Timashev PS, Kagan VE, Badylak SF. Lipidomics and RNA sequencing reveal a novel subpopulation of nanovesicle within extracellular matrix. Biomaterials. 2020;6: eaay4361.
127. Huleihel L, Hussey GS, Naranjo JD, Zhang L, Dziki JL, Turner NJ, Stolz DB, Badylak SF. Matrix-bound nanovesicles within ECM bioscaffolds. Sci Adv. 2016;2:e1600502.
128. Lv LL, Cao YH, Liu D, Xu M, Liu H, Tang RN, Ma KL, Liu BC. Isolation and quantification of microRNAs from urinary exosomes/microvesicles for biomarker discovery. Int J Biol Sci. 2013;9:1021–31.
129. Zarovni N, Corrado A, Guazzi P, Zocco D, Lari E, Radano G, Muhhina J, Fondelli C, Gavrilova J, Chiesi A. Integrated isolation and quantitative analysis of exosome shuttled proteins and nucleic acids using immunocapture approaches. Methods. 2015;87:46–58.

130. Zara M, Guidetti GF, Camera M, Canobbio I, Amadio P, Torti M, Tremoli E, Barbieri SS. Biology and role of extracellular vesicles (EVs) in the pathogenesis of thrombosis. Int J Mol Sci. 2019;20:2840.
131. van der Pol E, Boing AN, Harrison P, Sturk A, Nieuwland R. Classification, functions, and clinical relevance of extracellular vesicles. Pharmacol Rev. 2012;64:676–705.
132. Thery C, Boussac M, Veron P, Ricciardi-Castagnoli P, Raposo G, Garin J, Amigorena S. Proteomic analysis of dendritic cell-derived exosomes: a secreted subcellular compartment distinct from apoptotic vesicles. J Immunol. 2001;166:7309–18.
133. van Balkom BWM, de Jong OG, Smits M, Brummelman J, den Ouden K, de Bree PM, van Eijndhoven MAJ, Pegtel DM, Stoorvogel W, Wurdinger T, Verhaar MC. Endothelial cells require miR-214 to secrete exosomes that suppress senescence and induce angiogenesis in human and mouse endothelial cells. Blood. 2013;121:3997–4006.
134. Cantaluppi V, Biancone L, Figliolini F, Beltramo S, Medica D, Deregibus MC, Galimi F, Romagnoli R, Salizzoni M, Tetta C, Segoloni GP, Camussi G. Microvesicles derived from endothelial progenitor cells enhance neoangiogenesis of human pancreatic islets. Cell Transplant. 2012;21:1305–20.
135. Zhang B, Yin YJ, Lai RC, Tan SS, Choo ABH, Lim SK. Mesenchymal stem cells secrete immunologically active exosomes. Stem Cells Dev. 2014;23:1233–44.
136. Distler JHW, Jungel A, Huber LC, Seemayer CA, Reich CF, Gay RE, Michel BA, Fontana A, Gay S, Pisetsky DS, Distler O. The induction of matrix metalloproteinase and cytokine expression in synovial fibroblasts stimulated with immune cell microparticles. Proc Natl Acad Sci U S A. 2005;102:2892–7.
137. Rome S. Biological properties of plant-derived extracellular vesicles. Food Funct. 2019;10:529–38.
138. Teodori L, Costa A, Marzio R, Perniconi B, Coletti D, Adamo S, Gupta B, Tarnok A. Native extracellular matrix: a new scaffolding platform for repair of damaged muscle. Front Physiol. 2014;5:218.
139. Reing JE, Zhang L, Myers-Irvin J, Cordero KE, Freytes DO, Heber-Katz E, Bedelbaeva K, McIntosh D, Dewilde A, Braunhut SJ, Badylak SF. Degradation products of extracellular matrix affect cell migration and proliferation. Tissue Eng A. 2009;15:605–14.
140. Pasquinelli AE, Reinhart BJ, Slack F, Martindale MQ, Kuroda MI, Maller B, Hayward DC, Ball EE, Degnan B, Muller P, Spring J, Srinivasan A, Fishman M, Finnerty J, Corbo J, Levine M, Leahy P, Davidson E, Ruvkun G. Conservation of the sequence and temporal expression of let-7 heterochronic regulatory RNA. Nature. 2000;408:86–9.
141. Lagos-Quintana M, Rauhut R, Meyer J, Borkhardt A, Tuschl T. New microRNAs from mouse and human. RNA. 2003;9:175–9.
142. Lopera HM, Griffiths LG. Antigen removal process preserves function of small diameter venous valved conduits, whereas SDS-decellularization results in significant valvular insufficiency. Acta Biomater. 2020;107:115–28.
143. Takagi K, Fukunaga S, Nishi A, Shojima T, Yoshikawa K, Hori H, Akashi H, Aoyagi S. In vivo recellularization of plain decellularized xenografts with specific cell characterization in the systemic circulation: histological and immunohistochemical study. Artif Organs. 2006;30:233–41.
144. Robertson MJ, Dries-Devlin JL, Kren SM, Burchfield JS, Taylor DA. Optimizing recellularization of whole decellularized heart extracellular matrix. PLoS One. 2014;9:e90406.
145. De Waele J, Reekmans K, Daans J, Goossens H, Berneman Z, Ponsaerts P. 3D culture of murine neural stem cells on decellularized mouse brain sections. Biomaterials. 2015;41:122–31.
146. Navarro-Tableros V, Sanchez MBH, Figliolini F, Romagnoli R, Tetta C, Camussi G. Recellularization of rat liver scaffolds by human liver stem cells. Tissue Eng A. 2015;21:1929–39.

147. Pati F, Jang J, Ha DH, Won Kim S, Rhie JW, Shim JH, Kim DH, Cho DW. Printing three-dimensional tissue analogues with decellularized extracellular matrix bioink. Nat Commun. 2014;5:3935.
148. Lee H, Han W, Kim H, Ha DH, Jang J, Kim BS, Cho DW. Development of liver decellularized extracellular matrix bioink for three-dimensional cell printing-based liver tissue engineering. Biomacromolecules. 2017;18:1229–37.
149. Valentin JE, Stewart-Akers AM, Gilbert TW, Badylak SF. Macrophage participation in the degradation and remodeling of extracellular matrix scaffolds. Tissue Eng A. 2009;15:1687–94.
150. Lu P, Takai K, Weaver VM, Werb Z. Extracellular matrix degradation and remodeling in development and disease. Cold Spring Harb Perspect Biol. 2011;3:a005058.
151. Record RD, Hillegonds D, Simmons C, Tullius R, Rickey FA, Elmore D, Badylak SF. In vivo degradation of 14C-labeled small intestinal submucosa (SIS) when used for urinary bladder repair. Biomaterials. 2001;22:2653–9.
152. Brinker MR, O'Connor DP. Exchange nailing of ununited fractures. J Bone Joint Surg. 2007;89:177–88.
153. Alberti KA, Xu Q. Biocompatibility and degradation of tendon-derived scaffolds. Regen Biomater. 2016;3:1–11.
154. Agrawal V, Kelly J, Tottey S, Daly KA, Johnson SA, Siu BF, Reing J, Badylak SF. An isolated cryptic peptide influences osteogenesis and bone remodeling in an adult mammalian model of digit amputation. Tissue Eng A. 2011;17:3033–44.
155. Cao Z, Dou C, Dong S. Scaffolding biomaterials for cartilage regeneration. J Nanomater. 2014;2014:1–8.
156. Brown BN, Londono R, Tottey S, Zhang L, Kukla KA, Wolf MT, Daly KA, Reing JE, Badylak SF. Macrophage phenotype as a predictor of constructive remodeling following the implantation of biologically derived surgical mesh materials. Acta Biomater. 2012;8:978–87.
157. Agrawal V, Tottey S, Johnson SA, Freund JM, Siu BF, Badylak SF. Recruitment of progenitor cells by an extracellular matrix cryptic peptide in a mouse model of digit amputation. Tissue Eng A. 2011;17:2435–43.
158. Davis GE, Bayless KJ, Davis MJ, Meininger GA. Regulation of tissue injury responses by the exposure of matricryptic sites within extracellular matrix molecules. Am J Pathol. 2000;156:1489–98.
159. Ghuman H, Mauney C, Donnelly J, Massensini AR, Badylak SF, Modo M. Biodegradation of ECM hydrogel promotes endogenous brain tissue restoration in a rat model of stroke. Acta Biomater. 2018;80:66–84.
160. Badylak SF. The extracellular matrix as a scaffold for tissue reconstruction. Semin Cell Dev Biol. 2002;13:377–83.
161. Musarò A. The basis of muscle regeneration. Adv Biol. 2014;2014:612471.
162. Schiaffino S, Pereira MG, Ciciliot S, Rovere-Querini P. Regulatory T cells and skeletal muscle regeneration. FEBS J. 2017;284:517–24.
163. Huard J, Li Y, Fu FH. Current concepts review - muscle injuries and repair: current trends in research. J Bone Joint Surg. 2002;84A:822–32.
164. Tidball JG, Villalta SA. Regulatory interactions between muscle and the immune system during muscle regeneration. Am J Physiol. 2010;298:R1173–87.
165. Wozniak AC, Kong JM, Bock E, Pilipowicz O, Anderson JE. Signaling satellite-cell activation in skeletal muscle: markers, models, stretch, and potential alternate pathways. Muscle Nerve. 2005;31:283–300.
166. Charge SBP, Rudnicki MA. Cellular and molecular regulation of muscle regeneration. Physiol Rev. 2004;84:209–38.
167. Hawke TJ, Garry DJ. Myogenic satellite cells: physiology to molecular biology. J Appl Physiol. 2001;91:534–51.
168. Kharraz Y, Guerra J, Mann CJ, Serrano AL, Munoz-Canoves P. Macrophage plasticity and the role of inflammation in skeletal muscle repair. Mediat Inflamm. 2013;2013:491497.

169. Saclier M, Cuvellier S, Magnan M, Mounier R, Chazaud B. Monocyte/macrophage interactions with myogenic precursor cells during skeletal muscle regeneration. FEBS J. 2013;280:4118–30.
170. Tidball JG, Dorshkind K, Wehling-Henricks M. Shared signaling systems in myeloid cell-mediated muscle regeneration. Development. 2014;141:1184–96.
171. Badylak SF, Dziki JL, Sicari BM, Ambrosio F, Boninger ML. Mechanisms by which acellular biologic scaffolds promote functional skeletal muscle restoration. Biomaterials. 2016;103:128–36.
172. Brown BN, Badylak SF. Expanded applications, shifting paradigms and an improved understanding of host-biomaterial interactions. Acta Biomater. 2013;9:4948–55.
173. Wong ML, Griffiths LG. Immunogenicity in xenogeneic scaffold generation: antigen removal vs. decellularization. Acta Biomater. 2014;10:1806–16.
174. Mosser DM, Edwards JP. Exploring the full spectrum of macrophage activation. Nat Rev Immunol. 2008;8:958–69.
175. Mantovani A, Biswas SK, Galdiero MR, Sica A, Locati M. Macrophage plasticity and polarization in tissue repair and remodelling. J Pathol. 2013;229:176–85.
176. Keane TJ, Londono R, Turner NJ, Badylak SF. Consequences of ineffective decellularization of biologic scaffolds on the host response. Biomaterials. 2012;33:1771–81.
177. Sicari BM, Dziki JL, Siu BF, Medberry CJ, Dearth CL, Badylak SF. The promotion of a constructive macrophage phenotype by solubilized extracellular matrix. Biomaterials. 2014;35:8605–12.
178. Slivka PF, Dearth CL, Keane TJ, Meng FW, Medberry CJ, Riggio RT, Reing JE, Badylak SF. Fractionation of an ECM hydrogel into structural and soluble components reveals distinctive roles in regulating macrophage behavior. Biomater Sci. 2014;2:1521–34.
179. Allman AJ, McPherson TB, Badylak SF, Merrill LC, Kallakury B, Sheehan C, Raeder RH, Metzger DW. Xenogeneic extracellular matrix grafts elicit a TH2-restricted immune response. Transplantation. 2001;71:1631–40.
180. Deng M, Tan J, Hu C, Hou T, Peng W, Liu J, Yu B, Dai Q, Zhou J, Yang Y, Dong R, Ruan C, Dong S, Xu J. Modification of PLGA Scaffold by MSC-derived extracellular matrix combats macrophage inflammation to initiate bone regeneration via TGF-beta-induced protein. Adv Healthc Mater. 2020;2020:e2000353.
181. Brennan EP, Reing J, Chew D, Myers-Irvin JM, Young EJ, Badylak SF. Antibacterial activity within degradation products of biological scaffolds composed of extracellular matrix. Tissue Eng. 2006;12:2949–55.
182. Gilbert TW, Freund JM, Badylak SF. Quantification of DNA in biologic scaffold materials. J Surg Res. 2009;152:135–9.
183. Sadtler K, Allen BW, Estrellas K, Housseau F, Pardoll DM, Elisseeff JH. The Scaffold immune microenvironment: biomaterial-mediated immune polarization in traumatic and nontraumatic applications. Tissue Eng A. 2017;23:1044–53.
184. Keane TJ, Dziki J, Sobieski E, Smoulder A, Castleton A, Turner N, White LJ, Badylak SF. Restoring mucosal barrier function and modifying macrophage phenotype with an extracellular matrix hydrogel: potential therapy for ulcerative colitis. J Crohn's Colitis. 2017;11:360–8.
185. Dziki JL, Sicari BM, Wolf MT, Cramer MC, Badylak SF. Immunomodulation and mobilization of progenitor cells by extracellular matrix bioscaffolds for volumetric muscle loss treatment. Tissue Eng A. 2016;22:1129–39.
186. Wang Y, Bao J, Wu X, Wu Q, Li Y, Zhou Y, Li L, Bu H. Genipin crosslinking reduced the immunogenicity of xenogeneic decellularized porcine whole-liver matrices through regulation of immune cell proliferation and polarization. Sci Rep. 2016;6:24779.
187. Wu RX, He XT, Zhu JH, Yin Y, Li X, Liu X, Chen FM. Modulating macrophage responses to promote tissue regeneration by changing the formulation of bone extracellular matrix from filler particles to gel bioscaffolds. Mater Sci Eng C. 2019;101:330–40.

188. Adams GB, Scadden DT. A niche opportunity for stem cell therapeutics. Gene Ther. 2008;15:96–9.
189. Schulz C, von Andrian UH, Massberg S. Hematopoietic stem and progenitor cells: their mobilization and homing to bone marrow and peripheral tissue. Immunol Res. 2009;44:160–8.
190. Yang ZJ, Xu SL, Chen B, Zhang SL, Zhang YL, Wei W, Ma DC, Wang LS, Zhu TB, Li CJ, Wang H, Cao KJ, Gao W, Huang J. Clinical and experimental pharmacology and physiology. J Mater Chem B. 2009;36:790–6.
191. Brown BN, Badylak SF. Extracellular matrix as an inductive scaffold for functional tissue reconstruction. Transl Res. 2014;163:268–85.
192. Mauney J, Olsen BR, Volloch V. Matrix remodeling as stem cell recruitment event: a novel in vitro model for homing of human bone marrow stromal cells to the site of injury shows crucial role of extracellular collagen matrix. Matrix Biol. 2010;29:657–63.
193. Kenichiro Tashiro GCS, Weeks B, Sasakig M, Martinn GR, Kleinman HK, Yamada Y. A synthetic peptide containing the IKVAV sequence from the a chain of laminin mediates cell attachment, migration, and neurite outgrowth. J Biol Chem. 1989;264:16174–82.
194. Hu XX, He PP, Qi GB, Gao YJ, Lin YX, Yang C, Yang PP, Hao HX, Wang L, Wang H. Transformable nanomaterials as an artificial extracellular matrix for inhibiting tumor invasion and metastasis. ACS Nano. 2017;11:4086–96.
195. Dettin M, Bagno A, Gambaretto R, Iucci G, Conconi MT, Tuccitto N, Menti AM, Grandi C, Di Bello C, Licciardello A, Polzonetti G. Covalent surface modification of titanium oxide with different adhesive peptides: surface characterization and osteoblast-like cell adhesion. J Biomed Mater Res A. 2009;90A:35–45.
196. Maquart F-X, Pasco S, Ramont L, Hornebeck W, Monboisse J-C. An introduction to matrikines: extracellular matrix-derived peptides which regulate cell activity: Implication in tumor invasion. Crit Rev Oncol Hematol. 2004;49:199–202.
197. Freytes DO, Martin J, Velankar SS, Lee AS, Badylak SF. Preparation and rheological characterization of a gel form of the porcine urinary bladder matrix. Biomaterials. 2008;29:1630–7.
198. Badylak SF, Park K, Peppas N, McCabe G, Yoder M. Marrow-derived cells populate scaffolds composed of xenogeneic extracellular matrix. Exp Hematol. 2001;29:1310–8.
199. Adair-Kirk TL, Senior RM. Fragments of extracellular matrix as mediators of inflammation. Int J Biochem Cell Biol. 2008;40:1101–10.
200. Hynes RO. The extracellular matrix: not just pretty fibrils. Science. 2009;326:1216–9.
201. Sottile J. Regulation of angiogenesis by extracellular matrix. Biochim Biophys Acta. 2004;1654:13–22.
202. Shepherd BR, Enis DR, Wang F, Suarez Y, Pober JS, Schechner JS. Vascularization and engraftment of a human skin substitute using circulating progenitor cell-derived endothelial cells. FASEB J. 2006;20:1739–41.
203. Agrawal V, Siu BF, Chao H, Hirschi KK, Raborn E, Johnson SA, Tottey S, Hurley KB, Medberry CJ, Badylak SF. Partial characterization of the Sox2+ cell population in an adult murine model of digit amputation. Tissue Eng A. 2012;18:1454–63.
204. Tottey S, Corselli M, Jeffries EM, Londono R, Peault B, Badylak SF. Extracellular matrix degradation products and low-oxygen conditions enhance the regenerative potential of perivascular stem cells. Tissue Eng A. 2011;17:37–44.
205. Tottey S, Johnson SA, Crapo PM, Reing JE, Zhang L, Jiang H, Medberry CJ, Reines B, Badylak SF. The effect of source animal age upon extracellular matrix scaffold properties. Biomaterials. 2011;32:128–36.
206. Wolf MT, Daly KA, Reing JE, Badylak SF. Biologic scaffold composed of skeletal muscle extracellular matrix. Biomaterials. 2012;33:2916–25.
207. Cooper DK. A brief history of cross-species organ transplantation. PRO. 2012;25:49–57.
208. Robb KP, Shridhar A, Flynn LE. Decellularized matrices as cell-instructive scaffolds to guide tissue-specific regeneration. ACS Biomater Sci Eng. 2018;4:3627–43.

209. He F, Pei M. Extracellular matrix enhances differentiation of adipose stem cells from infrapatellar fat pad toward chondrogenesis. J Tissue Eng Regen Med. 2013;7:73–84.
210. Ng SL, Narayanan K, Gao S, Wan AC. Lineage restricted progenitors for the repopulation of decellularized heart. Biomaterials. 2011;32:7571–80.
211. Cortiella J, Niles J, Cantu A, Brettler A, Pham A, Vargas G, Winston S, Wang J, Walls S, Nichols JE. Influence of acellular natural lung matrix on murine embryonic stem cell differentiation and tissue formation. Tissue Eng A. 2010;16:2565–80.
212. Cheung HK, Han TT, Marecak DM, Watkins JF, Amsden BG, Flynn LE. Composite hydrogel scaffolds incorporating decellularized adipose tissue for soft tissue engineering with adipose-derived stem cells. Biomaterials. 2014;35:1914–23.
213. Zhang X, Dong J. Direct comparison of different coating matrix on the hepatic differentiation from adipose-derived stem cells. Biochem Biophys Res Commun. 2015;456:938–44.
214. Rothrauff BB, Yang G, Tuan RS. Tissue-specific bioactivity of soluble tendon-derived and cartilage-derived extracellular matrices on adult mesenchymal stem cells. Stem Cell Res Ther. 2017;8:133.
215. Tan MY, Zhi W, Wei RQ, Huang YC, Zhou KP, Tan B, Deng L, Luo JC, Li XQ, Xie HQ, Yang ZM. Repair of infarcted myocardium using mesenchymal stem cell seeded small intestinal submucosa in rabbits. Biomaterials. 2009;30:3234–40.
216. Fan MR, Gong M, Da LC, Bai L, Li XQ, Chen KF, Li-Ling J, Yang ZM, Xie HQ. Tissue engineered esophagus scaffold constructed with porcine small intestinal submucosa and synthetic polymers. Biomed Mater. 2014;9:015012.
217. Poghosyan T, Gaujoux S, Vanneaux V, Bruneval P, Domet T, Lecourt S, Jarraya M, Sfeir R, Larghero J, Cattan P. In vitro development and characterization of a tissue-engineered conduit resembling esophageal wall using human and pig skeletal myoblast, oral epithelial cells, and biologic scaffolds. Tissue Eng A. 2013;19:2242–52.
218. Tan B, Wang M, Chen X, Hou J, Chen X, Wang Y, Li-Ling J, Xie H. Tissue engineered esophagus by copper--small intestinal submucosa graft for esophageal repair in a canine model. Sci China Life Sci. 2014;57:248–55.
219. Sievert KD. Off-shelf commercially available acellular collagen matrix SIS (R) by cook for urethral reconstruction. Eur Urol Suppl. 2005;4:242.
220. Liu Y, Ma W, Liu B, Wang Y, Chu J, Xiong G, Shen L, Long C, Lin T, He D, Butnaru D, Alexey L, Zhang Y, Zhang D, Wei G. Urethral reconstruction with autologous urine-derived stem cells seeded in three-dimensional porous small intestinal submucosa in a rabbit model. Stem Cell Res Ther. 2017;8:63.
221. Sievert KD, et al. Homologous bladder acellular matrix graft (BAMG) in comparison to homologous small intestine submucosa (SIS) for the reconstruction of the canine bladder: in vivo functional and histologic evaluation. Eur Urol Suppl. 2005;4:206.
222. Kajbafzadeh AM, Khorramirouz R, Sabetkish S, Ataei Talebi M, Akbarzadeh A, Keihani S. In vivo regeneration of bladder muscular wall using decellularized colon matrix: an experimental study. Pediatr Surg Int. 2016;32:615–22.
223. Sievert KD, Wefer J, Bakircioglu ME, Nunes L, Dahiya R, Tanagho EA. Heterologous acellular matrix graft for reconstruction of the rabbit urethra: histological and functional evaluation. J Urol. 2001;165:2096–102.
224. Kajbafzadeh AM, Khorramirouz R, Masoumi A, Keihani S, Nabavizadeh B. Decellularized human fetal intestine as a bioscaffold for regeneration of the rabbit bladder submucosa. J Pediatr Surg. 2018;53:1781–8.
225. Badylak S, Kokini K, Tullius B, Simmons-Byrd A, Morff R. Morphologic study of small intestinal submucosa as a body wall repair device. J Surg Res. 2002;103:190–202.
226. Song Z, Peng Z, Liu Z, Yang J, Tang R, Gu Y. Reconstruction of abdominal wall musculofascial defects with small intestinal submucosa scaffolds seeded with tenocytes in rats. Tissue Eng A. 2013;19:1543–53.

227. Zhang J, Wang GY, Xiao YP, Fan LY, Wang Q. The biomechanical behavior and host response to porcine-derived small intestine submucosa, pericardium and dermal matrix acellular grafts in a rat abdominal defect model. Biomaterials. 2011;32:7086–95.
228. He SK, Guo JH, Wang ZL, Zhang Y, Tu YH, Wu SZ, Huang FG, Xie HQ. Efficacy and safety of small intestinal submucosa in dural defect repair in a canine model. Mater Sci Eng C. 2017;73:267–74.
229. Chen MK, Badylak SF. Small bowel tissue engineering using small intestinal submucosa as a scaffold. J Surg Res. 2001;99:352–8.
230. Wang M, Li YQ, Cao J, Gong M, Zhang Y, Chen X, Tian MX, Xie HQ. Accelerating effects of genipin-crosslinked small intestinal submucosa for defected gastric mucosa repair. J Mater Chem B. 2017;5:7059–71.
231. Gu Y, Dai K. Substitution of porcine small intestinal submucosa for rabbit Achilles tendon, an experimental study. Zhonghua Yi Xue Za Zhi. 2002;82:1279–82.
232. Bertone AL, Goin S, Kamei SJ, Mattoon JS, Litsky AS, Weisbrode SE, Clarke RB, Plouhar PL, Kaeding CC. Metacarpophalangeal collateral ligament reconstruction using small intestinal submucosa in an equine model. J Biomed Mater Res A. 2008;84:219–29.
233. Corno AF, Smith P, Bezuska L, Mimic B, Decellularized Porcine I. Small intestine sub-mucosa patch suitable for aortic arch repair? Front Pediatr. 2018;6:149.
234. Cao G, Huang Y, Li K, Fan Y, Xie H, Li X. Small intestinal submucosa: superiority, limitations and solutions, and its potential to address bottlenecks in tissue repair. J Mater Chem B. 2019;7:5038–55.
235. Padalino MA, Castellani C, Dedja A, Fedrigo M, Vida VL, Thiene G, Stellin G, Angelini A. Extracellular matrix graft for vascular reconstructive surgery: evidence of autologous regeneration of the neoaorta in a murine model. Eur J Cardiothorac Surg. 2012;42:e128–35.
236. Parmaksiz M, Dogan A, Odabas S, Elcin AE, Elcin YM. Clinical applications of decellularized extracellular matrices for tissue engineering and regenerative medicine. Biomed Mater. 2016;11:022003.
237. Feng X, Shen R, Tan J, Chen X, Pan Y, Ruan S, Zhang F, Lin Z, Zeng Y, Wang X, Lin Y, Wu Q. The study of inhibiting systematic inflammatory response syndrome by applying xenogenic (porcine) acellular dermal matrix on second-degree burns. Burns. 2007;33:477–9.
238. Callcut RA, Schurr MJ, Sloan M, Faucher LD. Clinical experience with alloderm: a one-staged composite dermal/epidermal replacement utilizing processed cadaver dermis and thin autografts. Burns. 2006;32:583–8.
239. Dieterich M, Faridi A. Biological matrices and synthetic meshes used in implant-based breast reconstruction - a review of products available in Germany. Geburtshilfe Frauenheilkd. 2013;73:1100–6.
240. Valdatta L, Cattaneo AG, Pellegatta I, Scamoni S, Minuti A, Cherubino M. Acellular dermal matrices and radiotherapy in breast reconstruction: a systematic review and meta-analysis of the literature. Plast Surg Int. 2014;2014:472604.
241. Salzberg CA. Nonexpansive immediate breast reconstruction using human acellular tissue matrix graft (AlloDerm). Ann Plast Surg. 2006;57:1–5.
242. Chun YS, Verma K, Rosen H, Lipsitz S, Morris D, Kenney P, Eriksson E. Implant-based breast reconstruction using acellular dermal matrix and the risk of postoperative complications. Plast Reconstr Surg. 2010;125:429–36.
243. Vos JD, Latev MD, Labadie RF, Cohen SM, Werkhaven JA, Haynes DS. Use of AlloDerm in type I tympanoplasty: a comparison with native tissue grafts. Laryngoscope. 2005;115:1599–602.
244. Hernandez SC, Sibley H, Fink DS, Kunduk M, Schexnaildre M, Kakade A, McWhorter AJ. Injection laryngoplasty using micronized acellular dermis for vocal fold paralysis: long-term voice outcomes. Otolaryngol Head Neck Surg. 2016;154:892–7.
245. Taylor JB, Gerlach RC, Herold RW, Bisch FC, Dixon DR. A modified tensionless gingival grafting technique using acellular dermal matrix. Int J Periodont Restorat Dentist. 2010;30:513–21.

246. Agarwal C, Kumar BT, Mehta DS. An acellular dermal matrix allograft (Alloderm((R))) for increasing keratinized attached gingiva: a case series. J Indian Soc Periodontol. 2015;19:216–20.
247. Snyder SJ, Arnoczky SP, Bond JL, Dopirak R. Histologic evaluation of a biopsy specimen obtained 3 months after rotator cuff augmentation with GraftJacket Matrix. Arthroscopy. 2009;25:329–33.
248. Mohsina A, Kumar N, Sharma AK, Mishra B, Mathew DD, Remya V, Shrivastava S, Negi M, Kritaniya D, Tamil Mahan P, Maiti SK, Shrivastava S, Singh KP. Bioengineered acellular dermal matrices for the repair of abdominal wall defects in rats. Hernia. 2015;19:219–29.
249. Gruskin E, Doll BA, Futrell FW, Schmitz JP, Hollinger JO. Demineralized bone matrix in bone repair: history and use. Adv Drug Deliv Rev. 2012;64:1063–77.
250. Luo X, Kulig KM, Finkelstein EB, Nicholson MF, Liu XH, Goldman SM, Vacanti JP, Grottkau BE, Pomerantseva I, Sundback CA, Neville CM. In vitro evaluation of decellularized ECM-derived surgical scaffold biomaterials. J Biomed Mater Res Pt B. 2017;105:585–93.
251. Hoganson DM, O'Doherty EM, Owens GE, Harilal DO, Goldman SM, Bowley CM, Neville CM, Kronengold RT, Vacanti JP. The retention of extracellular matrix proteins and angiogenic and mitogenic cytokines in a decellularized porcine dermis. Biomaterials. 2010;31:6730–7.
252. Hoganson DM, Meppelink AM, Hinkel CJ, Goldman SM, Liu XH, Nunley RM, Gaut JP, Vacanti JP. Differentiation of human bone marrow mesenchymal stem cells on decellularized extracellular matrix materials. J Biomed Mater Res A. 2014;102:2875–83.
253. Lai PH, Chang Y, Chen SC, Wang CC, Liang HC, Chang WC, Sung HW. Acellular biological tissues containing inherent glycosaminoglycans for loading basic fibroblast growth factor promote angiogenesis and tissue regeneration. Tissue Eng. 2006;12:2499–508.
254. Cui H, Chai Y, Yu Y. Progress in developing decellularized bioscaffolds for enhancing skin construction. J Biomed Mater Res A. 2019;107:1849–59.
255. Datta P, Chatterjee J, Dhara S. Phosphate functionalized and lactic acid containing graft copolymer: synthesis and evaluation as biomaterial for bone tissue engineering applications. J Biomater Sci. 2013;24:696–713.
256. Lee MS, Lee DH, Jeon J, Tae G, Shin YM, Yang HS. Biofabrication and application of decellularized bone extracellular matrix for effective bone regeneration. J Ind Eng Chem. 2020;83:323–32.
257. Ye Y, Pang Y, Zhang Z, Wu C, Jin J, Su M, Pan J, Liu Y, Chen L, Jin K. Decellularized periosteum-covered chitosan globule composite for bone regeneration in rabbit femur condyle bone defects. Macromol Biosci. 2018;18:e1700424.
258. Liu Q, Hatta T, Qi J, Liu H, Thoreson AR, Amadio PC, Moran SL, Steinmann SP, Gingery A, Zhao C. Novel engineered tendon-fibrocartilage-bone composite with cyclic tension for rotator cuff repair. J Tissue Eng Regen Med. 2018;12:1690–701.
259. Chen G, Lv Y. Decellularized bone matrix scaffold for bone regeneration. In: Turksen K, editor. Decellularized scaffolds and organogenesis: methods and protocols. New York: Springer; 2018. p. 239–54.
260. Heine J, Schmiedl A, Cebotari S, Karck M, Mertsching H, Haverich A, Kallenbach K. Tissue engineering human small-caliber autologous vessels using a xenogenous decellularized connective tissue matrix approach: preclinical comparative biomechanical studies. Artif Organs. 2011;35:930–40.
261. Lee JS, Lee K, Moon SH, Chung HM, Lee JH, Um SH, Kim DI, Cho SW. Mussel-inspired cell-adhesion peptide modification for enhanced endothelialization of decellularized blood vessels. Macromol Biosci. 2014;14:1181–9.
262. Ozeki M, Narita Y, Kagami H, Ohmiya N, Itoh A, Hirooka Y, Niwa Y, Ueda M, Goto H. Evaluation of decellularized esophagus as a scaffold for cultured esophageal epithelial cells. J Biomed Mater Res A. 2006;79:771–8.
263. Kimicata M, Allbritton-King JD, Navarro J, Santoro M, Inoue T, Hibino N, Fisher JP. Assessment of decellularized pericardial extracellular matrix and poly(propylene fumarate) biohybrid for small-diameter vascular graft applications. Acta Biomater. 2020;110:68–81.

264. Oswal D, Korossis S, Mirsadraee S, Wilcox H, Watterson K, Fisher J, Ingham E. Biomechanical characterization of decellularized and cross-linked bovine pericardium. J Heart Valve Dis. 2007;16:165–74.
265. Wu J, Brazile B, McMahan SR, Liao J, Hong Y. Heart valve tissue-derived hydrogels: preparation and characterization of mitral valve chordae, aortic valve, and mitral valve gels. J Biomed Mater Res B Appl Biomater. 2019;107:1732–40.
266. Weber B, Dijkman PE, Scherman J, Sanders B, Emmert MY, Grunenfelder J, Verbeek R, Bracher M, Black M, Franz T, Kortsmit J, Modregger P, Peter S, Stampanoni M, Robert J, Kehl D, van Doeselaar M, Schweiger M, Brokopp CE, Walchli T, Falk V, Zilla P, Driessen-Mol A, Baaijens FP, Hoerstrup SP. Off-the-shelf human decellularized tissue-engineered heart valves in a non-human primate model. Biomaterials. 2013;34:7269–80.
267. Pang K, Du L, Wu X. A rabbit anterior cornea replacement derived from acellular porcine cornea matrix, epithelial cells and keratocytes. Biomaterials. 2010;31:7257–65.
268. Hashimoto Y, Funamoto S, Sasaki S, Honda T, Hattori S, Nam K, Kimura T, Mochizuki M, Fujisato T, Kobayashi H, Kishida A. Preparation and characterization of decellularized cornea using high-hydrostatic pressurization for corneal tissue engineering. Biomaterials. 2010;31:3941–8.
269. Proulx S, Audet C, Uwamaliya J, Deschambeault A, Carrier P, Giasson CJ, Brunette I, Germain L. Tissue engineering of feline corneal endothelium using a devitalized human cornea as carrier. Tissue Eng A. 2009;15:1709–18.
270. Kim H, Park MN, Kim J, Jang J, Kim HK, Cho DW. Characterization of cornea-specific bioink: high transparency, improved in vivo safety. J Tissue Eng. 2019;10:2041731418823382.
271. Rothrauff BB, Shimomura K, Gottardi R, Alexander PG, Tuan RS. Anatomical region-dependent enhancement of 3-dimensional chondrogenic differentiation of human mesenchymal stem cells by soluble meniscus extracellular matrix. Acta Biomater. 2017;49:140–51.
272. Leor J, Amsalem Y, Cohen S. Cells, scaffolds, and molecules for myocardial tissue engineering. Pharmacol Ther. 2005;105:151–63.
273. Tan QW, Zhang Y, Luo JC, Zhang D, Xiong BJ, Yang JQ, Xie HQ, Lv Q. Hydrogel derived from decellularized porcine adipose tissue as a promising biomaterial for soft tissue augmentation. J Biomed Mater Res A. 2017;105:1756–64.
274. Okazaki H, Igarashi M, Nishi M, Tajima M, Sekiya M, Okazaki S, Yahagi N, Ohashi K, Tsukamoto K, Amemiya-Kudo M, Matsuzaka T, Shimano H, Yamada N, Aoki J, Morikawa R, Takanezawa Y, Arai H, Nagai R, Kadowaki T, Osuga J, Ishibashi S. Identification of a novel member of the carboxylesterase family that hydrolyzes triacylglycerol: a potential role in adipocyte lipolysis. Diabetes. 2006;55:2091–7.
275. Crapo PM, Medberry CJ, Reing JE, Tottey S, van der Merwe Y, Jones KE, Badylak SF. Biologic scaffolds composed of central nervous system extracellular matrix. Biomaterials. 2012;33:3539–47.
276. Hong P, Bezuhly M, Graham ME, Gratzer PF. Efficient decellularization of rabbit trachea to generate a tissue engineering scaffold biomatrix. Int J Pediatr Otorhinolaryngol. 2018;112:67–74.
277. Giraldo-Gomez DM, Garcia-Lopez SJ, Tamay-de-Dios L, Sanchez-Sanchez R, Villalba-Caloca J, Sotres-Vega A, Del Prado-Audelo ML, Gomez-Lizarraga KK, Garciadiego-Cazares-D, Pina-Barba MC. Fast cyclical-decellularized trachea as a natural 3D scaffold for organ engineering. Mater Sci Eng C. 2019;105:110142.
278. Porzionato A, Sfriso MM, Pontini A, Macchi V, Petrelli L, Pavan PG, Natali AN, Bassetto F, Vindigni V, De Caro R. Decellularized human skeletal muscle as biologic scaffold for reconstructive surgery. Int J Mol Sci. 2015;16:14808–31.
279. Baiguera S, Del Gaudio C, Lucatelli E, Kuevda E, Boieri M, Mazzanti B, Bianco A, Macchiarini P. Electrospun gelatin scaffolds incorporating rat decellularized brain extracellular matrix for neural tissue engineering. Biomaterials. 2014;35:1205–14.

280. Mellott AJ, Shinogle HE, Nelson-Brantley JG, Detamore MS, Staecker H. Exploiting decellularized cochleae as scaffolds for inner ear tissue engineering. Stem Cell Res Ther. 2017;8:41.
281. Mirzaeian L, Eivazkhani F, Hezavehei M, Moini A, Esfandiari F, Valojerdi MR, Fathi R. Optimizing the cell seeding protocol to human decellularized ovarian scaffold: application of dynamic system for bio-engineering. Cell J. 2020;22:227–35.
282. Hassanpour A, Talaei-Khozani T, Kargar-Abarghouei E, Razban V, Vojdani Z. Decellularized human ovarian scaffold based on a sodium lauryl ester sulfate (SLES)-treated protocol, as a natural three-dimensional scaffold for construction of bioengineered ovaries. Stem Cell Res Ther. 2018;9:252.
283. Henning NF, LeDuc RD, Even KA, Laronda MM. Proteomic analyses of decellularized porcine ovaries identified new matrisome proteins and spatial differences across and within ovarian compartments. Sci Rep. 2019;9:20001.
284. Miyazaki K, Maruyama T. Partial regeneration and reconstruction of the rat uterus through recellularization of a decellularized uterine matrix. Biomaterials. 2014;35:8791–800.
285. Hiraoka T, Hirota Y, Saito-Fujita T, Matsuo M, Egashira M, Matsumoto L, Haraguchi H, Dey SK, Furukawa KS, Fujii T, Osuga Y. STAT3 accelerates uterine epithelial regeneration in a mouse model of decellularized uterine matrix transplantation. JCI Insight. 2016;1:e87591.
286. Shakouri-Motlagh A, O'Connor AJ, Kalionis B, Heath DE. Improved ex vivo expansion of mesenchymal stem cells on solubilized acellular fetal membranes. J Biomed Mater Res A. 2019;107:232–42.
287. Guruswamy Damodaran R, Vermette P. Decellularized pancreas as a native extracellular matrix scaffold for pancreatic islet seeding and culture. J Tissue Eng Regen Med. 2018;12:1230–7.
288. Mestan K, Xin H, Su E. Vascular endothelial growth factor A administration rescues fetoplacental endothelial cell defects seen in severe fetal growth restriction. Placenta. 2016;45:67.
289. Schneider KH, Aigner P, Holnthoner W, Monforte X, Nurnberger S, Runzler D, Redl H, Teuschl AH. Decellularized human placenta chorion matrix as a favorable source of small-diameter vascular grafts. Acta Biomater. 2016;29:125–34.
290. Flynn L, Semple JL, Woodhouse KA. Decellularized placental matrices for adipose tissue engineering. J Biomed Mater Res A. 2006;79:359–69.
291. Mahmoudi-Rad M, Abolhasani E, Moravvej H, Mahmoudi-Rad N, Mirdamadi Y. Acellular amniotic membrane: an appropriate scaffold for fibroblast proliferation. Clin Exp Dermatol. 2013;38:646–51.
292. Balland O, Poinsard AS, Famose F, Goulle F, Isard PF, Mathieson I, Dulaurent T. Use of a porcine urinary bladder acellular matrix for corneal reconstruction in dogs and cats. Vet Ophthalmol. 2016;19:454–63.
293. Fu H, Teng L, Bai R, Deng C, Lv G, Chen J. Application of acellular intima from porcine thoracic aorta in full-thickness skin wound healing in a rat model. Mater Sci Eng C Mater Biol Appl. 2017;71:1135–44.
294. Sakakibara S, Ishida Y, Hashikawa K, Yamaoka T, Terashi H. Intima/medulla reconstruction and vascular contraction-relaxation recovery for acellular small diameter vessels prepared by hyperosmotic electrolyte solution treatment. J Artif Organs. 2014;17:169–77.
295. Syed O, Kim JH, Keskin-Erdogan Z, Day RM, El-Fiqi A, Kim HW, Knowles JC. SIS/aligned fibre scaffold designed to meet layered oesophageal tissue complexity and properties. Acta Biomater. 2019;99:181–95.
296. Ghassemi T, Saghatolslami N, Matin MM, Gheshlaghi R, Moradi A. CNT-decellularized cartilage hybrids for tissue engineering applications. Biomed Mater. 2017;12:065008.
297. Gupta SK, Dinda AK, Potdar PD, Mishra NC. Modification of decellularized goat-lung scaffold with chitosan/nanohydroxyapatite composite for bone tissue engineering applications. Biomed Res Int. 2013;2013:651945.

298. Iijima M, Aubin H, Steinbrink M, Schiffer F, Assmann A, Weisel RD, Matsui Y, Li RK, Lichtenberg A, Akhyari P. Bioactive coating of decellularized vascular grafts with a temperature-sensitive VEGF-conjugated hydrogel accelerates autologous endothelialization in vivo. J Tissue Eng Regen Med. 2018;12:e513–22.
299. Marinval N, Morenc M, Labour MN, Samotus A, Mzyk A, Ollivier V, Maire M, Jesse K, Bassand K, Niemiec-Cyganek A, Haddad O, Jacob MP, Chaubet F, Charnaux N, Wilczek P, Hlawaty H. Fucoidan/VEGF-based surface modification of decellularized pulmonary heart valve improves the antithrombotic and re-endothelialization potential of bioprostheses. Biomaterials. 2018;172:14–29.
300. Yang Y, Lei D, Zou H, Huang S, Yang Q, Li S, Qing FL, Ye X, You Z, Zhao Q. Hybrid electrospun rapamycin-loaded small-diameter decellularized vascular grafts effectively inhibit intimal hyperplasia. Acta Biomater. 2019;97:321–32.
301. Khang G, Rhee JM, Shin P, Kim IY, Lee B, Lee SJ, Lee YM, Lee HB, Lee I. Preparation and characterization of small intestine submucosa powder impregnated poly(L-lactide) scaffolds: The application for tissue engineered bone and cartilage. Macromol Res. 2002;10:158–67.
302. Da L, Gong M, Chen A, Zhang Y, Huang Y, Guo Z, Li S, Li-Ling J, Zhang L, Xie H. Composite elastomeric polyurethane scaffolds incorporating small intestinal submucosa for soft tissue engineering. Acta Biomater. 2017;59:45–57.
303. Zhang Q, Qian C, Xiao W, Zhu H, Guo J, Ge Z, Cui W. Development of a visible light, cross-linked GelMA hydrogel containing decellularized human amniotic particles as a soft tissue replacement for oral mucosa repair. RSC Adv. 2019;9:18344–52.
304. Choi JS, Yang HJ, Kim BS, Kim JD, Kim JY, Yoo B, Park K, Lee HY, Cho YW. Human extracellular matrix (ECM) powders for injectable cell delivery and adipose tissue engineering. J Control Release. 2009;139:2–7.
305. Choi JS, Kim BS, Kim JY, Kim JD, Choi YC, Yang HJ, Park K, Lee HY, Cho YW. Decellularized extracellular matrix derived from human adipose tissue as a potential scaffold for allograft tissue engineering. J Biomed Mater Res A. 2011;97:292–9.
306. Nakamura S, Ijima H. Solubilized matrix derived from decellularized liver as a growth factor-immobilizable scaffold for hepatocyte culture. J Biosci Bioeng. 2013;116:746–53.
307. Tabuchi M, Negishi J, Yamashita A, Higami T, Kishida A, Funamoto S. Effect of decellularized tissue powders on a rat model of acute myocardial infarction. Mater Sci Eng C. 2015;56:494–500.
308. Peterson B, Whang PG, Iglesias R, Wang JC, Lieberman JR. Osteoinductivity of commercially available demineralized bone matrix. Preparations in a spine fusion model. J Bone Joint Surg. 2004;86:2243–50.
309. Sclafani AP, Romo T, Jacono AA, McCormick S, Cocker R, Parker A. Evaluation of acellular dermal graft in sheet (AlloDerm) and injectable (micronized AlloDerm) forms for soft tissue augmentation. Clinical observations and histological analysis. Arch Facial Plast Surg. 2000;2:130–6.
310. Pearl AW, Woo P, Ostrowski R, Mojica J, Mandell DL, Costantino P. A preliminary report on micronized alloderm injection laryngoplasty. Laryngoscope. 2002;112:990–6.
311. Spang MT, Christman KL. Extracellular matrix hydrogel therapies: in vivo applications and development. Acta Biomater. 2018;68:1–14.
312. Saldin LT, Cramer MC, Velankar SS, White LJ, Badylak SF. Extracellular matrix hydrogels from decellularized tissues: structure and function. Acta Biomater. 2017;49:1–15.
313. Johnson TD, Dequach JA, Gaetani R, Ungerleider J, Elhag D, Nigam V, Behfar A, Christman KL. Human versus porcine tissue sourcing for an injectable myocardial matrix hydrogel. Biomater Sci. 2014;2014:60283D.
314. Flemming RG, Murphy CJ, Abrams GA, Goodman SL, Nealey PF. Effects of synthetic micro- and nano-structured surfaces on cell behavior. Biomaterials. 1999;20:573–88.
315. Singelyn JM, Sundaramurthy P, Johnson TD, Schup-Magoffin PJ, Hu DP, Faulk DM, Wang J, Mayle KM, Bartels K, Salvatore M, Kinsey AM, Demaria AN, Dib N, Christman KL. Catheter-deliverable hydrogel derived from decellularized ventricular extracellular matrix

increases endogenous cardiomyocytes and preserves cardiac function post-myocardial infarction. J Am Coll Cardiol. 2012;59:751–63.
316. Agarwal T, Narayan R, Maji S, Ghosh SK, Maiti TK. Decellularized caprine liver extracellular matrix as a 2D substrate coating and 3D hydrogel platform for vascularized liver tissue engineering. J Tissue Eng Regen Med. 2018;12:e1678–90.
317. Paduano F, Marrelli M, White LJ, Shakesheff KM, Tatullo M. Odontogenic differentiation of human dental pulp stem cells on hydrogel scaffolds derived from decellularized bone extracellular matrix and collagen type I. PLoS One. 2016;11:e0148225.
318. Johnson TD, Lin SY, Christman KL. Tailoring material properties of a nanofibrous extracellular matrix derived hydrogel. Nanotechnology. 2011;22:494015.
319. Massensini AR, Ghuman H, Saldin LT, Medberry CJ, Keane TJ, Nicholls FJ, Velankar SS, Badylak SF, Modo M. Concentration-dependent rheological properties of ECM hydrogel for intracerebral delivery to a stroke cavity. Acta Biomater. 2015;27:116–30.
320. La W-G, Jang J, Kim BS, Lee MS, Cho D-W, Yang HS. Systemically replicated organic and inorganic bony microenvironment for new bone formation generated by a 3D printing technology. RSC Adv. 2016;6:11546–53.
321. Faulk DM, Londono R, Wolf MT, Ranallo CA, Carruthers CA, Wildemann JD, Dearth CL, Badylak SF. ECM hydrogel coating mitigates the chronic inflammatory response to polypropylene mesh. Biomaterials. 2014;35:8585–95.
322. Wolf MT, Carruthers CA, Dearth CL, Crapo PM, Huber A, Burnsed OA, Londono R, Johnson SA, Daly KA, Stahl EC, Freund JM, Medberry CJ, Carey LE, Nieponice A, Amoroso NJ, Badylak SF. Polypropylene surgical mesh coated with extracellular matrix mitigates the host foreign body response. J Biomed Mater Res A. 2014;102:234–46.
323. Zhang Y, He Y, Bharadwaj S, Hammam N, Carnagey K, Myers R, Atala A, Van Dyke M. Tissue-specific extracellular matrix coatings for the promotion of cell proliferation and maintenance of cell phenotype. Biomaterials. 2009;30:4021–8.
324. Kim JY, Ahn G, Kim C, Lee JS, Lee IG, An SH, Yun WS, Kim SY, Shim JH. Synergistic effects of beta tri-calcium phosphate and porcine-derived decellularized bone extracellular matrix in 3D-printed polycaprolactone scaffold on bone regeneration. Macromol Biosci. 2018;18:e1800025.
325. Mollica PA, Booth-Creech EN, Reid JA, Zamponi M, Sullivan SM, Palmer XL, Sachs PC, Bruno RD. 3D bioprinted mammary organoids and tumoroids in human mammary derived ECM hydrogels. Acta Biomater. 2019;95:201–13.
326. Das S, Kim SW, Choi YJ, Lee S, Lee SH, Kong JS, Park HJ, Cho DW, Jang J. Decellularized extracellular matrix bioinks and the external stimuli to enhance cardiac tissue development in vitro. Acta Biomater. 2019;95:188–200.
327. Xu L, Huang Y, Wang D, Zhu S, Wang Z, Yang Y, Guo Y. Reseeding endothelial cells with fibroblasts to improve the re-endothelialization of pancreatic acellular scaffolds. J Mater Sci Mater Med. 2019;30:85.
328. Xiao J, Duan H, Liu Z, Wu Z, Lan Y, Zhang W, Li C, Chen F, Zhou Q, Wang X, Huang J, Wang Z. Construction of the recellularized corneal stroma using porous acellular corneal scaffold. Biomaterials. 2011;32:6962–71.
329. Ingram JH, Korossis S, Howling G, Fisher J, Ingham E. The use of ultrasonication to aid recellularization of acellular natural tissue scaffolds for use in anterior cruciate ligament reconstruction. Tissue Eng. 2007;13:1561–72.
330. Crabbe A, Liu Y, Sarker SF, Bonenfant NR, Barrila J, Borg ZD, Lee JJ, Weiss DJ, Nickerson CA. Recellularization of decellularized lung scaffolds is enhanced by dynamic suspension culture. PLoS One. 2015;10:e0126846.
331. Cuzzone DA, Albano NJ, Aschen SZ, Ghanta S, Mehrara BJ. Decellularized lymph nodes as scaffolds for tissue engineered lymph nodes. Lymphat Res Biol. 2015;13:186–94.
332. Welman T, Michel S, Segaren N, Shanmugarajah K. Bioengineering for organ transplantation: progress and challenges. Bioengineered. 2015;6:257–61.

333. Barrère F, Mahmood TA, de Groot K, van Blitterswijk CA. Advanced biomaterials for skeletal tissue regeneration: Instructive and smart functions. Mater Sci Eng. 2008;59:38–71.
334. Kiani M, Abbasi M, Ahmadi M, Salehi B. Organ transplantation in Iran; current state and challenges with a view on ethical consideration. J Clin Med. 2018;7:45.
335. Schiele NR, Koppes RA, Chrisey DB, Corr DT. Engineering cellular fibers for musculoskeletal soft tissues using directed self-assembly. Tissue Eng A. 2013;19:1223–32.
336. Hoque ME, Chuan YL, Pashby I. Extrusion based rapid prototyping technique: an advanced platform for tissue engineering scaffold fabrication. Biopolymers. 2012;97:83–93.
337. Willemse J, Verstegen MMA, Vermeulen A, Schurink IJ, Roest HP, van der Laan LJW, de Jonge J. Fast, robust and effective decellularization of whole human livers using mild detergents and pressure controlled perfusion. Mater Sci Eng C. 2020;108:110200.
338. Hillebrandt KH, Everwien H, Haep N, Keshi E, Pratschke J, Sauer IM. Strategies based on organ decellularization and recellularization. Transpl Int. 2019;32:571–85.
339. Peloso A, Dhal A, Zambon JP, Li P, Orlando G, Atala A, Soker S. Current achievements and future perspectives in whole-organ bioengineering. Stem Cell Res Ther. 2015;6:107.
340. Alexanian RA, Mahapatra K, Lang D, Vaidyanathan R, Markandeya YS, Gill RK, Zhai AJ, Dhillon A, Lea MR, Abozeid S, Schmuck EG, Raval AN, Eckhardt LL, Glukhov AV, Lalit PA, Kamp TJ. Induced cardiac progenitor cells repopulate decellularized mouse heart scaffolds and differentiate to generate cardiac tissue. Biochim Biophys Acta, Mol Cell Res. 2020;1867:118559.
341. Nonaka PN, Campillo N, Uriarte JJ, Garreta E, Melo E, de Oliveira LV, Navajas D, Farre R. Effects of freezing/thawing on the mechanical properties of decellularized lungs. J Biomed Mater Res A. 2014;102:413–9.
342. Godin LM, Sandri BJ, Wagner DE, Meyer CM, Price AP, Akinnola I, Weiss DJ, Panoskaltsis-Mortari A. Decreased laminin expression by human lung epithelial cells and fibroblasts cultured in acellular lung scaffolds from aged mice. PLoS One. 2016;11:e0150966.
343. Stahl EC, Bonvillain RW, Skillen CD, Burger BL, Hara H, Lee W, Trygg CB, Didier PJ, Grasperge BF, Pashos NC, Bunnell BA, Bianchi J, Ayares DL, Guthrie KI, Brown BN, Petersen TH. Evaluation of the host immune response to decellularized lung scaffolds derived from alpha-Gal knockout pigs in a non-human primate model. Biomaterials. 2018;187:93–104.
344. Nguyen DT, O'Hara M, Graneli C, Hicks R, Miliotis T, Nystrom AC, Hansson S, Davidsson P, Gan LM, Magnone MC, Althage M, Heydarkhan-Hagvall S. Humanizing miniature hearts through 4-flow cannulation perfusion decellularization and recellularization. Sci Rep. 2018;8:7458.
345. Ott HC, Matthiesen TS, Goh SK, Black LD, Kren SM, Netoff TI, Taylor DA. Perfusion-decellularized matrix: using nature's platform to engineer a bioartificial heart. Nat Med. 2008;14:213–21.
346. Yang W, Xia R, Zhang Y, Zhang H, Bai L. Decellularized liver scaffold for liver regeneration. In: Turksen K, editor. Decellularized scaffolds and organogenesis: methods and protocols. New York: Springer; 2018. p. 11–23.
347. Hassanein W, Cimeno A, Werdesheim A, Buckingham B, Harrison J, Uluer MC, Khalifeh A, Rivera-Pratt C, Klepfer S, Woodall JD, Dhru U, Bromberg E, Parsell D, Drachenberg C, Barth RN, LaMattina JC. Liver scaffolds support survival and metabolic function of multilineage neonatal allogenic cells. Tissue Eng A. 2018;24:786–93.
348. Xu T, Zhu M, Guo Y, Wu D, Huang Y, Fan X, Zhu S, Lin C, Li X, Lu J, Zhu H, Zhou P, Lu Y, Wang Z. Three-dimensional culture of mouse pancreatic islet on a liver-derived perfusion-decellularized bioscaffold for potential clinical application. J Biomater Appl. 2015;30:379–87.
349. Totonelli G, Maghsoudlou P, Garriboli M, Riegler J, Orlando G, Burns AJ, Sebire NJ, Smith VV, Fishman JM, Ghionzoli M, Turmaine M, Birchall MA, Atala A, Soker S, Lythgoe MF, Seifalian A, Pierro A, Eaton S, De Coppi P. A rat decellularized small bowel scaffold that preserves villus-crypt architecture for intestinal regeneration. Biomaterials. 2012;33:3401–10.

350. Guan Y, Liu S, Liu Y, Sun C, Cheng G, Luan Y, Li K, Wang J, Xie X, Zhao S. Porcine kidneys as a source of ECM scaffold for kidney regeneration. Mater Sci Eng C. 2015;56:451–6.
351. Nakayama KH, Batchelder CA, Lee CI, Tarantal AF. Decellularized rhesus monkey kidney as a three-dimensional scaffold for renal tissue engineering. Tissue Eng A. 2010;16:2207–16.
352. Song JJ, Guyette JP, Gilpin SE, Gonzalez G, Vacanti JP, Ott HC. Regeneration and experimental orthotopic transplantation of a bioengineered kidney. Nat Med. 2013;19:646–51.
353. Mirmalek-Sani SH, Orlando G, McQuilling JP, Pareta R, Mack DL, Salvatori M, Farney AC, Stratta RJ, Atala A, Opara EC, Soker S. Porcine pancreas extracellular matrix as a platform for endocrine pancreas bioengineering. Biomaterials. 2013;34:5488–95.
354. Baptista PM, Siddiqui MM, Lozier G, Rodriguez SR, Atala A, Soker S. The use of whole organ decellularization for the generation of a vascularized liver organoid. Hepatology. 2011;53:604–17.
355. Debnath T, Mallarpu CS, Chelluri LK. Development of bioengineered organ using biological acellular rat liver scaffold and hepatocytes. Organogenesis. 2020;2020:1–12.
356. Su J, Satchell SC, Shah RN, Wertheim JA. Kidney decellularized extracellular matrix hydrogels: rheological characterization and human glomerular endothelial cell response to encapsulation. J Biomed Mater Res A. 2018;106:2448–62.
357. Akbarzadeh A, Khorramirouz R, Ghorbani F, Beigi RSH, Hashemi J, Kajbafzadeh AM. Preparation and characterization of human size whole heart for organ engineering: scaffold microangiographic imaging. Regen Med. 2019;14:939–54.
358. Hussein KH, Park K-M, Kang K-S, Woo H-M. Heparin-gelatin mixture improves vascular reconstruction efficiency and hepatic function in bioengineered livers. Acta Biomater. 2016;38:82–93.
359. Hussein KH, Saleh T, Ahmed E, Kwak HH, Park KM, Yang SR, Kang BJ, Choi KY, Kang KS, Woo HM. Biocompatibility and hemocompatibility of efficiently decellularized whole porcine kidney for tissue engineering. J Biomed Mater Res A. 2018;106:2034–47.

Chapter 2
The Decellularization of Tissues

Guangxiu Cao and Xiaoming Li

Abstract Decellularized tissues have a wide range of applications as implantable biomaterials and/or bioscaffolds for tissue repair and show good clinical performance. Decellularized tissue characteristics, such as its shape, structure, mechanical properties, and biological activity, are strongly influenced by the decellularization methods. Physical, chemical, and biological methods or a combination of these approaches can be used to lyse cells and then rinse to remove cell residues. Making informed choices about the agents and techniques used in the decellularization can prepare the decellularized tissue preserving extracellular matrix (ECM) integrity and bioactivity. An overview of traditional and emerging decellularization methods, the effective decellularization index, and the subsequent cleaning and sterilization are presented herein. Meanwhile, the effect of ineffective decellularization and the effect of decellularization process on the performance of decellularized tissues are discussed.

Keywords Physical methods · Chemical methods · Biological methods · Removal of immunogenicity

2.1 Description of Decellularization Protocols

Decellularization was the process of removing all cells and cellular components from a tissue while reserving the extracellular matrix (ECM) structure and proteins and obtaining a biomechanically adequate scaffold (Fig. 2.1) [1]. Decellularized tissues had been extensively applied in tissue engineering and regenerative medicine [2, 3]. The decellularization of tissues provides broad opportunities for multifunctional tools in the field of tissue engineering. In particular, the challenging

G. Cao · X. Li (✉)
Key Laboratory for Biomechanics and Mechanobiology of Ministry of Education, Beijing Advanced Innovation Center for Biomedical Engineering, School of Biological Science and Medical Engineering, Beihang University, Beijing, China
e-mail: lixm@buaa.edu.cn; x.m.li@hotmail.com

X. Li, H. Xie (eds.), *Decellularized Materials*,
https://doi.org/10.1007/978-981-33-6962-7_2

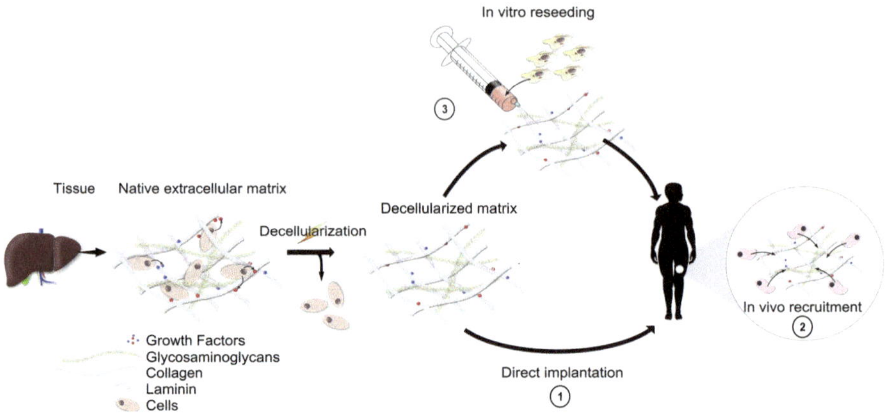

Fig. 2.1 The concept of tissue decellularization. The resulted ECM could be directly transplanted into the patient's body (1), completely dependent on the ability to direct resident cells to repair endogenous tissue (2). Alternatively, the ECM could be "seeded" with cells that "prime" the matrix (for example, to promote its remodeling or vascularization) and/or "get primed" toward specific functions (for example, proliferation or differentiation) (3). (Adapted with permission from [1]. Copyright 2013 Elsevier Ltd)

creation of decellularized scaffolds composed of retained tissue-specific ECM provides a key application area. The importance of bioscaffolds with natural structure of ECM is based on their similarity with in vivo conditions to maintain the mechanical, functional, and biocompatibility of cells while enabling cell attachment, proliferation, and differentiation. Compared with synthetic engineered tissues, the advantages of decellularized tissues were that the native matrix ultrastructure and intrinsic biological cues including growth factors, cytokines, and glycosaminoglycans (GAGs) could be preserved [4, 5]. After implantation, the cells repopulate in the matrix. The matrix degrades over time and is reshaped by proteins secreted by transplanted or ingrown cells. The native three-dimensional (3D) structure of the acellular ECM possesses a unique protein composition, mechanical properties, and bioactive molecular niche, providing great potential for replacing artificial scaffolds and using acellular scaffolds for tissue regeneration. Acellular ECM facilitates constructive host tissue remodeling rather than scar tissue formation, which has great benefits and avoids long-term complications and overall costs. Decellularized tissue can be applied for medical intervention directly or be refilled with cells prior to use. Lately, the decellularized tissue has been further processed to generate ECM slurry, hydrogel, coating, or bio-ink for 3D printing.

A variety of decellularization methods had been established, which varied with the origin, density, thickness, and biologic properties of the tissue [6]. Decellularization protocols were mainly divided into physics, chemistry, biology, or a combination of these methods [7, 8], which generally rely on lysis of the cell membrane for removing all the cellular contents from the tissue. Choosing an appropriate decellularization method is very important to retain the biological activity of the ECM. The cells in the tissue can be dissociated by mechanical grinding combined with enzymes, or chemical agents, which help cells remove from the underlying basement membrane. Although most studies of decellularized tissues report successful removal of cells, the final ECM structure still depends on precise methods. Before starting to prepare decellularized tissues, researchers require to consider which decellularization method is appropriate for their goals. Considering the application and properties of decellularized tissues, it is necessary to choose a suitable decellularization method.

The method of removing all visible cellular materials can produce bioscaffolds being safe for implantation. Many natural ECM materials have been approved by regulatory agencies and can be used in human patients, such as dermis, small intestinal submucosa, and heart valves. There are more and more biological scaffolds applied in tissue repair, which renders the continuous development of decellularization programs a clinically relevant and vital work. Decellularization treatment has been broadly applied to reduce potential immunogenicity and enhance the anticalcification properties of biologically derived materials. The decellularization protocol has been described for almost all tissues. The decellularization method should be adjusted to the target tissue. It is common to describe different protocols with various detergents for decellularization of the same tissue. Techniques applied to characterize decellularized tissues include histological staining to determine cellular and nuclear residues, as well as visualization of ECM architecture, quantification of residual DNA and nuclear fragments, mechanical testing, and other material characterization analysis.

2.1.1 Physical Methods

Decellularization is the process of removing natural cells from the tissue, leaving the structure of the ECM, while retaining the mechanics and biological activity of the tissue. In order to successfully decellularize tissues, decades of research have been conducted. Many factors (e.g. the cell density, thickness, and lipid content of the original tissue) may affect cell removal. Thus, it is vital to determine which method or combination is most appropriate for a particular tissue type. Physical decellularization protocols could destroy ECM structure and change ECM mechanics. Physical, chemical, and enzymatic approaches are combined to realize a good decellularization effect. Physical protocols that can promote decellularization of tissues include freeze-thaw, ultrasound, pressure gradient, supercritical fluid extraction, electroporation, immersion and agitation, and so on [9, 10].

2.1.1.1 Freeze-Thaw

Many decellularization approaches have been developed to remove cells as much as possible. A freeze-thaw process has been performed to achieve tissue decellularization without compromising mechanical properties. The freeze-thaw

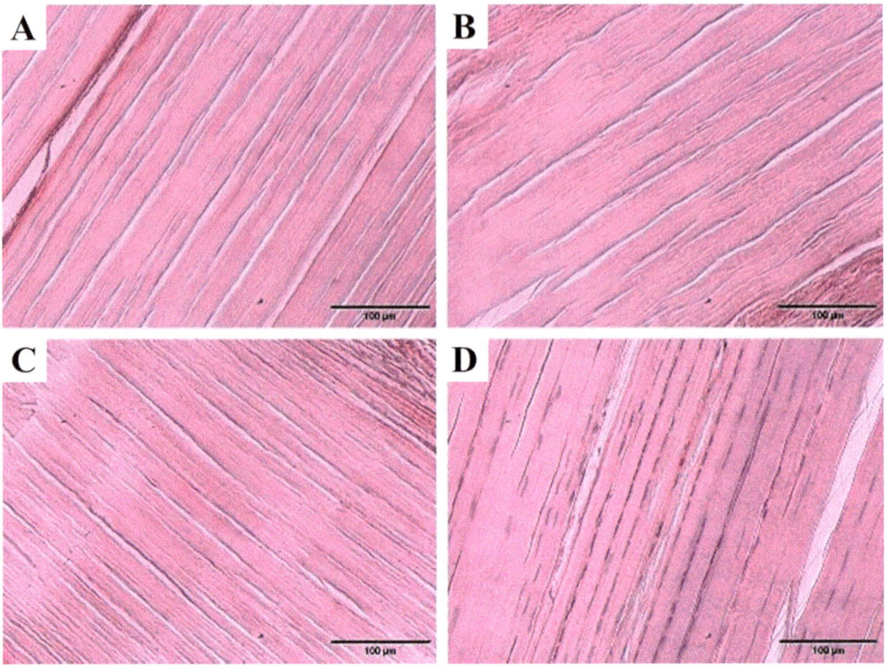

Fig. 2.2 Histological evaluation of decellularization. Hematoxylin and eosin (H&E) stainings of tendon at (**a**) group 1 (auto-protocol 1), (**b**) group 2 (auto-protocol 2), (**c**) group 3 (manual protocol) show a significant reduction in nuclei compared to tendon samples of (**d**) the internal controls (no decellularization). (Adapted with permission from [15]. Copyright 2017 Springer Ltd)

method can be used to lyse the cells in the tissue. A single freeze-thaw cycle could reduce adverse immune responses such as leukocyte infiltration in vascular ECM scaffolds [11]. Multiple freeze-thaw cycles might be used in the decellularization process, and the loss of ECM protein in the tissue would not be significantly increased [12]. Freeze-thaw method can produce less damage to the structure of ECM, so it should only be used when this effect is acceptable in the final ECM product. For load-bearing and mechanically strong tissues, freeze-thaw processing has minimal impact on mechanical properties. Wang et al. showed that an ideal cell-free and myelin-free scaffold material could be obtained by extracting the sciatic nerve with freeze-thawing and enzyme method [13]. Moreover, the decellularized nerve scaffold was not immunogenic and retained the mechanical properties of normal sciatic nerve. The study by Burk et al. showed that when treated with freeze-thawing and Triton X-100, the resident cells in the decellularization of the superficial flexor tendon samples were reduced by 99% [14]. Research by Ross et al. proved that automatic and manual freeze-thaw cycles had the same high efficiency in decellularization of large tendon samples (Fig. 2.2) [15].

2.1.1.2 Ultrasound

The unique characteristics of ultrasonic treatment are short treatment time, cell membrane rupture, and uniform treatment. Subsequent washing of the treated tissue with sodium dodecyl sulfate (SDS) detergent will result in a complete decellularized tissue in a short time. A large number of studies have shown that ultrasonic treatment technology could comprehensively establish an effective decellularization treatment method [16–18]. Azhim found that ultrasonic power had potential to increase the efficiency of meniscus decellularization [19]. In another study, they reported that ultrasonic treatment could be used to decellularize aortic tissue to prepare biological scaffolds [20]. Ultrasound technology was applied in the SDS solution to produce the appropriate effect, thereby performing an efficient decellularization process during the treatment.

2.1.1.3 Pressure Gradient

The pressure gradient caused in the whole tissue can be applied to complement the enzyme treatment, so that the structure is well preserved. Montoya et al. reported that compared with conventional stirring, after umbilical vein convection flow decellularization, DNA was qualitatively eliminated, and phospholipids were reduced by 150%. Both methods used the same volume of 20% acetone and 60% ethanol solution [21]. The 5 mmHg tested is the lowest transmural pressure, which can produce accumulated protein extracts and an ECM modulus with no significant difference from natural tissues or agitated decellularized ECM. Pressure gradients could realize the decellularization of bladder tissue by integrating immersion and circulating pressure within the bladder. Collagen I and IV, GAG, laminin, and some intracellular proteins were retained.

Hydrostatic pressure is more effective at removing cells from corneal tissue than detergents or enzymes, although the high pressure formation of ice crystals may destroy the ECM ultrastructure. The increase in temperature prevents the formation of ice crystals, but due to the increase in entropy, ECM may be destroyed, and this increase can be alleviated by colloids (such as dextran). The high hydrostatic pressure (HHP) approach is a unique physical method that can destroy cell membranes above 2000 atm, microbial membranes above 6000 atm, and virus membranes above 9000 atm. The unique feature of the HHP method is the uniform processing and short processing time. Sasaki et al. reported that the HHP method could be used to produce decellularized tissues without any chemical reagents [22]. Research by Negishi et al. showed that small diameter vascular grafts made using HHP could decellularize porcine radial arteries [23]. Prasertsung et al. developed a new method to prepare acellular dermis (ADM) using regular pressurized technology with enzymatic treatment [24]. The results showed that the application of regular pressurized technology and enzymatic treatment had great potential to become a novel approach for producing ADM for skin tissue engineering.

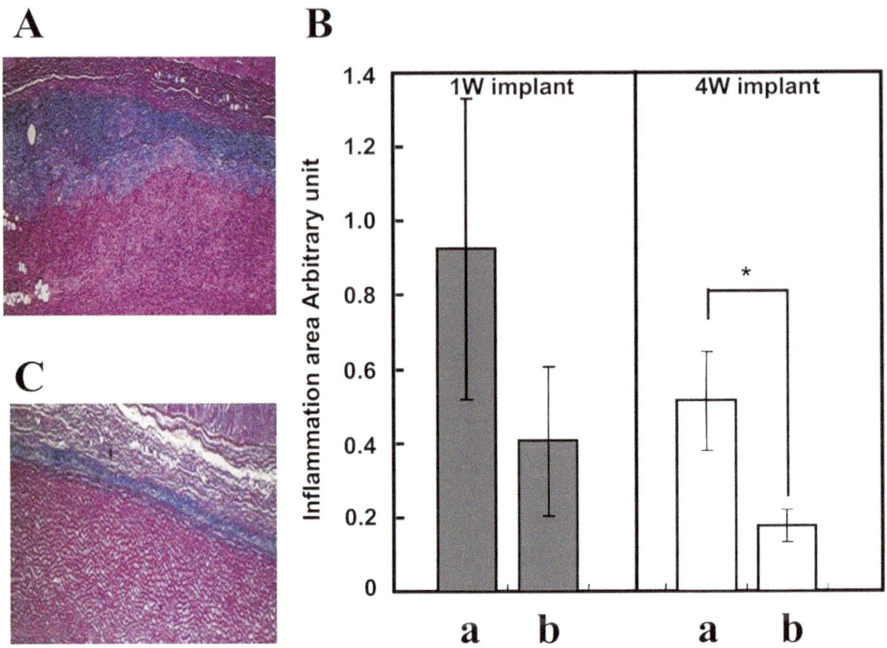

Fig. 2.3 Histological assessment of xenotransplantation research and immune response after xenotransplantation. (**a**) Natural aortic tissue. (**b**) Aortic tissue decellularized using the HHP method. (**c**) Immune response of implanted decellularized aortic tissue. a. Non-treated. b. HHP treated by condition II. $*p < 0.05$. (Adapted with permission from [25]. Copyright 2010 Elsevier Ltd)

The HHP can disrupt the cells inside the tissue, and a simple washing process can eliminate the cell debris, producing clean, decellularized tissue. Funamoto et al. reported that the HHP could be used for the decellularization of porcine aortic blood vessel [25]. HHP treatment did not change the mechanical properties of the decellularized tissue. Xenogenic transplant experiments confirmed the reduction of inflammation in acellular tissues (Fig. 2.3). Allogeneic transplantation studies displayed that decellularized blood vessels could withstand arterial blood pressure, and there was no clot formation on the lumen surface. Moreover, cell infiltration into the blood vessel wall was observed at 4 weeks after implantation, indicating that HHP therapy could be broadly used as a high-quality decellularization protocol.

2.1.1.4 Supercritical Fluid Extraction

Supercritical fluid extraction in the decellularization uses inert substances, such as carbon dioxide, which have low viscosity and high transmission properties. When the controlled rate of supercritical fluid is similar to the critical point drying, cellular components are removed from the tissue, thereby minimizing the damage to the

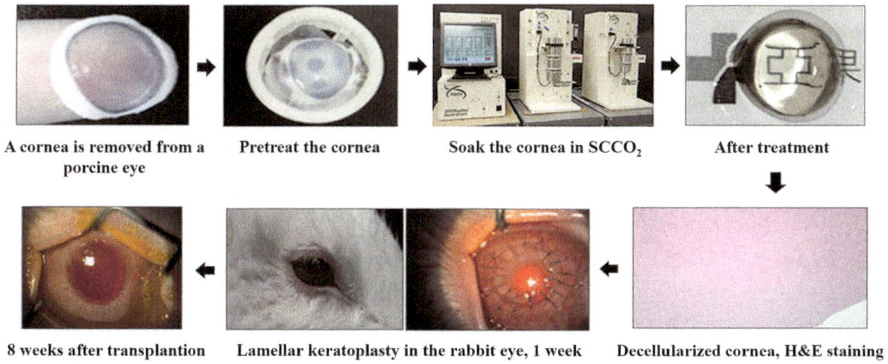

Fig. 2.4 Preparation process of decellularized scaffolds for corneal tissue engineering using $SCCO_2$ extraction protocol. (Adapted with permission from [26]. Copyright 2017 Elsevier Ltd)

mechanical properties of the tissue ECM. The prepared decellularized tissue is dry, which facilitates the storage and avoids lyophilization.

During the decellularization of tissues, the high pressure of the fluid bursts the cells and rapid depressurization can be effective in removing the cells from the tissues. Huang et al. proposed to use supercritical carbon dioxide ($SCCO_2$) extraction technology to prepare cell-free scaffolds for corneal tissue engineering (Fig. 2.4) [26]. Sawada et al. reported that the DNA in the aortic tissue was qualitatively eliminated after 15 min of treatment with $SCCO_2$ and ethanol [27]. This developed $SCCO_2$/entrainer system could be used to prepare acellular artificial tissues. Zambon et al. reported that dry acellular esophageal matrix could be prepared by $SCCO_2$ [28]. Guler et al. developed $SCCO_2$-assisted decellularization trials of cornea and aorta tissues and found that $SCCO_2$ decellularization could offer markedly reduced treatment times, complete decellularization, and retained ECM structure [29]. A novel $SCCO_2$-based decellularization method was developed by Casali and colleagues to maintain the hydration and mechanical properties of the scaffold [30].

2.1.1.5 Electroporation

Electroporation can also be used as a method for tissue decellularization. Weak electrical pulses could disrupt the transmembrane potential on the cell membrane, thereby mediating the formation of micropores [31]. The formation of micropores would destroy the stability of the cells and eventually lead to cell death [32]. Zager et al. evaluated and optimized non-thermal irreversible electroporation (NTIRE) protocols for in vivo myocardial decellularization [33]. The study by Wyman and colleagues demonstrated that decellularizing vaginal scaffolds using NTIRE of vaginal tissue was a step toward improving vaginal tissue grafts [34]. Phillips et al. used NTIRE to decellularize rabbit carotid arteries in vivo [35]. The study showed that after about three days of NTIRE treatment, the cell residue was basically removed from the tissue. Selecting appropriate parameters during the electroporation treatment could reduce heat generation and preserve the integrity and morphology of the ECM, allowing the host cell to be recellularized.

More importantly, since it has been proposed that the mechanism of cell removal is immune-mediated, the process of decellularization needs to occur in vivo, which greatly limits the potential applications. No completely decellularized ECM scaffold can promote a different host response when compared with those after cell clearance.

2.1.1.6 Immersion and Agitation

Another common decellularization method for tissues was to immerse them in a decellularization agent under stirring [11, 36, 37]. Agitation can promote the transport of decellularized solution to cells and remove cell debris from the tissue. The immersion and agitation approaches for the decellularization had been reported, involving various tissues such as skeletal muscle (Fig. 2.5) [38], tendon [39], peripheral nerve [40], spinal cord [41], cartilage/meniscus [42], dermis [43], urinary

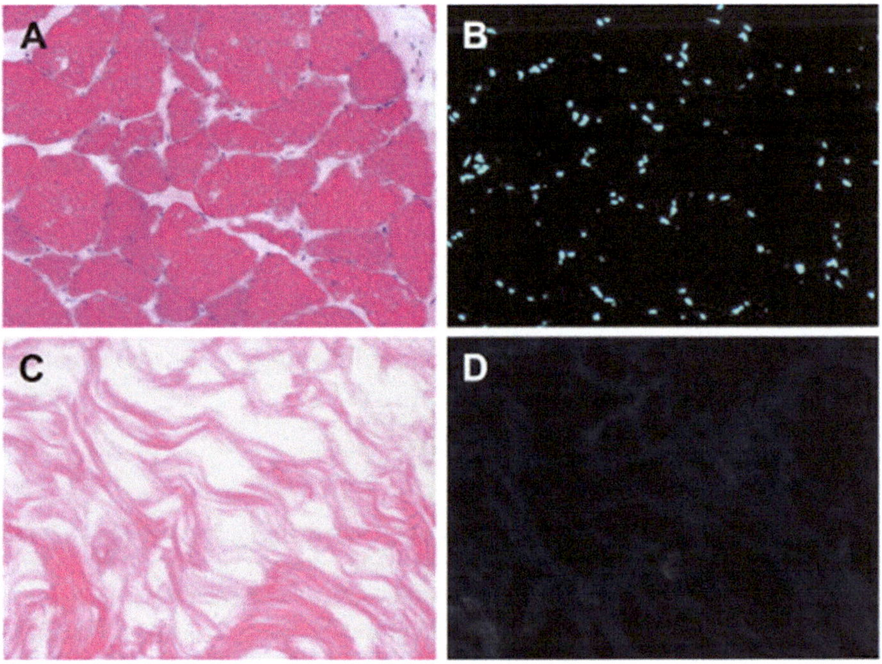

Fig. 2.5 The decellularization of skeletal muscle using the immersion and agitation approaches. H&E (**a-c**) or DAPI (**b**, **d**) staining of skeletal muscle tissue. Before decellularization, the muscle fibers and abundant nuclei of the cells could be seen. After decellularization, cellular material and nuclei were absent, while ECM remained. (Adapted with permission from [38]. Copyright 2009 Elsevier Ltd)

bladder [44], etc. The duration of the protocol is related to tissue density and thickness, the intensity of agitation, and the cleaning agent used.

Agitating is the most common way to decellularize thin tissues without proper vasculature. It promotes chemical exposure and destroys cells. Thin tissues (e.g. small intestine submucosa and bladder) after a short period of peracetic acid (PAA) agitation can be decellularized. Dense tissues (e.g. trachea, tendons, and dermis) need prolonged agitation regimens, lasting from days to months, and are frequently exposed to a combination of enzyme solutions, alcohol, and detergents.

2.1.2 Chemical Methods

Chemical treatments used alkalis, acids, detergents, hypotonic and hypertonic solutions, alcohol, and other solvents to damage bonds between intercellular and extracellular connections and cell membranes [45]. Although most studies on acellular tissues have reported successful removal of cells, the ECM structures, physical properties, and biological activity vary depending on precise methods. Alkali and acid catalyze the hydrolytic degradation of nucleic acids, cytoplasmic components, and biomolecules. Hypertonic solution can separate DNA from protein, and hypotonic solution can easily result in cell lysis through osmosis, but the matrix molecule and structure change little. Non-ionic, ionic, and zwitterionic detergents could dissolve cell membranes and separate DNA from proteins, thereby removing cellular components from tissues, but they might destroy and separate proteins in ECM [46]. The use of alcohol promotes lipid degreasing and causes tissue dehydration, leading to cell lysis. In addition, the chemicals used for decellularization, especially surfactants, may be difficult to completely eliminate from the ECM, causing poor material and cell interactions.

2.1.2.1 Alkaline and Acid Treatments

Acidic and alkaline substances can react with proteins and denature them, thereby dissolving cell components and changing nucleic acids, thereby rupturing cells. They are not selective and therefore also change the ECM composition, especially growth factors, GAG, and collagen. Chromosomal and plasmid DNA could be denatured by alkaline treatments [47–49]. Sheridan et al. reported that the use of 0.5M NaOH and ultrasonic treatment can arise the degradation of the small collagen fibers in decellularized tissue to improve access of cells during the recellularization [50]. The use of calcium oxide could significantly reduce GAG content and change the viscoelasticity of pericardial tissue [51]. The alkali treatment in the decellularization process can also lead to the loss of bioactive factors of the dermal ECM construct.

Acid can separate DNA from ECM and denature proteins in ECM, including GAG, growth factors, and collagen. The acetic acid and deoxycholic acid were the commonly used acids for decellularization [52, 53]. When using acid for decellularization, it is vital to optimize the exposure time and dosage. The 0.1% PAA could remove nucleic acids and have the least impact on ECM structure and composition. 0.1% (v/v) PAA combined with suitable mechanical approaches could decellularize thin tissues thoroughly. However, it has been shown that acetic acid can cause damage and remove collagen from the ECM and accordingly reduce the construct strength.

2.1.2.2 Non-ionic Detergents

The functional conformation of proteins in the tissues could be maintained after treatment with non-ionic detergents. Triton X-100 has been one of the extensively studied non-ionic detergents for decellularization methods. Triton X-100 can be employed to remove cell residues in valve conduits, accompanied by loss of ECM protein, and reduced adverse immune reactions in vivo. However, whether Triton X-100 can successfully achieve effective decellularization may be caused by differences in the composition and structure of the source tissue. For instance, Grauss et al. [54] reported little or no removal of rat aortic valve treated with 1–5% Triton X-100 (Fig. 2.6), while Liao et al. [55] reported that the use of 1% Triton X-100 effectively decellularized porcine aortic valve. The study by Luo et al. indicated that the combined use of Triton X-100 and sodium deoxycholate might be an appropriate solution for decellularization of heart valves [56]. These results indicated that the difference in tissue cellularity and density needs to be suitable for the decellularization process of the target tissue. Although generally considered as the detergent of choice for decellularization protocols [57], like other decellularization agents, Triton X-100 can change the structural characteristics of ECM [58].

2.1.2.3 Ionic Detergents

SDS and sodium deoxycholate are the most generally used ionic detergents. SDS is a powerful cleaning agent that can remove almost all tissue content except collagen. Moreover, SDS was effective in removing cellular components from tissues [59], because it could disrupt cell-to-cell and cell-to-ECM interactions, break protein–protein interactions, and emulsify phospholipids [60]. Although SDS can successfully remove unwanted natural components from tissues, it may destroy structural and signaling proteins. The effect of SDS decellularization is related to exposure time and concentration. The most common decellularization concentration of SDS is 0.1%. However, some studies have used other concentrations of SDS for decellularization, which can help to remove cells. High concentration or prolonged exposure to SDS solution may cause changes in the structure of ECM, such as fiber breakage, protein denaturation, and damage to the basement membrane. Compared with other detergents, SDS can more completely remove nuclear residues and cytoplasmic proteins. However, it seems that SDS does not remove collagen from

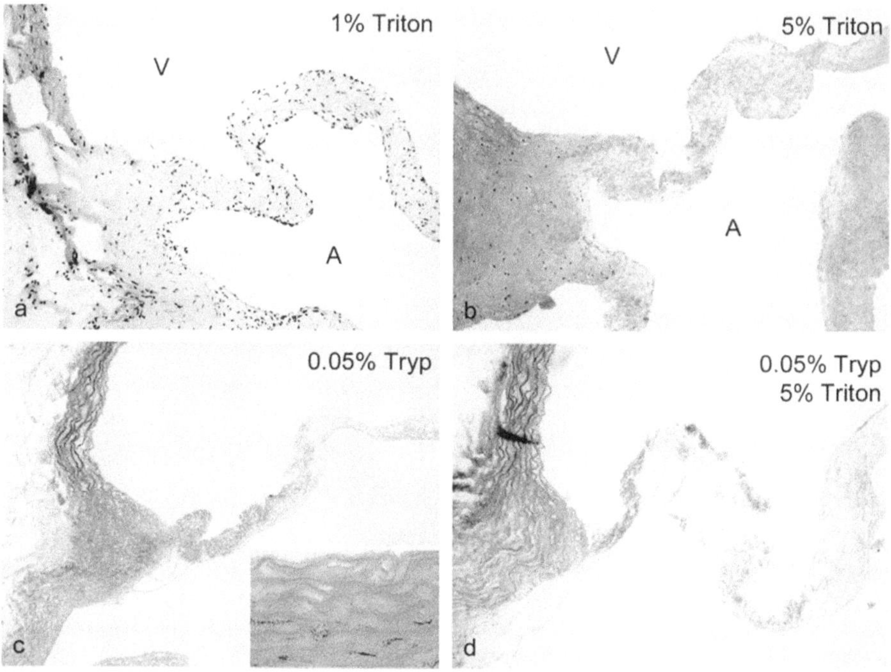

Fig. 2.6 The decellularization of aortic valve with different detergents treatment. The various described protocols that had been tested to decellularize the aortic valve did not remove cells at all from the rat valve (**a**), or remnants of cells found in the leaflets (**b**) or in the aortic wall (**c**). Severe damage to the leaflets could be observed in the trypsin-containing protocol (**c**, **d**). Arterial (A) and ventricular (V) side of the valve. (Adapted with permission from [54]. Copyright 2003 Elsevier Ltd)

the tissue. Wilshaw et al. reported that 0.03% (w/v) SDS could successfully remove the cellular components in human amniotic membrane [61]. Bolland et al. successfully developed an acellular porcine bladder matrix by immersing in 0.1% (w/v) SDS and nuclease enzymes [62]. Zhang et al. prepared acellular porcine corneal matrix with 0.5% SDS solution [63]. Singelyn et al. reported that SDS could be used to isolate ventricular ECM from fresh porcine ventricular myocardium (Fig. 2.7) [64]. After about 4–5 days in 1% SDS, the cellular material was effectively removed, resulting in white translucent ventricular ECM. The H&E sections of the acellular ECM confirmed that the cells had been removed and there was no nucleus. The cardiac ECM was then lyophilized and ground into powder and finally dissolved by enzyme digestion, so that it could be injected in subsequent studies.

Sodium deoxycholate is effective for removal of cell residues. There is no report using sodium deoxycholate alone for tissue decellularization. It was reported that the combination of zwitterionic detergents and Triton X-200 could produce the decellularized neural ECM. The ion detergent treatment is easy to use in decellularization of tissues, but this method can cost long treatment time and lead to changes in mechanical properties.

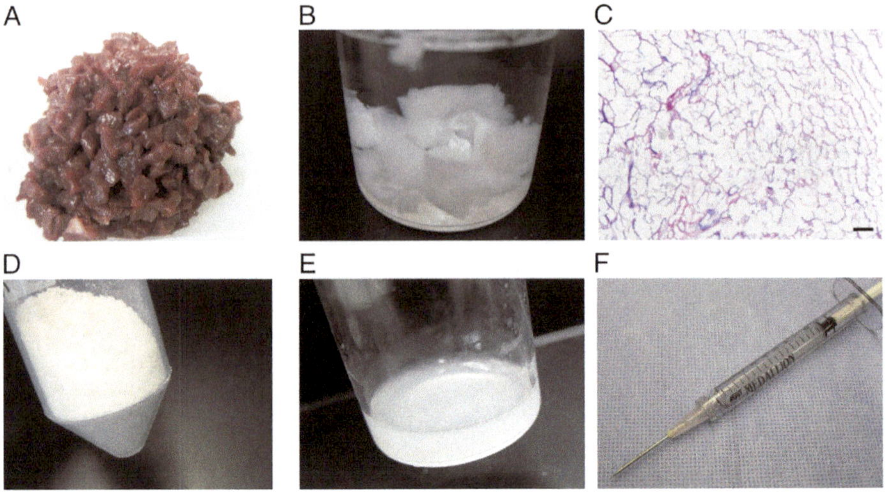

Fig. 2.7 The use of SDS to isolate ventricular ECM from fresh porcine ventricular myocardium. Slice (**a**) and decellularization (**b**) of the porcine ventricular myocardium using SDS. The cell removal of tissue section demonstrated by H&E staining (**c**). Prepared powder of decellularized ECM by milling (**d**). Solubilization by enzymatic digestion (**e**) and injection via syringe (**f**). (Adapted with permission from [64]. Copyright 2012 Elsevier Ltd)

2.1.2.4 Zwitterionic Detergents

Zwitterionic detergents have both non-ionic and ionic detergent properties and can be applied for thinner tissues that require gentle treatment. During decellularization, the electrical charge in detergents can protect natural state of the protein. Compared with non-ionic detergents, zwitterionic detergents are prone to denature proteins [65–67]. The 3-[(3-cholamidopropyl) dimethylammonio]-1-propanesulfonate (CHAPS) had been used for vascular decellularization research, and sulfobetaine-10 (SB-10) and sulfobetaine-6 (SB-16) could be applied for neural decellularization. The arterial tissue decellularized with CHAPS had natural elastin and collagen morphology histologically, and the collagen content was close to that of natural arteries. CHAPS decellularization could significantly reduce the burst pressure of the tissue. SB-10 and SB-16 combined with ionic detergent Triton X-200 could be used to decellularize peripheral nerves. This combination therapy had less harmful effects on nerve ECM, compared with the combination therapy of sodium deoxycholate and Triton X-100. Various detergents were used to evaluate the use of detergents to retain the bladder basement membrane complex (BMC) (Fig. 2.8). Surprisingly, 8 mM CHAPS disrupted the collagen network. CHAPS-treated bladder and 1% SDS-treated bladder had similar degree of damage to the integrity of the basement membrane [57].

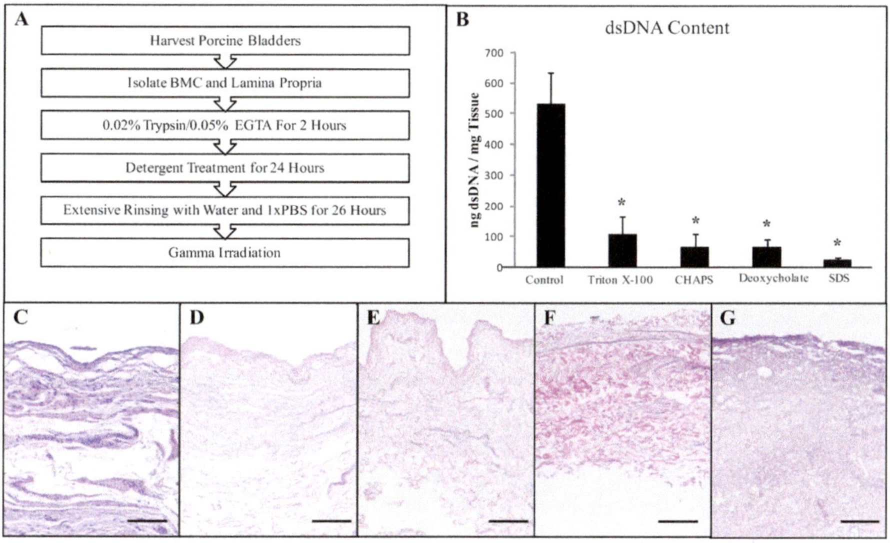

Fig. 2.8 The use of various detergents to retain the bladder basement membrane complex (BMC). (**a**) Preparation process of the BMC scaffold. (**b**) Compared to those treated with type I water, the dsDNA quantification results of the scaffolds treated with various detergents displayed obvious removal. H&E stainings of BMC scaffolds obtained with various detergents (**c-g**). (Adapted with permission from [57]. Copyright 2014 Elsevier Ltd)

2.1.2.5 Tri(n-butyl)phosphate

Tri(n-butyl)phosphate (TBP) can disrupt protein–protein interactions [68]. The study by Deeken et al. found that 1% TBP could effectively remove all cell nuclei, preserving the ECM strength, biocompatibility of the original tissue, and resistance to enzymatic degradation, the results of which showed that 1% TBP was an appropriate decellularization treatment for porcine diaphragm tendon [69]. For tendon and ligament tissues, some studies reported that TBP was comparable to SDS for cell removal ability in decellularization protocols [70, 71]. TBP might be a promising option that warranted further research for other tissues.

2.1.2.6 Hypotonic and Hypertonic Treatments

Hypertonic solutions could dissociate DNA from proteins, and hypotonic solution could easily cause cell lysis through simple osmosis, but the matrix molecule and structure changed little [72, 73]. Alternate treatment of hypotonic and hypertonic solutions has the greatest effect. Generally, hypotonic and hypertonic treatments are combined with other treatments because they can lyse cells but cannot completely remove cellular components from tissues. Hypotonic and hypertonic treatments combined with enzymatic treatment and mechanical agitation have decellularized the vocal cords of pigs. This treatment retained the 3D network of matrix proteins with intact basement membrane structure. The acellular scaffold promoted the adhesion, proliferation, and ECM production of human vocal cord fibroblasts [74].

2.1.2.7 Polyethylene Glycol

Polyethylene glycol (PEG) was a biocompatible amphoteric polymer that could be used to induce cell fusion and decellularization [75]. PEG could strengthen the interaction between the hydrophobic part of the molecule [76], which could destroy proteins and lipids on the cell membrane. Changes in the cell membrane cause a series of changes in cell permeability and osmotic pressure, eventually destroying and clearing the cells. However, PEG alone cannot completely remove cells and nuclei and needs to be used together with other reagents. In the process of decellularization, physical methods such as mechanical stirring or gently squeezing the tissue with a glass rod [77] could achieve better results. Compared with the detergent method or trypsin method, PEG is better for devascularization of blood vessels, and PEG is not a detergent, has less cytotoxicity, does not require extensive washing, and is safer for transplantation applications.

2.1.2.8 Alcohols and Acetone

If the cell membrane is permeabilized, the polar hydroxyl group of the alcohol can diffuse into the cell, where the alcohol replaces the water in the cell, and the cell is lysed by dehydration. Ethanol or methanol can be used as the final cleaning solution to remove residual nucleic acid from the tissue. The phospholipids in the tissues contributed to the calcification and destruction of the prosthesis and could be extracted with alcohol [78, 79]. In fact, alcohols are more effective than lipases in the removal of lipids from tissues. Isopropanol and ethanol could delipidize cornea and adipose tissue [80, 81]. However, Levy et al. [82] reported that ethanol treatment of tissue could alter the structure of collagen. Lumpkins et al. [83] achieved the decellularization of the porcine temporomandibular joint disk by using a mixture of ethanol and acetone. During decellularization, lipids can be removed by using acetone. However, like alcohol, acetone was a tissue fixative [84] and damaged the ultrastructure of the ECM [85, 86].

2.1.3 Biological Methods

Depending on the nature of the decellularization agent, biological methods included enzymatic and non-enzymatic agents [87]. The enzymes, generally used in decellularization, include trypsin, nuclease, dispase, lipase, and α-galactosidase, which can provide the specificity for removing cell residues or unwanted ECM components. Chelating agents, one kind of non-enzymatic agents, can disrupt the attachment of cells to fibronectin and collagen on the receptor by chelating divalent cations, thereby helping cells to dissociate from ECM proteins. Ethylenediaminetetraacetic acid (EDTA) and ethylene glycol tetraacetic acid (EGTA) are commonly employed chelating agents. They help the subtle disruption of protein–protein interactions, thereby removing cells from the ECM.

In the process of cell lysis resulted from decellularization agents, the intracellular proteases could be released, which might adversely affect ECM [88]. Serine protease inhibitors are usually employed to prevent the interaction/prolonged exposure of ECM and intracellular proteases, thereby preserving the ultrastructure of ECM. The application of streptokinase in the washing step can promote decellularization, but the actual mechanism behind it remains unclear. By using antibiotics and antifungal drugs in the decellularization system, microbial contamination in the decellularization of chemical agents can be avoided. Amphotericin B, sodium azide, streptomycin, and penicillin can be employed to prevent microbial contamination, but this may lead to a management barrier for decellularized scaffolds.

2.1.3.1 Nuclease

DNase and RNase can hydrolyze deoxyribonucleotide and ribonucleotide chains, respectively. DNase and RNase treatment combined with other methods can promote the removal of residual DNA/RNA in tissues. Generally, if the decellularization cannot be completed with detergent alone, these enzyme reagents are added to the detergent treatment to help remove residual DNA [89, 90]. Similarly, using a combination of freeze-thaw and washing solutions in hypotonic buffer and SDS in hypotonic buffer plus protease inhibitors, the porcine superflexor tendons were decellularized and then treated with nuclease [91].

2.1.3.2 Trypsin

Trypsin is one of the most commonly employed proteolytic enzymes in the process of decellularization. Trypsin cleaves peptides at the C-terminus of lysine and arginine residues. ECM proteins (such as collagen) had limited resistance to trypsin cleavage [92], so the duration of trypsin treatment became very vital in trypsin involving decellularization. Compared with detergents, the speed of removing cells by using trypsin is slower, but the preservation effect of GAG content is better. The destruction of ECM could lead to changes in mechanical properties [93]. Since trypsin can be effectively applied to destroy the ultrastructure of tissues and enhance the permeability of the subsequent decellularization matrix, it is possible to expose the tissue to trypsin at an early stage to enhance the permeability of the subsequent decellularization agent, especially when used in the complete removal of nucleus from the dense tissue. Chelating agents (such as EDTA, ethylenediaminetetraacetic acid) and an environment of pH $= 8$ at 37 °C are essential to achieve the maximum enzyme activity of trypsin. Schenke-Layland et al. found that by placing it in a trypsin-EDTA solution and incubating at 37 °C, then washing with PBS, the pulmonary valve could almost completely eliminate cells [94].

2.1.3.3 Dispase

Using dispase alone can only remove some cells of the tissue, and mechanical abrasion may be required to completely remove cells. Dispase can cut fibronectin, type IV collagen, and to a lesser extent type I collagen. The decellularization effect of dispase treatment was better than that of trypsin treatment, and trypsin could retain more ECM [24]. After subcutaneous implantation, there existed better cell infiltration in the tissues treated with dispase. Thicker tissues (such as the dermis) could be treated with dispase or trypsin combined with detergents [95]. The dispersion activity is inhibited by EDTA and EGTA, but not by serum.

2.1.3.4 Lipase

Lipase is an enzyme that can catalyze the hydrolysis of lipids. Most lipases act at specific positions on the glycerol backbone of lipid substrates. However, it was usually not possible to remove all lipids by lipase [96]. A decellularization method involving multiple approaches, such as freeze-thaw cycles, stirring, EDTA, deoxyribonuclease, ribonuclease, trypsin, lipase, ethanol, and 2-propanol, was used to obtain ECM scaffolds [97]. The decellularization method was proven to remove cellular material and maintain the ECM structure of the omentum.

2.1.3.5 α-Galactosidase

α-Galactosidase could be employed to inhibit the cell surface antigen galactose-α-1,3-galactose (Gal epitope) of acellular tissues [98]. Gal epitope could cause human xenogeneic rejection [99] and existed in small amounts in decellularized tissues [100]. In view of the mild immune response, it is unclear whether decellularization with α-galactosidase is needed in simple tissues. In addition, α-galactosidase can decrease the immunogenicity of ECM in vitro and has no adverse influence on remodeling of heterologous ECM in vivo.

2.1.3.6 Chelating Agents

Chelating agents could help cells dissociate from ECM proteins by chelating metal ions [101]. Chelating agents might cause the subtle disruption of protein–protein interactions through the same mechanism [102]. Chelating agents alone were not sufficient to remove surface cells even under agitation [103], so they were usually combined with enzymes such as trypsin (Fig. 2.9) [104] or detergent [73]. The

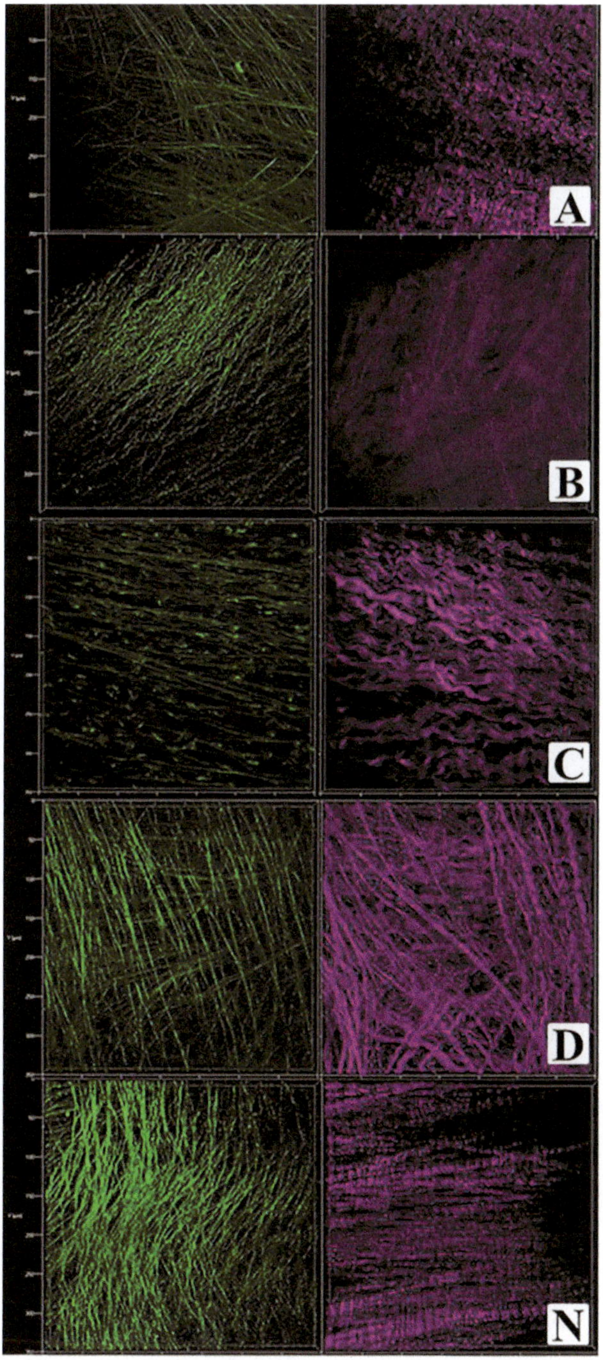

Fig. 2.9 Illustration of porcine pulmonary valve leaflets in a two-photon laser scanning microscope. The rows A, B, C, D of the images are the groups decellularized with sodium deoxycholate, SDS, trypsin/EDTA, trypsin–triton–nuclease, respectively, and the row N represents the undecellularized natural group. (Adapted with permission from [104]. Copyright 2010 Elsevier Ltd)

combined effect of chelating agents and simple hypertonic or hypotonic solutions on decellularization was unclear [11].

2.1.4 New Methods

2.1.4.1 Apoptosis

When optimizing tissue decellularization methods, the main purpose is the retainment of tissue structure. Despite this, many current solutions still rely on detergents, which help extract cellular components but also destroy tissue structure. The apoptosis induction has great potential in the decellularization of tissues with no use of stimulating agents. Through apoptosis, the cell separated from the ECM, degraded its protein and nucleic acid, and isolated its content into small fragments (as shown in Fig. 2.10) for easy removal. Therefore, induction of apoptosis might be an ideal method to produce intact cell-free tissues.

Apoptosis has been used as a means of preparing decellularized tissues and offers an alternative to traditionally used decellularization methods. Bourgine et al. first proposed the use of apoptosis for decellularization [1]. It had been proven to be a feasible approach to generate de novo matrix from cells designed to respond to pro-apoptotic signals [105, 106]. Cornelison et al. reported that there were no remaining chemicals in apoptotic acellular nerve grafts, which had immune tolerance in a rat model. Notably, this new decellularization way retained the natural nerve structure and was superior to the method of lysing cells with water. The structure of the treated tissues was almost the same as that of untreated peripheral nerves [107].

2.1.4.2 Serum

The use of the recipient's serum could create a decellularized tissue with high biocompatibility [108]. By immersing the heterogeneous tissue in serum that had been conditioned to activate the DNase I and complement system, its cellular components could be removed. Compared with the SDS-treated graft, the serum-treated graft retained the natural structure of its ECM (Fig. 2.11). When subcutaneously implanted into isogenic inbred rats, the recipient serum-treated grafts had fewer immune rejections than SDS-treated grafts.

Serum related to nucleic acid fragments could help remove them from tissues, but it could not remove certain immunogenic components, such as phospholipids [109]. In addition, the disadvantage of xenogeneic serum is the introduction of immunogenic components that may be associated with ECM, thereby enhancing downstream adverse host reactions. Therefore, the method of using serum in preparing ECM for clinical application is limited.

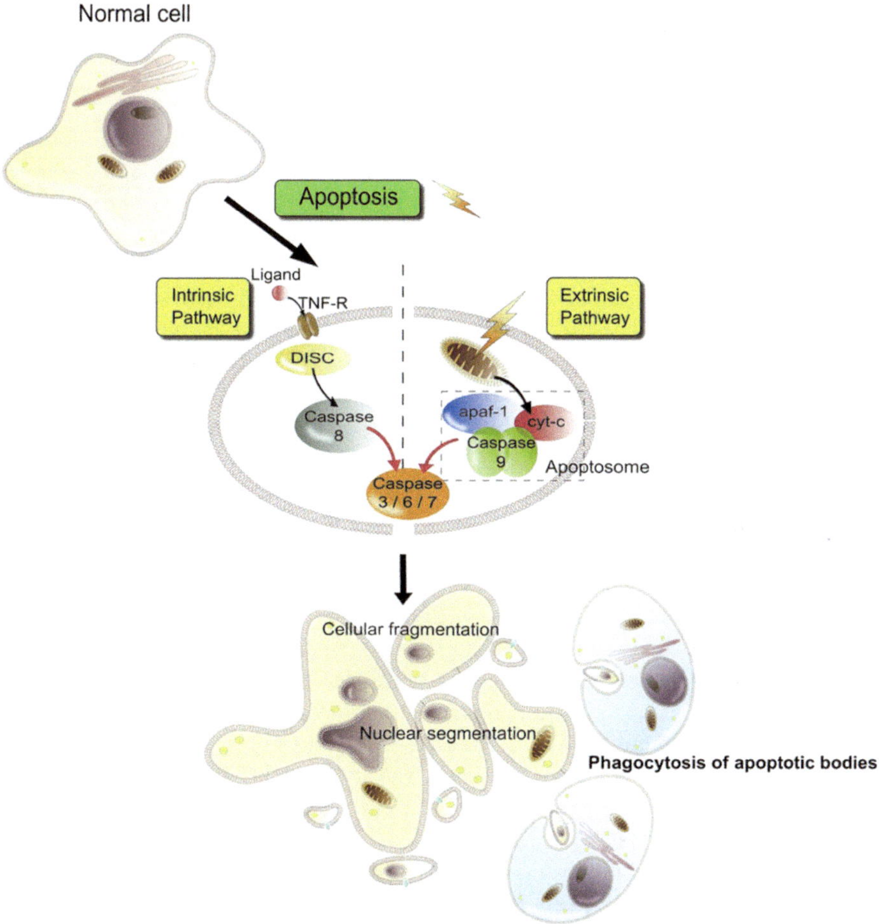

Fig. 2.10 Apoptosis could cause cell death. The apoptotic program's activation through internal or external pathways involves a series of molecular events that cause cells to coagulate and destroy into small apoptotic bodies. (Adapted with permission from [1]. Copyright 2013 Elsevier Ltd)

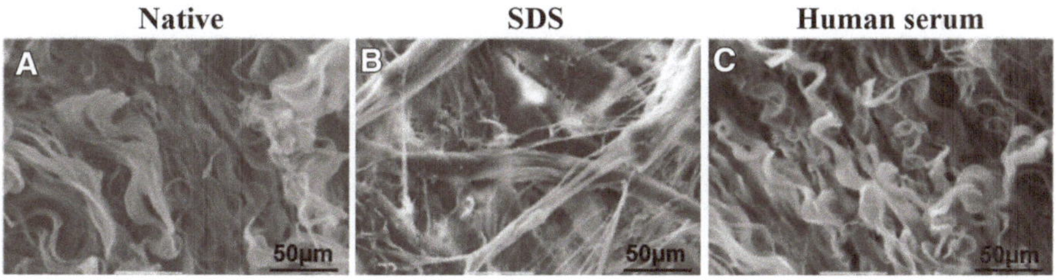

Fig. 2.11 The fiber structure of the outer membrane surface. Smooth wavy fibers of natural tissue (**a**), stretched fibers after SDS treatment (**b**), compared to natural tissue, there was no significant change after serum treatment (**c**). (Adapted with permission from [108]. Copyright 2015 Springer Ltd)

2.1.5 Combinatorial Methods

Because each of the above technologies has its advantages and disadvantages, in order to retain the required characteristics in the engineering tissue, several technologies have been combined to complement each other. The most powerful decellularization protocol is the combined use of physical, chemical, and enzymatic ways. The decellularization method usually starts by the use of physical treatment or ionic solution to lyse the cell membrane, then the use of enzyme treatment to separate the cell components from the ECM, the use of detergents to dissolve the cytoplasm and nuclear cell components. Finally, the cell debris were removed from the tissue. In order to increase the effectiveness of decellularization, these steps can be combined with mechanical stirring. For example, mechanical methods usually do little damage to the tissue structure. Sonication and mechanical agitation can be used simultaneously with chemical treatments to help cell lysis and remove cell debris. The use of a magnetic stir plate, an orbital oscillator, or a low profile roller can produce mechanical agitation. Research has not been conducted to detect the best sonication amplitude or frequency for cell destruction. In these processes, the optimum speed, reagent volume, and mechanical stirring time depend on the tissue types. Conversely, low concentrations of surfactants or enzymes employed alone may not remove cell debris completely. Combining the two treatment methods in a multi-step process can produce an acellular ECM suitable for its specific application. In order to suit a specific tissue, it is necessary to optimize each component parameter. These tissue-specific agreements can therefore be more effective. Generally, the mildest method can be used to produce cell-free ECM without destroying its functional and structural components. The trypsin/EDTA enzymatic treatment can be added before detergent treatment to help break the bond between the ECM and cell membrane.

At present, it is very important to explore effective decellularization schemes for natural tissues with better ECM preservation and good biocompatibility. The combined method is a method to achieve effective tissue decellularization for successful tissue regeneration applications. For example, Yang et al. proved that the decellularization approach combined with detergent and enzyme extraction could achieve complete decellularization and maintain the tissue structure and mechanical function of the bovine pericardium [93]. By combining traditional physical, chemical, and enzymatic pathways with light-emitting diodes, a decellularized tracheal scaffold was obtained [110]. An effective program was established to decellularize the bovine pericardium with a combination of chemical detergents (1% Triton X-100 and 0.5% sodium deoxycholate) and physical methods (freeze-thaw cycles) [111]. Multi-step decellularization procedures are required for thick tissues (such as fat and adipose tissues) that are needed in reconstructive surgery. Choi et al. developed a decellularization protocol for extracting an intact ECM from porcine adipose tissue by using chemical and enzymatic treatments [112]. In another study, adipose tissue was successfully decellularized by using a method that included freeze-thaw cycles, agitation, and washing with salt solution of different concentrations [113]. The samples were then treated with trypsin/EDTA, washed, and immersed in isopropanol. Finally, the samples were treated in Triton X-100, then washed in ultrapure water, and then washed in PBS. The study by Baiguera et al. displayed that the use of a modified detergent-enzymatic method could achieve complete tissue decellularization without losing most of the brain matrix [114]. Gillies et al. reported a decellularization method that utilized the destruction of actin treated with latrunculin B, cell lysis caused by osmotic shock, myosin depolymerization by exposure to high ionic strength salt solution, and DNase I treatment for removal of residual DNA from skeletal muscle. This approach could successfully remove the muscle fibers and degrade the sarcomere components in the muscle tissue without changing the ECM mechanical properties or structure [39].

It has been proven that multi-step chemical, enzymatic, and mechanical methods are not conducive to the properties of certain tissues. In the protocol, the different hypotonic buffers were employed in the treatment of pig cartilage disks [115]. Then, under the action of hyaluronidase, the porosity of the tissue was increased, and freeze-thaw and nuclease digestion were alternately performed. In order to further increase the porosity, the tissue was also ultrasonically treated with sodium hydroxide. To help remove DNA and subsequent recellularization, it is desirable to create more porous scaffolds. This method could effectively remove most cells and DNA and maintain collagen and ultrastructure of the tissues. However, as the porosity increased, some mechanical properties decreased.

2.2 Evaluation of Decellularization

2.2.1 Establishing Metrics for Effective Decellularization

The acellular matrix had no antigenicity and inflammation that caused tissue rejection [116], and the success of the implanted acellular ECM was attributed to the molecular clues provided by components of ECM. The complete acellular ECM serves as a reservoir for many molecular components and growth factors existed in natural tissues. Reducing the immunogenicity of the scaffold is one of the most key requirements for decellularization. This particular aspect is essential to prevent the use of decellularized ECM as a scaffold in clinical applications. If it is not completely decellularized, the cellular immunogenic components existed in the tissue may cause adverse host reactions. The increase in the clinical use of ECM-based scaffolds has prompted the need to establish decellularization guidelines. The most important thing is to determine the criteria for effective decellularization to guarantee long-term use of decellularized tissue. Standard evaluation approaches of decellularized ECM include histological analysis, immunohistochemistry, DNA quantification, SDS polyacrylamide gel electrophoresis (SDS-PAGE), mechanical property, among others.

Tissue section staining and DNA quantification are usually carried out to assess cell removal [12]. Although it is impossible for any decellularization protocol to completely remove all cell residues, it is possible to quantitatively analyze the remaining cellular components (e.g. DNA, phospholipids). Currently, no quantitative indicators have been proposed to assess the efficacy of decellularization protocols. A study correlated the quantitative degree of decellularization with the results of host remodeling [117]. A large number of studies have displayed that the ECM can facilitate constructive tissue remodeling. The effect of decellularization is one of the factors found to affect the remodeling of biological scaffolds.

Commercially available scaffolds are regulated by the U.S. Food and Drug Administration (FDA) and must comply with sterility guidelines and endotoxin levels. Endotoxin is particularly plentiful in the gut that is the source of tissue for small intestine submucosa. 0.5 EU/mL is the FDA standard for the biologic scaffold elution, while the standard for cerebrospinal devices is 0.06 EU/mL. Although the exact level of endotoxin concentration required to induce adverse reactions is not yet known, the immune response of dermal matrix incorporated 20 times the FDA limit to devices below FDA standards was similar, which indicated that the endotoxin standards might fall below the levels needed to elicit an acute proinflammatory response [118]. Although there are the guidelines for some indicators, the FDA has no guidelines related to the residual cell content in commercially available acellular tissues.

The criteria mentioned above may be too strict, sufficient, or too loose; these standards may not be applicable to all source tissues for preparing such bioscaffolds. It is not clear, but the threshold level of cellular content that may induce an inflammatory host response varies depending on the anatomical location. Mitochondria had evolved from bacteria and had a molecular pattern related to damage. If there was a certain threshold level or higher, it might cause a pre-inflammatory host response [119]. With a deeper understanding of the link between host response and cellular components, standards may need to be revised or supplemented.

Although these criteria can serve as useful guidelines, it can be considered to define the denucleation of tissues. These standards use DNA to replace other intracellular or membrane-like molecules, assuming they have the same effect after being removed. Because some decellularized ECM were derived from xenogeneic sources, it had always been considered that the existence of DNA was worrying, and there was concern that (so far unproven) DNA could be combined into recipient cells [120]. Except for Gal epitopes, the persistence of cellular antigens has not been fully explored. In order to prevent immune rejection, the natural antigens in the scaffold must be removed [121]. Although these were immunogenic molecules that adversely affected the host, ECM structural proteins had shown potential immunogenicity [122]. Therefore, although these proteins are required to retain the mechanical and structural properties of the constructs, they may be specific to the natural organism. Other methods must be adopted to reduce its immunogenicity, especially the use of autologous cells for recellularizing the ECM to prevent host rejection.

2.2.2 *Evaluation Methods*

Decellularized tissues have been studied extensively and have been used clinically. The research scope of decellularized tissue is expanding. ECM is not only a scaffold for cells, but also a control center for cell functions, as well as a niche for more complex biological functions. A method for decellularization of any tissue of interest must be developed through systematic, controlled experiments, and extensive evaluations must be conducted to ensure the effectiveness of cell removal, the effect of treatment on the proteins of ECM, and the response of cells to scaffolds. The effectiveness of decellularization was assessed by several methods [115, 123], such as histological analysis, immunohistochemistry, DNA quantification, SDS-PAGE, and mechanical testing. In addition, propidium iodide and PicoGreen detection methods can provide quantitative data about the presence of DNA in the sample.

Following decellularization, it is necessary to detect whether the crucial components of the tissue are still preserved in the decellularized ECM or not. Mechanical testing of ECM after decellularization can provide insight into the existence and integrity of proteins in the scaffold.

2.2.2.1 Histological Analysis

Several methods can be used to assess the efficiency of ECM decellularization and preservation. Histological stains could be used to detect cellular and nuclear remnants, as well as to visualize ECM architecture (Fig. 2.12) [124]. H&E staining was a standard technique for basic histological examination and analysis [70]. Grauss et al.

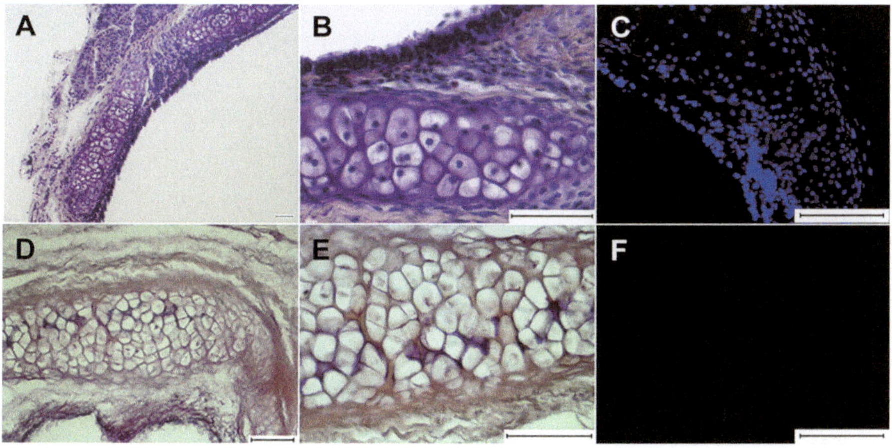

Fig. 2.12 Characterization of decellularized tracheal matrix. H&E (**a, b, d, e**), and DAPI (**c, f**) staining of native (**a-c**) and decellularized (**d-f**) tracheas. (Adapted with permission from [124]. Copyright 2014 Elsevier Ltd)

reported that DNase and RNase could produce a completely acellular material by using H&E staining. The qualitative determination of nuclear components could be assessed by histological staining [125]. Higuita et al. investigated the effect of antigen removal (AR) method or SDS decellularization on cellularity of bovine saphenous vein (SV) scaffolds using histologic assay. Qualitative histological analysis displayed both AR and SDS decellularization generated acellular SV scaffolds as determined by complete absence of visible nuclei on H&E-stained sections [126].

2.2.2.2 Immunohistochemistry

Immunohistochemical methods could be used for specific intracellular proteins, such as actin and vimentin [70]. In addition, immunohistochemistry was also applied to detect the presence of extracellular proteins [127]. The study by Flynn et al. reported that immunohistochemical staining localized the basement membrane components laminin and type IV collagen in the acellular matrix (Fig. 2.13) [96].

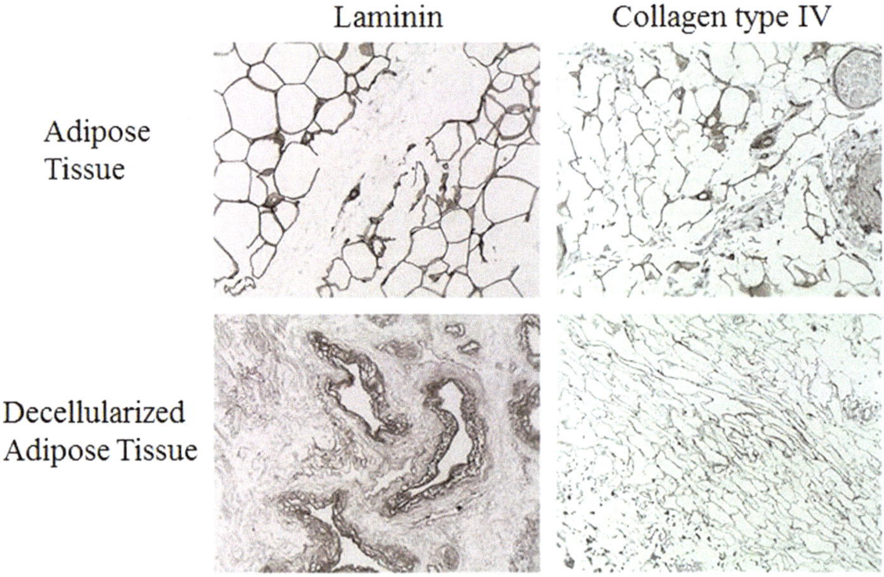

Fig. 2.13 Immunohistochemical stained images of normal human white fat and acellular adipose tissue matrix. Immunostaining of laminin and type IV collagen indicated that both basement membrane components were located in the network area of the acellular adipose tissue scaffold and along the lumens of the acellular vascular structure. (Adapted with permission from [96]. Copyright 2010 Elsevier Ltd)

2.2.2.3 DNA Assay

The DNA could be checked by staining the sample with DAPI or Hoechst [128, 129]. DNA probe technology could be used to detect whether there was DNA in decellularized tissue [130]. Electron microscopy or polymerase chain reaction (PCR) protocols are possible, but due to the technical complexity and expense of the daily operations of these technologies, they are usually not applied to check for the presence of residual cytoplasmic debris or nuclear material.

DNA quantification could be determined by spectrophotometric analysis, measuring DNA concentration by means of fluorometric stains, which were incorporated into the structure of DNA [131]. Stained DNA is then exposed to a specific UV radiation wavelength and the resulting emitted radiation rate is directly proportional to the concentration of nucleic acids in the sample. It was important t o qıantify the cell residues in the decellularized scaffold, such as DNA, mitochondria, or membrane-related molecules through commercial assay methods [132]. For example, the DNA intercalators and gel electrophoresis can be used for DNA quantification. Kral et al. developed methods that could provide quantitative analysis about the DNA in the sample using propidium iodide and PicoGreen [133]. Youngstrom et al. showed that the fluorescent dye Quant-iT PicoGreen was a single extraction technique based on ethanol to quantitatively measure the DNA content in the papain scaffold digestion solution [134].

2.2.2.4 SDS-PAGE

Electrophoresis was a basic separative method for proteins, using the distribution of molecules on a synthetic matrix based on their molecular weight and charge in electrical field as an analysis tool [135]. SDS-PAGE is the most commonly used gel electrophoresis technique used for proteins. The composition of the ECM-derived hydrogel could be analyzed by SDS-PAGE and Western blotting (Fig. 2.14) [136]. SDS-PAGE showed that collagen chains β, α1, and α2 existed under all conditions. One study focused on decellularization of skin, SDS-PAGE was used as an evaluation tool and the results showed that decellularization was effective since collagen did not show any alterations.

2.2.2.5 Mechanical Testing

The mechanical properties of decellularized tissues were very important factors in tissue engineering [137]. In regenerating tissues through decellularization, maintaining the mechanical properties of natural tissues was essential to ensure proper functionality [44]. The main properties of interest include modulus of elasticity, modulus of viscosity, tensile, and yield strength. However, the most critical attributes ultimately depend on the nature of the functions required by the tissue. In

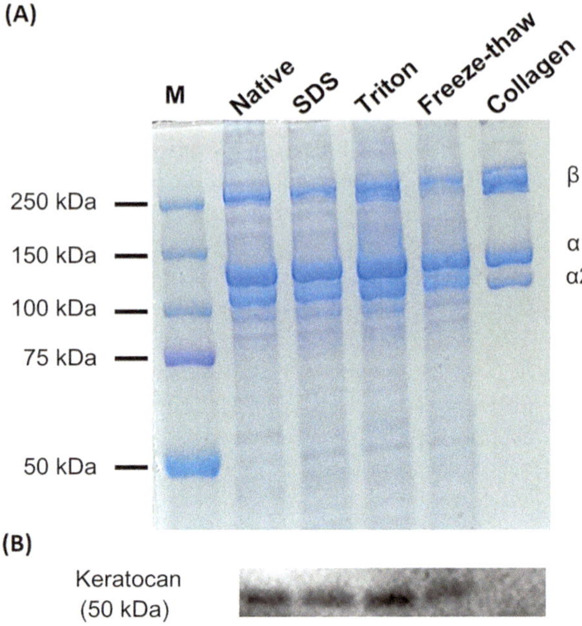

Fig. 2.14 SDS-PAGE and Western blotting to analyze the composition of the ECM-derived hydrogel. (**a**) Biochemical composition of ECM-derived hydrogel by SDS-PAGE (7%); (**b**) Western blot against keratocan. *M* molecular weight ladder. (Adapted with permission from [136]. Copyright 2019 Nature Ltd)

addition, the anisotropic or isotropic characteristics of the tissue must be adjusted, because they usually determined the direction of repopulation: this was the case with cardiomyocytes in myocardial regeneration [138]. These properties are mainly controlled by the ECM structural proteins. Each decellularization approach will have a different effect on these proteins, so the technology should be selected based on the tissue biomechanics required for proper function. The biomechanical properties of the decellularized ECM scaffold are also very important factors in tissue engineering. Acellular xenogeneic or allogeneic matrices have been used in tissue engineering heart valves, because the biomechanical properties of the valve can be preserved by the best decellularization method, which can remove cells while keeping the matrix intact.

With the verification of complete cell removal, the impact of decellularization on the mechanical properties of the ECM scaffolds is of interest. The study by Courtman et al. showed that the prepared decellularized pericardial matrix using a four-step detergent extraction and enzymatic digestion method had similar mechanical properties to natural tissues in addition to slightly increased stress relaxation [139]. The study by Samouillan et al. indicated that the method of decellularization could influence mechanical properties of acellular tissues [140]. The biomechanical tests showed that decellularization approach reported by Baiguera et al. did not led to any influence on tracheal mechanical properties [141]. Choi et al. reported that the decellularization of the human corneal stroma reliably removed cellular components from these tissues while retaining the ECM structure and sufficient mechanical properties (Fig. 2.15) [48].

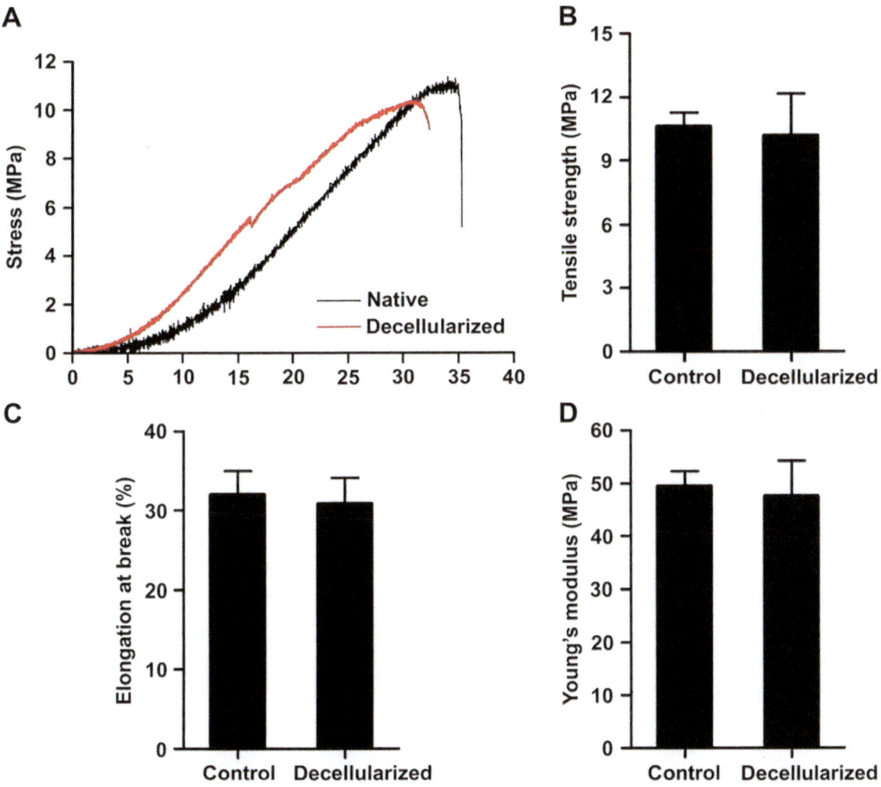

Fig. 2.15 Mechanical properties of acellular corneal stroma. The mechanical tester was used to analyze the mechanical properties of natural and decellularized human corneal stroma including stress–strain curves (**a**), tensile strength (**b**), elongation at break (**c**), and Young's modulus (**d**). (Adapted with permission from [48]. Copyright 2010 Elsevier Ltd)

2.2.3 *Effect of Ineffective Decellularization*

Effective tissue decellularization is the proper balance between cell removal and maintaining the integrity of ECM. In order to simplify the cell removal process, excess tissue should be removed before using the decellularization agent. There are numerous issues such as immunological rejection, calcification, degeneration, compliance mismatch, and zoonosis, resulting from the ineffective decellularization. When cells are destroyed during decellularization, they release proteases, which may destroy the innate ECM structure. Thus, in order to inhibit the activity of proteases, it is feasible to add protease inhibitors (aprotinin, leupeptin, and phenylmethylsulfonyl fluoride (PMSF)) or buffer solutions. The protease activity inhibition can be controlled by adjusting the time and temperature of decellularization.

The host tissue response after implantation of scaffolds in vivo depended on the efficiency of decellularization [142]. It is usually difficult to achieve the complete decellularization of the tissue, because some decellularized scaffolds retain cell residuals that can cause adverse host reactions in vivo (such as chronic inflammation). In addition, due to the necessity to reduce the adverse immune response generated by the allogeneic or heterogeneous receptors of the ECM, effective removal of intracellular components and epitopes related to tissue cell membranes has become a crucial matter. The chemical agents are mostly preferred for decellularization because of their approved effectiveness upon cell removal. Besides the effective removal, most of the chemical agents have adverse effects on ECM structure by irreversibly disrupting protein–protein interactions or undesired removal of vital ECM components such as elastin, collagen, and GAGs. Notably excessive exposure to the chemical agents results in an ECM with poor structural properties. Therefore, there is an unmet need for a fast and effective methodology that can provide a preserved ECM structure. In the decellularization process, the most commonly employed to remove cellular components from tissues was surfactants such as SDS. However, residual surfactants may be cytotoxic in vivo and are reported to affect remodeling after implantation. Moreover, the decellularization with surfactants may destroy important ECM structures that allow the decellularized tissue to act as a scaffold for cells. Although enzymes possess the specificity for removing undesirable components, enzyme residues may impair recellularization or cause an immune response. The use of ethanol can facilitate protein precipitation. Therefore, its use is limited to the elimination of phospholipids from tissues, because the phospholipids remained in tissues could lead to calcification.

If the tissue is not decellularized properly and/or overuse of chemicals, the original ECM structure may be lost, and the ECM protein present in the tissue may also be lost in large amounts. As mentioned earlier, a balance must be struck between removing cellular material and damaging the matrix. Decellularization agents with the ability to remove cell residues may destroy the collagen structure and eliminate growth factors and other key ECM components. However, a strong inflammatory response can be caused by insufficient decellularization. Small intestinal submucosa ECM produced by phosphate-buffered saline washes, and PAA treatment of different time had been compared by Keane et al. [115]. The results indicated that the degree of decellularization was related to host tissue response, the absence of seroma formation, and less swelling of surrounding tissues, as measured by reduced DNA content. However, DNA content is not the only determining factor of host response. In order to study how the scaffold remodeling and host's immune response are influenced by residual DNA and other cellular remnants, more research will be needed.

Studies on ineffective decellularization methods have shown that materials containing large amounts of residual DNA have a proinflammatory response. Moreover, a small but growing number of studies had revealed that acellular scaffolds might have inhibitory effects on cell proliferation and even exhibit cytotoxic effects [143]. Researchers had attributed these negative effects to various factors, such as residual sterilization chemicals, residual detergents, and matrix structure or biochemical changes owing to decellularization [144]. Complete decellularization of the tissue is still essential to minimize or avoid adverse immune responses to foreign cell antigens, which can cause the calcification and trigger adverse immune responses following implantation. Unfortunately, a complete acellular xenogeneic scaffold can sometimes still cause foreign body-type inflammation. In addition to the complete removal of cells and cell debris, there is an additional consideration that decellularization methods may have an adverse effect on the residual ECM. For example, enzyme treatment can degrade ECM components, chemical treatment can denature or selectively remove ECM components, and physical decellularization approaches can destroy the ECM structure and change the ECM mechanism.

It had been suggested that the SDS remaining in ECM biomaterials had serious cytotoxicity, which might be the reason for preventing the cellular ingrowth [13]. Rieder et al. [89] reported that the use of SDS decellularized scaffolds could not be recellularized, and the scaffold decellularized with sodium deoxycholate and tert-octylphenyl-polyoxyethylene could enable host recellularization. The residual detergent content in the washing solution was studied after decellularization with SDS or a mixture of SDS and sodium deoxycholate [145]. The results showed that a detergent concentration of less than 50 mg/L had no influence on the receptiveness of the scaffold to reseeding with cells. The combination of SDS and sodium deoxycholate was easier to flush out of the tissue.

Under acidic conditions, the methylene blue and SDS can produce blue compounds. This colorimetric assay can be used to quantify the residual SDS in the ECM. Wang et al. reported that the residual SDS fell below 10 ug/mg dry weight after SDS perfusion, and then washing could almost completely remove the residual SDS. Sansoto et al. [146] evaluated the influence of the SDS on decellularizing uterine tissue compared with high hydrostatic pressure. Their method realized effective decellularization. The samples decellularized using high hydrostatic pressure could stimulate the migration of cells into the scaffold, while cells did not migrate into the acellular scaffolds produced by SDS decellularization.

The decellularization method can significantly affect the host response to the biological scaffold. These effects may still be due to the difference in structure and composition caused by the decellularization method, rather than the residual detergent in the scaffold by SDS treatment. The study showed that the viability of the cells and level of repopulation were the same regardless of the washing method and the level of SDS in the scaffold [147].

2.3 Subsequent Cleaning and Sterilization

2.3.1 Subsequent Cleaning of Decellularized Tissues

Physical methods are employed in combination with chemical and biological agents to lyse cells and then rinse to remove cell residues. The decellularization of tissues can be optimized by choosing appropriate agents and techniques. Generally, the tissue of interest is briefly exposed to a chemical or detergent solution and then rinsed, combined with some physical treatments, e.g. agitation, freeze-thaw, or manual scraping to promote the removal of cellular components. After decellularization, all remaining chemicals must be removed to avoid host tissues from adversely responding to the chemicals. There is a need to develop methods to quantify the remaining chemicals in decellularized scaffolds. Similarly, some of the processes already described above include enzymes (i.e. DNase, RNase, trypsin) that are usually derived from bovine sources. Although these enzymes can shorten the length of the fragment and subsequently prevent an obvious immunogenic response, they are almost unable to separate the fragment from the ECM. Conversely, due to the heterogeneous source of nucleases, there is a risk of triggering an immune response. Extensive cleaning procedures are usually followed to remove chemicals, DNase and RNase residues.

After decellularization, it is crucial to flush away remaining chemicals from the ECM, especially cleaning agents that penetrate into thick tissues. Even thin tissues need multiple stirrings to completely remove the detergent. Regardless of the decellularization protocol used, the final step almost always involved extensive washing, usually with deionized water or PBS [148]. According to the report by Singelyn et al., an aliquot of the decellularized ECM was washed overnight with deionized water after completion of decellularization before lyophilization [64]. Chen et al. prepared the decellularized fibrocartilaginous matrix graft. The fibrocartilage tissues were immersed in 0.1% SDS for 12 h at 4 °C with gentle agitation and then rinsed with PBS three times (8 h/each time). Then the tissues were digested with nuclease solution at 37 °C for 12 h with agitation and then rinsed with PBS three times at 4 °C with gentle agitation (6 h/each time) [149]. Cornelison et al. examined several wash buffers and conditions to achieve effective apoptotic body removal (Fig. 2.16) [107]. It is worth remembering that any residual chemical reagents in the ECM after decellularization might have cytotoxic effects on colonizing cells. In order to completely remove these residues, several PBS washing cycles were used.

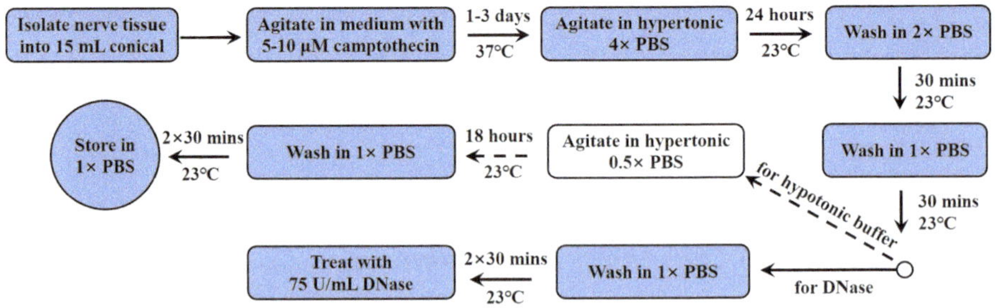

Fig. 2.16 Flow chart of the generalized process of apoptosis decellularization. The dashed arrows represent steps excluded from the final process of nerve tissue. (Adapted with permission from [107]. Copyright 2018 Elsevier Ltd)

If the residue is still present during the rinsing step, then the chemicals remaining in the tissue will determine the response of the cells. Therefore, an assay, which can accurately determine the presence of chemical substances that are preferably used in a quantitative manner in any protocol is needed. Alternatively, it would be very attractive to develop new methods for tissue decellularization that do not require chemical residues.

2.3.2 Regulatory Requirements for Sterilization

Biological scaffolds of decellularized tissues are usually regulated by the FDA as a medical device. This classification needs that the method of terminal sterilization of ECM products must comply with guidelines on bacterial load. The influence of terminal sterilization on the structure of ECM is a special concern for ECM. All biological scaffolds must undergo sterility testing before they can be approved for clinical use. The microbial sterilization verification is to evaluate the microbial lethality of sterilization and guarantee that the residual level meets the requirements. There exist other key requirements for the sterilization protocol used for decellularized tissues. First of all, sterilization should not elicit major changes in the decellularized tissues that could cause adverse response in vivo. Second, sterilization should not obviously change the mechanical properties of decellularized tissues that could impair the function of the ECM.

2.3.3 Sterilization of Decellularized Tissues

Before implantation or in vitro use, it is necessary to sterilize the ECM scaffold. In the process of tissue collection, processing, and decellularization, the resulting tissue may be contaminated by fungi, bacteria, and other organisms. Thus, these decellularized tissues require to be sterilized before application. There existed various sterilization methods including γ-irradiation, ethylene oxide, $SCCO_2$, etc. [150]. However, each method has its own beneficial or harmful effects.

The clinical application of acellular tissue scaffolds would require sterilization of the donor scaffold. For instance, the acellular dermal matrix used for breast reconstruction and abdominal wall repair needed to be sterilized prior to use. The sterilization process of the decellularized scaffold mainly eliminates endotoxins, complete viral and bacterial DNA. These toxins may cause harmful inflammation after the scaffold is implanted. Before implanting any decellularized tissue, a final sterilization method is required. The way should be tested according to the standards applicable in the medical material to guarantee that its structure and function are maintained. For this reason, the suitable sterilization method is supposed to eliminate all biological burdens in biological materials and, at the same time, prevent major changes in biological and mechanical properties [151].

2.3.3.1 γ-Irradiation

For more than a half-century, γ-irradiation has been employed as a low-temperature sterilization way. γ-Irradiation is a photon emitted by a 60 Co light source, without charge or mass. The wavelength of γ-irradiation is lower than 1010 m, the frequency is higher than 1018 Hz, and the energy is 100 keV. Compared with e-beam, γ-irradiation has greater penetrating power. This is a safe, efficient sterilization of various medical products (e.g. transplanted tissues, disposable medical equipment, medicines, medical devices). γ-Irradiation attacks DNA molecules to kill microorganisms. There are no chemical residues by using γ-sterilization that does not generate radioactivity. Mirazul Islam et al. assessed the effects of radiation on the optical, ultrastructure, biological, and mechanical properties of the decellularized porcine cornea (DPC) [152]. Transmission electron microscopy showed that the decellularized porcine cornea (G-DPC) irradiated with gamma rays could retain its structural integrity (Fig. 2.17). In addition, radiation did not decrease the optical properties of the tissue. When exposed to plasma, neither DPC nor G-DPC resulted in further activation of the complement system compared with natural porcine cornea. Although DPC was mechanically equivalent to natural tissues, γ-ray irradiation could improve tissue hydrophobicity, mechanical strength, and resistance to enzymatic degradation. Despite these changes, human corneal epithelial, stromal, endothelial, and hybrid neuroblastoma cells still grew and differentiated on DPC and G-DPC. Therefore, γ-ray irradiation could achieve effective tissue disinfection without affecting the key characteristics essential for the survival of corneal transplantation. However, Somers et al. [153] reported that γ-irradiation could change the structure of decellularized valves. The change included cross-linking, molecular fragmentation, and degradation of protein materials through peptide chain scission, resulting in significant changes in mechanical properties. This unfavorable structural change contributed to decreased cell adhesion.

γ-Irradiation is effective for terminal sterilization of tissues because it has the effect of resisting the spread of viral and bacterial diseases. However, γ-irradiation had a dose-dependent harmful effect on the biological and mechanical properties of tissues, especially in the high range ($>$25 kGy) [154, 155]. Sun et al. found that γ-ray irradiation could cause serious damage to the natural dermal ECM (Fig. 2.18). Even at a moderate dose, γ-ray irradiation could elicit ECM fragmentation and might severely damage the stability of collagen molecules [156–158].

2.3.3.2 Ethylene Oxide Exposure

Ethylene oxide has bactericidal, spore-killing, and virus-killing effects and has been broadly employed as a low-temperature disinfectant since the 1950s. Ethylene oxide is a colorless gas that is explosive and flammable. The bactericidal activity of ethylene oxide was the result of DNA, RNA, and the alkylation of protein in microorganisms, which prevented normal cell replication and metabolism, thereby

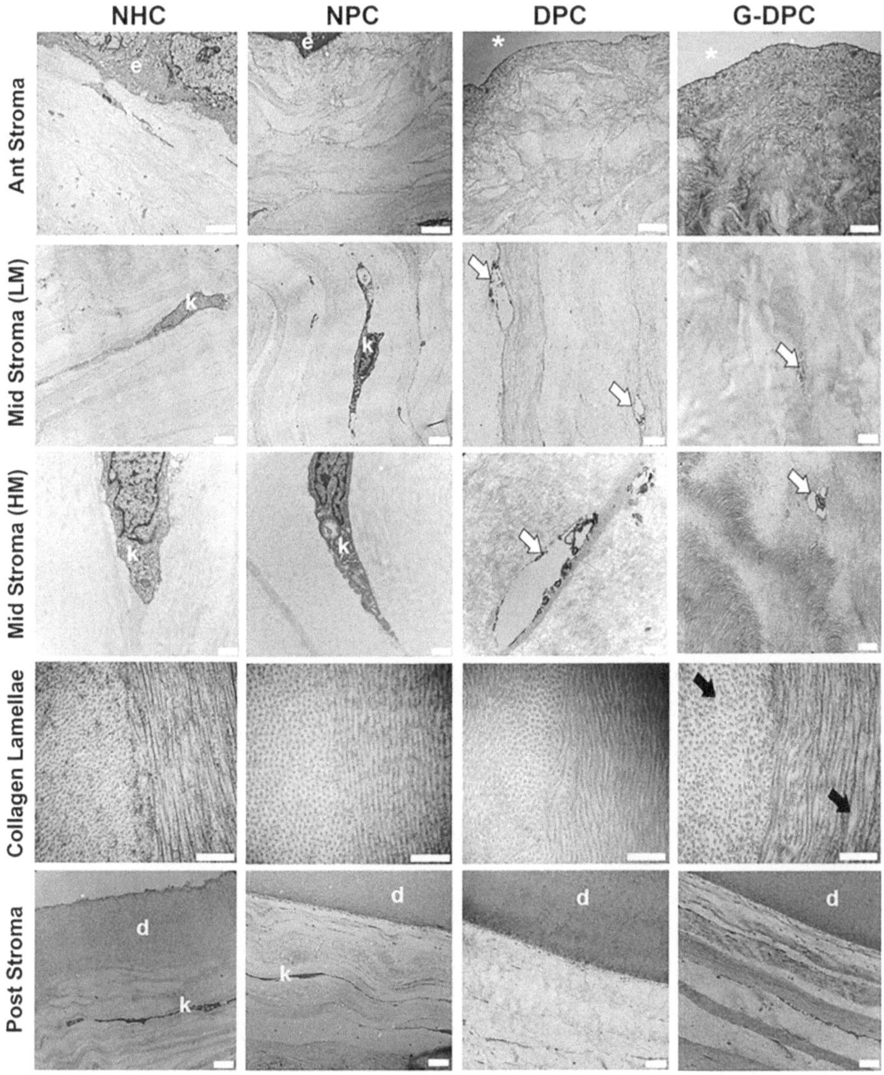

Fig. 2.17 Transmission electron micrographs of native human cornea (NHC), natural porcine cornea (NPC), decellularized porcine cornea (DPC), and gamma-irradiated decellularized porcine cornea (G-DPC). (Adapted with permission from [152]. Copyright 2019 Elsevier Ltd)

making the affected microorganisms unable to survive [159]. The high reactivity and high diffusibility of ethylene oxide play vital roles in the inactivation of microorganisms. In the process of ethylene oxide sterilization, the main factors affecting the lethality of sterilization include humidity, temperature, exposure time, and ethylene oxide concentration.

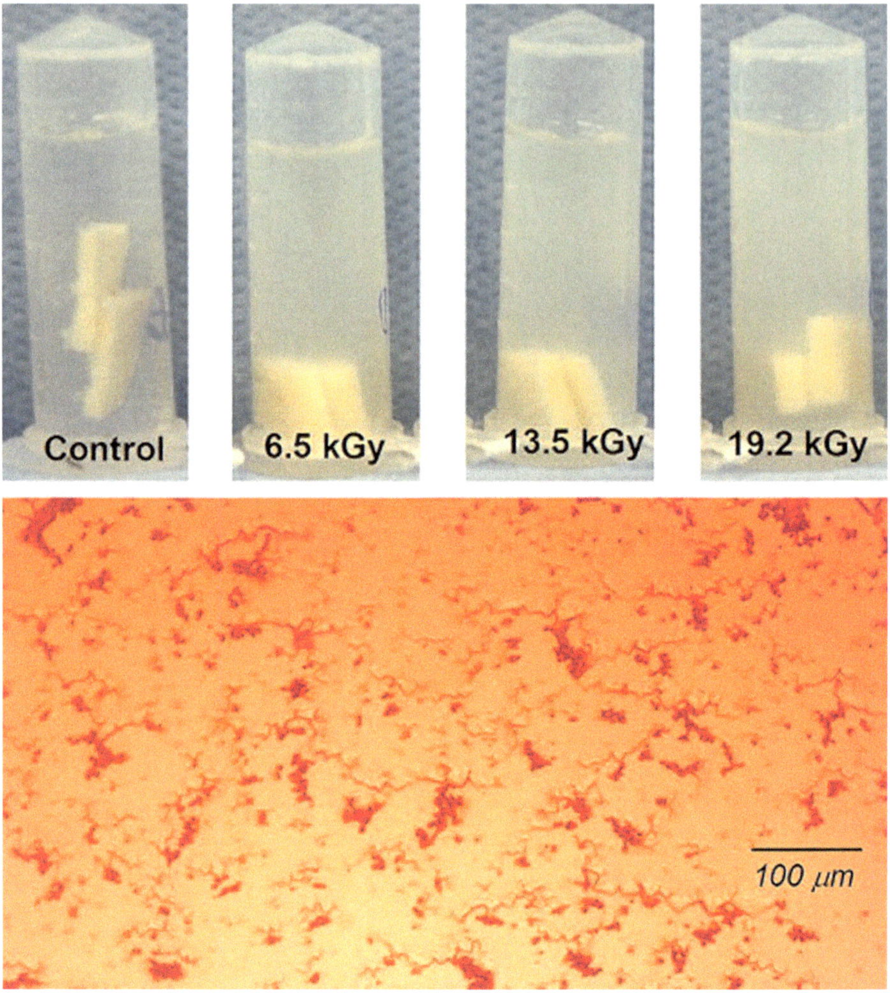

Fig. 2.18 Fragmentation of human acellular dermal matrix caused by γ-irradiation. At 37 °C, tiny tissue fragments were leached from the rehydrated and swollen tissue matrix into PBS (pH 7.5). The sirius red reagent in picric acid was employed to stain the tissue fragments. (Adapted with permission from [156]. Copyright 2008 Elsevier Ltd)

Although ethylene oxide has toxicity and safety issues, it is the acceptable sterilization way for certain sensitive decellularized tissues. Compared to radiation and heat sterilization, which have harmful influence on materials, ethylene oxide is compatible with many materials, and its permeability makes ethylene oxide sterilization the most appropriate treatment for most moisture or heat-sensitive products. It had been demonstrated that ethylene oxide sterilization could maintain the 3D architecture and topographical features of the original small intestinal submucosa (Fig. 2.19) [160]. Ethylene oxide treatment could alkylate proteins [161] or change

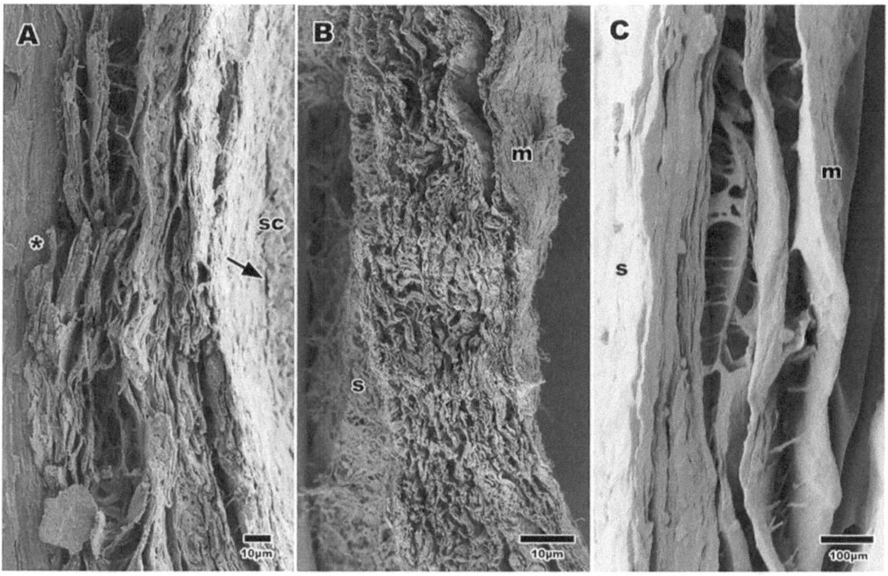

Fig. 2.19 The maintenance of 3D architecture and topographical features of small intestinal submucosa using ethylene oxide sterilization. Scanning electron micrographs of cross-sections of porcine intestine with unprocessed small intestinal submucosa in situ (**a**), small intestinal submucosa with PAA sterilization (**b**), and small intestinal submucosa with ethylene oxide sterilization (**c**) at magnifications of 500, 100, and 1000×, respectively. (Adapted with permission from [160]. Copyright 2007 Springer Ltd)

the mechanical properties of ECM [162, 163]. Ethylene oxide sterilization could leave behind residues such as mutagen and carcinogen, which might lead to undesirable host immune responses [164]. Ethylene oxide sterilization might result in a certain degree of damage to the dermal structure, but if incorporated with a glycerol treatment in advance, the damage is minimal. Thus, it should be considered carefully before the use of the ethylene oxide.

2.3.3.3 Supercritical Carbon Dioxide

$SCCO_2$ is known to have a sterilization effect. It was first discovered in the 1950s that dense carbon dioxide has a bactericidal effect [165]. The supercritical state is the only physical state defined as the "critical point" under a specific pressure and temperature combination. $SCCO_2$ has liquid and gaseous physical states, which has the viscosity of liquid and the transmission efficiency of gas that can realize high-permeability and efficient delivery. The special properties of $SCCO_2$ make it an attractive sterilization choice. Good diffusion properties of $SCCO_2$ allow it to penetrate deeply into the material. Additionally, it is non-toxic and can be readily removed by decompression and degassing. The efficiency of $SCCO_2$ on

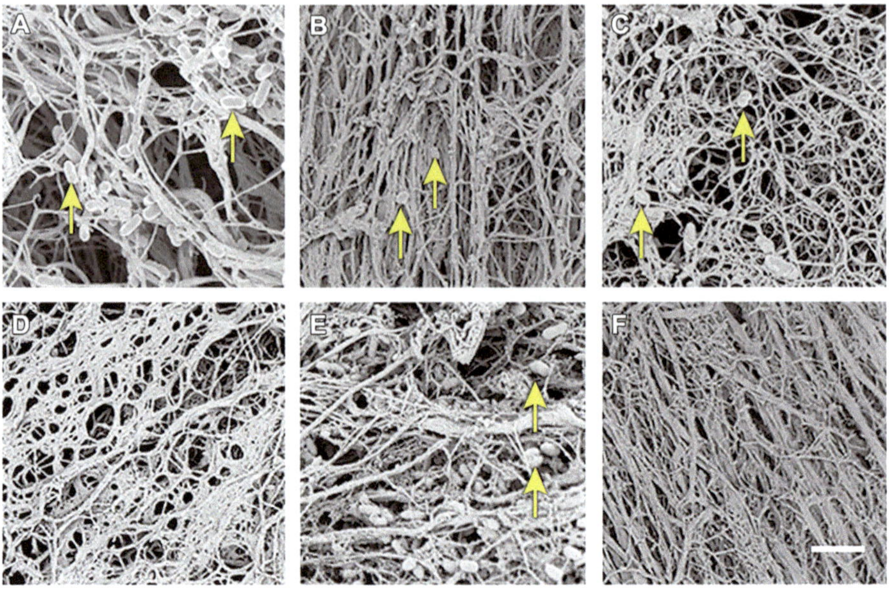

Fig. 2.20 The complete sterilization of the samples after treatment with $SCCO_2$. Scanning electron micrographs of the sinuses after sterilization by several methods. (**a**) The presence of microbes (arrows) in the non-sterile acellular sinus. (**b**) The disability to remove microbes (arrows) by sterilization with electrolyzed water. (**c**) The obvious microbes observed after sterilization with γ-radiation. (**d**) No microbes after sterilization with the ethanol and PAA. (**e**) The presence of microbes (arrows) indicating no complete sterilization of sinus tissue after treatment with liquid hydrogen peroxide. (**f**) The presence of no microbes indicating the complete sterilization of sinus tissue after treatment with $SCCO_2$. (Adapted with permission from [166]. Copyright 2017 Elsevier Ltd)

microorganisms is related to some parameters, including treatment time, pressure, and temperature.

$SCCO_2$ treatment is compatible with various decellularized tissues, including acellular dermal matrix and bone, and there are no toxic residues in the tissues. Studies had shown that $SCCO_2$-PAA sterilization had little effect on the biomechanical and biochemical properties of the acellular dermal matrix [150]. Hennessy found that the $SCCO_2$ sterilization method could effectively provide 100% sterility for the samples (Fig. 2.20) [166]. It neither damaged nor crosslinked the tissue. Compared with valves treated with other technologies, the valves treated with $SCCO_2$ had higher cusp tensile properties. The sterilization by $SCCO_2$ showed the excellent sterility and integrity in the decellularized valves. This sterilization method might be expected to be used for other decellularized soft tissues.

Supercritical fluid technology had been used in many various fields especially in sterilization of biomaterials and medical devices [167, 168]. $SCCO_2$ has been widely employed in sterilization of food and pharmaceuticals [169]. Recently, $SCCO_2$ has been studied as a way for sterilization of biologic scaffolds. $SCCO_2$ was compatible with biological materials and left no toxic residues within the treated material

[170]. $SCCO_2$ sterilization of dermal ECM had been studied and caused only minor alterations in mechanical properties. Additional research will be needed to detect the efficacy of $SCCO_2$ as a sterilization protocol for ECM obtained from other tissues.

2.4 Conclusions and Future Considerations

The decellularized tissues have the potential benefits for reconstruction of damaged or missing tissues. There are many exciting research avenues, especially related to the progress of tissue decellularization. It seems obvious that the method of removing all of the visible cellular material produces a bioscaffold material that is safe for implantation. The protocols of the tissue decellularization, subsequent cleaning and terminal sterilization described herein can provide guidelines in the preparation of effective decellularized tissues.

The main goal of decellularization is to preserve the 3D structure of ECM, which is composed of structural and functional proteins, growth factors, GAGs, etc. ECM is an immunologically inert scaffold that can retain the natural blood vessels and neural networks of the source material and can be used for tissue engineering. The efficiency of the decellularization process depends on several factors including tissue organization and density, biological composition, decellularization method, and targeted clinical application. The combination of physical, chemical, and biological methods can provide better results of cell removal. A major concern is the development of an optimal decellularization protocol, and it continues to exist in tissue repair.

Future research related to ECM technology may be carried out in multiple aspects. First, and most importantly, a more thorough basic understanding of how tissue structure affects the decellularization is needed. This will enable the selection of appropriate decellularization protocols for specific tissue types and reduce reliance on empirical methods. In addition, the development of improved decellularization technology would be beneficial. At present, most technologies either destroy the ECM structure or delete specific ECM components. A promising method to improve decellularization is based on the use of $SCCO_2$, which may be an active area of future research.

References

1. Bourgine PE, Pippenger BE, Todorov A, Tchang L, Martin I. Tissue decellularization by activation of programmed cell death. Biomaterials. 2013;34(26):6099–108.
2. Cheung DY, Duan B, Butcher JT. Current progress in tissue engineering of heart valves: multiscale problems, multiscale solutions. Expert Opin Biol Ther. 2015;15(8):1155–72.
3. Yu Y, Cui H, Zhang C, Zhang D, Yin J, Wen G, Chai Y. Human nail bed extracellular matrix facilitates bone regeneration via macrophage polarization mediated by the JAK2/STAT3 pathway. J Mater Chem B. 2020;8:4067–79.

4. Cerqueira SR, Lee YS, Cornelison RC, Mertz MW, Wachs RA, Schmidt CE, Bunge MB. Decellularized peripheral nerve supports Schwann cell transplants and axon growth following spinal cord injury. Biomaterials. 2018;177:176–85.
5. Mao AS, Mooney DJ. Regenerative medicine: current therapies and future directions. Proc Natl Acad Sci U S A. 2015;112(47):14452–9.
6. Oberwallner B, Brodarac A, Choi YH, Saric T, Anic P, Morawietz L, Stamm C. Preparation of cardiac extracellular matrix scaffolds by decellularization of human myocardium. J Biomed Mater Res A. 2014;102(9):3263–72.
7. Heath DE. A review of decellularized extracellular matrix biomaterials for regenerative engineering applications. Regenerat Eng Transl Med. 2019;5:155–66.
8. Shakouri-Motlagh A, Khanabdali R, Heath DE, Kalionis B. The application of decellularized human term fetal membranes in tissue engineering and regenerative medicine (TERM). Placenta. 2017;59:124–30.
9. Shakouri-Motlagh A, O'Connor AJ, Brennecke SP, Kalionis B, Heath DE. Native and solubilized decellularized extracellular matrix: a critical assessment of their potential for improving the expansion of mesenchymal stem cells. Acta Biomater. 2017;55:1–12.
10. Norzarini A, Kitajima T, Feng Z, Sha'ban M, Azhim A. Characterization based on biomechanical properties for meniscus scaffolds by sonication decellularization treatment. J Biomater Tiss Eng. 2017;7(3):223–32.
11. Lehr EJ, Rayat GR, Chiu B, Churchill T, McGann LE, Coe JY, Ross DB. Decellularization reduces immunogenicity of sheep pulmonary artery vascular patches. J Thorac Cardiovasc Surg. 2011;141(4):1056–62.
12. Crapo PM, Gilbert TW, Badylak SF. An overview of tissue and whole organ decellularization processes. Biomaterials. 2011;32(12):3233–43.
13. Wang Q, Zhang C, Zhang L, Guo W, Feng G, Zhou S, Zhang Y, Tian T, Li Z, Huang F. The preparation and comparison of decellularized nerve scaffold of tissue engineering. J Biomed Mater Res A. 2014;102(12):4301–8.
14. Burk J, Erbe I, Berner D, Kacza J, Kasper C, Pfeiffer B, Winter K, Brehm W. Freeze-thaw cycles enhance decellularization of large tendons. Tissue Eng Part C Methods. 2014;20(4):276–84.
15. Roth SP, Glauche SM, Plenge A, Erbe I, Heller S, Burk J. Automated freeze-thaw cycles for decellularization of tendon tissue-a pilot study. BMC Biotechnol. 2017;17:13.
16. Azhim A, Yamagami K, Muramatsu K, Morimoto Y, Tanaka M. The use of sonication treatment to completely decellularize blood arteries: A pilot study. Conf Proc IEEE Eng Med Biol Soc. 2011;2011:2468–71.
17. Fischer PD, Narayanan K, Liang MD. The use of high-frequency ultrasound for the dissection of small-diameter blood vessels and nerves. Ann Plast Surg. 1992;28(4):326–30.
18. Mardhiyah A, Sha'Ban M, Azhim A. Evaluation of histological and biomechanical properties on engineered meniscus tissues using sonication decellularization. Conf Proc IEEE Eng Med Biol Soc. 2017;2017:2064–7.
19. Azhim A, Takahashi T, Muramatsu K, Morimoto Y, Tanaka M. Decellularization of meniscal tissue using ultrasound chemical process for tissue-engineered scaffold applications. World Congr Biomech. 2010;2010:915–8.
20. Azhim A, Syazwani N, Morimoto Y, Furukawa KS, Ushida T. The use of sonication treatment to decellularize aortic tissues for preparation of bioscaffolds. J Biomater Appl. 2014;29(1):130–41.
21. Montoya CV, McFetridge PS. Preparation of ex vivo-based biomaterials using convective flow decellularization. Tissue Eng Part C Methods. 2009;15(2):191–200.
22. Sasaki S, Funamoto S, Hashimoto Y, Kimura T, Honda T, Hattori S, Kobayashi H, Kishida A, Mochizuki M. In vivo evaluation of a novel scaffold for artificial corneas prepared by using ultrahigh hydrostatic pressure to decellularize porcine corneas. Mol Vis. 2009;15:2022–8.
23. Negishi J, Funamoto S, Kimura T, Nam K, Higami T, Kishida A. Porcine radial artery decellularization by high hydrostatic pressure. J Tissue Eng Regen Med. 2015;9(11):E144–51.

24. Prasertsung I, Kanokpanont S, Bunaprasert T, Thanakit V, Damrongsakkul S. Development of acellular dermis from porcine skin using periodic pressurized technique. J Biomed Mater Res B Appl Biomater. 2008;85B(1):210–9.
25. Funamoto S, Nam K, Kimura T, Murakoshi A, Hashimoto Y, Niwaya K, Kitamura S, Fujisato T, Kishida A. The use of high-hydrostatic pressure treatment to decellularize blood vessels. Biomaterials. 2010;31(13):3590–5.
26. Huang YH, Tseng FW, Chang WH, Peng IC, Hsieh DJ, Wu SW, Yeh ML. Preparation of acellular scaffold for corneal tissue engineering by supercritical carbon dioxide extraction technology. Acta Biomater. 2017;58:238–43.
27. Sawada K, Terada D, Yamaoka T, Kitamura S, Fujisato T. Cell removal with supercritical carbon dioxide for acellular artificial tissue. J Chem Technol Biotechnol. 2008;83(6):943–9.
28. Zambon A, Vetralla M, Urbani L, Pantano MF, Ferrentino G, Pozzobon M, Pugno NM, Coppi PD, Elvassore N, Spilimbergo S. Dry acellular oesophageal matrix prepared by supercritical carbon dioxide. J Supercrit Fluids. 2016;115:33–41.
29. Guler S, Aslan B, Hosseinian P, Aydin HM. Supercritical carbon dioxide (sc-CO_2) assisted decellularization of aorta and cornea. Tissue Eng Part C Methods. 2017;23(9):540–7.
30. Casali DM, Handleton RM, Matthews MA. A novel supercritical CO_2-based decellularization method for maintaining scaffold hydration and mechanical properties. J Supercrit Fluids. 2018;131:72–81.
31. Sano MB, Neal RE, Garcia PA, Gerber D, Robertson J, Davalos RV. Towards the creation of decellularized organ constructs using irreversible electroporation and active mechanical perfusion. Biomed Eng Online. 2010;9(1):83.
32. Lee RC. Cell injury by electric forces. Ann N Y Acad Sci. 2006;1066(1):85–91.
33. Zager Y, Kain D, Landa N, Leor J, Maor E. Optimization of irreversible electroporation protocols for in-vivo myocardial decellularization. PLoS One. 2016;11(11):e0165475.
34. Wyman O, Hakim J, Dickinson ME. Decellularizing vaginal scaffolds using non-thermal irreversible electroporation of vaginal tissue-A step toward improving vaginal tissue grafts. J Pediatr Adolesc Gynecol. 2018;31(2):210.
35. Phillips M, Maor E, Rubinsky B. Nonthermal irreversible electroporation for tissue decellularization. J Biomech Eng. 2010;132(9):091003.
36. Meyer SR, Chiu B, Churchill TA. Comparison of aortic valve allograft decellularization techniques in the rat. J Biomed Mater Res A. 2006;79A(2):254–62.
37. Tudorache I, Cebotari S, Sturz G, Kirsch L, Hurschler C, Hilfiker A, Haverich A, Lichtenberg A. Tissue engineering of heart valves: biomechanical and morphological properties of decellularized heart valves. J Heart Valve Dis. 2007;16(5):567–73.
38. Stern MM, Myers RL, Hammam N, Stern KA, Eberli D, Kritchevsky SB, Soker S, Van Dyke M. The influence of extracellular matrix derived from skeletal muscle tissue on the proliferation and differentiation of myogenic progenitor cells ex vivo. Biomaterials. 2009;30 (12):2393–9.
39. Gillies AR, Smith LR, Lieber RL, Varghese S. Method for decellularizing skeletal muscle without detergents or proteolytic enzymes. Tissue Eng Part C Methods. 2011;17(4):383–9.
40. Karabekmez FE, Duymaz A, Moran SL. Early clinical outcomes with the use of decellularized nerve allograft for repair of sensory defects within the hand. Hand. 2009;4(3):245–9.
41. Guo SZ, Ren XJ, Wu B, Jiang T. Preparation of the acellular scaffold of the spinal cord and the study of biocompatibility. Spinal Cord. 2010;48(7):576–81.
42. Elder BD, Kim DH, Athanasiou KA. Developing an articular cartilage decellularization process toward facet joint cartilage replacement. Neurosurgery. 2010;66(4):722–7.
43. Reing JE, Brown BN, Daly KA, Freund JM, Gilbert TW, Hsiong SX, Huber A, Kullas KE, Tottey S, Wolf MT, Badylak SF. The effects of processing methods upon mechanical and biologic properties of porcine dermal extracellular matrix scaffolds. Biomaterials. 2010;31 (33):8626–33.
44. Freytes DO, Stoner RM, Badylak SF. Uniaxial and biaxial properties of terminally sterilized porcine urinary bladder matrix scaffolds. J Biomed Mater Res B. 2008;84B(2):408–14.

45. Nakamura N, Kimura T, Kishida A. Overview of the development, applications, and future perspectives of decellularized tissues and organs. ACS Biomater Sci Eng. 2016;3(7):1236–44.
46. White LJ, Taylor AJ, Faulk DM, Keane TJ, Saldin LT, Reing JE, Swinehart IT, Turner NJ, Ratner BD, Badylak SF. The impact of detergents on the tissue decellularization process: a ToF-SIMS study. Acta Biomater. 2017;50:207–19.
47. Gilbert TW, Wognum S, Joyce EM, Freytes DO, Sacks MS, Badylak SF. Collagen fiber alignment and biaxial mechanical behavior of porcine urinary bladder derived extracellular matrix. Biomaterials. 2008;29(36):4775–82.
48. Choi JS, Williams JK, Greven M, Walter KA, Laber PW, Khang G, Soker S. Bioengineering endothelialized neo-corneas using donor-derived corneal endothelial cells and decellularized corneal stroma. Biomaterials. 2010;31(26):6738–45.
49. Keane TJ, Swinehart IT, Badylak SF. Methods of tissue decellularization used for preparation of biologic scaffolds and in vivo relevance. Methods. 2015;84:25–34.
50. Sheridan WS, Duffy GP, Murphy BP. Mechanical characterization of a customized decellularized scaffold for vascular tissue engineering. J Mech Behav Biomed Mater. 2012;8:58–70.
51. Mendoza-Novelo B, Avila EE, Cauich-Rodríguez JV, Jorge-Herrero E, Rojo FJ, Guinea GV, Mata-Mata JL. Decellularization of pericardial tissue and its impact on tensile viscoelasticity and glycosaminoglycan content. Acta Biomater. 2011;7(3):1241–8.
52. Dong X, Wei X, Yi W, Gu C, Kang X, Liu Y, Li Q, Yi D. RGD-modified acellular bovine pericardium as a bioprosthetic scaffold for tissue engineering. J Mater Sci Mater Med. 2009;20 (11):2327–36.
53. Ozeki M, Narita Y, Kagami H, Ohmiya N, Itoh A, Hirooka Y, Niwa Y, Ueda M, Goto H. Evaluation of decellularized esophagus as a scaffold for cultured esophageal epithelial cells. J Biomed Mater Res A. 2006;79(4):771–8.
54. Grauss RW, Hazekamp MG, Van Vliet S, Gittenberger-de Groot AC, DeRuiter MC. Decellularization of rat aortic valve allografts reduces leaflet destruction and extracellular matrix remodeling. J Thorac Cardiovasc Surg. 2003;126(6):2003–10.
55. Liao J, Joyce EM, Sacks MS. Effects of decellularization on the mechanical and structural properties of the porcine aortic valve leaflet. Biomaterials. 2008;29(8):1065–74.
56. Luo Y, Lou D, Ma L, Gao C. Optimizing detergent concentration and processing time to balance the decellularization efficiency and properties of bioprosthetic heart valves. J Biomed Mater Res A. 2019;107(1):1–9.
57. Faulk DM, Carruthers CA, Warner HJ, Kramer CR, Reing JE, Zhang L, D'Amore A, Badylak SF. The effect of detergents on the basement membrane complex of a biologic scaffold material. Acta Biomater. 2014;10(1):183–93.
58. Vavken P, Joshi S, Murray MM. TRITON-X is most effective among three decellularization agents for ACL tissue engineering. J Orthop Res. 2009;27(12):1612–8.
59. Cartmell JS, Dunn MG. Effect of chemical treatments on tendon cellularity and mechanical properties. J Biomed Mater Res. 2000;49(1):134–40.
60. Roderjan JG, Noronha L, Stimamiglio MA, Correa A, Leitolis A, Bueno RRL, Costa FDA. Structural assessments in decellularized extracellular matrix of porcine semilunar heart valves: Evaluation of cell niches. Xenotransplantation. 2019;26(3):e12503.
61. Wilshaw SP, Kearney JN, Fisher J, Ingham E. Production of an acellular amniotic membrane matrix for use in tissue engineering. Tissue Eng. 2006;12(8):2117–29.
62. Bolland F, Korossis S, Wilshaw SP, Ingham E, Fisher J, Kearney JN, Southgate J. Development and characterisation of a full-thickness acellular porcine bladder matrix for tissue engineering. Biomaterials. 2007;28(6):1061–70.
63. Zhang J, Zhang CW, Du LQ, Wu XY. Acellular porcine corneal matrix as a carrier scaffold for cultivating human corneal epithelial cells and fibroblasts in vitro. Int J Ophthalmol. 2016;9 (1):1–8.
64. Singelyn JM, Sundaramurthy P, Johnson TD, Schup-Magoffin PJ, Hu DP, Faulk DM, Wang J, Mayle KM, Bartels K, Salvatore M, Kinsey AM, Demaria AN, Dib N, Christman

KL. Catheter-deliverable hydrogel derived from decellularized ventricular extracellular matrix increases endogenous cardiomyocytes and preserves cardiac function post-myocardial infarction. J Am Coll Cardiol. 2012;59(8):751–63.
65. Gilbert TW, Sellaro TL, Badylak SF. Decellularization of tissues and organs. Biomaterials. 2006;27(19):3675–83.
66. Hudson TW, Liu SY, Schmidt CE. Engineering an improved acellular nerve graft via optimized chemical processing. Tissue Eng. 2004;10(9-10):1346–58.
67. Du L, Wu X, Pang K, Yang Y. Histological evaluation and biomechanical characterisation of an acellular porcine cornea scaffold. Br J Ophthalmol. 2011;95(3):410–4.
68. Horowitz B, Bonomo R, Prince AM, Chin SN, Brotman B, Shulman RW. Solvent/detergent-treated plasma: a virus-inactivated substitute for fresh frozen plasma. Blood. 1992;79 (3):826–31.
69. Deeken CR, White AK, Bachman SL, Ramshaw BJ, Cleveland DS, Loy TS, Grant SA. Method of preparing a decellularized porcine tendon using tributyl phosphate. J Biomed Mater Res B Appl Biomater. 2011;96(2):199–206.
70. Woods T, Gratzer PF. Effectiveness of three extraction techniques in the development of a decellularized bone-anterior cruciate ligament-bone graft. Biomaterials. 2005;26 (35):7339–49.
71. Cartmell JS, Dunn MG. Development of cell-seeded patellar tendon allografts for anterior cruciate ligament reconstruction. Tissue Eng. 2004;10(7-8):1065–75.
72. Rana D, Zreiqat H, Benkirane-Jessel N, Ramakrishna S, Ramalingam M. Development of decellularized scaffolds for stem cell-driven tissue engineering. J Tissue Eng Regen Med. 2017;11(4):942–65.
73. Yang B, Zhang Y, Zhou L, Sun Z, Zheng J, Chen Y, Dai Y. Development of a porcine bladder acellular matrix with well-preserved extracellular bioactive factors for tissue engineering. Tissue Eng Part C Methods. 2010;16(5):1201–11.
74. Xu CC, Chan RW, Tirunagari N. A biodegradable, acellular xenogeneic scaffold for regeneration of the vocal fold lamina propria. Tissue Eng. 2007;13(3):551–66.
75. Ota T, Taketani S, Iwai S, Miyagawa S, Furuta M, Hara M, Uchimura E, Okita Y, Sawa Y. Novel method of decellularization of porcine valves using polyethylene glycol and gamma irradiation. Ann Thorac Surg. 2007;83(4):1501–7.
76. Vergara A, Paduano L, Sartorio R. Mechanism of protein-poly(ethylene glycol) interaction from a diffusive point of view. Macromolecules. 2002;35(4):1389–98.
77. Uchimura E, Sawa Y, Taketani S, Yamanaka Y, Hara M, Matsuda H, Miyake J. Novel method of preparing acellular cardiovascular grafts by decellularization with poly(ethylene glycol). J Biomed Mater Res A. 2003;67(3):834–7.
78. Levy RJ, Vyavahare N, Ogle M, Ashworth P, Bianco R, Schoen FJ. Inhibition of cusp and aortic wall calcification in ethanol- and aluminum-treated bioprosthetic heart valves in sheep: background, mechanisms, and synergism. J Heart Valve Dis. 2003;12(2):209–16.
79. Dunmore-Buyze J, Boughner DR, Macris N, Vesely I. A comparison of macroscopic lipid content within porcine pulmonary and aortic valves. Implications for bioprosthetic valves. J Thorac Cardiovasc Surg. 1995;110(6):1756–61.
80. Brown BN, Freund JM, Han L, Rubin JP, Reing JE, Jeffries EM, Wolf MT, Tottey S, Barnes CA, Ratner BD, Badylak SF. Comparison of three methods for the derivation of a biologic scaffold composed of adipose tissue extracellular matrix. Tissue Eng Part C Methods. 2011;17 (4):411–21.
81. Márquez SP, Martínez VS, Ambrose WM, Wang J, Gantxegui NG, Schein O, Elisseeff J. Decellularization of bovine corneas for tissue engineering applications. Acta Biomater. 2009;5(6):1839–47.
82. Levy RJ, Vyavahare N, Ogle M, Ashworth P, Bianco R, Schoen FJ. Inhibition of cusp and aortic wall calcification in ethanol-and aluminum-treated bioprosthetic heart valves in sheep: background, mechanisms, and synergism. J Heart Valve Dis. 2003;12(2):209–16.

83. Lumpkins SB, Pierre N, McFetridge PS. A mechanical evaluation of three decellularization methods in the design of a xenogeneic scaffold for tissue engineering the temporomandibular joint disc. Acta Biomater. 2008;4(4):808–16.
84. Jamur MC, Oliver C. Cell fixatives for immunostaining. Methods Mol Biol. 2010;588:55–61.
85. Cole MB. Alteration of cartilage matrix morphology with histological processing. J Microsc. 1984;133(Pt 2):129–40.
86. Gorschewsky O, Puetz A, Riechert K, Klakow A, Becker R. Quantitative analysis of biochemical characteristics of bone-patellar tendon-bone allografts. Biomed Mater Eng. 2005;15 (6):403–11.
87. Fu RH, Wang YC, Liu SP, Shih TR, Lin HL, Chen YM, Sung JH, Lu CH, Wei JR, Wang ZW, Huang SJ, Tsai CH, Shyu WC, Lin SZ. Decellularization and recellularization technologies in tissue engineering. Cell Transplant. 2014;23(4):621–30.
88. Gupta SK, Mishra NC, Dhasmana A. Decellularization methods for scaffold fabrication. Methods Mol Biol. 2017;1577:1–10.
89. Rieder E, Kasimir MT, Silberhumer G, Seebacher G, Wolner E, Simon P, Weigel G. Decellularization protocols of porcine heart valves differ importantly in efficiency of cell removal and susceptibility of the matrix to recellularization with human vascular cells. J Thorac Cardiovasc Surg. 2004;127(2):399–405.
90. Grauss RW, Hazekamp MG, Oppenhuizen F, van Munsteren CJ, Gittenberger-de Groot AC, DeRuiter MC. Histological evaluation of decellularised porcine aortic valves: matrix changes due to different decellularisation methods. Eur J Cardio Thorac. 2005;27(4):566–71.
91. Jones G, Herbert A, Berry H, Edwards JH, Fisher J, Ingham E. Decellularisation and characterisation of porcine superflexor tendon: a potential anterior cruciate ligament replacement. Tissue Eng Part A. 2017;23(3-4):124–34.
92. Waldrop FS, Puchtler H, Meloan SN, Younker TD. Histochemical investigations of different types of collagen. Acta Histochem Suppl. 1980;21:23–31.
93. Yang M, Chen CZ, Wang XN, Zhu YB, Gu YJ. Favorable effects of the detergent and enzyme extraction method for preparing decellularized bovine pericardium scaffold for tissue engineered heart valves. J Biomed Mater Res B Appl Biomater. 2009;91(1):354–61.
94. Schenke-Layland K, Vasilevski O, Opitz F, König K, Riemann I, Halbhuber KJ, Wahlers T, Stock UA. Impact of decellularization of xenogeneic tissue on extracellular matrix integrity for tissue engineering of heart valves. J Struct Biol. 2003;143(3):201–8.
95. Chen RN, Ho HO, Tsai YT, Sheu MT. Process development of an acellular dermal matrix (ADM) for biomedical applications. Biomaterials. 2004;25(13):2679–86.
96. Flynn LE. The use of decellularized adipose tissue to provide an inductive microenvironment for the adipogenic differentiation of human adipose-derived stem cells. Biomaterials. 2010;31 (17):4715–24.
97. Porzionato A, Sfriso MM, Macchi V, Rambaldo A, Lago G, Lancerotto L, Vindigni V, De Caro R. Decellularized omentum as novel biologic scaffold for reconstructive surgery and regenerative medicine. Eur J Histochem. 2013;57(1):e4.
98. Xu H, Wan H, Zuo W, Sun W, Owens RT, Harper JR, Ayares DL, McQuillan DJ. A porcine-derived acellular dermal scaffold that supports soft tissue regeneration: Removal of terminal Galactose-α-(1,3)-Galactose and retention of matrix structure. Tissue Eng Part A. 2009;15 (7):1807–19.
99. Galili U. The α-gal epitope (Galα1-3Galβ1-4GlcNAc-R) in xenotransplantation. Biochimie. 2001;83(7):557–63.
100. Daly KA, Stewart-Akers AM, Hara H, Ezzelarab M, Long C, Cordero K, Johnson SA, Ayares D, Cooper DKC, Badylak SF. Effect of the αGal epitope on the response to small intestinal submucosa extracellular matrix in a nonhuman primate model. Tissue Eng Part A. 2009;15(12):3877–88.
101. Gailit J, Ruoslahti E. Regulation of the fibronectin receptor affinity by divalent cations. J Biol Chem. 1988;263(26):12927–32.

102. Maurer P, Hohenester E. Structural and functional aspects of calcium binding in extracellular matrix proteins. Matrix Biol. 1997;15(8-9):569–80.
103. Hopkinson A, Shanmuganathan VA, Gray T, Yeung AM, Lowe J, James DK, Dua HS. Optimization of amniotic membrane (AM) denuding for tissue engineering. Tissue Eng Part C Methods. 2008;14(4):371–81.
104. Zhou J, Fritze O, Schleicher M, Wendel HP, Schenke-Layland K, Harasztosi C, Hu S, Stock UA. Impact of heart valve decellularization on 3-D ultrastructure, immunogenicity and thrombogenicity. Biomaterials. 2010;31(9):2549–54.
105. Bourgine PE, Scotti C, Pigeot S, Tchang LA, Todorov A, Martin I. Osteoinductivity of engineered cartilaginous templates devitalized by inducible apoptosis. Proc Natl Acad Sci U S A. 2014;111(49):17426–31.
106. Bourgine PE, Gaudiello E, Pippenger B, Jaquiéry C. Engineered extracellular matrices as biomaterials of tunable composition and function. Adv Funct Mater. 2017;27(7):1605486.
107. Cornelison RC, Wellman SM, Park JH, Porvasnik SL, Song YH, Wachs RA, Schmidt CE. Development of an apoptosis-assisted decellularization method for maximal preservation of nerve tissue structure. Acta Biomater. 2018;77:116–26.
108. Ishino N, Fujisato T. Decellularization of porcine carotid by the recipient's serum and evaluation of its biocompatibility using a rat autograft model. J Artif Organs. 2015;18:136–42.
109. Gui L, Chan SA, Breuer CK, Niklason LE. Novel utilization of serum in tissue decellularization. Tissue Eng Part C Methods. 2010;16(2):173–84.
110. Evaristo TC, CruzAlves FCM, Moroz A, Mion W, Acorci-Valério MJ, Felisbino SL, Rossi-Ferreira R, Ruiz Júnior RL, Deffune E. Light-emitting diode effects on combined decellularization of tracheae. A novel approach to obtain biological scaffolds. Acta Cir Bras. 2014;29(8):485–92.
111. Li N, Li Y, Gong D, Xia C, Liu X, Xu Z. Efficient decellularization for bovine pericardium with extracellular matrix preservation and good biocompatibility. Interact Cardiovasc Thorac Surg. 2018;26(5):768–76.
112. Choi YC, Choi JS, Kim BS, Kim JD, Yoon HI, Cho YW. Decellularized extracellular matrix derived from porcine adipose tissue as a xenogeneic biomaterial for tissue engineering. Tissue Eng Part C Methods. 2012;18(11):866–76.
113. Wang L, Johnson JA, Zhang Q, Beahm EK. Combining decellularized human adipose tissue extracellular matrix and adipose-derived stem cells for adipose tissue engineering. Acta Biomater. 2013;9(11):8921–31.
114. Baiguera S, Del Gaudio C, Lucatelli E, Kuevda E, Boieri M, Mazzanti B, Bianco A, Macchiarini P. Electrospun gelatin scaffolds incorporating rat decellularized brain extracellular matrix for neural tissue engineering. Biomaterials. 2014;35(4):1205–14.
115. Luo L, Eswaramoorthy R, Mulhall KJ, Kelly DJ. Decellularization of porcine articular cartilage explants and their subsequent repopulation with human chondroprogenitor cells. J Mech Behav Biomed. 2015;55:21–31.
116. Ketchedjian A, Jones AL, Krueger P, Robinson E, Crouch K, Wolfinbarger L, Hopkins R. Recellularization of decellularized allograft scaffolds in ovine great vessel reconstructions. Ann Thorac Surg. 2005;79(3):888–96.
117. Keane TJ, Londono R, Turner NJ, Badylak SF. Consequences of ineffective decellularization of biologic scaffolds on the host response. Biomaterials. 2012;33(6):1771–81.
118. Daly KA, Liu S, Agrawal V, Brown BN, Huber A, Johnson SA, Reing J, Sicari B, Wolf M, Zhang X, Badylak SF. The host response to endotoxin-contaminated dermal matrix. Tissue Eng Part A. 2012;18(11-12):1293–303.
119. Zhang Q, Raoof M, Chen Y, Sumi Y, Sursal T, Junger W, Brohi K, Itagaki K, Hauser CJ. Circulating mitochondrial DAMPs cause inflammatory responses to injury. Nature. 2010;464(7285):104–7.
120. Zheng MH, Chen J, Kirilak Y, Willers C, Xu J, Wood D. Porcine small intestine submucosa (SIS) is not an acellular collagenous matrix and contains porcine DNA: possible implications in human implantation. J Biomed Mater Res B Appl Biomater. 2005;73:61–7.

121. Wong ML, Griffiths LG. Immunogenicity in xenogeneic scaffold generation: antigen removal vs. decellularization. Acta Biomater. 2014;10(5):1806–16.
122. Boeer U, Buettner FFR, Klingenberg M, Antonopoulos GC, Meyer H, Haverich A, Wilhelmi M. Immunogenicity of intensively decellularized equine carotid arteries is conferred by the extracellular matrix protein collagen type VI. PLoS One. 2014;9(8):e105964.
123. Parmaksiz M, Elcin AE, Elcin YM. Decellularization of bovine small intestinal submucosa and its use for the healing of a critical-sized full-thickness skin defect, alone and in combination with stem cells, in a small rodent model. J Tissue Eng. 2017;11(6):1754–65.
124. Baiguera S, Del Gaudio C, Kuevda E, Gonfiotti A, Bianco A, Macchiarini P. Dynamic decellularization and cross-linking of rat tracheal matrix. Biomaterials. 2014;35(24):6344–50.
125. Remlinger NT, Czajka CA, Juhas ME, Vorp DA, Stolz DB, Badylak SF, Gilbert S, Gilbert TW. Hydrated xenogeneic decellularized tracheal matrix as a scaffold for tracheal reconstruction. Biomaterials. 2010;31(13):3520–6.
126. Lopera HM, Griffiths LG. Antigen removal process preserves function of small diameter venous valved conduits, whereas SDS-decellularization results in significant valvular insufficiency. Acta Biomater. 2020;107:115–28.
127. Xu J, Liu S, Wang S, Qiu P, Chen P, Lin X, Fang X. Decellularised nucleus pulposus as a potential biologic scaffold for disc tissue engineering. Mater Sci Eng C. 2019;99:1213–25.
128. Kakkar R, Grover SR. Theoretical study of molecular recognition by Hoechst 33258 derivatives. J Biomol Struct Dyn. 2005;23:37–47.
129. Quintana JR, Lipanov AA, Dickerson RE. Low-temperature crystallographic analyses of the binding of Hoechst 33258 to the doublehelical DNA dodecamer C-G-C-G-A-A-T-T-C-G-C-G. Biochemistry. 1991;30:10294–306.
130. Jackson DW, Simon TM. Donor cell survival and repopulation after intraarticular transplantation of tendon and ligament allografts. Microsc Res Tech. 2002;58:25–33.
131. Weis C, Littlefield N, Jackson CD. A direct fluorometric assay for DNA in alkaline sucrose gradients using Hoechst 33258 dye. Am Biotechnol Lab. 1993;11(11):50–2.
132. Syed O, Walters NJ, Day RM, Kim HW, Knowles JC. Evaluation of decellularization protocols for production of tubular small intestine submucosa scaffolds for use in oesophageal tissue engineering. Acta Biomater. 2014;10(12):5043–54.
133. Kral T, Widerak K, Langner M, Hof M. Propidium iodide and PicoGreen as dyes for the DNA fluorescence correlation spectroscopy measurements. J Fluoresc. 2005;15(2):179–83.
134. Youngstrom DW, Barrett JG, Jose RR, Kaplan DL. Functional characterization of detergent-decellularized equine tendon extracellular matrix for tissue engineering applications. PLoS One. 2013;8(5):e64151.
135. Dorri Y. Two-dimensional gel electrophoresis: vertical isoelectric focusing. Methods Mol Biol. 2012;869:235–46.
136. Fernández-Pérez J, Ahearne M. The impact of decellularization methods on extracellular matrix derived hydrogels. Sci Rep. 2019;9:14933.
137. Rodriguez-Rodriguez VE, Martínez-González B, Quiroga-Garza A, Reyes-Hernández CG, Fuente-Villarreal D, Garza-Castro O, Guzmán-López S, Elizondo-Omana RE. Human umbilical vessels: choosing the optimal decellularization method. ASAIO J. 2018;64(5):575–80.
138. Wang F, Guan J. Cellular cardiomyoplasty and cardiac tissue engineering for myocardial therapy. Adv Drug Deliv Rev. 2010;62(7-8):784–97.
139. Courtman DW, Pereira CA, Kashef V, McComb D, Lee JM, Wilson GJ. Development of a pericardial acellular matrix biomaterial: biochemical and mechanical effects of cell extraction. J Biomed Mater Res. 1994;28(6):655–66.
140. Samouillan V, Dandurand-Lods J, Lamure A, Maurel E, Lacabanne C, Gerosa G, Venturini A, Casarotto D, Gherardini L, Spina M. Thermal analysis characterization of aortic tissues for cardiac valve bioprostheses. J Biomed Mater Res. 1999;46(4):531–8.
141. Baiguera S, Jungebluth P, Burns A, Mavilia C, Haag J, Coppi PD, Macchiarini P. Tissue engineered human tracheas for in vivo implantation. Biomaterials. 2010;31(34):8931–8.

142. Badylak SF. Decellularized allogeneic and xenogeneic tissue as a bioscaffold for regenerative medicine: factors that influence the host response. Ann Biomed Eng. 2014;42(7):1517–27.
143. Luo X, Kulig KM, Finkelstein EB, Nicholson MF, Liu XH, Goldman SM, Vacanti JP, Grottkau BE, Pomerantseva I, Sundback CA, Neville CM. In vitro evaluation of decellularized ECM-derived surgical scaffold biomaterials. J Biomed Mater Res B Appl Biomater. 2017;105 (3):585–93.
144. Morris AH, Chang J, Kyriakides TR. Inadequate processing of decellularized dermal matrix reduces cell viability, in vitro, and increases apoptosis and acute inflammation, in vivo. Biores Open Access. 2016;5(1):177–87.
145. Cebotari S, Tudorache I, Jaekel T, Hilfiker A, Dorfman S, Ternes W, Haverich A, Lichtenberg A. Detergent decellularization of heart valves for tissue engineering: toxicological effects of residual detergents on human endothelial cells. Artif Organs. 2010;34(3):206–10.
146. Santoso EG, Yoshida K, Hirota Y, Aizawa M, Yoshino O, Kishida A, Osuga Y, Saito S, Ushida T, Furukawa KS. Application of detergents or high hydrostatic pressure as decellularization processes in uterine tissues and their subsequent effects on in vivo uterine regeneration in murine models. PLoS One. 2014;9(7):e103201.
147. Gratzer PF, Harrison RD, Woods T. Matrix alteration and not residual sodium dodecyl sulfate cytotoxicity affects the cellular repopulation of a decellularized matrix. Tissue Eng. 2006;12 (10):2975–83.
148. Brown BN, Buckenmeyer MJ, Prest TA. Preparation of decellularized biological scaffolds for 3D cell culture. Methods Mol Biol. 1612;2017:15–27.
149. Chen C, Chen Y, Li M, Xiao H, Shi Q, Zhang T, Li X, Zhao C, Hu J, Lu H. Functional decellularized fibrocartilaginous matrix graft for rotator cuff enthesis regeneration: a novel technique to avoid in-vitro loading of cells. Biomaterials. 2020;250:119996. https://doi.org/10.1016/j.biomaterials.2020.119996.
150. Qiu QQ, Leamy P, Brittingham J, Pomerleau J, Kabaria N, Connor J. Inactivation of bacterial spores and viruses in biological material using supercritical carbon dioxide with sterilant. J Biomed Mater Res B Appl Biomater. 2009;91B(2):572–8.
151. Fidalgo C, Iop L, Sciro M, Harder M, Mavrilas D, Korossis S, Bagno A, Palù G, Aguiari P, Gerosa G. A sterilization method for decellularized xenogeneic cardiovascular scaffolds. Acta Biomater. 2018;67:282–94.
152. Mirazul Islam M, Sharifi R, Mamodaly S, Islam R, Nahra D, Abusamra DB, Hui RC, Adibnia Y, Goulamaly M, Paschalis EI, Cruzat A, Kong J, Nilsson PH, Argeso P, Mollnes TE, Chodosh J, Dohlman CH, Gonzalez-Andrades M. Effects of gamma radiation sterilization on the structural and biological properties of decellularized corneal xenografts. Acta Biomater. 2019;96:330–44.
153. Somers P, Cuvelier CA, Somer FD, Cornelissen M, Cox E, Verloo M, Chiers K, van Nooten G. Gamma radiation alters the ultrastructure in tissue-engineered heart valve scaffolds. Tissue Eng Part A. 2009;15(11):3597–604.
154. Nguyen H, Morgan DAF, Forwood MR. Sterilization of allograft bone: is 25 kGy the gold standard for gamma irradiation. Cell Tissue Bank. 2007;8(2):81–91.
155. Gouk SS, Lim TM, Teoh SH, Sun WQ. Alterations of human acellular tissue matrix by gamma irradiation: Histology, biomechanical property, stability, in vitro cell repopulation, and remodeling. J Biomed Mater Res B Appl Biomater. 2008;84B(1):205–17.
156. Sun WQ, Leung P. Calorimetric study of extracellular tissue matrix degradation and instability after gamma irradiation. Acta Biomater. 2008;4(4):817–26.
157. Moreau MF, Gallois Y, Baslé MF, Chappard D. Gamma irradiation of human bone allografts alters medullary lipids and releases toxic compounds for osteoblast-like cells. Biomaterials. 2000;21(4):369–76.
158. Matuska AM, McFetridge PS. The effect of terminal sterilization on structural and biophysical properties of a decellularized collagen-based scaffold; implications for stem cell adhesion. J Biomed Mater Res B Appl Biomater. 2015;103(2):397–406.

159. Mendes GCC, Brandao TR, Silva CL, Portugal P. Ethylene oxide sterilization of medical devices: a review. Am J Infect Control. 2007;35(9):574–81.
160. Hodde J, Janis A, Ernst D, Zopf D, Sherman D, Johnson C. Effects of sterilization on an extracellular matrix scaffold: part I. Composition and matrix architecture. J Mater Sci. 2007;18 (4):537–43.
161. Dellarco VL, Generoso WM, Sega GA, Fowle JR, Jacobson-Kram D. Review of the mutagenicity of ethylene oxide. Environ Mol Mutagen. 1990;16(2):85–103.
162. Rosario DJ, Reilly GC, Ali SE, Glover M, Bullock AJ, Macneil S. Decellularization and sterilization of porcine urinary bladder matrix for tissue engineering in the lower urinary tract. Regen Med. 2008;3(2):145–56.
163. Jackson DW, Grood ES, Wilcox P, Butler DL, Simon TM, Holden JP. The effects of processing techniques on the mechanical properties of bone-anterior cruciate ligament-bone allografts: an experimental study in goats. Am J Sports Med. 1988;16(2):101–5.
164. Jackson DW, Windler GE, Simon TM. Intraarticular reaction associated with the use of freeze-dried, ethylene oxide-sterilized bone-patella tendon-bone allografts in the reconstruction of the anterior cruciate ligament. Am J Sports Med. 1990;18(1):1–10.
165. Fraser D. Bursting bacteria by release of gas pressure. Nature. 1951;167(4236):33–4.
166. Hennessy RS, Jana S, Tefft BJ, Helder MR, Young MD, Hennessy RR, Stoyles NJ, Lerman A. Supercritical carbon dioxide-based sterilization of decellularized heart valves. JACC. 2017;2(1):71–84.
167. Checinska A, Fruth IA, Green TL, Crawford RL, Paszczynski AJ. Sterilization of biological pathogens using supercritical fluid carbon dioxide containing water and hydrogen peroxide. J Microbiol Methods. 2011;87(1):70–5.
168. Balestrini JL, Liu A, Gard AL, Huie J, Blatt KM, Schwan J, Zhao L, Broekelmann TJ, Mecham RP, Wilcox EC, Niklason LE. Sterilization of lung matrices by supercritical carbon dioxide. Tissue Eng Part C Methods. 2016;22(3):260–9.
169. Rizvi SSH, Sikin AM. Recent patents on the sterilization of food and biomaterials by supercritical fluids. Recent Pat Food Nutr Agric. 2011;3(3):212–25.
170. Donati I, Benincasa M, Foulc MP, Turco G, Toppazzini M, Solinas D, Spilimbergo S, Kikic I, Paoletti S. Terminal sterilization of BisGMA-TEGDMA thermoset materials and their bioactive surfaces by supercritical CO_2. Biomacromolecules. 2012;13(4):1152–60.

Chapter 3
Different Forms of Decellularized Tissues and Their Characteristics, Applications in Tissue Repair as Well as Performance Optimization

Lincui Da, Xiongxin Lei, Yuting Song, Yizhou Huang, and Huiqi Xie

Abstract Decellularized tissues, which could maintain cell phenotype or promote stem cell differentiation into specific tissue cells, play a central role in promoting the reconstruction of functional tissue/organ. Tissue ECM in diverse formats has been developed for tissue repair and regeneration, which can generally be divided into five categories: (1) Scaffolds that preserve the morphological structure of original tissues. (2) Powders obtained by cryogenic grinding. (3) Hydrogels. (4) Coatings. (5) 3D-printing. This chapter will elaborate the preparation methods for creating decellularized tissue scaffolds with the above formats. What is more, the characteristics of these materials will be studied, and the respective applications and the efforts in performance optimization in tissue repair will also be introduced.

Keywords Decellularization · Formats of tissue extracellular matrix · Hydrogels · 3D-printing

Decellularized tissues, whose ultrastructure, biological characteristics, and biomechanical properties were similar to natural tissue, play a pivotal role in damaged tissue/organ functional regeneration [1–3]. These tissue ECM could be obtained from different animals by using the decellularization methods introduced in Chap. 2 to remove cell-related immunogenicity. It is composed of structural protein (such as collagen and elastin), cell adhesion protein (such as laminin and fibronectin), polysaccharide (such as glycosaminoglycans and proteoglycans), growth factor,

---

L. Da
Reproductive Center, Fujian Provincial Maternity and Children's Hospital, Affiliated Hospital of Fujian Medical University, Fuzhou, China

X. Lei · Y. Song · Y. Huang · H. Xie (✉)
Laboratory of Stem Cell and Tissue Engineering, Orthopedic Research Institute, State Key Laboratory of Biotherapy and Cancer Center, West China Hospital, Sichuan University and Collaborative Innovation Center of Biotherapy, Chengdu, China
e-mail: xiehuiqi@scu.edu.cn

X. Li, H. Xie (eds.), *Decellularized Materials*,
https://doi.org/10.1007/978-981-33-6962-7_3

and signal molecule whose specific composition is varied with the tissues [4]. Decellularized tissues could maintain cell phenotype or promote stem cell differentiation into specific tissue cells. Related researches demonstrated that adipose stem cells can differentiate into adipocytes on adipose-derived decellular matrix, bone marrow mesenchymal stem cells can differentiate into chondrocytes on cartilage-derived decellular matrix, and stem cells can differentiate into neural cells on brain derived decellular matrix [5–9].

Diverse formats of tissue ECM have been developed for tissue repair and regeneration, which can generally be divided into five categories: (1) Scaffolds in sheet and tubular. (2) Powders obtained by cryogenic grinding. (3) Hydrogels. (4) Coatings. (5) 3D-printing.

This chapter will elaborate the preparation methods for creating decellularized tissue scaffolds with the above formats, which have been derived from tissues of cartilage and osseous, connective tissue, muscle, central nervous system, eye and ear, circulatory system, skin, digestive system, respiratory system, urinary and reproductive system. What is more, the characteristics of these materials will be studied, and the respective applications and the efforts in performance optimization in tissue repair will also be introduced.

3.1 Scaffolds in Sheet and Tubular

Microarchitectures of the scaffolds have a great influence on the morphology, behavior, and function of cells and regulation of immune responses to the scaffolds [1, 10–14]. For example, cells seeded on a flat surface will fuse into cell sheets on this surface, while cells seeded on the surface of 3D irregular porous scaffolds will further grow into the pore of the scaffolds and may ultimately form tissue-engineered tissue [15]. Keeping the tissues intact during the acellular process, which will reserve the microscopic and ultrastructural features along with the biochemical composition of ECM for cell growth, is particularly useful when the purpose is to engineer transplantable tissues of the same position of the body [16].

An extraordinary variety of decellularized tissues scaffolds obtained directly after acellular disposal (e.g. acellular dermal matrix (ADM), acellular amniotic membrane (AAM), and small intestinal submucosa (SIS), etc.) are promised to repair the defect of various tissues including but not limited to: skin, tendon, bladder, nerve, esophagus, vascular, and bone, which are often in forms of sheets or tubes and have been extensively investigated in recent decades [17–23].

The ADM patch, a decellularized tissue derived from split-thickness skin, is a good example of these scaffolds. The ADM that is capable of inhibiting contraction of wound and affording skin with mechanical support has been approved by the State Food and Drug Administration and was available to large numbers of patients in clinical [24]. In the study of Srivastava et al, 0.005- or 0.017 inch-thick ADMs were successfully fabricated and proven effective as an artificial skin to repair the damage of rat full-thickness skin defects (225 mm^2) [25]. The interfibrillar spaces between

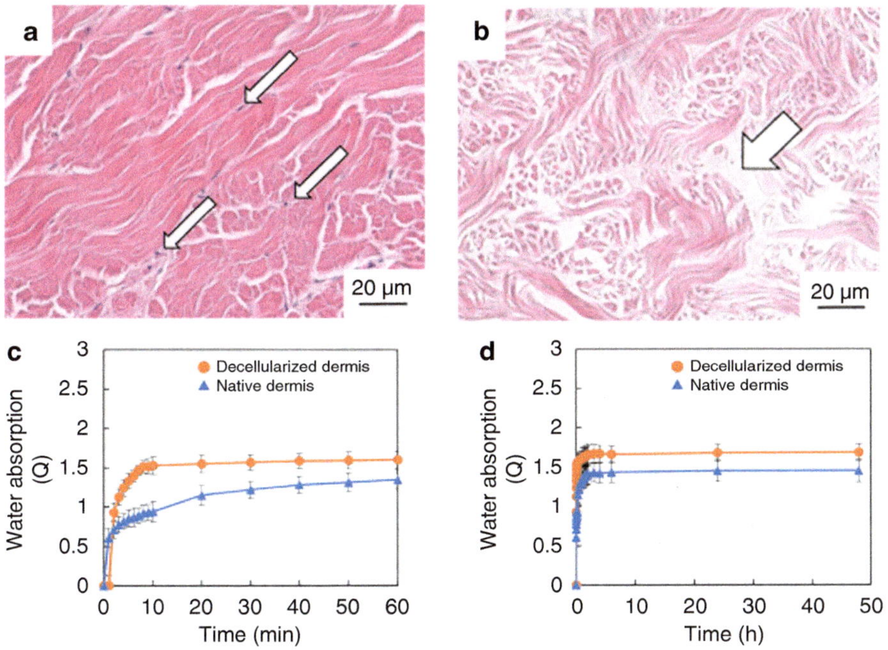

Fig. 3.1 One research characterized the water-absorbing quality of ADM decellularized by sodium dodecyl sulfate solution. Hematoxylin-eosin stained pictures showed that the larger interfibrillar spaces (thick arrow) have been found in the decellularized dermis group (**b**) than that in native dermis (**a**), which may affect water absorption and lead to a better water absorptivity of dermis (**c**, **d**). Reproduced with permission [26], copyright 2018

collagen fibers in the decellularized dermis group are much larger than that in native dermis, which leads to a better water absorptivity of dermis (Fig. 3.1) [26].

Alloderm is an ADM product invented by LifeCell incorporation in the USA. Askari et al. applied meshed AlloDerm (0.50–1.05 mm thick) together with thin split-thickness skin grafts (0.305 mm) to the open wound in nine patients undergoing release of burn contractures of the hand [27]. Patients were followed-up for 10–25 months, and all of them retained at least 83% of the corrected range of motion and at least 89% of correction at each webspace [27]. An "ultrathin" sheet of AlloDerm was applied by Patel and Cervino to manage recurrent keloid scars of 1–6 cm, which only causes low recurrence [28]. Vascularization of ADM film could be further improved when it was placed in a solution containing siRNA-mediated inhibition of prolyl hydroxylase domain-2 [29].

What is more, meshing of the ADM, placing an ultrathin meshed, and expanding autogenous split-thickness skin grafts on top are often involved in a one-stage procedure of burn treatment [30–32]. Meshed ratio is generally at 1:1. The interstices of the meshed ADM may provide a channel for the plasmatic imbibition process, thus support the overlying skin graft during revascularization [33]. In addition, meshing can also enhance the integration, reduce drain time and seroma in ADM assisted direct-to-implant breast reconstruction. A retrospective comparative analysis of women undergoing one-stage or two-stage immediate breast reconstruction with meshed and nonmeshed bovine ADM has been studied by Lotan et al. [34]. Results showed that significantly fewer major seromas, fewer total hematomas, fewer total infections, and fewer drain removal time had been observed in meshed ADM group when compared with a control group undergoing the same procedure with nonmeshed acellular dermal matrix [34].

Besides, efforts for further performance optimization of the ADM, such as adding coatings, crosslinking, stacking multiple layers, have been employed to modify the mechanical and degradation performance of the scaffolds [35–37]. In order to increase the scaffolds stability for better tissue regeneration, researchers fabricated silk fibroin protein modified ADM by coating the ADM with silk fibroin protein in different concentrations (5, 10, 15%) [36]. The biodegradation study revealed enzymatic condition [36]. In the study of Ma et al. in 2020, ADM was crosslinked with oxidized chitosan oligosaccharide (OCOS), carbodiimide (EDC), and glutaraldehyde (GA) [35]. Results indicated that the tensile strength of ADM crosslinked by OCOS was inferior to ADM crosslinked by EDC and GA. Further, the ADM crosslinked by EDC improved mechanical properties, antidegradation capability, and thermal stability of the scaffolds [35].

However, tissue regeneration is a process often beyond several weeks, and one time exogenous growth factors is insufficient to meet the requirements. In order to seek effective approaches to solve the problem, fibroblast growth factor-basic (bFGF) or bone morphogenetic protein-2 (BMP-2) has been loaded on ADM and assessed their differential effects during bone regeneration procedure in animal models [38]. Results showed that more amount of both new bone formation and expression of osteopontin in the groups of ADM loaded with bFGF or BMP-2 were observed than those in ADM alone [38].

AAM, a thin tough membrane normally 20–500 μm in thickness, is derived from the innermost layer of the embryolemma without blood vessels, nerves, and lymphatics [39]. Because of its ability to anti-inflammatory, antibiosis, pro-epithelialization, and reduce scarring, inflammation and tissue adhesion, the AAM, which often in patch form, has been applied as a biodegradable and bioactive scaffold for a variety of complex tissue regeneration [40–44]. AAM decellularized by using surfactant, lipase, and DNAase has been applied to repair full-thickness skin deficiency [45]. The biological repairing mechanism may be that the AAM can enhance the secretion of vascular endothelial growth factor and alpha-smooth muscle actin while reducing the expression of transforming growth factor beta-1 at early stage and then promoted skin defects regeneration and reduced scar formation [45]. What is more, in the study of Ehsan et al., human fetal fibroblasts achieved from the skin of fetus after spontaneous pregnancy termination have been seeded on AAM to fabricate a tissue-engineered skin substitute [46]. AAM is also effective in the treatment of chronic wounds. Wu et al. found AAM could be an innovative way to treat venous ulcers [47]. After grafted it onto venous ulcers in four patients, a significant reduction in ulcer size and depth with newborn granulation tissue has been occurred in these four cases [47]. Moreover, AAM or/and chorion membrane could be used in improving healing of diabetic foot ulcerations [48–51]. AAM also could remarkably promote healing of chronic wounds in patients with epidermolysis bullosa [52, 53]. Recently, the AAM patching, whose stiffness was more proper than that of plastic surfaces, was found to have positive effects on cardiomyocytes maturation [54]. AAM has also been found to have favorable functional and anatomical outcomes in treating ophthalmic diseases (e.g. recurrent macular hole) [55]. In the study of Caporossi et al., they implanted AAM into the macular hole in 10 patients [56]. Patients have been followed-up for 6 months with final retinal reattachment, macular hole closure, and best corrected visual acuity significantly improved [56]. Sung et al. found deep anterior lamellar keratoplasty with AAM patch could activate fibroblast and facilitate epithelialization [57]. AAM also has a series of studies and applications in pericardium substitute, reconstructive urology, and tissue regeneration within orthopedics [58, 59]. AAM could also be an osteoinductive biomaterial [60]. Li et al. invented a lyophilized multilayered AAM scaffold to repair rat tibia defect, which showed great efficiency in inhibiting fibrous tissue ingrowth, promoting bone-to-implant connection, and inducing the growth and maturation of massive bone [61].

However, the poor mechanical property as well as rapid degradation rate has hinder the AAM widespread clinic use. To overcome these problems, an integrated bilayer nanofiber-AAM film composed of an electrospinning nanofiber net sheet bonded to AAM sheet through interfacial conjugation has been fabricated by Liu et al. in 2018 (Fig. 3.2a) [62]. The graph of the morphology of the nanofiber-AAM with and without compression taken by scanning electron microscopy has been shown in Fig. 3.2b [62]. As shown in Fig. 3.2c, d, the ultimate tensile strength and toughness of nanofiber-AAM prepared from poly(ε-caprolactone) (PCL), poly(D, L-lactide-co-glycolide) (PLA), and poly(lactic acid) (PLGA) were obviously higher than those of rehydrated AAM, indicating that these integrated bilayer method significantly enhanced the physical properties of AAM [62]. And data of suture retention strengths have shown that the measured failure forces of nanofiber-AAM prepared from PCL, PLGA, and PLA were 14.7, 13.2, and 10.5 times higher than that of AAM alone, respectively (Fig. 3.2e) [62]. One year later, the research team constructed the composite PCL/AAM membrane by using method mentioned above and tested it in comparison to AAM alone in a rabbit limbal stem cell deficiency corneal epithelial defect model [63]. This PCL/AAM membrane not only showed better stability and slower degradation than AAM, but also maintained the pro-regenerative and immunomodulatory properties of AAM, thus reducing inflammation and neovascularization and promoting the re-epithelialization, which demonstrates the translational potential of their PCL/AAM membrane for ocular surface damage treatment [63].

Renewed cellularization with autologous or allogeneic cells on acellular scaffolds is called "recellularization" [64]. Recellularization of the decellularized tissues may not only bring back functions to the bioengineered tissues, but also elevate the durability of the decellularized tissues [65–68]. Zhou et al. seeded AAM with induced pluripotent stem cells-derived CD200+/ITGA6+ epithelial stem cells to develop a substitute for skin defects repairing [69]. After transplanting the cell–AAM construct onto skin defects of nude mice, the injured skin has been

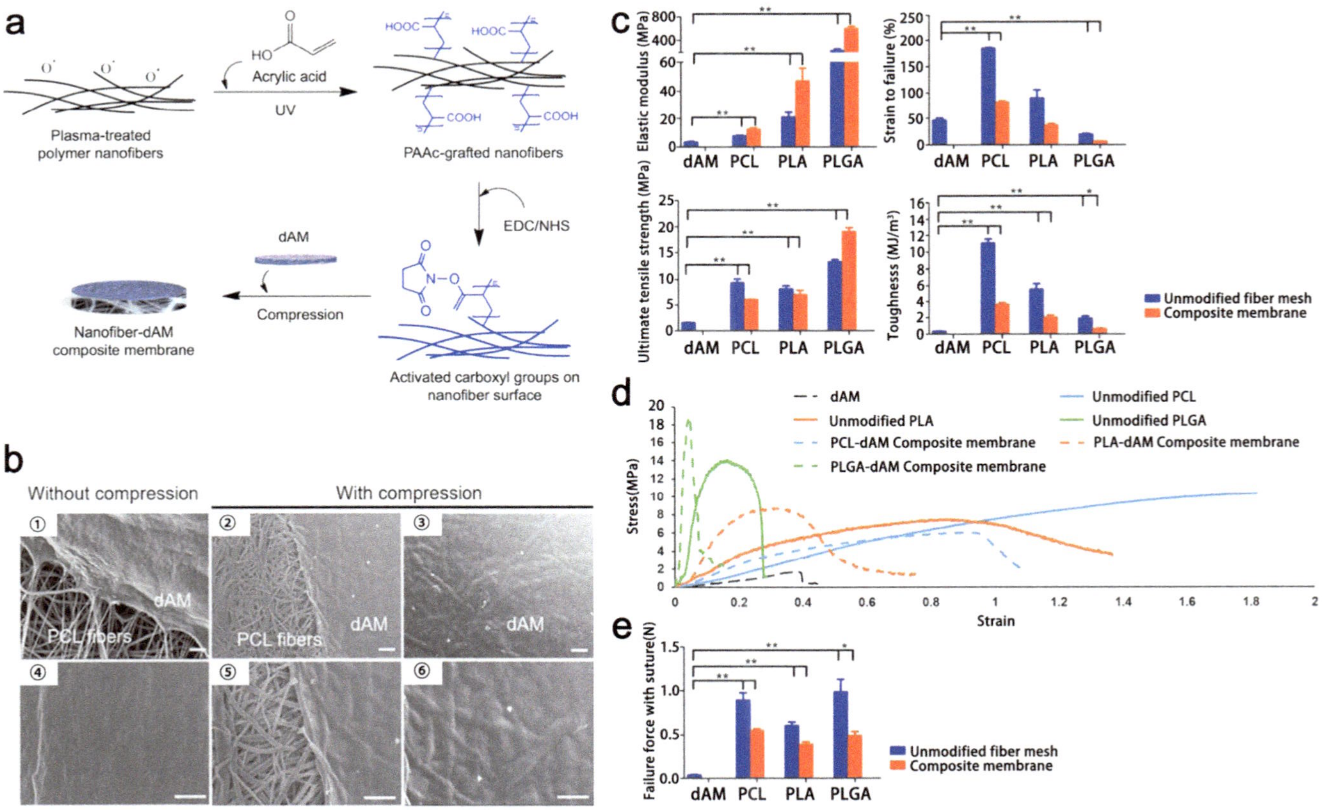

Fig. 3.2 One study example of improving mechanical property of AAM by adding synthetic materials, showing that introducing poly(ε-caprolactone) (PCL), poly(D,L-lactide-co-glycolide) (PLA), and poly(lactic acid) (PLGA) to AAM (prepared by grafting electrospinning synthetic nanofiber net sheet with poly

(acrylic acid) (PAAc), crosslinking with AAM, and compression process (**a**)) significantly enhanced the physical properties of AAM. The morphologies of the nanofiber-AAM with and without compression were different. Scanning electron microscopy images (**b**) show that the conjugation of nanofiber-AAM films after compression (②, ③, ⑤, ⑥) was more thorough with features of nanofiber visible on the AAM side, while the conjugation of nanofiber-AAM films without compression (①, ④) was not uniform with some regions of the AAM detached. Mechanical property results (**c–e**) showed that the elastic modulus (**c**), ultimate tensile strength (**c**), toughness (**c**), and the measured failure forces (**e**) of nanofiber-AAM prepared from poly(ε-caprolactone) (PCL), poly(D,L-lactide-co-glycolide) (PLA), and poly (lactic acid) (PLGA) were obviously higher than those of rehydrated AAM which had moderate stretchability (**d**) (*$p < 0.05$, **$p < 0.01$). Reproduced with permission [62], copyright 2018

reconstructed with newly hair follicles and interfollicular epidermis formation in the injured regions [69]. Besides, bone marrow derived mesenchymal stem cells were also seeded on AAM with better therapeutic effect on the osteochondral defect than that of the AAM group ($p < 0.05$) [70]. In a liver regeneration study, a cell–AAM patch composed of allogenic mesenchymal stem cells and AAM has been transplanted on a remnant liver [71]. Interestingly, this cell–AAM patch can not only promote tissue regeneration and survival rate, but also modulate the lipid metabolism in hepatocytes [71].

The decellularized SIS sheet, a translucent, flexible, and resilient film with distinct ultrastructural differences prepared by opening along the longitudinal axis is another established and broadly applied acellular matrices [72, 73]. Dry SIS has a porous microstructure with normally 10–50 μm in single sheet of thickness [74, 75]. While SIS was used in flat shape, its dimensions and mechanic performance may vary by the donor age and the size of the obtained intestinal tube [76]. After implanted in vivo, SIS was degraded 40–60% after 4 weeks and completely degraded after 3 months [77–79].

Nowadays, SIS patch has been used for the repair of a variety of tissues including the lower urinary tract, esophagus, myocardium, cornea, bone regeneration, abdominal wall, hernia, carotid, and musculotendinous tissues [75, 80–90]. SIS patch is frequently applied in rebuilding urogenital tissue. 12 months after a single layer of SIS patch (2×5 cm) sewn to partial cystectomy bladder, newly formed detrusor and serosal layers, vessel, and nerve were observed in regenerated bladder wall without inflammation [91]. In a case report, Xu et al. repaired ureteral stenoses with SIS, and mucosa covered the surfaces of the original patches with no apparent stenoses after 2 months of operation, indicating that SIS is a feasible material to repair ureters stenosis [82]. SIS also is a frequently used biomaterial in hernia repair [92]. The remodeled tissue after SIS treatment healed more strongly than the normal tissue [93]. Recently, SIS has become a potential candidate biomaterial for heart repair. Tan MY et al. found that the dimension and contractile function of left ventricular in rabbit myocardial infarction models could be significantly improved by transplanting 5-bromo-2-deoxyuridine-labeled mesenchymal stem cells (MSCs)-SIS construct into infarctive region [94]. What is more, the capillary density of infarctive region and myocardial pathological changes were also improved, suggesting that this cell-SIS construct has an excellent foreground in chronic myocardial infarction treatment [94].

Nevertheless, SIS may show insufficient mechanical support especially for high strain environments or very deep lesions. Multilayer scaffolds is an effective way to provide better structural support and greater degradation resistant capacity [95]. The mechanical strength of the dry SIS from monolayer to four-layer was 20.6 MPa, 24.7 MPa, 31.2 MPa, and 36.7 MPa [74]. Similar trends were observed in suture retention, with the suture retention of dry SIS from monolayer to four-layer were 0.5 N, 1.3 N, 2.1 N, and 2.8 N, respectively [74]. Significantly less degradation was found in three-layer (49%) and four-layer (47%) SIS than that of monolayer (82%) and two-layer SIS (75%), indicating that the increase of the layers will likely increase the degradation resistant capacity of the scaffolds [74]. The efficacy of four-layer SIS in treatment of deep corneal lesions, which has the advantage in mechanic support and corneal transparency, has been demonstrated by Barachetti et al. [86]. However, increasing the number of layers of SIS may enact an elevated inflammatory response. Xia et al. found that the eosinophil and neutrophil infiltrations were more elevated and lasted in four-layered SIS than that of monolayer SIS with comparable impermeability [81]. In addition, fabricating composite patches composed of SIS and synthetic materials may be another promising option. SIS-polypropylene mesh (PPM) patches have been used as patches for the abdominal wall defects in adult beagle dogs [88]. At 2 weeks after scaffolds implantation, the tensile strength of SIS-PPM was considerably higher than those of SIS ($p < 0.05$) [88]. Besides, modification of SIS by crosslinking method has also been introduced in some researches to improve the mechanics performance and degradation resistant capacity. Recently, genipin, a biocompatible naturally occurring product with crosslinking capability, has been used to optimize the performance of SIS with concentrations of 0.05%, 0.1%, 0.3%, and 0.5% for better gastric mucosa repairing (Fig. 3.3a) [96]. As shown in Fig. 3.3b, along with the increase of genipin concentration, the ultimate tensile strength of genipin-crosslinked SIS (GP-CR SIS) was significantly increased ($P < 0.05$) [96]. However, no significant difference was observed between GP-CR SIS and SIS in stiffness, and no significant difference was observed between GP-CR SIS before and 8 weeks after immersing in simulated gastric fluid (Fig. 3.3c) [96]. In addition, GP-CR SIS scaffolds crosslinked in different concentrations of genipin kept intact after simulated gastric fluid immersion for 8 weeks, indicating the degradation resistant performance has been enhanced after crosslinking (Fig. 3.3d) [96].

Furthermore, decellularized tissue scaffolds in sheet can also be fabricated by fixing the tissue on a cryostat with Tissue-Tek® optimal cutting temperature compound and cut it into slices into thicknesses as designed and followed with decellularization procedure [97, 98]. Decellularized tendon slice (DTS) was one representative example for it [91, 99–103]. Both in terms of better nutrient supply and cellular infiltration, the thickness of tendon slices that are suitable for tendon repairment is one of the key manufacturing parameters. In the study of Qin et al. in 2012, tendon slices (100, 200, 300, 400, and 500 μm in thickness) were fabricated by longitudinally slicing the bundles of Achilles tendon using cryo-sectioning technique mentioned above [104]. They found that the tendon slices equal to or greater than 300 mm in thickness have similar Young's modulus and ultimate tensile strength as compared to the intact tendon bundle, and without any structural damage after 21,000 cycle fatigue testing under a strain of 5% or less, which indicated that the tendon slices equal to or greater than 300 mm in thickness would be preferable for tendon tissue engineering from a mechanical strength perspective [104]. Along with the parameter of thickness was determined, the team developed the DTS of 300 μm thickness by removing the cells and DNA using repetitive freeze/thaw and nuclease treatment [100]. The ultimate tensile strength, percent strain, elastic modulus, and stiffness of the DTS of 300 μm were 29.68 ± 6.73 MPa, $19.16\% \pm 4.58\%$, 210.68 ± 46.43 MPa, and 27.40 ± 8.66 N mm^{-1}, respectively [100]. In addition to retain tendon ECM mechanical properties, experimental results of ELISA analysis

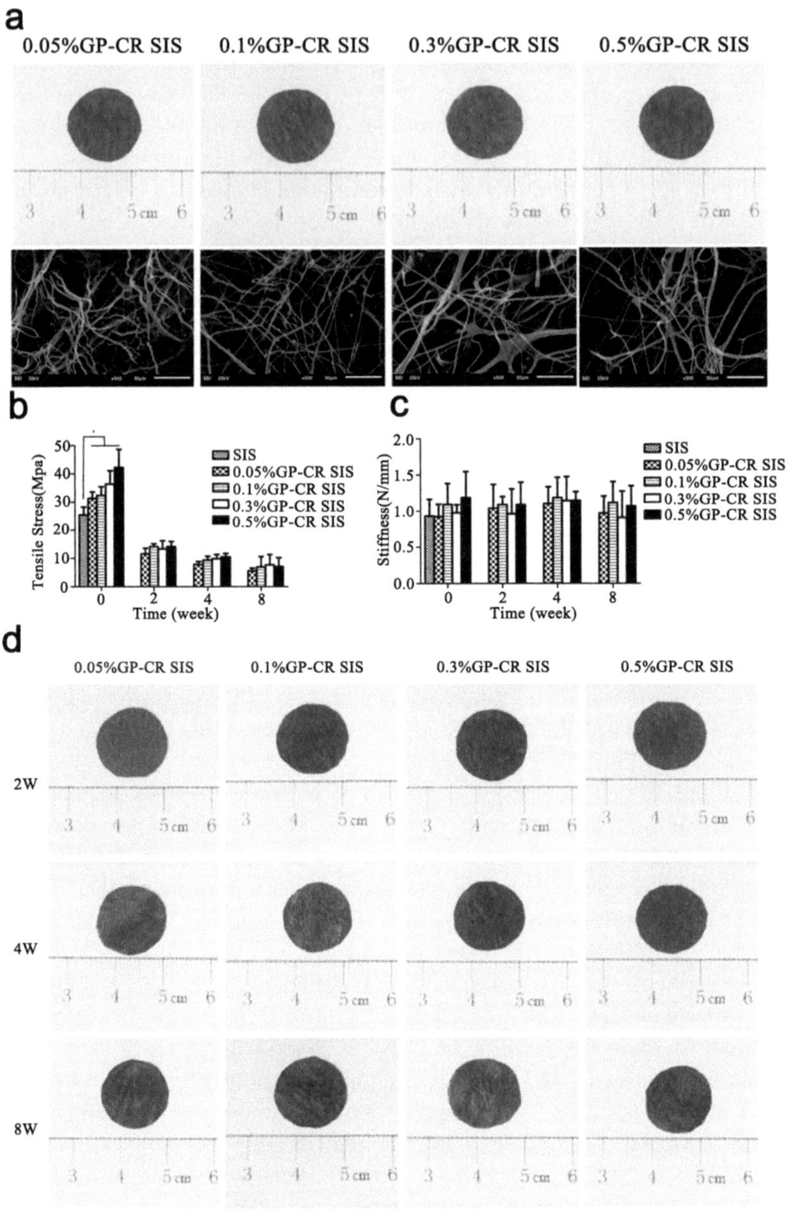

Fig. 3.3 One study example of improving physical properties of SIS by using crosslinking methods, showing that usage of genipin as crosslinker has significantly enhanced the ultimate tensile strength and the gastric juice resistance degradation ability of SIS. Macro-observation (**a**, left) and scanning electron microscopy images of GP-CR SIS (**a**, right) showed the morphologies of SIS after crosslinking with genipin in concentrations of 0.05%, 0.1%, 0.3%, and 0.5%. Then, the ultimate tensile strength and stiffness of GP-CR SIS before and after simulated gastric fluid treatment at different points in time were determined. Results showed that the ultimate tensile strength of GP-CR SIS (**b**) was significantly increased compared with SIS ($*P < 0.05$) and showed

showed that proteoglycans and growth factors (>93%) were inherent in tendon ECM, which may facilitate cell attachment and repopulation in vitro, were also preserved in the DTS of 300 μm in thickness, indicating that the DTS of 300 μm in thickness may have promising applications in tendon repair [100]. After in-depth researches for 3 years, the team found that when tendon-derived stem cells or bone marrow derived stem cells (BMSC) were seeded on the DTS, the stem cells showed alignment along the direction of the DTS fibrous with highly cell viability (Fig. 3.4a) [105]. Encouragingly, the expression of TNMD and THBS4 in BMSC seeded DTS is significantly higher than the pure BMSC group at all time points ($p < 0.01$), indicating that the DTS may provide an inductive microenvironment for the tenogenic differentiation of rat BMSC based on the tendon-specific gene expression (Fig. 3.4b) [105]. In order to construct a functional engineered tendon patch to improve clinical outcomes, Qin et al. then seeded DTS with BMSC and subjected to cyclic mechanical stimulation for 7 days (Fig. 3.4c) [106]. The cyclically stretched tendon constructs exhibit significantly higher expression of tenomodulin and decorin than those of the unstrained control group ($P < 0.05$), with similar ultimate tensile strength and stiffness compared to the unstrained control group ($P > 0.05$), which indicate that the cyclically stretched tendon constructs could be served as a scaffold to improve the surgical repair of rotator cuff tears (Fig. 3.4d) [106].

Meanwhile, decellularized multilayer sliced tendon scaffold has also been fabricated and proved to provide a microenvironment for the differentiation of BMSC into the tenogenic lineage [107]. Further, Alberti et al. immersed the stack multilayer DTS in 5% glutaraldehyde in PBS for 30 min to enhance its mechanical properties [108]. The ultimate tensile strength of crosslinked stack multilayer DTS is more than 20-fold over non-crosslinked samples, with a maximum strength of 13.59 ± 1.35 MPa [108]. And the modulus of crosslinked stack multilayer DTS is nearly 50-fold over non-crosslinked samples, with a maximum modulus of 145.54 ± 18.72 MPa [108]. What is more, rotating adjacent layers of the stacked DTS also enable the improvement of transversely isotropic mechanical properties [108].

In addition to sheet form, tube was another commonly used form of decellularized tissue scaffolds which reserved the microscopic and ultrastructural features along with the biochemical composition of ECM [109, 110].

In the field of tubular tissue regeneration, the structural requirement of scaffolds for nervous system injury repair is relatively higher. To be specific, for the sake of bridging the nerve gap under tension-free suturing and avoiding any compression of the regenerating tissue, appropriate dimensions is an important parameter of the

Fig. 3.3 (continued) a dose-dependent effect. Although no significant difference was observed between GP-CR SIS and SIS in stiffness (**b**), and no significant difference was observed between GP-CR SIS in stiffness and in ultimate tensile strength (**b**, **c**) before and 8 weeks after immersing in simulated gastric fluid, usage of genipin as crosslinker was proved to keep the structure of SIS intact after simulated gastric fluid immersion for 8 weeks (**b–d**). Reproduced with permission [96], copyright 2018

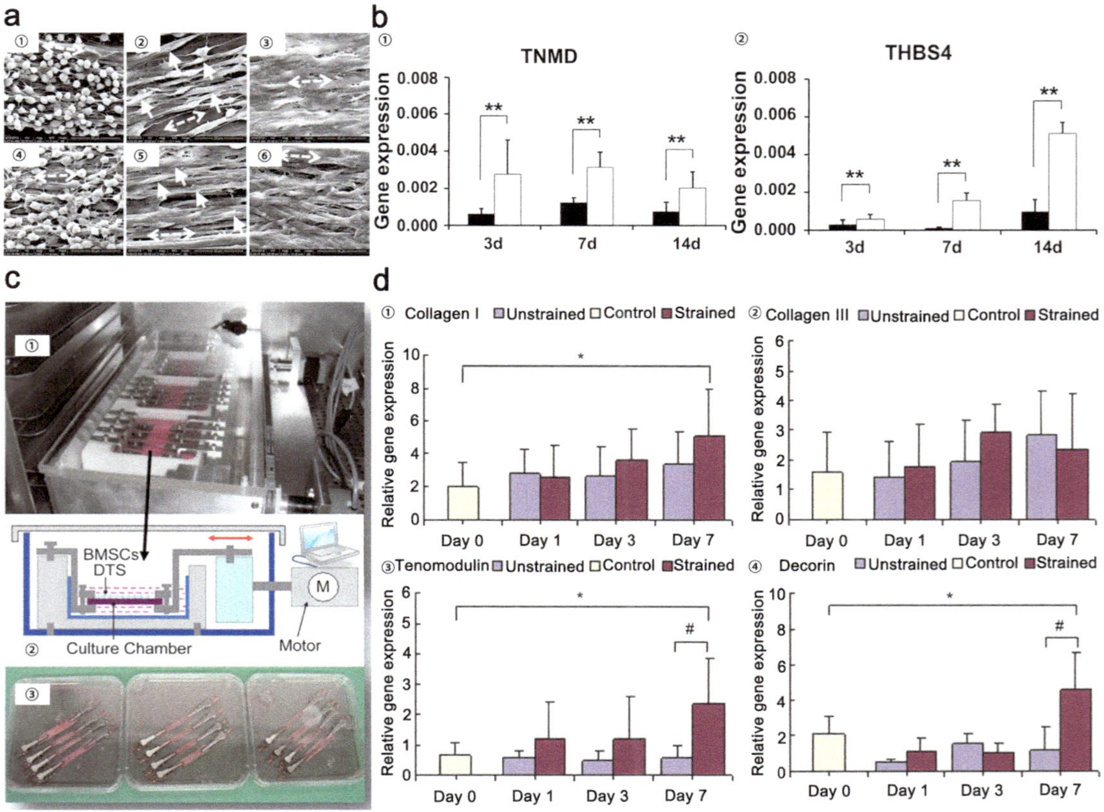

Fig. 3.4 One study example for decellularized tissue scaffolds in sheet fabricated by using cryo-sectioning technique. DTS tendon slices have been fabricated by longitudinally slicing the bundles of Achilles tendon and followed with freeze/thaw and nuclease treatment to serve as a scaffold to improve the surgical

repair of rotator cuff tears. Tendon-derived stem cells or bone marrow derived stem cells (BMSC) were seeded on the DTS. Scanning electron microscopy images (**a**) showed that both tendon-derived stem cells (①-③) and BMSC (④-⑥) were distribution and grow along the direction of the DTS fibrous. Quantitative real-time PCR analysis results (**b**) showed that the expression of TNMD and THBS4 in BMSC seeded on DTS is significantly higher than that in the pure BMSC group at all time points ($^{**}p < 0.01$). After subjecting the BMSC seeded DTS to cyclic mechanical stimulation provided by a custom mechanical stimulation device (**c**) for 7 days, gene expressions results (**d**) showed that significantly higher expression of tenomodulin and decorin has been observed in the cyclically mechanical stimulated tendon constructs than those in the unstrained control group ($^{\#}P < 0.05$). Reproduced with permission [105, 106], copyright 2015

scaffolds [111, 112]. Acellular nerve matrix (ANM), which contains Schwann cell basement membrane tube, nerve bundle membrane matrix, and outer membrane matrix obtained after removing myelin, axon, and cell components in nerve tissue, has been evaluated as an alternative means in treatment of nerve defect recently [111, 113–118]. The degree of decellularization and demyelination, the integrity of nerve fiber tubes and laminin activity that may affect the axon regeneration are the evaluation indexes of ANM [119]. Multiple sources of ANM have been fabricated through a variety of decellularization methods [120, 121]. Acellular nerve segment (7 mm) has been prepared from the rats femoral nerve by means of repeated freezing and thawing for five times [122]. Results show that the basal lamina and endoneurial ECM contain a large number of molecules that promote axon growth and Schwann cell migration, and the growth of AMN is related to the migration of Schwann cells from proximal and distal stumps [122]. He and the team developed an AMN graft from human peripheral nerves, which has been proved by monkey experiments to be effective in restoring radial nerve function, and it can also have the effect of digital nerve repair in clinic [123–125]. In 2019, ANM derived from decellularized porcine nerves, whose cellular components have been successfully eliminated while many neural ECM components and bioactive molecules have been retained, has been developed by using chemical decellularization methods [20].

Mechanical properties of acellular rat sciatic nerves have been calculated by Borschel et al. [126]. The average ultimate strain, Young's moduli, and normalized work to failure for ANM were 0.480 ± 0.117, 576 ± 160 kPa, and 0.35 ± 0.14 N, respectively [126]. Remarkably, nerve allograft preserved by freeze may impair functional recovery. In the study of reconstruction of a 3-cm peroneal nerve gap in New Zealand White rabbits, Bulstra et al. found that the cold-preserving is better for the functional recovery of ANM which is similar to the gold standard autograft [127]. Due to the loss of neurotrophic factors and angiogenic factors during decellularization, the nerve regeneration efficiency and functional recovery have declined, especially for long-segment nerve repair (>3 cm) [128–132]. Studies have shown that the aperture diameter of basal membrane conduit that less than 10 μm may not conducive to the migration of cells, and deficiency porosity of ANM is likely to restrict the functionality of decellularization reagent and the exchange of nutrients [133, 134]. To overcome this problem, unidirectional freeze-drying and axial puncture have been used to develop ANM with large channels [135]. In addition, zwitterionic detergent 3-[(3-cholamidopropyl) dimethylammonio]-1-propanesulfonate together with low concentrations of Triton X-100 has been introduced as acellular chemical reagents for better preservation of nerve ECM bioactive components [135]. The multichannel ANM demonstrated outstanding capability in promoting Schwann cells proliferation and penetration [135]. After transferring it to the 10 mm rat sciatic nerve defects, multichannel ANM group showed greater ability to guide regenerative nerve fibers through the defective segment and restore the target innervation, thus implementing better function recovery of muscle and motor than conventional acellular scaffolds, which demonstrated that this ANM offers a new vision for future neural regeneration [135].

Acellular trachea is a promising tubular scaffold in trachea-tissue engineering. Zhang and colleagues found the detergent-enzymatic treatment could effectively remove antigenicity and preserved structural integrity and adequate mechanical strength. Moreover, it would not induce significant inflammatory reaction or foreign-body reactions in vivo [136, 137]. Macchiarini et al. reported clinical effects of acellular trachea fabricated by removing cells and major histocompatibility complex antigens from human trachea [138]. Over 5 years of follow-up, the implanted is well recellularized and vascularized and remained open over its whole length [139]. A child, the first pediatric patient, received an acellular trachea seeded stem cell treatment. Through 2–4 years follow-up, the epithelium was complete restoration, with squamoid, respiratory type epithelium, and scanty ciliated cells observed by biopsy [140, 141].

Moreover, decellularized tissue scaffolds in form of tube can also be obtained by suturing or pasting sheet-type scaffolds, which provides more choices of scaffolds for tubular tissue repair and reconstruction. The tubular SIS has applied for intestinal, carotid artery, tubular tricuspid valve, and esophageal tissue repair [142–146]. Similar with the problems met in SIS patch, tubular SIS also shows insufficient mechanical support and high degradation rate. After implanting the tubular SIS between isolated jejunal segments in animal model for 8 weeks, the tubular monolayer SIS and tubular four-layer SIS were contracted to 50% and 29% of its initial length, respectively [142]. Although the contract degree in tubular four-layer SIS was below that in tubular monolayer SIS, both of them cannot support significant regeneration of intestinal [142]. In the study of Fan et al., synthetic polyesters biomaterial poly(3-hydroxybutyrate-co-3-hydroxyhexanoate) (PHBHHx) and poly (lactide-co-glycolide) (PLGA) in various ratios (3:7, 5:5, and 7:3) were instigated to elevate the mechanic and degradation performance of tubular multilayer SIS [147]. Particularly, the peak load of SIS/PHBHHx-PLGA in ratio of 5:5 was much higher than that of other group and was close to those of esophageal tissue [147]. Furthermore, the biological degradation performance of SIS/PHBHHx-PLGA in ratio of 5:5 ($34.8 \pm 2.8\%$) by week 12 immersed in simulated body fluid is more proper than that of 7:3 in ratio ($49.6 \pm 5.9\%$) and pure SIS, indicating that this SIS/PHBHHx-PLGA scaffolds may provide a new method for esophageal defects reparation [147].

3.2 Extracellular Matrix Powders

ECM creates a microenvironment for cell adhesion, migration, differentiation, and proliferation, which can promote tissue remodeling and regulate gene expression [148]. It is of great significance that the way in which cells interact with extracellular matrix determines cell fate and sustains cell phenotypes [149]. ECM materials, such as acellular organs and plaques (acellular small intestinal submucosa), retaining their original structure, can provide an intact vascular system, accurate derived tissue model, and maintain good mechanical properties [150]. Although the structure and composition of ECM are relatively complete during the preparation process, while their original structure limited the adaptation of their tissue conformation, so that their clinical application is limited to a certain extent. It is necessary to consider the structure, composition, and source of the ECM used when inducing right tissue formation at a certain position of injury [151]. In practical applications, the morphology of ECM-based materials is able to be modulated to suit the expected usage by adjusting its shape. ECM can retain its original tissue conformation, or it can be processed to form specific configuration required. Powder materials have many excellent properties, such as good bulk density, fluidity, formability, and compressibility, which make powder materials have great application prospects.

ECM powder and its derivatives (ECM suspension and gel) can fill the damage areas and conform to fit the contours of natural tissue. Because of its conformational adaptation and solubility, the powder can be minimally implanted, such as injection. It is different from ECM scaffolds or patches and sheets that have a fixed conformation and are easy to contract, ECM particles can maintain a specific shape that is adopted after dissolution [152]. The dispersion and solubility of the powder give it a spatial advantage and the ability to combine with other molecules efficiently. Therefore, the powder can be used as a carrier to deliver cells, growth factors, drugs, and so on. The variable characters of ECM powder allow them to be used in many fields, such as 3D bioprinting and the treatment of various tissue injuries. ECM powder and its derived materials seem to provide a wide range of application potential in tissue engineering and regenerative materials, there are still many challenges in production and majorization, which limit the wider application of ECM powder.

ECM powder can be obtained by pulverizing ECM, and the powder can be used in particle or liquid and hydrogel dissolved enzymatically. The preparation of ECM powder involves multiple processes, which may change its biological integrity. The ECM processing process may affect its component content, mechanical strength, microstructure and ultrastructure. composition content, mechanical strength, ultrastructure, and microstructure [153]. Undoubtedly, the active ingredients of ECM obtained by chemical reagents or enzyme treatment may be destroyed in each step, and, in return, the process may finally alter the host response after implantation in the body [154]. Although ECM powder is widely used, standard guide and reasonable methods of powder manufacturing and processing have not yet been determined. The preparation and modification pretense a challenge to the extended use of ECM powder. The use of ECM powders and its derived constructs are need more exploration. The challenges of extending their applications in tissue engineering and regenerative medicine require special attention.

3.2.1 Preparation Methods and Characteristics

The origin and composition of ECM and the method of powder manufacture may lead to differences in particle size distribution and shape [156].

The preparation process of ECM powder and its derivative constructs is of great significance to its effective use and is worthy of further study. Generally, the preparation of powdered ECM usually involves separating ECM from fat, connective, and other tissues, and then performing operations such as decellularization, freezing, lyophilization, crushing, and grinding [153, 154].

Decellularization is the key and necessary course to prepare the medical devices based on ECM, and decellularization reagents usually involve detergents, organic solvents, and enzymatic solutions. Each of the methods used in the process may have influences on the content, ultrastructure, and mechanical properties of the materials, including the host response to the material [157, 158].

Usually, removing the tunica muscularis externa and tunica submucosa layers is the first step to prepare decellularized matrix of tubular organs, [159]. Then the retained tissues were washed with 0.1% peracetic acid solution for 2 h, rinsed in phosphate buffered saline and distilled water subsequently in order to sterilize the material and remove any cellular remnants [159].

Lyophilized ECM can be further processed into powders according to actual needs. Analyze the particle size distribution by sonic sifting and laser diffraction. Sonic sifting is the separation of powders by size through a range of vertically stacked grading screens. The powder passes through the sieve under the action of a mechanical agitation along the vertical axis of the chimney and the plane of the sieve. Usually, the diameter range of particle size is mostly from 25 to 200 μm. References [159, 160] show the process of acellular matrix of various tissues and organs, and the process of preparing ECM powder and its derivatives constructs [159].

Jakus et al. fabricated a series of "tissue papers" by mixing powders of tissues ECM derived from heart, kidney, liver, muscle, ovaries, and uterine horns with polylactic-co-glycolic acid and then air-drying [159]. The "tissue papers," whose microstructural characteristics, physical, and mechanical properties were distinct, can support human mesenchymal stem cells adhesion, viability, and proliferation over 4 weeks [159]. This process, however, can hardly meet the requirement for preparing specimen thicker than several millimeters. Da et al. described a method to develop three-dimensional porous composites by mixing water-based polyurethane and tissue ECM in nontoxic aqueous solvent and lyophilization, which can take the shape of any mold easily [161]. The obtained composites, shown as in Fig. 3.5, showed a bimodal pore structure with the discrete voids (400–1000 μm in diameters) encircled by the pores (20–250 μm in diameters) with ECM powders dispersed across the matrix of polyurethane homogeneously [161]. Other results revealed that the produced composites were characterized by high bioactivity and resilience and could be used for soft tissue engineering applications [161].

Particle forms make the topical delivery of the ECM suspension through minimally invasive injection or fabrication of 3D scaffolds or patches/sheets by the method of crosslinking, then freeze-dried, or compacted [154]. Especially, granularity, surface area, dosage form, and the form of liquid carrier are the primary related variables affecting the convenience and usability while the material was

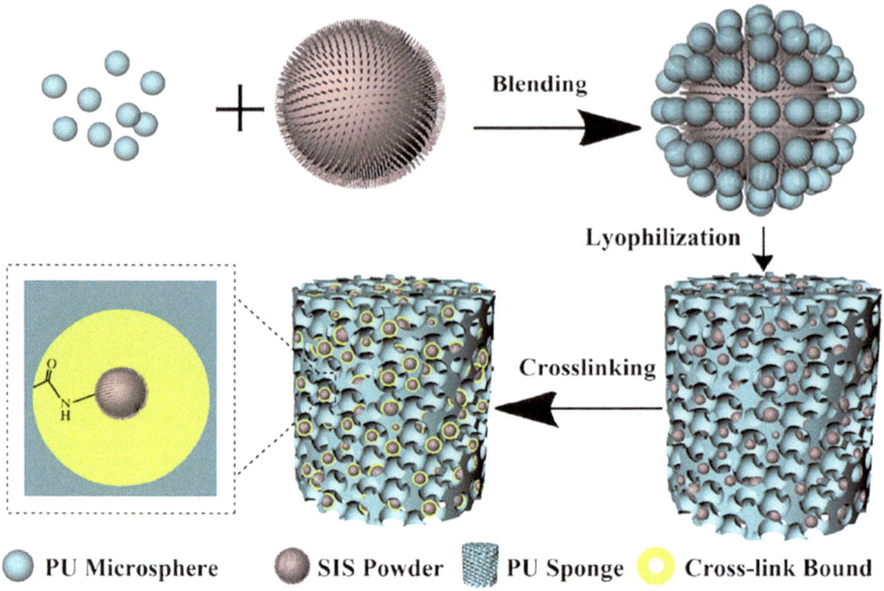

Fig. 3.5 One study example of improving bioactivity and elasticity of SIS by combining SIS with water-based polyurethane for an expanded application in soft tissue engineering. The picture showed the protocol of synthesis of PU and PU/SIS composites, mainly including two steps: (**a**) synthesis of PU via bulk polymerization of PTMG 1000, IPDI, and DMBA species. (**b**) Preparation of the crosslinked PU/SIS composites. Reproduced with permission [161], copyright 2017

conveyed [154]. The way to prepare ECM powder are various. Here, Gilbert et al. reported two ways were used to prepare UBM particle [156].

The first way involves freeze-drying the sterilized material, which is first cut into small pieces and then immersed in liquid nitrogen for further crushing. Then use Waring agitator to grind the quick-frozen materials into small pieces in order to obtain particles small enough to be put into the rotary knife grinder. Restrict the collected powder size within 250 mm by the 60-mesh.

The second way was to immerse the disinfected material in a 30% (w/v) NaCl solution for 5 min at first, then was snap-frozen in liquid nitrogen and lyophilized to remove residual water and crystalline sodium chloride, material was then crushed as described in the first method. Immersed the tissue into NaCl, it can be anticipated that the embedded salt crystals will make the material rupture into homogeneous particles. Then suspend the particles in deionized water and centrifuge at 1000 rpm for 5 min and repeat three times to remove NaCl. Snap-frozen and lyophilized again, the obtained powder was pour into the rotary knife mill to grind into uniform particles.

The powder size was measured by delivering the dried particles through the way of the HeNe laser beam by laser diffraction method [162, 163]. During the detection process, each particle can diffract light, and the diffraction angle is inversely proportional to particle size. The diffraction pattern was measured by the detector array, the particle size was evaluated by the observed angle and intensity. To analyze the ultrastructure of ECM particles by SEM. In short, attach a thin layer of powder coated with platinum on the conductive adhesive tape. The ultrastructure of the samples was detected at magnifications under appropriate multiple [156].

3.2.2 Optimization and Applications

Factors should be considered, including the amount and concentration, particle size and morphology, solubility, and the crosslinking to optimize of ECM powder. Especially, these methods and projects should be reproducible if these materials are to be used in clinical [155].

3.2.3 Usage Concentration

The amount and concentration of ECM in the liquid or gel will affect the chemical and physical properties of these materials, such as the porosity, compressive modulus, gelation threshold, and rheology of the constructs. Also, the amount of powder implanted may influence the tissue remodeling process. In addition, it is worth noting that excessive use of ECM can lead to ECM deposition in the body, which can lead to tissue fibrosis [155]. So it is essential to optimize the appropriate concentrations of ECM used in regeneration and repair of specific tissues.

Usually, constructs with a low ECM concentration showed highly porous after lyophilization, but exhibited the tendency to degradation in vivo and low mechanical property. Augmenting the concentration of the powder may decrease the porosity but can significantly extend the degradation time and present stronger contraction ability [164].

However, the concentration of ECM powder is usually determined according to the specific conditions of use. Lower ECM concentration is beneficial to cell adhesion and proliferation, while higher ECM concentration stays longer in vivo. Therefore, it is necessary to balance the biological activity of scaffolds against the effectiveness of filling tissue defect areas [165].

The concentration of ECM also affects the properties of the rheological properties and viscosity of the hydrogel. Hydrogels with higher ECM concentrations have lower viscosity and greater elastic modulus, while hydrogels of lower concentrations presented higher rheological properties and lower shear properties. However, the pore size of the constructs does not appear to be directly related to the powder concentration used, depending on the ECM type. The concentration of urinary bladder matrix (UBM) has no effect on the porosity of hydrogel; however, increasing concentration of dermal ECM would reduce the pore size of hydrogel. So the amount and concentration of ECM are related to its type [166].

Obviously, the higher the concentration of ECM powder, the greater the viscosity after gelation, preventing therapeutic injection and requiring larger needle lumen in clinical [166].

3.2.4 *Particle Morphology*

Despite the same biological constituent, ECM powder can exert various biological effects in vivo due to the distinct conformation. Size and morphology of the particles can lead to different biological behavior of cells in terms of proliferation, differentiation, tissue development, and so on [165, 167, 168]. The preparation processes and applications of ECM powder are diverse, leading to the wide particle size distribution, and the gauges of the needle used in clinical are also different (Fig. 3.6) [169]. For instance, the type of crusher and sieve, the size of the cutter, and the drying degree of the sample all affect the particle size distribution after ECM grinding. In addition, the residual moisture in the powder tends to agglomerate and agglomerate, resulting in poor dispersion during dissolution [170]. Although particles with different shapes and sizes are excellent matrixes for cell adhesion and growth [171]. Homogeneous particles are reported to better regulate tissue development in the body. For instance, powers with different particle sizes used in skin tissue engineering heterogeneity can testify meshes with gaps which lead to form noncontiguous tissue cosmetically unesthetic [169]. Usually, the diameter range of particle size is mostly from 25 to 200 μm [159, 160]. So, further research is necessary to demonstrate whether the bioengineering of specific tissues needs specific particle sizes.

Electrospinning is a variable way to control fiber building and properties at the nano and micron levels [8, 10]. Yoon et al. have previously prepared aligned polycaprolactone/small intestinal submucosa fibrous webs by electrospinning under electric voltage of 17 kV and syringe pumping rate of 3 ml/h [160]. The orthotropic mechanical behavior of polycaprolactone/small intestinal submucosa fibrous webs was superior than the fibrous webs without ECM powders [160]. Adding of tissue ECM powder may increase electrical conductivity and decrease surface tension, and thus lead to more finely fibers relative to that of fibers without tissue ECM [172]. After studying the SEM images, it was found that randomly arranged fiber decreased the charge density on the surface and increased with the concentration of ECM [172]. Furthermore, the elastic modulus, maximum stress, and the break strain of the fiber mats are related to its alignment and molecular orientation closely [172]. Mechanical properties of poly(ester urethane) urea (PEUU)/ADM electrospun patch have been evaluated by Hong et al. [173] The obtained results revealed that materials tensile strengths for PEUU/ADM with the ratios of 67/33, 72/28, and 80/20 were within 80–187 kPa on the vertical axis and 41–91 kPa on the circumferential axis with 645–938% breaking strains [173].

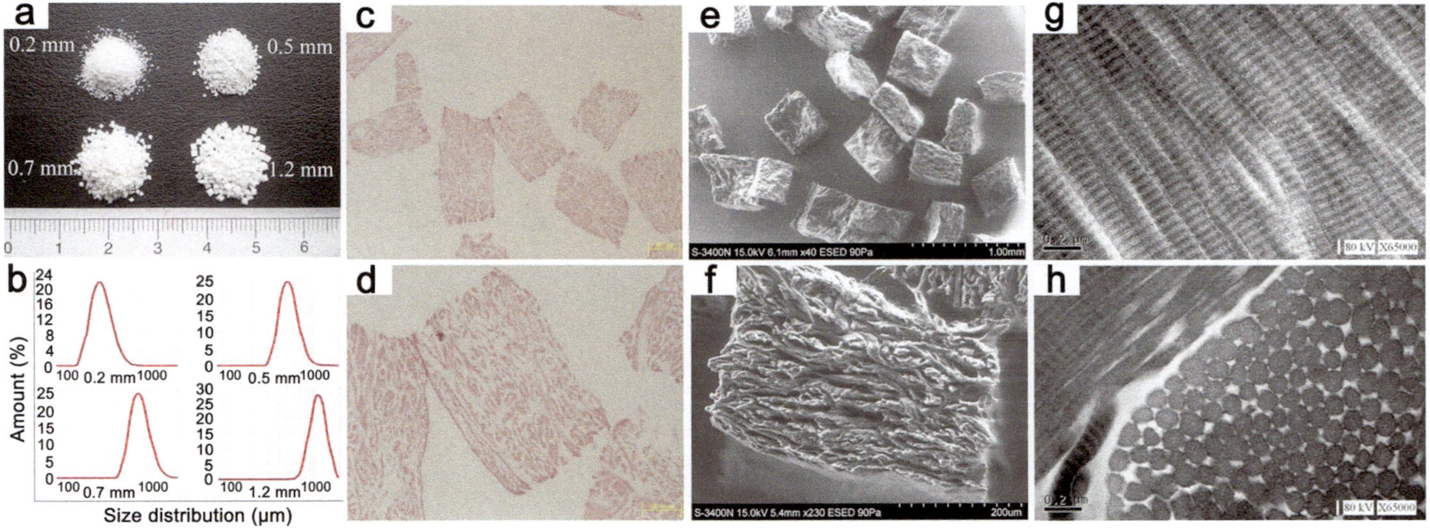

Fig. 3.6 The study example of the morphology and particle size distribution of particulate acellular dermal matrix (PADM) with different particle sizes, and the evaluation of histological morphology of PADM, as well as the pore structure and arrangement of collagen fibers in freeze-dried state. Overview (**a**) and size distribution (**b**) of PADM powders of four gauges (0.2, 0.5, 0.7, 1.2 mm) in macroscopic, H&E staining (**c**, **d**), and SEM micrographs (**e**, **f**) of PADM in microcosmic, and TEM micrographs of PADM (**g**, **h**) in ultra-microcosmic have been presented above. These results indicated that proper distribution of PADM particle size which is prerequisite for its application in filling materials and injectable materials wide could be achieved by selecting appropriate preparation processes. Reproduced with permission [169], copyright 2012

3.2.5 Sterilization Methods

ECM biomaterials need to be thoroughly disinfected prior to use in clinical. Sterilization would be performed in a manner that is least destructive and most effective. Many bactericidal technologies and reagent are used, such as gamma irradiation, e-Beam, glutaraldehyde, ethylene oxide, peracetic acid (PAA) [174].

The selection of sterilization methods and processes adheres to the principle that they are effective and do not destroy the original mechanical properties and biological integrity of biomaterials. In general, it is difficult to production: high temperatures denature the proteins [175], irradiation can weaken the growth factors [176], and soluble reagents may penetrate the matrix uneffectively [175].

Specifically, the choice of sterilization method of ECM needs further consideration of the ECM state, as it affects the sterilization effect, namely whether ECM is hydrated or freeze-dried [176]. For instance, to sterilize lyophilized ECM using ethylene oxide or electron beam would destroy the triple helix structure of collagen, reducing its mechanical properties and resistance to enzymatic hydrolysis. However, the same approach does not affect the properties of hydrated ECM [177].

Some mild forms of sterilization, such as PAA and ethanol, do little damage to ECM properties, but have limited bactericidal effects. In addition, the solution bactericidal form is not applicable to powder, which may change particle size and shape due to dissolution or aggregation [178]. It is necessary to further study the best sterilization method of ECM particles in order to popularize it in clinical application.

3.2.6 Solubilization Properties

ECM powder is often dissolved by strong acid, alkali, and enzyme to get the homogenous liquids or gels. Strong acid and base seriously destroy the structural integrity of ECM and make it lose its biological function. A certain concentration of sodium hydroxide or hydrochloric acid solutions is feasible. Pepsin is an enzyme commonly used to digest ECM, which requires a suitable acidic environment. So hydrochloric acid or acetic acid is generally used to enhance its solubility. Acidic solubilized ECM should be neutralized to physiological pH when used in vivo. Otherwise, the acidic microenvironment created by ECM prevents vascularization and induces inflammation, granulation, and fibrosis [179].

In addition, inadequate or incomplete digestion will result in the precipitation of particulate residues from the solution and the construction will have heterogeneous components [170].

3.2.7 *Crosslinking Methods*

The powder or its derived structures such as lyophilized scaffolds, liquids, and hydrogels are short of sufficient mechanical strength. So, it is often crosslinked to intensify their structure and resist degradation, but the chemicals crosslinker may lead to poor biocompatibility [180].

Moreover, crosslinking can conjugate additional molecules (e.g. drugs, fluorescent markers) to ECM constructs. Crosslinkers like formaldehyde, glutaraldehyde, genipin, carbodiimide 1-ethyl-3-(3-dimethylaminopropyl) carbodiimide hydrochloride (EDC), and N-hydroxysuccinimide (NHS) are commonly used crosslinking agent to protein. The theory of reaction is mainly to bond with amino and carboxyl groups in ECM molecules through intermolecular reaction. However, the residual crosslinking agent will reduce the biocompatibility of the material [181].

3.2.8 *The Applications of ECM Powder in Tissue Engineering*

ECM powder has the potential to be used in various fields of tissue engineering, shown as Fig. 3.7.

3.2.8.1 ECM Powder as Carrier

ECM powders provide unique surface morphology and adjustable granularity as well as bioactive components such as cells and growth factors. Moreover, powder of ECM can be combined with pharmacologic drugs to serve as drug delivery system.

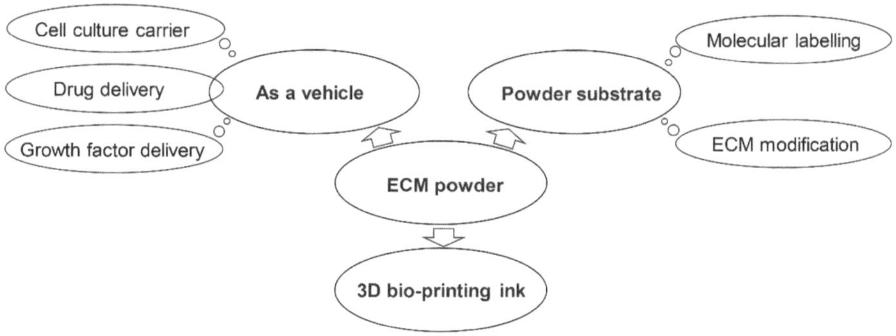

Fig. 3.7 Here are examples of various scenarios in the application of ECM powder in tissue engineering, mainly including using as delivery media, substrates, 3D bioprinting, etc.

3.2.8.2 Cell Culture Carrier

ECM powder or its derived dissolved powder can be used as a carrier for 3D cell culture or as a carrier for delivery of minimally invasive cells at certain sites. 3D cell culture has been obtained in spheroids, sponges, hydrogels, and acellular organs, which may not allow cell distribution or nutrient diffusion uniformly. In addition, acellular biofilms or scaffolds cannot be injected and do not meet in vivo filling requirements.

ECM is processed into cell-sized particles that can let the cells to uniformly bind and attach [155]. ECM particles have a large ratio of surface area to volume, which maximizes the adhesion ratio between cells and ECM. Different performances of ECM particles, such as size, concentration, and dispersity regulate the correlation cell–matrix and cell–cell.

ECM power has already been used as cell carriers. The cells are able to be inoculated into the powder or its lysate directly [171]. A cell–ECM system is exceptionally superiority in application, because it can deliver cells via injection. It has been reported that human fat-derived stem cell–ECM powder injection transplantation is superior to autologous fat transplantation in that its size does not decrease over time. However, adverse results from autologous grafts containing natural extracellular matrix suggest that the use of lysed specific extracellular matrix as a cell carrier has an advantage over uncrushed cell delivery [182].

Cell culture in ECM lysate also promotes the differentiation of progenitor cells. Fong et al. found that induced pluripotent stem cells (iPSCs) in cardiac ECM lysates could differentiate into mature cardiac myocytes, while iPSc cells generally tend to differentiate into cardiomyocytes with fetal phenotype, suggesting that ECM lysis solution may provide a microenvironment, making progenitor cells differentiate ultimately [183]. In addition, cells encapsulated by soluble ECM powder extracted from the native environment can increase their levels of proliferation and differentiation and maintain specific cell phenotypes [184]. In cell–ECM system, ECM provides a three-dimensional platform for cell attachment and a microenvironment for cell proliferation and differentiation, where the correlation cell–matrix and cell–cell lead to specific combination of cells such as spheres. It is important for ECM powders to have the ability to induce the formation of cellular aggregates because aggregates can act as a tissue unit. About articular cartilage repair, Yin Hui et al. found that the functional "Microsoft bone" aggregates were formed by mesenchymal stem cells and chondrogenic ECM powder, which could induce natural stem cells and ECM particles to construct and form hyaline articular cartilage. Obviously, ECM powder provided a particulate carrier for cells to attach and was beneficial to the maintenance of chondrocyte phenotype in the experiment, which tended to disappear over time. The polymers of the cell–ECM system could be delivered arthroscopically to the position of injury, achieving a minimally invasive approach to cartilage repair by one step [185].

3.2.8.3 Drug Delivery Vehicle

Effective drug delivery is important for successful drug therapy. Polymer constructs and nanomaterial delivery systems have shown their potential as drug carriers [186]. But ECM powder offers a more tunable and accurate way to deliver drug. It can be mixed with drugs homogeneously, and targeted at specific position in vivo through fixed-point injection.

The drug bioavailability could be easily regulated by controlling the ratio of ECM powder to drug in the system. Besides, ECM powder also has certain ability in sustaining and controlling release of drugs [187].

3.2.8.4 Powder Substrate

ECM mainly consists of protein in which amino acid contains lots of active groups such as amino and carboxyl. So, powder is easily to be modified by interacting with other molecules. Powder can act as a conjugate of which it can be compounded with synthetic polymers to improve the biocompatibility of polymer materials while enhancing the mechanical properties of ECM-based composites. Generally, ECM powder can be regarded as substrates to which polymers as well as molecules are conjugated.

ECM powder serves as a superior substrate for bioconjugate reactions due to its active groups which can reacts with a variety of chemical reagents, especially crosslinkers like formaldehyde, glutaraldehyde, genipin, carbodiimide 1-ethyl-3-(3-dimethylaminopropyl) carbodiimide hydrochloride (EDC), N-hydroxysuccinimide (NHS), etc. [188] Specially, fluorescent labeling are usually used to label biological molecules, fluorescent molecules, and ECM powder can be bioconjugated by N-hydroxysuccinimide (NHS) and other linkers as well, for example, conjugate fluorescent Cy3 molecules with matrix collagen in ECM [189]. Notably, fluorescently labeled solubilized powder in real time can serve as a monitor for material degradation upon minimally invasive injection, from which the degradation of samples can be presented.

3.2.8.5 ECM Modification

Synthetic materials are subject to many restrictions in clinical applications due to their poor biocompatibility. Synthetic polymers conjugated with biomaterials are widely used in tissue engineering and wound repair due to their compatibility with biomaterials and the mechanical properties. ECM powder mixed with synthetic polymers was used to prepare more biocompatible constructs in research and clinical application. It is reported that dermal powder solubilized has already been employed to coat polypropylene-based surgery instruments at a hydrogel form.

Polypropylene, a kind of thermoplastic synthetic resin, can resist to chemical, heat, high-strength, and good high wear, which has been widely used in medical equipment, chemical containers, food, pharmaceutical packaging, and other fields. Polypropylene, a nondegradable polymer, may cause foreign body response and serious inflammation when used as medical instruments. ECM-based composite device, however, can ease inflammation and the deposition of fibrosis and scar tissue subsequently to a certain extent [190].

Being modificated by ECM hydrogel, the mechanical properties of surgical devices have not been compromised because of lacking mechanical strength. In one research, ECM modified surgical devices can support the weight of bearing loads in the process of repairing ventral hernias, inguinal hernias, and pelvic organ prolapse effectively [190, 191]. D'Amore et al. found that polyurethane cardiac patches covered with soluble cardiac ECM powder can effectively prevent the maladjustive remolding after cardiac infarct and reduce left ventricular compliance ultimately, promote angiogenesis, and prevent left ventricular from dilatate, scar, and transform thin.

Moreover, solubilized ECM can also be used to modify scaffolds like synthetic-polymer scaffolds as well as metal scaffolds. It has been reported that polylactic acid-glycolic acid (PLGA) modified by powder of the small intestinal submucosa (SIS) is applied for the regeneration of the nucleus pulposus of the intervertebral disc.

Together with synthetic polymers, ECM powders are advantageous to design hybrid composites with tunable mechanical and degradation properties, highly biomimetic properties, and biological activity. These biomaterials confirmed prospect performance properties for the regeneration of different tissues, such as nerves, bone, and abdominal walls [8, 161, 172, 192, 193]. Recommended solvent for developing composites with ECM powders are 2-butoxyethanol (0.3 g per 0.35 g ECM powders), dibutyl phthalate (0.15 g per 0.35 g ECM powders), dichloromethane (3 mL for every 0.35 g of ECM powders), and water (10 mL for every 0.3 g of ECM powders) [159, 161]. Physical methods such as air-drying, lyophilization, and electrospinning could be utilized for scaffolds shaping with controlled architecture and mechanical properties [8, 159, 161, 172, 194].

PLGA devices may yield acidic degradation products in body when used in tissue engineering lonely, however, when it is modified by SIS-ECM, its biocompatibility increases. The cells cultured in SIS-PLGA composite scaffold presented more superior cellular adhesion property compared with the cells cultured in PLGA scaffold [195].

Furthermore, Kang et al. discovered that chitosan by adding bone ECM powder exhibited better bone regeneration effects compared to that of chitosan scaffold alone when used to fill bone defect area [196]. Collagen in ECM has the special triple helix structure which can induce calcium ion deposition and accelerate bone healing when used as bone filled graft [197].

3.2.8.6 3D Bioprinting Ink

3D bioprinting is a bio-manufacture technology in which 3D construct was fabricated by layer-by-layer assembly and was used extensively in biomedicine field as well as tissue engineering [198]. Specifically, 3D printing was mainly determined by the "ink" that was used. While ECM-based powder "ink" has been commonly used to fabricate complex constructs for application in tissue engineering [199].

To some extent, dissolved ECM powder and cell suspension are all "bioink" and can be combined together through duplex printing. In the printed constructs, ECM provided a microenvironment for cell to attach and grow, however, ECM protected cell from mechanical damage during printing process. The colloidal ECM after dissolution has moderate viscosity, which can protect cells from shear stress, which may lead to cell apoptosis [200]. For the use of ECM in 3D printing, we will introduce in detail in the following chapters.

3.2.8.7 Clinical Application of Powder-Based ECM

ECM materials have various applications in the clinical requirements, including wound healing, surgical closure, tissue reinforcement, and defect tissue reconstruction [201].

It is not like cell therapy, ECM-based therapies mainly exert a regulatory role in the microenvironment between the injured and healthy tissues, promoting the regeneration of blood vessels and nerves across the injured area [202]. Shooter GK et al. reported that ECM has effect in healing refractory wounds that were incurable with conventional methods and repairing patients whose wounds delay to heal, such as elder people, diabetic people, etc. [203] Therefore, the ECM has benefits in helping maintaining wound closure and supporting surgical anastomosis. ECM powder has a prospect broad clinical application because it could be adjusted to various shapes at solubilized form and can be even delivered via minimally invasive injection. The clinical application of ECM-based materials is summarized in Table 3.1.

In clinic, ECM powders were often coordinate with ECM patch in treatment. ECM powder is used for filling, while ECM patch is used to seal the wound [216].

ECM powder can be used not only in forms of dry powder but also in forms of suspension or gels. ECM dry powder is used to fulfill open or deep surgery, while the liquid and gels can be applied to minimally invasive injection in clinical to promote tissue repair and regeneration [150]. According to the previous reports, the injectable ECM hydrogel has been used in regeneration of tendons, cartilage, and skeletal muscles, the treatment of tendinopathies, muscle loss, cartilage degeneration, and the reparation of meniscus successfully [168, 222–224].

Various kinds of ECM powders have been injected directly for tissue repair. Powders made from urinary bladder matrix, with glycerin used as a carrier to increase viscosity of the system, have been used an injectable materials successfully

Table 3.1 Clinical application of ECM-based materials

Forms	Applications	Sources	Reference
Scaffold	Treatment of trachea defects	Porcine cartilage powder	[204]
Patch	Muscle loss repairing	Porcine urinary bladder, small intestinal submucosa	[205]
	Septal ulceration	Acellular human dermal allograft	[206]
	Diabetic foot ulcers, radiation wounds repairing, diabetic, decubitus ulcerations	Porcine small intestinal submucosa	[207–209]
	Cartilage defects filling	Bovine cartilage	[210]
	Esophagojejunal repairing	Porcine urinary bladder	[211]
	Hernia repairing	Porcine urinary bladder	[212]
	Covering surgical trauma	Porcine urinary bladder	[213–216]
Powder	Injectable cell delivery, adipose tissue engineering	Human adipose tissue	[182]
	Synovial sarcoma, seminoma, sarcoma of the sacrum, diabetic, decubitus ulcerations, anal fistulas, pilonidal disease	Porcine urinary bladder	[208, 209, 217–219]
	Full-filling wound	Porcine urinary bladder, small intestine submucosa, and liver (LECM)	[216]
	3D bioprinting	Porcine adipose, cartilage, and heart	[148, 200, 220, 221]

to treat the acquired urinary incontinence in preclinical researches [225]. Another example of the application of ECM powders was to inject micronized ADM into laryngeal tissue, but tissue augmentation did not occur due to the uncontrollable degradability in vivo [226].

In the case of ischemic myocardial injury, myocardial infarction leads to upregulation of metalloproteinases which result in degrading of native ECM, while the self-produced adaptive reconstruction eventually leads to fibrosis and scar formation.

ECM particle injection offers potential in reconstructing ischemic myocardial injury and avoids adaptive reconstruction after ischemic injury which may ultimately lead to fibrosis and scar formation. A study showed that intramyocardial injection of particulate cardiac acellular matrix could promote the development of new myocardial tissue, angiogenesis, thicken ventricular wall and improve cardiac function [227]. Also, ECM-based materials have already been made into patches and intravenous injection liquid, while these forms of ECM biomaterials are not suitable for the management of internal organ injure such as myocardial infarction because they do not infiltrate the myocardium directly [228]. Solubilized ECM powder, however, was injected via the tissue (epicardium) to the damaged site (ischemia).

The form of ECM can be regulated as the change of temperature, concentration, and the pH of solution. Some ECM-based temperature-sensitive and pH-sensitive hydrogels are gelled based on changes in temperature and pH [73]. In stroke models, soluble ECM was injected into brain in which ECM particle emulsions formed a hydrogel at body temperature [229]. The original form of the ECM material is liquid, which is easy for injection. After being gelled at body temperature, it can be combined with the surrounding tissues of the damaged area to play the role of filling and supporting. In addition, transforming into gel form can prevent the liquid from dissipating when the solubilized ECM powder is injected into the stroke cavity, which ensures the ECM remains inside the lesion. Injection of urinary bladder matrix in this manner has successfully remodeled pathology brain and enhanced the endogenous repairing of surrounding penumbra [229]. The same way has also been applied for tendon regeneration, in which the liquid tendon ECM was injected percutaneously into tendon injury area and gelled at body temperature to achieve tendon repair [222].

ECM powder has great application potential in tissue engineering and regenerative medicine. Compared to scaffolds and patches forms, ECM powder overcomes determined structure and can be packed, spread, sprayed, or solubilized for minimally invasive delivery by injection. The dry or soluble ECM powder has good dispersion effect or fluidity, can be used for wound covering and filling, and can conform to the contours of the tissue. However, easy shrinkage, poor shaping ability, and rapid degradation are also problems faced in the use of ECM-based powder materials.

The powder has a large specific surface area, which can provide an ideal platform for cell adhesion and growth, sufficient diffusion of nutrients, and reagents incorporation, allowing 3D cell culture and bioconjugate with growth factors, drugs, and molecular markers. In addition, composed of collagen, ECM powder can be easily modified and usually conjugated with polymer materials to improve their biocompatibility. Moreover, synthetic devices modified with ECM have significantly reduced inflammation and foreign-body reactions caused by materials implanted in the body.

Also, ECM powder derived bioink can enhance biocompatibility of 3D constructs in 3D bioprinting compared to that of synthetic materials, and the stabilized method of 3D constructs is different from the method that gel and scaffold used, of which the crosslinking was by the chemical agent on internal way, which might release in the body over time. Whereas, by external crosslinking, the 3D constructs can not only maintain the forms of it, but also avoid the direct contact between the cells inside the printed constructs and the chemical crosslinking agent, reducing the toxicity to the cells.

The implantation method of ECM is also different from that of scaffold. The ECM powder was invaded by injection, while the implantation of ECM scaffold was through surgical procedures. Through the injection method, we can not only achieve the purpose of minimally invasive, but also target the damaged area, which are hard for achieving for scaffold that needs to be implanted through surgical procedures.

ECM powder provides a promising treatment strategy for currently intractable diseases in tissue regeneration. Adjunctive therapy with ECM powder, sheet or patch has successfully used in multiple disease models, and results show that specific advantages have been founded in ECM powder.

However, the use of acellular matrix powder is both good and bad. The advantage of ECM powder and its soluble derivatives is that it can be prepared into various form of surgical dressing according to clinic needs, such as patches/sheets, scaffolds, gels, etc. However, in the process of preparing ECM powder, the original 3D structure was destroyed and part of the original composition was lost, which compromised the biological integrity of the native ECM. Thus, ideally biomimetic of the inherent property of native ECM can not be completely achieved in ECM powder or powder derivative due to the losing of their original structures.

ECM powder preparation is not easy, because there are a series of problems faced in the preparation process. Freeze-dried ECM is easy to absorb moisture, thus affecting trituration. At present, there are mainly two kinds of crushing methods commonly used, one is crushing by blades, and the other is by collision. But no matter which method is used, there are problems of incomplete crushing, low yield, and agglomeration after crushing. Usually, screens with different apertures are commonly used for multiple screening to obtain ECM powder with the required particle size. Solubilization with acidic, so that the ECM solubilized derivatives need to be neutralized before implantation in the body to prevent inflammation. It is necessary to address these key issues in order to make ECM powder more widely used.

ECM powder is regarded as a biological therapy with greater potential if we know more about the preparation of ECM and its particle form. Further research on the properties of ECM powder provides the possibility of tissue-specific damage repair.

3.3 Extracellular Matrix Hydrogels

Hydrogel is a three-dimensional polymer network or peptide-based gel with hydrophilic groups that flow freely at ambient temperature [230–232]. The polymer of hydrogel network can be divided into synthetic polymer and natural polymer (includes polysaccharides, collagen, fibronectin, fibrin, and silk protein) [233].

3.3.1 Preparation Methods

Although methods for hydrogel preparation varied with the characteristics that need in the specific application, two key steps are still necessary in the process of ECM hydrogel: (1) decellularized tissue powders treated by enzymes, acid, urea, or some other solution to prepare homogeneous ECM pre-gel; (2) methods such as adjusting the temperature, ionic strength, and pH or adding crosslinking agent have been used to prepare hydrogel [234–238]. Gelation of the pre-gel may contribute to the entropy generation in the collagen self-polymerization process contained in ECM [234]. By this means, hydrogels derived from various tissues, such as bladder [239], small intestinal submucosa [73], heart [240], liver [241], kidney [242], meniscus

[223, 243], pancreas [244–246], uterus [247], nervous [113, 248], adipose [5, 236, 249] are available now and are summarized in Table 3.2. Figure 3.8 showed the general process of ECM (SIS) hydrogel preparation.

Ideally, purification is able to destroy or remove low molecular weight peptides and the active growth factors of ECM should be minimized or avoid. Recently, Wang et al. describe a way to prepare hydrogel derived from with no purification steps [73]. First of all, lyophilized small intestinal submucosa (SIS) powder filtered through an 80 mesh screen was digested by porcine pepsin in 0.01M HCl solution at for 48 h at 25 °C. Then, freeze dry SIS digest solution again and cut it into flocculent pieces. Finally, the pieces of SIS matrix form a gel after dissolved in phosphate buffered saline (PBS) and raised the pH and temperature to physiological levels.

3.3.2 Characteristics

3.3.2.1 Microstructure

The original structure of decellularized tissues has been destroyed along with the decellularization and the dissolution process of tissue ECM. And a new three-dimensional porous structure self-assembly by the collagen contained in the ECM has formed during gelation [234]. Tissue source of ECM, conditions of gelation, and concentration of ECM proteins were main factors that affect the hydrogel structure. Furthermore, the impact of gelation conditions or ECM proteins concentration on the size of pore of the hydrogel are bigger than that of tissue source of ECM [113, 251].

3.3.2.2 Rheological Properties

Rheological properties of ECM pre-gel are important criteria that evaluate whether this gel could be used in minimally invasive surgery, which determines the injectability and whether a hydrogel can be formed at physiological temperature and then retained at a specific surgical site.

Wang et al. demonstrated the rheological properties of SIS hydrogel [73]. While the SIS pre-gel solution is kept at 10 °C, it has low storage modulus (G') and loss modulus (G'') values as a liquid [73]. After the temperature rises to 37 °C, the G' and the G'' were increased [73]. Simultaneously, the gelation rate of the pre-gel solution was increased with the increasing of SIS concentration, which makes it possible to regulate the rheologic properties meet the requirements of various applications [73].

Similar with the SIS hydrogel, both G' and G'' of decellularized urinary bladder solution changed with time and became gel phase when temperature raises from 15 °C to 37 °C for about 8–12 min [237]. The G' is greater than G'', suggesting that this hydrogel executes as a solid material after self-assembly [237].

Table 3.2 Preparation of dECM hydrogels from different tissue sources

ECM origin	Preparation technology	Applications/Results	Reference
Bladder	Frozen, pulverize; pepsin, HCl, PBS, NaOH	Bioactive factors preserve well, support the adherence and growth of PIEC cells	[239]
Small intestinal submucosa	Mechanical remove, methanol, frozen, chloroform, 0.05% trypsin, 0.05% ethylenediamine tetraacetic acid, 0.5% SDS, 9% NaCl, 0.1% NaOH, 20% ethanol; 1 mg/ml porcine pepsin, HCl, PBS	Promoted neovascularization	[73]
Heart	Perfusion, 1% SDS, 1% Triton X-100; slice, pepsin-HCl, NaOH, PBS	Supported the proliferation and cardiomyogenic differentiation of BADSCs; promote myocardial regeneration	[240]
Liver	Chopped, HBSS, 2% Triton X-100, 0.05 mM EDTA, 0.025% ammonium hydroxide, dialysis; 10% pepsin, 0.01M HCl	It is superior to collagen in 2D and 3D culture and tissue engineering of hepatocytes; supported the development of microvasculature and vasculogenesis	[241]
Kidney	Minced, rinsed, 0.1% sodium dodecyl sulfate, frozen; milled, 1 mg/mL pepsin, 0.01M HCl, frozen, 1.0M NaOH, PBS	Exhibited high cell viability and proliferation over a one-week culture period	[242]
Meniscus	Pulverize, 1% SDS/PBS, frozen; pepsin/0.01M HCl, 0.1M NaOH, PBS, froze, freeze-dried, sputter coated	Good injectability and good cellular compatibility; upregulated expression of fibrochondrogenic markers, transmission cells	[223, 243]
Pancreas	Frozen, ground, NaCl, HCl, 10 mg/mL pepsin	Supported long-term islet culture and preservation of glucose stimulated insulin release	[244–246]
Uterus	Perfusion, PBS, SDS, Triton X-100, 2 g/mL DNase; milled, pepsin	Possibly be used as part of a treatment of Asherman's syndrome and endometrial atrophy in the future	[247]
Nervous	125 mM sulfobetaine-10 and 0.14% Triton X-200/0.6 mM SB-16, PBS, 200 μL 0.02 U/mL chondroitinase; 1 mg/mL porcine pepsin, 0.01M HCl, NaOH, PBS	Use after contusion spinal cord injury	[113, 248]
Adipose	1% sodium dodecyl sulfate, 2.5 mM sodium deoxycholate added with 500 U lipase and colipase, in deionized (DI) water with 5000 I.U. penicillin and 5 mg/mL streptomycin, 100% isopropanol, milled; 0.1M HCl with 2 mg/mL pepsin, 1M NaOH, 12 mg/mL PBS	Stimulating subcutaneous adipose regeneration, restore function to adipose deficits; reconstructive surgery such as craniomaxillofacial surgeries, trauma, and other corrective procedures	[5, 236, 249]

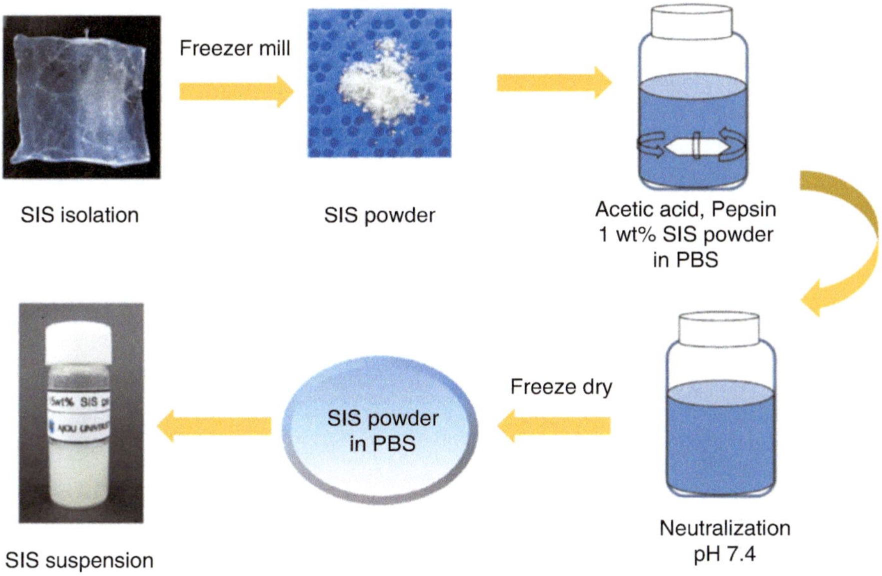

Fig. 3.8 Digesting SIS film or powder into SIS suspension or gel can achieve the transformation of SIS as a filling material to an injectable material. Here is one example of schematic diagram for the preparation of the ECM (SIS) suspension, mainly including crushing, enzymatic lysis, freeze-drying, reconstitution, etc. Reproduced with permission [250], copyright 2011

3.3.3 Applications and Efforts in Performance Optimization

Hydrogels are widely being applied in various fields, which are usually used as food additives, pharmaceuticals, biomedical implants, tissue engineering and regenerative medicines, diagnostics, cellular immobility, cell encapsulation, etc. (Fig. 3.9) [252]. Hydrogel form can further extend the application of decellularized tissue by delivering it into the irregularly shaped anatomic sites via minimally invasive methods and fill defects with any shape and size, which is more readily than a suspension of powders [154].

Injectable hydrogels have two forms: in situ forming (formation after injection in physiological conditions) and preformed hydrogels (shear-thinning hydrogel) [253, 254]. Injection hydrogels are perfect for filling defects and simply encapsulating cells, reducing surgical invasion, infections, scarring, and high cost during surgery [254, 255]. Injectable ECM hydrogel loaded with biphasic calcium phosphate powder (BCP) was used to fill the irregular bone defects and could promote bone formation [256]. There are research found that ECM hydrogel was implanted into the sub-acute stroke cavity to make it stay in the lesion cavity for a long time, so as to hence stabilize the stroke environment and reduce the lesion volume [173]. The use of tiny needles in transplant programs can reduce damage to brain tissue [257].

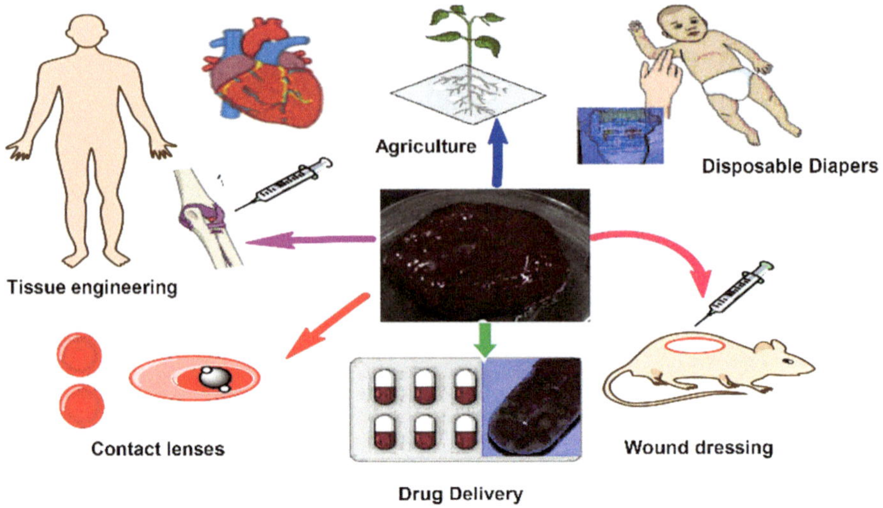

Fig. 3.9 Hydrogel is a kind of materials with a wide range of applications in different fields. Here are exemplified the application of hydrogels in medical engineering, agriculture, daily use, medicine, cell biology, and other fields. Reproduced with permission [252], copyright 2020

Although satisfactory prognosis has been obtained after ECM hydrogel injection in repair of damaged tissues, its therapeutic effect is still unstable because of the differences of the composition, mechanical properties, biological characteristics, and preparation process of hydrogels derived from different tissues.

At present, many strategies including encapsule cells or bioactive factors, crosslinking, and creating composites have introduced in performance optimization for hydrogels [243, 250, 258–260].

The delivery systems, functionalization of hydrogels with cells or bioactive factors, have been already demonstrated to offer an additional design capability of scaffold function. Chronic tendon injury can be repaired by surgery, but slow tissue healing would lead to poor functional recovery, adding platelet-rich plasma (PRP) and adipose-derived stem cells (ASCs) to the tendon hydrogel can promote tendon repair process [259]. The dense ECM of meniscus can obstruct cellular infiltration and integration, hydrogel prepared by decellularized bovine meniscus combined with human mesenchymal stem cells (hMSCs) was injected into the damaged part, which can promote the repair of meniscus and the growth of cells [243]. Studie also show that mixing mesenchymal stem cells (MSCs) into SIS hydrogel could promote scarless vocal fold healing [260].

After adding bovine serum albumin (BSA), the SIS suspension could form gel after injection. This hydrogel can be used as a delivery system, and it was found that BSA could be released continuously for 30 days in vivo [250]. Moreover, heparin-binding growth factor can be immobilized to decellularized porcine pericardia hydrogel by sulfated glycosaminoglycans (GAGs) to promote cellular infiltration and acute neovascularization in ischemic regions [258].

Crosslinking, which can enhance the mechanical property and degradation property of ECM, is another modification strategy to hydrogels [261, 262]. Crosslinking can control the degradation of extracellular matrix and increase the persistence of adipose transfer [5]. Pericardial matrix hydrogel (PMNT) doped with carbodihydrazide (CDH) functionalized multiwall carbon nanotubes (CDH-MWCNT) could increase electrical conductivity and mechanical properties [263]. Hydrogel crosslinked to EDAC (1-ethyl-3-3-dimethylaminopropyl carbodiimide) could reinforce antidegradation property and improve the mechanical properties [255].

In addition, manipulating hydrogels into composite materials is also a valuable approach to improve the ECM properties of cell delivery, mechanical, and degradation. ECM hydrogel/PEUU fiber composites with complex lamellar structure containing elastic were created by concurrent gel electrospray/polymer electrospinning, which could increase the mechanical properties on the basis of good biocompatibility and bioactivity of hydrogels (Fig. 3.10) [264]. Decellularized adipose tissue (DAT) can also be used as bioactive matrix for in situ polymerization with synthetic hydrogels, which could be conducive to the delivery of adipose-derived stem cells (ASCs) and promote the viability and adipogenic differentiation [265, 266].

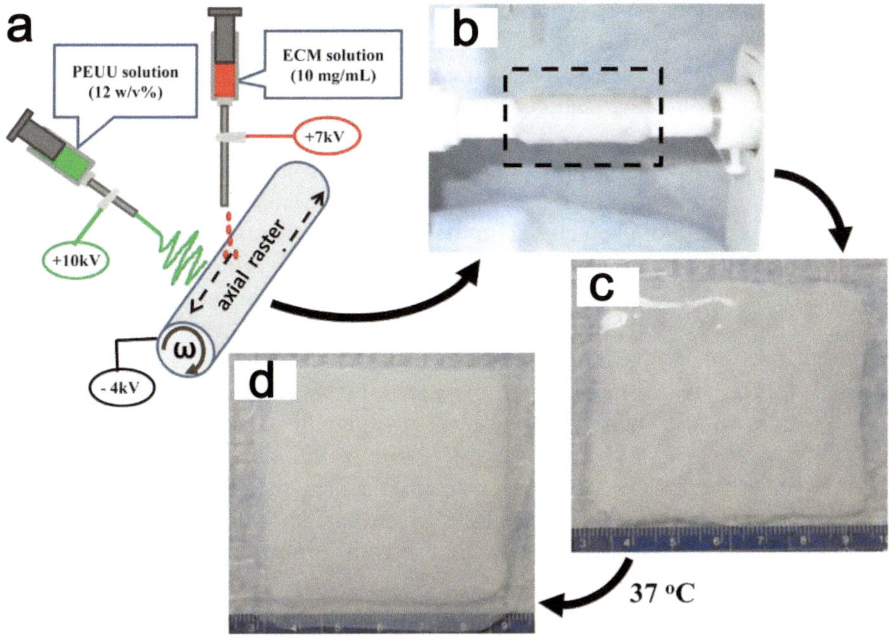

Fig. 3.10 The solubilization or gelation can expand the application field of ECM. Liquefied or gelled ECM can be combined with other materials according to requirements. One application of ECM solution is the concurrent electrospin/electrospray technique for PEUU/dECM biohybrid scaffold fabrication. Reproduced with permission [264], copyright 2011

Moreover, mixing acellular matrix with natural hydrogels can also improve the function of hydrogels. The addition of high content of native heart matrix to collagen hydrogel can improve the contractile function of cardiac cells [267]. Rat decellularized adipose tissue (DAT) mixed with silk fibroin hydrogel can accelerate vascularization [268]. In addition, nanomaterials embedded matrix of the hydrogel can also improve its properties [253].

3.4 Extracellular Matrix Coating

ECM at the interface between the material and the tissue promotes the fusion of the implant material and the tissue by changing the characteristics of the material surface to reduce the inflammatory response and infection [269, 270].

3.4.1 Preparation Methods and Characteristics

ECM coatings are created by adding the ECM hydrogels on the surface of scaffolds or tissue culture plate directly or dried through open-air drying or freeze-drying [235, 271–274]. Concurrent electrospin/electrospray technique is also common method used for fabrication of biomedical devices. The key material properties of ECM coatings are coating stability following rinsing, modulus, and water content [275, 276].

The balance between modulus and water content is controlled by setting or curing time, which also determines the stability of redissolution. The properties, water content, and stability of the material can be adjusted by treatment time, temperature, or concentration [275]. The concentration of ECM can be adjusted to show appropriate surface properties [277]. Moreover, the desorption or rearrangement of the coating should also be maintained stable [275]. The biological activity of ECM may depend largely on the specific ECM source, how and when the ECM is digested, and the type of cell being loaded. For example, the attachment, proliferation, and adipogenesis of ASC cultured on a DAT coating digestion with the glycolytic enzyme α-amylase would be enhanced [276]. In addition, the microcosm and topography of the extracellular matrix coating can regulate the morphometry and phenotype of human corneal endothelial cells and can adjust topography to enhance the performance of cells according to the components that constitute the carrier of transplanted cells [278].

Remarkably, biological characteristics of ECM coatings may be affected when the preparation processes were changed. For example, Shridhar and other scholars found that the cell attachment, proliferation, and adipogenesis of human adipose-derived stem/stromal cells were enhanced on the $\boldsymbol{\alpha}$-amylase-digested decellularized adipose tissue coatings whose bioactivity was better conserved, compared with the proteolytic enzyme pepsin-digested decellularized adipose tissue coatings [276].

3.4.2 Applications and Efforts in Performance Optimization

ECM coatings, which have been shown to provide important tissue-specific cues, are a favorable platform in vitro for culture of cells (e.g. Schwann cells, fibroblast, and chondrocytes) based on their larger quantities and more easily to retain cells than sections [223, 237]. In that case, ECM coatings taken from various tissues are expected to build models for pathology, tissue repair mechanism, and cell biology researches.

Firstly, ECM coatings derived from diseased tissues can be used to recreate the special microenvironment in vitro and then establish a model for exploring the cellular dysfunction mechanisms during disease progression [279]. Some researchers studied the metabolic mechanism of adipocyte by seeding cells onto the ECM coatings derived from adipose tissue with diabetes or without diabetes [280]. Results indicated that ECM coatings from adipose tissue without diabetes rescued metabolic dysfunction in diabetes adipocytes, whereas ECM coatings from adipose tissue with diabetes imparted features of metabolic dysfunction to non-diabetic adipocytes [280].

The investigator can use appropriate methods to enhance the tissue specificity of the coating, bringing it closer to the mature ECM microenvironment in vivo [281]. Agarwal et al. found that CLECM can provide a biochemical microenvironment that supports the differentiation of HepG2 cells. 2D cell culture on a CLECM coated polystyrene plate revealed enhanced secretion of albumin and the synthesis of urea, glycogen, and GAGs [241]. The survival rate and hepatogenic differentiation rate of bone marrow mesenchymal stem cells coated with liver ECM cultured on a plate were significantly increased, and the liver ECM coating could have induced additional metabolic functions [271].

Secondly, ECM coatings can also be used to provide plenty of 3D environment and space for stem cells to maintain the stemness and proliferation. Studies on the proliferation and differentiation of human renal cells using extracellular matrix (ECM) coatings have shown that ECMs coatings consisting of basement membrane components, laminin, or collagen IV can maintain differentiated epithelium for up to 3 weeks [282].

Whether the microenvironment of a given tissue can enable stem cells maturation, and directional differentiate to particular adult cells have also been investigated by using ECM coatings derived from several tissues. Wang et al. investigated the ability of liver ECM in facilitating the bone marrow derived mesenchymal stem cells (BMSC) hepatic differentiation and hepatocyte-specific functions in vitro [271]. After 28-day culture, BMSC on the decellularized liver hydrogel showed hepatocyte-like morphology with metabolic functions induced [271]. Both synthesis of albumin, urea, as well as antialpha-fetoprotein and expression of certain hepatic markers were enhanced when cells were cultured on ECM [271]. Liver ECM can provide a suitable biochemical, biophysical, and structural microenvironment for the survival, maturation, and differentiation of hepatocytes, and it can upregulate the hepatic gene expression of human adipogenic stem cells and enhance the

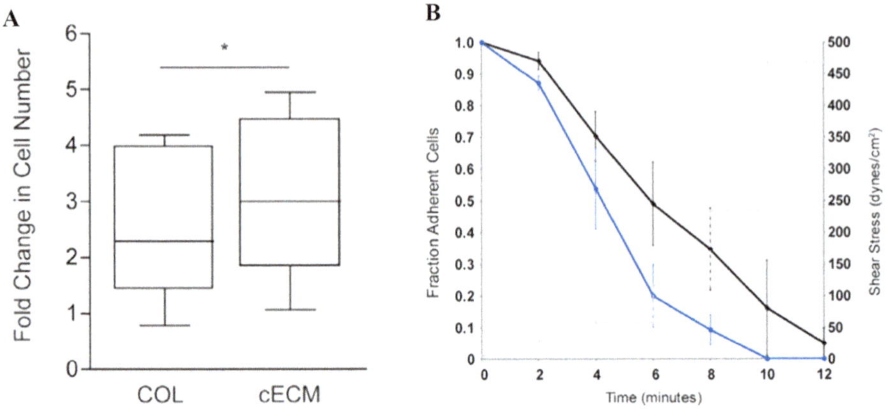

Fig. 3.11 The main component of ECM is collagen. What is more, ECM not only contains collagen, but also contains other components (integrin, cytokines, etc.) that benefit to cell adhesion and proliferation. Here is one example of comparing the influences of cECM and COL on CPC proliferation and adhesion CPC. After 48 h culture, results of fold change in cell number (**a**) showed that proliferation of the CPC seeded on cECM was significantly increased than that of CPC seeded on COL($*p < 0.05$). *COL* collagen, *cECM* cardiac decellularized extracellular matrix. Further, to determine whether CPC adhered more strongly on cECM or on COL, CPC adherence was determined by microfluidic adhesion assay. The grouped data (**b**) showed that CPC adhere more tightly to cECM (black) as compared to COL (blue). These data show that cECM is a better substrate for CPC proliferation as compared to COL. Reproduced with permission [283], copyright 2012

differentiation of liver cells [277]. DeQuach et al. developed an acellular muscle matrix coating that retained ECM proteins, peptides, and GAGs to promote committed muscle progenitor differentiation and in vitro maturation of stem cells [281]. Moreover, decellularized cardiac ECM coated plate can promote the proliferation and adhesion of cardiac progenitor cells (CPC) and reduce cell apoptosis (Fig. 3.11) [283].

These results, which provide a theoretical basis, may be helpful for scientists to explore the possible mechanisms in tissue repair. What is more, tissue-derived ECM can be also coated around the backbones of synthetic scaffolds to provide a biomimetic environment for better host response and integration (Fig. 3.12). ECM extracted from the Wharton's jelly can promote the adhesion and proliferation of human mesenchymal stem cells and mature endothelial cells and can be used as a coating in vascular tissue engineering [284].

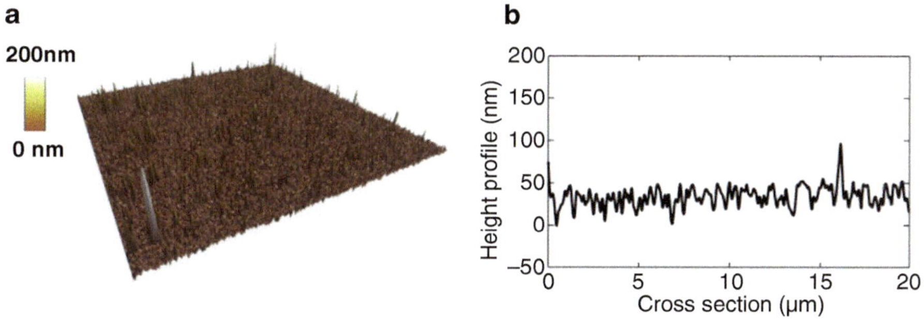

Fig. 3.12 ECM powder can be used as a coating material after solubilization, which can improve the biocompatibility of metals, polymers, etc. Here is one study example of tissue-derived ECM being coated around the backbones of synthetic scaffolds to provide a biomimetic environment for better host response and integration. Figure shows the characterization of WJ-ECM derived coating. Atomic force microscopy image of WJ-ECMB (**a**) checking the surface roughness of the substrate modified with ECM. Cross-section thickness of WJ-ECM coating (**b**) revealing a measured thickness of 45 ± 12 nm and an even distribution. Reproduced with permission [284], copyright 2017

3.5 3D Printing Tissue ECM Scaffolds

Placing all the components at the specified location in the 3D scaffolds to generate structures that resemble native tissues has promising implications in tissue engineering. In tissue engineering, 3D printing is a biofabrication strategy that can accurately control the position and placement of bioinks (e.g. biomaterials, cells, growth factors, etc.) with hierarchical architecture according to the model of artificial tissue or organ designed by researchers, so that the engineering tissues have formed with native-like characteristics [285–287]. Designable ability of the bioprinted scaffolds is superior to the products produced by traditional material molding methods such as fiber bonding, phase separation, gas foaming, and so on [288–291]. Tissue ECM scaffolds derived from the nature have great potential to serve as bioinks for advanced tissue fabrication benefits by their biocompatible characteristics [292]. In this section, the 3D printing technology particularly with decellularized tissues as bioink, including bioink fabrication, common bioprinting techniques, significance considerations of bioink, and the recent applications and efforts in performance optimization, has been introduced in detail.

3.5.1 Preparation Methods and Characteristics

The ECM scaffolds derived from human or animal tissue can be made into solution or hydrogel and used as bioinks to print tissue-specific scaffolds [220, 293–299]. Skin-derived ECM bioink, for instance, could be manufactured by dissolving the lyophilized decellularized skin tissue in pepsin-acetic acid solution and adjusting the pH of ECM solution to 7.4 before use [300]. Other tissue-specific bioinks, such as cornea, adipose, tendon, heart, and cartilage-derived ECM, have already been developed either [301–304]. The ECM bioink may facilitate stem cells to proliferate and differentiate into particular cell types. Lee at al. developed a liver ECM bioink with excellent print capabilities for 3D printing, which exhibits higher stem cell differentiation potency and higher secreted albumin levels and urea secretion rate of human hepatocellular carcinoma cells over other commercially available bioinks such as collagen bioinks [305]. In addition, Jinah et al. developed a 3D pre-vascularization stem cell patch using the extracellular matrix of decellularized heart tissue and mesenchymal stem cells as bioink, which was proved to promote angiogenesis and enhance the therapeutic efficacy for cardiac repair [306]. Remarkably, tissue ECM bioinks derived from porcine, which has high breeding potential, rapid growth, and short reproductive maturity, are easily available and are becoming one of the most popular bioink formulations [307]. Choi et al. found that the ECM bioink derived from porcine skeletal muscle could provide a myogenic environment to myoblasts, promote the formation and maturation of myotubes, and respond well to electrical stimulation [308]. However, porcine ECM has some problems, for example, porcine ECM is affected by endogenous retrovirus. Although it does not self-replicate under physiological conditions, transmission of endogenous retrovirus was observed when pig cells were injected into immunocompromised mice [309, 310].

The process of bioprinting was controlled by software. The 3D image data of tissue ECM can either be manually designed or acquired from 3D patient-specific digital model generated by medical images obtained through technology including computed tomography scanning, magnetic resonance imaging, optical microscopy, etc. [311–313] Tissue ECM bioink was then deposited in the correct xyz coordinates by 3D bioprinter through a syringe controlled b y omputer [314]. The entire 3D bioprinting process is depicted in Fig. 3.13.

Currently, extrusion-based and droplet-based bioprinting techniques are commonly applied in the 3D bioprinting of ECM [301, 315–318]. Extrusion-based bioprinting, which originates from fused deposition modeling printing technology, is a needle-syringe-type system and an automated robotic system combination that performs to extrusion and bioprinting, respectively [319, 320]. The needle-syringe-type system can be driven by a pneumatic-, mechanical-, electromagnetic-based system to deposit bioinks [296, 321–324]. To be specific, pneumatic-driven extrusion-based bioprinting is a convenient and feasible method that extrudes bioinks from a syringe and nozzle though air pressure at a controllable rate by using a valve-based or a valve-free configuration [325]. Mechanical-driven extrusion-based bioprinting is a technology that extrudes bioinks though deposition forces provided by screw- and/or piston-driven configurations [326]. This system can drive larger pressure drop from syringe to the nozzle which is beneficial to print higher viscosity bioinks. However, the large driving forces may cause cytomembrane fragmentate in the printing process and thus may not be applicable to print bioinks containing living cells [316]. Electromagnetic-driven extrusion-based bioprinting is a system

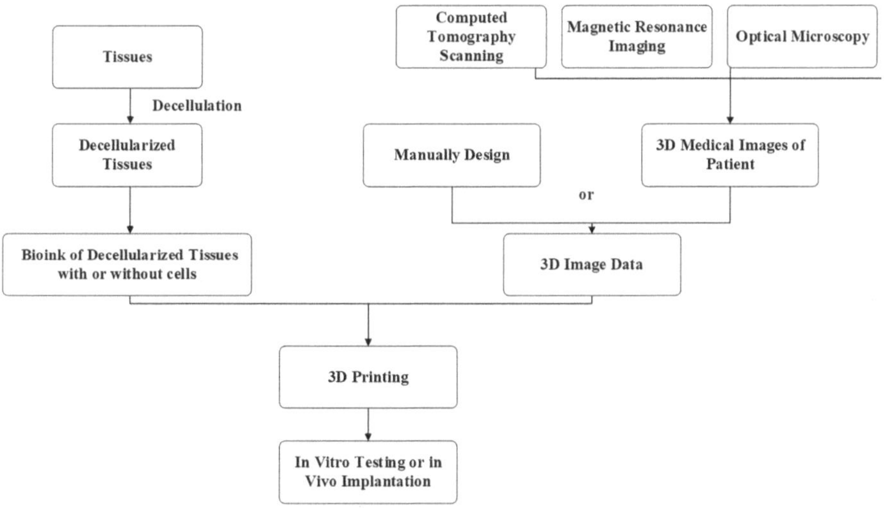

Fig. 3.13 A flow diagram for 3D bioprinting. After importing manually designed image data or patient-specific 3D image data into software, tissue ECM bioink was deposited in the correct xyz coordinates and then printed out the designed shapes through a syringe controlled by the software

that extrudes bioinks though electrical pulses produced by canceling the magnetic pull force [319]. Meanwhile, droplet-based bioprinting is another commonly used technique which can deposit preset volume of bioink drops onto a targeted spatial position to form three-dimensional shapes in accordance with the image data accepted from digital signal [296, 327]. Droplet-based bioprinting methods include the following contents: inkjet bioprinting, acoustic bioprinting, and micro-valve bioprinting [328–330]. Inkjet bioprinting can be further classified into continuous-inkjet bioprinting, drop-on-demand bioprinting, and electrohydrodynamic jet bioprinting. And physical means, such as gas pressure, gravity, and hydrodynamics characteristics of inks, have been utilized to discharge droplets [331]. Whereas, in acoustic bioprinting, droplets were produced by gentle acoustic field which protects the bioinks from detrimental stressors (e.g. heat, large voltage, and high pressure, etc.) during droplet ejection [332]. However, it also has some flaws. For instance, the movement of printhead and/or substrate during printing may bring undesirable acoustic interference and led to uncontrol of ink-jet process, and viscous bioinks are not capable of being printed by using this methods [333]. In addition to inkjet bioprinting and acoustic bioprinting, the micro-valve bioprinting, which generates droplets by an electromechanical valve, is favorable for printing various biomaterials and cells [331, 334]. However, the larger droplets lead to lower resolution of micro-valve bioprinters than other droplet-based bioprinting methods [331].

The significance considerations of the ECM inks are bioprintability, mechanical and structural integrity, degradability, swelling properties, and contractile properties [335–337]. Among them, bioprintability is dominated by the rheological property and crosslinking mechanism of ECM during and after the bioprinting process, the requirements of which are varied with the printing techniques used [293, 331, 336]. The rheological property of bioinks, which is crucial to the distribution and survival of cells during extrusion process, the printability and printing parameters, and the preservation of printed module shape, is commonly output as solution viscosity, gelation kinetics, and the modulus of the crosslinked solution [323]. Notably, the viscosity is proportion to the concentration for bioinks. Shear sweep analysis and frequency sweep can be used as measurement methods for viscosity and linear viscoelastic region testing. Extrusion-based bioprinting can print bioinks with wide range of fluid viscosities (30-6 $\times$ 10^7 mPa s) [336]. Though bioinks with higher viscosity gain better printed structure fidelity, high viscosity may not be suitable for the even distribution of cells and may have negative effect on cell functionalities [338, 339]. While with regard to droplet-based bioprinting, bioinks with viscosities $<$10 mPa s are more appropriate [336]. Further, the bioprinted scaffolds should provide proper strength and structural support for cells proliferation, cells differentiation, and tissues reconstruction [340]. In that case, the mechanical properties of the bioink, which mainly depend upon the concentration and the intrinsic properties of the materials, should be emphasized. Despite the probable advantages of supporting cell proliferation in a lower concentration, its mechanical properties would be poor and thus result in the collapse of the printed scaffolds [316].

Meanwhile, excessive strength, which might have an adverse effect on cellular activities, is also not desirable [316]. Besides, printing parameters may affect the results of printed cells. For example, when using myocyte coated extracellular matrix bioink for 3D printing, the cellular alignment of the constructs depends on the linewidth of muscle construct [308]. Almost 50% of acellular matrix usually degrades in a month [306]. What is more, the degradation rate of ECM bioink is governed by the concentration, the crosslinking mechanism, its intrinsic properties and/or external additive [293, 335, 341]. Keeping consistent between the degradation rate of bioinks and the regeneration rate of tissue in vivo is an ideal situation in regenerative medicine. Other essential characteristics of the printed tissue ECM scaffolds are swelling and contractile properties. Both overly swelling and excessively contractile may result in the deformation of designed structure. In addition, contraction may also close the channel which is essential for nutrient delivery and cell migration [342].

3.5.2 Applications and Efforts in Performance Optimization

In recent years, ECM scaffolds prepared by 3D bioprinting have been successfully applied for fabricating 3D tissue models, organ-on-a-chip models, tissue and organ constructs, and minitissues [300, 331, 343–352]. 3D tissue models and organ-on-a-chip models have advantages of building proper biomicroenvironment in disease modeling and in vitro drug testing than animal models and 2D culture models [331]. Mollica and his coworkers describe mammary-specific in vitro 3D models fabricated with ECM from decellularized breast tissue by 3D bioprinting platform

[353]. Results demonstrate that the mammary-specific in vitro 3D models, whose unique profiles of structural and signaling affect variety of cellular processes, allows for growth of organoids/tumoroids, offering suitable systems for epithelial/cancer cell researches [353]. Skin ECM bioink is also useful as 3D full-thickness (epidermis/dermis) human skin models which lead to less contraction as compared to the collagen skin tissue. In addition, bioinks can provide physical stability and tissue-specific microenvironment to promote organized epidermis differentiation. EPCs-laden 3D-printed skin patch with ASCs can rapidly form in vitro pre-vascularization on day 3 [300].

Different types of tissue or organ constructs, such as cardiac tissue, blood vessel, cornea, muscle, esophageal and liver, have been developed for different applications [303, 308, 354–357]. For instance, Das's group engineered a heart tissue with function characteristics and structure features similar to native tissue [354]. Using a 3DXPrinter, neonatal rat cardiomyocytes were bioprinted within porcine heart tissue-derived ECM and cultured statically or dynamically [354]. Afterwards, immunostaining and qRT-PCR were performed to detect cardiac-specific proteins expression and relative transcription factors [354]. After 14 days of culture, the maturation of cardiomyocytes in heart tissue-derived ECM was enhanced, while the collagen group under similar culture conditions was not [354]. Recently, a cell/drug loaded engineered blood vessel has been printed with a bioink composed of endothelial progenitor cells (EPCs), atorvastatin-loaded poly (lactic-co-glycolic) acid microspheres, decellularized vascular tissue and alginate [356]. Therapeutic efficacy has been evaluated in a model of athymic nude mice with hindlimb ischemia. Results revealed that the cell/drug-laden bio-blood-vessel not only promotes neovascularization but also remarkable salvage of ischemic limbs [356]. Bioprinting of endotracheal endothelial and fibroblasts cells from tracheal ECM can form a natural vascular network and offer a natural niche for endotheliocyte repositioning and angiogenesis [348]. What is more, human turbinate-derived MSC has been wrapped in the corneal derived ECM (CodECM) for bioprinting [303]. It was found that the stem cells differentiated into keratinocytes only within the CodECM bioink [303]. In addition, the content of collagen and GAGs in CodECM is similar to that of natural cornea [303].

However, since the ECM bioinks lost the mechanical and structural integrity during preparation process, their mechanical properties are very weak [319, 358]. The general improvement scheme is to use supporting materials (e.g. polycaprolactone, poly (ethylene/vinyl acetate), or Pluronic F-127) to keep the ECM structure not collapsed. However, the mechanical property of the scaffolds produced in this way may mismatch with the native tissue, and its degradation time may be much longer than a natural biomaterial system [220, 355]. Nevertheless, the mechanical performance of the scaffolds or constructs fabricated by 3D-printing remains far below that of native comparisons. In order to eliminate unwanted supporting materials, various methods, including combining with crosslinkable hydrogels, chemical modification of the ECM bioinks, and/or crosslinking, were investigated to enhance the mechanical performance of the ECM bioinks [316, 359]. For composite ECM/crosslinkable hydrogels bioinks, the role of ECM is to provide tissue-specific components, and the role of crosslinkable hydrogels is to enhance the mechanical properties [325]. Jung CS and his coworkers chose silk fibroin to optimize the printability of cartilage acellular matrix benefit by its physically crosslink property and controllable viscosity [304]. Furthermore, cartilage acellular matrix-silk scaffolds were crosslinked with 1-ethyl-3-(-3-dimethylaminopropyl) carbodiimide/N-hydroxysuccinimide in water (EDC-W) or in 80% methanol (EDC-M), or methanol only for 12 h [304]. Results showed that the mechanical strength of EDC-M-treated scaffolds is much greater than that of methanol- or EDC-W-crosslinked scaffolds, indicating that the mechanical properties of cartilage acellular matrix-silk scaffolds can be adjusted by utilizing different crosslinking methods [304]. Liver-, cardiac-, and skeletal muscle tissue-derived ECM bioinks have been prepared by Skardal A et al. [360] And the HyStem-HP hydrogel kits, which were used to produce crosslinkable hydrogel, have been combined with the tissue-specific ECM abovementioned to provide soft and extrudable ECM-containing hydrogel that supported extrusion bioprinting [360]. The HyStem-HP hydrogel kits are comprised of thiolated gelatin, thiolated hyaluronic acid, acrylate-based crosslinkers, and alkyne-based crosslinkers [360]. What is more, unmodified gelatin and hyaluronic acid have been added to the bioinks to improve its shear thinning and extrusion properties [360]. Stage 1 crosslinking has been allowed to occur to form thiol-acrylate bonds when bioinks were loaded into printer cartridges of 3D bioprinting platform [360]. In this stage, the gel is soft and can be extruded smoothly (Fig. 3.14a) [360]. Afterwards, ultraviolet light was administrated to the printed scaffolds to initiate stage 2 crosslinking procedure during which a near-instantaneous thiol-alkyne polymerization reaction occurred [360]. In this stage, stiffness is significantly increased (Fig. 3.14a) [360]. What is more, a wide range of final stiffness could be reached by manipulating the concentrations, geometries, and molecular weights in stage 2 (Fig. 3.14b) [360]. It means that final construct with mechanical properties similar to the tissues in the body could be achieved by regulating the molecule structure of the crosslinker [360].

Although the mechanical properties and printability have been improved, the above approaches still have limitations in fabricating printed cell–ECM constructs because the EDC and the UV systems may induce increased cell death or stress. In order to address this, gelatin methacrylate, which has methacrylate groups so that it can undergo radical polymerization under a white light system, has been used to mix with cardiac extracellular matrix and human cardiac progenitor cells to fabricate bioprinted cardiac patches for cardiac defects repairing [355]. What is more, acellular ECM can also be dissolved in an acid solution and chemically modified as a photocrosslinkable hydrogel by methacrylation to fabricate an ECM bioink with tunable biomechanical properties [361]. Another common method to make ECM bioinks crosslinkable was to add photoinitiator (e.g. riboflavin, lithium phenyl-2,4,6, trimethylbenzoylphosphinate, etc.) directly [359, 362]. Jang et al. successfully used riboflavin as photosensitizers and ultraviolet A as light source, to develop ECM-containing bioinks for 3D-printing. After photocrosslinking, the hardness of ECM bioink was greatly improved [363]. Further, this printed bioconstruct was proved could support high cell viability, and promote proliferation and

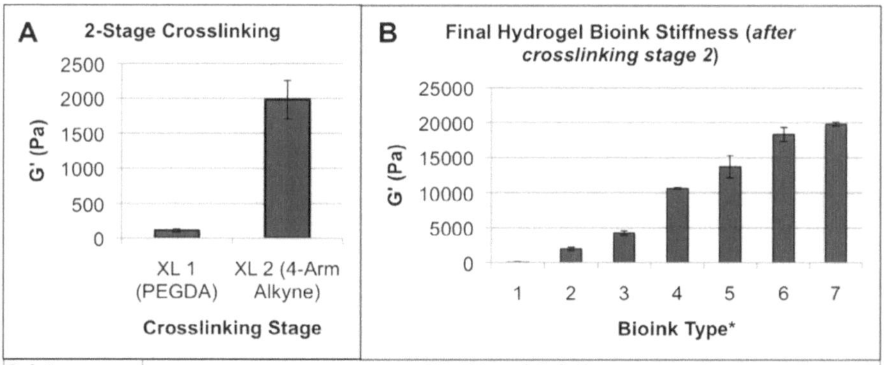

Gel #	1% HA and Gelatin	
	Crosslinker 1	Crosslinker 2
1	4% PEGDA	-
2	4% PEGDA	4% 4-Arm Alkyne
3	4% 4-Arm Acrylate, 20 kDa	10% 8-Arm Alkyne, 10 kDa
	2% HA and Gelatin	
4	8% PEGDA	8% 8-Arm Alkyne, 10 kDa
5	8% 8-arm Alkyne, 10 kDa	-
6	8% PEGDA	16% 8-arm Alkyne, 10 kDa
7	8% 4-Arm Acrylate, 10 kDa	20% 8-Arm Alkyne, 10 kDa

Fig. 3.14 One study example to produce crosslinkable ECM bioinks. In this study, liver-, cardiac-, and skeletal muscle tissue-derived ECM bioinks have been combined with the HyStem-HP hydrogel kits to produce crosslinkable hydrogel. Stiffness results of the crosslinkable ECM bioinks at different crosslinking stage (**a**) showed that after stage 1 crosslinking the ECM bioinks is soft, while after stage 2 crosslinking the stiffness is significantly increased. Stiffness results of the crosslinkable ECM bioinks with varied formulations (**b**) showed that the final stiffness after stage 2 crosslinking would be varied with the bioink formulations. Reproduced with permission [360], copyright 2015

differentiation of cells [363]. Ma et al. developed the photocrosslinkable liver ECM ink and printed it into liver lobule structure by using digital light processing-based 3D bioprinting [241]. This scaffold can not only support cell activity, but also provides a stable physiological and mechanical environment [241].

3.6 Conclusions

Decellularized tissue is one of the major promising biomaterials which promote the reconstruction of functional tissue/organ. Diverse formats (e.g. sheet and tubular, powders, hydrogels, coatings, and 3D-printing) have been introduced in decellularized tissue fabrication to meet different application requirements. Among them, scaffolds in sheets or tubes, whose microscopic and ultrastructural features has been reserved, have an edge on other forms in orthotopic tissue transplantation, including but not limited to: skin, tendon, bladder, nerve, esophagus, vascular, and bone. Secondly, powders and hydrogels show advantage in filling the irregular damage. Besides, ECM coatings are favorable to provide a platform in vitro for culture of cells or coat around the backbones of synthetic scaffolds to provide a biomimetic environment. Finally, the 3D printing technology, which can accurately place bioinks (e.g. biomaterials, cells, growth factors, etc.) to generate structures as designed to engineer native-like tissues, has inspired high hopes by biomedical researchers, while future studies are still needed in bioinks optimum and complex structures printing. In general, we are confident that decellularized tissue with diverse formats will achieve new breakthroughs in precision regenerative therapy.

References

1. Taylor DA, Sampaio LC, Ferdous Z, Gobin AS, Taite LJ. Decellularized matrices in regenerative medicine. Acta Biomater. 2018;74:74–89.
2. Atala A. Tissue engineering of reproductive tissues and organs. Fertil Steril. 2012;98(1):21–9.
3. Shakouri-Motlagh A, O'Connor AJ, Brennecke SP, Kalionis B, Heath DE. Native and solubilized decellularized extracellular matrix: a critical assessment of their potential for improving the expansion of mesenchymal stem cells. Acta Biomater. 2017;55:1–12.
4. Keane TJ, Badylak SF. Biomaterials for tissue engineering applications. Semin Pediatr Surg. 2014;23(3):112–8.
5. Wu I, Nahas Z, Kimmerling KA, Rosson GD, Elisseeff JH. An injectable adipose matrix for soft-tissue reconstruction. Plast Reconstr Surg. 2012;129(6):1247–57.
6. Ingram JH, Korossis S, Howling G, Fisher J, Ingham E. The use of ultrasonication to aid recellularization of acellular natural tissue scaffolds for use in anterior cruciate ligament reconstruction. Tissue Eng. 2007;13(7):1561–72.
7. Crapo PM, Tottey S, Slivka PF, Badylak SF. Effects of biologic scaffolds on human stem cells and implications for CNS tissue engineering. Tissue Eng Part A. 2014;20(1-2):313–23.
8. Baiguera S, Del Gaudio C, Lucatelli E, Kuevda E, Boieri M, Mazzanti B, Bianco A, Macchiarini P. Electrospun gelatin scaffolds incorporating rat decellularized brain extracellular matrix for neural tissue engineering. Biomaterials. 2014;35(4):1205–14.
9. Lu H, Hoshiba T, Kawazoe N, Chen G. Autologous extracellular matrix scaffolds for tissue engineering. Biomaterials. 2011;32(10):2489–99.
10. Guilak F, Butler DL, Goldstein SA, Baaijens FP. Biomechanics and mechanobiology in functional tissue engineering. J Biomech. 2014;47(9):1933–40.
11. Sellaro TL, Ravindra AK, Stolz DB, Badylak SF. Maintenance of hepatic sinusoidal endothelial cell phenotype in vitro using organ-specific extracellular matrix scaffolds. Tissue Eng. 2007;13(9):2301–10.
12. Kimura T, Kondo M, Hashimoto Y, Fujisato T, Nakamura N, Kishida A. Surface topography of PDMS replica transferred from various decellularized aortic lumens affects cellular orientation. ACS Biomater Sci Eng. 2019;5(11):5721–6.
13. Lamers E, Walboomers XF, Domanski M, te Riet J, van Delft FC, Luttge R, Winnubst LA, Gardeniers HJ, Jansen JA. The influence of nanoscale grooved substrates on osteoblast behavior and extracellular matrix deposition. Biomaterials. 2010;31(12):3307–16.
14. Culenova M, Bakos D, Ziaran S, Bodnarova S, Varga I, Danisovic L. Bioengineered scaffolds as substitutes for grafts for urethra reconstruction. Materials. 2019;12(20):3449.
15. Kajbafzadeh AM, Abbasioun R, Sabetkish S, Sabetkish N, Rahmani P, Tavakkolitabassi K, Arshadi H. Future prospects for human tissue engineered urethra transplantation: decellularization and recellularization-based urethra regeneration. Ann Biomed Eng. 2017;45(7):1795–806.

16. Arnalich-Montiel F, Moratilla A, Fuentes-Julian S, Aparicio V, Cadenas Martin M, Peh G, Mehta JS, Adnan K, Porrua L, Perez-Sarriegui A, De Miguel MP. Treatment of corneal endothelial damage in a rabbit model with a bioengineered graft using human decellularized corneal lamina and cultured human corneal endothelium. PLoS One. 2019;14(11):e0225480.
17. Milan PB, Lotfibakhshaiesh N, Joghataie MT, Ai J, Pazouki A, Kaplan DL, Kargozar S, Amini N, Hamblin MR, Mozafari M, Samadikuchaksaraei A. Accelerated wound healing in a diabetic rat model using decellularized dermal matrix and human umbilical cord perivascular cells. Acta Biomater. 2016;45:234–46.
18. Lu CC, Zhang T, Amadio PC, An KN, Moran SL, Gingery A, Zhao C. Lateral slit delivery of bone marrow stromal cells enhances regeneration in the decellularized allograft flexor tendon. J Orthop Transl. 2019;19:58–67.
19. Zhou L, Xia J, Wang P, Jia R, Zheng J, Yao X, Chen Y, Dai Y, Yang B. Autologous smooth muscle progenitor cells enhance regeneration of tissue-engineered bladder. Tissue Eng Part A. 2018;24(13-14):1066–81.
20. Li T, Sui Z, Matsuno A, Ten H, Oyama K, Ito A, Jiang H, Ren X, Javed R, Zhang L, Ao Q. Fabrication and evaluation of a xenogeneic decellularized nerve-derived material: preclinical studies of a new strategy for nerve repair. Neurotherapeutics. 2019;89:115–25.
21. Cai Z, Gu Y, Cheng J, Li J, Xu Z, Xing Y, Wang C, Wang Z. Decellularization, cross-linking and heparin immobilization of porcine carotid arteries for tissue engineering vascular grafts. Cell Tissue Bank. 2019;20(4):569–78.
22. Chen K, Lin X, Zhang Q, Ni J, Li J, Xiao J, Wang Y, Ye Y, Chen L, Jin K, Chen L. Decellularized periosteum as a potential biologic scaffold for bone tissue engineering. Acta Biomater. 2015;19:46–55.
23. Londono R, Badylak SF. Regenerative medicine strategies for esophageal repair. Tissue Eng Part B Rev. 2015;21(4):393–410.
24. Sorkin M, Qi J, Kim HM, Hamill JB, Kozlow JH, Pusic AL, Wilkins EG. Acellular dermal matrix in immediate expander/implant breast reconstruction: a multicenter assessment of risks and benefits. Plast Reconstr Surg. 2017;140(6):1091–100.
25. Basonbul RA, Cohen MS. Use of porcine small intestinal submucosa for pediatric endoscopic tympanic membrane repair. World J Otorhinolaryngol. 2017;3(3):142–7.
26. Zhang Y, Iwata T, Nam K, Kimura T, Wu P, Nakamura N, Hashimoto Y, Kishida A. Water absorption by decellularized dermis. Heliyon. 2018;4(4):e00600.
27. Askari M, Cohen MJ, Grossman PH, Kulber DA. The use of acellular dermal matrix in release of burn contracture scars in the hand. Plast Reconstr Surg. 2011;127(4):1593–9.
28. Patel NP, Lawrence Cervino A. Keloid treatment: is there a role for acellular human dermis (Alloderm)? J Plast Reconstr Aesthet Surg. 2010;63(8):1344–8.
29. Vandegrift MT, Szpalski C, Knobel D, Weinstein A, Ham M, Ezeamuzie O, Warren SM, Saadeh PB. Acellular dermal matrix-based gene therapy augments graft incorporation. J Surg Res. 2015;195(1):360–7.
30. Stapleton PA, McBride CR, Yi J, Nurkiewicz TR. Uterine microvascular sensitivity to nanomaterial inhalation: an in vivo assessment. Toxicol Appl Pharmacol. 2015;288(3):420–8.
31. Callcut RA, Schurr MJ, Sloan M, Faucher LD. Clinical experience with alloderm: a one-staged composite dermal/epidermal replacement utilizing processed cadaver dermis and thin autografts. Burns. 2006;32(5):583–8.
32. Gore DC. Utility of acellular allograft dermis in the care of elderly burn patients. J Surg Res. 2005;125(1):37–41.
33. Wainwright DJ, Bury SB. Acellular dermal matrix in the management of the burn patient. Aesthet Surg J. 2011;31(7 Suppl):13S–23S.
34. Maisel Lotan A, Ben Yehuda D, Allweis TM, Scheflan M. Comparative study of meshed and nonmeshed acellular dermal matrix in immediate breast reconstruction. Plast Reconstr Surg. 2019;144(5):1045–53.

35. Ma P, Wang Y, Li B, Hou H. Cross-linking effects of carbodiimide, oxidized chitosan oligosaccharide and glutaraldehyde on acellular dermal matrix of basa fish (Pangasius bocourti). Int J Biol Macromol. 2020;164:677–86.
36. Dhasmana A, Singh L, Roy P, Mishra NC. Silk fibroin protein modified acellular dermal matrix for tissue repairing and regeneration. Mater Sci Eng C Mater Biol Appl. 2019;97:313–24.
37. Mirzaei-Parsa MJ, Ghanbari H, Alipoor B, Tavakoli A, Najafabadi MRH, Faridi-Majidi R. Nanofiber-acellular dermal matrix as a bilayer scaffold containing mesenchymal stem cell for healing of full-thickness skin wounds. Cell Tissue Res. 2019;375(3):709–21.
38. Du M, Zhu T, Duan X, Ge S, Li N, Sun Q, Yang P. Acellular dermal matrix loading with bFGF achieves similar acceleration of bone regeneration to BMP-2 via differential effects on recruitment, proliferation and sustained osteodifferentiation of mesenchymal stem cells. Mater Sci Eng C Mater Biol Appl. 2017;70(Pt 1):62–70.
39. Wilshaw S-P, et al. Production of an acellular amniotic membrane matrix for use in tissue engineering. Tissue Eng. 2006;12(8):2117–29.
40. Huddleston HP, Cohn MR, Haunschild ED, Wong SE, Farr J, Yanke AB. Amniotic product treatments: clinical and basic science evidence. Curr Rev Musculoskelet Med. 2020;13 (2):148–54.
41. Niknejad H, et al. Properties of the amniotic membrane for potential use in tissue engineering. Eur Cell Mater. 2008;15:88–99.
42. He H, Li W, Tseng DY, Zhang S, Chen SY, Day AJ, Tseng SC. Biochemical characterization and function of complexes formed by hyaluronan and the heavy chains of inter-alpha-inhibitor (HC*HA) purified from extracts of human amniotic membrane. J Biol Chem. 2009;284 (30):20136–46.
43. Soluble factors of amnion-derived cells in treatment of inflammatory and fibrotic pathologies. pdf.
44. Su YN, Zhao DY, Li YH, Yu TQ, Sun H, Wu XY, Zhou XQ, Li J. Human amniotic membrane allograft, a novel treatment for chronic diabetic foot ulcers: a systematic review and meta-analysis of randomised controlled trials. Int Wound J. 2020;17(3):753–64.
45. Song M, Wang W, Ye Q, Bu S, Shen Z, Zhu Y. The repairing of full-thickness skin deficiency and its biological mechanism using decellularized human amniotic membrane as the wound dressing. Mater Sci Eng C Mater Biol Appl. 2017;77:739–47.
46. E. T, B. B, N. A. Amniotic membrane seeded fetal fibroblasts as skin substitute for wound regeneration. Methods Mol Biol. 2019;1879:211–9.
47. Wu Z, et al. Human acellular amniotic membrane is adopted to treat venous ulcers. Exp Ther Med. 2018;16(2):1285–9.
48. Shah AP. Using amniotic membrane allografts in the treatment of neuropathic foot ulcers. J Am Podiatr Med Assoc. 2014;104(2):198–202.
49. Abdo RJ. Treatment of diabetic foot ulcers with dehydrated amniotic membrane allograft: a prospective case series. J Wound Care. 2016;25(Suppl 7):S4–9.
50. DiDomenico LA, Orgill DP, Galiano RD, Serena TE, Carter MJ, Kaufman JP, Young NJ, Jacobs AM, Zelen CM. Use of an aseptically processed, dehydrated human amnion and chorion membrane improves likelihood and rate of healing in chronic diabetic foot ulcers: a prospective, randomised, multi-centre clinical trial in 80 patients. Int Wound J. 2018;15 (6):950–7.
51. DiDomenico LA, Orgill DP, Galiano RD, Serena TE, Carter MJ, Kaufman JP, Young NJ, Zelen CM. Aseptically processed placental membrane improves healing of diabetic foot ulcerations: prospective, randomized clinical trial. Plast Reconstr Surg Glob Open. 2016;4 (10):e1095.
52. Lo V, Lara-Corrales I, Stuparich A, Pope E. Amniotic membrane grafting in patients with epidermolysis bullosa with chronic wounds. J Am Acad Dermatol. 2010;62(6):1038–44.

53. Hasegawa T, Mizoguchi M, Haruna K, Mizuno Y, Muramatsu S, Suga Y, Ogawa H, Ikeda S. Amnia for intractable skin ulcers with recessive dystrophic epidermolysis bullosa: report of three cases. J Dermatol. 2007;34(5):328–32.
54. Parveen S, Singh SP, Panicker MM, Gupta PK. Amniotic membrane as novel scaffold for human iPSC-derived cardiomyogenesis, in vitro cellular & developmental biology. Animal. 2019;55(4):272–84.
55. Rizzo S. A human amniotic membrane plug to promote retinal breaks repair and recurrent macular hole closure. Retina. 2019;39(Suppl 1):S95–S103.
56. Caporossi T, Angelis L, Pacini B, Tartaro R, Finocchio L, Barca F, Rizzo S. A human amniotic membrane plug to manage high myopic macular hole associated with retinal detachment. Acta Ophthalmol. 2019;98(2):e252–6.
57. Wee SW, Choi SU, Kim JC. Deep anterior lamellar keratoplasty using irradiated acellular cornea with amniotic membrane transplantation for intractable ocular surface diseases. Korean J Ophthalmol. 2015;29(2):79–85.
58. Francisco JC, Correa Cunha R, Cardoso MA, Baggio Simeoni R, Mogharbel BF, Picharski GL, Dziedzic DSM, Guarita-Souza LC, Carvalho KA. Decellularized amniotic membrane scaffold as a pericardial substitute: an in vivo study. Transplant Proc. 2016;48(8):2845–9.
59. Ramuta TŽ, Kreft ME. Human amniotic membrane and amniotic membrane–derived cells: how far are we from their use in regenerative and reconstructive urology? Cell Transplant. 2017;27(1):77–92.
60. Tang K, et al. Human acellular amniotic membrane: a potential osteoinductive biomaterial for bone regeneration. J Biomater Appl. 2018;32(6):754–64.
61. Li W, et al. Investigating the potential of amnion-based scaffolds as a barrier membrane for guided bone regeneration. Langmuir. 2015;31(31):8642–53.
62. Liu H, Zhou Z, Lin H, Wu J, Ginn B, Choi JS, Jiang X, Chung L, Elisseeff JH, Yiu S, Mao HQ. Synthetic nanofiber-reinforced amniotic membrane via interfacial bonding. ACS Appl Mater Interfaces. 2018;10(17):14559–69.
63. Zhou Z, Long D, Hsu CC, Liu H, Chen L, Slavin B, Lin H, Li X, Tang J, Yiu S, Tuffaha S, Mao HQ. Nanofiber-reinforced decellularized amniotic membrane improves limbal stem cell transplantation in a rabbit model of corneal epithelial defect. Acta Biomater. 2019;97:310–20.
64. Fu RH, Wang YC, Liu SP, Shih TR, Lin HL, Chen YM, Sung JH, Lu CH, Wei JR, Wang ZW, Huang SJ, Tsai CH, Shyu WC, Lin SZ. Decellularization and recellularization technologies in tissue engineering. Cell Transplant. 2014;23(4-5):621–30.
65. Paulo Zambon J, Atala A, Yoo JJ. Methods to generate tissue-derived constructs for regenerative medicine applications. Methods. 2019;171:3–10.
66. Font Tellado S, Balmayor ER, Van Griensven M. Strategies to engineer tendon/ligament-to-bone interface: biomaterials, cells and growth factors. Adv Drug Deliv Rev. 2015;94:126–40.
67. Steffens D, Braghirolli DI, Maurmann N, Pranke P. Update on the main use of biomaterials and techniques associated with tissue engineering. Drug Discov Today. 2018;23(8):1474–88.
68. Al Haddad E, LaPar DJ, Dayton J, Stephens EH, Bacha E. Complete atrioventricular canal repair with a decellularized porcine small intestinal submucosa patch. Congenit Heart Dis. 2018;13(6):997–1004.
69. Zhou H, et al. Feasibility of repairing full-thickness skin defects by iPSC-derived epithelial stem cells seeded on a human acellular amniotic membrane. Stem Cell Res Ther. 2019;10(1):155.
70. Zhang Z, et al. Amniotic membrane-derived stem cells help repair osteochondral defect in a weight-bearing area in rabbits. Exp Ther Med. 2017;14(1):187–92.
71. A. D, C. D, C. G, Mesenchymal stem cells seeded on a human amniotic membrane improve liver regeneration and mouse survival after extended hepatectomy.
72. Andree B, Bar A, Haverich A, Hilfiker A. Small intestinal submucosa segments as matrix for tissue engineering: review. Tissue Eng. 2013;19(4):279–91.

73. Wang W, Zhang X, Chao NN, Qin TW, Ding W, Zhang Y, Sang JW, Luo JC. Preparation and characterization of pro-angiogenic gel derived from small intestinal submucosa. Acta Biomater. 2016;29:135–48.
74. Shi L, et al. Biochemical and biomechanical characterization of porcine small intestinal submucosa (SIS): a mini review. Int J Burn Trauma. 2013;3(4):173–9.
75. Zhang X, Fang Z, Cho E, Huang K, Zhao J, Jiang J, Huangfu X. Use of a novel, reinforced, low immunogenic, porcine small intestine submucosa patch to repair a supraspinatus tendon defect in a rabbit model. Biomed Res Int. 2019;2019:9346567.
76. Tottey S, Johnson SA, Crapo PM, Reing JE, Zhang L, Jiang H, Medberry CJ, Reines B, Badylak SF. The effect of source animal age upon extracellular matrix scaffold properties. Biomaterials. 2011;32(1):128–36.
77. Gilbert TW, Stewart-Akers AM, Badylak SF. A quantitative method for evaluating the degradation of biologic scaffold materials. Biomaterials. 2007;28(2):147–50.
78. Carey LE, Dearth CL, Johnson SA, Londono R, Medberry CJ, Daly KA, Badylak SF. In vivo degradation of 14C-labeled porcine dermis biologic scaffold. Biomaterials. 2014;35 (29):8297–304.
79. Gilbert TW, Stewart-Akers AM, Simmons-Byrd A, Badylak SF. Degradation and remodeling of small intestinal submucosa in canine Achilles tendon repair. J Bone Joint Surg Am. 2007;89 (3):621–30.
80. Liang X, Zhang L, Lv Y. Treatment of total pelvic organ prolapse using a whole biological patch: a pilot study of 17 patients. J Obstet Gynaecol Res. 2017;43(1):164–72.
81. Xia D, Yang Q, Fung KM, Towner RA, Smith N, Saunders D, Greenwood-Van Meerveld B, Kropp BP, Madihally SV, Lin HK. Immunomodulatory response of layered small intestinal submucosa in a rat bladder regeneration model. J Biomed Mater Res B Appl Biomater. 2018;107(6):1960–9.
82. Xu Q, Chen C, Xu Z, Chen F, Yu Y, Hong X, Xu S, Chen J, Ding Q, Chen H. Ureteral reconstruction with decellularized small intestinal submucosa matrix for ureteral stricture: a preliminary report of two cases. Asian J Urol. 2020;7(1):51–5.
83. Tan B, Wang M, Chen X, Hou J, Chen X, Wang Y, Li-Ling J, Xie H. Tissue engineered esophagus by copper—small intestinal submucosa graft for esophageal repair in a canine model. Sci China Life Sci. 2014;57(2):248–55.
84. Rosen M, Roselli EE, Faber C, Ratliff NB, Ponsky JL, Nicholas M, Smedira G. Small intestinal submucosa intracardiac patch: an experimental study. Surg Innov. 2005;12:227–31.
85. Woo JS, Fishbein MC, Reemtsen B. Histologic examination of decellularized porcine intestinal submucosa extracellular matrix (CorMatrix) in pediatric congenital heart surgery. Cardiovasc Pathol. 2016;25(1):12–7.
86. Barachetti L, Zanni M, Stefanello D, Rampazzo A. Use of four- layer porcine small intestinal submucosa alone as a scaffold for the treatment of deep corneal defects in dogs and cats: preliminary results. Vet Rec. 2019;186(19):e28.
87. Wang S, Wu W, Liu Y, Wang X, Tang L, You P, Han J, Li B, Zhang Y, Wang M. Bone augmentation of peri-implant dehiscence defects using multilaminated small intestinal submucosa as a barrier membrane: an experimental study in dogs. Biomed Res Int. 2019;2019:8962730.
88. Wang Z-L, Wu S-Z, Li Z-F, Guo J-H, Zhang Y, Pi J-K, Hu J-G, Yang X-J, Huang F-G, Xie H-Q. Comparison of small intestinal submucosa and polypropylene mesh for abdominal wall defect repair. J Biomater Sci Polym Ed. 2018;29(6):663–82.
89. Franklin M, Russek K. Use of porcine small intestine submucosa as a prosthetic material for laparoscopic hernia repair in infected and potentially contaminated fields: long-term follow-up assessment. Surg Endosc. 2010;25(5):1693–4.
90. Weber SS, Annenberg AJ, Wright CB, Braverman TS, Mesh CL. Early pseudoaneurysm degeneration in biologic extracellular matrix patch for carotid repair. J Vasc Surg. 2014;59 (4):1116–8.

91. Ayyildiz A, Akgul KT, Huri E, Nuhoglu B, Kilicoglu B, Ustun H, Gurdal M, Germiyanoglu C. Use of porcine small intestinal submucosa in bladder augmentation in rabbit: long-term histological outcome. ANZ J Surg. 2008;78(1-2):82–6.
92. Korwar V, Adjepong S, Pattar J, Sigurdsson A. Biological mesh repair of paraesophageal hernia: an analysis of our outcomes. J Laparoendosc Adv Surg Tech A. 2019;29(11):1446–50.
93. Chang CG, Thackeray L. Laparoscopic hiatal hernia repair in 221 patients: outcomes and experience. JSLS. 2016;20(1):e2015.
94. Tan MY, Zhi W, Wei RQ, Huang YC, Zhou KP, Tan B, Deng L, Luo JC, Li XQ, Xie HQ, Yang ZM. Repair of infarcted myocardium using mesenchymal stem cell seeded small intestinal submucosa in rabbits. Biomaterials. 2009;30(19):3234–40.
95. Wang B, Zhang W, Shan CX, Liu S, Jiang ZG, Qiu M. Long-term outcomes of cruroplasty reinforcement with composite versus biologic mesh for gastroesophageal reflux disease. Surg Endosc. 2016;30(7):2865–72.
96. Wang M, Li Y-Q, Cao J, Gong M, Zhang Y, Chen X, Tian M-X, Xie H-Q. Accelerating effects of genipin-crosslinked small intestinal submucosa for defected gastric mucosa repair. J Mater Chem B. 2017;5(34):7059–71.
97. Dai X, Xu Q. Nanostructured substrate fabricated by sectioning tendon using a microtome for tissue engineering. Nanotechnology. 2011;22(49):494008.
98. Dai X, Schalek R, Xu Q. Staining and etching: a simple method to fabricate inorganic nanostructures from tissue slices. Adv Mater. 2012;24(3):370–4.
99. Burk J, Erbe I, Berner D, Kacza J, Kasper C, Pfeiffer B, Winter K, Brehm W. Freeze-thaw cycles enhance decellularization of large tendons. Tissue Eng. 2014;20(4):276–84.
100. Ning LJ, Zhang Y, Chen XH, Luo JC, Li XQ, Yang ZM, Qin TW. Preparation and characterization of decellularized tendon slices for tendon tissue engineering. J Biomed Mater Res A. 2012;100(6):1448–56.
101. Deeken CR, White AK, Bachman SL, Ramshaw BJ, Cleveland DS, Loy TS, Grant SA. Method of preparing a decellularized porcine tendon using tributyl phosphate. J Biomed Mater Res B Appl Biomater. 2011;96(2):199–206.
102. Tong WY, Shen W, Yeung CW, Zhao Y, Cheng SH, Chu PK, Chan D, Chan GC, Cheung KM, Yeung KW, Lam YW. Functional replication of the tendon tissue microenvironment by a bioimprinted substrate and the support of tenocytic differentiation of mesenchymal stem cells. Biomaterials. 2012;33(31):7686–98.
103. Alberti KA, Xu Q. Slicing, stacking and rolling: fabrication of nanostructured collagen constructs from tendon sections. Adv Healthc Mater. 2013;2(6):817–21.
104. Qin TW, Chen Q, Sun YL, Steinmann SP, Amadio PC, An KN, Zhao C. Mechanical characteristics of native tendon slices for tissue engineering scaffold. J Biomed Mater Res B Appl Biomater. 2012;100(3):752–8.
105. The utilization of decellularized tendon slices to provide an inductive microenvironment for the proliferation and tenogenic differentiation of stem cells.
106. Qin TW, Sun YL, Thoreson AR, Steinmann SP, Amadio PC, An KN, Zhao C. Effect of mechanical stimulation on bone marrow stromal cell-seeded tendon slice constructs: a potential engineered tendon patch for rotator cuff repair. Biomaterials. 2015;51:43–50.
107. Omae H, Zhao C, Sun YL, An KN, Amadio PC. Multilayer tendon slices seeded with bone marrow stromal cells: a novel composite for tendon engineering. J Orthop Res. 2009;27 (7):937–42.
108. Alberti KA, Sun JY, Illeperuma WR, Suo Z, Xu Q. Laminar tendon composites with enhanced mechanical properties. J Mater Sci. 2015;50(6):2616–25.
109. Wang F, Maeda Y, Zachar V, Ansari T, Emmersen J. Regeneration of the oesophageal muscle layer from oesophagus acellular matrix scaffold using adipose-derived stem cells. Biochem Biophys Res Commun. 2018;503(1):271–7.
110. Rothrauff BB, Tuan RS. Decellularized bone extracellular matrix in skeletal tissue engineering. Biochem Soc Trans. 2020;48(3):755–64.

111. Carriel V, Alaminos M, Garzon I, Campos A, Cornelissen M. Tissue engineering of the peripheral nervous system. Expert Rev Neurother. 2014;14(3):301–18.
112. Kubiak CA, Kung TA, Brown DL, Cederna PS, Kemp SWP. State-of-the-art techniques in treating peripheral nerve injury. Plast Reconstr Surg. 2018;141(3):702–10.
113. Lin T, Liu S, Chen S, Qiu S, Rao Z, Liu J, Zhu S, Yan L, Mao H, Zhu Q, Quan D, Liu X. Hydrogel derived from porcine decellularized nerve tissue as a promising biomaterial for repairing peripheral nerve defects. Acta Biomater. 2018;73:326–38.
114. Kasper M, Deister C, Beck F, Schmidt CE. Bench-to-bedside lessons learned: commercialization of an acellular nerve graft. Adv Healthc Mater. 2020;9(16):e2000174.
115. Shin YH, Park SY, Kim JK. Comparison of systematically combined detergent and nuclease-based decellularization methods for acellular nerve graft: an ex vivo characterization and in vivo evaluation. J Tissue Eng Regen Med. 2019;13(7):1241–52.
116. Neubauer D, Graham JB, Muir D. Chondroitinase treatment increases the effective length of acellular nerve grafts. Exp Neurol. 2007;207(1):163–70.
117. Buckenmeyer MJ, Meder TJ, Prest TA, Brown BN. Decellularization techniques and their applications for the repair and regeneration of the nervous system. Methods. 2020;171:41–61.
118. Wieringa PA, Goncalves de Pinho AR, Micera S, van Wezel RJA, Moroni L. Biomimetic architectures for peripheral nerve repair: a review of biofabrication strategies. Adv Healthc Mater. 2018;7(8):e1701164.
119. Sun M, Wang X, Zhao B. Quality estimation and influence factors of the larger chemically acellular nerve allografts in vitro. Zhongguo Xiu Fu Chong Jian Wai Ke Za Zhi. 2006;20(8):779–82.
120. Philips C, Cornelissen M, Carriel V. Evaluation methods as quality control in the generation of decellularized peripheral nerve allografts. J Neural Eng. 2018;15(2):021003.
121. Cornelison RC, Wellman SM, Park JH, Porvasnik SL, Song YH, Wachs RA, Schmidt CE. Development of an apoptosis-assisted decellularization method for maximal preservation of nerve tissue structure. Acta Biomater. 2018;77:116–26.
122. Kubek T, Ghalib N, Dubovy P. Endoneurial extracellular matrix influences regeneration and maturation of motor nerve axons--a model of acellular nerve graft. Neurosci Lett. 2011;496(2):75–9.
123. He B, Zhu Q, Chai Y, Ding X, Tang J, Gu L, Xiang J, Yang Y, Zhu J, Liu X. Safety and efficacy evaluation of a human acellular nerve graft as a digital nerve scaffold: a prospective, multicentre controlled clinical trial. J Tissue Eng Regen Med. 2015;9(3):286–95.
124. Wang D, Liu XL, Zhu JK, Hu J, Jiang L, Zhang Y, Yang LM, Wang HG, Zhu QT, Yi JH, Xi TF. Repairing large radial nerve defects by acellular nerve allografts seeded with autologous bone marrow stromal cells in a monkey model. J Neurotrauma. 2010;27(10):1935–43.
125. Wang D, Liu XL, Zhu JK, Jiang L, Hu J, Zhang Y, Yang LM, Wang HG, Yi JH. Bridging small-gap peripheral nerve defects using acellular nerve allograft implanted with autologous bone marrow stromal cells in primates. Brain Res. 2008;1188:44–53.
126. Borschel GH, Kia KF, Kuzon WM, Dennis RG. Mechanical properties of acellular peripheral nerve. J Surg Res. 2003;114(2):133–9.
127. Bulstra LF, Hundepool CA, Friedrich PF, Bishop AT, Hovius SER, Shin AY. Functional outcome after reconstruction of a long nerve gap in rabbits using optimized decellularized nerve allografts. Plast Reconstr Surg. 2020;145(6):1442–50.
128. Farber SJ, Hoben GM, Hunter DA, Yan Y, Johnson PJ, Mackinnon SE, Wood MD. Vascularization is delayed in long nerve constructs compared with nerve grafts. Muscle Nerve. 2016;54(2):319–21.
129. Brooks DN, Weber RV, Chao JD, Rinker BD, Zoldos J, Robichaux MR, Ruggeri SB, Anderson KA, Bonatz EE, Wisotsky SM, Cho MS, Wilson C, Cooper EO, Ingari JV, Safa B, Parrett BM, Buncke GM. Processed nerve allografts for peripheral nerve reconstruction: a multicenter study of utilization and outcomes in sensory, mixed, and motor nerve reconstructions. Microsurgery. 2012;32(1):1–14.

130. Moore AM, MacEwan M, Santosa KB, Chenard KE, Ray WZ, Hunter DA, Mackinnon SE, Johnson PJ. Acellular nerve allografts in peripheral nerve regeneration: a comparative study. Muscle Nerve. 2011;44(2):221–34.
131. Pan D, Hunter DA, Schellhardt L, Jo S, Santosa KB, Larson EL, Fuchs AG, Snyder-Warwick AK, Mackinnon SE, Wood MD. The accumulation of T cells within acellular nerve allografts is length-dependent and critical for nerve regeneration. Exp Neurol. 2019;318:216–31.
132. Hoben GM, Ee X, Schellhardt L, Yan Y, Hunter DA, Moore AM, Snyder-Warwick AK, Stewart S, Mackinnon SE, Wood MD. Increasing nerve autograft length increases senescence and reduces regeneration. Plast Reconstr Surg. 2018;142(4):952–61.
133. Wang ZZ, Sakiyama-Elbert SE. Matrices, scaffolds & carriers for cell delivery in nerve regeneration. Exp Neurol. 2019;319:112837.
134. Joung D, Lavoie NS, Guo SZ, Park SH, Parr AM, McAlpine MC. 3D printed neural regeneration devices. Adv Funct Mater. 2020;30(1):1906237.
135. Yu T, Wen L, He J, Xu Y, Li T, Wang W, Ma Y, Ahmad MA, Tian X, Fan J, Wang X, Hagiwara H, Ao Q. Fabrication and evaluation of an optimized acellular nerve allograft with multiple axial channels. Acta Biomater. 2020;115:235–49.
136. Zang M, Zhang Q, Chang EI, Mathur AB, Yu P. Decellularized tracheal matrix scaffold for tissue engineering. Plast Reconstr Surg. 2012;130(3):532–40.
137. Zang M, Zhang Q, Chang EI, Mathur AB, Yu P. Decellularized tracheal matrix scaffold for tracheal tissue engineering: in vivo host response. Plast Reconstr Surg. 2013;132(4):549e–59e.
138. Dodson A, Bellini S, Parnigotto PP, Dickinson SC, Hollander AP, Mantero S, Conconi MT, Birchall MA. Clinical transplantation of a tissue-engineered airway. Lancet. 2008;372 (9655):2023–30.
139. Otti AG, Jaus MO, Barale SBD, Comin C, Lavorini F, Fontana G, Sibila O, Rombola G, Jungebluth P, Macchiarini P. The first tissueengineered airway transplantation: 5-year follow-up results. Lancet. 2014;383(9913):238–44.
140. Elliott MJ, Coppi PD, Speggiorin S, Roebuck D, et al. Stem-cell-based, tissue engineered tracheal replacement in a child: a 2-year follow-up study. Lancet. 2012;380(9846):994–1000.
141. Hamilton NJ, Kanani M, Roebuck DJ, Hewitt RJ, Cetto R, Culme-Seymour EJ, Toll E, Bates AJ, Comerford AP, McLaren CA, Butler CR, Crowley C, McIntyre D, Sebire NJ, Janes SM, O'Callaghan C, Mason C, De Coppi P, Lowdell MW, Elliott MJ, Birchall MA. Tissue-engineered tracheal replacement in a child: a 4-year follow-up study. Am J Transplant. 2015;15(10):2750–7.
142. Lee M, Chang PC, Dunn JC. Evaluation of small intestinal submucosa as scaffolds for intestinal tissue engineering. J Surg Res. 2008;147(2):168–71.
143. Pavcnik D, Obermiller J, Uchida BT, Van Alstine W, Edwards JM, Landry GJ, Kaufman JA, Keller FS, Rosch J. Angiographic evaluation of carotid artery grafting with prefabricated small-diameter, small-intestinal submucosa grafts in sheep. Cardiovasc Intervent Radiol. 2009;32(1):106–13.
144. Zafar F, Hinton RB, Moore RA, Baker RS, Bryant R, Narmoneva DA, Taylor MD, Morales DL. Physiological growth, remodeling potential, and preserved function of a novel bioprosthetic tricuspid valve: tubular bioprosthesis made of small intestinal submucosa-derived extracellular matrix. J Am Coll Cardiol. 2015;66(8):877–88.
145. Syed O, Walters NJ, Day RM, Kim HW, Knowles JC. Evaluation of decellularization protocols for production of tubular small intestine submucosa scaffolds for use in oesophageal tissue engineering. Acta Biomater. 2014;10(12):5043–54.
146. Liu Y, Ma W, Liu B, Wang Y, Chu J, Xiong G, Shen L, Long C, Lin T, He D, Butnaru D, Alexey L, Zhang Y, Zhang D, Wei G. Urethral reconstruction with autologous urine-derived stem cells seeded in three-dimensional porous small intestinal submucosa in a rabbit model. Stem Cell Res Ther. 2017;8(1):63.
147. Fan MR, Gong M, Da LC, Bai L, Li XQ, Chen KF, Li-Ling J, Yang ZM, Xie HQ. Tissue engineered esophagus scaffold constructed with porcine small intestinal submucosa and synthetic polymers. Biomed Mater. 2014;9(1):015012.

148. Brunello G, Sivolella S, Meneghello R, Ferroni L, Gardin C, Piattelli A, Zavan B, Bressan E. Powder-based 3D printing for bone tissue engineering. Biotechnol Adv. 2016;34:740–53.
149. Ahmed M, Ffrench-Constant C. Extracellular matrix regulation of stem cell behavior. Curr Stem Cell Rep. 2016;2(3):197–206.
150. Bloom JP, Ott HC. Regenerative medicine applications in organ transplantation. Ann Surg. 2015;262(6):1166.
151. Londono R, Badylak SF. Biologic scaffolds for regenerative medicine: mechanisms of in vivo remodeling. Ann Biomed Eng. 2015;43(3):577–92.
152. Visscher DO, Bos EJ, Peeters M, Kuzmin NV, Groot ML, Helder MN, van Zuijlen PPM. Cartilage tissue engineering: preventing tissue scaffold contraction using a 3D-printed polymeric cage. Tissue Eng Part C Methods. 2016;22(6):573–84.
153. Reing JE, Brown BN, Daly KA, Freund JM, Gilbert TW, Hsiong SX, Huber A, Kullas KE, Tottey S, Wolf MT, Badylak SF. The effects of processing methods upon mechanical and biologic properties of porcine dermal extracellular matrix scaffolds. Biomaterials. 2010;31(33):8626–33.
154. Badylak SF, Freytes DO, Gilbert TW. Reprint of: Extracellular matrix as a biological scaffold material: Structure and function. Acta Biomater. 2015;23(Suppl):S17–26.
155. Edgar L, Altamimi A, Garcia Sanchez M, Tamburrinia R, Asthana A, Gazia C, Orlando G. Utility of extracellular matrix powders in tissue engineering. Organogenesis. 2018;14(4):172–86.
156. Gilbert TW, Stolz DB, Biancaniello F, Simmons-Byrd A, Badylak SF. Production and characterization of ECM powder: implications for tissue engineering applications. Biomaterials. 2005;26(12):1431–5.
157. Valentin JE, Badylak JS, McCabe GP, Badylak SF. Extracellular matrix bioscaffolds for orthopaedic applications - a comparative histologic study. J Bone Joint Surg. 2006;88A(12):2673–86.
158. Badylak SF, Valentin JE, Ravindra AK, McCabe GP, Stewart-Akers AM. Macrophage phenotype as a determinant of biologic scaffold remodeling. Tissue Eng A. 2008;14(11):1835–42.
159. Jakus AE, Laronda MM, Rashedi AS, Robinson CM, Lee C, Jordan SW, Orwig KE, Woodruff TK, Shah RN. "Tissue papers" from organ-specific decellularized extracellular matrices. Adv Funct Mater. 2017;27(34):1700992.
160. Yoon H, Kim G. Micro/nanofibrous scaffolds electrospun from PCL and small intestinal submucosa. J Biomater Sci Polym Ed. 2010;21(5):553–62.
161. Da L, Gong M, Chen A, Zhang Y, Huang Y, Guo Z, Li S, Li-Ling J, Zhang L, Xie H. Composite elastomeric polyurethane scaffolds incorporating small intestinal submucosa for soft tissue engineering. Acta Biomater. 2017;59:45–57.
162. Barber D, Keuter J, Kravig K. A logical stepwise approach to laser diffraction particle size distribution analysis methods development and validation. Pharm Dev Technol. 1998;3(2):153.
163. Kleinebudde P, Jumaa M, Saleh FE. Influence of degree of polymerization on behavior of cellulose during homogenization and extrusion/spheronization. AAPS PharmSci. 2000;2(3):18–27.
164. Chang CH, Chen CC, Liao CH, Lin FH, Hsu YM, Fang HW. Human acellular cartilage matrix powders as a biological scaffold for cartilage tissue engineering with synovium-derived mesenchymal stem cells. J Biomed Mater Res A. 2014;102(7):2248–57.
165. Almeida HV, Cunniffe GM, Vinardell T, Buckley CT, O'Brien FJ, Kelly DJ. Coupling freshly isolated CD44(+) infrapatellar fat pad-derived stromal cells with a TGF-β3 eluting cartilage ECM-derived scaffold as a single-stage strategy for promoting chondrogenesis. Adv Healthc Mater. 2015;4:1043–53.
166. Wolf MT, Daly KA, Brennan-Pierce EP, Johnson SA, Carruthers CA, D'Amore A, Nagarkar SP, Velankar SS, Badylak SF. A hydrogel derived from decellularized dermal extracellular matrix. Biomaterials. 2012;33(29):7028–38.

167. Thitiset T, Damrongsakkul S, Bunaprasert T, Leeanansaksiri W, Honsawek S. Development of collagen/demineralized bone powder scaffolds and periosteum-derived cells for bone tissue engineering application. Int J Mol Sci. 2013;14(1):2056–71.
168. Almeida HV, Eswaramoorthy R, Cunniffe GM, Buckley CT, O'Brien FJ, Kelly D. Fibrin hydrogels functionalized with cartilage extracellular matrix and incorporating freshly isolated stromal cells as an injectable for cartilage regeneration. Acta Biomater. 2016;36:55–62.
169. Haibin Z, Daizhi P, Bixiang Z, Xiaoling L, Yong W, Lihua W, Xin Z, Jing L. Regeneration of mature dermis by transplanted particulate acellular dermal matrix in a rat model of skin defect wound. J Mater Sci Mater Med. 2012;23(12):2933–44.
170. Seif-Naraghi S, Singelyn J, Dequach J, Schup-Magoffin P, Christman K. Fabrication of biologically derived injectable materials for myocardial tissue engineering. J Vis Exp. 2010;46:2109.
171. Cha PF, Gao JH, Chen Y, Lu F. Construction of scaffold with human extracellular matrix from adipose tissue. Chin J Plast Surg. 2012;28(1):55.
172. Electrospun micro/nanofibrous conduits composed of poly(e-caprolactone) and small intestine submucosa powder for nerve tissue regeneration.
173. Ghuman H, Gerwig M, Nicholls FJ, Liu JR, Donnelly J, Badylak SF, Modo M. Long-term retention of ECM hydrogel after implantation into a sub-acute stroke cavity reduces lesion volume. Acta Biomater. 2017;63:50–63.
174. Edgar L, Pu T, Porter B, Aziz JM, La Pointe C, Asthana A, Orlando G. Regenerative medicine, organ bioengineering and transplantation. Br J Surg. 2020;107(7):793–800.
175. Minae A, Kihwan K, Junbeom P, Dong-Ryeol R, Jung AS, Jihee Lee K, Ji Ha C, Eun-Mi P, Kyung Eun L, Minna W, Minsuk K. Extracellular matrix-derived extracellular vesicles promote cardiomyocyte growth and electrical activity in engineered cardiac atria. Biomaterials. 2017;146:49–59.
176. Dearth CL, Keane TJ, Carruthers CA, Reing JE, Huleihel L, Ranallo CA, Kollar EW, Badylak SF. The effect of terminal sterilization on the material properties and in vivo remodeling of a porcine dermal biologic scaffold. Acta Biomater. 2016;33:78–87.
177. Benedikt LP, Gabriel SP, Braden CF, Jakob TS, Joshua K, Michael LH, Martha MM. Effect of low-temperature ethylene oxide and electron beam sterilization on the in vitro and in vivo function of reconstituted extracellular matrix-derived scaffolds. J Biomater Appl. 2015;30:435–49.
178. Andrea MM, Peter SM. The effect of terminal sterilization on structural and biophysical properties of a decellularized collagen-based scaffold; implications for stem cell adhesion. J Biomed Mater Res B Appl Biomater. 2014;103(2):397–406.
179. Lauren E, Kyle MN, Theresa W, Riccardo T, Ravi K, Giuseppe O. Heterogeneity of scaffold biomaterials in tissue engineering. Materials. 2016;9(5):332.
180. Beck EC, Barragan M, Tadros MH, Gehrke SH, Detamore MS. Approaching the compressive modulus of articular cartilage with a decellularized cartilage-based hydrogel. Acta Biomater. 2016;38:94–105.
181. Parenteau-Bareil R, Gauvin R, Berthod F. Collagen-based biomaterials for tissue engineering applications. Materials. 2010;3(3):1863–87.
182. Choi JS, Yang H-J, Kim BS, Kim JD, Kim JY, Yoo B, Park K, Lee HY, Cho YW. Human extracellular matrix (ECM) powders for injectable cell delivery and adipose tissue engineering. J Control Release. 2009;139(1):2–7.
183. Fong AH, Romero-Lopez M, Heylman CM, Keating M, Tran D, Sobrino A, Tran AQ, Pham HH, Fimbres C, Gershon PD, Botvinick EL, George SC, Hughes CCW. Three-dimensional adult cardiac extracellular matrix promotes maturation of human induced pluripotent stem cell-derived cardiomyocytes. Tissue Eng A. 2016;22(15-16):1016–25.
184. Zhang Y, He Y, Bharadwaj S, Hammam N, Carnagey K, Myers R, Atala A, Van Dyke M. Tissue-specific extracellular matrix coatings for the promotion of cell proliferation and maintenance of cell phenotype. Biomaterials. 2009;30(23-24):4021–8.

185. Yin H, Wang Y, Sun Z, Sun X, Xu Y, Li P, Meng H, Yu X, Xiao B, Fan T. Induction of mesenchymal stem cell chondrogenic differentiation and functional cartilage microtissue formation for in vivo cartilage regeneration by cartilage extracellular matrix-derived particles. Acta Biomater. 2016;33:96–109.
186. Garg T, Singh O, Arora S, Murthy RSR. Scaffold: a novel carrier for cell and drug delivery. Crit Rev Ther Drug Carrier Syst. 2012;29(1):1–63.
187. Kwon JS, Yoon SM, Shim SW, Park JH, Min KJ, Oh HJ, Kim JH, Kim YJ, Yoon JJ, Choi BH. Injectable extracellular matrix hydrogel developed using porcine articular cartilage. Int J Pharm. 2013;454(1):183–91.
188. Kim HJ, Lee S, Yun H-W, Yin XY, Kim SH, Choi BH, Kim YJ, Kim MS, Min B-H. In vivo degradation profile of porcine cartilage-derived extracellular matrix powder scaffolds using a non-invasive fluorescence imaging method. J Biomater Sci. 2016;27(2):177–90.
189. Madler S, Bich C, Touboul D, Zenobi R. Chemical cross-linking with NHS esters: a systematic study on amino acid reactivities. J Mass Spectrom. 2009;44(5):694–706.
190. Wolf MT, Carruthers CA, Dearth CL, Crapo PM, Huber A, Burnsed OA, Londono R, Johnson SA, Daly KA, Stahl EC, Freund JM, Medberry CJ, Carey LE, Nieponice A, Amoroso NJ, Badylak SF. Polypropylene surgical mesh coated with extracellular matrix mitigates the host foreign body response. J Biomed Mater Res A. 2014;102(1):234–46.
191. Faulk DM, Londono R, Wolf MT, Ranallo CA, Carruthers CA, Wildemann JD, Dearth CL, Badylak SF. ECM hydrogel coating mitigates the chronic inflammatory response to polypropylene mesh. Biomaterials. 2014;35(30):8585–95.
192. Debels H, Gerrand YW, Poon CJ, Abberton KM, Morrison WA, Mitchell GM. An adipogenic gel for surgical reconstruction of the subcutaneous fat layer in a rat model. J Tissue Eng Regen Med. 2017;11(4):1230–41.
193. Lee SJ, Lee IW, Lee YM, Lee HB, Khang G. Macroporous biodegradable natural/synthetic hybrid scaffolds as small intestine submucosa impregnated poly(D,L-lactide-co-glycolide) for tissue-engineered bone. J Biomater Sci Polym Ed. 2004;15(8):1003–17.
194. Wang B, Lv X, Li Z, Zhang M, Yao J, Sheng N, Lu M, Wang H, Chen S. Urethra-inspired biomimetic scaffold: a therapeutic strategy to promote angiogenesis for urethral regeneration in a rabbit model. Acta Biomater. 2020;102:247–58.
195. Kim SH, Song JE, Lee D, Khang G. Development of poly(lactide-co-glycolide) scaffold-impregnated small intestinal submucosa with pores that stimulate extracellular matrix production in disc regeneration. J Tissue Eng Regen Med. 2014;8(4):279–90.
196. Kang X, Zhao Z, Wu X, Shen Q, Wang Z, Kang Y, Xing Z, Zhang T. Experimental study on chitosan/allogeneic bone powder composite porous scaffold to repair bone defects in rats. Chin J Reparat Reconstruct Surg. 2016;30(3):298–302.
197. Lei X, Gao J, Xing F, Zhang Y, Ma Y, Zhang G. Comparative evaluation of the physicochemical properties of nano-hydroxyapatite/collagen and natural bone ceramic/collagen scaffolds and their osteogenesis-promoting effect on MC3T3-E1 cells. Regenerat Biomater. 2019;6(6):361–71.
198. Gao G, Cui X. Three-dimensional bioprinting in tissue engineering and regenerative medicine. Biotechnol Lett. 2016;38(2):203–11.
199. Alhnan MA, Okwuosa TC, Sadia M, Wan K-W, Ahmed W, Arafat B. Emergence of 3D printed dosage forms: opportunities and challenges. Pharm Res. 2016;33(8):1817–32.
200. Pati F, Jang J, Ha D-H, Kim SW, Rhie J-W, Shim J-H, Kim D-H, Cho D-W. Printing three-dimensional tissue analogues with decellularized extracellular matrix bioink. Nat Commun. 2014;5:3935.
201. Chaudhary C, Garg T. Scaffolds: a novel carrier and potential wound healer. Crit Rev Ther Drug Carrier Syst. 2015;32(4):277–321.
202. Turner NJ, Yates AJ Jr, Weber DJ, Qureshi IR, Stolz DB, Gilbert TW, Badylak SF. Xenogeneic extracellular matrix as an inductive scaffold for regeneration of a functioning musculotendinous junction. Tissue Eng A. 2010;16(11):3309–17.

203. Shooter GK, Van Lonkhuyzen DR, Croll TI, Cao Y, Xie Y, Broadbent JA, Stupar D, Lynam EC, Upton Z. A pre-clinical functional assessment of an acellular scaffold intended for the treatment of hard-to-heal wounds. Int Wound J. 2015;12(2):160–8.
204. Shin YS, Choi JW, Park J-K, Kim YS, Yang SS, Min B-H, Kim C-H. Tissue-engineered tracheal reconstruction using mesenchymal stem cells seeded on a porcine cartilage powder scaffold. Ann Biomed Eng. 2015;43(4):1003–13.
205. Han N, Yabroudi MA, Stearns-Reider K, Helkowski W, Sicari BM, Rubin JP, Badylak SF, Boninger ML, Ambrosio F. Electrodiagnostic evaluation of individuals implanted with extracellular matrix for the treatment of volumetric muscle injury: case series. Phys Ther. 2016;96 (4):540–9.
206. Liu Y-CC, Chhabra N, Houser SM. Novel treatment of a septal ulceration using an extracellular matrix scaffold (septal ulceration treatment using ECM). Am J Otolaryngol. 2016;37 (3):195–8.
207. Lev-Tov H, Li C-S, Dahle S, Isseroff RR. Cellular versus acellular matrix devices in treatment of diabetic foot ulcers: study protocol for a comparative efficacy randomized controlled trial. Trials. 2013;14:8.
208. Rommer EA, Peric M, Wong A. Urinary bladder matrix for the treatment of recalcitrant nonhealing radiation wounds. Adv Skin Wound Care. 2013;26(10):450–5.
209. LeCheminant J, Field C. Porcine urinary bladder matrix: a retrospective study and establishment of protocol. J Wound Care. 2012;21(10):476–82.
210. Jia S, Zhang T, Xiong Z, Pan W, Liu J, Sun W. In vivo evaluation of a novel oriented scaffold-BMSC construct for enhancing full-thickness articular cartilage repair in a rabbit model. PLoS One. 2015;10(12):e0145667.
211. Afaneh C, Abelson J, Schattner M, Janjigian YY, Ilson D, Yoon SS, Strong VE. Esophageal reinforcement with an extracellular scaffold during total gastrectomy for gastric cancer. Ann Surg Oncol. 2015;22(4):1252–7.
212. Martin Riganti J, Ciotola F, Amenabar A, Craiem D, Graf S, Badaloni A, Gilbert TW, Nieponice A. Urinary bladder matrix scaffolds strengthen esophageal hiatus repair. J Surg Res. 2016;204(2):344–50.
213. Abu Saleh WK, Al Jabbari O, Bruckner BA, Reardon MJ. Case report: A rare case of left atrial hemangioma: surgical resection and reconstruction. Methodist Debakey Cardiovasc J. 2016;12(1):51–4.
214. Abu Saleh WK, Al Jabbari O, Ramlawi B, Bruckner BA, Loebe M, Reardon MJ. Case report: Cardiac tumor resection and repair with porcine xenograft. Methodist Debakey Cardiovasc J. 2016;12(2):116–8.
215. Abu Saleh WK, Al Jabbari O, Ramlawi B, Bruckner BA, Loebe M, Reardon MJ. Right atrial tumor resection and reconstruction with use of an acellular porcine bladder membrane. Tex Heart Inst J. 2016;43(2):175–7.
216. Kruper GJ, Vandegriend ZP, Lin H-S, Zuliani GF. Salvage of failed local and regional flaps with porcine urinary bladder extracellular matrix aided tissue regeneration. Case Rep Otolaryngol. 2013;2013:917183.
217. Iorio T, Blumberg D. Short-term results of treating primary and recurrent anal fistulas with a novel extracellular matrix derived from porcine urinary bladder. Am Surg. 2015;81 (5):498–502.
218. Dorman RM, Bass KD. Novel use of porcine urinary bladder matrix for pediatric pilonidal wound care: preliminary experience. Pediatr Surg Int. 2016;32(10):997–1002.
219. Frykberg RG, Cazzell SM, Arroyo-Rivera J, Tallis A, Reyzelman AM, Saba F, Warren L, Stouch BC, Gilbert TW. Evaluation of tissue engineering products for the management of neuropathic diabetic foot ulcers: an interim analysis. J Wound Care. 2016;25(7):S18–25.
220. Abaci A, Guvendiren M. Designing decellularized extracellular matrix-based bioinks for 3D bioprinting. Adv Healthc Mater. 2020;9:e2000734.
221. Ghorbani F, Li D, Ni S, Zhou Y, Yu B. 3D printing of acellular scaffolds for bone defect regeneration: a review. Mater Today Commun. 2020;22:100979.

222. Farnebo S, Woon CYL, Schmitt T, Joubert L-M, Kim M, Hung P, Chang J. Design and characterization of an injectable tendon hydrogel: a novel scaffold for guided tissue regeneration in the musculoskeletal system. Tissue Eng A. 2014;20(9-10):1550–61.
223. Wu J, Ding Q, Dutta A, Wang Y, Huang Y-H, Weng H, Tang L, Hong Y. An injectable extracellular matrix derived hydrogel for meniscus repair and regeneration. Acta Biomater. 2015;16:49–59.
224. Fu Y, Fan X, Tian C, Luo J, Zhang Y, Deng L, Qin T, Lv Q. Decellularization of porcine skeletal muscle extracellular matrix for the formulation of a matrix hydrogel: a preliminary study. J Cell Mol Med. 2016;20(4):740–9.
225. Response of endothelial cells to decellularized extracellular matrix deposited by bone marrow mesenchymal stem cells.
226. Lundy DS, Casiano RR, McClinton ME, Xue JW. Early results of transcutaneous injection laryngoplasty with micronized acellular dermis versus type-I thyroplasty for glottic incompetence dysphonia due to unilateral vocal fold paralysis. J Voice. 2003;17(4):589–95.
227. Soucy KG, Smith EF, Monreal G, Rokosh G, Keller BB, Yuan F, Matheny RG, Fallon AM, Lewis BC, Sherwood LC, Sobieski MA, Giridharan GA, Koenig SC, Slaughter MS. Feasibility study of particulate extracellular matrix (P-ECM) and left ventricular assist device (HVAD) therapy in chronic ischemic heart failure bovine model. ASAIO J. 2015;61(2):161–9.
228. Slaughter MS, Soucy KG, Matheny RG, Lewis BC, Hennick MF, Choi Y, Monreal G, Sobieski MA, Giridharan GA, Koenig SC. Development of an extracellular matrix delivery system for effective intramyocardial injection in ischemic tissue. ASAIO J. 2014;60(6):730–6.
229. Massensini AR, Ghuman H, Saldin LT, Medberry CJ, Keane TJ, Nicholls FJ, Velankar SS, Badylak SF, Modo M. Concentration-dependent rheological properties of ECM hydrogel for intracerebral delivery to a stroke cavity. Acta Biomater. 2015;27:116–30.
230. Ullah F, Othman MB, Javed F, Ahmad Z, Md Akil H. Classification, processing and application of hydrogels: a review. Mater Sci Eng C Mater Biol Appl. 2015;57:414–33.
231. Ruel-Gariepy E, Leroux JC. In situ-forming hydrogels--review of temperature-sensitive systems. Eur J Pharm Biopharm. 2004;58(2):409–26.
232. Wichterle O, Lím D. Hydrophilic gels for biological use. Nature. 1960;185(4706):117–8.
233. Pacifici A, Laino L, Gargari M, Guzzo F, Velandia Luz A, Polimeni A, Pacifici L. Decellularized hydrogels in bone tissue engineering: a topical review. Int J Med Sci. 2018;15(5):492–7.
234. Ciuffreda MC, Malpasso G, Chokoza C, Bezuidenhout D, Goetsch KP, Mura M, Pisano F, Davies NH, Gnecchi M. Synthetic extracellular matrix mimic hydrogel improves efficacy of mesenchymal stromal cell therapy for ischemic cardiomyopathy. Acta Biomater. 2018;70:71–83.
235. Bi H, Ming L, Cheng R, Luo H, Zhang Y, Jin Y. Liver extracellular matrix promotes BM-MSCs hepatic differentiation and reversal of liver fibrosis through activation of integrin pathway. J Tissue Eng Regen Med. 2017;11(10):2685–98.
236. Adam Young D, Bajaj V, Christman KL. Award winner for outstanding research in the PhD category, 2014 Society for Biomaterials annual meeting and exposition, Denver, Colorado, April 16-19, 2014: decellularized adipose matrix hydrogels stimulate in vivo neovascularization and adipose formation. J Biomed Mater Res A. 2014;102(6):1641–51.
237. Freytes DO, Martin J, Velankar SS, Lee AS, Badylak SF. Preparation and rheological characterization of a gel form of the porcine urinary bladder matrix. Biomaterials. 2008;29(11):1630–7.
238. Saheli M, Sepantafar M, Pournasr B, Farzaneh Z, Vosough M, Piryaei A, Baharvand H. Three-dimensional liver-derived extracellular matrix hydrogel promotes liver organoids function. J Cell Biochem. 2018;119(6):4320–33.
239. Jiang D, Huang J, Shao H, Hu X, Song L, Zhang Y. Characterization of bladder acellular matrix hydrogel with inherent bioactive factors. Mater Sci Eng C Mater Biol Appl. 2017;77:184–9.

240. Bai R, Tian L, Li Y, Zhang J, Wei Y, Jin Z, Liu Z, Liu H. Combining ECM hydrogels of cardiac bioactivity with stem cells of high cardiomyogenic potential for myocardial repair. Stem Cells Int. 2019;2019:6708435.
241. Agarwal T, Narayan R, Maji S, Ghosh SK, Maiti TK. Decellularized caprine liver extracellular matrix as a 2D substrate coating and 3D hydrogel platform for vascularized liver tissue engineering. J Tissue Eng Regen Med. 2018;12(3):e1678–90.
242. Su J, Satchell SC, Shah RN, Wertheim JA. Kidney decellularized extracellular matrix hydrogels: Rheological characterization and human glomerular endothelial cell response to encapsulation. J Biomed Mater Res A. 2018;106(9):2448–62.
243. Yuan X, Wei Y, Villasante A, Ng JJD, Arkonac DE, Chao PG, Vunjak-Novakovic G. Stem cell delivery in tissue-specific hydrogel enabled meniscal repair in an orthotopic rat model. Biomaterials. 2017;132:59–71.
244. Sackett SD, Tremmel DM, Ma F, Feeney AK, Maguire RM, Brown ME, Zhou Y, Li X, O'Brien C, Li L, Burlingham WJ, Odorico JS. Extracellular matrix scaffold and hydrogel derived from decellularized and delipidized human pancreas. Sci Rep. 2018;8(1):10452.
245. Jiang K, Chaimov D, Patel SN, Liang JP, Wiggins SC, Samojlik MM, Rubiano A, Simmons CS, Stabler CL. 3-D physiomimetic extracellular matrix hydrogels provide a supportive microenvironment for rodent and human islet culture. Biomaterials. 2019;198:37–48.
246. Gaetani R, Aouad S, Demaddalena LL, Straessle H, Dzieciatkowska M, Wortham M, Bender HR, Nguyen-Ngoc KV, Schmid-Schoenbein GW, George SC, Hughes CCW, Sander M, Hansen KC, Christman KL. Evaluation of different decellularization protocols on the generation of pancreas-derived hydrogels. Tissue Eng. 2018;24(12):697–708.
247. C. H, C. I, P. A. Bioengineering strategies of the uterus towards improving current investigative models and female reproductive health. Facts Views Vis Obgyn. 2019;11:1.
248. Cornelison RC, Gonzalez-Rothi EJ, Porvasnik SL, Wellman SM, Park JH, Fuller DD, Schmidt CE. Injectable hydrogels of optimized acellular nerve for injection in the injured spinal cord. Biomed Mater. 2018;13(3):034110.
249. Poon CJ, Cotta MVPE, Sinha S, Palmer JA, Woods AA, Morrison WA, Abberton KM. Preparation of an adipogenic hydrogel from subcutaneous adipose tissue. Acta Biomater. 2013;9(3):5609–20.
250. Kang KN, Kim Y, Yoon SM, Kwon JS, Seo HW, Kim ES, Lee B, Kim JH, Min BH, Lee HB, Kim MS. In vivo release of bovine serum albumin from an injectable small intestinal submucosa gel. Int J Pharm. 2011;420(2):266–73.
251. Ungerleider JL, Johnson TD, Rao N, Christman KL. Fabrication and characterization of injectable hydrogels derived from decellularized skeletal and cardiac muscle. Methods. 2015;84:53–9.
252. Sharma S, Tiwari S. A review on biomacromolecular hydrogel classification and its applications. Int J Biol Macromol. 2020;162:737–47.
253. Asadi N, Alizadeh E, Salehi R, Khalandi B, Davaran S, Akbarzadeh A. Nanocomposite hydrogels for cartilage tissue engineering: a review. Artif Cells Nanomed Biotechnol. 2018;46(3):465–71.
254. Singh NK, Lee DS. In situ gelling pH- and temperature-sensitive biodegradable block copolymer hydrogels for drug delivery. J Control Release. 2014;193:214–27.
255. Gao S, Yuan Z, Guo W, Chen M, Liu S, Xi T, Guo Q. Comparison of glutaraldehyde and carbodiimides to crosslink tissue engineering scaffolds fabricated by decellularized porcine menisci. Mater Sci Eng C Mater Biol Appl. 2017;71:891–900.
256. Ventura RD, Padalhin AR, Kim B, Park M, Lee BT. Evaluation of bone regeneration potential of injectable extracellular matrix (ECM) from porcine dermis loaded with biphasic calcium phosphate (BCP) powder. Mater Sci Eng C Mater Biol Appl. 2020;110:110663.
257. Bible E, Chau DY, Alexander MR, Price J, Shakesheff KM, Modo M. Attachment of stem cells to scaffold particles for intra-cerebral transplantation. Nat Protoc. 2009;4(10):1440–53.

258. Seif-Naraghi SB, Horn D, Schup-Magoffin PJ, Christman KL. Injectable extracellular matrix derived hydrogel provides a platform for enhanced retention and delivery of a heparin-binding growth factor. Acta Biomater. 2012;8(10):3695–703.
259. Chiou GJ, Crowe C, McGoldrick R, Hui K, Pham H, Chang J. Optimization of an injectable tendon hydrogel: the effects of platelet-rich plasma and adipose-derived stem cells on tendon healing in vivo. Tissue Eng Part A. 2015;21(9-10):1579–86.
260. Choi JW, Park JK, Chang JW, Kim DY, Kim MS, Shin YS, Kim C-H. Small intestine submucosa and mesenchymal stem cells composite gel for scarless vocal fold regeneration. Biomaterials. 2014;35(18):4911–8.
261. Mercuri JJ, Patnaik S, Dion G, Gill SS, Liao J, Simionescu DT. Regenerative potential of decellularized porcine nucleus pulposus hydrogel scaffolds: stem cell differentiation, matrix remodeling, and biocompatibility studies. Tissue Eng Part A. 2013;19(7-8):952–66.
262. Rajabi-Zeleti S, Jalili-Firoozinezhad S, Azarnia M, Khayyatan F, Vahdat S, Nikeghbalian S, Khademhosseini A, Baharvand H, Aghdami N. The behavior of cardiac progenitor cells on macroporous pericardium-derived scaffolds. Biomaterials. 2014;35(3):970–82.
263. Roshanbinfar K, Hilborn J, Varghese OP, Oommen OP. Injectable and thermoresponsive pericardial matrix derived conductive scaffold for cardiac tissue engineering. RSC Adv. 2017;7(51):31980–8.
264. Mechanical properties and in vivo behavior of a biodegradable synthetic polymer microfiber - extracellular matrix hydrogel biohybrid scaffold.
265. Brown CF, Yan J, Han TT, Marecak DM, Amsden BG, Flynn LE. Effect of decellularized adipose tissue particle size and cell density on adipose-derived stem cell proliferation and adipogenic differentiation in composite methacrylated chondroitin sulphate hydrogels. Biomed Mater. 2015;10(4):045010.
266. Cheung HK, Han TT, Marecak DM, Watkins JF, Amsden BG, Flynn LE. Composite hydrogel scaffolds incorporating decellularized adipose tissue for soft tissue engineering with adipose-derived stem cells. Biomaterials. 2014;35(6):1914–23.
267. Duan Y, Liu Z, O'Neill J, Wan LQ, Freytes DO, Vunjak-Novakovic G. Hybrid gel composed of native heart matrix and collagen induces cardiac differentiation of human embryonic stem cells without supplemental growth factors. J Cardiovasc Transl Res. 2011;4(5):605–15.
268. Kayabolen A, Keskin D, Aykan A, Karslioglu Y, Zor F, Tezcaner A. Native extracellular matrix/fibroin hydrogels for adipose tissue engineering with enhanced vascularization. Biomed Mater. 2017;12(3):035007.
269. Tobin EJ. Recent coating developments for combination devices in orthopedic and dental applications: a literature review. Adv Drug Deliv Rev. 2017;112:88–100.
270. Xing H, Lee H, Luo L, Kyriakides TR. Extracellular matrix-derived biomaterials in engineering cell function. Biotechnol Adv. 2019;42:107421.
271. Wang B, Li W, Dean D, Mishra MK, Wekesa KS. Enhanced hepatogenic differentiation of bone marrow derived mesenchymal stem cells on liver ECM hydrogel. J Biomed Mater Res A. 2018;106(3):829–38.
272. Chowdhury F, Na S, Li D, Poh YC, Tanaka TS, Wang F, Wang N. Material properties of the cell dictate stress-induced spreading and differentiation in embryonic stem cells. Nat Mater. 2010;9(1):82–8.
273. Kollmer M, Keskar V, Hauk TG, Collins JM, Russell B, Gemeinhart RA. Stem cell-derived extracellular matrix enables survival and multilineage differentiation within superporous hydrogels. Biomacromolecules. 2012;13(4):963–73.
274. Zia S, Mozafari M, Natasha G, Tan A, Cui Z, Seifalian AM. Hearts beating through decellularized scaffolds: whole-organ engineering for cardiac regeneration and transplantation. Crit Rev Biotechnol. 2016;36(4):705–15.
275. Landry MJ, Rollet FG, Kennedy TE, Barrett CJ. Layers and multilayers of self-assembled polymers: tunable engineered extracellular matrix coatings for neural cell growth. Langmuir. 2018;34(30):8709–30.

276. Shridhar A, Lam AYL, Sun Y, Simmons CA, Gillies ER, Flynn LE. Culture on tissue-specific coatings derived from alpha-amylase-digested decellularized adipose tissue enhances the proliferation and adipogenic differentiation of human adipose-derived stromal cells. Biotechnol J. 2019;15:e1900118.
277. Lee JS, Shin J, Park HM, Kim YG, Kim BG, Oh JW, Cho SW. Liver extracellular matrix providing dual functions of two-dimensional substrate coating and three-dimensional injectable hydrogel platform for liver tissue engineering. Biomacromolecules. 2014;15(1):206–18.
278. Koo S, Muhammad R, Peh GS, Mehta JS, Yim EK. Micro- and nanotopography with extracellular matrix coating modulate human corneal endothelial cell behavior. Acta Biomater. 2014;10(5):1975–84.
279. Frank LA, Onzi GR, Morawski AS, Pohlmann AR, Guterres SS, Contri RV. Chitosan as a coating material for nanoparticles intended for biomedical applications. React Funct Polym. 2020;147:104459.
280. Baker NA, Muir LA, Washabaugh AR, Neeley CK, Chen SY, Flesher CG, Vorwald J, Finks JF, Ghaferi AA, Mulholland MW, Varban OA, Lumeng CN, O'Rourke RW. Diabetes-specific regulation of adipocyte metabolism by the adipose tissue extracellular matrix. J Clin Endocrinol Metab. 2017;102(3):1032–43.
281. DeQuach JA, Mezzano V, Miglani A, Lange S, Keller GM, Sheikh F, Christman KL. Simple and high yielding method for preparing tissue specific extracellular matrix coatings for cell culture. PLoS One. 2010;5(9):e13039.
282. Zhang H, Tasnim F, Ying JY, Zink D. The impact of extracellular matrix coatings on the performance of human renal cells applied in bioartificial kidneys. Biomaterials. 2009;30 (15):2899–911.
283. French KM, Boopathy AV, DeQuach JA, Chingozha L, Lu H, Christman KL, Davis ME. A naturally derived cardiac extracellular matrix enhances cardiac progenitor cell behavior in vitro. Acta Biomater. 2012;8(12):4357–64.
284. Dan P, Velot E, Francius G, Menu P, Decot V. Human-derived extracellular matrix from Wharton's jelly: an untapped substrate to build up a standardized and homogeneous coating for vascular engineering. Acta Biomater. 2017;48:227–37.
285. Jeon O, Bin Lee Y, Hinton TJ, Feinberg AW, Alsberg E. Cryopreserved cell-laden alginate microgel bioink for 3D bioprinting of living tissues. Mater Today Chem. 2019;12:61–70.
286. Park J, Lee SJ, Chung S, Lee JH, Kim WD, Lee JY, Park SA. Cell-laden 3D bioprinting hydrogel matrix depending on different compositions for soft tissue engineering: Characterization and evaluation. Mater Sci Eng C Mater Biol Appl. 2017;71:678–84.
287. Gao G, Schilling AF, Yonezawa T, Wang J, Dai G, Cui X. Bioactive nanoparticles stimulate bone tissue formation in bioprinted three-dimensional scaffold and human mesenchymal stem cells. Biotechnol J. 2014;9(10):1304–11.
288. Jia W, Gungor-Ozkerim PS, Zhang YS, Yue K, Zhu K, Liu W, Pi Q, Byambaa B, Dokmeci MR, Shin SR, Khademhosseini A. Direct 3D bioprinting of perfusable vascular constructs using a blend bioink. Biomaterials. 2016;106:58–68.
289. Pati F, Song TH, Rijal G, Jang J, Kim SW, Cho DW. Ornamenting 3D printed scaffolds with cell-laid extracellular matrix for bone tissue regeneration. Biomaterials. 2015;37:230–41.
290. Levato R, Webb WR, Otto IA, Mensinga A, Zhang Y, van Rijen M, van Weeren R, Khan IM, Malda J. The bio in the ink: cartilage regeneration with bioprintable hydrogels and articular cartilage-derived progenitor cells. Acta Biomater. 2017;61:41–53.
291. Heinrich MA, Liu W, Jimenez A, Yang J, Akpek A, Liu X, Pi Q, Mu X, Hu N, Schiffelers RM, Prakash J, Xie J, Zhang YS. 3D Bioprinting: from benches to translational applications. Small. 2019;15(23):e1805510.
292. Ramos T, Moroni L. Tissue engineering and regenerative medicine 2019: the role of biofabrication-a year in review. Tissue Eng. 2020;26(2):91–106.
293. Hospodiuk M, Dey M, Sosnoski D, Ozbolat IT. The bioink: a comprehensive review on bioprintable materials. Biotechnol Adv. 2017;35(2):217–39.

294. Gungor-Ozkerim PS, Inci I, Zhang YS, Khademhosseini A, Dokmeci MR. Bioinks for 3D bioprinting: an overview. Biomater Sci. 2018;6(5):915–46.
295. Touri M, Kabirian F, Saadati M, Ramakrishna S, Mozafari M. Additive manufacturing of biomaterials – the evolution of rapid prototyping. Adv Eng Mater. 2019;21(2):1800511.
296. Cui H, Nowicki M, Fisher JP, Zhang LG. 3D bioprinting for organ regeneration. Adv Healthc Mater. 2017;6(1):201601118.
297. Pati F, Ha DH, Jang J, Han HH, Rhie JW, Cho DW. Biomimetic 3D tissue printing for soft tissue regeneration. Biomaterials. 2015;62:164–75.
298. Jiang L, Wang Y, Liu Z, Ma C, Yan H, Xu N, Gang F, Wang X, Zhao L, Sun X. Three-dimensional printing and injectable conductive hydrogels for tissue engineering application. Tissue Eng. 2019;25(5):398–411.
299. Fedorovich NE, Alblas J, de Wijn JR, Hennink WE, Verbout AJ, Dhert WJA. Hydrogels as extracellular matrices for skeletal tissue engineering: state-of-the-art and novel application in organ printing. Tissue Eng. 2007;13(8):1905–25.
300. Kim BS, Kwon YW, Kong JS, Park GT, Gao G, Han W, Kim MB, Lee H, Kim JH, Cho DW. 3D cell printing of in vitro stabilized skin model and in vivo pre-vascularized skin patch using tissue-specific extracellular matrix bioink: a step towards advanced skin tissue engineering. Biomaterials. 2018;168:38–53.
301. Pati F, Jang J, Ha DH, Won Kim S, Rhie JW, Shim JH, Kim DH, Cho DW. Printing three-dimensional tissue analogues with decellularized extracellular matrix bioink. Nat Commun. 2014;5:3935.
302. Toprakhisar B, Nadernezhad A, Bakirci E, Khani N, Skvortsov GA, Koc B. Development of bioink from decellularized tendon extracellular matrix for 3D bioprinting. Macromol Biosci. 2018;18(10):e1800024.
303. Kim H, Park MN, Kim J, Jang J, Kim HK, Cho DW. Characterization of cornea-specific bioink: high transparency, improved in vivo safety. J Tissue Eng. 2019;10:2041731418823382.
304. Jung CS, Kim BK, Lee J, Min BH, Park SH. Development of printable natural cartilage matrix bioink for 3D printing of irregular tissue shape. Tissue Eng Regen Med. 2018;15(2):155–62.
305. Lee H, Han W, Kim H, Ha DH, Jang J, Kim BS, Cho DW. Development of liver decellularized extracellular matrix bioink for three-dimensional cell printing-based liver tissue engineering. Biomacromolecules. 2017;18(4):1229–37.
306. Jang J, Park HJ, Kim SW, Kim H, Park JY, Na SJ, Kim HJ, Park MN, Choi SH, Park SH, Kim SW, Kwon SM, Kim PJ, Cho DW. 3D printed complex tissue construct using stem cell-laden decellularized extracellular matrix bioinks for cardiac repair. Biomaterials. 2017;112:264–74.
307. Dzobo K, Motaung K, Adesida A. Recent trends in decellularized extracellular matrix bioinks for 3D printing: an updated review. Int J Mol Sci. 2019;20(18):4628.
308. Choi YJ, Kim TG, Jeong J, Yi HG, Park JW, Hwang W, Cho DW. 3D cell printing of functional skeletal muscle constructs using skeletal muscle-derived bioink. Adv Healthc Mater. 2016;5(20):2636–45.
309. Choudhury D, Tun HW, Wang T, Naing MW. Organ-derived decellularized extracellular matrix: a game changer for bioink manufacturing? Trends Biotechnol. 2018;36(8):787–805.
310. van der Laan LJ, Lockey CG, Frasier FS, Wilson CA, Onions DE, Hering BJ, Long Z, Otto E, Torbett BE, Salomon DR. Infection by porcine endogenous retrovirus after islet xenotransplantation in SCID mice. Nature. 2000;7:4076800.
311. Vijayavenkataraman S, Yan WC, Lu WF, Wang CH, Fuh JYH. 3D bioprinting of tissues and organs for regenerative medicine. Adv Drug Deliv Rev. 2018;132:296–332.
312. Alonzo M, AnilKumar S, Roman B, Tasnim N, Joddar B. 3D Bioprinting of cardiac tissue and cardiac stem cell therapy. Transl Res. 2019;211:64–83.
313. Lee W, Hong Y, Dai G. 3D bioprinting of vascular conduits for pediatric congenital heart repairs. Transl Res. 2019;211:35–45.

314. Jessop ZM, Al-Sabah A, Gardiner MD, Combellack E, Hawkins K, Whitaker IS. 3D bioprinting for reconstructive surgery: principles, applications and challenges. J Plast Reconstr Aesthet Surg. 2017;70(9):1155–70.
315. Hann SY, Cui H, Esworthy T, Miao S, Zhou X, Lee SJ, Fisher JP, Zhang LG. Recent advances in 3D printing: vascular network for tissue and organ regeneration. Transl Res. 2019;211:46–63.
316. Ning L, Chen X. A brief review of extrusion-based tissue scaffold bio-printing. Biotechnol J. 2017;12(8):91–7.
317. Ostrovidov S, Salehi S, Costantini M, Suthiwanich K, Ebrahimi M, Sadeghian RB, Fujie T, Shi X, Cannata S, Gargioli C, Tamayol A, Dokmeci MR, Orive G, Swieszkowski W, Khademhosseini A. 3D bioprinting in skeletal muscle tissue engineering. Small. 2019;15 (24):e1805530.
318. Leberfinger AN, Dinda S, Wu Y, Koduru SV, Ozbolat V, Ravnic DJ, Ozbolat IT. Bioprinting functional tissues. Acta Biomater. 2019;95:32–49.
319. Ozbolat IT, Hospodiuk M. Current advances and future perspectives in extrusion-based bioprinting. Biomaterials. 2016;76:321–43.
320. Shafiee A, Atala A. Printing technologies for medical applications. Trends Mol Med. 2016;22 (3):254–65.
321. Zein I, et al. Fused deposition modeling of novel scaffold architectures for tissue engineering applications. Biomaterials. 2002;23:1169–85.
322. Wiggenhauser PS, Schwarz S, Koerber L, Hoffmann TK, Rotter N. Addition of decellularized extracellular matrix of porcine nasal cartilage improves cartilage regenerative capacities of PCL-based scaffolds in vitro. J Mater Sci Mater Med. 2019;30(11):121.
323. Matai I, Kaur G, Seyedsalehi A, McClinton A, Laurencin CT. Progress in 3D bioprinting technology for tissue/organ regenerative engineering. Biomaterials. 2020;226:119536.
324. Liu W, Zhong Z, Hu N, Zhou Y, Maggio L, Miri AK, Fragasso A, Jin X, Khademhosseini A, Zhang YS. Coaxial extrusion bioprinting of 3D microfibrous constructs with cell-favorable gelatin methacryloyl microenvironments. Biofabrication. 2018;10(2):024102.
325. Hiller T, Berg J, Elomaa L, Rohrs V, Ullah I, Schaar K, Dietrich AC, Al-Zeer MA, Kurtz A, Hocke AC, Hippenstiel S, Fechner H, Weinhart M, Kurreck J. Generation of a 3D liver model comprising human extracellular matrix in an alginate/gelatin-based bioink by extrusion bioprinting for infection and transduction studies. Int J Mol Sci. 2018;19(10):3129.
326. Visser J, et al. Biofabrication of multi-material anatomically shaped tissue constructs. Biofabrication. 2013;5(3):035007.
327. Yilmaz B, Tahmasebifar A, Baran ET. Bioprinting technologies in tissue engineering. Adv Biochem Eng Biotechnol. 2020;171:279–319.
328. Xu T, Binder KW, Albanna MZ, Dice D, Zhao W, Yoo JJ, Atala A. Hybrid printing of mechanically and biologically improved constructs for cartilage tissue engineering applications. Biofabrication. 2013;5(1):015001.
329. Faulkner-Jones A, Greenhough S, King JA, Gardner J, Courtney A, Shu W. Development of a valve-based cell printer for the formation of human embryonic stem cell spheroid aggregates. Biofabrication. 2013;5(1):015013.
330. Eagles PA, et al. Electrohydrodynamic jetting of mouse neuronal cells. Biochem J. 2006;394 (Pt 2):375–8.
331. Gudapati H, Dey M, Ozbolat I. A comprehensive review on droplet-based bioprinting: past, present and future. Biomaterials. 2016;102:20–42.
332. Demirci U, Montesano G. Single cell epitaxy by acoustic picolitre droplets. Lab Chip. 2007;7 (9):1139–45.
333. Ferris CJ, Gilmore KJ, Beirne S, McCallum D, Wallace GG, Panhuis M. Bio-ink for on-demand printing of living cells. Biomater Sci. 2013;1(2):224–30.
334. Ng WL, Lee JM, Yeong WY, Win Naing M. Microvalve-based bioprinting - process, bio-inks and applications. Biomater Sci. 2017;5(4):632–47.

335. Kim BS, Kim H, Gao G, Jang J, Cho DW. Decellularized extracellular matrix: a step towards the next generation source for bioink manufacturing. Biofabrication. 2017;9(3):034104.
336. Holzl K, Lin S, Tytgat L, Van Vlierberghe S, Gu L, Ovsianikov A. Bioink properties before, during and after 3D bioprinting. Biofabrication. 2016;8(3):032002.
337. Williams D, Thayer P, Martinez H, Gatenholm E, Khademhosseini A. A perspective on the physical, mechanical and biological specifications of bioinks and the development of functional tissues in 3D bioprinting. Bioprinting. 2018;9:19–36.
338. Kim MH, Lee YW, Jung WK, Oh J, Nam SY. Enhanced rheological behaviors of alginate hydrogels with carrageenan for extrusion-based bioprinting. J Mech Behav Biomed Mater. 2019;98:187–94.
339. Wang X, Rijff BL, Khang G. A building-block approach to 3D printing a multichannel, organ-regenerative scaffold. J Tissue Eng Regen Med. 2017;11(5):1403–11.
340. Ma X, Yu C, Wang P, Xu W, Wan X, Lai CSE, Liu J, Koroleva-Maharajh A, Chen S. Rapid 3D bioprinting of decellularized extracellular matrix with regionally varied mechanical properties and biomimetic microarchitecture. Biomaterials. 2018;185:310–21.
341. Kabirian F, Mozafari M. Decellularized ECM-derived bioinks: prospects for the future. Methods. 2020;171:108–18.
342. Murphy SV, Atala A. 3D bioprinting of tissues and organs. Nat Biotechnol. 2014;32 (8):773–85.
343. Gu Q, Tomaskovic-Crook E, Lozano R, Chen Y, Kapsa RM, Zhou Q, Wallace GG, Crook JM. Functional 3D neural mini-tissues from printed gel-based bioink and human neural stem cells. Adv Healthc Mater. 2016;5(12):1429–38.
344. Matsusaki M, Sakaue K, Kadowaki K, Akashi M. Three-dimensional human tissue chips fabricated by rapid and automatic inkjet cell printing. Adv Healthc Mater. 2013;2(4):534–9.
345. Peng W, Unutmaz D, Ozbolat IT. Bioprinting towards physiologically relevant tissue models for pharmaceutics. Trends Biotechnol. 2016;34(9):722–32.
346. Xia Z, Jin S, Ye K. Tissue and organ 3D bioprinting. SLAS Technol. 2018;23(4):301–14.
347. Barrs RW, Jia J, Silver SE, Yost M, Mei Y. Biomaterials for bioprinting microvasculature. Chem Rev. 2020;19:10887–949.
348. Park JY, Ryu H, Lee B, Ha DH, Ahn M, Kim S, Kim JY, Jeon NL, Cho DW. Development of a functional airway-on-a-chip by 3D cell printing. Biofabrication. 2018;11(1):015002.
349. Singh S, Choudhury D, Yu F, Mironov V, Naing MW. In situ bioprinting - bioprinting from benchside to bedside? Acta Biomater. 2020;101:14–25.
350. Liu J, Miller K, Ma X, Dewan S, Lawrence N, Whang G, Chung P, McCulloch AD, Chen S. Direct 3D bioprinting of cardiac micro-tissues mimicking native myocardium. Biomaterials. 2020;256:120204.
351. Yu F, Choudhury D. Microfluidic bioprinting for organ-on-a-chip models. Drug Discov Today. 2019;24(6):1248–57.
352. Kim J, Shim IK, Hwang DG, Lee YN, Kim M, Kim H, Kim SW, Lee S, Kim SC, Cho DW, Jang J. 3D cell printing of islet-laden pancreatic tissue-derived extracellular matrix bioink constructs for enhancing pancreatic functions. J Mater Chem B. 2019;7(10):1773–81.
353. Mollica PA, Booth-Creech EN, Reid JA, Zamponi M, Sullivan SM, Palmer XL, Sachs PC, Bruno RD. 3D bioprinted mammary organoids and tumoroids in human mammary derived ECM hydrogels. Acta Biomater. 2019;95:201–13.
354. Das S, Kim SW, Choi YJ, Lee S, Lee SH, Kong JS, Park HJ, Cho DW, Jang J. Decellularized extracellular matrix bioinks and the external stimuli to enhance cardiac tissue development in vitro. Acta Biomater. 2019;95:188–200.
355. Bejleri D, Streeter BW, Nachlas ALY, Brown ME, Gaetani R, Christman KL, Davis ME. A bioprinted cardiac patch composed of cardiac-specific extracellular matrix and progenitor cells for heart repair. Adv Healthc Mater. 2018;7(23):e1800672.
356. Gao G, Lee JH, Jang J, Lee DH, Kong J-S, Kim BS, Choi Y-J, Jang WB, Hong YJ, Kwon S-M, Cho D-W. Tissue engineered bio-blood-vessels constructed using a tissue-specific bioink and 3D coaxial cell printing technique: a novel therapy for ischemic disease. Adv Funct Mater. 2017;27(33):1700798.

357. Nam H, Jeong HJ, Jo Y, Lee JY, Ha DH, Kim JH, Chung JH, Cho YS, Cho DW, Lee SJ, Jang J. Multi-layered free-form 3D cell-printed tubular construct with decellularized inner and outer esophageal tissue-derived bioinks. Sci Rep. 2020;10(1):7255.
358. Croce S, Peloso A, Zoro T, Avanzini MA, Cobianchi L. A hepatic scaffold from decellularized liver tissue: food for thought. Biomolecules. 2019;9(12):813.
359. Serna JA, et al. Formulation and characterization of a SIS-based photocrosslinkable bioink. Polymers. 2019;11:569.
360. Skardal A, Devarasetty M, Kang HW, Mead I, Bishop C, Shupe T, Lee SJ, Jackson J, Yoo J, Soker S, Atala A. A hydrogel bioink toolkit for mimicking native tissue biochemical and mechanical properties in bioprinted tissue constructs. Acta Biomater. 2015;25:24–34.
361. Ali M, Pr AK, Yoo JJ, Zahran F, Atala A, Lee SJ. A photo-crosslinkable kidney ECM-derived bioink accelerates renal tissue formation. Adv Healthc Mater. 2019;8(7):e1800992.
362. Yu C, Ma X, Zhu W, Wang P, Miller KL, Stupin J, Koroleva-Maharajh A, Hairabedian A, Chen S. Scanningless and continuous 3D bioprinting of human tissues with decellularized extracellular matrix. Biomaterials. 2019;194:1–13.
363. Jang J, Kim TG, Kim BS, Kim SW, Kwon SM, Cho DW. Tailoring mechanical properties of decellularized extracellular matrix bioink by vitamin B2-induced photo-crosslinking. Acta Biomater. 2016;33:88–95.

Chapter 4
Applications of Decellularized Materials for Tissue Repair

Bo Liu, Xuewei Bi, Yuqi He, and Xiaoming Li

Abstract The pursuit of tissue engineering and regenerative medicine is to regenerate functional tissues and organs to replace damaged or diseased tissue. As a basic element of tissue engineering, scaffolds provide three-dimensional structure for cell growth while maintaining cell function, so the selection of scaffold materials is of great importance. Natural extracellular matrix is a potential therapeutic material because it is a naturally structured non-cellular tissue component with low immunogenicity, high biological activity, and can provide a unique microenvironment for cells. In addition, acellular materials can not only provide chemical and mechanical signals that determine the fate of cells, but also act as components of newly formed tissues, promoting damage repair. This chapter will focus on the repair ability of acellular materials in different tissues, and provide reference for the experimental research and development of different tissue repair materials and their clinical application.

Keywords Tissue engineering · Decellularized materials · Tissue regeneration

4.1 Introduction

Reconstruction of tissue defects represents a constant clinical challenge because tens of thousands of transplants procedures are performed each year. In this context, limitations of conventional treatment for trauma, cancer, congenital defects, and other diseases necessitate novel strategies to regenerate diseased or damaged tissue. It is well known that acellular matrix not only has remarkable biocompatibility, but also is a reservoir of various molecules that mediate the structure and/or biological characteristics of tissues since is capable of regulating intracellular signals and

B. Liu · X. Bi · Y. He · X. Li (✉)
Key Laboratory for Biomechanics and Mechanobiology of Ministry of Education, Beijing Advanced Innovation Center for Biomedical Engineering, School of Biological Science and Medical Engineering, Beihang University, Beijing, China
e-mail: lixm@buaa.edu.cn; x.m.li@hotmail.com

X. Li, H. Xie (eds.), *Decellularized Materials*,
https://doi.org/10.1007/978-981-33-6962-7_4

promoting tissue regeneration. In addition, as the material is treated with acellular technology, the occurrence of immune rejection can be largely avoided, providing a stable microenvironment for tissue growth and development. Various traditional treatment strategies, including autograft, allograft and other alternative materials, which have their own benefits and drawbacks. Although autografts still remain the "gold standard" for transplantation, the acellular materials introduced here break through the limitation of donor availability, avoid the occurrence of donor complications, meet the high requirement of interaction between biological artificial devices and natural tissues, which show great potential in clinical application.

4.2 Decellularized Materials for Musculoskeletal Tissue Engineering

4.2.1 Bone Tissue Engineering

4.2.1.1 Structure and Characteristics of Bone

Bone is a sophisticated composite on different hierarchical levels, as shown in Fig. 4.1, which a dynamic organization can respond to different daily mechanical loads [1]. From a structural perspective, more than 206 different bones make up the skeleton, from the long bones of the limbs, the short bones of the wrist and ankle, the flat bones of the sternum and skull, to the irregular bones of the pelvis and vertebrae. Herein, bone tissue mainly consists of two parts, arranged either in a compact pattern (cortical bone) or a trabecular pattern (cancellous bone) [2]. The combination of a dense shear stress resistant shell and a honeycomb internal structure [3] prevents buckling and has excellent bending resistance. By weight, approximately 60% of bones are composed of minerals (calcium and phosphate), 10–20% water, and 20–30% protein. About 90% of the protein content is type I collagen. According to the literature review, bone ECM has more than 200 different types of non-collagen matrix proteins, including fibronectin, osteonectin, osteocalcin, osteopontin, etc. [4] This combination of collagen matrix and calcium phosphate mineral [5] results in high mechanical properties and ductility. It helps to enrich the signal to regulate the cell behavior that interacts with the resident cells in the extracellular environment [1, 6, 7].

4.2.1.2 Requirements for Bone Tissue Engineering Materials

It is well known that bone tissue can not only provide protection and support for internal organs and muscle attachment points, but also participate in the metabolism of calcium and phosphorus, with strong function of repair and reconstruction. Reconstruction of skeletal defects represents a constant clinical challenge because of often limitations in conventional therapies including autograft, allograft, metal, and other alternative materials, for the treatment of trauma, cancer, congenital

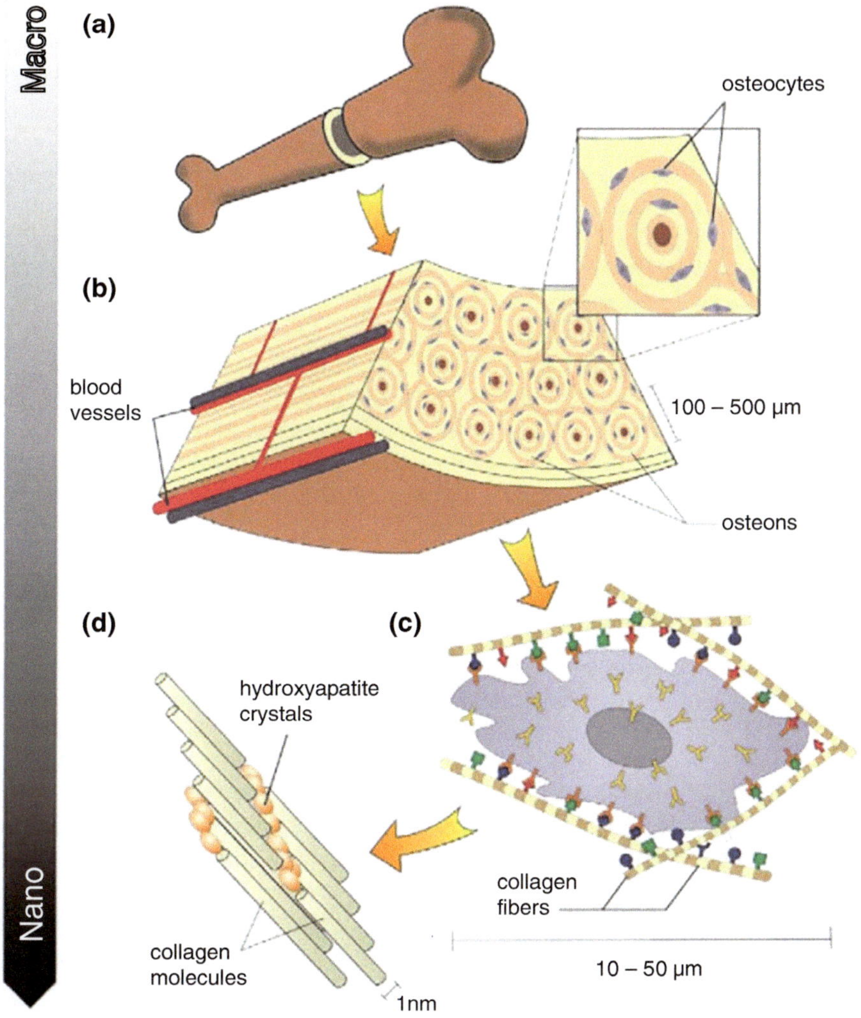

Fig. 4.1 Grading of bone on scales of different lengths. (**a**) A strongly calcified outer dense layer of bone (**b**). (**c**) The resident cells in a cluster of membrane receptors (**d**) The nanostructures of the surrounding extracellular matrix. (Reproduced with permission from [22]. Copyright 2008 Elsevier Ltd.)

defects, and other bone diseases [8–10]. The maturity of tissue engineering technology offers a promising future for repairing large bone defect [1]. Bone tissue engineering is a dynamic and complex process, normally utilizing synergistic combination of biomaterials and cell therapy [11]. Specifically, the cells were planted on the prefabricated 3D scaffold, and then the bone defect site was transplanted with cell hybrid material. As the gradual degradation of the biomaterials, new bone tissue is constantly generated, enabling the defect could be completely repaired [12]. In order to manufacture scaffolds suitable for bone tissue growth, the first step is to be clear bone formation comprising a complex sequence of events [13]. That is to say, engineering of a bone tissue requires a scaffold with specific properties conducive to bone regeneration both in vitro and in vivo, such as biocompatibility, suitable degradability and mechanical properties and osteoconductivity [8]. Currently, the bone repair materials have been investigated including bioceramics [14–16], metals [17, 18], and polymers [19, 20]. However, within the context of rapid development of modern regenerative medicine, tissue engineering approaches being developed by investigators satisfactory results have been obtained [21–23]. In such these tissue engineering approaches, the degree of success often depends on the properties of the material. In the case of bone, materials should preferably be both osteoinductive, osteoconductive, and capable of osseointegration [21]. To be more specific: (a) A 3D and highly porous interconnected network to maintain in vivo tissue in-growth, flow transport of nutrients and metabolic waste; (b) The porosity, Young's modulus, and solubility to match the tissue, a biocompatible and bioresorbable composition with degradation rates matching the new tissue formation rate; (c) A suitable surface chemistry for cell attachment, proliferation, and differentiation; and (d) Mechanical properties to match those of the tissue at the site of implantation [24–26]. Decellularized tissue explant is conducive to tissue growth by providing a diverse range of ECM molecules with tissue-specific orientation [1].

Applications of Decellularized Materials in Bone
Recent research has revealed many unique features of ECM that seem to play a key role in tissue regeneration. For instance, ECM could trigger specific cellular functions such as adhesion, migration, proliferation, and differentiation [27–29]. ECM is also a reservoir of growth factors including transforming growth factor (TGF)-b1and bone morphogenetic proteins (BMPs) which assists in tissue repair and remodeling [30, 31]. Besides its chemical composition, ECM mechanical properties are important in mediating and determining stem cell differentiation. Similarly, the crucial role of ECM in bone regeneration is also demonstrated by the fact that decellularized allogeneic bone grafts often exhibit clinical manifestations comparable to autologous bone grafts [26].

Bone matrix is composed of organic (40%) and inorganic compounds (60%) [32]. Type I collagen is the main organic component, approximately (90%) and mineral binding non-collagenous proteins, and the inorganic component mainly contains hydroxyapatite and a small amount of trace elements [33–35]. The organic matrix and inorganic matrix together provide mechanical properties for daily activities [1]. Due to the anticipated host immune response on the remaining bone marrow cells, many researchers have used decellularization techniques to eliminate antigenic cells to reduce the immunogenicity of these materials [36]. Furthermore, decellularization techniques do not destroy the ability of bone ECM scaffolds for bone reconstruction [37–39]. The acellular bone described in this article is widely used as a bone tissue engineering scaffold in clinical settings due to its three-dimensional (3D) structure and properties similarity to the native bone matrix.

A number of studies have demonstrated the potential of decellularized bone matrix to support and guide the osteogenic differentiation of MSC [37], ESC [40], induced pluripotent stem cells [41, 42]. In addition, 3D-printed hybrid decellularized bone ECM/PCL scaffolds promote healing in murine defects compared to PCL [43], proving that decellularized bone matrix is an excellent bone repair material. At present, cell-loaded bone ECM graft has been used widely for various reconstructive surgery. For instance, seeding rat MSCs in decellularized bone matrix can stimulate bone regeneration of calvarial defect in rat [44]. Clinically, decellularized bone ECM seeded with autogenous bone marrow cells has been successfully used in reconstruction segmental long bone defects. In a recent report, a hybrid acellular material was studied to treat nonunion fractures of the distal tibia by incorporating decellularized bovine trabecular disc with MSCs. The results showed that callus formed around the graft and natural bone after 6 weeks operation [45].

Another decellularized material, the dermis, has unique advantages in bone repair. Anatomically, the dermis is divided into two distinct layers: the upper (papillary) dermis and deep (reticular) dermis. The former is rich in thin collagen fibers, whereas the latter is much thicker and rich in dense irregular collagen fibers [46]. Moreover, dermal ECM is composed of type I and III dermal collagen [47] and elastic fibers and proteoglycans [48], thus providing the characteristic strength and elasticity of the dermis. Today, pre-clinical and clinical applications are leaning towards the use of decellularized dermal ECM due to all major histocompatibility Class II antigens have been removed, make it immunologically inert, and it preserves the structural proteins of dermal ECM, resulting in scaffolds that are well tolerated by the host upon implantation in vivo [49]. Some studies found that decellularized dermal ECM has been shown to support bone formation in bone defect dogs and this matrix is also potentially superior to promote bone regeneration in critical size defects in rat skull when seeded with canine ASCs [46, 50]. This is because decellularized dermal ECM possesses osteoconductivity and the binding site for ASCs, enabling ASCs to attach and osteogenic differentiation [51]. decellularized dermal ECM combined with platelet-rich plasma(PRP) to form a blood derivative rich in growth factors [52], which showed significant therapeutic effects. Jeong et al. found that the combination of the two improved bone regeneration in rabbit calvarial defects is better than acellular dermal ECM or PRP, respectively, the PRP with ADM could promote mature bone formation throughout the defect (Fig. 4.2) [53]. Considering above studies, we thought that this material has already been shown to be efficacious in the treatment of calvarial bone defect, as a substrate material for bone regeneration, and has shown promise in plastic surgery.

As mentioned earlier, decellularized tissues have structural and mechanical properties that similar to natural tissues, which represent a new technology for scaffolds for bone tissue engineering. ECM scaffolds derived from bone and skin have been supported bone formation in animal models, and osteoinductive behavior. However, prior to clinical settings of decellularized tissues, further studies are needed to be explored on the efficacy of acellular therapy in various tissues. Once

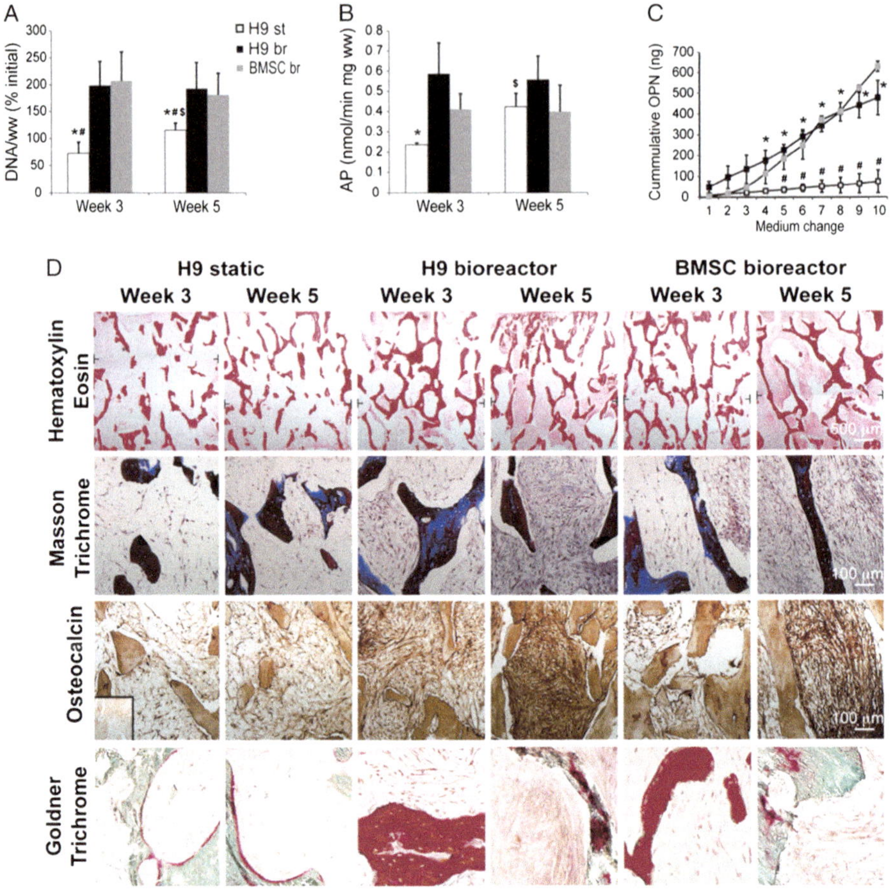

Fig. 4.2 Decellularized trabecular bone used for bone tissue engineering. (Reproduced by permission from [1], Copyright 2014, Elsevier.)

decellularized constructs are available for clinical use, they will provide orthopedic surgeons with an additional tool for the repair of musculoskeletal tissue.

4.2.2 Cartilage Tissue Engineering

4.2.2.1 Structure and Characteristics of Cartilage

Cartilage is a specialized dense connective tissue and almost without vascular, neural, and lymphatic [54, 55]. In the fetus and young age, cartilage tissue was widely distributed and was gradually replaced by bone tissue. In the adults, only

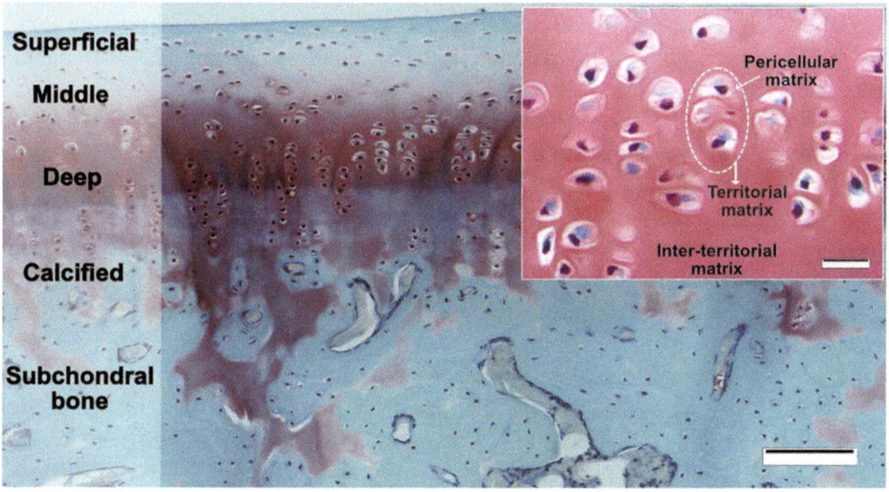

Fig. 4.3 Safranin O/Fast Green stained image of articular cartilage structure of rabbit knee joint shows low cell density. Scale bar: 100 μm. In the small image, the different regions of the cartilage ECM are indicated. Scale bar: 20 μm. (Reproduced by permission from [68], Copyright 2018, Elsevier.)

some cartilage is scattered, mainly exists in the articular surface of the bone, costal cartilage, trachea, auricle, intervertebral disc, etc., and its effect varies depending on the location [56–58]. Cartilage ECM is mainly a collagen network [59–62], another major component of cartilage is proteoglycan, which imparts elasticity and expansion properties to cartilage [63, 64]. Different cartilage has distinct properties due to its different collagen and glycosaminoglycan (GAG) composition [65].

According to the ECM composition, cartilage can be classified into three types: elastic cartilage (if elastic fibers are present in the ECM), fibrous cartilage (if the matrix is rich in collagenous fibers), and hyaline cartilage (if the matrix is predominantly composed of GAGs). The latter, also known as articular cartilage, which is most studied in tissue engineering [66]. Hyaline cartilage is a tissue rich in ECM. The main components of ECM include proteoglycan aggrecan, which is rich in glycosaminoglycan (GAG) and collagen II. There are chondrocytes in cartilage ECM, which comprise only 1–5% of total cartilage volume (Fig. 4.3) [67, 68]. Due to the lower cell content, the regeneration capacity of cartilage is low. Therefore, taste requires regenerative therapy to promote cartilage formation.

4.2.2.2 Requirements for Cartilage Tissue Engineering Materials

Due to the complex natural tissue structure of cartilage, it is a reasonable choice to prepare ideal cartilage repair scaffold to imitate natural cartilage extracellular matrix [69, 70]. Cartilage tissue engineering is designed specifically for the treatment of cartilage damage, in which several key factors must be considered: seeding cells, growth factors, scaffolds, and their mechanical and biological properties [71]. Scaffold selection plays a key role in cartilage tissue engineering [72]. In general, good biocompatibility and biomechanical properties, proper degradation time, and three-dimensional (3D) nanoporous structures are the aspects that an ideal scaffold must have [73]. In addition, the scaffold specifically used for cartilage regeneration should promote cell migration, induce new ECM assembly, and have appropriate mechanical properties to withstand the physiological load required by the human body during exercise [74]. Looking forward to the function of this microenvironment to seek inspiration has provided insight into how to improve the materials used in tissue regeneration. Numerous synthesized and natural materials have been used as scaffolds for cartilage regeneration [75]. Although synthetic materials have adjustable mechanical and biochemical properties, they cannot completely mimic the various interactions that occur in tissues [76, 77]. The advantages of natural materials as tissue engineering scaffolds for cartilage repair are non-toxic and biodegradable profiles, generally high degrees of porosity and biological properties are similar to those of native tissues [69, 78, 79]. A potential therapeutic natural material is the extracellular matrix (ECM), which is a non-cellular component of tissue that can provide structural support and biochemical clues to determine the fate of cells. Decellularized ECM (dECM) from various tissues has been used in tissue engineering and achieved gratifying results [80–83]. For example, dECM derived from cartilage tissue has been extensively studied as a cartilage engineering scaffold for its inherent composition, unique structure, and micromechanical properties [84–87].

4.2.2.3 Applications of Decellularized Materials in Cartilage

Studies have shown that scaffolds derived from naturally decellularized cartilage ECM do not contain allogeneic or xenogeneic cell antigens, while retaining most ECM structural and functional proteins [88, 89]. In addition, cartilage ECM-derived scaffolds have been shown to retain the main components of natural cartilage, including glycosaminoglycans (GAG) and type II collagen, which have good biocompatibility and support chondrocytes and mesenchymal stem cells (MSCs) attachment and proliferation and differentiation [90]. Therefore, due to the heterogeneous structure, and the biochemical and biomechanical properties of natural articular cartilage, a decellularized allogeneic and xenogeneic articular cartilage matrix (CECM) scaffold is proposed as ideal cartilage regeneration scaffolds. The main CECM scaffolds currently used in cartilage tissue engineering are porcine or human derived. The latest strategy for cartilage repair relies on cell-based therapies which involving biological materials that have been implanted with chondrocytes or progenitor cells, which is a promising method for repairing cartilage defects and studying the formation of cartilage tissue in tissue engineering [91].

Chondrocytes, the resident cells of cartilage, produce components of ECM and therefore represent the logically selected cells in cartilage tissue engineering. Chondrocytes help maintain and remodel cartilage matrix tissue by carefully balancing catabolism and anabolic processes [76]. Adult chondrocytes have been isolated from various sources are used in cartilage tissue engineering [92]. Many studies have focused on the use of chondrocytes as a viable source of cartilage repair.

Cheng et al. studied the porous CECM [93], they seed human or porcine articular chondrocytes on porous CECM. After 28 days culture, histology and immunohistochemistry showed abundant cartilage-specific macromolecule deposition. In another study, Bomhard et al. further proved that CECM was favored for possible clinical application [94]. Additionally, according to the result of Gong et al., the authors discovered CECM with chondrocytes could promote the formation of mature cartilage after 12 weeks of implantation [95]. Li et al. inoculated rabbit-derived chondrocytes on a laser-modified CECM scaffold and cultured in vitro for 8 weeks. They found that the cartilage-like tissue was formed from cell-scaffold constructs within the micropores and on the surface of scaffold. In vivo, the cell-scaffold constructs formed ivory-white new cartilage with high level of DNA and cartilage matrix components (Fig. 4.4) [96]. Furthermore, after the transplantation into rabbit knee articular cartilage defects, the cell-scaffold constructs induced neocartilage and improved structural restoration in 8 weeks. Jin et al. also proved that chondrocytes-scaffold constructs repaired the osteochondral defects in vivo [97]. Subsequently, Nie et al. researched the chondrocyte-scaffold constructs in porcine knee joints. Histology, biochemical assays, DNA microarrays, and mechanical tests demonstrated the fine morphology, composition, phenotype, microstructure, and mechanical properties of the regenerated hyaline-like cartilaginous neo-tissue [98]. Similar studies were performed by Bomhard and colleagues, who found that CECM scaffolds seeded with chondrocytes have the ability to repair the injured cartilage defect.

Recent evidence showed that dECMs combined with chondrocyte/stem cells (C-dECMs) benefited cartilage regeneration by promoting cell proliferation and differentiation. Recently, pluripotent bone marrow mesenchymal stem cells (MSCs) have aroused great interest in cartilage tissue engineering because of their abundant sources, easy isolation, and high expansion ability in vitro, among them bone marrow and adipose tissue-derived MSCs play a significant role in cartilage tissue engineering [99, 100]. MSCs are often incorporated into scaffolds for cartilage tissue engineering [101], CECM is associated with inducing the chondrogenesis differentiation of MSCs which is important for cartilage repair, through its various components and topographies. MSCs have the ability to differentiate into chondrocytes [102]. Recently, due to the rapid proliferation ability, pluripotency, and relatively easy isolation of bone marrow mesenchymal stem cells (BMSCs), they have proved to be promising candidates for cartilage regeneration [103, 104]. Using tissue engineering methods, the combination of BMSC and dECM scaffold has achieved the repair of cartilage defects [105]. Lu et al. developed a scaffold from human CECM scaffold, which is a natural, cell-free, three-dimensional interconnected porous. In vitro studies have shown that the scaffold could promote the ability of BMSCs to proliferate and differentiate into chondrocytes after 21 days of culture. In rat model, the scaffold loaded with BMSCs stimulated the formation of cartilage-like tissue [106]. Similar results were obtained by Xue et al.,

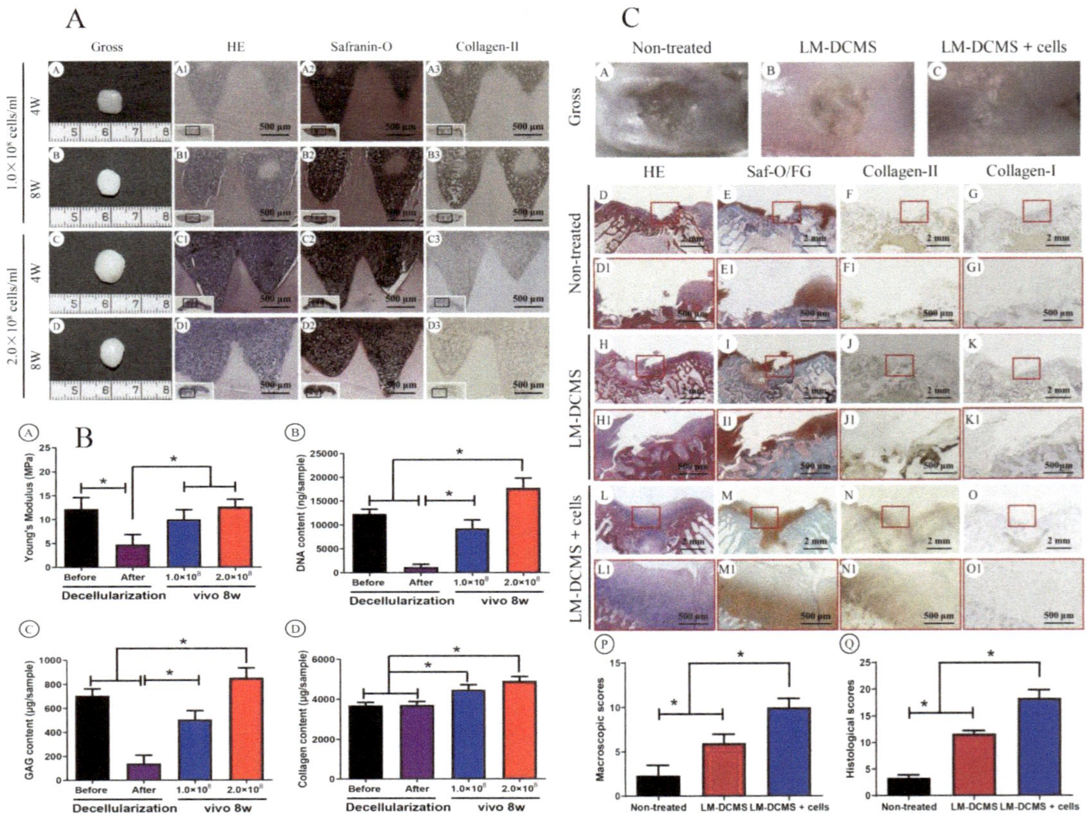

Fig. 4.4 In vivo regeneration of cartilage in tissue-engineered constructs. (**a**) Gross view, HE staining, safranin-O staining and collagen II staining showed the constructs composed of laser-modified decellularized cartilage matrix scaffolds (LM-DCMS) with chondrocytes seeded with the cell density of 1.0×10^8 cells/ml or 2.0×10^8 cells/ml at 4 and 8 weeks. (**b**) Biomechanical and biochemical tests of in vivo neocartilage at 8 weeks. (A) Young's modulus; (B) DNA content; (C) GAG content; and (D) total collagen content. $*P < 0.05$; (**c**) Comparison of microscopic images and histological evaluation of cartilage defects at 8 weeks

postoperatively. (A–C) Macrophotographs of the cartilage defects in the non-treated group, LM-DCMS without cells group and autologous chondrocytes-seeded LM-DCMS group at 8 weeks post-surgery. (D, D1) HE histological images: untreated group, (H, H1) LM-DCMS without cells group and (L, L1) chondrocyte-seeded LM-DCMS group at 8 weeks. Safranin-O/fast green staining: (E, E1) untreated group, (I, I1) LM-DCMS without cells group, and (M, M1) chondrocyte-seeded LM-DCMS group at 8 weeks. Collagen II staining (F, F1): untreated group, (J, J1) LM-DCMS without cells group; (N, N1) chondrocyte-seeded LM-DCMS group at 8 weeks. Collagen I staining: (G, G1) untreated group; (K, K1) LM-DCMS without cells group; (O, O1) chondrocyte-seeded LM-DCMS group at 8 weeks. (P) Results of ICRS scoring system of gross macroscopic scoring and histological scoring. (Q) Results of O'Driscoll scoring system of histological scoring. $*P < 0.05$. (Reproduced by permission from [96], Copyright 2019, Elsevier.)

they seeded BMSCs on CECM that were then stacked layer-by-layer to form BMSC-ACS constructs. After 4 weeks of culture, cartilage-like tissues were formed [107]. Yang et al. constructed CECM scaffold carrying BMSCs that was well tolerated when implanted into rats with cartilage defects in rabbit knees and improved cartilage repair [108].

Apart from co-culture with BMSCs, the researchers inoculated adipose-derived stem cells (ASCs) on porous dECM scaffolds derived from adult pig articular cartilages. Biochemistry analysis showed that the cartilage formation of ASC implanted in the scaffold. Further examination of histology and immunohistochemistry revealed that cartilage-specific ECM components, especially type II collagen, were produced in large quantities after 4 or 6 weeks of culture. These results indicated that the porous scaffold derived from articular cartilage has the ability to induce ASC cartilage differentiation [109]. Similar studies were performed by Luo and colleagues, who found that CECM scaffold seeded with infrapatellar fat pad derived stem cells (FPSCs) has the ability to produce a matrix rich in proteoglycans and collagen [110]. In another study, Kang et al. further explored the feasibility of application of ASCs loaded CECM scaffold for cartilage defect repair in rabbits. The results showed that in all defects were fully filled with repair tissue [111].

Except for seeding cell on the scaffold, many studies have shown that growth factors can be added to the CECM scaffold. Studies have reported that dECM serves as vectors for growth factors to provide the possibility of growth factor delivery is emerging as a new way to enhance intrinsic mechanisms of cartilage repair [112–114]. It is generally known that transforming growth factor-β (TGF-β) family members play a key role in driving MSCs cartilage formation, which have a natural affinity for components of ECM (such as proteoglycans) [115–118]. This indicated that the cartilage ECM-derived scaffold can be employed as a delivery system of natural growth factor. Kelly et al. used CECM to create a scaffold that can deliver growth factors to promote the generation of human infrapatellar-fat-pad-derived stem cells (FPSCs) into cartilage. They proved that under the stimulation of transforming growth factor TGF-β3, CECM-derived scaffolds can promote FPSCs to form cartilage, moreover, ECM-derived scaffold with TGF-β3 loaded can induce highly chondrogenesis than TGF-β3 in media, which optimized for cartilage tissue engineering applications. In addition, they proved that the most part of TGF-β3 that loaded onto the scaffold was controlledly-release over the first 10 days culture, with the observation of comparable long-term chondrogenesis (Fig. 4.5) [119].

4.2.3 Ligament and Tendon Tissue Engineering

4.2.3.1 Structure and Characteristics of Tendon and Ligament

Tendons and ligaments are bundles of dense connective tissue that mediate normal joint movement and stability and transmit mechanical forces between muscles and bones [120]. Tendons attach muscles to bones, while ligaments secure bone elements to each other and stabilize bone joints. The composition of both ligament and

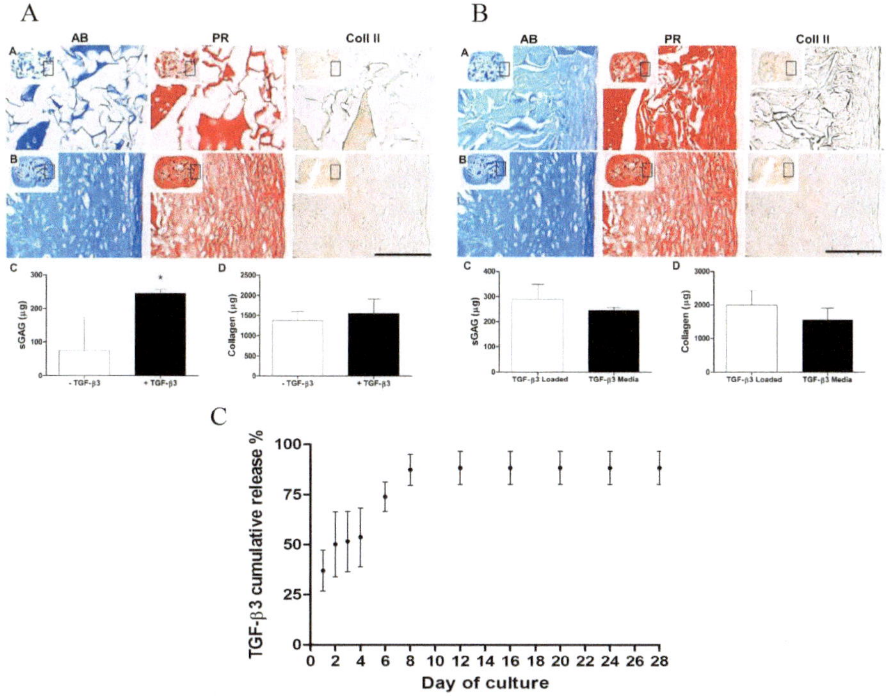

Fig. 4.5 An image shows promotion chondrogenesis due to controlled release of TGF-β3 from cartilage extra cellular matrix derived scaffolds. (**a**) Exogenously supplied TGF-β3 induced robust chondrogenesis. After 28 days of culture, Alcian blue (AB), picro-sirius red (PR) and collagen type II (Coll II) staining of ECM-derived scaffold histological sections, (A) without TGF-β3 supplementation; (B) with TGF-β3 supplementation. The supplemented group had higher sGAG and collagen accumulation (scale bar: 100 μm). Biochemical assays result for ECM-derived scaffold with and without TGF-β3 supplementation seeded with fat pad derived stem cells (FPSC) of human infrapatellar. (C) sGAG and (D) Collagen content ($n = 4$, $^{*}p < 0.05$); (**b**) ECM-derived scaffold with TGF-β3 loaded can induce highly chondrogenesis. After 28 days of culture, Alcian blue (AB), picro-sirius red (PR), and collagen type II (Coll II) staining of ECM-derived scaffold loaded with TGF-β3 and TGF-β3 in media, (scale bar: 100 μm). (A) TGF-β3 loaded; (B) TGF-β3 in media. The sGAG (C) and collagen (D) accumulation are similar for both groups ($n = 4$); (**c**) Release profile of TGF-β3. ELISA results showed that TGF-β3 was released into the culture medium from the ECM-derived scaffold loaded with TGF-β3 ($n = 3$). The cumulative release value is expressed as a percentage of the initial amount of TGF-β3 loaded into the stent. (Reproduced by permission from [119], Copyright 2014, Elsevier.)

tendon is similar, and they are composed of fibroblasts (termed tenocytes and ligamentocytes, respectively) and collagenous fibers that are crosslinked to one another [121–123]. The tendon is a bundle of uniaxial dense collagen fiber connective tissue connecting the skeletal muscle and bone. It is a small elastic and gliovascular tissue. It is used to transmit the force generated by the contraction of the muscles and draw the bone to produce movement. Tendon itself does not have the ability to contract, but has a strong ability to withstand pressure, tension, and friction. Therefore, mechanical stimulation must have a great influence on the biological characteristics of tendon cells [124]. The ligaments are mostly located around the joint (extracapsular ligament) or in the joint cavity (intracapsular ligament). They extend in parallel and have strong tensile strength and certain elasticity. Ligaments can be divided into two categories: elastic connective tissue and irregular dense connective tissue interwoven with collagen fibers. The function of the ligament is to strengthen the joint, maintain the stability of the joint movement, and limit its activities beyond the physiological range. If the ligament bends beyond its physiological range (such as a sprain), it may cause the ligament to stretch or break [125, 126]. Because of their structural roles, injuries to tendons and ligaments are extremely common and often debilitating. Moreover, tendons and ligaments have poor regeneration capacity because of their low cell density and nutrient and oxygen requirements [127]. Thus, a fundamental question in tendons and ligaments tissue engineering is how these connective tissue structures develop in the correct locations and acquire the strength necessary to translate contractions of muscles into skeletal movements [128].

4.2.3.2 Requirements for Tendon and Ligament Tissue Engineering Materials

Scaffolds for tendon and ligament engineering require material with excellent combination of biocompatibility, biodegradability, and suitable mechanical property. Specifically, the scaffold acts as a temporary template of the native ECM [129, 130] and will gradually degrade with the neo-tissue resynthesized [131, 132]. In addition, scaffolds should have appropriate morphology and porosity to facilitate the cell behavior, such as adhesion, proliferation, differentiation, etc [133]. Besides, the scaffold should mimic the properties of the native tendon and ligament tissue, such as proper topography, geometry, and porosity to recreate the native microenvironment and aid the cell adhesion, growth, and differentiation of the populating cells [133]. In fact, biomaterials, synthetic or natural materials, have been designed to treat, enhance or replace tendon and ligament tissues that perform the intended functions of the human body [134]. Among them, naturally derived scaffolds have important advantages including biocompatibility, preservation of the extracellular matrix protein, and biodegradation that allows for maintenance of tensile properties. One potential naturally derived scaffold that has been provoked is decellularization of whole tissue, which has been considered as the most suitable scaffold for ligament/tendon replacement because of their complex structure is similar with tendons and ligaments.

4.2.3.3 Applications of Decellularized Materials in Tendon and Ligament

Currently, the widely used decellularized tissue in the repair of tendon and ligament tissue is the porcine small intestinal submucosa (SIS). SIS has good mechanical

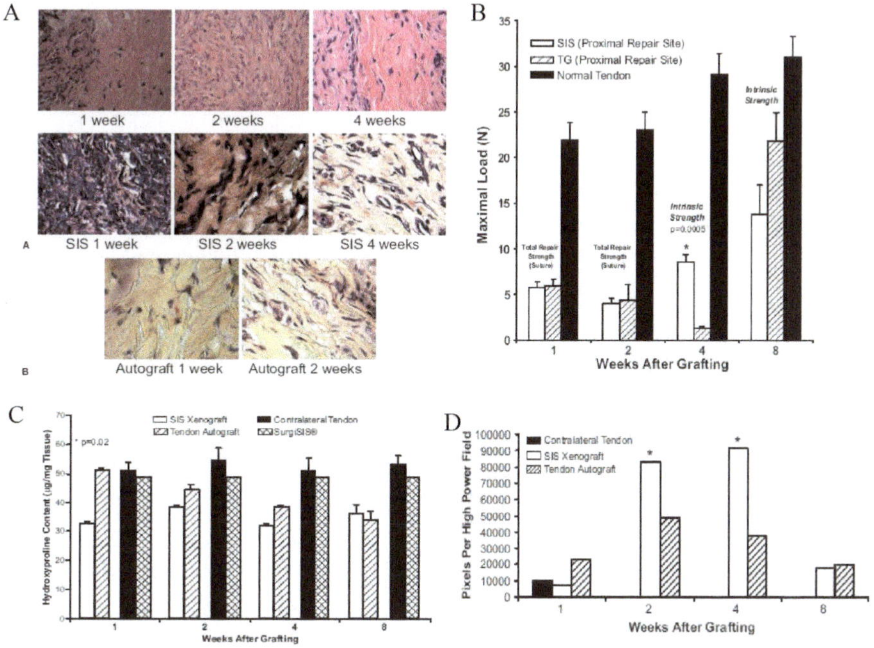

Fig. 4.6 SIS promotes flexor tendon regeneration by elevated TGF-β1 expression. (**a**) A representative hematoxylin-eosin section of the distal repair site of the SIS in the 2nd week and improved resolution in the 4th week. B Movat staining showing increased cellularity and immature collagen at 1 and 2 weeks. (**b**) Tensile test showing that SIS have greater strength compared with native tendon repairs at 4 weeks ($p = 0.0005$) (Values expressed as mean ± SEM). (**c**) Hydroxyproline content test showing that there is a significant difference between porcine SIS and tendon autograft at 1 week ($p = 0.02$) but no significant difference thereafter. (Values expressed as mean ± SEM); (**d**) Immunohistochemical staining showed that the TGF-β1 in the SIS graft than in autograft tendon at 2 and 4 weeks after implantation. ($*p < 0.05$, values expressed as mean ± SEM). (Reproduced by permission from [138], Copyright 2008, Elsevier.)

properties, tissue compatibility, and low immunogenicity. It is a natural extracellular matrix material and is often used as a good scaffold material for cellular tissue repair and tissue engineering reconstruction [135, 136]. As early as 1995, Badylak et al. evaluated the tissue response of SIS used to repair Achilles tendon defects in dogs. The results showed that the new tendon remodeled by SIS 12 weeks after the operation was similar to the normal tendon on the opposite side. In addition, Yan et al. studied the effect of SIS for Achilles tendon in rabbit model. Histological and maximum load tests proved that the SIS completely replaced the host tendon, and its tissue structure and mechanical properties were extremely close to that of normal tendon after 16 weeks implantation [137]. Furthermore, the SIS as a potential human flexor tendon graft substitute was compared with tendon autografts. Results showed that the strength of them is comparable. Moreover, earlier neovascularization, increased TGF-1 levels, and increased collagen deposition, as well as higher intrinsic repair strength made SIS a promising one flexor tendon graft replacement [138] (Fig. 4.6).

Recently, application of cell-matrix composites with an appropriate combination of cells and biochemical cues have shown to affect the process of tendon healing. Human dermal fibroblasts [139] and pluripotent stem cells [140] have been successfully used to recellularizations of tendon grafts. Ideally, resident tendon fibroblasts are the ideal candidates for their regeneration since this type of cell allows reproducing the collagenous matrix, the destruction of matrix would decrease the pre-implant mechanical strength of the graft [141]. Tischer et al. collected the semitendon of New Zealand White (NZW) rabbits, and then injected the cultured dermal fibroblasts into the tendon. These constructs were further cultured for 4, 7, or 14 days under the same culture conditions. Biomechanical and histological test results showed that the maximum breaking load of the tendon was comparable to that of the natural tendon. Histologically, immunohistochemical staining of positive collagen I showed the metabolic activity of seed cells [142]. However, the location and the limited quantity and proliferation potential of these fibroblasts restricted their use. With the advancement of stem cell technology, pluripotent and multipotent stem cells for ligament/tendon tissue engineering, include embryonic stem cells (ESCs), bone marrow mesenchymal stem cells (BMSCs), induced pluripotent stem cells (iPSC) and adipose-tissue-derived mesenchymal stem cells (AMSCs), have been more and more used.

Rotator cuff tear is a common tendon injury disease, which can cause severe pain and impair active movement. The incidence of rotator cuff rupture exceeds 54% in older people [143]. Rupture of the rotator cuff usually requires surgical repair to relieve pain and restore shoulder function. However, traditional surgical repair usually fails because the spontaneous healing efficiency of the tendon–bone interface is very low [144]. Although various tissue engineering strategies have been developed to reconstruct the rotator cuff and promote the formation of the fibrocartilage interface [145], studies have shown that nearly half of surgical rotator cuffs cannot achieve the tendon–bone interface. Furthermore, they also cannot promote biological healing [146]. Therefore, there is an urgent need to develop promising materials and tissue engineering strategies to regenerate fibrocartilage at the tendon–bone interface. Recently, the use in the repair of rotator cuffs is proposed due to its no immunogenicity while retaining the integrity of the extracellular matrix, thereby forming a collagen scaffold with good characteristics [147]. In a study, a human acellular dermal matrix graft is compared with autologous tendon in a full-thickness infraspinatus tendon tear dog model. Mechanical tests showed that the initial mechanical strength of the two materials is similar. Over time, the mechanical strength of the control group remained unchanged, while the experimental group gradually increased. By 12 weeks, the strength of the experimental repair was equal to that of the control. Histological tests indicated that cells infiltrated the control and experimental specimens by 6 weeks. At 6 months, the control and experimental specimens were similar to normal structures [148]. In another study, Xu et al. conducted a study to repair and augment the excision defect of the supraspinatus tendon of the rotator cuff in the green monkeys by using porcine acellular dermal

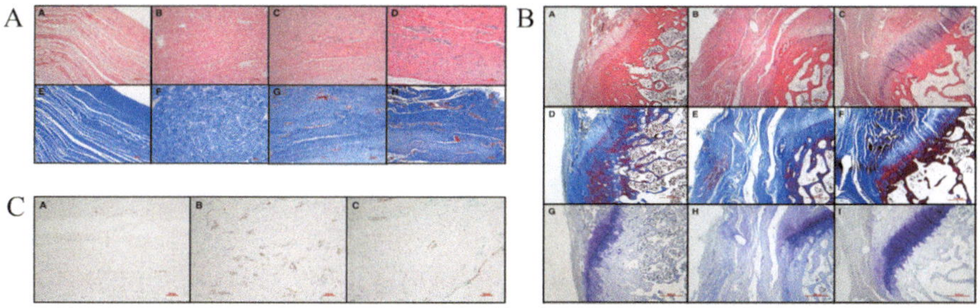

Fig. 4.7 Porcine acellular dermal promote rotator cuff repair. (**a**) Representative H&E and Masson trichrome staining slides (×100), (A, E) native unoperated supraspinatus tendon, (B, F) Conexa graft at time 0, (C, G) 3 months' and 6 months' implant duration (D, H). (**b**) Representative H&E-stained (A-C), Masson trichrome–stained (D-F), and toluidine blue–stained (G-I) histology slides (×100), (A, D, G) native unoperated supraspinatus tendon, (B, E, H) Conexa implanted in African green monkey rotator cuff repair site for 3 months and 6 months (C, F, I). (**c**) Representative muscle cell actin immunohistochemical staining (×100). (A) Native unoperated supraspinatus tendon, (B) Conexa implanted in African green monkey rotator cuff repair site for 3 months and (C) 6 months. (Reproduced by permission from [149], Copyright 2012, Elsevier.)

matrix. Gross examination of grafted scaffold showed the appearance of integration with tendon. Histologically, scaffold appeared to have remodeled to tendon-like architecture, with parallel alignment of collagen fibers and homogeneous distribution of fibroblast cells in the direction of force after 3 months implantation [149] (Fig. 4.7).

Anterior cruciate ligament (ACL) is a common knee ligament, and 400,000 patients in the USA receive ACL reconstruction each year. The transplantation options for ACL reconstruction include autologous transplantation and allogeneic transplantation [150]. Allografts and autografts will ideally produce tissues that are histologically and biochemically similar to natural ACL [151]. Despite the advantages of allograft and autograft, tissue engineering strategies for ACL reconstruction rely primarily on biodegradable polymer scaffolds. However, the current biodegradable stent initially does not possess the mechanical properties necessary for reconstruction, and is related to the potential release of toxic by-products during the degradation process. Therefore, the focus has shifted to the use of naturally derived scaffold materials for ACL reconstruction. The naturally derived that currently used to ACL reconstruction is decellularized freeze-dried human Achilles tendon tissues. Whitlock et al. evaluated the properties of scaffold derived from freeze-dried human Achilles tendon. The experimental results showed that the scaffold had good cell compatibility in vitro. During the tensile test, compared with the allograft, the elongation of the scaffold increased significantly and no other significant differences in tensile properties were observed in vitro or in vivo. Furthermore, in vivo experiments, the number of infiltrating host cells and the depth of infiltration of these cells were significantly greater [152]. Except for decellularized human Achilles tendon tissues, there has been reported that SIS successfully used for reconstruction of lateral collateral ligament, by comparing with sham-treated and control groups, the results prove that SIS surgical repair eliminates joint slack during surgery.

SIS-treated joints had significantly less relaxation at 8 weeks than sham surgery and showed a gradual increase in repair tissue density and fiber arrangement, which was significantly larger than sham-treated ligaments. Furthermore, SIS repaired ligaments tend to have greater peak stress [153].

4.3 Decellularized Materials for Skin Tissue Engineering

4.3.1 Structure and Characteristics of Skin

Skin, including countless sensors, glands, channels, and pores, is an important interface between the human body and the environment, enabling the organism to feel pain, regulating body temperature and humidity, etc. It is the body's first protective barrier against external factors [154]. As described previously, skin must be able to withstand some degree of damage and recover on its own, while mitigating tear propagation that may occur during growth, movement, and injuries to its integrity [155]. Hence, understanding the composition of the skin and the functions is important for application of artificial skin grafts.

Traditionally, adult skin consists of three main regions, and it described by the schematic in Fig. 4.8. Its structure mainly includes: a keratinized, strati filed epidermis, an underlying layer of collagen-rich dermal connective tissue, and subcutaneous adipose tissue (hypodermis) [156]. Thus, skin can be considered a layered composite material [157], each layer fulfills a certain function. It is well known that the epidermis (50–150 μm in thickness [158, 159] for humans) is the external

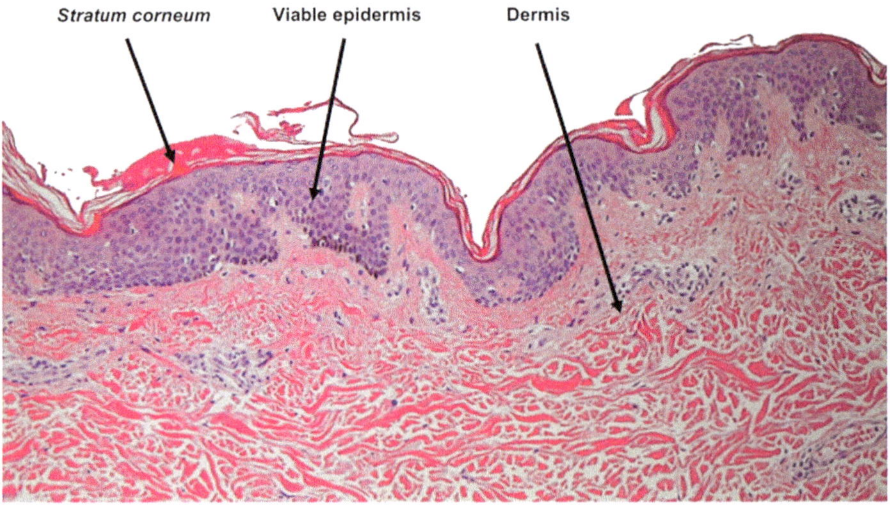

Fig. 4.8 Histological sections of the human dorsal skin, stained with hematoxylin and eosin. (Reproduced with permission from [162]. Copyright 2015, Elsevier.)

layer of the skin, constructed from packed sheets of keratinocytes, primarily acting as a barrier and exchange interface with the exterior [160]. The epidermis is built up on top of the stratum basale, a meshed network of the proteins collagen IV and laminin [161]. Keratinocytes then migrate outward and form the hardest keratinized layer [162]. The dermis, which is composed of collagen and some elastin and glycosaminoglycans [161, 163], supports the epidermis by providing nutrients to the epidermis and structural support. The reticular dermis is composed of various sensors, blood vessels, and glands and provides mechanical properties for the skin through the layered dense arrangement of wavy collagen fibers [161, 163, 164]. The extracellular matrix (ECM) of the inner dermis layer generates a negatively charged, hydrophilic, and dynamic space capable of hydration of the tissue and dispersing forces [165, 166]. Fibroblasts present the major cell type of the dermis and play a major role in wound healing since they are capable of producing remodeling enzymes such as collagenases and proteases. The hypodermis (or subcutaneous layer) is the innermost layer of skin, consisting mostly of adipose tissue, in which adipocytes form clumps called lobules, surrounded by a network of fibrous collagen. It contributes mostly to mechanical as well as thermoregulatory properties of the skin [161, 163]. Its loose connective tissue structure also enables it to minimize friction with neighboring muscle tissue. And its thickness varies greatly depending on the region of the body and the subject.

4.3.2 Requirements for Skin Tissue Engineering Materials

Skin defects can result in various factors, such as burns, trauma, surgery, and chronic ulcers [167]. When a large area of injury occurs, it is difficult to restore the original shape and function of the skin spontaneously [168]. Accordingly, it is needed to use skin transplantations to cover these wounds. Generally, the gold standard for skin regeneration is autograft [169, 170], but there are limitations such as in the donor-site availability and complications [171–173]. The development of a bioartificial skin would be beneficial to the treatment of patients with various skin-related disorders, but this approach also has to meet high demands with respect to the interaction of the bioartificial devices and the human natural tissues. In particular, full-thickness skin substitutes should consider the structural and functional properties and requirements of both epidermis and dermis, so as to better imitate the natural skin [174, 175]. Consequently, artificial skin substitute with biomimetic design shows great potential in skin tissue engineering.

Great efforts have been made to develop dermal and epidermal replacements in order to better treat skin defects, overcome poor skin quality and subsequent scarring, especially for extensive burn wounds [176, 177]. It should be noted that in the presence of a dermal substrate, the epidermis of an adult can regenerate rapidly, but in severe burns dermis regeneration may be slow or nonexistent [178]. Consequently, dermal regeneration should be the first step to speed up the formation of new skin. In summary, therefore, optimal skin substitutes must contain several outstanding properties for clinical use [179]:

a. Protect the wound from infection, further damage, and loss of fluid;
b. Avoid immunoreactive processes, as well as provide a stable and biodegradable template for the synthesis of normal skin over the wound;
c. Reduce pain, and allow ease of handling and resist tear forces;
d. Either host or enable the influx of cells that will function as dermal cells, producing dermal tissue rather than scar tissue.

There are types of scaffold matrices for skin defects repair, such as autologous, natural, synthetic, degradable or non-degradable materials, and so forth, with both advantages and drawbacks. Compared with natural scaffolds, synthetic scaffolds, and cell sheets, decellularized biomaterials show significant potential clinical value in covering skin wound defects [180]. Because the artificial organs and tissues produced by decellularization method have no unfavorable immune response [181], and decellularized materials show their inherent advantages such as a relatively integrated natural structure, excellent biocompatibility, and desirable biological activity that could release of growth factors in an ideal spatiotemporally controlled method [182]. Several studies have shown that tissue scaffolds and decellularized organ have ECM structures, components, and various cytokines [183]. The protein-based decellularized materials provide mechanical strength and biological activity, which are important for successful tissue repair and regeneration [166, 184]. During the last decade, many achievements have been made in the research of cell-free biomaterials for investigations on organ and tissue substitutes [185]. Ott and his coworkers successfully engineered kidneys [186, 187], lungs [188, 189], and limbs [190] using decellularized scaffolds. Moreover, Badylak and coworkers investigated decellularized esophagus [191], spinal cords [192], cardiac patches [193], brains, and skeletal muscles [194–196]. In addition to the above decellularized organs, many researchers have concentrated on developing decellularized scaffolds as skin substitutes, dermis, amniotic membrane, the intestinal tissues, peritoneum, and skin flaps have been studied extensively.

4.3.3 Applications of Decellularized Materials in Skin

The dermis, located between the epidermis and the subcutaneous tissue, has high physical strength and flexibility as well as extensive blood vessels. Loewe proved that the dermis could improve the tenacity of grafted skin, inhibit granulation tissue, and prevent scar formation, around 1913. The ADM is the remaining ECM of the dermal and basement membranes once the cellular components have been removed, which serves multiple functions to facilitate wound healing. So far, experiment on animals and preliminary clinical trials have shown that acellular dermal matrix (ADM), with beneficial ECM skin microstructure, can be used effectively to repair skin defects. Since the 1990s, ADM has been produced and developed into types of commercialized products. Typically, LifeCell Corporation developed AlloDerm regenerative tissue matrix in the 1990s and it was approved by the U.S. Food and Drug Administration (FDA) for the replacement of damaged integumental tissue. AlloDerm, derived from allogenic human cadaveric dermis, had ability of promoting cell adhesion in vitro [197, 198]. Although previous researches have shown that dermal grafts obtained from human subjects have a significant therapeutic effect in treating burns and full-thickness skin defects [199], the high cost, limited availability, and the risk of disease transmission limit their application [200]. Consequently, ADMs derived from the skins of goats, fish, porcine, and ostriches have been studied and yielded satisfactory results with respect to wound healing [201–203]. In particularly, dermis obtained from porcine has been commercialized, such as Medeor Matrix, Permacol, and Strattice. Medeor matrix retained more transforming growth factor (TGF)-β and vascular endothelial growth factor (VEGF), which has positive effects on the proliferation and migration of cells. Moreover, compared with other acellular dermal products, the adhesion of this material to cells was obviously enhanced. With its better crosslinking properties, Permacol exhibits high durability and stability and can be used to reconstruct the skin wall. These surgical implants provide lasting structural stability to guarantee the completeness of the graft and the duration of wound coverage during the healing process in vivo. However, the use of crosslinking agents can lead to poor cell adhesion [196]. In general, the stiffness of the ADM may affect wound healing. Han et al. found that epidermis treated with a softer ADM derived from younger mice exhibits better collagen density than that derived from older mice, and the stiffness of the new tissue was closer to the normal tissue [204]. Decellularized dermis contains advantageous dermal ECM that has achieved remarkable clinical achievement in the treatment of skin lesions. However, poor penetration of cells limits its applicability. Another shortcoming may be the implantation of dermal ECM scaffolds in vivo that can lead to an increase in the pro-inflammatory M1-like macrophage phenotype [205]. Thus, further in vitro and in vivo studies need to be performed in order to the improvement of clinical applications.

Mesothelial tissues, found surrounding the abdominal cavity, chest cavity, and heart, have a strong, thin layer of ECM that supports mesothelial cells capable of rapid healing. In addition, mesothelium tissue can produce various types of ECM proteins, such as collagen (type I, III, and IV), fibronectin, and laminin [206]. If this thin and strong membrane retains its inherent strength and assumed significant biological activity as an acellular matrix material, it may have surgical and regenerative medicine applications [207]. Herein, we focus on the applications of the decellularized pericardium and decellularized peritoneum. After removal of cellular components from these two materials, components of ECM are usually maintained and are quite intact [207], likewise, cytokines have also been shown to be preserved [208, 209]. It has also been reported that decellularized peritoneum has a short half-life in vivo, combined with promoting wound healing cytokines, which can accelerate wound healing of full-thickness wounds. It has also been reported that with a short half-life in vivo and healing cytokines, the decellularized peritoneum can accelerate the healing of full-thickness wounds (Fig. 4.9) [210, 211].

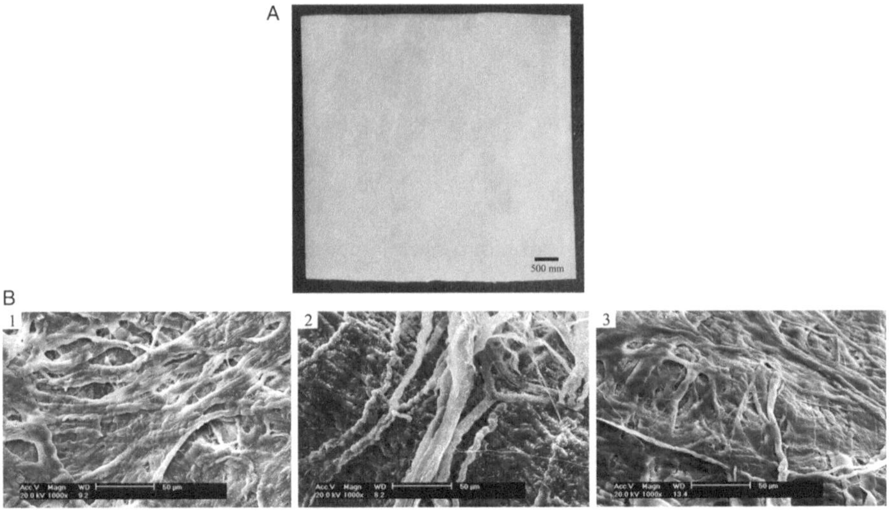

Fig. 4.9 Scanning electron microscope (SEM) was used to examine the scaffold. (**a**) An acellular scaffold with a smooth surface. (**b**) The acellular scaffolds were examined by scanning electron microscopy. The outer membrane surface of the scaffold is relatively smooth. (Reproduced with permission from [211]. Copyright 2014, Elsevier.)

Decellularized mesothelium has shown remarkable promise as surgical implants and scaffolds for regenerative medicine. Commercially available porcine mesothelium graft, Meso BioMatrix Surgical Mesh is produced by DSM's proprietary Optrix process, which retains the natural structure of the biological matrix. In vitro evaluation shows Meso BioMatrix that supports cell proliferation and promotes cell migration and chemotactic signals [212]. In vivo, it can facilitate infiltration of fibroblasts and macrophages [213]. In vivo response to decellularized mesothelium scaffolds: Capella-Monsonís et al. compared the biochemical and biological properties of the only two commercially available porcine mesothelium grafts [Meso Biomatrix® and Puracol® Ultra ECM (PM-MB and PM-PC)] to traditionally use wound healing grafts, which these mesothelium grafts have shown efficiency in the skin wound defects repair or skin wound healing enhancement [214]. Another decellularized mesothelial product derived from bovine pericardium is Veritas Collagen Matrix. Numerous studies have shown that Veritas stimulates faster soft tissue remodeling compared to other acellular dermal alternatives, resulting in more complete remodeling within a month [215–217]. However, there is no report that it can repair skin wound defects. In short, the decellularized mesothelium contributes to faster tissue remodeling. There are still many shortcomings to be solved, including the high degradation rate in the body, poor biomechanical properties, and the absence of skin formation during skin repair.

One of the most studied decellularized matrices for a variety of biomedical and tissue-engineering applications is derived from the intestine [218, 219], including ileum [220], jejunum [221], and small intestinal submucosa (SIS) [222]. Rebuild distinct organs using decellularized intestinal scaffolds had been reported in many studies, such as cornea [223], urethra [224], vagina [225], lung [226], and heart [227]. The decellularized intestine is a vital platform for angiogenesis [228] and repair of the abdominal wall [229]. This material may have applications in wound coverage. Human fibroblasts, keratinocytes, and human microvascular endothelial cells were seeded for the first time on a cell-free intestinal scaffold by Groeber et al. [230]. After 14 days of co-culture, they can clearly observe the formation of the skin barrier. In addition, by seeding adipose-derived stem cells, recellularization of cell-free SIS also improved angiogenesis and wound healing [231].

Since the decellularization of porcine small intestine submucosa (SIS) by using well known chemical detergents was first proposed in 1995 [232, 233], the number of decellularization studies have increased dramatically, especially in the field of tissue repair. SIS, a natural three-dimensional (3D) form with a biocompatible and regenerative structure, as well as continuous dynamic remodeling after implantation meet the requirements of damaged tissue repair [80, 234, 235] is a promising biomaterial for tissue repair. In addition, SIS scaffolds preserve higher amounts of active molecules, modulate intracellular signaling to support the body's own tissue repair mechanisms and eliminate the risk of adverse effects such as immune-mediated inflammatory reactions, increasing the possibility of the clinical use of decellularized tissue [236–238]. Recent researches demonstrated that SIS-based biomaterials, either alone or crosslinking and combining one or more distinct types of materials with complementary features, can overcome some of the problems in skin injury repair by regulating cellular phenotypes, inhibiting local inflammation, and increasing wound healing rate [239, 240].

OaSIS Wound Matrix (Cook Biotech Inc., West Lafayette, IN), a trilayered natural matrix, has smooth surfaces and no opening pores. This material is decellularized and can be applied for the management of wounds [212, 241–243], including venous ulcers, diabetic ulcers, trauma wounds, pressure ulcers, vascular ulcers, and full-thickness wounds. In a clinical trial of 120 patients with chronic leg ulcer, the rate of complete healing after 12 weeks was significantly higher in the SIS group than in the standard treatment group. The results of six months' follow-up findings showed no recurrent ulcers in SIS-treated patients, supporting the applicability of SIS products in the treatment of chronic trauma [244]. Moreover, in 2008, researchers designed functional wound dressing by SIS membrane with SIS cryogel. It has a high potential to treat full-thickness wound owing to sufficient mechanical properties, appropriate anti-infective behavior as well as remarkable biocompatibility [245]. Another SIS-based multilayered matrix material-SurgiSIS, 4-Layer Tissue Graft (Cook Surgical, Bloomington, IN), has also been applications in skin wound repair in a clinical case report, which has a normal healing process [246–248]. While the SIS offers promising consequences in research studies, it should be pointed out that rapid degradation leading to mechanical property loss and spectrum of natural occurring heterogeneity limit its application, which should to overcome these problems.

The human amniotic membrane (HAM) is made of a thick basement membrane and an avascular matrix. It is considered suitable for clinical application as an attractive skin repair material due to critical anti-inflammatory and anti-scarring properties. HAM has been reported to cover ulcerated skin and burned wounds have marked effects on pain reduction [249, 250]. Amniotic membrane transplantation as a permanent or temporary graft can promote epithelial healing [251]. After decellularization, cell-free HAM preserved all major components of the matrix. Other excellent properties are not significantly reduced [252]. Cell-free HAM has been shown to be used as a feeder layer for certain stem cells that induce neuronal differentiation. What's more, acellular HAM scaffold could promote epithelialization, prevent scarring, exhibit antimicrobial and antifibroblastic effects [252–254]. Living skin equivalents (LSEs) can be constructed using acellular HAM by activating platelet-rich plasma and seeding fibroblasts on the chorionic side and keratinocytes on the epithelial side, and other methods [255–258]. The cultured HAM-LSE is strong enough and has a layered epidermis. After being transplanted into full-thickness skin wounds of nude mice, it showed good morphology and mechanical properties [258]. According to reports, HAM-LSM can induce epidermis and skin regeneration in short time [259]. However, it should be noted that the reduction of anti-scarring components and anti-inflammatory as well as secondary contamination during preparation process can reduce their advantage to some extent. In addition, Ambiodry is a commercially available cell-free HAM allogeneic transplant believed to have clinical applications in ocular surface reconstruction [260], but not suitable for skin defects, which requires further research and evaluation.

The flap is made from healthy skin and fat or skin, fat and muscle. Clinically, acellular skin flaps are typically used to cover severe wounds. Therefore, the study of the decellularization of the skin flap is very important. The perfusion method has been used to successfully fabricate decellularized epigastric flaps with an average size of 5×3.5 cm^2 in 2010, by Henderson et al. [261]. Subsequently, the researchers conducted extensive research on infusion decellularization and developed an optimized protocol (Fig. 4.10) [262], which has been successfully reported in a miniature pig model [263]. In cell culture experiments, it has been shown that decellularized skin flaps can promote the survival and proliferation of fibroblasts and fatty stem cells. In addition, by inoculating endothelial cells from human umbilical veins, re-endothelialization of the acellular flaps has been achieved [264]. In the nude mouse model, a vascularized skin flap is inoculated with stem cells derived from adipose, then anastomosed with the recipient area of the wound defect. In a short time, the functional circulation maintains a balanced blood flow. Three months after the operation, the skin flaps were remodeled and fatty tissue evidently formed [264].

The above studies have shown that although functional circulation can be reconstructed in the decellularized flap, the non-functional aorta limits the success of transplantation of biomaterial constructs. As a result, more research is needed to confirm the efficacy of acellular flaps, especially in terms of clinical manifestations. In general, the application of acellular biomaterials in skin wound repair shows promise although there are still many challenges to overcome.

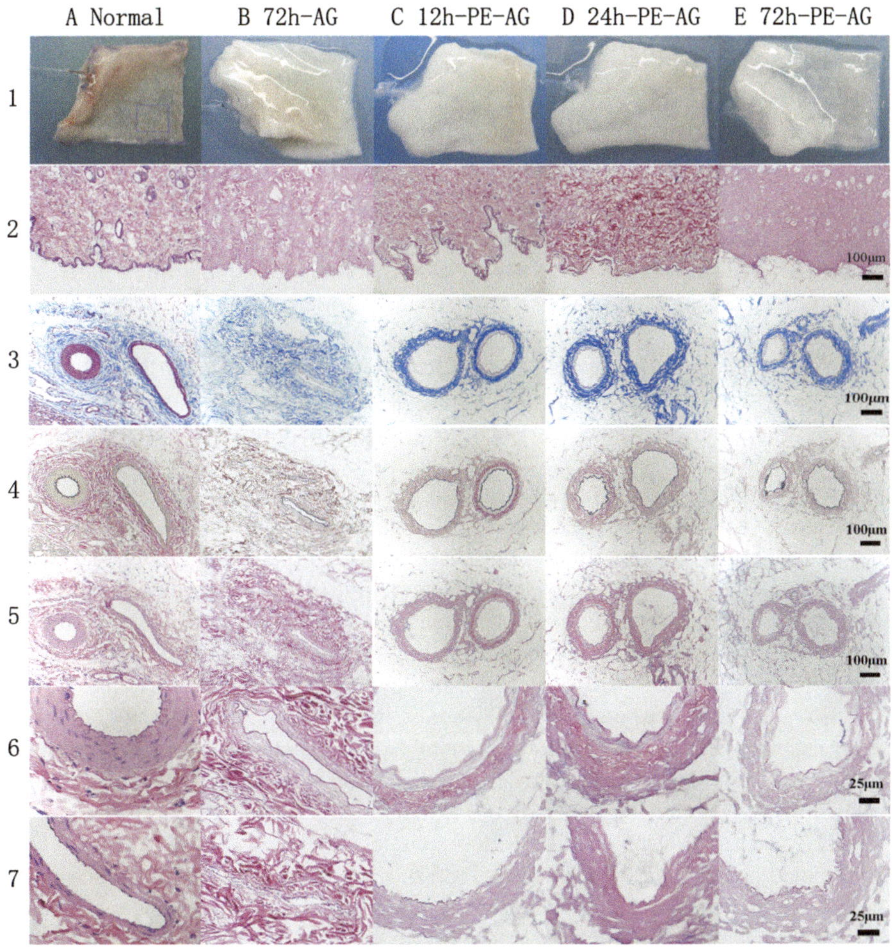

Fig. 4.10 Gross observation and histological evaluation of normal flap and decellularized flaps with different methods and treatment times. Normal (**a**); 72h-AG (**b**); 12h-PE-AG (**c**); 24h-PE-AG (**d**); 72h-PE-AG (**e**). Row 1, General observation of flap. Row 2, H&E staining of the superficial skin structure in a magnification of ×100. Row 3, Masson trichrome staining of the artery and vein in a magnification of ×100. Row 4, Verhoeff's Van Gieson E staining of the artery and vein in a magnification of ×100. Row 5, H&E staining of the artery and vein in a magnification of ×100. Row 6, H&E staining of the artery in a magnification of ×400. Row 7, H&E staining of the vein in a magnification of ×400. (Reproduced with permission from [262]. Copyright 2015, Wolters Kluwer Health, Inc.)

4.4 Decellularized Materials for Bladder Tissue Engineering

4.4.1 Structure and Characteristics of Bladder

Bladder is a complex organ with a sac-like structure sitting in the pelvis that functions as a reservoir for storing urine produced in the kidneys. Histologically, the bladder wall is composed mainly of mucosa, submucosa, and muscularis. The normally functioning bladder consists of two main parts: a flexible muscle wall and a fully specialized urothelium, involved in various physiological functions and pathological development, and carefully organized under the control of the autonomic and sensory nerves. The urethral epithelium is a transitional epithelium derived from the endoderm, which is necessary to prevent hypertonic urine from passing through the blood and exchanging toxic metabolites. It is an impermeable blood-urinary barrier that protects the urethra against high urinary reflux, pressure, and infections, which can also act as a large capacity low pressure liquid storage reservoir [265]. About 90% of the urethral epithelium is made up of basal cells, which have mitotically activity, and 5% of the urethral epithelium is made up of intermediate and surface cells located in the upper basal layer and the luminal layer, respectively. In homeostasis, adult urothelial cells are in a static state and the cell turnover rate is approximately 40 weeks. However, the damaged of urothelium due to acute injury, toxins, or urinary tract pathology will induce the process of rapid exfoliation and regeneration within 72 h [266]. Individually, each layer of the bladder plays a specific role. The lamina propria accommodates the vascular system, provides the necessary oxygen and nutrients to the urinary epithelium, and anchors it to the bladder muscles [267, 268]. Detrusor facilitates low pressure urine storage and coordinated contraction of urine output [269]. In general, they are all necessary for the unique and essential function of the bladder. Due to the special properties of smooth muscle fibers, blood vessels and nerves in the bladder, repeated and coordinated bladder contractions can be performed. In addition, here is no impact on the storage capacity of the bladder or the protection of the upper airways of these contractions, which is why when this organ is damaged, all of these characteristics should be taken into account when attempting to replace it, in order to reproduce or recreate them individually for an overall functional recovery [266].

4.4.2 Requirements for Bladder Tissue Engineering Materials

Numerous conditions related to loss or lack of bladder ability, such as trauma, inflammation, infection, cancer, or iatrogenic injury, should require reconstructive procedures to reduce the risk of such injury. In addition, congenital or neurological diseases can also cause a gradual loss of bladder function, leading to urinary

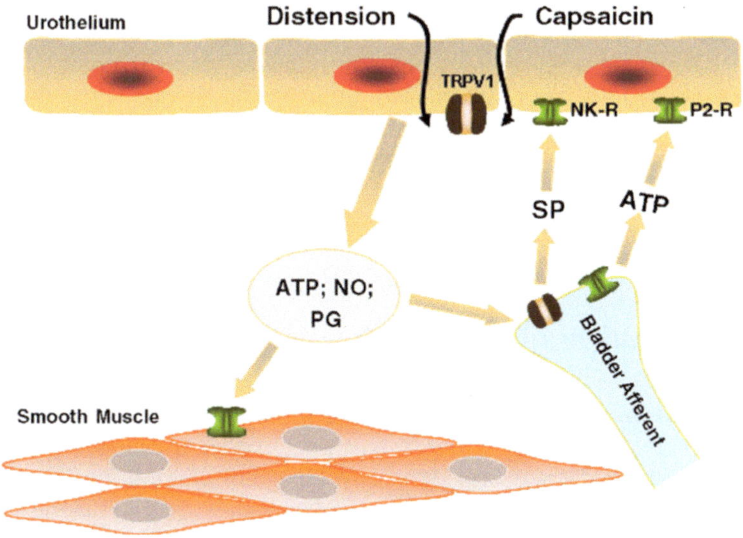

Fig. 4.11 The schematic diagram depicts the possible interactions between the bladder afferent nerve, urinary epithelial cells, and the bladder smooth muscle. (Reproduced with permission from [275]. Copyright 2006, Elsevier.)

incontinence or kidney failure [270]. Traditionally, expanding bladder capacity and reducing bladder pressure are the standard surgical method to treat diseased bladder [271]. However, this technique can result in related complications. Tissue engineering of urinary bladder has been developed as an attractive alternative to autologous enhancement in the treatment of bladder abnormalities [272, 273]. The bladder is an inherently multilayered mechanical organ, which the cell converts mechanical stimulation into chemical activity at the cellular level, mechanical stimulation strongly influences the development of new tissue [274–277] (Fig. 4.11). The importance of replicating the target tissue or organ from a functional standpoint means that more attention to fulfill biomechanical requirement [278], especially if the tissue or organ such as bladder is subjected to frequent mechanical loading and unloading. Limited vascular supply after bladder implantation remains a major challenge, thereby special attention should be paid to material selection [279, 280]. In addition, the ability of diseased bladder cells to colonize scaffolds successfully is often impaired, and their safety in transplantation may be limited by certain disease states, such as bladder cancer or spina bifida [265, 281]. In summary, the complex functions of the bladder create unique challenges for bioengineering tissue. Tissue bladder engineering requires ideally designed bladder scaffolds, especially the selection of suitable biocompatible materials to withstand the mechanical force required to fill and empty the bladder [276]. Polymeric biomaterials used to make bladder tissue must meet certain requirements to provide structural support for several different cell layers and prevent the hollow organ from prematurely collapsing. The ideal biomaterial luminal side should provide a suitable surface for the attachment of epithelial cells of the urethra as a barrier to prevent urine from entering surrounding tissues and allow rapid innervation and blood vessel formation [282]. Designed to avoid the host's harmful immune response, which can affect the function and durability of the allograft, to support the transplanted bladder, it is necessary to design a bladder to support transplantation and avoid harmful host immune responses that may affect the function and durability of allografts [283, 284]. Therefore, these complex requirements still require comprehensive research to achieve suitable bladder materials for clinical treatment.

4.4.3 Applications of Decellularized Materials in Bladder

Tissue scaffolds from natural acellular have been shown to be a valuable election for regeneration of tissue and whole organ, providing exceeding functional results than other synthetic biomaterials tested [285, 286]. Generally, decellularized scaffolds are derived from pig organs because the anatomical and biological similarities between pig and human bladders make porcine bladder cells a competent model for tissue engineering strategies [81, 287]. Natural acellular matrix used to repair bladder tissue is extracted from bladder [208] or small intestinal mucosal [265] tissues by mechanical or chemical acellular processes. Since the composition and structure of ECM are unique to individual tissues, in-situ derived tissues may have advantages. Therefore, BAM is thought to contain more biological clues about the bladder than SIS [208, 288]. These preserved bioactive factors regulate cell adhesion, proliferation, migration, and differentiation, and they are thought to have great potential for tissue repair.

Liu et al. found that BAM porous scaffolds derived from the bladder submucosa (BSM) (Fig. 4.12) can be used for bladder regeneration, recellularized by human bladder cells and subsequently by adjusting porosity of the scaffolds to facilitate the penetration of cellular matrix and promote cell growth in vivo [289].

In bladder reconstruction, BAM and SIS are either used as grafts alone or seeded with UCs and SMCs in a synergistic manner [276, 290, 291]. After transplantation into different animal models, regeneration of the urothelium and muscle layers of the urethra was observed, with some vascularization and innervation [290, 292]. However, the successful use of acellular matrix bladder regeneration seems to depend on the rate of graft revascularization and the degree of primary bladder injury. In small animals such as rats, although both seed and non-seed matrices have been successfully used [293] for bladder repair, similar implants have failed in larger models like pigs and dogs. Graft contractions due to insufficient vascularization, incomplete or disordered smooth muscle development, and rapid growth of fibroblasts are considered to be the causes of graft failure [291, 294]. Therefore, the success of bladder reconstruction project requires the pre-implantation of scaffolds to rapid establish vascular growth, support cell survival and bladder wall regeneration.

Colon is commonly considered as an alternative material for bladder augmentation due to the large lumen and remarkable mechanical properties. However, rich

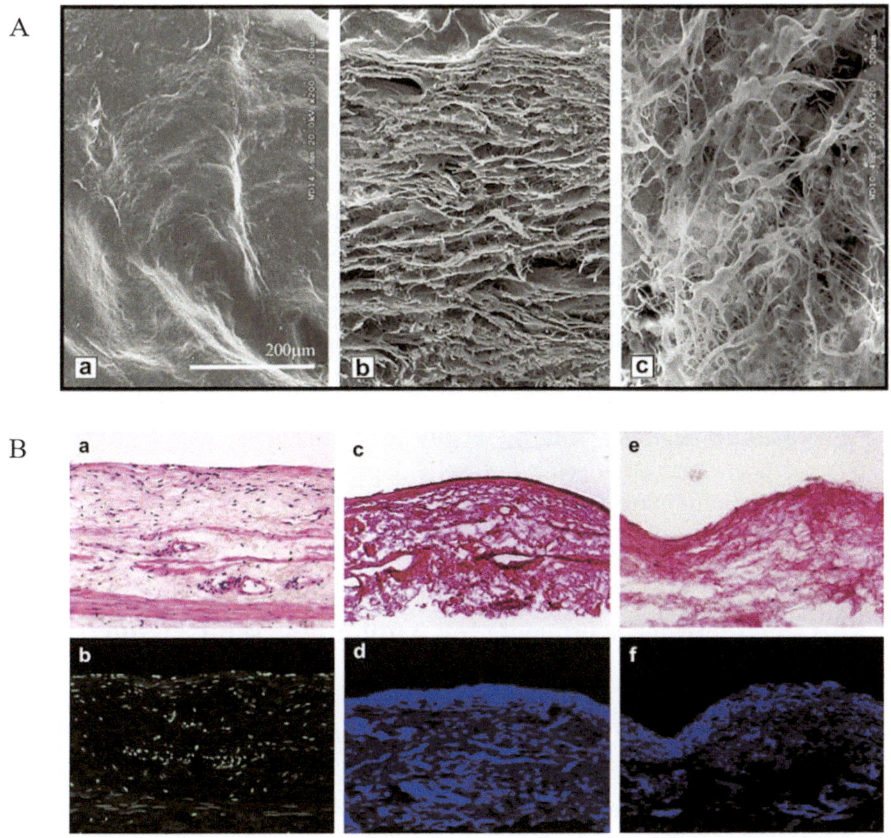

Fig. 4.12 The picture shows the microstructure of fresh and decellularized BSM through scanning electron microscopy and histological examination, including pores and residual cellular material. (**a**): Scanning electron micrographs (SEM) of fresh bladder submucosa (BSM) at 1000× magnification: surface (a) and cross section (b, c). The scale bar indicates 200 μm. (**b**) Cells and nuclei were visible in fresh BSM tissue (A, B), decreased after decellularization alone (C, D), and completely disappeared after desellation and 5% PAA treatment (E, F).The microstructure of BSM explains the increase of porosity after PAA treatment (E, F). (Reproduced with permission from [289]. Copyright 2009, Elsevier.)

bacterial flora, more mucus production, and the risk of long-term malignancy make it an imperfect bladder alternative [295]. When acellular, the above shortcomings are no longer concerned and also maintain appropriate mechanical properties [296], which could theoretically be a promising scaffold for bladder expansion. A recent study suggests that acellular colon tissue provides a new non-immunogenic scaffold that effectively retains its biomechanical properties and may be a suitable candidate for experimental model transplantation [296]. The histological and immunological results of this study verified that full-thickness bladder regeneration occurs in the acellular matrix in a rat model, suggesting that decellularized colon may provide a viable material for bladder augmentation in rats, providing a new idea for future applications of this natural collagen scaffold.

In summary, the integrity of ECM (including structural proteins) in the acellular colon is an important factor in the successful application of these scaffolds as grafts. Therefore, the preparation of small intestinal submucosa (SIS) and acellular matrix bladder grafts (BAMG) also limited the application of these heterogeneous materials [297, 298]. In addition, non-gastrointestinal alternatives like omentum, fascia, placenta, and skin can also applied for bladder repair, but their applicability is not ideal due to complications and limited donor availability [295, 299]. Thus, finding a better bladder repair material with satisfactory long-term results remains a challenge. The tissue engineering of the bladder structure into the human body and fully functioning physiologically is a complex process, and many limitations need to be addressed before this treatment becomes a clinical reality. We believe that in the near future, tissue-engineered scaffolds with the ability to control topography and deliver a variety of nerve and angiogenic factors will provide a potential option for the construction of physiologically functional bladders.

4.5 Decellularized Materials for Abdominal Wall Tissue Engineering

4.5.1 Structure and Characteristics of Abdominal Wall

The abdominal wall is composed of muscle and fascia and other soft tissues except for the posterior spine as a scaffold. It is divided into anterior abdominal wall and posterior abdominal wall by the extension of the posterior axillary line on both sides [300]. The anterior abdominal wall not only protects the internal organs, but also maintains the stability of the position of the internal organs and maintains the intra-abdominal pressure. The posterior function is central to voluntary control of coughing, micturition, and defecation [301]. Abdominal wall defect is one of the common diseases in surgery, which mostly occurs on the anterior lateral wall of the abdomen. The main pathogenic factors are wound infection, trauma, abdominal wall tumor resection, congenital malformation, etc. Incisional hernia and abdominal wall tumor resection after abdominal surgery are the most common causes of abdominal wall defects. The goals of abdominal wall reconstruction are: (1) reconstruction of abdominal wall muscle fascia and soft tissue structural integrity; (2) prevention of the increase of intra-abdominal pressure; (3) restoration of the complete function of the abdominal wall. That is: the reconstructed abdominal wall must not only provide mechanical support with sufficient strength to protect the contents of the abdominal cavity, but also make the abdominal wall covered with myofascial tissue with blood vessels and nerves [302].

4.5.2 Requirements for Abdominal Wall Tissue Engineering Materials

Reconstruction of the abdominal wall is important for large resections from surgery. Abdominal wall defects can be caused by many factors such as trauma, previous surgery, infection, and tumor resection [303, 304]. The defects of abdominal wall lead to the loss of abdominal wall structure, loss of domains, and infection. The purpose of abdominal wall reconstruction is to protect abdominal contents and restore functional support [305]. Complex abdominal wall reconstruction is a demanding procedure, which requires the correct choice of prosthesis. The ideal surgical mesh material for abdominal wall hernia repair should have a certain degree of flexibility, provide mechanical support, and allow tissue to grow inward while degrading, avoiding adhesion, pockets or erosion of nearby tissues [306–308]. It was first reported in 1958 that the use of mesh materials for hernia repair was preliminary, and then many other mesh products with different properties were developed [309]. Synthetic materials have been widely used for abdominal wall reconstruction and can reduce hernia and repair the abdominal wall. However, these materials can cause serious complications, including infections at the surgical site, delayed wound healing, skin rupture, seroma collection, and fistula formation [310–312]. Limitations of synthetic biomaterials for abdominal wall repair have led investigators to seek naturally derived matrices. Currently, biological scaffold materials from human cadavers or other animal sources decellularized tissue utilized for surgical treatment of abdominal wall defects have become increasingly, such as human acellular dermal matrix (HADM), small intestine submucosa (SIS), and so on [234, 313–316]. These biomaterials have been shown to facilitate the constructive remodeling of many different tissues in both pre-clinical animal studies and in human clinical applications. These materials mainly consist of a complex mixture of molecules that mediate structural and or biological properties. It has been reported that these materials also could effectively promoted cellular infiltration, new extracellular matrix deposition, and neovascularization [317]. In addition, these materials could become indistinguishable from native tissues. These characteristics might provide distinct advantages over synthetic meshes, especially in overcoming the postoperative complications, such as stiffness of abdominal wall, adhesions, chronic inflammation, foreign body reaction, and so on.

4.5.3 Applications of Decellularized Materials in Abdominal Wall

In 2003, acellular dermal matrix (ADM) was proposed as a substitute for abdominal wall reconstruction, which showed that it could enhance the fusion with the surrounding tissues, and has less tendency to infection, erosion, extrusion, adhesion formation, and rejection [318–320]. Human-derived acellular dermis (HADM) is a graft successfully used for abdominal wall repair. The treatment method of HADM will not destroy or change the structural structure of the dermal matrix, and this process makes the material easy to be accepted by the human body without immune or obvious foreign body reaction, thus easily supporting cell reproduction and revascularization. Animal studies have confirmed that HADM has the ability to rapidly rebuild blood vessels in the abdominal wall, and also has the ability to resist visceral adhesions [320]. Study has proved that human-derived acellular dermis (HADM) has better repair ability than pig-derived acellular dermis (PADM). Ngo et al. established an open hernia in the rabbit model, and the comparison of the two ADM at week 4 and week 20 showed that the fibroblast infiltration in the HADM group was significantly greater than the PADM group, and the HADM tissue was inward. Growth is significantly faster (Fig. 4.13) [321]. In a study conducted by Eberli et al., HADM was used in a rabbit abdominal hernia repair model. A quarter of the abdominal wall was replaced with each biological material. After 24 weeks, the overall and histological evaluation of the samples taken from these implants was performed. The results showed that HADM maintained their strength in the rabbit hernia repair model, and no hernia formation or intestinal adhesions occurred. Macro and micro evaluations show excellent integration and organization formation. It is further indicated that the HADM is suitable for abdominal hernia repair [317]. In addition, it has been reported that even when the placed HADM directly on the internal organs and when exposed to and/or contaminated with bacteria in the surgical area, large, complex abdominal wall defects can be successfully reconstructed [316, 322]. However, due to the degradation and remodeling caused by collagenase and matrix metalloproteinase (MMP), HADM grafts have poor long-term durability and lose mechanical strength and integrity after implantation. The premature failure of HADM grafts and insufficient deposition of functional repair tissue can lead to swelling, cracking, and regeneration. Chemical crosslinking has been used to increase durability of acellular matrix. However, such crosslinking alters the ECM structure and may inhibit cellular infiltration, revascularization, and matrix remodeling, adversely affecting graft incorporation with host tissues and limiting its application [323]. Hence, at present, it is a concern to fabricate a biological graft material with sufficient biological properties to resist the risk of early infections and durable mechanical properties. Sahoo et al. produced fiber-reinforced HADM grafts. The experiments evaluated the fatigue biomechanics of fiber-reinforced HADM grafts after 8-hour enzymatic degradation in vitro. Results showed that r-HADM still retains key sub-failure parameters and cyclic expansion strain, no sample will fail prematurely. These data indicate that fiber reinforcement improves biomechanical properties and imparts mechanical durability to fiber-reinforced HADM during enzymatic degradation. The results of the study indicate that fiber reinforcement may be a strategy to reduce the loss of mechanical properties of HADM grafts after implantation in vivo [324].

However, some researchers indicated that the dermis does not have the ability to promote muscle regeneration needed to restore abdominal wall function. In contrast, small intestinal submucosa (SIS) grafts have been shown to promote muscle generation in a volumetric muscle loss model. Moreover, some reports valuated that

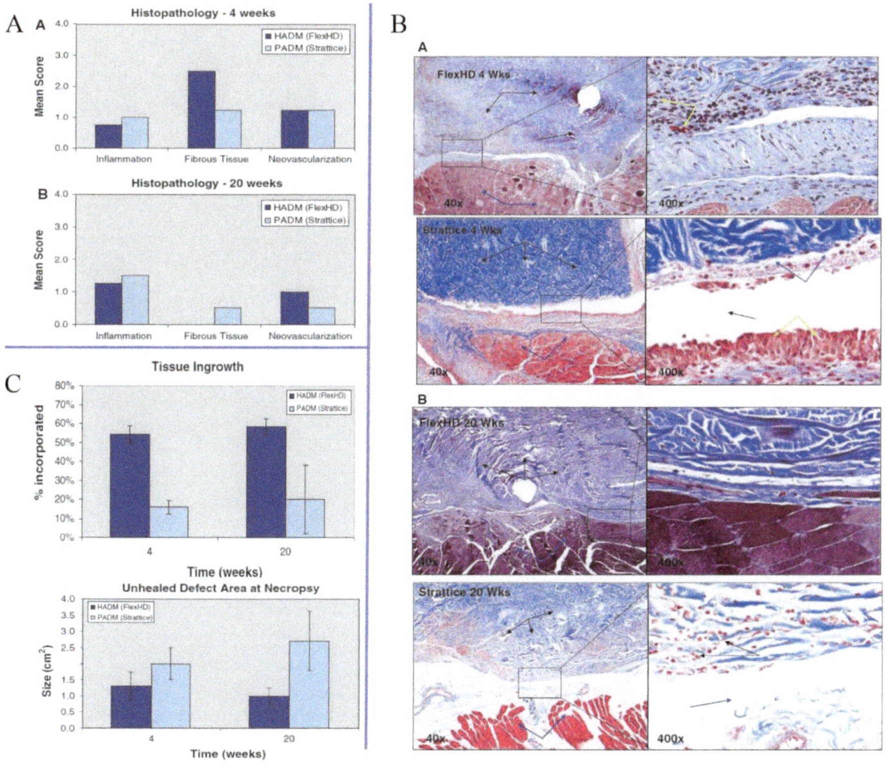

Fig. 4.13 An image shows the comparison of the effectiveness of HADM and PADM in the repair of incisional abdominal wall hernias in a rabbit model. (**a**) Implants histopathological evaluation. Created an incisional hernia defect in rabbit abdomen, and repair them with HADM or PADM after 3 weeks following the primary surgery. (**b**) Representative photomicrographs. (A) After 4 weeks post-implantation, photomicrograph of FlexHD and Strattice hernia repair implants. Photos are from the interface area between the implant (black arrows) and the host skeletal muscle (blue arrows). (B) At 20 weeks post-implantation, photomicrographs of FlexHD and Strattice hernia repair implants. Photos are from the interface area between the implant (black arrows) and the host skeletal muscle (blue arrows). (Reproduced by permission from [321], Copyright 2012, Springer.)

compared with ADM grafts, SIS is absorbable, induces less inflammation, and is more biocompatible, and therefore might be more useful and suitable for closure of abdominal wall defects [325, 326]. Liu et al. compared the repair effects of ADM and SIS in the rat model of full-thickness abdominal wall defects. They examined the thickness intensity of hernia, infection, adhesion, contraction, and new tissue at defect sites. It was found that ADM induces less tissue adhesion and slow thickness increase, while SIS undergoes better host integration [327] (Fig. 4.14), moreover, SIS has continuous dynamic remodeling of after implantation of damaged tissue [328]. Research by Jiang et al. showed that SIS has good biocompatibility and causes less inflammation, which indicates that SIS may be suitable for closing abdominal wall defects [326].

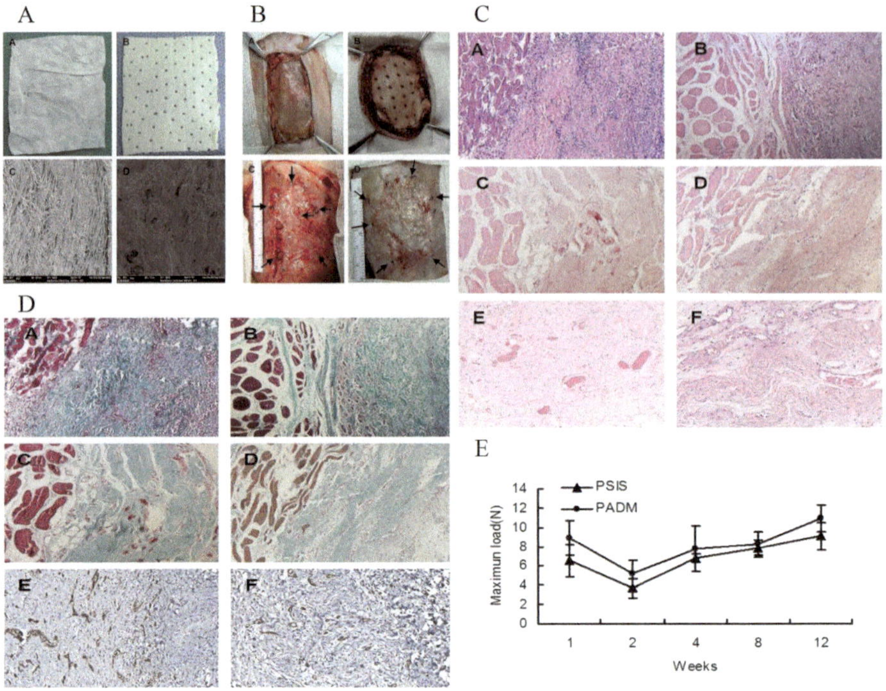

Fig. 4.14 This image illuminated structures, macroscopic observation, histological observations, and mechanical properties of PSIS and PADM scaffold at 12 weeks after implantation. The implants contraction around the sutures (arrows) repairing the abdominal wall defects at 12 weeks after implantation. (**a**) (A) PSIS scaffold. (B) PADM scaffold. (C) SEM photographs of PSIS, ×500 and D) PADM, ×500. (**b**) (A) PSIS implant. (B) PADM implant. (C) Contracted PSIS implant. (D) Contracted PADM implant. (**c**) Histologic observation of the PSIS and PADM explants (H & E staining). (A) PSIS at Week 1 (×100). (B) PADM at Week 1 (×100). (C) PSIS at Week 12 (×100). (D) PADM at Week 12 (×100). (E) PSIS at Week 12 (×200). (F) PADM at Week 12 (×200). (**d**) Histologic observation of the explants of PSIS and PADM (Masson trichrome staining and Immunostaining). A–D: Masson trichrome staining; E–F: Immunostaining. (A) PSIS at Week 1 (×100). (B) PADM at Week 1 (×100). (C) PSIS at Week 12 (×100). (D) PADM at Week12 (×100). (E) PSIS at Week 12 (×200). (F) PADM at Week 12 (×200). (**e**) The bond strength of PSIS and PADM implants during the experiment (Reproduced by permission from [327], Copyright 2011, Elsevier.)

However, related research showed that after implantation, the degradation rate of the biological material is too fast to match the neo-tissue regeneration. In order to further improve the effect of abdominal wall repair, many methods to modify SIS have been developed [329]. Wang et al. fabricated SIS + PPM patches for repairing lateral ventral wall defects in adult dogs, they demonstrated that composite patch could be successfully applied to repair abdominal defect [330]. Dreger et al. reported that SIS-ECM coated with L-valine-co-L-phenylalanine poly (ester urea) (PEU) copolymer film can enhance the mechanical properties of SIS-ECM burst test without increasing the hardness, this indicates that the composite membrane may be a potential candidate for hernia repair.

In addition, many studies have proved that the combination of SIS and growth factors or stem cells is more conducive to tissue repair. Liu et al constructed SIS-DNPs/VEGF-PLGA-SIS (SVDPS) composite scaffold, the in vivo study showed that the SVDPS composite scaffold promoted higher early therapeutic neovascularization within 2 weeks post-surgery by continuously releasing biologically active VEGF [331]. In another study, bFGF-coated gelatin hydrogel-coated SIS was used to construct a new biomaterial to repair abdominal wall defects, indicating that the biomaterial can promote remodeling and regenerate host tissue to reconstruct abdominal wall defects [332]. In a recent study, Klinger and colleagues developed a SIS graft implanted by ASCs, the study indicates that the scaffold promotes cell infiltration and new blood vessel formation, and integration with the abdominal wall tissue, which demonstrates the feasibility of using SIS grafts implanted by ASCs as a valuable tool for abdominal wall reconstruction [229]. In another study, SIS scaffolds modified with silver nanoparticles manufactured by self-assembly showed significantly enhanced antibacterial properties, excellent biomechanical properties, and biosafety in repairing contaminated abdominal wall defects [333].

4.6 Decellularized Materials for Cardiovascular Tissue Engineering

4.6.1 Structure and Characteristics of Cardiovascular

Cardiovascular tissue is the human body's "circulatory system," which is a closed circulation channel composed of heart, arteries, capillaries, and veins. The heart is the dynamic organ of the vascular system. It is in the chest cavity and between the two lungs. It has a structure that conducts excitement and regulates the beat of the rhythm. Blood vessels are the channels through which blood flows. In the human body, they are divided into arteries, veins and capillaries. The blood vessels that transport blood originates from the heart to the various organs of the body are called arteries, and the arteries are branched gradually. The capillaries converge into a vein and return to the heart. Blood flows through it and supplies oxygen, various nutrients, hormones, etc. to organs and tissues, and then transports the wastes metabolized by the tissues to the excretory organs to maintain the steady state of the body's environment, the progress of metabolism and maintain normal life activities [334, 335].

4.6.2 Requirements for Cardiovascular Tissue Engineering Materials

Cardiovascular disease (CVD) represents one of the leading causes of death worldwide [336]. Researchers are paying special attention to the new treatment methods provided by cardiovascular disease, regenerative medicine. As a promising regeneration strategy, cell therapies using a variety of stem cells (i.e., cardiac stem/progenitor cells, stem cells derived from bone marrow and adipose tissue, induced pluripotent stem cells) have been extensively studied [337]. However, some significant limitations such as a low transplantation rate and a low survival rate of stem/progenitor cells after transplantation restrict their clinical application [338]. To improve the regeneration potential of stem cells, combining them with appropriate scaffolds can represent an improved method of treatment. Recently, as scaffolds for heart tissue repair, many biomaterials, from synthetic polymers [339, 340] to natural matrices [341], have been studied. Among them, decellularized pericardium and myocardium have aroused incremental interest in application of cardiac tissue engineering.

4.6.3 Applications of Decellularized Materials in Cardiovascular

In the heart, ECM is responsible for various pathophysiological reactions including fibrosis, inflammation, angiogenesis, etc. [342]. As people gradually realize the meaning of using tissue-specific matrix to produce biological heart scaffolds, the decellularized tissue of rat or porcine-derived myocardial tissue has been widely used [343]. In the past 10 years, great efforts have been made to obtain decellularized human cardiac muscle with high regeneration capacity. Improving wall stress by applying a myocardial patch on the epicardial surface or injecting it directly into the myocardium aims to replace gangrenous myocardial cells and/or repair stained extracellular matrix [344]. Myocardial ECM patch can promote the survival and implantation of cell components as a cell delivery platform, and can guide their differentiation to replace necrotic tissue [345].

Acellular myocardial sheets retained the features of a native tissue. After decellularization processing, the cardiac ECM sheets can be processed into a microparticle powder and then self-assemble into cardiac ECM hydrogel with saved bioactivity. These ECM forms have the prospective of in vitro and in vivo purpose for cardiac regeneration [342]. The characteristics of cardiac ECM support cells have been demonstrated by supporting the growth of mouse HL-1 cardiomyocytes. In a study, ECM particles are suspended on culture medium or embedded in hydrogel to cover the surface of the culture plate, and HL-1 cells are cultured on it. Result showed that HL-1 cells had strong metabolic activity and proliferation ability [346]. Furthermore, it has been proved that the cardiac ECM can also support the multiplication of human myocardial fibroblasts and epicardial progenitor cells, and has certain immunomodulatory effects on human immune cells [347].

Johnson and colleagues also made an injectable form of human myocardial matrix and the study highlighted the significant differences between patients in the study [348]. In addition, Wang and colleagues evaluated the immune response to the injection of human-derived myocardial ECM hydrogels in subcutaneous tissues of mice. In vivo studies have shown that this animal model is of great importance in studying the biocompatibility of biomaterials and promoting the quality of remodeling before clinical application. To be of clinical value, the cell-free myocardial matrix must display biocompatibility and non-immunogenic characteristics [349]. Based on this research, Godier-Furnémont and colleagues created a new type of composite scaffold that embedded human mesenchymal stem cells in a fibrin hydrogel and then loaded it onto decellularized myocardium. Fibrin allows cells to retain and emit local signals, while acellular myocardium provides structural and mechanical support [335]. Another reformatory composite myocardial-based scaffold was prepared by Guhathakurta et al. [350], who fabricated the electrospun poly-(L)lactic acid/polycaprolactone/collagen nanofibers coated with acellular cardiac ECM and subsequently injected cord blood mononuclear cells, the engineered myocardial patches displayed troponin I and cardiac myosin expression in vitro culture. By implanting mesenchymal stem cells from the bone marrow, filling of the acellular myocardial scaffolds has been successfully achieved. By implanting bone marrow mesenchymal stem cells, human endothelial cells (HUVEC), human cardiac progenitor cells, and cardiomyocytes into the myocardial matrix fused into the fibers over time, the cell-free myocardial scaffold has been successfully repopulation with newborn myocardium [344], myocardial fiber differentiation, progenitor cells of cardiac and smooth muscle cell lines were also obtained [351], and cardiomyocytes derived from induced pluripotent stem cells produced contracting slices spontaneously (Fig. 4.15) [352]. Overall, the study of myocardial ECM in recellularization shows that the myocardial ECM is beneficial to cell adhesion, proliferation, and differentiation, which may be related to the retained three-dimensional structure and biological information with it. In addition, it has been proved by subcutaneous implantation experiments in rats that myocardium fragments can promote $CD163^{+}$ macrophage response, which is same with the reconstruction and remodeling of implants. A larger model of myocardial infarction was reported by Guhathakurta et al., they implanted a myocardial scaffolding covered with nanofibers and umbilical cord blood mononuclear cells in sheep with ischemic myocardium, and found that the patch was completely integrated into the host myocardial tissue.

Recently, acellular pericardium has been used as an effective scaffold for bioprosthetic cardiac valves and patches in cardiac surgery. Human-derived pericardium can better simulate the natural extracellular matrix, stimulate the cell response, and integrate with the host. Based on this direction, many authors are devoted to the research on the application of human-derived acellular pericardium to cardiovascular tissue engineering. It has been reported that human-derived pericardium can be obtained from cadavers or from patients undergoing cardiothoracic surgery. The methods for decellularization of the pericardium mainly include detergent-enzymatic

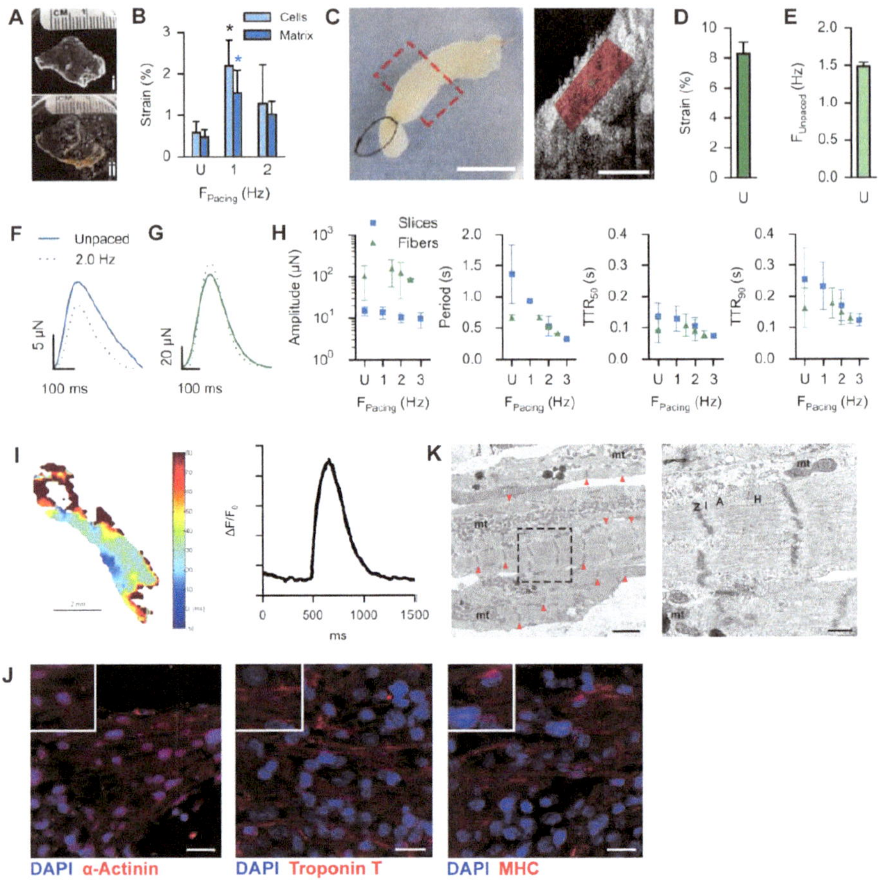

Fig. 4.15 Generation of functional tissue structure based on decellularized human myocardial slices and fibers and measurement of strain analysis, force curves in spontaneously contracting cardiac tissues and action potentials at multiple points. (**a**) Cardiac matrix slices: decellularized human cardiac matrix slice (200 μm) in a culture well (1), with induced pluripotent stem cell (iPS)–derived cardiomyocytes seeded (2) for 120-d culture. Ruler ticks are 1 mm. (**b**) The regional strain of iPS-derived cardiomyocytes and the underlying cardiac matrix were compared to spontaneous contraction (U; $n = 5$ beating regions) with electrical stimulation at 1 Hz ($n = 4$) and 2 Hz ($n = 4$)). Error bars, SD. Black and blue asterisks (*) indicate $P < 0.05$ compared with unstimulated cells and matrix, respectively. (**c**) Left, Macrophotography of seeded heart fibers. Scale bar, 5 mm. (**d**) HDM calculates the surface strain under non-spaced conditions (n = 4 fibers). Error bars, SD. (**e**, **f**) represents the force curve of a single area of contraction in a myocardial slice. (**g**) The force curve representing a single reimplanted cardiac fiber shows the force modulation under non-interval (solid line) or pacing (dotted line) conditions. (**h**) The average magnitude of the force curve captured from the cross section, the period, the time to 50% relaxation (TTR50) and the time to 90% relaxation (TTR90) (gray squares; $n = 9, 9, 9$ and 3, no interval, 1 2 and 3 Hz) and optical fibers (red triangles; $n = 4, 3, 4$ and 2 without spacing of 1.5, 2 and 2.5 Hz, respectively) and optical fibers under stimulation conditions (U). Error bars, SD. (**i**) Conductivity diagram of the seeded heart fibers (left) and example of action potential (right). (**j**) Immunofluorescence staining was performed on the reimplanted cardiac fibers to detect cardiac markers α-actin, troponin T and myosin heavy chain (MHC). Scale bars, 25 μm. (**k**) Recellularized cardiac fibers transmission electron micrograph. (Reproduced by permission from [352], Copyright 2016, Elsevier.)

treatment, SDS-based decellularization method, and traditional chemical reagent methods [353, 354]. The results show that all these decellularization methods can completely remove cell surface antigens and will not affect the ECM structure. Importantly, it is retaining the original mechanical properties, which is helpful for the pericardium to play a certain role under high pressure when implanted in the heart or large vessels [355]. Finally, expose and extraction cytotoxicity tests were conducted to evaluate the effect of any remaining decellularizing agent on viability and proliferation of cells. In contact with the decellularized scaffold extract, human fibroblasts and human epithelial cells were cultured successfully, which indicates that the chemical/biological decellularizing agent has been reduced to non-toxic levels [356]. Interestingly, cell-free pericardium was also prepared in the form of injection matrix or microporous gel by freeze-drying and pepsin digestion. The manufacture of 3D microporous scaffolds derived from pericardium can obtain better cellular reactions in the cardiac progenitor cells proliferation, viability, migration, and differentiation [357]. In addition, the decellularized human pericardium was loaded with progenitor cells derived from porcine mediastinal adipose tissue and self-assembled RAD16-1 peptide. This can generate the ideal 3D network inside the scaffold, thereby promoting the proliferation and differentiation of endothelial cell lineage [358]. For in vivo experiments, subcellular decellularized pericardium implantation in mouse models showed decreased cytotoxicity, reduced immunogenicity, and improved graft integration compared to fresh/frozen or glutaraldehyde-fixed implanted tissue and lack of calcification [359]. A study of implanting scaffolds loaded with cardiac progenitor cells into subcutaneous tissues of mice was also conducted to evaluate the angiogenesis and differentiation characteristics of pericardial substitutes in vivo, results showed that the transplanted constructs have a weaker immune response and improved angiogenesis and differentiation of cardiomyocytes. In addition, they injected a thin acellular pericardium into the left ventricular wall of healthy rats and found that it formed a fibrous and porous scaffold in the body. This proof suggests that pericardial matrix gel can be used as a structural and biochemical scaffold to support infiltration and revascularization of destroyed myocardial tissue to control remodeling after myocardial infarction. In addition, chemically decellularized pericardial membranes are implanted in the abdominal aorta of rats or carotid/aortic [353, 360] in the form of vascular membranes, and are tested to regenerate the arterial wall in vivo. These intravascular studies have highlighted the graft's super integration ability and low immunogenicity. Finally, by implanting an innovative ECM/hydrogel RAD16-I composite pericardial bioprosthesis loaded with porcine adipose stem cells in a porcine myocardial infarction model in the ischemic heart region, the prosthesis was found to have a higher biological phase. Compatibility and mobility have also been observed in damaged myocardial cell bioprostheses. In addition, the scaffold comprehensive with the basic myocardium showed the ability of re-colonization, new blood vessel formation, and nerve germination. Overall, decellularized pericardial constructs can significantly reduce infarct size and improve heart function, reduce inflammation, and alter collagen deposition [361].

In cardiology, end-stage valvular heart disease is an increasingly common disease that can only be treated by artificial repair [362]. Currently, there are several commercial heart valves, but their clinical use has serious drawbacks. First, mechanical grafts have thrombosis and require lifelong anticoagulation. In addition, although heterogeneous or allogeneic bioprostheses can ensure excellent hemodynamic behavior, but durability will be reduced owing to tissue degradation [363]. In both cases, valve replacement does not permit somatic cells to grow or reshape after implantation. The ideal valve replacement should have immunogenic and thrombotic properties, as well as the potential for remodeling, regeneration, and growth. Recently, aortic heart valves and lung heart valves have been decellularized to produce allogeneic human heart valves, which have attracted widespread attention [364, 365].

For heart valves, structures are required to ensure enough mechanical functions immediately after implantation to ensure patient survival. Therefore, for successful clinical use, heart valve replacements must mimic and maintain the functional and mechanical properties of natural tissues. On this basis, mechanical measurements were performed on the decellularized aortic valve and the pulmonary valve, which showed that the decellularization method has no significant effect on tissue resistance or biomechanics [366, 367]. Based on the extract and expose cytotoxicity test, the acellular heart valve is not toxic when contacted with 3T3 mouse fibroblasts and baby hamster kidney cells [368]. In vitro migration experiments showed that the decellularization process decreased the human monocytes migration to heart valve tissue, implied a strong reduction in humoral response in implantation in vivo [369].

Though the rapid development of cardiac tissue engineering investigation, in vitro refilling of decellularized valve catheters is still lacking. Under dynamic flow conditions, human endothelial cells derived from the umbilical cord vein or great saphenous vein were cultured on the luminal surface of the aorta and lung acellular valve, showing the confluence of living cells expressing specific endothelial lineage marker. Once the cell-free aortic valve roof and wall are implanted, human bone marrow mesenchymal stem cells can maintain the viability of the culture, thus confirming the non-cytotoxic effect of decellularization [370]. Most importantly, when human bone marrow mesenchymal stem cells are used to repopulate decellularized pulmonary valve leaflets, they rebuild the endothelial lining and differentiate into myofibroblasts, fibroblasts, and smooth muscle cells. In addition, the valve layers (fibrous and ventricular) both affect the reproliferative potential of human bone marrow MSC differently, and the 3D diffusion and differentiation on the surface of the ventricle are higher [371]. Finally, heart mesenchymal stromal cells were cultured on the cell-free aortic valve matrix, adhered and proliferated to the stent, and expressed the mature endothelial marker von Willebrand factor. In addition, the nurtured expression of matrix metalloproteinase 2 demonstrates the capabilities of these cells to transfer from the surface of the sheet structure to the inner layer, where a reduced efficient population is indeed observed [372]. Using an acellular aortic valve matrix, Koening and his colleagues framed and fabricated a new type of decellularized and replanted bioprosthesis made up of biological and synthetic components. In this creative structure, the allogeneic aortic valve provides the tip and the polyurethane patch forms the wall. By dynamically implanting fibroblasts and endothelial cells derived from the human saphenous vein and re-implanting the biological hybrid aortic valve, a confluent and uniform layer of cells is obtained over the entire stent surface[373].

4.7 Others

4.7.1 Trachea Tissue Engineering

4.7.1.1 Structure and Characteristics of Trachea Tissue

The trachea is made of cartilage, muscle, connective tissue, and mucous. In trachea, the collagen fibers are not clear orientated: in the thin layer close to the lumen, collagen fibers are oriented either circumferentially (along the cartilage length) or longitudinally (along the axis of the airway), while in the deeper layer, they are oriented less regularly with a circumferential tendency. Elastic fibers are abundant and elaborated organized: there are discrete longitudinal elastic bundles in the submucosa of the tracheal membranous wall and diffuse longitudinal elastic fibers in the non-membranous parts, deep to the cartilage rings [374].

4.7.1.2 Requirements of Trachea Tissue Engineering

The reconstruction of the trachea has encouraged more than a century of research. To date, using prosthetic materials, autologous and alloprosthetic replacements have failed to provide functional solutions for the treatment of oblong tracheal defects. The replacement of solid or non-porous prosthesis cannot be absorbed by local tissues, and has problems of infection, displacement, migration, compression, and stenosis, and it has not been widely used. Various regenerative medicine methods for airway replacement have been proposed, ranging from collagen scaffolds supported by silicone scaffolds to cartilage tubes made of silicone in vitro culture methods [375, 376]. Similarly, due to incomplete epithelialization accompanied by stenosis or lack of mechanical integrity leading to tracheal softening, these strategies are not sufficient to replace the trachea. Tracheal replacements must be rigid in the lateral direction, but flexible in the longitudinal direction, and must have a surface lining containing ciliated respiratory epithelium. In addition, the catheter must first be airtight and must be integrated into adjacent tissues to avoid chronic inflammation, granulation tissue, infection, and erosion. In addition, the materials used for tracheal replacement must be airtight, biocompatible, non-immunogenic, and toxic. Ideally, they should provide or promote resurfacing of the epithelial surface to avoid stricture or late flexion. They must be permanently constructed. Finally, for a method to be practically considered, the technique of construction or insertion of the conduit must be surgically straightforward and the results predictably successful.

4.7.1.3 Applications of Decellularized Materials for Trachea Tissue Engineering

Starting with the success of decellularized tissues and organ-derived bioscaffolds, people began to pay attention to the fact that decellularized tracheal matrix can be used to achieve functional tracheal replacement. To date, different acellular matrix and decellularization protocols have been studied to obtain suitable acellular tracheal grafts. DM is a simple and effective method of bioengineering tracheal matrix, characterized by retaining its natural structural integrity, lack of immunogenicity [377]. We obtained a decellularized pig jejunum segment, in which autologous costal chondrocytes, smooth muscle cells, airway epithelial cells, and endothelial progenitor cells were inoculated and a cellular tracheal functional element formation [378]. In addition, the porcine tracheal matrix decellularized by the same method can support the in vitro adhesion of chondrocytes and tracheal epithelial cells in the ear [379]. Jungebluth et al. investigated the regeneration of decellularized pig trachea in vivo. The decellularized porcine tracheal scaffold is regulated with monocytes and growth and differentiation factors to promote differentiation of stem/progenitor cells and the regenerative process Fig. 4.16. After 2 weeks in the postoperative period, the trachea was almost normal, with airway epithelium and double cartilage, no mechanical difference compared with natural tissue [380]. There is no cell matrix under the intestinal mucosa of the small intestine. Regardless of the use of stents, chondral flaps, or allogenic cartilage membranes, moderate success can be achieved on small defects (less than half the circumference) with sufficient structural support, but they cannot be fully recovered. Moreover, the function of trachea tissue will be impaired to a certain extent [381–383].

4.7.2 Gastrointestinal Tissue Engineering

4.7.2.1 Structure and Characteristics of Gastrointestinal

The gastrointestinal tract (GI) is a continuous tubular system that extends from the mouth to the anus. It can be divided into the upper and lower digestive tracts and consists of the following major parts: the mouth, esophagus, stomach, small intestine, and large intestine [384, 385]. They alternately contract and relax, the same process as bowel motility, helps propel and digest food. Normally, the digested material is absorbed through the epithelium into the bloodstream to provide nourishment, while the indigestible and unabsorbed material is excreted as waste through the anus. Initially, the gastrointestinal tract is a simple and uniform embryonic tube, which subsequently differentiates along the anterior and posterior axes to form complex organs in the anterior, midgut, and posterior intestinal regions [386]. The histological layer of the gastrointestinal tract is derived from three types of derm: endoderm, mesoderm, and ectoderm, each of which constitutes its own cell number and ECM component, of which most of the ECM in the gastrointestinal tract come

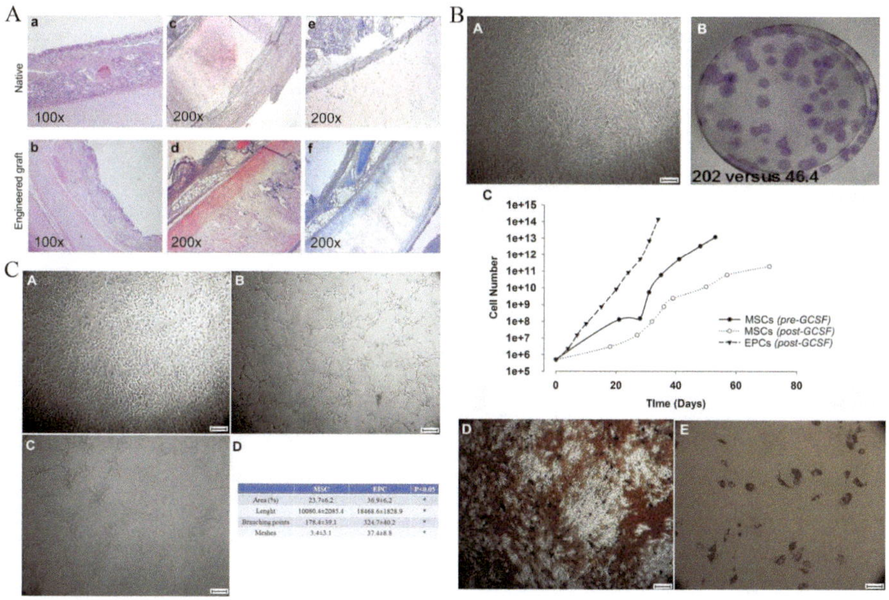

Fig. 4.16 Tracheal tissue regeneration. (**a**) Tracheal tissue. Representative sections of H&E staining (a, b) collagen II stained cartilage (c, d) and cytokeratin stained epithelial cells (e, f) of native pig trachea (a, c, e) and the in vivo tissue-engineered tracheal graft (b, d, f). (**b**) MSC cultures. (A) Microscopy showing attached spindle-shaped cells (B). (C) Proliferation curves of MSCs and EPCs expanded from samples obtained pre- and postoperative. (D) Alizarin Red S staining. (E) Oil red O staining. Scale bar: 100 μm. (**c**) EPC cultures. Microscopy showing (A) EPC-like cells; the arrangement of EPCs (B) and of MSCs (C) after being cultured on Matrigel for 9 h. (D) Quantitative analysis of dimensional and topological parameters of cell capillary-like mesh (*$p < 0.05$ with respect to EPCs). Scale bar: 100 μm. (Reproduced by permission from [380], Copyright 2012, Elsevier.)

from the mesoderm [387, 388]. To better simulate human gastrointestinal tissue, construction strategies need to consider not only the composition of ECM, but also the selection of suitable scaffold structures (e.g., tubular grafts, laminates, or hydrogels). Figure 4.17 shows the various ECM stent configurations and their applications in the gastrointestinal tract.

4.7.2.2 Requirements for Gastrointestinal Tissue Engineering Materials

Gastrointestinal injury may be caused by a variety of factors, such as injury [389, 390], inflammation [391–393], and neoplasia [394–397]. Some diseases often change the dynamics of the entire gastrointestinal tract, while surgical intervention cannot provide long-term solutions [398–403]. Biological scaffolds composed of extracellular matrix are expected to serve as repair materials, providing opportunities for functional remodeling of damaged or diseased gastrointestinal

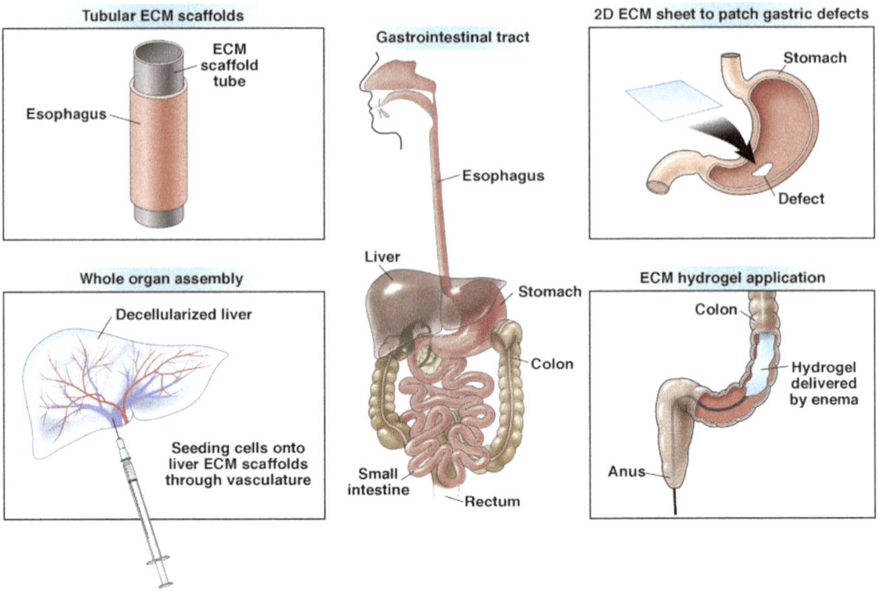

Fig. 4.17 ECM scaffold configurations and their use in the GI tract. (Reproduced by permission from [388], Copyright 2017, Elsevier.)

tissues such as congenital or acquired defects, inflammatory bowel disease, and tumors. Because the gastrointestinal tract is a continuous tubular structure, perforated stents are commonly used for gastrointestinal tissue engineering. Clinical studies are currently under way to test different biomaterials suitable for gastric tissue engineering [404, 405]. The porosity of the material can regulate the smooth arrangement, contraction, and relaxation of muscle cells, which requires special attention. In addition, the scaffold must support neuronal differences, which are a key to regulate peristalsis. Depending on the field of application, biomaterials may include natural polymers, synthetic polymers, or a combination of both [406–409]. Although progress has been made in the manufacture of scaffolds, their mechanical properties and degradation kinetics still limit the steps of their design [410–415], especially the gastrointestinal tract, which requires elastic mechanical regulation. The extracellular matrix, which includes collagen, laminin, and other proteoglycan, effectively regulates a range of cellular activities. Thus, in the design of scaffolds, the addition of ECM components should be fully considered to promote cell–cell and cell–scaffold interaction. Moreover, studies have found that glycosaminoglycan crosslinking with scaffolds can improve cell attachment, migration, proliferation, and differentiation in many applications [416–418]. It is therefore advantageous to use acellular matrix as a scaffold because it preserves extracellular matrix components while eliminating immune responses.

4.7.2.3 Applications of Decellularized Materials in Gastrointestinal

Strategies for the repair and regeneration of gastrointestinal tissue are complex and uniquely challenging for the digestive tract, including extreme pH changes that require a continuous nutrient supply host and require direct peristalsis along the gastrointestinal tract [419]. Extracellular matrix (ECM) is an ideal therapeutic matrix for functional remodeling of impaired gastrointestinal (GI) tissue although additional work is required before ECM bioscaffolds see widespread clinical application, promising results have been reported with the use of ECM scaffold approaches for GI tissue repair and regeneration.

Currently, ECM stents have been reported as promising for gastrointestinal tissue repair and regeneration, such as the use of ECM bioflutter to promote the construction and functional regeneration of gastrointestinal tissue. The use of acellular ECM biological scaffolds has the many advantages. First, the degradation products of acellular ECM can promote the differentiation of stem cells [420–423].Secondly, it can reduce the host inflammatory response, which can not only promote the formation of new blood vessels, but also fully integrate with the surrounding tissues [424–427].Therefore, ECM biological scaffolds provide morphogenesis clues that affect the recruitment, survival, proliferation, migration, polarization and differentiation of recipient cells [428–431], accelerating tissue growth and remodeling.

4.7.3 Esophagus Tissue Engineering

4.7.3.1 Structure and Characteristics of Esophagus

The esophagus extends from the pharynx to the stomach along the muscular canal. It regulates the transport of food to the stomach through a series of coordinated and rhythmic peristaltic waves. The lumen along the esophagus contains fine epithelium, and the muscles are divided into two parts: the upper esophagus consists of striated muscles and the lower esophagus of smooth muscles. In order to ensure normal swallowing and peristalsis, both types of muscle are regulated by endogenous and exogenous neural pathways [432].

4.7.3.2 Requirements for Esophagus Tissue Engineering Materials

Any disease that affects one or more of these ingredients affects the transport of food through the esophagus. Impaired peristaltic function in patients with esophageal atresia and tracheoesophageal fistula can lead to esophageal stenosis and gastro-esophageal reflux disease [432]. Other forms of disease include esophageal cancer, which also has a significant impact on quality of life. Partial defects in the esophagus can have devastating consequences for both children and adults, so different esophagus tissue engineering approaches attempt to reverse life-threatening problems.

Because the esophagus has many functions, bioengineered tissue must mimic the structure and function of the body in order to achieve long-term success. To design a tissue-engineered esophagus, the following factors must be considered. First, the modified esophagus must have a hollow tubular structure similar to the original esophagus. Second, the artificial esophagus must be capable of normal swallowing and peristalsis, and the restoration of the gravity property of the esophagus and the coordinated rhythmic activity should be considered. Third, it is desirable to design the esophagus to contain important mediators necessary for the growth and differentiation of different cell types, such as epithelial cells. Evaluation of the composition of the basement membrane of the esophagus provides a starting point for the design of tissue-engineered scaffolds [433].

4.7.3.3 Applications of Decellularized Materials in Esophagus

Due to the unique advantages of natural materials, its application in esophageal tissue engineering has been studied. For example, the OptiMa is collagen scaffold successfully constructed smooth muscle tissue [434]. Decellularized matrices have also gained widespread attention, the different sources of tissue such as the esophagus, stomach, bladder matrix substrate, and small intestine submucosa have been applied in esophageal reconstruction, in addition, the ovine of the esophagus material was decellularized and resulted in matrices with rough surface of substrate, which can be used in the esophageal tissue engineering [435]. Extracellular matrix was extracted from pig bladder and implanted into dog model to reconstruct esophagus. The implanted biological scaffold could not only fuse with natural muscle tissue, but also form functional esophagus with similar mechanical properties to natural tissue [403]. Tan et al. isolated bone marrow mesenchymal stem cells (BMSCs) from dogs and inoculated them into the small intestinal submucosa (SIS). A histological and immunohistochemical evaluation, the inoculated scaffolds promoted angiogenesis and muscle regeneration [436].

With the continuous development of science and technology, the field of gastrointestinal regeneration is no longer in its infancy, but the complexity of the gastrointestinal tract brings more and more challenges. Although the above acellular materials have achieved some success in this field, further clinical studies are needed to continuously enrich improve their repair ability.

4.7.4 Nerve Tissue Engineering

4.7.4.1 Structure and Characteristics of Nerve

Most of the central nervous system injuries are caused by traumatic injuries, accounting for nearly 80%, and other diseases account for the remaining 20% [437]. Spinal cord injury (SCI) caused by vehicle and fall accidents accounts for nearly 70% of all SCI cases in the USA [438]. Peripheral nervous system injury (PNI) is the result of the interruption of the complex feedback network from peripheral organs to the central nervous system.

4.7.4.2 Requirements for Nerve Tissue Engineering Materials

Recently, the use of tissue-engineered catheters to repair nerve damage has shown therapeutic potential. The nerve conduit works by enclosing the proximal and distal ends of the damaged nerve in a cylindrical shell, providing an environment conducive to nerve regeneration. To optimize nerve regeneration, catheters must conform to certain design principles: (a) catheters must minimize immunogenic and inflammatory responses that impair nerve regeneration; (b) Be biodegradable and strong enough to support surgical procedures (e.g. sutures), prevent collapse, and withstand the stress of surrounding tissues, and should also facilitate the effective exchange of metabolic nutrients and wastes, while preventing the invasion of cells and biological materials that may interfere with nerve regeneration activities; (c) The surface of the material in the catheter must be guided by the principle of optimal promotion of nerve regeneration in order to promote the regeneration of the ordered structure within the nerve. FDA has approved biodegradable and biocompatible nerve catheters for the repair of short nerve defects in the human body [438], but these catheters only partially meet the above design principles. Therefore, after nerve repair with these products, motor function recovery is incomplete, and the maximum repair length is limited within 4 cm [439–442]. In addition to the above shortcomings, nerve regeneration achieved in this way may also be accompanied by obvious fibrosis, with a layer of myofibroblasts surrounding the nerve pressing, thus hindering recovery [443, 444].

4.7.4.3 Applications of Decellularized Materials in Nerve

Nerve damage after trauma or surgery can lead to severe morbidity and disability. Despite advances in microsurgery, recovery from nerve injury is often disappointing. The traditional "gold standard" treatment for nerve defects is autologous nerve transplantation. However, due to the limited donor collection area and the different size of the graft and recipient nerves, it cannot be used in a large number of treatments. To overcome these shortcomings, various artificial and biological nerve conduits have been developed with varying degrees of success. Compared with other strategies, acellular materials have been widely concerned for their low immunogenicity, promotion of cell re-proliferation, and promotion of angiogenesis [445, 446]. Each targeted therapy has its own challenges based on the differences in the microenvironment inherent in CNS tissue defects. Zhang et al. used SDS to remove paravertebral muscle tissue to remove cell debris [447], and the results showed that this method could promote the parallel and linear germination of axons in the lesion site, and improve the survival rate of axons. In addition, Crapo et al. and Tukmachev et al. demonstrated that acellular materials were superior to other repair materials [192, 448]. Tukmachev et al. found that SC dECM and UBM both increased vascular density, neuronal expansion, and macrophage polarization in spinal cord injury in rats [448]. DECM for PNS regeneration most applications attempt to improve nerve transplantation by generating acellular allografts [449–451] or xenografts [452]. Others have used decellularized nerve [453] or bladder [454] sources to make hydrogels or wraps to enhance existing repair procedures.

Currently, only a few neuro-specific dECM products have entered clinical trials. One of these products is Avance® nerve graft from AxoGen, an acellular allograft used to treat PNI. In addition to two ongoing studies, AxoGen has completed two clinical studies of the product [455]. In addition, a number of clinical studies were conducted [456–458]. Other AxoGen products, such as AxoGuard neuroprotectors, have shown good clinical efficacy in the initial treatment of recurrent cubic tube syndrome [459].

More and more diseases are treated with dECM and its derivatives because of their inherent ability to promote constructive remodeling of tissues [460, 461]. Various treatments are used in the acellular process [462]. The unique materials of the remaining acellular scaffolds provide an enabling environment for the promotion of post-injury repair and reconstruction [463, 464], especially to support functional repair of the nervous system. In one study, [465] a biodegradable collagen nerve conduit containing acellular sciatic nerve matrix was fabricated and compared with a silicone catheter used for peripheral nerve regeneration in rats. Enzyme-linked immunosorbent assay (ELISA) confirmed that the collagen-based catheter contained nerve growth factor, brain-derived neurotrophic factor and laminin. Compared with the sham operation control group and the silicone catheter group, the motor and sensory nerve functions of the rats with sciatic nerve defect implanted with collagen tube were significantly improved, and the nerve regeneration ability was significantly enhanced.

The dECM can also be used as an alternative therapy to stimulate peripheral nerve regeneration. A significant advantage of using acellular nerve grafts is the avoidance of donor-site disease [466, 467]. In clinical practice, acellular allografts can improve the motor and sensory functions of small nerve spaces (mean about 3.2 cm) [468, 469]. However, there is currently no better method to obtain an ideal graft, and this finding highlights the importance of selecting an appropriate acellular method to optimize peripheral nerve injury recovery.

References

1. Cheng CW, Solorio LD, Alsberg E. Decellularized tissue and cell-derived extracellular matrices as scaffolds for orthopedic tissue engineering. Biotechnol Adv. 2014;32:462–84.
2. Guarino V, Causa F, Ambrosio L. Bioactive scaffolds for bone and ligament tissue. Expert Rev Med Devices. 2007;4:405–18.
3. Gibson LJ. Biomechanics of cellular solids. J Biomech. 2005;38:377–99.

4. Young MF. Bone matrix proteins: their function, regulation, and relationship to osteoporosis. Osteoporos Int. 2003;14(Suppl 3):S35–42.
5. Wegst UGK, Ashby MF. The mechanical efficiency of natural materials. Philos Mag. 2004;84:2167–86.
6. Ning CQ. Biomaterials for bone tissue engineering. In: Biomechanics and biomaterials in orthopedics. New York: Springer; 2016. p. 35–57.
7. Hutmacher DW, Schantz JT, Lam CX, Tan KC, Lim TC. State of the art and future directions of scaffold-based bone engineering from a biomaterials perspective. J Tissue Eng Regen Med. 2007;1:245–60.
8. Murphy WL, Kohn DH, Mooney DJ. Growth of continuous bonelike mineral within porous poly(lactide-co-glycolide) scaffolds in vitro. J Biomed Mater Res. 2000;50:50–8.
9. Gazdag AR, Lane JM, Glaser D, Forster RA. Alternatives to autogenous bone graft: efficacy and indications. J Am Acad Orthop Surg. 1995;3:1–8.
10. Jeong HJ, Gwak SJ, Seo KD, Lee S, Yun JH, Cho YS, Lee SJ. Fabrication of three-dimensional composite scaffold for simultaneous alveolar bone regeneration in dental implant installation. Int J Mol Sci. 2020;21:1863.
11. Li X, Wang L, Fan Y, Feng Q, Cui FZ, Watari F. Nanostructured scaffolds for bone tissue engineering. J Biomed Mater Res A. 2013;101:2424–35.
12. Hench L, Potak L. Third-generation biomedical materials. Science. 2002;8:1014–7.
13. Bruder SP, Jaiswal N, Ricalton NS, Mosca JD, Kraus KH. Mesenchymal stem cells in osteobiology and applied bone regeneration. Clin Orthop Relat Res. 1998;355:S247–56.
14. Vallet-Regí M, Romero AM, Ragel CV, LeGeros RZ. XRD, SEM-EDS, and FTIR studies of in vitro growth of an apatite-like layer on sol-gel glasses. J Biomed Mater Res. 1999;44:416–21.
15. Li PJ, Nakanishi K, Kokubo T. Induction and morphology of hydroxyapatite, precipitated from metastable simulated body fluids on sol-gel prepared silica. Biomaterials. 1993;14:963–8.
16. Abe Y, Kokubo T, Yamamuro T. Apatite coating on ceramics, metals and polymers utilizing a biological process. J Mater Sci Mater Med. 1990;1:233–8.
17. Campbell AA, Fryxell GE, Linehan JC, Graff GL. Surface-induced mineralization: a new method for producing calcium phosphate coatings. J Biomed Mater Res. 1996;32:111–8.
18. Wen HB, Wijn JRD, Cui FZ, Groot KD. Preparation of calcium phosphate coatings on titanium implant materials by simple chemistry. J Biomed Mater Res. 1998;41:227–36.
19. Tretinnikov ON, Kato K, Ikada Y. In vitro hydroxyapatite deposition onto a film surface-grated with organophosphate polymer. J Biomed Mater Res. 1994;28:1365–73.
20. Tanahashi M, Yao T, Kokubo T. Apatite coated on organic polymers by biomimetic process: improvement in adhesion to substrate by HCl treatment. J Mater Sci Mater Med. 1995;6:319–26.
21. Putnam AJ, Mooney DJ. Tissue engineering using synthetic extracellular matrices. Nat Med. 1996;2:824–6.
22. Stevens MM. Biomaterials for bone tissue engineering. Mater Today. 2008;11:18–25.
23. Minabe M. A critical review of the biologic rationale for guided tissue regeneration. J Periodontol. 1991;62:171–9.
24. Susmita B, Mangal R, Amit B. A recent advances in bone tissue engineering scaffolds. Trends Biotechnol. 2012;30:546–54.
25. Byrne DP, Lacroix D, Planell JA, Kelly DJ, Prendergast PJ. Simulation of tissue differentiation in a scaffold as a function of porosity, Young's modulus and dissolution rate: application of mechanobiological models in tissue engineering. Biomaterials. 2007;28:5544–54.
26. Mansour A, Mezour MA, Badran Z, Tamimi F. Extracellular matrices for bone regeneration: a literature review. Tissue Eng Part A. 2017;23:1436–51.
27. Zeitouni S, Krause U, Clough BH, Halderman H, Falster A, Blalock DT, Chaput CD, Sampson HW, Gregory CA. Human mesenchymal stem cell derived matrices for enhanced osteoregeneration. Sci Transl Med. 2012;4:132ra155.

28. Loo Y, Goktas M, Tekinay AB, Guler MO, Hauser CA, Mitraki A. Self-assembled proteins and peptides as scaffolds for tissue regeneration. Adv Healthc Mater. 2015;4:2557–86.
29. Legate KR, Wickstrom SA, Fassler R. Genetic and cell biological analysis of integrin outside-in signaling. Genes Dev. 2009;23:397–418.
30. Mathews S, Mathew SA, Gupta PK, Bhonde R, Totey S. Glycosaminoglycans enhance osteoblast differentiation of bone marrow derived human mesenchymal stem cells. J Tissue Eng Regen Med. 2014;8:143–52.
31. Adele LB. Bone composition: relationship to bone fragility and antiosteoporotic drug effects. BoneKEy Rep. 2013;2:447.
32. Clarke B. Normal bone anatomy and physiology. Clin J Am Soc Nephrol. 2008;3(Suppl 3): S131–9.
33. Viguet-Carrin S, Garnero P, Delmas PD. The role of collagen in bone strength. Osteoporos Int. 2006;17:319–36.
34. Ao HY, Xie YT, Yang SB, Wu XD, Li K, Zheng XB, Tang TT. Covalently immobilised type I collagen facilitates osteoconduction and osseointegration of titanium coating implants. J Orthop Transl. 2016;5:16–25.
35. Weatherholt AM, Fuchs RK, Warden SJ. Specialized connective tissue: bone, the structural framework of the upper extremity. J Hand Ther. 2012;25:123–31. quiz 132
36. Campana V, Milano G, Pagano E, Barba M, Cicione C, Salonna G, Lattanzi W, Logroscino G. Bone substitutes in orthopaedic surgery: from basic science to clinical practice. J Mater Sci Mater Med. 2014;25:2445–61.
37. Hashimoto Y, Funamoto S, Kimura T. The effect of decellularized bone/bone marrow produced by high-hydrostatic pressurization on the osteogenic differentiation of mesenchymal stem cells. Biomaterials. 2011;32:7060–7.
38. Gerhardt LC, Widdows KL, Erol MM. Neocellularization and neovascularization of nanosized bioactive glass-coated decellularized trabecular bone scaffolds. J Biomed Mater Res A. 2013;101:827–41.
39. Ventura RD, Padalhin AR, Min Y-K, Lee B-T. Bone regeneration using hydroxyapatite sponge scaffolds with in vivo deposited extracellular matrix. Tissue Eng A. 2015;21:2649–61.
40. Marcos-Campos I, Marolt D, Petridis P, Bhumiratana S, Schmidt D, Vunjak-Novakovic G. Bone scaffold architecture modulates the development of mineralized bone matrix 2155 by human embryonic stem cells. Biomaterials. 2012;33:8329–42.
41. de Peppo GM, Marcos-Campos I, Kahler DJ, Alsalman D, Shang L, Vunjak-Novakovic G, Marolt D. Engineering bone tissue substitutes from human induced pluripotent stem cells. Proc Natl Acad Sci U S A. 2013;110:8680–5.
42. Mirjam F, Warren L, Grayson D, Marolt J, Gimble M, Nevenka K, Gordana V. Bone grafts engineered from human adipose-derived stem cells in perfusion bioreactor culture. Tissue Eng. 2010;16:179–89.
43. Hung BP, Naved BA, Nyberg EL, Dias M, Holmes CA, Elisseeff JH, Dorafshar AH, Grayson WL. Three-dimensional printing of bone extracellular matrix for craniofacial regeneration. ACS Biomater Sci Eng. 2016;2:1806–16.
44. Lee DJ, Diachina S, Lee YT, Zhao L, Zou R, Tang N, Han H, Chen X, Ko CC. Decellularized bone matrix grafts for calvaria regeneration. J Tissue Eng. 2016;7:2041731416680306.
45. Hesse E, Kluge G, Atfi A, Correa D, Haasper C, Berding G, Shin HO, Viering J, Langer F, Vogt PM, Krettek C, Jagodzinski M. Repair of a segmental long bone defect in human by implantation of a novel multiple disc graft. Bone. 2010;46:1457–63.
46. Borges GJ, Novaes AB Jr, Grisi MF, Palioto DB, Taba M Jr, de Souza SL. Acellular dermis as a dermal matrix of tissue engineered skin substitute for burns treatment. Clin Oral Implants Res. 2009;20:1105–15.
47. Uitto J. Connective tissue biochemistry of the aging dermis. Age-related alterations in collagen and elastin. Clin Geriatr Med. 1989;5:127–47.
48. Smith MM, Melrose J. Proteoglycans in normal and healing skin. Adv Wound Care. 2015;4:152–73.

49. Medalie DA, Tompkins RG, Morgan JR. Evaluation of acellular human dermis as a dermal analog in a composite skin graft. ASAIO J. 1996;42:M455.
50. Kim HJ, Park SS, Oh SY, Kim H, Kweon OK, Woo HM, Kim WH. Effect of acellular dermal matrix as a delivery carrier of adipose-derived mesenchymal stem cells on bone regeneration. J Biomed Mater Res B Appl Biomater. 2012;100:1645–53.
51. Gimble J, Guilak F. Adipose-derived adult stem cells: isolation, characterization, and differentiation potential. Cytotherapy. 2003;5:362–9.
52. Smith SE, Roukis TS. Bone and wound healing augmentation with platelet-rich plasma. Clin Pediatr Med Surg. 2009;26:559–88.
53. Jeong WH, Roh TS, Kim YS, Kang EH, Jung BK, Lee WJ, Lew DH, Yun IS. Acceleration of osteogenesis by platelet-rich plasma with acellular dermal matrix in a calvarial defect model. Childs Nerv Syst. 2016;32:1653–9.
54. Sun Y, Yan LQ, Chen S, Pei M. Functionality of decellularized matrix in cartilage regeneration: a comparison of tissue versus cell sources. Acta Biomater. 2018;74:56–73.
55. De Mori A, Pena Fernandez M, Blunn G, Tozzi G, Roldo M. 3D printing and electrospinning of composite hydrogels for cartilage and bone tissue engineering. Polymers. 2018;10(3):285.
56. Liu M, Zeng X, Ma C, Yi H, Ali Z, Mou XB, Li S, Deng Y, He NY. Injectable hydrogels for cartilage and bone tissue engineering. Bone Res. 2017;5:17014.
57. Bian LM, Hou C, Tous E, Rai R, Mauck RL, Burdick JA. The influence of hyaluronic acid hydrogel crosslinking density and macromolecular diffusivity on human MSC chondrogenesis and hypertrophy. Biomaterials. 2013;34(2):413–21.
58. Kosiur JR, Collins RA. Weight-bearing compared with non-weight-bearing following osteochondral autograft transfer for small defects in weight-bearing areas in the femoral articular cartilage of the knee. J Bone Joint Surg Am. 2014;96:16.
59. Gentili C, Cancedda R. Cartilage and bone extracellular matrix. Curr Pharm Des. 2009;15 (12):1334–48.
60. Sophia Fox AJ, Bedi A, Rodeo SA. The basic science of articular cartilage: structure, composition, and function. Sports Health. 2009;1(6):461–8.
61. Chen JL, Duan L, Zhu W, Xiong J, Wang D. Extracellular matrix production in vitro in cartilage tissue engineering. J Transl Med. 2014;12:88.
62. Hardin JA, Cobelli N, Santambrogio L. Consequences of metabolic and oxidative modifications of cartilage tissue. Nat Rev Rheumatol. 2015;11(9):521–9.
63. Maldonado M, Nam J. The role of changes in extracellular matrix of cartilage in the presence of inflammation on the pathology of osteoarthritis. Biomed Res Int. 2013;2013:284873.
64. Osiecka-Iwan A, Hyc A, Radomska-Lesniewska DM, Rymarczyk A, Skopinski P. Antigenic and immunogenic properties of chondrocytes. Implications for chondrocyte therapeutic transplantation and pathogenesis of inflammatory and degenerative joint diseases. Cent Eur J Immunol. 2018;43(2):209–19.
65. Almarza AJ, Athanasiou KA. Design characteristics for the tissue engineering of cartilaginous tissues. Ann Biomed Eng. 2004;32(1):2–17.
66. O'Connell G, Garcia J, Amir J. 3D bioprinting: new directions in articular cartilage tissue engineering. ACS Biomater Sci Eng. 2017;3(11):2657–68.
67. Akkiraju H, Nohe A. Role of chondrocytes in cartilage formation, progression of osteoarthritis and cartilage regeneration. J Dev Biol. 2015;3(4):177–92.
68. Armientoa AR, Stoddartab MJ, Alinia M, Eglina D. Biomaterials for articular cartilage tissue engineering: learning from biology. Acta Biomater. 2018;65:1–20.
69. Cheng G, Davoudi Z, Xing X, Yu X, Cheng X, Li ZB, Deng HB, Wang Q. Advanced silk fibroin biomaterials for cartilage regeneration. ACS Biomater Sci Eng. 2018;4(8):2704–15.
70. Basiri A, Farokhi M, Azami M, Ebrahimi-Barough S, Mohamadnia A, Rashtbar M, Hasanzadeh E, Mahmoodi N, Baghaban Eslaminejad M, Ai J. A silk fibroin/decellularized extract of Wharton's jelly hydrogel intended for cartilage tissue engineering. Prog Biomater. 2019;8(1):31–42.

71. Bhardwaj N, Nguyen QT, Chen AC, Kaplan DL, Sah RL, Kundu SC. Potential of 3-D tissue constructs engineered from bovine chondrocytes/silk fibroin-chitosan for in vitro cartilage tissue engineering. Biomaterials. 2011;32(25):5773–81.
72. Gardner OFW, Musumeci G, Neumann AJ, Eglin D, Archer CW, Alini M, Stoddart MJ. Asymmetrical seeding of MSCs into fibrin-poly (ester-urethane) scaffolds and its effect on mechanically induced chondrogenesis. J Tissue Eng Regen Med. 2017;11(10):2912–21.
73. Bhattacharjee M, Coburn J, Centola M, Murab S, Barbero A, Kaplan DL, Martin I, Ghosh S. Tissue engineering strategies to study cartilage development, degeneration and regeneration. Adv Drug Deliv Rev. 2015;84:107–22.
74. Fazal N, Latief N. Bombyx mori derived scaffolds and their use in cartilage regeneration: a systematic review. Osteoarthr Cartil. 2018;26(12):1583–94.
75. Liao JF, Shi K, Ding QX, Qu Y, Luo F, Qian ZY. Recent developments in scaffold-guided cartilage tissue regeneration. J Biomed Nanotechnol. 2014;10(10):3085–104.
76. Liao JH, Guo XA, Grande-Allen KJ, Kasper FK, Mikos AG. Bioactive polymer/extracellular matrix scaffolds fabricated with a flow perfusion bioreactor for cartilage tissue engineering. Biomaterials. 2010;31(34):8911–20.
77. Madry H, Kaul G, Zurakowski D, Vunjak-Novakovic G, Cucchiarini M. Cartilage constructs engineered from chondrocytes overexpressing Igf-I improve the repair of osteochondral defects in a rabbit model. Eur Cell Mater. 2013;25:229–47.
78. Szychlinska MA, Stoddart MJ, D'Amora U, Ambrosio L, Alini M, Musumeci G. Mesenchymal stem cell-based cartilage regeneration approach and cell senescence: can we manipulate cell aging and function? Tissue Eng Part B. 2017;23(6):529–39.
79. Shin SC, Park HY, Shin N, Jung DW, Kwon HK, Kim JM, Wang SG, Lee JC, Sung ES, Park GC, Lee BJ. Evaluation of decellularized xenogenic porcine auricular cartilage as a novel biocompatible filler. J Biomed Mater Res B. 2018;106(7):2708–15.
80. Gilbert TW, Sellaro TL, Badylak SF. Decellularization of tissues and organs. Biomaterials. 2006;27(19):3675–83.
81. Crapo PM, Gilbert TW, Badylak SF. An overview of tissue and whole organ decellularization processes. Biomaterials. 2011;32(12):3233–43.
82. Zhang YY, He YJ, Bharadwaj S, Hammam N, Carnagey K, Myers R, Atala A, Van Dyke M. Tissue-specific extracellular matrix coatings for the promotion of cell proliferation and maintenance of cell phenotype. Biomaterials. 2009;30(23-24):4021–8.
83. Benders KEM, van Weeren PR, Badylak SF, Saris DBF, Dhert WJA, Malda J. Extracellular matrix scaffolds for cartilage and bone regeneration. Trends Biotechnol. 2013;31(3):169–76.
84. Kiyotake EA, Beck EC, Detamore MS. Cartilage extracellular matrix as a biomaterial for cartilage regeneration. Ann N Y Acad Sci. 2016;1383:139–59.
85. Dikina AD, Almeida HV, Cao M, Kelly DJ, Alsberg E. Scaffolds derived from ECM produced by chondrogenically induced human MSC condensates support human MSC chondrogenesis. ACS Biomater Sci Eng. 2017;3(7):1426–36.
86. He F, Chen XD, Pei M. Reconstruction of an In vitro tissue-specific microenvironment to rejuvenate synovium-derived stem cells for cartilage tissue engineering. Tissue Eng Pt A. 2009;15(12):3809–21.
87. Zhang WX, Zhu Y, Li J, Guo QY, Peng J, Liu SC, Yang JH, Wang Y. Cell-derived extracellular matrix: basic characteristics and current applications in orthopedic tissue engineering. Tissue Eng Part B. 2016;22(3):193–207.
88. Kim YS, Majid M, Melchiorri AJ, Mikos AG. Applications of decellularized extracellular matrix in bone and cartilage tissue engineering. Bioeng Transl Med. 2019;4(1):83–95.
89. Jia SJ, Liu L, Pan WM, Meng GL, Duan CG, Zhang LQ, Xiong Z, Liu J. Oriented cartilage extracellular matrix-derived scaffold for cartilage tissue engineering. J Biosci Bioeng. 2012;113(5):647–53.
90. Jin CZ, Choi BH, Park SR, Min BH. Cartilage engineering using cell-derived extracellular matrix scaffold in vitro. J Biomed Mater Res A. 2010;92a(4):1567–77.

91. Bhardwaj N, Devi D, Mandal BB. Tissue-engineered cartilage: the crossroads of biomaterials, cells and stimulating factors. Macromol Biosci. 2015;15(2):153–82.
92. Zeng L, Chen XF, Zhang Q, Yu F, Li YL, Yao YC. Redifferentiation of dedifferentiated chondrocytes in a novel three-dimensional microcavitary hydrogel. J Biomed Mater Res A. 2015;103(5):1693–702.
93. Cheng NC, Estes BT, Young TH, Guilak F. Engineered cartilage using primary chondrocytes cultured in a porous cartilage-derived matrix. Regen Med. 2011;6(1):81–93.
94. von Bomhard A, Elsaesser A, Riepl R, Pippich K, Faust J, Schwarz S, Koerber L, Breiter R, Rotter N. Cartilage regeneration using decellularized cartilage matrix: long-term comparison of subcutaneous and intranasal placement in a rabbit model. J Cranio-Maxillofac Surg. 2019;47(4):682–94.
95. Gong YY, Xue JX, Zhang WJ, Zhou GD, Liu W, Cao YL. A sandwich model for engineering cartilage with acellular cartilage sheets and chondrocytes. Biomaterials. 2011;32(9):2265–73.
96. Li YQ, Xu Y, Liu YQ, Wang ZX, Chen WM, Duan L, Gu DY. Decellularized cartilage matrix scaffolds with laser-machined micropores for cartilage regeneration and articular cartilage repair. Mat Sci Eng C Mater. 2019;105:110139.
97. Jin CZ, Cho JH, Choi BH, Wang LM, Kim MS, Park SR, Yun JH, Oh HJ, Min BH. The maturity of tissue-engineered cartilage in vitro affects the repairability for osteochondral defect. Tissue Eng Pt A. 2011;17(23-24):3057–65.
98. Nie XL, Chuah YJ, Zhu WZ, He PF, Peck Y, Wang DA. Decellularized tissue engineered hyaline cartilage graft for articular cartilage repair. Biomaterials. 2020;235:119821.
99. Lehmann J, Nurnberger S, Narcisi R, Stok S, van der Eerden BCJ, Koevoet WJLM, Kops N, ten Berge D, van Osch GJ. Recellularization of auricular cartilage via elastase-generated channels. Biofabrication. 2019;11(3):035012.
100. Ibsirlioglu T, Elcin AE, Elcin YM. Decellularized biological scaffold and stem cells from autologous human adipose tissue for cartilage tissue engineering. Methods. 2020;171:97–107.
101. Sawatjui N, Limpaiboon T, Schrobback K, Klein T. Biomimetic scaffolds and dynamic compression enhance the properties of chondrocyte- and MSC-based tissue-engineered cartilage. J Tissue Eng Regen Med. 2018;12(5):1220–9.
102. Jaipaew J, Wangkulangkul P, Meesane J, Raungrut P, Puttawibul P. Mimicked cartilage scaffolds of silk fibroin/hyaluronic acid with stem cells for osteoarthritis surgery: morphological, mechanical, and physical clues. Mater Sci Eng C Mater Biol Appl. 2016;64:173–82.
103. Song L, Baksh D, Tuan RS. Mesenchymal stem cell-based cartilage tissue engineering: cells, scaffold and biology. Cytotherapy. 2004;6(6):596–601.
104. Jorgensen C, Gordeladze J, Noel D. Tissue engineering through autologous mesenchymal stem cells. Curr Opin Biotechnol. 2004;15(5):406–10.
105. Zhou GD, Liu W, Cui L, Wang XY, Liu TY, Cao YL. Repair of porcine articular osteochondral defects in non-weightbearing areas with autologous bone marrow stromal cells. Tissue Eng. 2006;12(11):3209–21.
106. Yang Q, Peng J, Lu S, Huang J, Yao J, Yang F, Wang S, Wang A, Xu W, Guo Q, Zhang L. A cartilage ECM-derived 3-D porous acellular matrix scaffold for in vivo cartilage tissue engineering with PKH26-labeled chondrogenic bone marrow-derived mesenchymal stem cells. Tissue Eng Pt A. 2008;14(5):800.
107. Xue JX, Gong YY, Zhou GD, Liu W, Cao Y, Zhang WJ. Chondrogenic differentiation of bone marrow-derived mesenchymal stem cells induced by acellular cartilage sheets. Biomaterials. 2012;33(24):5832–40.
108. Yang ZQ, Shi YY, Wei XC, He JR, Yang SH, Dickson G, Tang JQ, Xiang JJ, Song C, Li G. Fabrication and repair of cartilage defects with a novel acellular cartilage matrix scaffold. Tissue Eng Part C Med. 2010;16(5):865–76.
109. Cheng NC, Estes BT, Awad HA, Guilak F. Chondrogenic differentiation of adipose-derived adult stem cells by a porous scaffold derived from native articular cartilage extracellular matrix. Tissue Eng Pt A. 2009;15(2):231–41.

110. Luo L, Chu JYJ, Eswaramoorthy R, Mulhall KJ, Kelly DJ. Engineering tissues that mimic the zonal nature of articular cartilage using decellularized cartilage explants seeded with adult stem cells. ACS Biomater Sci Eng. 2017;3(9):1933–43.
111. Kang HJ, Peng J, Lu SB, Liu SY, Zhang L, Huang JX, Sui X, Zhao B, Wang AY, Xu WJ, Luo ZJ, Guo QY. In vivo cartilage repair using adipose-derived stem cell-loaded decellularized cartilage ECM scaffolds. J Tissue Eng Regen Med. 2014;8(6):442–53.
112. Cucchiarini M, Madry H. Biomaterial-guided delivery of gene vectors for targeted articular cartilage repair. Nat Rev Rheumatol. 2019;15(1):18–29.
113. Badylak SF. The extracellular matrix as a biologic scaffold material. Biomaterials. 2007;28 (25):3587–93.
114. Ekenseair AK, Kasper FK, Mikos AG. Perspectives on the interface of drug delivery and tissue engineering. Adv Drug Deliv Rev. 2013;65(1):89–92.
115. Yang Q, Teng BH, Wang LN, Li K, Xu C, Ma XL, Zhang Y, Kong DL, Wang LY, Zhao YH. Silk fibroin/cartilage extracellular matrix scaffolds with sequential delivery of TGF-beta3 for chondrogenic differentiation of adipose-derived stem cells. Int J Nanomedicine. 2017;12:6721–33.
116. Macri L, Silverstein D, Clark RA. Growth factor binding to the pericellular matrix and its importance in tissue engineering. Adv Drug Deliv Rev. 2007;59(13):1366–81.
117. Hennig T, Lorenz H, Thiel A, Goetzke K, Dickhut A, Geiger F, Richter W. Reduced chondrogenic potential of adipose tissue derived stromal cells correlates with an altered TGF beta receptor and BMP profile and is overcome by BMP-6. J Cell Physiol. 2007;211 (3):682–91.
118. Solorio LD, Vieregge EL, Dhami CD, Dang PN, Alsberg E. Engineered cartilage via self-assembled hMSC sheets with incorporated biodegradable gelatin microspheres releasing transforming growth factor-beta 1. J Control Release. 2012;158(2):224–32.
119. Almeida HV, Liu YR, Cunniffe GM, Mulhall KJ, Matsiko A, Buckley CT, O'Brien FJ, Kelly DJ. Controlled release of transforming growth factor-beta 3 from cartilage-extra-cellular-matrix-derived scaffolds to promote chondrogenesis of human-joint-tissue-derived stem cells. Acta Biomater. 2014;10(10):4400–9.
120. Lim WL, Liau LL, Ng MH, Chowdhury SR, Law JX. Current progress in tendon and ligament tissue engineering. Tissue Eng Regen Med. 2019;16(6):549–71.
121. Tozer S, Duprez D. Tendon and ligament: development, repair and disease. Birth Defects Res C Embryo Today. 2005;75(3):226–36.
122. Yang G, Rothrauff BB, Tuan RS. Tendon and ligament regeneration and repair: clinical relevance and developmental paradigm. Birth Defects Res C Embryo Today. 2013;99 (3):203–22.
123. Birch HL, Thorpe CT, Rumian AP. Specialisation of extracellular matrix for function in tendons and ligaments. Muscles Ligaments Tendons J. 2013;3(1):12–22.
124. Hodgson RJ, O'Connor PJ, Grainger AJ. Tendon and ligament imaging. Br J Radiol. 2012;85 (1016):1157–72.
125. Silva M, Ferreira FN, Alves NM, Paiva MC. Biodegradable polymer nanocomposites for ligament/tendon tissue engineering. J Nanobiotechnol. 2020;18(1):23.
126. Asahara H, Inui M, Lotz MK. Tendons and ligaments: connecting developmental biology to musculoskeletal disease pathogenesis. J Bone Miner Res. 2017;32(9):1773–82.
127. Kumai T, Yamada G, Takakura Y, Tohno Y, Benjamin M. Trace elements in human tendons and ligaments. Biol Trace Elem Res. 2006;114(1-3):151–61.
128. Kuo CK, Marturano JE, Tuan RS. Novel strategies in tendon and ligament tissue engineering: advanced biomaterials and regeneration motifs. Sports Med Arthrosc Rehabil Ther Technol. 2010;2:20.
129. Woo SL. Tissue engineering: use of scaffolds for ligament and tendon healing and regeneration. Knee Surg Sports Traumatol Arthrosc. 2009;17(6):559–60.
130. Li WJ, Laurencin CT, Caterson EJ, Tuan RS, Ko FK. Electrospun nanofibrous structure: a novel scaffold for tissue engineering. J Biomed Mater Res. 2002;60(4):613–21.

131. Leong NL, Petrigliano FA, McAllister DR. Current tissue engineering strategies in anterior cruciate ligament reconstruction. J Biomed Mater Res A. 2014;102(5):1614–24.
132. Wheelton A, Mace J, Khan WS, Anand S. Biomaterials and fabrication to optimise scaffold properties for musculoskeletal tissue engineering. Curr Stem Cell Res Ther. 2016;11 (7):578–84.
133. Smith BD, Grande DA. The current state of scaffolds for musculoskeletal regenerative applications. Nat Rev Rheumatol. 2015;11(4):213–22.
134. No YJ, Tarafder S, Reischl B, Ramaswamy Y, Dunstan CR, Friedrich O, Lee CH, Zreiqat H. High-strength fiber-reinforced composite hydrogel scaffolds as biosynthetic tendon graft material. ACS Biomater Sci Eng. 2020;6(4):1887–98.
135. Cao GX, Huang Y, Li K, Fan YB, Xie HQ, Li XM. Small intestinal submucosa: superiority, limitations and solutions, and its potential to address bottlenecks in tissue repair. J Mater Chem B. 2019;7(33):5038–55.
136. Wang M, Li YQ, Cao J, Gong M, Zhang Y, Chen X, Tian MX, Xie HQ. Accelerating effects of genipin-crosslinked small intestinal submucosa for defected gastric mucosa repair. J Mater Chem B. 2017;5(34):7059–71.
137. Gu Y, Dai K. Substitution of porcine small intestinal submucosa for rabbit Achilles tendon, an experimental study. Zhonghua Yi Xue Za Zhi. 2002;82(18):1279–82.
138. Murphy KD, Mushkudiani IA, Kao D, Levesque AY, Hawkins HK, Gould LJ. Successful incorporation of tissue-engineered porcine small-intestinal submucosa as substitute flexor tendon graft is mediated by elevated TGF-beta1 expression in the rabbit. J Hand Surg Am. 2008;33(7):1168–78.
139. Pridgen BC, Woon CYL, Kim M, Thorfinn J, Lindsey D, Pham H, Chang J. Flexor tendon tissue engineering: acellularization of human flexor tendons with preservation of biomechanical properties and biocompatibility. Tissue Eng Part C Med. 2011;17(8):819–28.
140. Lipner J, Shen H, Cavinatto L, Liu WY, Havlioglu N, Xia YN, Galatz LM, Thomopoulos S. In vivo evaluation of adipose-derived stromal cells delivered with a nanofiber scaffold for tendon-to-bone repair. Tissue Eng Pt A. 2015;21(21-22):2766–74.
141. Subramanian A, Schilling TF. Tendon development and musculoskeletal assembly: emerging roles for the extracellular matrix. Development. 2015;142(24):4191–204.
142. Tischer T, Vogt S, Aryee S, Steinhauser E, Adamczyk C, Milz S, Martinek V, Imhov AB. Tissue engineering of the anterior cruciate ligament: a new method using acellularized tendon allografts and autologous fibroblasts. Arch Orthop Traum. 2007;127(9):735–41.
143. Hsu J, Keener JD. Natural history of rotator cuff disease and implications on management. Oper Tech Orthop. 2015;25(1):2–9.
144. Aleem AW, Brophy RH. Outcomes of rotator cuff surgery: what does the evidence tell us? Clin Sports Med. 2012;31(4):665–74.
145. Han B, Jones IA, Yang Z, Fang W, Vangsness CT. Repair of rotator cuff tendon defects in aged rats using a growth factor injectable gel scaffold. Arthroscopy. 2020;36(3):629–37.
146. Maffulli N, Longo UG, Loppini M, Berton A, Spiezia F, Denaro V. Tissue engineering for rotator cuff repair: an evidence-based systematic review. Stem Cells Int. 2012;2012:418086.
147. Holton LH 3rd, Kim D, Silverman RP, Rodriguez ED, Singh N, Goldberg NH. Human acellular dermal matrix for repair of abdominal wall defects: review of clinical experience and experimental data. J Long-Term Eff Med Implants. 2005;15(5):547–58.
148. Adams JE, Zobitz ME, Reach JS, An KN, Steinmann SP. Rotator cuff repair using an acellular dermal matrix graft: an in vivo study in a canine model. Arthroscopy. 2006;22(7):700–9.
149. Xu H, Sandor M, Qi SJ, Lombardi J, Connor J, McQuillan DJ, Iannotti JP. Implantation of a porcine acellular dermal graft in a primate model of rotator cuff repair. J Shoulder Elb Surg. 2012;21(5):580–8.
150. Stuart D. Kinsella publication rates of podium versus poster presentations at the American Orthopaedic Society for Sports Medicine meetings: 2006-2010. Am J Sports Med. 2015;43 (5):1255–9.

151. Gulotta LV, Rodeo SA. Biology of autograft and allograft healing in anterior cruciate ligament reconstruction. Clin Sports Med. 2007;26(4):509–24.
152. Whitlock PW, Seyler TM, Parks GD, Ornelles DA, Smith TL, Van Dyke ME, Poehling GG. A novel process for optimizing musculoskeletal allograft tissue to improve safety, ultrastructural properties, and cell infiltration. J Bone Joint Surg Am. 2012;94a(16):1458–67.
153. Bertone AL, Goin S, Kamei SJ, Mattoon JS, Litsky AS, Weisbrode SE, Clarke RB, Plouhar PL, Kaeding CC. Metacarpophalangeal collateral ligament reconstruction using small intestinal submucosa in an equine model. J Biomed Mater Res A. 2008;84a(1):219–29.
154. Meskell M. Principles of anatomy and physiology. J Anat. 2010;217:631.
155. Pissarenko A, Meyers MA. The materials science of skin: analysis, characterization, and modeling. Prog Mater Sci. 2019;110:100634.
156. Keck M, Lumenta DB, Kamolz LP. Skin tissue engineering. In: Dermal replacements in general, burn, and plastic surgery. New York: Springer; 2013. p. 13–25.
157. Butler BJ, Boddy RL, Bo C, Arora H, Williams A, Proud WG, Brown KA. Composite nature of fresh skin revealed during compression. Bioinsp Biomim Nanobiomater. 2015;4:133–9.
158. Huzaira M, Rius F, Rajadhyaksha M. Topographic variations in normal skin, as viewed by in vivo reflectance confocal microscopy. J Investig Dermatol. 2001;116:846–52.
159. Sandby-Moller J, Poulsen T, Wulf HC. Epidermal thickness at different body sites: relationship to age, gender, pigmentation, blood content, skin type and smoking habits. Acta Derm Venereol. 2003;83:410–3.
160. Hendriks FM, Brokken D, Oomens CW, Bader DL, Baaijens FP. The relative contributions of different skin layers to the mechanical behavior of human skin in vivo using suction experiments. Med Eng Phys. 2006;28:259–66.
161. Jor JW, Parker MD, Taberner AJ, Nash MP, Nielsen PM. Computational and experimental characterization of skin mechanics: identifying current challenges and future directions. Wiley Interdiscip Rev Syst Biol Med. 2013;5:539–56.
162. Leyva-Mendivil MF, Page A, Bressloff NW, Limbert G. A mechanistic insight into the mechanical role of the stratum corneum during stretching and compression of the skin. J Mech Behav Biomed Mater. 2015;49:197–219.
163. Wilkes GL, Brown IA, Wildnauer RH. The biomechanical properties of skin. CRC Crit Rev Bioeng. 1973;1:453.
164. Limbert G. Mathematical and computational modelling of skin biophysics: a review. Proc Math Phys Eng Sci. 2017;473:20170257.
165. Bielefeld KA, Amini-Nik S, Whetstone H, Poon R, Youn A, Wang J, Alman BA. Fibronectin and -catenin act in a regulatory loop in dermal fibroblasts to modulate cutaneous healing. J Biol Chem. 2011;286:27687–97.
166. Avery D, Govindaraju P, Jacob M, Todd L, Monslow J, Pure E. Extracellular matrix and dermal fibroblast function in the healing wound. Matrix Biol. 2018;67:90–106.
167. Allgöwer M, Schoenenberger GA, Sparkes BG. Burning the largest immune organ. Burns. 1995;21:S7–47.
168. Manimalha B, Kumar TR, Babu M. Skin substitutes: a review. Burns. 2001;27:534–44.
169. Metcalfe AD, Ferguson MW. Bioengineering skin using mechanisms of regeneration and repair. Biomaterials. 2007;28:5100–13.
170. Arno AI, Amini-Nik S, Blit PH, Al-Shehab M, Belo C, Herer E, Tien C, Jeschke MG. Human Wharton's jelly mesenchymal stem cells promote skin wound healing through paracrine signaling. Stem Cell Res Ther. 2014;5:28.
171. Blais M, Parenteau-Bareil R, Cadau S, Berthod F. Concise review: tissue-engineered skin and nerve regeneration in burn. Stem Cells Transl Med. 2013;2:545–51.
172. Silvera-Tawil D, Rye D, Velonaki M. Artificial skin and tactile sensing for socially interactive robots: a review. Robot Auton Syst. 2015;63:230–43.
173. Sheikholeslam M, Wright MEE, Jeschke MG, Amini-Nik S. Biomaterials for skin substitutes. Adv Healthc Mater. 2018;7:1700897.

174. Jeschke MG, Sadri AR, Belo C, Amini-Nik S. A surgical device to study the efficacy of bioengineered skin substitutes in mice wound healing models. Tissue Eng Part C Methods. 2017;23:237–42.
175. Nie C, Yang D, Morris SF. Local delivery of adipose-derived stem cells via acellular dermal matrix as a scaffold: a new promising strategy to accelerate wound healing. Med Hypotheses. 2009;72:679–82.
176. Abdullahi A, Amini-Nik S, Jeschke MG. Animal models in burn research. Cell Mol Life Sci. 2014;71:3241–55.
177. Amini-Nik S, Yousuf Y, Jeschke MG. Scar management in burn injuries using drug delivery and molecular signaling: current treatments and future directions. Adv Drug Deliv Rev. 2018;123:135–54.
178. Chin CD, Khanna K, Sia SK. A microfabricated porous collagen-based scaffold as prototype for skin substitutes. Biomed Microdevices. 2008;10:459–67.
179. Hogrebe NJ, Reinhardt JW, Gooch KJ. Biomaterial microarchitecture: a potent regulator of individual cell behavior and multicellular organization. J Biomed Mater Res A. 2017;105:640–61.
180. Zhou H, You C, Wang X, Jin R, Wu P, Li Q, Han C. The progress and challenges for dermal regeneration in tissue engineering. J Biomed Mater Res A. 2017;105:1208–18.
181. Tatara AM, Mikos AG. Tissue engineering in orthopaedics. J Bone Joint Surg Am. 2016;98:1132–9.
182. James KS, Cornwell KC, Greenburg AG. Extracellular matrix bioscaffolds for orthopaedic applications. J Bone Joint Surg. 2010;92:1316.
183. Ricard-Blum S, Vallet SD. Fragments generated upon extracellular matrix remodeling: biological regulators and potential drugs. Matrix Biol. 2019;75-76:170–89.
184. Rilla K, Mustonen AM, Arasu UT, Harkonen K, Matilainen J, Nieminen P. Extracellular vesicles are integral and functional components of the extracellular matrix. Matrix Biol. 2019;75-76:201–19.
185. Kawasaki T, Kirita Y, Kami D, Kitani T, Ozaki C, Itakura Y, Toyoda M, Gojo S. Novel detergent for whole organ tissue engineering. J Biomed Mater Res A. 2015;103:3364–73.
186. Song JJ, Guyette JP, Gilpin SE, Gonzalez G, Vacanti JP, Ott HC. Regeneration and experimental orthotopic transplantation of a bioengineered kidney. Nat Med. 2013;19:646–51.
187. Yu YL, Shao YK, Ding YQ, Lin KZ, Chen B, Zhang HZ, Zhao LN, Wang ZB, Zhang JS, Tang ML, Mei J. Decellularized kidney scaffold-mediated renal regeneration. Biomaterials. 2014;35:6822–8.
188. Ott HC, Clippinger B, Conrad C, Schuetz C, Pomerantseva I, Ikonomou L, Kotton D, Vacanti JP. Regeneration and orthotopic transplantation of a bioartificial lung. Nat Med. 2010;16:927–33.
189. Ott HC, Matthiesen TS, Goh SK, Black LD, Kren SM, Netoff TI, Taylor DA. Perfusion-decellularized matrix: using nature's platform to engineer a bioartificial heart. Nat Med. 2008;14:213–21.
190. Jank BJ, Xiong L, Moser PT, Guyette JP, Ren X, Cetrulo CL, Leonard DA, Fernandez L, Fagan SP, Ott HC. Engineered composite tissue as a bioartificial limb graf. Biomaterials. 2015;61:246–56.
191. Keane TJ, DeWard A, Londono R, Saldin LT, Castleton AA, Carey L, Nieponice A, Lagasse E, Badylak SF. Tissue-specific effects of esophageal extracellular matrix. Tissue Eng Part A. 2015;21:2293–300.
192. Tukmachev D, Forostyak S, Koci Z, Zaviskova K, Vackova I, Vyborny K, Sandvig I, Sandvig A, Medberry CJ, Badylak SF, Sykova E, Kubinova S. Injectable extracellular matrix hydrogels as scaffolds for spinal cord injury repair. Tissue Eng Part A. 2016;22:306–17.
193. D'Amore A, Yoshizumi T, Luketich SK, Wolf MT, Gu X, Cammarata M, Hoff R, Badylak SF, Wagner WR. Bi-Layered polyurethane – extracellular matrix cardiac patch improves ischemic ventricular wall remodeling in a rat model. Biomaterials. 2016;107:1–14.

194. Wolf MT, Dearth CL, Sonnenberg SB, Loboa EG, Badylak SF. Naturally derived and synthetic scaffolds for skeletal muscle reconstruction. Adv Drug Deliv Rev. 2015;84:208–21.
195. Zhang J, Cheng WY, Hu ZQ, Turner NJ, Zhang L, Wang Q, Badylak SF. A panel data set on harvest and perfusion decellularization of porcine rectus abdominis. Data Brief. 2016;7:1375–82.
196. Cui H, Chai Y, Yu Y. Progress in developing decellularized bioscaffolds for enhancing skin construction. J Biomed Mater Res A. 2019;107:1849–59.
197. Ng KW, Khor HL, Hutmacher DW. In vitro characterization of natural and synthetic dermal matrices cultured with human dermal fibroblasts. Biomaterials. 2004;25:2807–18.
198. Ellis CV, Kulber DA. Acellular dermal matrices in hand reconstruction. Plast Reconstr Surg. 2012;130:256S–69S.
199. Levin F, Turbin RE, Langer PD. Acellular human dermal matrix as a skin substitute for reconstruction of large periocular cutaneous defects. Ophthalmic Plast Reconstr Surg. 2011;27:44–7.
200. Liu Z, Zhou Q, Zhu J, Xiao J, Wan P, Zhou C, Huang Z, Qiang N, Zhang W, Wu Z, Quan D, Wang Z. Using genipin-crosslinked acellular porcine corneal stroma for cosmetic corneal lens implants. Biomaterials. 2012;33:7336–46.
201. Zhang Z, Lv L, Mamat M, Chen Z, Liu L, Wang Z. Xenogenic (porcine) acellular dermal matrix is useful for the wound healing of severely damaged extremities. Exp Ther Med. 2014;7:621–4.
202. Baldursson BT, Kjartansson H, Konradsdottir F, Gudnason P, Sigurjonsson GF, Lund SH. Healing rate and autoimmune safety of full-thickness wounds treated with fish skin acellular dermal matrix versus porcine small-intestine submucosa: a noninferiority study. Int J Low Extrem Wounds. 2015;14:37–43.
203. Dogeon Y, Ji-Seon L, Young JS. Honey incorporated antibacterial acellular dermal matrix for full-thickness wound healing. Macromol Res. 2018;1:1–8.
204. Han X, Liu H, Chen M, Gong L, Pang H, Deng X, Jin Y. Acellular dermal matrix from one-day-old mouse skin on adult scarless cutaneous wound repair by second harmonic generation microscopic imaging. RSC Adv. 2016;6:71852–62.
205. Dziki JL, Wang DS, Pineda C, Sicari BM, Rausch T, Badylak SF. Solubilized extracellular matrix bioscaffolds derived from diverse source tissues differentially influence macrophage phenotype. J Biomed Mater Res A. 2017;105:138–47.
206. Oguzkurt P, Kayaselcuk F, Tuncer I, Alkan M, Hicsonmez A. Evaluation of extracellular matrix protein composition in sacs associated with undescended testis, hydrocele, inguinal hernia, and peritoneum. Urology. 2007;70:346–50.
207. Hoganson DM, Owens GE, O'Doherty EM, Bowley CM, Goldman SM, Harilal DO, Neville CM, Kronengold RT, Vacanti JP. Preserved extracellular matrix components and retained biological activity in decellularized porcine mesothelium. Biomaterials. 2010;31:6934–40.
208. Chun SY, Lim GJ, Kwon TG, Kwak EK, Kim BW, Atala A, Yoo JJ. Identification and characterization of bioactive factors in bladder submucosa matrix. Biomaterials. 2007;28:4251–6.
209. Hodde JP, Record RD, Lang HA, Badylak SF. Vascular endothelial growth factor in porcine-derived extracellular matrix. Endothel J Endothel Cell Res. 2001;8:11–24.
210. Wu Z, Tang Y, Fang H, Su Z, Xu B, Lin Y, Zhang P, Wei X. Decellularized scaffolds containing hyaluronic acid and EGF for promoting the recovery of skin wounds. J Mater Sci Mater Med. 2015;26:5322.
211. Su Z, Ma H, Wu Z, Zeng H, Li Z, Wang Y, Liu G, Xu B, Lin Y, Zhang P, Wei X. Enhancement of skin wound healing with decellularized scaffolds loaded with hyaluronic acid and epidermal growth factor. Mater Sci Eng C Mater Biol Appl. 2014;44:440–8.
212. Luo X, Kulig KM, Finkelstein EB, Nicholson MF, Liu XH, Goldman SM, Vacanti JP, Grottkau BE, Pomerantseva I, Sundback CA, Neville CM. In vitro evaluation of decellularized ECM-derived surgical scaffold biomaterials. J Biomed Mater Res B Appl Biomater. 2017;105:585–93.

213. Cronce MJ, Faulknor RA, Pomerantseva I, Liu XH, Goldman SM, Ekwueme EC, Mwizerwa O, Neville CM, Sundback CA. In vivo response to decellularized mesothelium scaffolds. J Biomed Mater Res B Appl Biomater. 2018;106:716–25.
214. Capella-Monsonis H, Tilbury MA, Wall JG, Zeugolis DI. Porcine mesothelium matrix as a biomaterial for wound healing applications. Mater Today Biol. 2020;7:100057.
215. Melman L, Jenkins ED, Hamilton NA, Bender LC, Brodt MD, Deeken CR, Greco SC, Frisella MM, Matthews BD. Early biocompatibility of crosslinked and non-crosslinked biologic meshes in a porcine model of ventral hernia repair. Hernia. 2011;15:157–64.
216. Deeken CR, Melman L, Jenkins ED, Greco SC, Frisella MM, Matthews BD. Histologic and biomechanical evaluation of crosslinked and non-crosslinked biologic meshes in a porcine model of ventral incisional hernia repair. J Am Coll Surg. 2011;212:880–8.
217. Singh H, Kumar N, Sharma AK, Kataria M, Munjal A, Kumar A, Dewangan R, Kumar V, Devarathnam J, Kumar S. Activity of MMP-9 after repair of abdominal wall defects with acellular and crosslinked bovine pericardium in rabbit. Int Wound J. 2014;11:5–13.
218. Ji Y, Zhou J, Sun T, Tang K, Xiong Z, Ren Z, Yao S, Chen K, Yang F, Zhu F, Guo X. Diverse preparation methods for small intestinal submucosa (SIS): decellularization, components, and structure. J Biomed Mater Res A. 2019;107:689–97.
219. Badylak SF. The extracellular matrix as a scaffold for tissue reconstruction. Cell Dev Biol. 2002;13:377–83.
220. Nowocin AK, Southgate A, Gabe SM, Ansari T. Biocompatibility and potential of decellularized porcine small intestine to support cellular attachment and growth. J Tissue Eng Regen Med. 2016;10:E23–33.
221. Scheller K, Dally I, Hartmann N, Munst B, Braspenning J, Walles H. Upcyte® microvascular endothelial cells repopulate decellularized scaffold. Tissue Eng Part C Methods. 2013;19:57–67.
222. Glynn JJ, Polsin EG, Hinds MT. Crosslinking decreases the hemocompatibility of decellularized, porcine small intestinal submucosa. Acta Biomater. 2015;14:96–103.
223. Oliveira AC, Garzon I, Ionescu AM, Carriel V, Cardona L, Gonzalez-Andrades M, Perez Mdel M, Alaminos M, Campos A. Evaluation of small intestine grafts decellularization methods for corneal tissue engineering. PLoS One. 2013;8:e66538.
224. Fiala R, Vidlar A, Vrtal R, Belej K, Student V. Porcine small intestinal submucosa graft for repair of anterior urethral strictures. Eur Urol. 2007;51:1702–8.
225. Lemos NL, Kamergorodsky G, Faria AL, Ribeiro PA, Auge AP, Aoki T. Small intestinal submucosa patch for extensive vaginal endometriosis resection. J Minim Invasive Gynecol. 2009;16:765–7.
226. Keckler SJ, Spilde TL, Peter SDS, Tsao K, Ostlie DJ. Treatment of bronchopleural fistula with small intestinal mucosa and fibrin glue sealant. Ann Thorac Surg. 2007;84:1383–6.
227. Woo JS, Fishbein MC, Reemtsen B. Histologic examination of decellularized porcine intestinal submucosa extracellular matrix (CorMatrix) in pediatric congenital heart surgery. Cardiovasc Pathol. 2016;25:12–7.
228. Dew L, English WR, Chong CK, MacNeil S. Investigating neovascularization in rat decellularized intestine: an in vitro platform for studying angiogenesis. Tissue Eng Part A. 2016;22:1317–26.
229. Klinger A, Kawata M, Villalobos M, Jones RB, Pike S, Wu N, Chang S, Zhang P, DiMuzio P, Vernengo J, Benvenuto P, Goldfarb RD, Hunter K, Liu Y, Carpenter JP, Tulenko TN. Living scaffolds: surgical repair using scaffolds seeded with human adipose-derived stem cells. Hernia. 2016;20:161–70.
230. Groeber F, Engelhardt L, Lange J, Kurdyn S, Schmid FF, Rucker C, Mielke S, Walles H. A first vascularized skin equivalent as an alternative to animal experimentation. ALTEX. 2016;33:415–22.
231. Liu S, Zhang H, Zhang X, Lu W, Huang X, Xie H, Zhou J, Wang W, Zhang Y, Liu Y, Deng Z, Jin Y. Synergistic angiogenesis promoting effects of extracellular matrix scaffolds and adipose-derived stem cells during wound repair. Tissue Eng Part A. 2011;17:725–39.

232. Badylak SF, Tullius R, Kokini K. The use of xenogeneic small intestinal submucosa as a biomaterial for Achilles tendon repair in a dog model. J Biomed Mater Res. 1995;29:977–85.
233. Badylak SF, Record R, Lindberg K, Hodde J, Park K. Small intestinal submucosa: a substrate for in vitro cell growth. J Biomater Sci Polym Ed. 1998;9:863–78.
234. Badylak SF. Xenogeneic extracellular matrix as a scaffold for tissue reconstruction. Transpl Immunol. 2004;12:367–77.
235. Rajabi-Zeleti S, Jalili-Firoozinezhad S, Azarnia M. Engineered whole organs and complex tissues. Biomaterials. 2014;35:970–82.
236. Da LC, Huang YZ, Xie HQ. Progress in development of bioderived materials for dermal wound healing. Regen Biomater. 2017;4:325–34.
237. Parmaksiz M, Dogan A, Odabas S, Elcin AE, Elcin YM. Clinical applications of decellularized extracellular matrices for tissue engineering and regenerative medicine. Biomed Mater. 2016;11:022003.
238. Hodde J, Janis A, Hiles M. Effects of sterilization on an extracellular matrix scaffold: part II. Bioactivity and matrix interaction. J Mater Sci Mater Med. 2007;18:545–50.
239. Badylak SF, Gilbert TW. Immune response to biologic scaffold materials. Semin Immunol. 2008;20:109–16.
240. Huleihel L, Dziki JL, Bartolacci JG, Rausch T, Scarritt ME, Cramer MC, Vorobyov T, LoPresti ST, Swineheart IT, White LJ, Brown BN, Badylak SF. Macrophage phenotype in response to ECM bioscaffolds. Semin Immunol. 2017;29:2–13.
241. Zeng R, Lin C, Lin Z. Approaches to cutaneous wound healing: basics and future directions. Cell Tissue Res. 2018;374:57–63.
242. Yeh DD, Nazarian RM, Demetri L, Mesar T, Dijkink S, Larentzakis A, Velmahos G, Sadik KW. Histopathological assessment of OASIS ultra on critical-sized wound healing: a pilot study. J Cutan Pathol. 2017;44:523–9.
243. Hodde J. Naturally occurring scaffolds for soft tissue repair and regeneration. Tissue Eng. 2002;8:295–308.
244. Seeger JM. Effectiveness of an extracellular matrix graft (OASIS Wound Matrix) in the treatment of chronic leg ulcers: a randomized clinical trial. Yearbook Surg. 2006;2006:391–2.
245. Wang L, Wang W, Liao J, Wang F, Jiang J, Cao C, Li S. Novel bilayer wound dressing composed of SIS membrane with SIS cryogel enhanced wound healing process. Mater Sci Eng C Mater Biol Appl. 2018;85:162–9.
246. Clough A, Ball J, Smith GS. Porcine small intestine submucosa matrix (Surgisis) for esophageal perforation. Ann Thorac Surg. 2011;91:e15–6.
247. Ginting N, Tremblay L, Kortbeek JB. Surgisis® in the management of the complex abdominal wall in trauma: a case series and review of the literature. Injury. 2010;41:970–3.
248. Brtil E, Brtil CP, Comandau DE. Perineal reconstruction with biologic graft vulvoplasty for verrucous carcinoma treated by repeated vulvar excisions: a case report. Romanian J Morphol Embryol. 2015;56:537–43.
249. Trelford JD, Trelford-Sauder M. The amnion in surgery, past and present. Am J Obstet Gynecol. 1979;134:833–45.
250. Simone HB. The grafting of preserved amniotic membrane to burned and ulcerated surfaces. Cell Tissue Bank. 2007;2007:77–84.
251. Tseng SCG, Espana EM, Kawakita T, Di Pascuale MA, Li W, He H, Liu T-S, Cho T-H, Gao Y-Y, Yeh L-K, Liu C-Y. How does amniotic membrane work? Ocul Surf. 2004;2:177–87.
252. Wilshaw SP, Kearney JN, Fisher J. Production of an acellular amniotic membrane matrix for use in tissue engineering. Tissue Eng. 2006;12:2117.
253. Chen YJ, Chung MC, Jane Yao CC, Huang CH, Chang HH, Jeng JH, Young TH. The effects of acellular amniotic membrane matrix on osteogenic differentiation and ERK1/2 signaling in human dental apical papilla cells. Biomaterials. 2012;33:455–63.
254. Wilshaw SP, Kearney J, Fisher J, Ingham E. Biocompatibility and potential of acellular human amniotic membrane to support the attachment and proliferation of allogeneic cells. Tissue Eng Part A. 2008;14:463–72.

255. Kshersagar J, Kshirsagar R, Desai S, Bohara R, Joshi M. Decellularized amnion scaffold with activated PRP: a new paradigm dressing material for burn wound healing. Cell Tissue Bank. 2018;19:423–36.
256. Yang L, Shirakata Y, Tokumaru S, Xiuju D, Tohyama M, Hanakawa Y, Hirakawa S, Sayama K, Hashimoto K. Living skin equivalents constructed using human amnions as a matrix. 107. New skin-equivalent model from de-epithelialized amnion membrane. J Dermatol Sci. 2009;56:188–95.
257. Yang L, Shirakata Y, Shudou M, Dai X, Tokumaru S, Hirakawa S, Sayama K, Hamuro J, Hashimoto K. New skin-equivalent model from de-epithelialized amnion membrane. Cell Tissue Res. 2006;326:69–77.
258. Kim SS, Song CK, Shon SK, Lee KY, Kim CH, Lee MJ, Wang L. Effects of amniotic membrane extract and deferoxamine on angiogenesis in wound healing: an in vivo model. Cell Tissue Res. 2009;336:59–66.
259. Momeni M, Zarehaghighi M, Hajimiri M, Khorasani G, Dinarvand R, Nekookar A, Sodeifi N, Khosravani P, Shayanasl N, Ebrahimi M. In vitro and in vivo investigation of a novel amniotic based chitosan dressing for wound healing. Wound Repair Regen. 2018;26:87–101.
260. Chuck RS, Graff JM, Bryant MR, Sweet PM. Biomechanical characterization of human amniotic membrane preparations for ocular surface reconstruction. Ophthalmic Res. 2004;36:341–8.
261. Henderson PW, Nagineni VV, Harper A, Bavinck N, Sohn AM, Krijgh DD, Jimenez N, Weinstein AL, Spector JA. Development of an acellular bioengineered matrix with a dominant vascular pedicle. J Surg Res. 2010;164:1–5.
262. Qu J, Van Hogezand RM, Zhao C, Kuo BJ, Carlsen BT. Decellularization of a fasciocutaneous flap for use as a perfusable scaffold. Ann Plast Surg. 2015;75:112–6.
263. Schmidt VJ, Wietbrock JO, Leibig N, Hernekamp JF, Henn D, Radu CA, Kneser U. Hemodynamically stimulated and in vivo generated axially vascularized soft tissue free flaps for closure of complex defects: Evaluation in a small animal model. J Tissue Eng Regen Med. 2018;12:622–32.
264. Zhang Q, Johnson JA, Dunne LW. Decellularized skin/adipose tissue flap matrix for engineering vascularized composite soft tissue flaps. Acta Biomater. 2016;35:166–84.
265. Jack GS, Zhang R, Lee M, Xu Y, Wu BM, Rodríguez LV. Urinary bladder smooth muscle engineered from adipose stem cells and a three dimensional synthetic composite. Biomaterials. 2009;30:3259–70.
266. McCloskey KD. Bladder interstitial cells: an updated review of current knowledge. Acta Physiol. 2013;207:7–15.
267. Carattino MD, Prakasam HS, Ruiz WG, Clayton DR, McGuire M, Gallo LI, Apodaca G. Bladder filling and voiding affect umbrella cell tight junction organization and function. Am J Physiol Renal Physiol. 2013;305:F1158–68.
268. Andersson KE, McCloskey KD. Lamina propria: the functional center of the bladder? Neurourol Urodyn. 2014;33:9–16.
269. Shah AP, Mevcha A, Wilby D, Alatsatianos A, Hardman JC, Jacques S, Wilton JC. Continence and micturition: an anatomical basis. Clin Anat. 2014;27:1275–83.
270. Wyndaele JJ. The management of neurogenic lower urinary tract dysfunction after spinal cord injury. Nat Rev Urol. 2016;13:705–14.
271. Atala A. Tissue engineering of human bladder. Br Med Bull. 2011;97:81–104.
272. Atala A, Bauer SB, Soker S, Yoo JJ, Retik AB. Tissue-engineered autologous bladders for patients needing cystoplasty. Lancet. 2006;367:1241–6.
273. Atala A. Bladder regeneration by tissue engineering. BJU Int. 2015;88:765–70.
274. Tiemessen D, de Jonge P, Daamen W, Feitz W, Geutjes P, Oosterwijk E. The effect of a cyclic uniaxial strain on urinary bladder cells. World J Urol. 2017;35:1531–9.
275. Birder LA. Urinary bladder urothelium: molecular sensors of chemical/thermal/mechanical stimuli. Vasc Pharmacol. 2006;45:221–6.

276. Farhat WA, Yeger H. Does mechanical stimulation have any role in urinary bladder tissue engineering? World J Urol. 2008;26:301–5.
277. Gill BC, Damaser MS, Chermansky CJ. Future perspectives in bladder tissue engineering. Curr Bladder Dysfunct Rep. 2014;10:443–8.
278. Guilak F, Butler DL, Goldstein SA, Baaijens FP. Biomechanics and mechanobiology in functional tissue engineering. J Biomech. 2014;47:1933–40.
279. Lin HK, Madihally SV, Palmer B, Frimberger D, Fung KM, Kropp BP. Biomatrices for bladder reconstruction. Adv Drug Deliv Rev. 2015;82-83:47–63.
280. Osborn SL, Kurzrock EA. Bioengineered bladder tissue—close but yet so far! J Urol. 2015;194:619–20.
281. Ajalloueian F, Lemon G, Hilborn J, Chronakis IS, Fossum M. Bladder biomechanics and the use of scaffolds for regenerative medicine in the urinary bladder. Nat Rev Urol. 2018;15:155–74.
282. Horst M, Madduri S, Gobet R, Sulser T, Milleret V, Hall H, Atala A, Eberli D. Engineering functional bladder tissues. J Tissue Eng Regen Med. 2013;7:515–22.
283. Wiles K, Fishman JM, De Coppi P, Birchall MA. The host immune response to tissue-engineered organs: current problems and future directions. Tissue Eng Part B Rev. 2016;22:208–19.
284. Wrenn SM, Weiss DJ. Whole-organ tissue engineering: no longer just a dream. Curr Pathobiol Rep. 2016;4:87–98.
285. Corona BT, Ward CL, Harrison BS, Christ GJ. Regenerative medicine: basic concepts, current status, and future applications. J Investig Med. 2010;58:849–58.
286. Simoes IN, Vale P, Soker S, Atala A, Keller D, Noiva R, Carvalho S, Peleteiro C, Cabral JM, Eberli D, da Silva CL, Baptista PM. Acellular urethra bioscaffold: decellularization of whole urethras for tissue engineering applications. Sci Rep. 2017;7:41934.
287. Turner AM, Subramaniam R, Thomas DF, Southgate J. Generation of a functional, differentiated porcine urothelial tissue in vitro. Eur Urol. 2008;54:1423–32.
288. Yang B, Zhang Y, Zhou L, Sun Z, Zheng J, Chen Y, Dai Y. Development of a porcine bladder acellular matrix with well-preserved extracellular bioactive factors for tissue engineering. Tissue Eng Part C Methods. 2010;16:1201–11.
289. Liu Y, Bharadwaj S, Lee SJ, Atala A, Zhang Y. Optimization of a natural collagen scaffold to aid cell-matrix penetration for urologic tissue engineering. Biomaterials. 2009;30:3865–73.
290. Oberpenning F, Meng J, Yoo JJ. De novo reconstitution of a functional mammalian urinary bladder by tissue engineering. Nat Biotechnol. 1999;17:149–55.
291. Zhang Y, Frimberger D, Cheng EY, Lin HK, Kropp BP. Challenges in a larger bladder replacement with cell-seeded and unseeded small intestinal submucosa grafts in a subtotal cystectomy model. BJU Int. 2006;98:1100–5.
292. Reddy PP, Barrieras DJ, Wilson G. Regeneration of functional bladder substitutes using large segment acellular matrix allografts in a porcine model. J Urol. 2000;164:936–41.
293. Sievert KD, Amend B, Stenzl A. Tissue engineering for the lower urinary tract: a review of a state of the art approach. Eur Urol. 2007;52:1580–9.
294. Brown A, Farhat W, Merguerian P. 22 week assessment of bladder acellular matrix as a bladder augmentation material in a porcine model. Biomaterials. 2002;23:2179–90.
295. Biers SM, Venn SN, Greenwell TJ. The past, present and future of augmentation cystoplasty. BJU Int. 2012;109:1280–93.
296. Kajbafzadeh AM, Khorramirouz R, Sabetkish S, Ataei Talebi M, Akbarzadeh A, Keihani S. Pediatr Surg Int. 2016;32:615–22.
297. Ayyildiz A, Akgul KT, Huri E, Nuhoglu B, Kilicoglu B, Ustun H, Gurdal M, Germiyanoglu C. Use of porcine small intestinal submucosa in bladder augmentation in rabbit: longterm histological outcome. ANZ J Surg. 2008;78:82–6.
298. Song L, Murphy SV, Yang B, Xu Y, Zhang Y, Atala A. Bladder acellular matrix and its application in bladder augmentation. Tissue Eng Part B Rev. 2014;20:163–72.

299. Elbahnasy A, Shalhav A, Hoenig D. Bladder wall substitution with synthetic and non-intestinal organic materials. J Urol. 1998;159:628–37.
300. Kapur SK, Butler CE. Lateral abdominal wall reconstruction. Semin Plast Surg. 2018;32 (3):141–6.
301. Hutan M, Bartko C, Majesky I, Prochotsky A, Sekac J, Skultety J. Reconstruction option of abdominal wounds with large tissue defects. BMC Surg. 2014;14:50.
302. Patel NG, Ratanshi I, Buchel EW. The best of abdominal wall reconstruction. Plast Reconstr Surg. 2018;141(1):113e–36e.
303. Rosen MJ, Krpata DM, Ermlich B, Blatnik JA. A 5-year clinical experience with single-staged repairs of infected and contaminated abdominal wall defects utilizing biologic mesh. Ann Surg. 2013;257(6):991–6.
304. Shahan CP, Stoikes NF, Webb DL, Voeller GR. Sutureless onlay hernia repair: a review of 97 patients. Surg Endosc. 2016;30(8):3256–61.
305. Rohrich RJ, Lowe JB, Hackney FL, Bowman JL, Hobar PC. An algorithm for abdominal wall reconstruction. Plast Reconstr Surg. 2000;105(1):202–16.
306. Dubay DA, Wang X, Adamson B, Kuzon WM, Dennis RG, Franz MG. Mesh incisional herniorrhaphy increases abdominal wall elastic properties: a mechanism for decreased hernia recurrences in comparison with suture repair. Surgery. 2006;140(1):14–24.
307. Bellows CF, Alder A, Helton WS. Abdominal wall reconstruction using biological tissue grafts: present status and future opportunities. Expert Rev Med Devices. 2006;3(5):657–75.
308. Smith SE, Cozad MJ, Grant DA, Ramshaw BJ, Grant SA. Materials characterization of explanted polypropylene hernia mesh: Patient factor correlation. J Biomater Appl. 2016;30 (7):1026–35.
309. Bringman S, Conze J, Cuccurullo D, Deprest J, Junge K, Klosterhalfen B, Parra-Davila E, Ramshaw B, Schumpelick V. Hernia repair: the search for ideal meshes. Hernia. 2010;14 (1):81–7.
310. Burger JWA, Luijendijk RW, Hop WCJ, Halm JA, Verdaasdonk EGG, Jeekel J. Long-term follow-up of a randomized controlled trial of suture versus mesh repair of incisional hernia. Ann Surg. 2004;240(4):578–83.
311. Buinewicz B, Rosen B. Acellular cadaveric dermis (AlloDerm): a new alternative for abdominal hernia repair. Ann Plast Surg. 2004;52(2):188–94.
312. Breuing K, Butler CE, Ferzoco S, Franz M, Hultman CS, Kilbridge JF, Rosen M, Silverman RP, Vargo D, Grp VHW. Incisional ventral hernias: review of the literature and recommendations regarding the grading and technique of repair. Surgery. 2010;148(3):544–58.
313. Margulies IG, Salzberg CA. The use of acellular dermal matrix in breast reconstruction: evolution of techniques over 2 decades. Gland Surg. 2019;8(1):3–10.
314. Burns NK, Jaffari MV, Rios CN, Mathur AB, Butler CE. Non-cross-linked porcine acellular dermal matrices for abdominal wall reconstruction. Plast Reconstr Surg. 2010;125(1):167–76.
315. Ghazi B, Deigni O, Yezhelyev M, Losken A. Current options in the management of complex abdominal wall defects. Ann Plast Surg. 2011;66(5):488–92.
316. Nemeth NL, Butler CE. Complex torso reconstruction with human acellular dermal matrix: long-term clinical follow-up. Plast Reconstr Surg. 2009;123(1):192–6.
317. Eberli D, Rodriguez S, Atala A, Yoo JJ. In vivo evaluation of acellular human dermis for abdominal wall repair. J Biomed Mater Res A. 2010;93a(4):1527–38.
318. Wang MG, Yang S, Cao Z, Hu SY. Application of acellular tissue matrix for enhancement of weak abdominal wall in animal model. Biomed Res Int. 2020;2020:3475289.
319. Sahoo S, Baker AR, Haskins IN, Krpata DM, Rosen MJ, Derwin KA. Assessment of human acellular dermis graft in porcine models for ventral hernia repair. Tissue Eng Part C Med. 2017;23(11):718–27.
320. Menon NG, Rodriguez ED, Byrnes CK, Girotto JA, Goldberg NH, Silverman RP. Revascularization of human acellular dermis in full-thickness abdominal wall reconstruction in the rabbit model. (vol 50, pg 523, 2002). Ann Plast Surg. 2003;51(2):228.

321. Ngo MD, Aberman HM, Hawes ML, Choi B, Gertzman AA. Evaluation of human acellular dermis versus porcine acellular dermis in an in vivo model for incisional hernia repair. Cell Tissue Bank. 2011;12(2):135–45.
322. Butler CE, Langstein HN, Kronowitz SJ. Pelvic, abdominal, and chest wall reconstruction with AlloDerm in patients at increased risk for mesh-related complications. Plast Reconstr Surg. 2005;116(5):1263–75.
323. Costa A, Naranjo JD, Londono R, Badylak SF. Biologic scaffolds. CSH Perspect Med. 2017;7 (9):a025676.
324. Sahoo S, DeLozier KR, Dumm RA, Rosen MJ, Derwin KA. Fiber-reinforced dermis graft for ventral hernia repair. J Mech Behav Biomed. 2014;34:320–9.
325. Ma JJ, Sahoo S, Baker AR, Derwin KA. Investigating muscle regeneration with a dermis/small intestinal submucosa scaffold in a rat full-thickness abdominal wall defect model. J Biomed Mater Res B. 2015;103(2):355–64.
326. Jiang WW, Zhang J, Lv XF, Lu CG, Chen H, Xu XQ, Tang WB. Use of small intestinal submucosal and acellular dermal matrix grafts in giant omphaloceles in neonates and a rabbit abdominal wall defect model. J Pediatr Surg. 2016;51(3):368–73.
327. Liu ZN, Tang R, Zhou ZY, Song ZC, Wang HC, Gu Y. Comparison of two porcine-derived materials for repairing abdominal wall defects in rats. PLoS One. 2011;6(5):e20520.
328. Palmer EM, Beilfuss BA, Nagai T, Semnani RT, Badylak SF, van Seventer GA. Human helper T cell activation and differentiation is suppressed by porcine small intestinal submucosa. Tissue Eng. 2002;8(5):893–900.
329. Cordero A, Hernandez-Gascon B, Pascual G, Bellon JM, Calvo B, Pena E. Biaxial mechanical evaluation of absorbable and nonabsorbable synthetic surgical meshes used for hernia repair: physiological loads modify anisotropy response. Ann Biomed Eng. 2016;44(7):2181–8.
330. Wang ZL, Wu SZ, Li ZF, Guo JH, Zhang Y, Pi JK, Hu JG, Yang XJ, Huang FG, Xie HQ. Comparison of small intestinal submucosa and polypropylene mesh for abdominal wall defect repair. J Biomat Sci Polym E. 2018;29(6):663–82.
331. Tang R, Wang X, Zhang HY, Liang X, Feng XY, Zhu XQ, Lu XW, Wu F, Liu ZN. Promoting early neovascularization of SIS-repaired abdominal wall by controlled release of bioactive VEGF. RSC Adv. 2018;8(9):4548–60.
332. Wang L, La DM, Yang B, Jiang ZP, Zhang YC, Zhou J, Lai W, Chen S. Reconstruction of abdominal wall defects using small intestinal submucosa coated with gelatin hydrogel incorporating basic fibroblast growth factor. Acta Cir Bras. 2014;29(4):252–60.
333. Zhou HY, Zhang J, Yan RL, Wang QA, Fan LY, Zhang Q, Wang WJ, Hu ZQ. Improving the antibacterial property of porcine small intestinal submucosa by nano-silver supplementation a promising biological material to address the need for contaminated defect repair. Ann Surg. 2011;253(5):1033–41.
334. Kankala RK, Zhu K, Sun XN, Liu CG, Wang SB, Chen AZ. Cardiac tissue engineering on the nanoscale. ACS Biomater Sci Eng. 2018;4(3):800–18.
335. Godier-Furnemont AFG, Martens TP, Koeckert MS, Wan L, Parks J, Arai K, Zhang GP, Hudson B, Homma S, Vunjak-Novakovic G. Composite scaffold provides a cell delivery platform for cardiovascular repair. Proc Natl Acad Sci U S A. 2011;108(19):7974–9.
336. Jones L. Heart Disease and Stroke Statistics-2009 update: a report from the American Heart Association Statistics Committee and Stroke Statistics Subcommittee (vol 119, pg e21, 2009). Circulation. 2011;124(16):E424.
337. Muller P, Lemcke H, David R. Stem cell therapy in heart diseases - cell types, mechanisms and improvement strategies. Cell Physiol Biochem. 2018;48(6):2607–55.
338. Perez-Estenaga I, Prosper F, Pelacho B. Allogeneic mesenchymal stem cells and biomaterials: the perfect match for cardiac repair? Int J Mol Sci. 2018;19(10):3236.
339. Hashizume R, Hong Y, Takanari K, Fujimoto KL, Tobita K, Wagner WR. The effect of polymer degradation time on functional outcomes of temporary elastic patch support in ischemic cardiomyopathy. Biomaterials. 2013;34(30):7353–63.

340. Lin YD, Ko MC, Wu ST, Li SF, Hu JF, Lai YJ, Harn HIC, Laio IC, Yeh ML, Yeh HI, Tang MJ, Chang KC, Su FC, Wei EIH, Lee ST, Chen JH, Hoffman AS, Wu WT, Hsieh PCH. A nanopatterned cell-seeded cardiac patch prevents electro-uncoupling and improves the therapeutic efficacy of cardiac repair. Biomater Sci. 2014;2(4):567–80.
341. Menasche P, Vanneaux V, Hagege A, Bel A, Cholley B, Cacciapuoti I, Parouchev A, Benhamouda N, Tachdjian G, Tosca L, Trouvin JH, Fabreguettes JR, Bellamy V, Guillemain R, Boissel CS, Tartour E, Desnos M, Larghero J. Human embryonic stem cell-derived cardiac progenitors for severe heart failure treatment: first clinical case report. Eur Heart J. 2015;36(30):2011–7.
342. Becker M, Maring JA, Oberwallner B, Kappler B, Klein O, Falk V, Stamm C. Processing of human cardiac tissue toward extracellular matrix self-assembling hydrogel for in vitro and in vivo applications. J Vis Exp. 2017;130:56419.
343. Moroni F, Mirabella T. Decellularized matrices for cardiovascular tissue engineering. Am J Stem Cells. 2014;3(1):1–20.
344. Sanchez PL, Fernandez-Santos ME, Costanza S, Climent AM, Moscoso I, Gonzalez-Nicolas MA, Sanz-Ruiz R, Rodriguez H, Kren SM, Garrido G, Escalante JL, Bermejo J, Elizaga J, Menarguez J, Yotti R, Perez del Villar C, Espinosa MA, Guillem MS, Willerson JT, Bernad A, Matesanz R, Taylor DA, Fernandez-Aviles F. Acellular human heart matrix: a critical step toward whole heart grafts. Biomaterials. 2015;61:279–89.
345. Oberwallner B, Brodarac A, Choi YH, Saric T, Anic P, Morawietz L, Stamm C. Preparation of cardiac extracellular matrix scaffolds by decellularization of human myocardium. J Biomed Mater Res A. 2014;102(9):3263–72.
346. Kappler B, Anic P, Becker M, Bader A, Klose K, Klein O, Oberwallner B, Choi YH, Falk V, Stamm C. The cytoprotective capacity of processed human cardiac extracellular matrix. J Mater Sci Mater Med. 2016;27(7):120.
347. Becker M, Maring JA, Schneider M, Martin AXH, Seifert M, Klein O, Braun T, Falk V, Stamm C. Towards a novel patch material for cardiac applications: tissue-specific extracellular matrix introduces essential key features to decellularized amniotic membrane. Int J Mol Sci. 2018;19(4):1032.
348. Johnson TD, Hill RC, Dzieciatkowska M, Nigam V, Behfar A, Christman KL, Hansen KC. Quantification of decellularized human myocardial matrix: a comparison of six patients. Proteomics Clin Appl. 2016;10(1):75–83.
349. Wang RM, Johnson TD, He JJ, Rong ZL, Wong M, Nigam V, Behfar A, Xu Y, Christman KL. Humanized mouse model for assessing the human immune response to xenogeneic and allogeneic decellularized biomaterials. Biomaterials. 2017;129:98–110.
350. Guhathakurta S, Mathapati S, Bishi DK, Rallapalli S, Cherian KM. Nanofiber-reinforced myocardial tissue-construct as ventricular assist device. Asian Cardiovasc Thorac Ann. 2014;22(8):935–43.
351. Di Meglio F, Nurzynska D, Romano V, Miraglia R, Belviso I, Sacco AM, Barbato V, Di Gennaro M, Granato G, Maiello C, Montagnani S, Castaldo C. Optimization of human myocardium decellularization method for the construction of implantable patches. Tissue Eng Part C Med. 2017;23(9):525–39.
352. Guyette JP, Charest JM, Mills RW, Jank BJ, Moser PT, Gilpin SE, Gershlak JR, Okamoto T, Gonzalez G, Milan DJ, Gaudette GR, Ott HC. Bioengineering human myocardium on native extracellular matrix. Circ Res. 2016;118(1):56–72.
353. van Steenberghe M, Schubert T, Xhema D, Bouzin C, Guiot Y, Duisit J, Abdelhamid K, Gianello P. Enhanced vascular regeneration with chemically/physically treated bovine/human pericardium in rodents. J Surg Res. 2018;222:167–79.
354. Madani MM. A clinical commentary on the article "patient-to-patient variability in autologous pericardial matrix scaffolds for cardiac repair". J Cardiovasc Transl. 2011;4(5):557–8.
355. Vinci MC, Tessitore G, Castiglioni L, Prandi F, Soncini M, Santoro R, Consolo F, Colazzo F, Micheli B, Sironi L, Polvani G, Pesce M. Mechanical compliance and immunological

compatibility of fixative-free decellularized/cryopreserved human pericardium. PLoS One. 2013;8(5):0064769.
356. Mirsadraee S, Wilcox HE, Korossis SA, Kearney JN, Watterson KG, Fisher J, Ingham E. Development and characterization of an acellular human pericardial matrix for tissue engineering. Tissue Eng. 2006;12(4):763–73.
357. Rajabi-Zeleti S, Jalili-Firoozinezhad S, Azarnia M, Khayyatan F, Vahdat S, Nikeghbalian S, Khademhosseini A, Baharvand H, Aghdami N. The behavior of cardiac progenitor cells on macroporous pericardium-derived scaffolds. Biomaterials. 2014;35(3):970–82.
358. Perea-Gil I, Galvez-Monton C, Prat-Vidal C, Jorba I, Segu-Verges C, Roura S, Soler-Botija C, Iborra-Egea O, Revuelta-Lopez E, Fernandez MA, Farre R, Navajas D, Bayes-Genis A. Head-to-head comparison of two engineered cardiac grafts for myocardial repair: from scaffold characterization to pre-clinical testing. Sci Rep. 2018;8(1):6708.
359. Mirsadraee S, Wilcox HE, Watterson KG, Kearney JN, Hunt J, Fisher J, Ingham E. Biocompatibility of acellular human pericardium. J Surg Res. 2007;143(2):407–14.
360. van Steenberghe M, Schubert T, Bouzin C, Caravaggio C, Guiot Y, Xhema D, Gianello P. Enhanced vascular biocompatibility and remodeling of decellularized and secured xenogeneic/allogeneic matrices in a porcine model. Eur Surg Res. 2018;59(1-2):58–71.
361. Galvez-Monton C, Bragos R, Soler-Botija C, Diaz-Guemes I, Prat-Vidal C, Crisostomo V, Sanchez-Margallo FM, Llucia-Valldeperas A, Bogonez-Franco P, Perea-Gil I, Roura S, Bayes-Genis A. Noninvasive assessment of an engineered bioactive graft in myocardial infarction: impact on cardiac function and scar healing. Stem Cells Transl Med. 2017;6 (2):647–55.
362. Cheung DY, Duan B, Butcher JT. Current progress in tissue engineering of heart valves: multiscale problems, multiscale solutions. Expert Opin Biol Ther. 2015;15(8):1155–72.
363. Naso F, Gandaglia A. Different approaches to heart valve decellularization: a comprehensive overview of the past 30 years. Xenotransplantation. 2018;25(1):e12354.
364. Cebotari S, Mertsching H, Kallenbach K, Kostin S, Repin O, Batrinac A, Kleczka C, Ciubotaru A, Haverich A. Construction of autologous human heart valves based on an acellular allograft matrix. Circulation. 2002;106(12 Suppl 1):I63–8.
365. Gerson CJ, Elkins RC, Goldstein S, Heacox AE. Structural integrity of collagen and elastin in SynerGraft(R) decellularized-cryopreserved human heart valves. Cryobiology. 2012;64 (1):33–42.
366. Jiao T, Clifton RJ, Converse GL, Hopkins RA. Measurements of the effects of decellularization on viscoelastic properties of tissues in ovine, baboon, and human heart valves. Tissue Eng Part A. 2012;18(3-4):423–31.
367. Weymann A, Schmack B, Okada T, Soos P, Istok R, Radovits T, Straub B, Barnucz E, Loganathan S, Patzold I, Chaimow N, Schies C, Korkmaz S, Tochtermann U, Karck M, Szabo G. Reendothelialization of human heart valve neoscaffolds using umbilical cord-derived endothelial cells. Circ J. 2013;77(1):207–16.
368. Vafaee T, Thomas D, Desai A, Jennings LM, Berry H, Rooney P, Kearney J, Fisher J, Ingham E. Decellularization of human donor aortic and pulmonary valved conduits using low concentration sodium dodecyl sulfate. J Tissue Eng Regen Med. 2018;12(2):E841–53.
369. Rieder E, Seebacher G, Kasimir MT, Eichmair E, Winter B, Dekan B, Wolner E, Simon P, Weigel G. Tissue engineering of heart valves - decellularized porcine and human valve scaffolds differ importantly in residual potential to attract monocytic cells. Circulation. 2005;111(21):2792–7.
370. Iop L, Paolin A, Aguiari P, Trojan D, Cogliati E, Gerosa G. Decellularized cryopreserved allografts as off-the-shelf allogeneic alternative for heart valve replacement: in vitro assessment before clinical translation. J Cardiovasc Transl. 2017;10(2):93–103.
371. Iop L, Renier V, Naso F, Piccoli M, Bonetti A, Gandaglia A, Pozzobon M, Paolin A, Ortolani F, Marchini M, Spina M, De Coppi P, Sartore S, Gerosa G. The influence of heart valve leaflet matrix characteristics on the interaction between human mesenchymal stem cells and decellularized scaffolds. Biomaterials. 2009;30(25):4104–16.

372. Dainese L, Guarino A, Burba I, Esposito G, Pompilio G, Polvani G, Rossini A. Heart valve engineering: decellularized aortic homograft seeded with human cardiac stromal cells. J Heart Valve Dis. 2012;21(1):125–34.
373. Koenig F, Lee JS, Akra B, Hollweck T, Wintermantel E, Hagl C, Thierfelder N. Is transcatheter aortic valve implantation of living tissue-engineered valves feasible? An in vitro evaluation utilizing a decellularized and reseeded biohybrid valve. Artif Organs. 2016;40(8):727–37.
374. Kamel KS, Beckert LE, Stringer MD. Novel insights into the elastic and muscular components of the human trachea. Clin Anat. 2009;22(6):689–97.
375. Kanzaki M, Yamato M, Hatakeyama H, Kohno C, Yang J, Umemoto T, Kikuchi A, Okano T, Onuki T. Tissue engineered epithelial cell sheets for the creation of a bioartificial trachea. Tissue Eng. 2006;12(5):1275–83.
376. Yamashita M, Kanemaru S, Hirano S, Magrufov A, Tamaki H, Tamura Y, Kishimoto M, Omori K, Nakamura T, Ito J. Tracheal regeneration after partial resection: a tissue engineering approach. Laryngoscope. 2007;117(3):497–502.
377. Jungebluth P, Go T, Asnaghi A, Bellini S, Martorell J, Calore C, Urbani L, Ostertag H, Mantero S, Conconi MT, Macchiarini P. Structural and morphologic evaluation of a novel detergent-enzymatic tissue-engineered tracheal tubular matrix. J Thorac Cardiovasc Surg. 2009;138(3):586–93.
378. Walles T, Giere B, Hofmann M, Schanz J, Hofmann F, Mertsching H, Macchiarini P. Experimental generation of a tissue-engineered functional and vascularized trachea. J Thorac Cardiovasc Surg. 2004;128(6):900–6.
379. Conconi MT, De Coppi P, Di Liddo R, Vigolo S, Zanon GF, Parnigotto PP, Nussdorfer GG. Tracheal matrices, obtained by a detergent-enzymatic method, support in vitro the adhesion of chondrocytes and tracheal epithelial cells. Transpl Int. 2005;18(6):727–34.
380. Jungebluth P, Bader A, Baiguera S, Moller S, Jaus M, Lim ML, Fried K, Kjartansdottir KR, Go T, Nave H, Harringer W, Lundin V, Teixeira AI, Macchiarini P. The concept of in vivo airway tissue engineering. Biomaterials. 2012;33(17):4319–26.
381. Park JW, Pavcnik D, Uchida BT, Timmermans H, Corless CL, Yamakado K, Yamada K, Keller FS, Rosch J. Small intestinal submucosa covered expandable Z stents for treatment of tracheal injury: an experimental pilot study in swine. J Vasc Interv Radiol. 2000;11(10):1325–30.
382. Gubbels SP, Richardson M, Trune D, Bascom DA, Wax MK. Tracheal reconstruction with porcine small intestine submucosa in a rabbit model. Otolaryng Head Neck. 2006;134(6):1028–35.
383. Zhang LF, Liu Z, Cui PC, Zhao DQ, Chen WX. SIS with tissue-cultured allogenic cartilages patch tracheoplasty in a rabbit model for tracheal defect. Acta Otolaryngol. 2007;127(6):631–6.
384. John R, Douglas S. Gastrointestinal anatomy and physiology. Gastroenterology. 2015;148:1481–2.
385. Zorn AM, Wells JM. Vertebrate endoderm development and organ formation. Annu Rev Cell Dev Biol. 2009;25:221–51.
386. Kuppan P, Sethuraman S, Krishnan UM. Interaction of human smooth muscle cells with nanofibrous scaffolds: effect of fiber orientation on cell adhesion, proliferation, and functional gene expression. J Biomed Mater Res A. 2015;103:2236–50.
387. Montgomery RK, Mulberg AE, Grand RJ. Development of the human gastrointestinal tract: twenty years of progress. Gastroenterology. 1999;116:702–31.
388. Hussey GS, Cramer MC, Badylak SF. Extracellular matrix bioscaffolds for building gastrointestinal tissue. Cell Mol Gastroenterol Hepatol. 2018;5(1):1–13.
389. Goke M, Podolsky DK. Regulation of the mucosal epithelial barrier. Baillieres Clin Gastroenterol. 1996;10:393–405.
390. Goke M, Zuk A, Podolsky DK. Regulation and function of extracellular matrix intestinal epithelial restitution in vitro. Am J Phys. 1996;271:G729–40.

391. Shimshoni D, Yablecovitch L, Baram L. ECM remodelling in IBD: innocent bystander or partner in crime? The emerging role of extracellular molecular events in sustaining intestinal inflammation. Gut. 2015;64:367–72.
392. Craig R, Amernic J. A privatization success story: accounting and narrative expression over time. Account Audit Account J. 2008;21:1085–115.
393. Speca S, Giusti I, Rieder F, Latella G. Cellular and molecular mechanisms of intestinal fibrosis. World J Gastroenterol. 2012;18:3635–61.
394. Teller IC, Beaulieu JF. Interactions between laminin and epithelial cells in intestinal health and disease. Expert Rev Mol Med. 2001;3:1–18.
395. Worthley DL, Giraud AS, Wang TC. The extracellular matrix in digestive cancer. Cancer Microenviron. 2010;3:177–85.
396. Verma S, Kesh K, Ganguly N, Jana S, Swarnakar S. Matrix metalloproteinases and gastrointestinal cancers: impacts of dietary antioxidants. World J Biol Chem. 2014;5:355–76.
397. Weaver AM, Yamaguchi H, Yoshida N, Takanashi M, Ito Y, Fukami K, Yanagihara K, Yashiro M, Sakai R. Stromal fibroblasts mediate extracellular matrix remodeling and invasion of scirrhous gastric carcinoma cells. PLoS ONE. 2014;9:e85485.
398. Levitt MA, Dickie B, Pena A. Evaluation and treatment of the patient with Hirschsprung disease who is not doing well after a pull-through procedure. Semin Pediatr Surg. 2010;19:146–53.
399. Beyer-Berjot L, Joly F, Maggiori L, Corcos O, Bouhnik Y, Bretagnol F, Panis Y. Segmental reversal of the small bowel can end permanent parenteral nutrition dependency. Ann Surg. 2012;256:739–44.
400. Sampietro GM, Corsi F, Maconi G, Ardizzone S, Frontali A, Corona A, Porro GB, Foschi D. Prospective study of long-term results and prognostic factors after conservative surgery for small bowel Crohn's disease. Clin Gastroenterol Hepatol. 2009;7:183–91.
401. Braghetto I, Korn O, Valladares H, Debandi A, Diaz JC, Brunet L. Segmental reversal of the small bowel can end permanent parenteral nutrition dependency: an experience of 38 adults with short bowel syndrome. Int Surg. 2011;96:95–103.
402. Songun I, Putter H, Kranenbarg EM-K, Sasako M, van de Velde CJH. Surgical treatment of gastric cancer: 15-year follow-up results of the randomised nationwide Dutch D1D2 trial. Lancet Oncol. 2010;11:439–49.
403. Pearson RK. Cetuximab and chemotherapy as initial treatment for metastatic colorectal cancer. Yearbook Gastroenterol. 2010;2010:91–3.
404. Zhou Y, Zhao Q, Wang M. Coaxial electrospray for cell encapsulation and delivery in cell-based tissue engineering. Front Bioeng Biotechnol. 2015;4:14–7.
405. Liu H, Li X, Zhou G, Fan H, Fan Y. Electro spun sulfate silk fibro in a no fibrous scaffolds for vascular tissue engineering. Biomaterials. 2011;32:3784–93.
406. Kim HI, Ishihara K, Lee S, Seo JH, Kim HY, Suh D, Kim MU, Konno T, Takai M, Seo JS. Tissue response to poly fu lactica ckd-based blend with phospholipid polymer for biodegradable cardiovasc uars tents. Biomaterials. 2011;32:2241–7.
407. Wang CC, Yang KC, Lin KH, Liu HC, Lin FH. A highly organized three-dimensional alginate scaffold for cartilage tissue engineering prepared by microfluidic technology. Biomaterials. 2011;32:7118–26.
408. Sreerekha PR, Menon D, Nair SV, Chennazhi KP. Fabrication of electrospun poly (lactide-co-glycolide)-fibrin multiscale scaffold for myocardial regeneration in vitro. Tissue Eng Part A. 2013;19:849–59.
409. Chen G, Xia Y, Lu X, Zhou X, Zhang F, Gu N. Effects of surface functionalization of PLGA membranes for guided bone regeneration on proliferation and behavior of osteoblasts. J Biomed Mater Res A. 2013;101:44–53.
410. Hong Y, Guan J, Fujimoto KL, Hashizume R, Pelinescu AL, Wagner WR. Tailoring the degradation kinetics of poly(ester carbonate urethane) Jure a thermoplastic elastomers for tissue engineering scaffolds. Biomaterials. 2010;31:4249–58.

411. Jeong CG, Hollister SJ. Mechanical, permeability, and degradation properties of 3D designed poly(1, 8octanediol-co-citrate) scaffolds for soft tissue engineering. J Biomed Mater Res B Appl Biomater. 2010;93:141–9.
412. Lu G, Sheng B, Wang G, Wei Y, Gong Y, Zhang X, Zhang L. Controlling the degradation of covalently cross-linked carboxymethyl chitosan utilizing bimodal molecular weight distribution. J Biomater Appl. 2009;23:435–51.
413. Pu XM, Yao QQ, Yang Y, Sun ZZ, Zhang QQ. Invitro degradation of three-dimensional chitosan/apatite composite rods prepared via insitu precipitation. Int J Biol Macromol. 2012;51:868–73.
414. White LJ, Hutter V, Tai H, Howdle SM, Shakesheff KM. The effect of processing variables on morphological and mechanical properties of supercritical CO_2 foamed scaffolds for tissue engineering. Acta Biomater. 2012;8:61–71.
415. Nguyen LH, Kudva AK, Saxena NS, Roy K. Engineering articular cartilage with spatially-varying matrix composition and mechanical properties from a single stemcell population using a multi-layered hydrogel. Biomaterials. 2011;32:6946–52.
416. Murphy CM, Haugh MG, O'Brien FJ. The effect of mean pore size on cell attachment, proliferation and migration in collagen-glycosaminoglycan scaffolds for bone tissue engineering. Biomaterials. 2010;31:461–6.
417. Harley BA, Kim HD, Zaman MH, Yannas IV, Lauffenburger DA, Gibson LJ. Microarchitecture of three-dimensional scaffolds influences cell migration behavior via junction interactions. Biophys J. 2008;95:4013–24.
418. Haugh MG, Murphy CM, McKiernan RC, Altenbuchner C, O'Brien FJ. Crosslinking and mechanical properties significantly influence cell attachment, proliferation, and migration within collagen glycosaminoglycan scaffolds. Tissue Eng Part A. 2011;17:1201–8.
419. Bitar KN, Zakhem E. Tissue engineering and regenerative medicine as applied to the gastrointestinal tract. Curr Opin Biotechnol. 2013;24:909–15.
420. Beattie AJ, Gilbert TW, Guyot JP, Yates AJ, Badylak SF. Chemoattraction of progenitor cells by remodeling extracellular matrix scaffolds. Tissue Eng Part A. 2009;15:1119–25.
421. Dziki JL, Sicari BM, Wolf MT, Cramer MC, Badylak SF. Immunomodulation and mobilization of progenitor cells by extracellular matrix bioscaffolds for volumetric muscle loss treatment. Tissue Eng Part A. 2016;22:1129–39.
422. Agrawal V, Tottey S, Johnson SA, Freund JM, Siu BF, Badylak SF. Recruitment of progenitor cells by an extracellular matrix cryptic peptide in a mouse model of digit amputation. Tissue Eng Part A. 2011;17:2435–43.
423. Crapo PM, Tottey S, Slivka PF, Badylak SF. Effects of biologic scaffolds on human stem cells and implications for CNS tissue engineering. Tissue Eng Part A. 2014;20:313–23.
424. Londono R, Badylak SF. Biologic scaffolds for regenerative medicine: mechanisms of in vivo remodeling. Ann Biomed Eng. 2015;43:577–92.
425. Galvez-Monton C, Fernandez-Figueras MT, Marti M, Soler-Botija C, Roura S, Perea-Gil I, Prat-Vidal C, Llucia-Valldeperas A, Raya A, Bayes-Genis A. Neoinnervation and neovascularization of acellular pericardial-derived scaffolds in myocardial infarcts. Stem Cell Res Ther. 2015;6:108.
426. Zhang X. Functional neovascularization in tissue engineering with porcine acellular dermal matrix and human umbilical vein endothelial cells. Tissue Eng Part C Methods. 2011;17:423–33.
427. Dziki J, Badylak S, Yabroudi M, Sicari B, Ambrosio F, Stearns K, Turner N, Wyse A, Boninger ML, Brown EHP, Rubin JP. An acellular biologic scaffold treatment for volumetric muscle loss: results of a 13-patient cohort study. NPJ Regen Med. 2016;1:16008.
428. Nelson CM, Bissell MJ. Of extracellular matrix, scaffolds, and signaling: tissue architecture regulates development, homeostasis. Annu Rev Cell Dev Biol. 2006;22:287–309.
429. Fata JE, Werb Z, Bissell MJ. Regulation of mammary gland branching morphogenesis by the extracellular matrix and its remodeling enzymes. Breast Cancer Res. 2004;6:1–11.

430. Huttenlocher A, Sandborg RP, Horwitz AF. Adhesion in cell migration. Curr Opin Cell Biol. 1995;7:697–706.
431. Cohen M, Joester D, Geiger B, Addadi L. Spatial and temporal sequence of events in cell adhesion: from molecular recognition to focal adhesion assembly. Chembiochem. 2004;5:1393–9.
432. Kovesi T, Rubin S. Long-term complications of congenital esophageal atresia and/or tracheoesophageal fistula. Chest. 2004;126:915–25.
433. Li Y, Zhu Y, Yu H, Chen L, Liu Y. Topographic characterization and protein quantification of esophageal basement membrane for scaffold design reference in tissue engineering. J Biomed Mater Res B Appl Biomater. 2012;100:265–73.
434. Badylak SF, Vorp DA, Spievack AR, Simmons-Byrd A, Hanke J, Freytes DO, Thapa A, Gilbert TW, Nieponice A. Esophageal reconstruction with ECM and muscle tissue in a dog model. J Surg Res. 2005;128:87–97.
435. Ackbar R, Ainoedhofer H, Gugatschka M, Saxena AK. Decellularized ovine esophageal mucosa for esophageal tissue engineering. Technol Health Care. 2012;20:215–23.
436. Tan B, Wei RQ, Tan MY, Luo JC, Deng L, Chen XH, Hou JL, Li XQ, Yang ZM, Xie HQ. Tissue engineered esophagus by mesenchymal stemcell seeding for esophageal repair in a canine model. J Surg Res. 2013;182:40–8.
437. Bauchet L, Lonjon N, Perrin FE, Gilbert C, Privat A, Fattal C. Strategies for spinal cord repair after injury: a review of the literature and information. Ann Phys Rehabil Med. 2009;52:330–51.
438. Richard S, Adrian P, Popescu A. Animal models in pressure ulcer research. J Spinal Cord Med. 2016;30:301–2.
439. Pabari A, Yang SY, Seifalian AM, Mosahebi A. Modern surgical management of peripheral nerve gap. J Plast Reconstr Aesthet Surg. 2010;63:1941–8.
440. Navissano M, Malan F, Carnino R, Battiston B. Neurotube for facial nerve repair. Microsurgery. 2005;25:268–71.
441. Weber RA, Breidenbach WC, Brown RE, Jabaley ME, Mass DP. A randomized prospective study of polyglycolic acid conduits for digital nerve reconstruction in humans. Plast Reconstr Surg. 2000;106:1036–45.
442. Battiston B, Geuna S, Ferrero M, Tos P. Nerve repair by means of tubulization: literature review and personal clinical experience comparing biological and synthetic conduits for sensory nerve repair. Microsurgery. 2005;25:258–67.
443. Chamberlain LJ, Yannas IV, Arrizabalaga A, Hsu HP, Norregaard TV, Spector M. Early peripheral nerve healing in collagen and silicone tube implants: myofibroblasts and the cellular response. Biomaterials. 1998;19:1393–403.
444. Merle M, Dellon AL, Campbell JN, Chang PS. Complications from silicon-polymer intubulation of nerves. Microsurgery. 1989;10:130–3.
445. Ribatti D, Conconi MT, Nico B, Baiguera S, Corsi P, Parnigotto PP, Nussdorfer GG. Angiogenic response induced by acellular brain scaffolds grafted onto the chick embryo chorioallantoic membrane. Brain Res. 2003;989:9–15.
446. Zhang XY, Xue H, Liu JM, Chen D. Chemically extracted acellular muscle: a new potential scaffold for spinal cord injury repair. J Biomed Mater Res A. 2012;100:578–87.
447. Crapo PM, Medberry CJ, Reing JE, Tottey S, van der Merwe Y, Jones KE, Badylak SF. Biologic scaffolds composed of central nervous system extracellular matrix. Biomaterials. 2012;33:3539–47.
448. Wang H, Lin XF, Wang LR, Lin YQ, Wang JT, Liu WY, Zhu GQ, Braddock M, Zhong M, Zheng MH. Decellularization technology in CNS tissue repair. Expert Rev Neurother. 2015;15:493–500.
449. Zilic L, Wilshaw SP, Haycock JW. Decellularisation and histological characterisation of porcine peripheral nerves. Biotechnol Bioeng. 2016;113(9):2041–53.

450. Meiqin C, Tengchao H, Bo H, Ying G. Role of demyelination efficiency within acellular nerve scaffolds during nerve regeneration across peripheral defects. Biomed Res Int. 2017;2017:4606387.
451. Wang Q, Zhang C, Zhang L, Guo W, Feng G, Zhou S, Zhang Y, Tian T, Li Z, Huang F. The preparation and comparison of decellularized nerve scaffold of tissue engineering. J Biomed Mater Res A. 2014;102:4301–8.
452. Prest TA, Yeager E, Lopresti ST, Zygeiyte E, Martin MJ, Dong L, Gibson A, Olutoye OO, Brown BN, Cheetham J. Nerve-specific, xenogeneic extracellular matrix hydrogel promotes recovery following peripheral nerve injury. J Biomed Mater Res A. 2018;106:450–9.
453. Ren T, Faust A, van der Merwe Y, Xiao B, Johnson S, Kandakatla A, Gorantla VS, Badylak SF, Washington KM, Steketee MB. Fetal extracellular matrix nerve wraps locally improve peripheral nerve remodeling after complete transection and direct repair in rat. Sci Rep. 2018;8:4474.
454. Means KR Jr, Rinker BD, Higgins JP, Payne SH Jr, Merrell GA, Wilgis EF. A multicenter, prospective, randomized, pilot study of outcomes for digital nerve repair in the hand using hollow conduit compared with processed allograft nerve. Hand. 2016;11:144–51.
455. Dellon AL. Management of peripheral nerve problems in the upper and lower extremity using quantitative sensory testing. Hand Clin. 1999;15:697–715.
456. Cho MS, Rinker BD, Weber RV, Chao JD, Ingari JV, Brooks D, Buncke GM. Functional outcome following nerve repair in the upper extremity using processed nerve allograft. J Hand Surg Am. 2012;37:2340–9.
457. Rinker BD, Ingari JV, Greenberg JA, Thayer WP, Safa B, Buncke GM. Outcomes of short-gap sensory nerve injuries reconstructed with processed nerve allografts from a multicenter registry study. J Reconstr Microsurg. 2015;31:384–90.
458. Papatheodorou LK, Williams BG, Sotereanos DG. Preliminary results of recurrent cubital tunnel syndrome treated with neurolysis and porcine extracellular matrix nerve wrap. J Hand Surg Am. 2015;40:987–92.
459. Badylak SF, Freytes DO, Gilbert TW. Extracellular matrix as a biological scaffold material: structure and function. Acta Biomater. 2009;5:1–13.
460. Jensen G, Morrill C, Huang Y. 3D tissue engineering, an emerging technique for pharmaceutical research. Acta Pharm Sin B. 2018;8:756–66.
461. Keane TJ, Swinehart IT, Badylak SF. Methods of tissue decellularization used for preparation of biologic scaffolds and in vivo relevance. Methods. 2015;84:25–34.
462. Maghsoudlou P, Totonelli G, Loukogeorgakis IG, Eaton SP, De S, Coppi PA. A decellularization methodology for the production of a natural acellular intestinal matrix. J Vis Exp. 2013;80:50658.
463. Meng F, Modo M, Badylak SF. Biologic scaffold for CNS repair. Regen Med. 2014;9:367–83.
464. Choi J, Kim JH, Jang JW, Kim HJ, Choi SH, Kwon SW. Decellularized sciatic nerve matrix as a biodegradable conduit for peripheral nerve regeneration. Neural Regen Res. 2018;13:1796–803.
465. Carriel V, Alaminos M, Garzon I, Campos A, Cornelissen M. Tissue engineering of the peripheral nervous system. Expert Rev Neurother. 2014;14:301–18.
466. Lee SK, Wolfe SW. Peripheral nerve injury and repair, peripheral nerve injury and repair. J Am Acad Orthop Surg. 2000;8:243–52.
467. Johnson PJ, Wood WD, Moore AM, Mackinnon SE. Tissue engineered constructs for peripheral nerve surgery. Eur Surg. 2013;45:122–35.
468. Tang P, Chauhan A. Decellular nerve allografts. J Am Acad Orthop Surg. 2015;23:641–7.
469. Moore AM, MacEwan M, Santosa KB, Chenard KE, Ray WZ, Hunter DA, Mackinnon SE, Johnson PJ. Acellular nerve allografts in peripheral nerve regeneration: a comparative study. Muscle Nerve. 2011;44:221–34.

Chapter 5
The Decellularization of Whole Organs

Yan Huang, Hangqi Yue, Zhongwei Lian, and Xiaoming Li

Abstract Since the development of tissue engineering, many mature decellularization schemes have been formed for different organs. The purpose of decellularization is to remove immunogenic components (such as alpha-Gal epitopes) and cellular components from the organ while retaining the inherent characteristics of natural extracellular matrix (ECM). The efficiency of removal of cells from the organs depends on the methods of decellularization. The commonly used decellularization methods can be divided into physical methods, chemical methods, and enzymatic hydrolysis methods according to the different reagents or conditions used. In the process of whole-organ decellularization, perfusion decellularization is currently the most widely used method and one of the safest and most effective decellularization methods. Each of these treatments affects the biochemical composition, tissue ultrastructure, mechanical behavior of the remaining ECM scaffold and detergent residue. This chapter presents an overview of the decellularization of different whole organs, including heart, lung, liver, kidney, pancreas, intestine, and others. Intended as a broad-ranging reference, this chapter will be of value for those concerned with the study of tissue engineering.

Keywords Decellularization · Organs · Extracellular matrix · Methods · Standards

Organ transplantation is the best way to treat end-stage organ diseases, but it is hindered by the problems of high prices and shortage of organs. In order to break this limitation, tissue engineering based on the organ decellularization scaffold has become the most promising effective method. The decellularized scaffold is composed of extracellular matrix (ECM) and has many unique advantages. Structurally, the decellularized scaffold can preserve the complete microstructure of the organ. In

Y. Huang (✉) · H. Yue · Z. Lian · X. Li
Key Laboratory for Biomechanics and Mechanobiology of Ministry of Education, Beijing Advanced Innovation Center for Biomedical Engineering, School of Biological Science and Medical Engineering, Beihang University, Beijing, China
e-mail: huangyan@buaa.edu.cn; lixm@buaa.edu.cn; x.m.li@hotmail.com

X. Li, H. Xie (eds.), *Decellularized Materials*,
https://doi.org/10.1007/978-981-33-6962-7_5

terms of source, it can ensure that organ scaffolds of appropriate sizes are obtained from appropriate species. In addition, the acellular scaffold retains many of the biological functions of natural ECM and avoids the risk of immune rejection [1]. Although in-depth research and certification are still required before being applied to humans, decellularized scaffolds have shown great potential in animal models.

5.1 Decellularization Method

The purpose of decellularization is to remove immunogenic components and cellular components from the organ while retaining the inherent characteristics of natural ECM. The selected organ determined the used acellular methods. The efficiency of removing cells from an organ depends on the organ itself and the choice of decellularization method. The commonly used decellularization methods can be divided into physical methods, chemical methods, and enzymatic hydrolysis methods according to the different reagents or conditions used, as shown in Table 5.1. During the decellularization process, the color of the organ will change, and eventually it will gradually become translucent, which is a direct result of the removal of cells. At the same time, this is also the most direct way to judge the effect of decellularization. The core step of decellularization is to destroy cell membranes, remove cellular components and nucleic acids.

Table 5.1 Summary of decellularization methods

Decellularization method	Reagents or conditions used	Types of decellularization Method
Freeze-thaw method	Low temperature conditions and cryo-protectants (5% trehalose)	Physical method
Stirring and shaking method	Shaking and stirring conditions	Physical method
Ultra-high pressure method	Gradient pressure condition	Physical method
Supercritical fluid decellularization method	Supercritical carbon dioxide	Physical method
Acid and alkaline reagent method	Peracetic acid, acetic acid and NaOH	Chemical method
Hypotonic and hypertonic reagent method	Hypotonic solution and hypertonic solution	Chemical method
Non-ionic detergents	TritonX-100	Chemical method
Ionic detergents	SDS, SDC, CHAPS	Chemical method
Chelating agent method	PEG, EDTA	Chemical method
Enzymatic hydrolysis method	Nuclease, trypsin, collagenase, and dispase	Enzyme decellularization

5.1.1 Physical Method

The main purpose of physical methods is to lyse cells. However, due to the large amount of biologically viscous substances contained in the components after cell lysis, it is difficult to remove cells from the decellularized scaffold by physical methods alone. Therefore, the physical method is usually only an auxiliary method, and generally needs to be used in conjunction with other methods to achieve the purpose of decellularization. In addition, physical methods will disrupt the ECM structure of the scaffold to a certain extent while lysing the cells. Therefore, the physical decellularization scheme is not suitable for organs with weak ECM structures.

Physical methods mainly include freeze-thaw, stirring and shaking, ultra-high pressure, and supercritical fluid method [2]. The most common physical method for decellularization is the freeze-thaw method. The principle of this method is based on the fact that liquid water will expand in volume after forming ice crystals. The enlarged ice crystals will lyse the cells, but may also damage the ECM structure, so it is usually necessary to add some cryoprotectants (such as 5% trehalose) during the use of this method. Therefore, adjusting the freeze-thaw cycle to reduce the damage to ECM is the research focus of this method.

Mechanical stirring and mechanical vibration are also commonly used auxiliary methods in the process of decellularization, which are used to improve the efficiency of decellularization and the effect of decellularization. Shaking or stirring the tissues or organs immersed in the decellularized chemical reagent, enzyme reagent or detergent not only can increase the contact between the tissue and the reagent, but also can destroy the cell structure by mechanical force to remove the cell. However, mechanical agitation and mechanical vibration may also cause damage to the ECM ultrastructure.

With the improvement of technical level, researchers have proposed a method of decellularization under ultra-high pressure. This method removes cell debris from the organ by applying gradient pressure on the organ to be decellularized. This method not only avoids the use of descaling agents, reduces the toxicity of decellularization scaffolds, but also improves the decellularization efficiency of decellularization reagents. In addition, this method can play a sterilizing function while removing cells. It has been proved that in high-pressure and low-temperature environments, collagen fibers can maintain aggregation and arrangement, and glycosaminoglycans (GAGs) can reduce loss. But the use cost of this method is too high.

In recent years, with the deepening of research on supercritical fluids, its high flux and low viscosity characteristics have successfully attracted the attention of researchers, and a supercritical fluid decellularization scheme has been successfully proposed [2]. Carbon dioxide (CO_2) can form supercritical fluids under certain conditions. It has the characteristics of being extremely easy to obtain, cheap, and inert, making it the reagent of choice for this scheme. Supercritical carbon dioxide ($SCCO_2$) as an ideal decellularization reagent can not only ensure the integrity of the

ECM scaffold microstructure to the greatest extent, but also maintain the stability of its mechanical properties. At the same time, $SCCO_2$ can also obtain scaffolds directly under dry conditions and reduce protein loss. In addition, $SCCO_2$ can also play a sterilizing role in the process of decellularization, which is an ideal sterilizing agent [3]. In addition to the above advantages, $SCCO_2$ is also very easy to extract, thereby avoiding the residue of toxic substances, and can significantly reduce the treatment time.

5.1.2 *Chemical Method*

The chemical method mainly achieves decellularization by changing the permeability of the cell membrane of the remaining cells in the scaffold. The chemical reagents used in the decellularization process mainly include acidic reagents, alkaline reagents, hypotonic and hypertonic solutions, detergents, and chelating agents. The chemical method is the most used method in the decellularization process, but different reagents have different characteristics. Usually, before decellularization, acetone or alcohol solution is used to dissolve lipids in order to remove cells better. In addition, when the whole organ is decellularized, the experimenter often controls the pressure or flow rate of the perfusion method to achieve uniform contact between the organ and the decellularization reagent [4, 5].

In the process of decellularization, acidic reagents or alkaline reagents are often used to dissolve cytoplasmic components or remove nucleic acids, but at the same time may adversely affect the mechanical structure of ECM [5]. Peracetic acid is less damaging to ECM, but it may hydrolyze biomolecules in organs and at the same time it cannot completely remove cells [6]. In addition, the combination of ethanol and peroxyacetic acid is often used as a sterilization reagent. Acetic acid has little effect on GAGs, but can decompose collagen, so it has an effect on the strength of the scaffold. The mechanical strength of the scaffold will be affected by alkaline reagents by reducing the cross-linking of matrix proteins. In addition, NaOH has been shown to have certain toxic and side effects on fibroblasts, but it can be reduced by increasing the washing time [7].

The hypotonic or hypertonic solution promotes cell expansion and lysis through the osmotic effect to achieve decellularization. In addition, the hypertonic solution can increase the solubility of the cellular DNA remaining in the scaffold, and then in the later stage, the scaffold can be washed to remove the residual nucleic acid in the scaffold. The hypotonic solution is usually used with detergents to enhance the effect of destroying cell membranes. But the remaining cell debris usually needs to be removed with other reagents.

Detergent is the most frequently used reagent in the decellularization process, and its use is also a very important step in the process. Detergents are more hydrophilic than biofilms, which allows them to dissolve biofilms and achieve the purpose of decellularization. According to the different charge types carried by detergents, they are usually divided into ionic detergents and non-ionic detergents. Different detergents have their own characteristics and are suitable for different various tissues or organs. However, in the process of use, attention should be paid to the ratio of ionic detergent to non-ionic detergent to achieve the best decellularization effect, while reducing the damage to the protein components in the scaffold.

Non-ionic detergents are not charged and have a relatively gentle effect on tissues. It hardly affects the collagen and protein in ECM, so it is the reagent of choice in the process of tissue decellularization. It achieves decellularization by disrupting the interaction between lipids and proteins or lipids. Therefore, the interaction between protein and protein is preserved, which helps to preserve the functional conformation of the protein. TritonX-100 is the most widely used non-ionic detergent, usually used for cell lysis [8]. It has the best decellularization effect in small intestine, but it has poor decellularization effect in heart and kidney [9].

Ionic detergents are charged, can destroy the nuclear membrane and cytoplasm, and have a more vigorous effect on tissues, so the decellularization effect is better than non-ionic detergents. But its strong decellularization ability can partly destroy the structure of proteins in ECM, resulting in the loss of matrix. The ionic detergents commonly used in the experiment are sodium dodecyl sulfate (SDS), sodium deoxycholate (SDC) and 3-[3-(cholamidopropyl) dimethylamino] propanesulfonic acid internal salt (CHAPS). SDS has anions, which can denature proteins by dissolving membrane proteins, thereby achieving decellularization. SDS is also recognized as a high-efficiency decellularization reagent, which has good decellularization effect in the heart, kidney, and aorta. But there are also studies that have shown that it has toxic side effects on cells. In addition, the SDS attached to the scaffold is difficult to remove, which not only increases the toxicity of the scaffold, but also reduces cell attachment and proliferation [10]. To remedy this problem, a mixed reagent of SDS and TritonX-100 was used to decellularize the kidneys and liver. The results showed that the structure of ECM was preserved. SDC is also a commonly used ionic detergent, which achieves decellularization by infiltrating cholate into the cell membrane, promoting cell membrane rupture and pore formation. SDC has shown excellent decellularization properties while still retaining the structure of the ECM scaffold [11]. The CHAPS is a commonly used zwitterionic detergent, which can directly destroy the lipid membrane to achieve decellularization. CHAPS has a relatively gentle effect on tissues and can retain the mechanical strength of the scaffold [12].

The chelating agents used in the decellularization process are mainly polyethylene glycol (PEG) and ethylenediaminetetraacetic acid (EDTA). They combined with Ca^{2+} and Fe^{2+} ions to separate the cells from the ECM. Chelating agents are often used in combination with trypsin or detergents during actual use. This is because the chelating agent alone does not remove the cells, and destroy the proteins in the scaffold.

5.1.3 Enzyme Decellularization

The enzymatic hydrolysis method mainly includes trypsin method, nuclease method, and the like. Proteases mainly include trypsin, α-galactosidase, collagenase, and dispase. Enzyme agents can destroy cell membranes, thereby weakening adhesion between cells, between cells and ECM, or between nucleic acids and ECM. It is often used in conjunction with chemical reagents to remove cellular debris. Because the enzyme reagents have low efficiency when used alone, it is not only difficult to completely remove the cells, but also has an adverse effect on the recellularization of the scaffold. In addition, if the enzymes act for too long or the concentration is too high, they will destroy the bioactive substances inside the ECM scaffold, such as aminodextran or collagen. Therefore, it is particularly important to choose the appropriate concentration of enzyme reagents, the appropriate temperature and time. The addition of selective protease inhibitors can also protect components in ECM.

5.1.4 Decellularized Perfusion

In the process of organ decellularization, perfusion decellularization method is the most used, and it is also one of the most safe and efficient decellularization methods. The perfusion decellularization method perfuses the decellularization reagent through the inherent vascular system network of the donor organ to achieve the purpose of removing cells and DNA and retaining the three-dimensional structure and composition of the organ ECM. The significance of the vascular network inside the organ is to minimize the distance between cells and oxygen, so it can also increase the contact between decellularization reagents and cells, making perfusion decellularization a highly effective decellularization method. The first appearance of perfusion decellularization method was to decellularize the heart and achieved good results. After this, more complex ECM scaffolds were gradually manufactured by perfusion decellularization, such as lung decellularized scaffolds, liver decellularized scaffolds, and kidney decellularized scaffolds.

The perfusion decellularization method requires perfusion of the decellularization agent through the artery or vein of the organ. After the reagent circulates in the organ for a week, the decellularized reagent is discharged through the vein or artery. Therefore, in this process, factors such as the composition of the decellularizing agent, the flow rate of the perfusion, and the pressure of the perfusion all affect the effect of the decellularization. In order to understand the influence of these factors on the perfusion decellularization method, researchers have been constantly exploring. In addition, factors such as the stability of the perfusion equipment and the ambient temperature at the time of perfusion will also affect the effect of decellularization and are factors that need to be considered. Perfusion and decellularization are usually performed at room temperature, but the structure of the ECM scaffold can be better protected at 4 °C; if the decellularization reagent contains enzyme reagents, it needs to be heated to 37 °C before perfusion [13].

In 2011, Baptista et al. proposed a method to manufacture acellular liver ECM scaffolds with a complete vascular network structure [14]. They perfused decellularized reagents into livers of different organisms through perfusion decellularization, successfully removing cellular components and retaining the vascular network. At the same time, it was proved that the reagent entering from the inlet can be dispersed into a capillary bed, and finally merged into an outlet blood vessel, which provides the basis for the construction of tissue engineering liver.

In 2012, Sullivan et al. used three different decellularization reagents, 0.25% SDS, 0.5% SDS, or 1% Triton X-100, to decellularize pig kidneys [15] in order to explore which decellularization reagent can more effectively retain the three-dimensional structure of the ECM scaffold. At the same time, it can ensure the efficiency of decellularization and the repeatability of the method. They also constructed a high-throughput decellularization system. The system is capable of infusing decellularization reagents from the arterial or portal system and outputting decellularization reagents from the venous system. The results show that all three reagents can achieve decellularization of organs and retain the vascular system. Among them, 0.5% SDS has the best decellularization effect and no cytotoxicity. The perfusion decellularization method developed by Sullivan et al. can also be used for other large organs, which provides a reference for subsequent perfusion decellularization research.

In the process of perfusion and decellularization, the uniformity is a factor that should be considered. In 2015, in order to reduce the influence of the poorly perfused area of the organ on the decellularization process, Hillebrandt et al. proposed that the liver be subjected to perfusion decellularization treatment under the pressure of oscillation [16]. They constructed a dedicated perfusion device to perfuse the liver from the hepatic artery and/or portal vein. Simultaneously imitating the breathing process, externally loaded with varying oscillation pressure. The results show that the perfusion through the hepatic artery is more uniform, and the oscillating pressure significantly improves the effect of perfusion decellularization and improves the performance of the rat liver decellularization scaffold.

Temperature is also important in perfusion decellularization. In 2017, Elebring et al. proposed an efficient cold perfusion technique for the pancreas [17]. At 4 °C, they used a decellularizing agent mixed with SD, TritonX-100, and DNase to perfuse the pancreas, successfully removing cellular material, and providing hope for the treatment of clinical diabetes.

Existing research results suggest that pulsatile flow can improve the effect of perfusion, because pulsatile perfusion can reduce the time that the organ contacts the decellularized reagent. To investigate the effect of pulsatile perfusion on the decellularization of rat heart, in 2018, Park et al. used energy equivalent pressure (EEP) and perfusion flow rate as parameters, using 1% SDS to perform 2 h pulsatile perfusion and 2 h non-pulsatile perfusion treatment on different rat hearts [18]. The results showed that under the same initial average perfusion pressure, the decellularization efficiency of the pulsatile perfusion group was higher, and the

EEP was significantly improved. The final result proves that pulsatile perfusion can promote the decellularization process.

In order to shorten the time consumed by perfusion decellularization, physical methods are often needed to assist. In 2019, in order to shorten the preparation time of renal decellularization scaffolds, Say et al. used ultrasound with different power to enhance the decellularization effect [19]. They chose to use 1% SDS for perfusion. At the same time, every 2 h of rest, 2 h of ultrasound with different power are loaded, and the cycle is repeated. It was found that the higher the power of ultrasound, the shorter the decellularization time required. And after loading ultrasound, it can significantly shorten the time consumed for perfusion. In 2020, Manalastas and others also explored the role of ultrasound in promoting perfusion decellularization and the effect of ultrasound on the ECM microstructure [20]. They explored the effects of these parameters on the perfusion time by applying different SDS concentrations, perfusion flow rates, and ultrasound system power to the pig kidneys. The results also found that ultrasound can significantly improve the decellularization efficiency and shorten the time it takes, but ultrasound will cause minimal to moderate damage to the integrity of the ECM scaffold microstructure. At the same time, the concentration of SDS and perfusion flow rate also play an important role in shortening the time spent for decellularization.

In 2020, Schilling et al. designed and manufactured a tissue infusion/perfusion device (TIPD) in combination with 3D printing technology and commercially available components to improve the repeatability of the decellularization process [21]. The device is a compact integrated system. In addition, users can define their own software functions to precisely control the changes of decellularization reagents during the perfusion process. The device has been applied to the decellularization of kidneys in all organs, and has successfully removed cellular components, retaining the ultrastructure of ECM. In the same year, in order to improve the efficiency of manufacturing human liver decellularization scaffolds, Willemse et al. proposed to control the pressure during the perfusion and designed a reactor, as shown in Fig. 5.1a [22]. They used TritonX-100 to decellularize the human liver. The results show that in the presence of pressure control, the decellularization process can be completed within 20 h. An almost transparent liver decellularization scaffold was obtained, cell components were completely removed, and the ultrastructure of ECM was also retained. Without pressure control, it takes 64–96 h to achieve complete decellularization. It was finally proved that pressure control can significantly shorten the decellularization time, and the increase in pressure and flow rate will not affect the ECM structure.

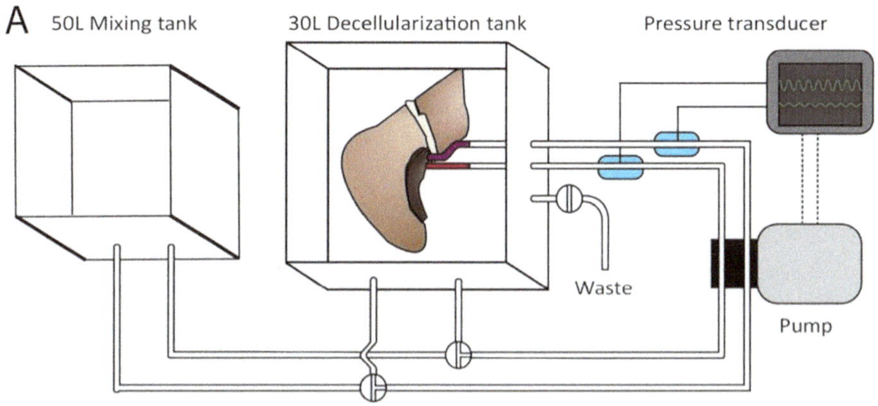

Fig. 5.1 Protocol and equipment for decellularization of human liver by pressure control: place the liver in the trough and connect it to the pump tube through the hepatic artery and portal vein. The pump tube is connected to the pressure sensor, and the pressure information will be transmitted to the double-head pump driver. Three-way valves are used to create closed or open loops. The 50 L mixing tank stores the decellularization reagent and can be replaced. Waste is discarded through a manual valve [22]

5.2 General Effective Standards for the Decellularization of Organs

The application of the decellularization scaffold is closely related to the nature of the decellularized scaffold. If the decellularization process is not completed, it will lead to severe immune rejection, and the decellularization process will also cause damage to the ECM, affecting its composition, ultrastructure, and mechanical properties [23, 24]. The ideal decellularization process not only removes cellular components, but also preserves the macro and micro structure of the organ as much as possible. The selection of the decellularization protocol should strike a balance between the removal of cellular components and the retention of extracellular matrix. Therefore, the existence of evaluation criteria for decellularized scaffolds is particularly important. The evaluation criteria are mainly divided into several aspects: cell composition, ECM structure, ECM composition, immunological analysis, mechanical strength, and detergent residue, as shown in Fig. 5.2.

5.2.1 Cell Composition

Cellular components are the main cause of the immune response. Therefore, its detection is crucial. The observation and determination of cellular components are mainly divided into three types. The simplest of these is to use H&E staining to observe the presence of nuclei in the decellularized scaffold. In addition, the other

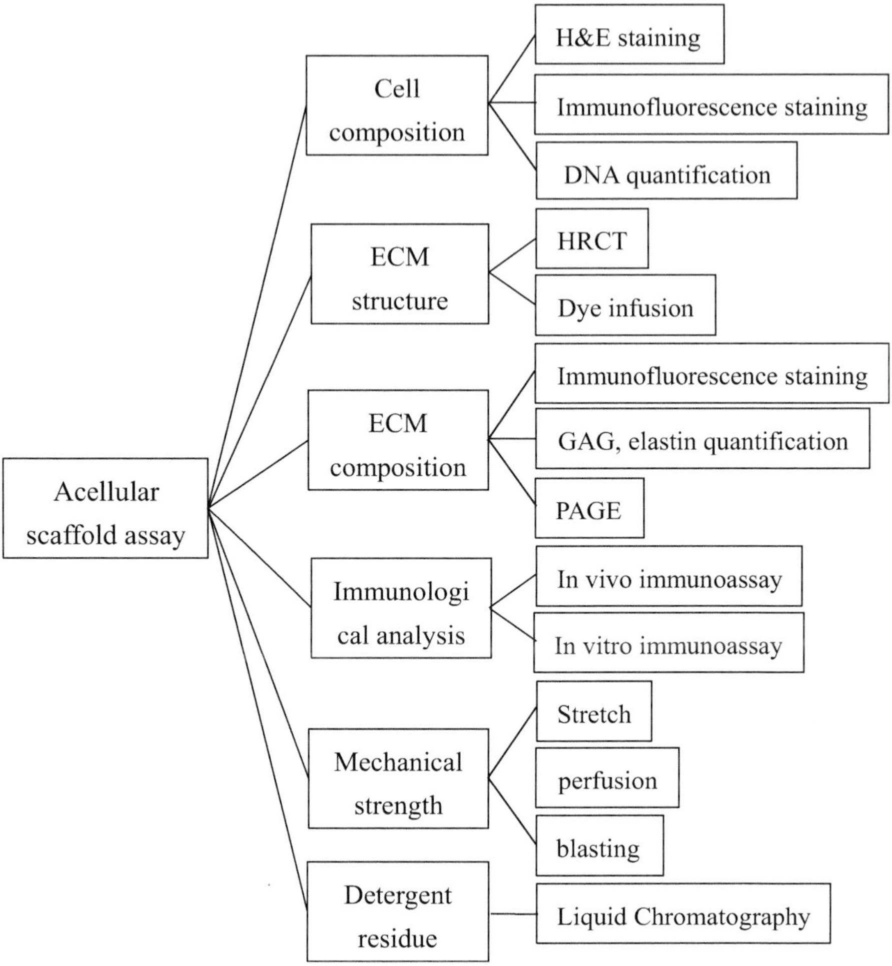

Fig. 5.2 Evaluation criteria for acellular scaffold

two more precise methods are: one is to perform immunofluorescence staining on the decellularized organ section to observe whether there is Nuclear DAPI staining. Another method is to quantify nucleic acid by DNA extraction kit, ultraviolet spectrophotometer, PCR experiment, and agarose gel electrophoresis. It is believed that when the DNA content of the scaffold is less than 50 ng per milligram of dry weight or the length of the existing DNA fragments is less than 200 bp, it is considered that the cellular components have been completely removed [25].

5.2.2 ECM Structure

The extracellular matrix is mainly composed of functional proteins, GAGs, and lipids. In addition to providing mechanical support and communication functions for cells, research also confirmed that ECM can specifically promote cell differentiation [26]. Therefore, maintaining the composition and structure of ECM is of great significance. Experiments usually use dye perfusion electron microscopy and high-resolution computed tomography (HRCT) to evaluate the ECM structure [27]. Due to the structural differences between different organs, there are many differences in the detection methods and standards of acellular scaffolds, which will be described in detail below.

5.2.3 ECM Composition

Acellular scaffolds mainly contain collagen, laminin, fibronectin, elastin, and GAGs. These proteins have important implications for the biological and biomechanical properties of the scaffold, and GAGs help maintain the binding of water and action and growth factors in the extracellular matrix. Therefore, the analysis and determination of the components of the decellularized scaffold is an important indicator of scaffold evaluation.

Experiments usually use immunofluorescence staining and polyacrylamide gel electrophoresis (PAGE) to determine protein content. There are also studies using liquid chromatography-tandem mass spectrometry for protein analysis, and the results are more accurate [28]. As for the measurement of GAGs, which are biological macromolecules, spectrophotometry is often used for measurement.

5.2.4 Immunological Analysis

Immunological analysis is divided into in vitro and in vivo analysis. In vitro analysis is similar to the measurement of extracellular matrix components. It is also measured by protein immunoassay. While in vivo analysis is performed by using the obtained decellularized scaffold for foreign body or heterogeneity transplant into the skin and observe its immune response [28, 29].

5.2.5 Mechanical Strength

The decellularized scaffold of the entire organ mainly has mechanical properties such as strain, elasticity, and stiffness. These mechanical properties have an important effect on subsequent cell retransplantation and transplantation, so they must be measured.

The determination of the mechanical properties of the decellularized scaffold mainly includes uniaxial or biaxial tensile test, perfusion pressure test, and burst pressure test. For example, a study conducted a biaxial mechanical test on natural heart, decellularized heart, and typical fibrin gel samples, plotted engineering stress–strain curves, and calculated tangent modulus [30]. There are also studies on the mechanical evaluation of decellularized scaffolds through expansion experiments [31]. With the development of science and technology, some studies have proposed the use of nuclear magnetic imaging technology to detect changes in the viscoelasticity of decellularized scaffolds [32].

5.2.6 Detergent Residue

The residual toxicity of the detergent will destroy the decellularized scaffold, be not conducive to subsequent recellularization, and lead to inflammation. Liquid chromatography is used to measure the residual toxicity of the detergent. It is generally believed that when the detergent residue is less than 50 mg/L (0.005%), it will no longer adversely affect the cells [33].

5.3 The Decellularization of Different Whole Organs

Since the development of tissue engineering, many mature decellularization schemes have been formed for different types of organs. However, the first step of most decellularization protocols is to start with freeze-thaw to promote cell lysis and at the same time ensure the stability of the mechanical structure of the ECM scaffold. Afterwards, the remaining cells and other components are removed by infusing chemical decellularization reagents. Finally, the obtained decellularized scaffold was evaluated. But no matter what kind of decellularization method or decellularization reagent, it may cause a certain degree of destruction or change of the decellularization scaffold. Therefore, when selecting a decellularization method, the standard should be to minimize the loss of ECM in the decellularization scaffold. At the same time, thrombosis or air embolism caused by bacterial contamination should be avoided as much as possible during subsequent processing. In addition, it is still necessary to study the optimal decellularization scheme for different organs in order to achieve the optimal decellularization scheme for specific ECM. At the same time, the impact of different decellularization methods on subsequent clinical applications also needs further study.

5.3.1 Heart

The heart, as an important vital organ of the vertebrate circulatory system, has a complicated chamber and blood vessel structure. In the physiological activities of the heart, heart valves, especially the aortic valve and mitral valve, which control the whole blood flow, play a vital role in the function of the heart. Today, cardiovascular disease has become one of the most common diseases in the world, and the global demand for heart and heart valve donors is huge [34]. Although heart transplantation is the only effective treatment for advanced heart failure, it still has many defects. Transplantation is not only limited by insufficient donors and immune rejection [35]. Even if the operation is successful, lifelong chronic immunosuppression can also induce diseases such as hypertension, diabetes, malignancy, and renal failure [36]. Therefore, finding more superior substitutes has become the research focus of the treatment of cardiovascular diseases.

Research on artificial hearts and artificial heart valves is similar. They initially used machinery to replace natural organs and tissues. The mechanical artificial valve was first invented in 1951 and was used in clinical practice in 1952. The artificial heart-lung machine was first used in open heart surgery in 1953. However, due to the complex organ structure and physiological process of the heart, compared with artificial valves, the research of artificial hearts lags far behind. The development of mechanical valves and biological valves of artificial valves is almost synchronized. The Starr-Edwards cage valve was first commercialized in 1960 [37], while the acellular scaffold was invented in 1984, and the interventional valve was invented in the 1990s, and was successfully used in surgery for the first time in 2000 [38]. In addition to obtaining more research results, the relatively simple structure of the heart valve also caused a very interesting phenomenon, that is, the first case of heart valve transplantation was successful in 1967 [38, 39], much later than the time when the artificial valve appeared. Unlike valves, artificial heart research has been focused on mechanical hearts for a long time. In 1957, artificial hearts made of polyethylene salts survived in dogs for 1.5 h. In 1982, the artificial heart was implanted into the human body for the first time. Although the patient can only survive for 112 days, it sets a milestone in artificial heart. By 2001, the artificial heart had been successfully applied to the human body. In 2004, the artificial heart named cardiowest became the world's first artificial heart officially put into clinical use. However, the mechanical heart lacks the biological characteristics of the natural heart, and, therefore, has many defects. Research and construction of artificial organs similar to natural hearts are of great significance, and decellularization technology provides new ideas for achieving this goal. The development process and main research results of the whole heart organ decellularization are shown in Fig. 5.3.

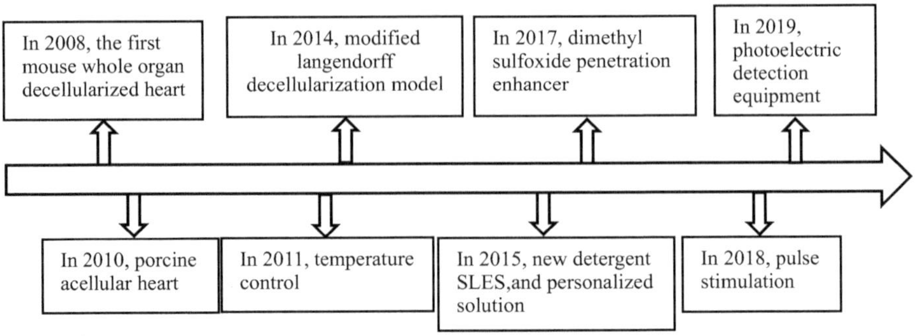

Fig. 5.3 Progress map of acellular heart research

In January 2008, 50 years after the first successful heart transplant operation, another pioneering work was realized that the whole heart was decellularized and recellularized, which was a milestone in creating biological artificial heart (BAH). For the first time, Ott et al. described the method of using PEG, Triton-X100, and SDS three different detergents on Langendorff equipment to reverse perfusion rat heart to obtain ECM. The study confirmed by H&E staining and immunofluorescence staining that the decellularized scaffold did not show nucleus, DAPI positive nucleus, cardiac α-myosin heavy chain, or sarcomere α-actin. Despite the loss of muscle fibers in the scaffold, immunofluorescence staining also revealed that not only collagen I and III, laminin and fibronectin of ECM, but also characteristic braids, coils, pillars, and dense epicardial fibers. In addition, the decellularization method retains the collagen and elastin fibers in the aortic wall, as well as the very important aortic valve leaflets. The experiment also used isobiaxial mechanical tests and red Mercox resin infusion to determine the mechanical properties and vascular structure of the decellularized heart. The study found that using SDS as a detergent can remove cell components better than PEG and Triton-X100. And it produces cell-free, perfusable vascular structures, effective cell-free valves, and complete chamber geometry. After the experimental decellularized scaffold was seeded in the heart or endothelial cells, the macroscopic contraction and pump function of the construct can be observed under physiological complexes and electrical stimulation (equivalent to 2% of adult or 25% of fetal heart function at 16 weeks) [31]. This experiment sets a precedent for the whole-organ decellularization technology and provides a systematic experimental method, and also provides hope for solving the heart transplantation problem.

After the success of the first whole heart decellularization, many scientists began to explore the possibilities of decellularization of whole organs. Rat is a commonly used and efficient experimental object, but its heart cannot be used as a transplant material for human. In fact, the size and structure of pig organs are closer to humans, which is a very suitable experimental material. But compared with the heart of rats, the heart of pigs is more difficult to operate. In 2010, some scientists used enzymatic, non-ionic detergents, ionic detergents, and acidic solutions together with hypotonic and hypertonic rinsing fluids to maximize the removal of cellular components and preserve ECM structure, and through the pulsatile retrograde aortic perfusion achieves complete decellularization of porcine heart [40]. The experiment used H&E staining, DAPI fluorescence staining, and DNA quantitative assessment of the cellular component removal level in the decellularized scaffold. ECM components were also assessed by Pentachrome staining, Herovici staining, mouse anti-human collagen IV antibody analysis, GAG assay, and elastin measurement. In addition, the research also used the load cell model and the ball burst compression cage to measure the biaxial burst strength, and found that the final biaxial breaking strength of the decellularized heart was almost the same as that of the natural tissue. The decellularization method proposed by this research expands the distribution of chemicals, while removing cellular components and retaining the ECM structure and mechanical properties, greatly reducing the time required for the decellularization process. This study provides specific steps for decellularization of a human-sized heart. It also showed a broader prospect for the construction of the heart of tissue engineering.

Although encouraging results have been reported for the decellularization of whole hearts in rats and pigs, artificial valves constructed using decellularization techniques have also achieved many research results [41–43]. However, the BAH still cannot show complete cardiac function. The main problem of the cardiac stent is the residual cell components caused by the decellularization process, the missing ECM components such as muscle fibers, GAGs, etc., and the ECM ultrastructure. The artificial valve also has the disadvantage of lack of durability, and the calcification or degeneration of the stent will occur in about 10 years. Studies have shown that lipids and cell debris play an important role in the calcification process [44]. Other studies have found that the decellularization process changes the pore size of heart valves [45]. Decellularization can also affect the microstructure and mechanical properties of the valve matrix, such as stiffness and tension [46]. Therefore, it is very important to improve and optimize the decellularization method. In 2011, scientists have improved the Langendorff decellularization model based on the perfusion decellularization scheme of ionic detergents to achieve effective temperature control, and quantify the quality of the decellularization process through histology and fluorescence microscopy, as well as histological examination of intracellular components. The content of multiple proteins proves the availability of this model [47]. Akhyari et al. used an automated model to control the temperature of the perfusant during the decellularization process, and compared four elution methods including SDS and trypsin, which further proved that the extracellular matrix components and cell washing methods are very strong relevance [48].

In 2012, there was a study to further improve the whole-organ decellularization method of the previous pig heart as is shown in Fig. 5.4a, b. The experiment used trypsin, Triton X-100, and sodium deoxycholate solution to perform decellularization through retrograde coronary perfusion, and used perfusion speed is more than 2 L/min for extended periods of time. Frequent solution replacement can reduce tissue contamination by cell debris and minimize the exposure time of each decellularization reagent to reduce the adverse effects on ECM [49]. The

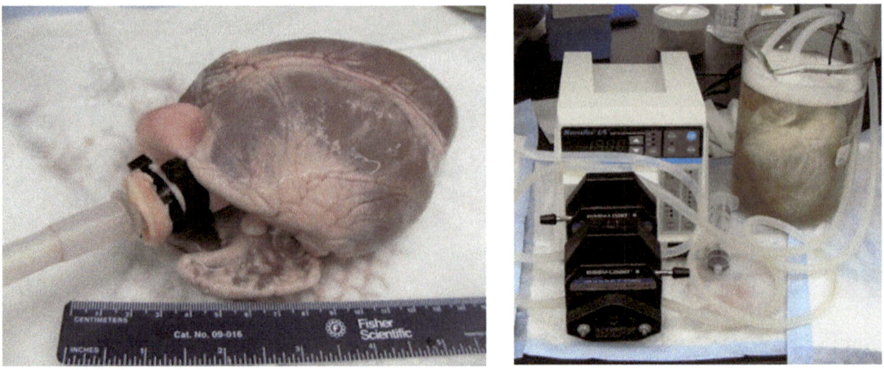

(a) (b)

Fig. 5.4 In one study, retrograde coronary artery decellularization was used to improve the acellularization protocol. (**a**) The barbed end of the tube is inserted into the aorta of the porcine heart. (**b**) Perfusion of porcine heart decellularization [49]

decellularized heart obtained from the experiment complies with the previously released decellularization standards.

A whole-organ decellularized heart needs to retain its complex vasculature, but the natural myocardium is a dense, high-vascular tissue, and each cell has almost capillaries to meet its high energy requirements, with a thickness of up to 1 cm. Therefore, decellularization of the heart requires complicated in vitro culture conditions. Compared to the decellularization of rat hearts, human-sized pig hearts lag far behind. In 2014, Weymann et al. used a modified Langendorff decellularization model that included a perfusion circuit and a pressure control module to perform antegrade coronary perfusion of pig hearts [50]. The model used a computer system to continuously record the filling pressure, used a heat exchanger to control the temperature, and interposed an air trap to ensure air-free perfusate. In the experiment, at a constant perfusion pressure of 100 mmHg, the heart was decellularized using SDS at 37 degrees Celsius. Through quantitative DNA analysis, protein determination, and mechanical pressure integrity measurement of the obtained decellularized heart, the results were compared with the control group and found that the DNA content was significantly reduced. It can be concluded that the improved decellularized model has higher practicality. In the same year, the effect of the decellularization process on human cardiac ECM was also reported. Experiments have shown that the scheme consisting of 6 h SDS and 3 h FBS can best preserve the ECM structure. This study expands the knowledge of decellularization of human heart [51].

The ideal whole-organ decellularization scheme should not only remove the cellular components to the greatest extent, but also retain the ECM components and their microstructure, and also avoid the adverse consequences of decellularization caused by inappropriate decellularization procedures. A large number of studies have shown that the existing common decellularization reagents including Triton X-100, SDS, trypsin, etc., have the disadvantages of insufficient removal of cellular components or damage to the ECM structure [52]. One of the most commonly used reagents for cardiac decellularization is SDS, which has a low cost and can completely remove genetic material. However, such chemicals can significantly alter ECM, including lowering glycans and cytokines. This damage may lead to poor cell implantation after tissue recellularization, and may lead to inflammation and coagulation during transplantation [53]. Other studies have found that SDS is related to fibrosis [54]. Solving this problem requires modification of the decellularization reagents. This process is also called pre-optimization. Pre-optimized protocols have been proposed for decellularized valves. Enzymes can be used as decellularized supplements to reduce the immunogenicity of the stent [55]. In addition, experiments have confirmed that protein solubilization can reduce protein residues [56]. And BAH has conducted many explorations in this direction. In 2015, scientists proposed to replace SDS with a novel detergent with weak chemical properties, sodium lauryl ether sulfate (SLES) [57]. After three cycles of freeze-thawing, the heart undergoes antegrade coronary perfusion and decellularization. The experiment evaluated the new detergent by analyzing the DNA content, cytokines, sugar content according to glycan, platelet adhesion degree, immunological analysis, microscopic observation, and lectin microarray analysis. The experiment also used LC-MS to test the residual detergent in the stent and found that the concentration of residual detergent in the samples treated by SDS or SLES was less than 0.001%, that is, far below 50 mg/L. The result found the total DNA content of the heart after 1% SDS and SLES treatment is less than 50 ng/mg dry tissue weight, which is considered to be the complete removal of genetic material. That is, SLES and SDS can provide the same cell component removal, but SLES takes longer. The experiment also found that although both reagents retained the macrostructure and composition of ECM, the SDS-treated heart not only lost the microstructure of ECM, but also greatly reduced the content of glycans. The research proves that the new anionic detergent can not only better maintain the integrity of ECM, but the heart it treats can almost avoid inflammation and thrombosis. This provides a new scheme for decellularization of the whole heart organ.

The whole-organ decellularization technology has made great progress, however, there are no methods known to individualize decellularization protocols while automatically ensuring a standard level of quality to minimize adverse effects on the resulting extracellular matrix. In 2015, Hülsmann et al. optimized the decellularization process by adding two components to the program. One was to monitore changes in the physical properties of the perfusate including the viscosity, protein, lipid, carbohydrate, and nucleic acid content during the decellularization of the rat heart. The other one was to observe the perfusion pressure, plot the RS-P curve, and measure the changes in properties during decellularization and after heart failure, such as dry weight and protein content. This study demonstrates for the first time the effect of biomass emissions on the perfusion-based decellularization variables for perfusate viscosity and pressure control during decellularization. This allows us to personalize the decellularization scheme into a single organ, thereby avoiding or minimizing the potential negative residence time of various molecules in the processing organ [58].

It is a challenge to maintain the integrity of the thinnest extracellular matrix in the process of perfusion and decellularization of structural heterogeneous organs such as the heart, and to completely remove cell debris from thicker areas. The high inflow rate required to maintain physiological pressure may swell or damage the thin tissue, but lower pressure will prolong the process and increase the possibility of contamination. Whole-organ decellularization of porcine heart is usually achieved using pressure control or perfusion flow rate control. However, a higher rate of perfusion will fill the left ventricle (LV), which may cause the aortic valve to deteriorate and reduce its ability to seal. Although aortic valve patency can maintain coronary perfusion and prevent retrograde filling of the left ventricle and atrium, it can in turn affect the mechanical integrity of the thin-walled atrium. This situation can be avoided by the heart's early preparation.

In 2017, a scientist used pig heart as experimental material to compare two novel retrograde decellularization methods: inverting the heart or venting the apex to decrease inflow rate. The experiment used hyperosmolar, hypotonic ionic solution, and SDS for decellularization perfusion, measured the flow dynamics through the aorta (Ao) and pulmonary artery (PA) under different Ao pressures, and evaluated the appearance of the heart and the turbidity of the effluent and coronary perfusion efficiency, as is shown in Fig. 5.5a. A rectangular image fit of the decellularized heart image is used to obtain the heart shape index. Using a nonlinear optical microscope, the microstructure of collagen and elastic fibers of the aortic valve tip was determined, as shown in Fig. 5.5b. The levels of DNA, GAGs, and residual detergent were also compared. The results of the study found that the inverted method showed higher coronary perfusion efficiency, more cell debris flow, higher collagen and elastin content, lower DNA content after decellularization, and better preserve the heart shape, which proves the inverted method was superior to the vented method [59, 60].

It is well known that excessive exposure to chemicals such as SDS during decellularization has the disadvantages such as structural changes of ECM or loss of some important components in ECM. In 2018, a scientist proposed to add dimethyl sulfoxide as a penetration enhancer to SDS to reduce the time required for decellularization by increasing the permeability of SDS [61]. The experimental results found that in the presence of DMSO, not only the permeability of SDS was improved, but almost all cells were removed from the aorta in only 3 h of exposure. And this method also increased the content of collagen fibers and GAGs in the scaffold, which showed that by adding DMSO can improve the decellularization efficiency and protect ECM components. This study provides an improved solution for decellularization of whole organs.

Many studies have shown that the mechanical stimulation produced by blood flow pulsation is involved in many physiological processes [62, 63]. In 2018, Park and others used SDS to decellularize rat hearts, using energy equivalent pressure (EEP) as a pulsation parameter to study the role of pulsatile flow in the decellularization process [18]. The experimental results found that compared with

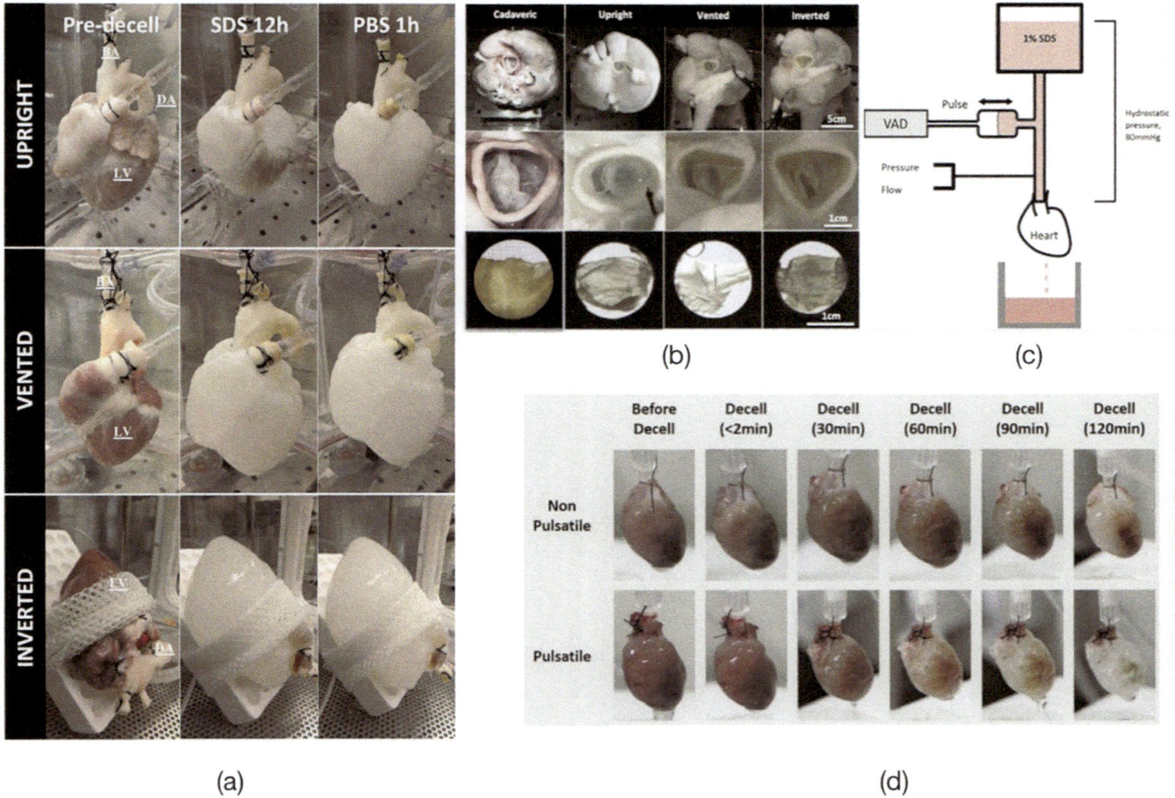

Fig. 5.5 To improve the acellular protocol, this study compared two methods of retrograde coronary artery decellularization by observing the heart and aortic valve. (**a**) Gross appearance of porcine hearts decellularized by using the conventional upright Langendorff method vented and inverted. Hearts are depicted before perfusion, after hypertonic, hypotonic, and SDS solutions, and after PBS washing for 1 h. All periods of perfusion occurred for the same duration under 60 mmHg aortic pressure. Descending aorta (DA); Left ventricle (LV). (**b**) The micrographs show three different methods to obtain the opacity difference

the control group, the EEP and perfusion flow of the beating group increased significantly, and the processed heart DNA content was lower, but the ECM content of the two was not significantly different. The study showed that pulsed stimulation is more beneficial to cardiac decellularization, as shown in Fig. 5.5c, d.

One of the main problems of whole-organ decellularization techniques today is the subjectivity of the final evaluation of its efficacy in various organs, which may cause the reproducibility of cell populations to be limited and the response to transplanted tissues to be worse [64]. In 2019, scientists developed a device consisting of a structured LED light source and a photodetector on an adjustable base to capture the relationship between transmitted light and residual DNA content during the perfusion of rat hearts. The study analyzes the optical distribution during the decellularization of the heart and hopes to use it as a method to evaluate the decellularization process of the organ [65]. The experimental results showed that it is possible to predict ineffective cell removal at the initial stage, regardless of the apparent transparency of the organ. This research showed the far-reaching prospects of decellularization of whole organs.

Since the twentieth century, people's exploration of the heart has continued to advance, and acellular heart valves have developed to this day and have been used in clinical practice. The optimization of the decellularization process is not limited to the pre-optimization mentioned above, but post-optimization methods including decellularized valve cross-linking and surface coating are also proposed [66, 67]. As a complex organ, the heart's decellularized scaffold must not only ensure the valve structure and mechanical properties, but also retain the vasculature, chamber structure, and ECM microstructure, which greatly increases the difficulty of constructing an artificial heart. Nevertheless, from the success of whole-organ decellularization in 2008 to the present, the exploration of decellularized hearts has also made great progress. Although it is still some time before clinical application due to many defects, it is certain that the waiting queue for transplantation will eventually disappear in the near future.

5.3.2 *Lung*

The lungs are an important part of the respiratory system as the respiratory organs of the human body. It is a highly complex organ composed of different cell types derived from the various germ layers of the early embryo [68]. Chronic lung disease, such as chronic obstructive pulmonary disease, is currently the third leading cause of

---

Fig. 5.5 (continued) between acellular heart aortic valve after the aortic root is cut off [59]. A similar study compared the decellularization process between simple perfusion and pulsatile perfusion. (**c**) A perfusion device that generates a pulsatile flow while measuring the pressure and flow of perfusion fluid. (**d**) Images of the decellularization process of the rat heart in the unbeatable and beatable groups [18]

death worldwide. Other chronic respiratory diseases, such as interstitial lung disease (ILD) or pulmonary hypertension, and genetic diseases, such as α1-antitrypsin disease and cystic fibrosis, are incurable and devastating lung diseases. Today, the treatment of lung diseases has become a global medical challenge. Although the rapid development of stem cell culture and lung tissue engineering technology provides a cure for chronic obstructive pulmonary disease and sarcoidosis that cause a certain lethality [60, 69], lung transplantation is still the only option in the late stage of the disease.

The right orthotopic right lung transplantation of dogs was successfully performed in 1950 for the first time, and the allogeneic lung transplantation of dogs was performed in 1954, which confirmed the feasibility of lung transplantation. Lung transplantation was first applied to humans in 1963, but patients survived only 18 days after surgery [70]. In 1979, a study found that the length of the lung bronchus was directly related to the comorbidity of bronchial anastomosis, and it was proposed to reduce the comorbidity by shortening the bronchus of the lung [71]. There are also studies confirming that nested bronchial anastomosis can reduce ischemic bronchial complications [72]. In 1983, another studies confirmed that the use of a large number of hormones seriously affects bronchial healing, and found a new anti-rejection inhibitor cyclosporine prime A which is used clinically [73]. With extensive research and exploration, the first unilateral lung transplantation was successfully performed in the same year. Since then, lung transplantation has been widely used. However, in addition to the severe shortage of donors, lung transplantation still has a high rate of acute and chronic rejection. Despite many efforts, the 5-year survival rate after surgery is only 50% [74]. For children with cystic fibrosis, surfactant dysfunction, pulmonary hypertension, or other lung anatomical defects or diseases that require lung transplantation, the problem of lung transplantation is more serious [75, 76]. Since there are no organs of matching size, lobectomy is often involved [77]. Excessive grafts can lead to complications, including impaired lung function, chest wall problems, and diaphragm damage [78].

In order to solve the problems of lung transplantation, scientists turned their attention to the construction of artificial lungs. As early as the 1940s, there was already a respirator called an iron lung. The artificial lungs used today, such as bubble oxygenators and membrane oxygenators, do not exhibit the biological characteristics of natural lungs. The emergence of whole-organ decellularization technology provides the possibility of constructing a natural organ transplant. The development history and main research results of acellular lung are shown in Fig. 5.6.

As a complex organ that undertakes gas exchange, the decellularization process must not only retain its vasculature, but the decellularized scaffold obtained must have a complete airway and alveolar structure to ensure its biological function. This is also the reason why the whole-organ decellularization of the lung is different from other organs. In 2010, 2 years after the first full-organ decellularized heart was achieved, Ott et al. used low-concentration SDS (0.1%) to perfusion and decellularize rat lungs to obtain a stent with vasculature, airway, and alveoli [79]. The decellularized lung was analyzed by DNA quantification, western blot

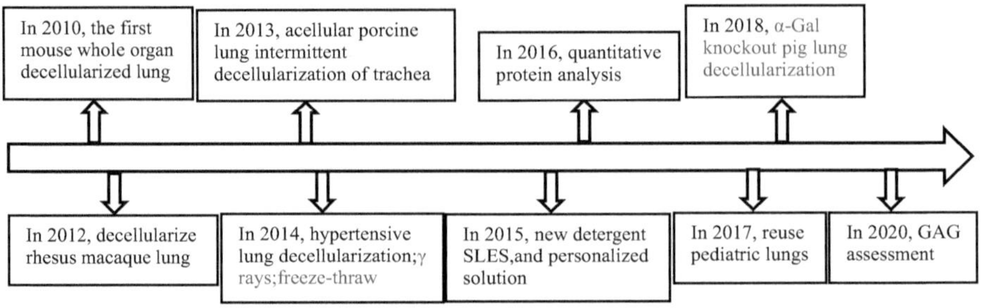

Fig. 5.6 Progress map of acellular lung research

analysis, and electron microscope observation. The results prove that the decellularization scheme can obtain decellularized scaffolds that remove cells and retain the microstructure of the lung, and only the residual cellular components are found in the tracheal cartilage ring. It was also found that compared with natural lung, the alveolar interval of decellularized lung was shortened, and compliance and lung volume were reduced. This study showed great prospects for the construction of biological artificial lungs. In the same year, there was a study using CHAPS and keeping the perfusion rate below 20 mmHg to decellularize rat lungs to obtain decellularized scaffolds that retain intact lung structure, arterial tree, and microvasculature [80]. The analysis of decellularized lung by microscopic computed tomography, immunofluorescence, and histochemical staining showed that the method not only removed cellular components but also retained cellular matrix components such as collagen, elastin, and laminin. The experiment also analyzed the ECM by transmission electron microscopy (TEM) and found that the decellularization method also restores the ultrastructure of the alveolar septum and the tiny blood vessels surrounding the alveoli. This research provides a new solution for lung decellularization.

Although previous studies have initially confirmed that the lung decellularization technology can be extended to human-sized lungs such as sheep, pigs, and primates, no complete plan has been proposed [79]. Rats are ideal experimental models, but they are not suitable for conversion to human clinical research. Compared with rats, non-human primate models have biological advantages very similar to humans. In 2012, a study was conducted to decellularize the heart and lung tissues of cynomolgus monkeys by infusing PBS containing 50 U/mL heparin and 5 μg/mL sodium nitroprusside into the pulmonary blood vessels through the pulmonary artery under the pressure of 25–30 cm H_2O through the gravity-driven flow. The left atrium is torn during this process to allow low resistance flow of lavage fluid from the pulmonary veins. Histological and protein analysis confirmed that this method can remove 92% of the DNA and retain the content of ECM components to the greatest extent to preserve the anatomy and ultrastructure of the lungs [81].

In addition to primates, pigs are also a common and effective experimental material. In 2013, there was a study using a combination of freezing and SDS washing to perfusion and decellularize pig lungs, bronchoscopy, multi-photon microscopy (MPM), the use of immunocytochemistry to evaluate ECM and the use of lung function test (PFT) assessment to assess the structural integrity of the AC lung stent. The study found that rapid thawing and slow injection of SDS helped to remove cellular components. The experimental results also showed that although the protein content of the decellularized lung of pigs changed, it did not affect lung function and cell adhesion [82]. In the same year, there was a study using three methods, including SDS and CHAPS, to decellularize thin sections of human and pig lungs. Their ECM composition, ultrastructure, and biomechanical changes were analyzed. Through a series of experiments such as DNA determination, histological analysis, protein content determination, scanning electron microscope observation, and immunochemical analysis, it was found that except for the acellular matrix of humans and pigs, except for the low elastin of pigs, the other ECM components are basically similar. The study also confirmed that CHAPS has a relatively good decellularization effect. The above research showed that pig lungs can provide materials for constructing grafts [83].

Although animal-derived lungs have the advantages of being readily available and well-studied, there are still some differences in extracellular matrix between different species. The regulatory signals provided by the extracellular matrix affect the basic cellular processes that control the lung, such as proliferation and migration [84], induction of metastasis or fibrosis activity [85], response to growth factors [86], and so on. Studies have compared the decellularized scaffolds of rats, pigs, primates, and humans through DNA determination, protein measurement, and mechanical evaluation. The experimental results showed that the lungs of humans and primates are more rigid and contain more elastin and less GAGs. During the process of cell inoculation, the adhesion of endothelial cells on the decellularized scaffolds of humans and primates was significantly enhanced [87].

Some scientists began to explore the possibility of expanding human lung decellularization. In 2013, there was a study that assessed the delay of autopsy, long-term storage (3 and 6 months). Two sterilization methods including irradiation or final washing of the lung structure and ECM of decellularized mice with peroxyacetic acid were used to explore the possibility of obtaining human lungs from autopsies. Although the experimental results indicate that delayed autopsy and long-term storage will seriously affect the effectiveness of decellularized scaffolds, as shown in Fig. 5.7a, the study provides new ideas for the source of decellularized materials [88]. In 2014, to explore whether the lungs of individuals with chronic lung disease can be used as decellularized materials, a study carried out decellularization of human lungs from normal patients and patients with chronic obstructive pulmonary disease (COPD), and analyzed the obtained decellularized scaffolds by histochemical and immune histochemical staining, electron microscopy, and mass spectrometry, as shown in Fig. 5.7b. The experimental results found that all the cell types inoculated into the decellularized emphysema lung were much lower than the normal lung, but the cell adhesion and proliferation on the soluble ECM homogenate coated on the tissue culture plate were not damaged, which confirmed the lungs of people with COPD lack ECM structures that maintain cell growth [89]. In the same year, a study used Triton X-100 to decellularize lungs from normal

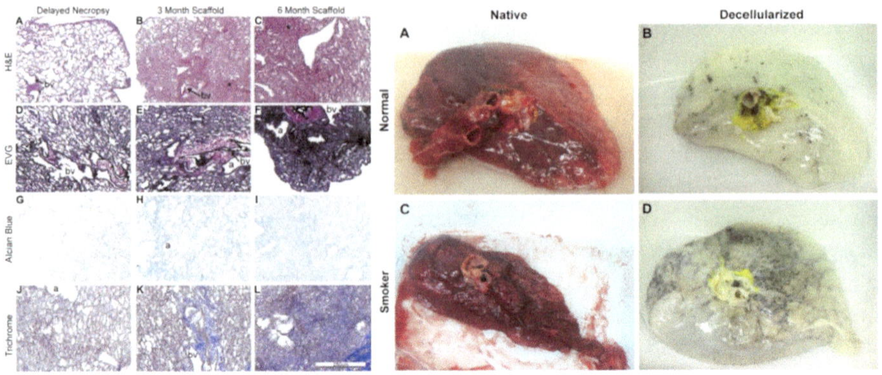

(a) (b)

Fig. 5.7 In order to expand the source of acellular lung, two studies have explored the effects of delayed autopsy and emphysema on acellular lung. (**a**) Effects of delayed autopsy on acellular lung matrix in mice [88]. (**b**) Morphological difference of acellular lung between normal and emphysema [89]

rats and porcine-induced pulmonary hypertension rats, and evaluated lung decellularized scaffolds through DNA quantification, western blotting, immunohistochemistry and proteomics. The experimental results showed that the lungs with hypertension can be decellularized normally and may be able to form healthy lung tissue. This study provides a possible solution to the shortage of lung transplant donors [90]. In 2017, there was a study on the use of decellularized scaffolds obtained from human lungs that did not meet the transplantation standards and the use of cells isolated from adult waste lungs as a source of cells for recellularization to construct bioengineered pediatric lungs. The experimental results showed that the artificial lungs contained well- developed blood vessels, airway epithelium, and lung tissue. It also confirmed the formation of alveolar-capillary connections. This research provides hope that small adult and child lung stents can be used to produce bioengineered lungs for pediatric patient transplantation [91]. Other scientists have explored the possibility of removing the xenogeneic lung immune response. In 2018, there was a study to decellularize and subcutaneously implant the α-Gal gene knockout pig lung, and evaluate its immune response by comparison with the wild type. The experimental results showed that the removal of Gal epitopes can potentially reduce the inflammatory response caused by xenotransplantation, which provides a solution to the problem of xenoimmunization [92].

Regardless of the research direction, the optimization of the decellularization process is very important. Although the decellularization protocols used in rats have been perfected, these protocols are not suitable for human-sized lungs. The prolonged decellularization process and inappropriate reagent selection can cause damage to the extracellular matrix. For example, SDS can destroy tissue ultrastructure and eliminate growth factors [93]. Therefore, it is of great significance to optimize the decellularization process. In 2013, a study compared the effects of different methods on the perfusion and decellularization of rats and sheep. The experiment used immunostaining analysis, DNA quantification, electron microscope observation, alveolar size analysis, and biomechanical testing to analyze the experimental results. The results showed that the trachea is the best entry point for lung decellularization, and intermittent decellularization in the trachea can not only greatly shorten the time required for decellularization, but also retain the extracellular matrix well [94]. In 2016, another study compared the effects of tracheal perfusion and pulmonary artery perfusion on the decellularization of rat lungs. The results showed that at high flow rates, pulmonary artery perfusion is more conducive to lung decellularization [95]. Sterilization is an important procedure for decellularized lungs to prevent the spread of viruses and bacteria before cellization, which can reduce the risk of infections such as HIV and hepatitis C. In 2014, a study used SDS to decellularize rat lungs and used it as an experimental object to analyze the effect of gamma ray sterilization on the mechanical properties of decellularized lungs. The experimental results showed that the radiation can change the mechanical properties of decellularized scaffolds [96]. In addition to sterilization, freezing is also a routine procedure for lung decellularization and storage. In the same year, another study also used SDS decellularized rat lung as an experimental material to analyze the effect of freezing and thawing on decellularized scaffolds. The results confirmed that this process only caused minor changes in the decellularized lung [97]. It is well known that an ideal decellularized organ should remove all cellular components that cause an immune response while preserving the extracellular matrix and organ structure. Nucleic acid has always been the main evaluation standard for decellularized scaffolds, and protein is also one of the most important factors. However, there is a lack of research on it. In 2016, a study used protein-type 13C-labeled peptides to perform more accurate quantitative analysis of extracellular matrix proteins in natural and decellularized lungs [98]. The experimental results showed the feasibility of the quantitative decellularization process, and also provide a certain theoretical basis for adjusting the lung decellularization process. In the same year, a study used mass spectrometry-based proteomics to analyze the residual protein of the human decellularized lung [29]. This experiment enhanced the understanding of the decellularized process and also provided a basis for quantifying the decellularized process. GAGs play a key role in cell growth, differentiation, and function, and this characteristic makes them play a key role in lung behavior [99, 100]. However, most GAGs are associated with cells and are removed during the decellularization process. In 2020, a study evaluated the content and function of GAGs in the human decellularized lung. The experimental results confirmed that the loss of GAG in the decellularized lung affected its function and matrix-related growth factors. This study further describes the biological properties of lung decellularized scaffolds and provides some theoretical support for the exploration of decellularized lungs [88].

The decellularized lung not only has profound significance for lung transplantation, but the lung decellularized scaffold can also be used to simulate asthma and other diseases, and plays an important role in the treatment of lung-related diseases

[101, 102]. After continuous exploration and development, the decellularized lung has achieved many results. Although it cannot be used in clinical practice, it is believed that with the continuous development of technology, the decellularized lung of the entire organ can solve various lung diseases in the future.

5.3.3 Liver

The liver is the largest gland in the body, divided into the right large lobe and the small left lobe [103]. As an important metabolic organ, it participates in vitamin metabolism, hormone metabolism, and also has the functions of bile secretion, enzyme activation, excretion, and detoxification. Loss of liver function can cause serious complications such as kidney failure, cerebral edema, and severe hypotension. Today, liver diseases including cirrhosis and hepatocellular carcinoma are still the main cause of global mortality [104]. From the first implementation of canine allogeneic liver transplantation in 1955, a large number of animal studies have been carried out to summarize the transplantation technology to provide experience for clinical use [105]. It was first applied to humans in 1963 [106]. Until now, liver transplantation has several limited operations, but this is the only way to treat end-stage liver disease [107]. The latest data shows that the demand for liver in some countries is almost three times that of transplantable donors. Many patients die while waiting for donors [108]. The long waiting queue indicates the urgent need for alternative therapies.

To solve the above problems, new treatment options including liver cell transplantation and engineered liver tissue have been proposed [109, 110]. However, these methods have certain drawbacks. For example, one method of hepatocyte transplantation is to inject primary hepatocytes into the liver parenchyma to restore liver activity and function, but this method has the problem of limited re-proliferation of transplanted cells, and the clinical application of hepatocyte transplantation is limited by the health of the graft and the frequent undetectable rejection [111]. Other studies have proposed bioartificial liver devices and liver-on-a-chip platforms based on regenerative medicine. The bioartificial liver devices can eliminate toxins accumulated in liver failure, and provide functional liver cell support in terms of synthesis and regulation functions [112]. The liver-on-a-chip as a microfluidic cell culture device can simulate the flow of physiological fluid and control nutrition distribution of substances and growth factors in cells [113]. However, these methods cannot replace liver transplantation, and can only be used as a temporary support to extend the waiting time for liver transplantation. The invention of whole-organ decellularization technology provides a new solution to the problem of liver transplantation. The main development studies of acellular liver are shown in Fig. 5.8.

In 2010, this was the first time that the entire organ of the liver had been decellularized. The experiment used SDS as a detergent, decellularized the rat liver through portal vein infusion, and evaluated the residual cellular components

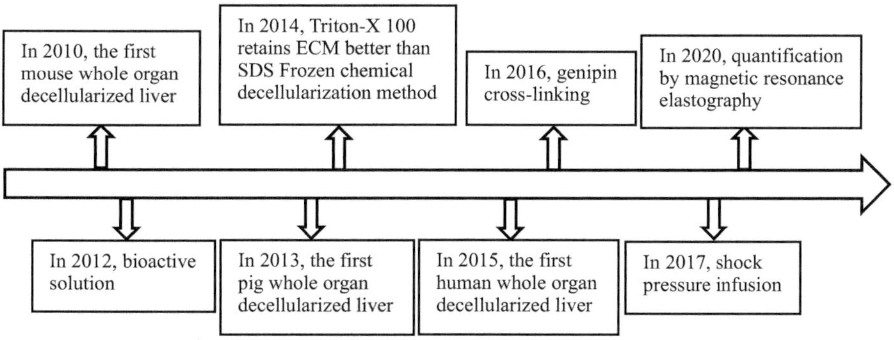

Fig. 5.8 Progress map of acellular liver

and the retention of ECM protein in the decellularized liver by immunostaining. The experimental results found that the content of the external matrix protein and laminin-β1 in the decellularized scaffold was similar to that of the natural liver. It was also found that the decellularized liver retained 100% fibrillar collagen and about 50% GAGs. In addition, the experiment also evaluated the retention of decellularized liver vascular structure through Allura Red dye perfusion and the construction of vascular casting molds. The experimental results confirmed that the decellularization method produced decellularized scaffolds that retained the natural extracellular matrix and vascular structure. The success of this research laid the foundation for exploring acellular liver [114]. In the same year, a study using Triton X-100 and 0.1% SDS to decellularize rat liver also successfully obtained acellular scaffold. In addition to further verifying the feasibility of decellularization of whole liver organs, this study also found some limitations such as perfusion flow rate is too slow to transport liver cells to the inside of the liver lobe, and relatively fast flow rate leads to a large number of intravascular thrombosis [115].

Although the decellularization method reported in previous studies can remove cellular components, strong ion detergents can cause damage to the extracellular matrix and microvascular structure. This poses a challenge for recellularization, which limits cell seeding due to its lack of intra-scaffold channels [116]. In 2011, there was research and development of a new type of scaffold that human liver cells can easily enter to refill the scaffold volume. The experiment used 1% Triton-X 100 and 0.1% ammonium hydroxide to decellularize the livers of rats and ferrets via portal vein perfusion. The obtained decellularized scaffold was evaluated by protein measurement, immunohistochemical analysis, and fluorescence microscope imaging technology. The results showed that the method better preserved the liver vascular tree structure [14]. In the same year, there was a study to improve the previous decellularization process. The experiment applied a biotonic active solution to the rat liver for portal vein perfusion, and successively perfused through the inferior vena cava and aorta using 1% Triton-X 100 and 1% SDS decellularization. The results showed that while removing cellular components and retaining extracellular matrix, the duration of decellularization was shortened to 60 min [117]. This further optimizes the liver decellularization process.

In 2013, a study compared the decellularization effects of Triton X-100 and SDS detergents. The experiment used rats as experimental materials for decellularization through portal vein perfusion, and performed histological analysis and immunofluorescence analysis of the obtained decellularized scaffolds, nucleic acid quantification, and protein determination. It was found that although both methods retained collagen, the elastin content in the decellularized liver treated with SDS and Triton X-100 fell to about 20% and 60% of the natural liver level, the GAGs content dropped to about 10% and 50%, and the HGF content dropped to about 20% and 60%. This indicates that Triton X-100 has better decellularization effect [118]. Although rodents are very good experimental models and already have a relatively complete decellularization protocol, they are not suitable for clinical applications. The existing protocol needs to be extended to the decellularization of human-sized livers. In the same year, it was published for the first time that the pig liver decellularization scaffold was successfully obtained, and the study used Triton X-100 and SDS to decellularize through pulmonary artery perfusion. The experiment evaluated the effect of scaffold decellularization through histological analysis, electron microscope observation, and subcutaneous implantation. It was found that the scheme retained the gallbladder and vascular structure and had low immunogenicity, which confirmed that the scheme can be applied to human decellularization of the liver [25].

In 2014, a study was conducted to obtain a decellularized scaffold using the method of freezing chemical decellularization. After the rat liver was frozen at −80 °C for 24 h, the cells were perfused through the portal vein using 1% Triton-X 100 and 0.1% ammonium hydroxide, as shown in Fig. 5.9a. The results proved that this method can retain the main components of extracellular matrix and vascular framework [119]. Although various liver whole-organ decellularization schemes have been proposed, there is little consensus on the best scheme. In the same year, a study compared the advantages and disadvantages of different decellularization time schemes of different detergents. The experiments were mainly divided into three groups: no detergent, ionic detergent (SDS), and non-ionic detergent (Triton-X 100). Staining evaluation and mechanical elasticity measurement evaluate the decellularized scaffold. As shown in Fig. 5.9b, it was found that time was not a key factor for decellularization, but the decellularization time should be shortened as much as possible to avoid structural collapse, and further proved that Triton-X 100 can retain the scaffold structure and mechanical properties better than SDS [120].

In 2015, there was a study using rats and sheep as experimental materials to compare the effects of SDS and Triton perfusion decellularization. After the decellularized scaffold is recellularized, xenogeneic and homologous transplantation is performed. The results of the study showed that the Triton method is a promising decellularization method that can retain the three-dimensional structure of the liver. The study also found that the same transplantation had better characteristics and a more organized structure than the xenotransplantation. This result poses a demand

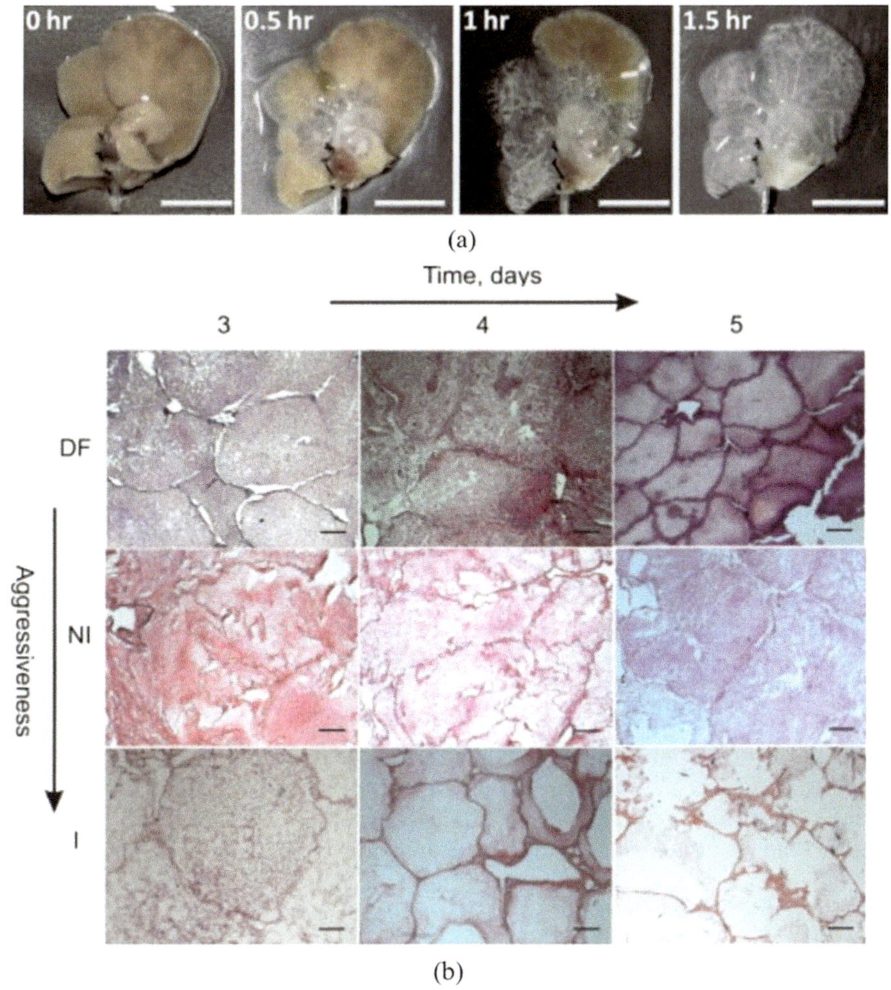

Fig. 5.9 In this study, acellular scaffolds were obtained by cryochemical decellularization (**a**) Decellularization process of mouse liver [119]. This study compared the effects of different acellular reagents and decellularization time on the scaffolds. (**b**) Pig liver changes with the time of immersion and stirring, and the change of liver matrix [120]

for human allotransplantation [27]. In the United Kingdom, over 40% of liver transplants are rejected due to prolonged ischemia time or comorbidities that are judged to exceed critical criteria [121]. This phenomenon presents a solution for allogeneic transplantation. In the same year, decellularization of human liver was reported for the first time. The experiment used cryochemistry to perform decellularization through retrograde perfusion of the hepatic venous system and the inferior vena cava, and proved the feasibility of this method for human liver by characterization of the decellularized stent. This study also provides new ideas for liver transplantation [32].

Human-sized acellular xenogeneic whole liver matrix is a reasonable material for the construction of artificial biological liver, but its transplantation has the problem of immune rejection. One way to reduce the immunogenicity of implants is to crosslink with fixatives. Cross-linking can increase mechanical strength and reduce the immunogenicity of the implant [122]. Glutaraldehyde is the most widely used reagent for cross-linking of xenograft tissues. However, despite the widespread use of glutaraldehyde, problems with calcification and cytotoxicity as well as sub-optimal growth and durability of glutaraldehyde-treated materials have been reported. Genipin, discovered in 1998 as a replacement for glutaraldehyde, has a good in vivo biocompatibility for the grafts it handles. However, the effects of these different cross-linking methods on biocompatibility and immune response have not been thoroughly studied, and the mechanism of occurrence is unclear. In 2016, there was a study using cryochemical methods to obtain acellular liver scaffolds through portal vein perfusion, and immunohistology to evaluate the pros and cons of the two cross-linking methods. The results confirmed that genipin cross-linking can reduce the immunogenicity of xenogenic decellularized whole liver scaffolds [123].

In 2017, a study used 1% Triton X-100 and 1% SDS as detergents to explore the decellularization effect on rat liver under shock pressure. The experiment compared four decellularization methods: through the portal vein or hepatic artery, or infusion with or without shock pressure. The results of the study found that decellularization through hepatic artery perfusion under shock pressure can efficiently remove cellular components within 3 h and retain extracellular matrix well. This study provides a gentle and effective method for obtaining liver decellularization scaffolds [124].

In 2020, a study compared the difference between using only the Triton-x-100 protocol and the Triton-x-100 and SDS combination method to decellularize human liver. The experiment evaluated the decellularization protocol by protein determination, nucleic acid quantification, and electron microscope observation. The results showed that the content of collagen and sGAG retained by the decellularized scaffolds obtained using the Triton-x-100 protocol alone was 1.5 times that of the Triton-x-100 and SDS combination protocol. It was also found that pressure control can greatly shorten the decellularization time without damaging the extracellular matrix [125]. In the same year, there was a study using the drive frequency of compact magnetic resonance elastography (MRE) to study the rat decellularized liver. The experiment found that the stiffness of liver tissue mainly depends on the content of cellular components, increase and decrease of viscous resistance in ECM. This study quantified changes in viscoelasticity during liver decellularization [33].

In the 10 years since the first successful acquisition of a decellularized scaffold for liver organs, after a long period of exploration, the decellularization technology for liver has made great progress. Although this technology cannot be applied clinically for now, it is believed that in the near future, liver transplantation will no longer have a long wait and severe immune rejection.

5.3.4 Kidney

A variety of cells, mainly cubic columnar epithelial cells, constitute the kidney. The kidney is the most important excretory and osmotic regulating organ in human and even mammalian bodies, and it is also one of the highly complex organs. Therefore, if kidney problems occur, it will seriously affect personal health and even life-threatening. Various stresses and oxidative damage can cause chronic kidney disease (CKD) [126]. Unfortunately, CKD is prevalent in adults. About 8–16% of people worldwide suffer from this disease, and CKD has become a global problem [127]. At present, the clinical treatment methods for CKD are mainly peritoneal dialysis, hemodialysis, and kidney transplantation [128], but the cost of treatment is severe and the patient's economic burden is seriously increased. The shortage of kidney sources and the mismatch between kidney transplant patients and donors make kidney transplantation, which is the most effective method for the treatment of end-stage renal disease, also face huge challenges [129]. However, compared with long-term dialysis treatment, kidney transplantation can improve the long-term survival rate and quality of life of patients [130]. In order to meet the growing kidney needs and reduce mortality and morbidity, a new set of solutions is urgently needed. Therefore, from the emergence of non-cellular whole kidney scaffolds that can support cell adhesion, growth, and differentiation, the decellularization technology with a new kidney as the core has begun to flourish. The development and progress of tissue engineering technology provides a solution to the problem of shortage of kidney sources, and has become a new technical means. The emergence of renal decellularization scaffolds provides unprecedented ideas for solving clinical problems, and also provides huge potential for the increase of donor organ banks for kidney transplantation.

After treatment with decellularization reagents, enzymes, or other cell lysis solutions, biologically derived ECM kidney decellularization scaffolds can be prepared. The ideal renal decellularization scaffold should have a unique vascular network and tissue-specific structure, which not only has the same mechanical strength as natural tissue, but also can provide cells with a suitable tissue framework for proliferation, differentiation, and replantation. Ultimately, it is guaranteed to be able to realize the construction of kidney engineered organs through recellularization technology. The kidney decellularized scaffold can still be regarded as a normal kidney, because it has similar characteristics to natural kidney [131]. Although the construction of kidney decellularization scaffold is an important method for preparing bioartificial kidneys, how to construct a suitable environment conducive to cell attachment need to be continuously explored. At present, it is already possible to obtain a relatively mature renal decellularized scaffold after being treated by decellularization technology, and its development is inseparable from its significant advantages and prospects. On the one hand, after decellularization, the kidney decellularization scaffold can not only retain the inherent three-dimensional structure of ECM, but also leave some biochemical components beneficial to the cells. On the other hand, the kidney decellularization scaffold not only has a complete

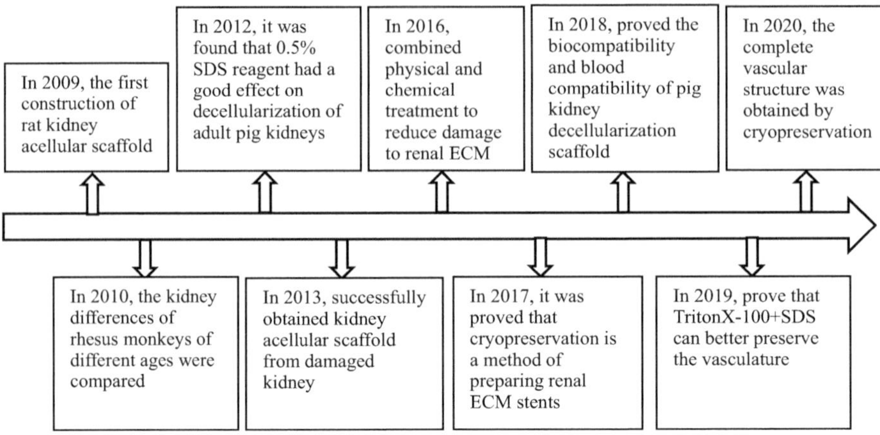

Fig. 5.10 A diagram of the research progress of acellular kidney

structure, but its high tissue specificity also becomes an ECM template for kidney transplantation. Renal decellularized scaffolds not only retain fibrin, but also soluble signals, including cytokines, growth factors, and chemokines. These signals have a huge impact on cell adhesion, migration, proliferation, and differentiation [132].

The research process of the whole-organ kidney decellularization scaffold is shown in Fig. 5.10. Since Ross and others first proposed the use of rat kidney as a template to make acellular renal scaffolds in 2009, researchers have gradually used pig or monkey kidneys for experiments [133]. They perfused rat kidneys with three different decellularization reagents, ionic, non-ionic detergent, and enzyme solution, for 5 days through blood vessels. The results showed that the ionic detergent SDC could not completely remove the cell debris, and the SDS solution could produce a complete decellularized scaffold. Their work laid the foundation for the establishment and optimization of the rodent kidney decellularization program.

Later, in order to systematically evaluate the preparation effect of rodent kidney acellular scaffolds, Caralt and others developed a histological scoring system [134]. In order to verify the scoring system, they compared three different methods of rat kidney decellularization. The final results show that treatment of rat kidneys with TritonX-100 reagent alone does not completely remove the cells, and the combination with SDS reagent or trypsin-EGTA reagent can effectively remove most of the cells. But trypsin-EGTA reagent will cause the deterioration of kidney decellularization scaffold structure and the loss of growth factors, so the decellularization effect is not as good as SDS reagent. It was finally proved that in order to ensure the integrity of the ECM structure and composition while removing the cells, the use of TritonX-100 and SDS is the most effective method.

In addition, physical decellularization methods are gradually emerging. In order to evaluate the effect of hypothermia on the kidney tissue structure, some researchers compared the difference between the kidney scaffold and the freshly isolated rat kidney after 3 months of low-temperature storage by different methods. They used a scanning electron microscope to observe the ultrastructure of different organs and found that their structures were effectively preserved. The results of immunohistochemical evaluation also showed that cryopreservation did not result in significant loss of collagen and laminin [135]. In 2020, Feng and others evaluated the integrity of the vessel structure of the kidney decellularized scaffold after cryopreservation in order to achieve the purpose of providing the kidney decellularized scaffold at any time [136]. They used a mixed reagent consisting of TritonX-100 and DNase to decellularize the rat kidney based on the perfusion method. Afterwards, the obtained scaffolds were pre-treated with different cryoprotectants and cryopreserved. Then the Micro-CT imaging method was used to confirm the vascular structure changes of the renal decellularization scaffold. The results are shown in Fig. 5.11a–e, which finally proved that the complete vascular structure was successfully preserved. Therefore, it is proved that cryopreservation is also an effective method for manufacturing functional renal ECM scaffolds.

Rhesus monkey kidneys are considered by researchers to be the most suitable material for studying kidney acellular scaffolds because they have characteristics

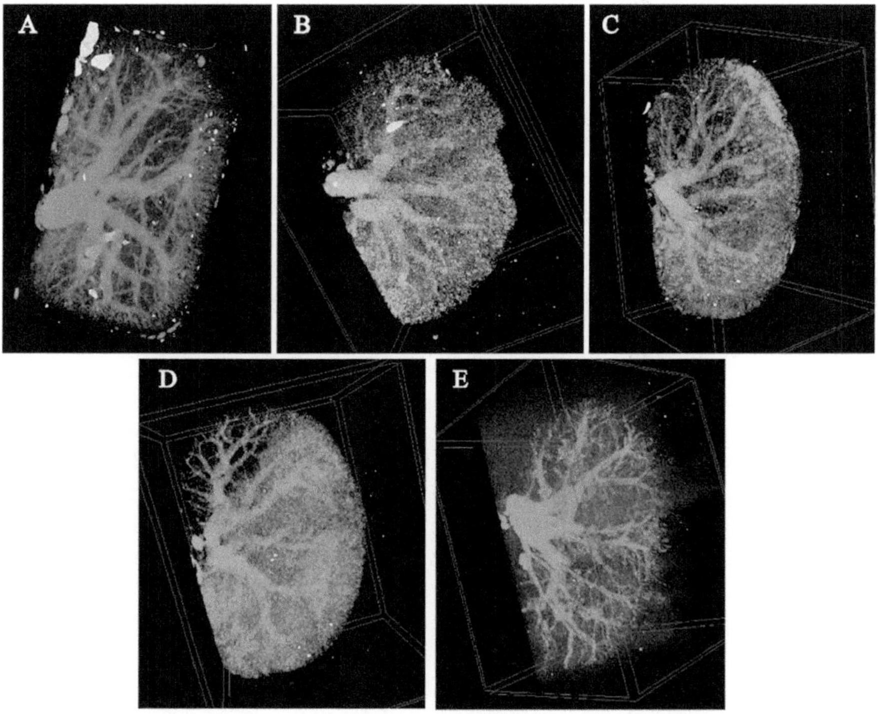

Fig. 5.11 After freezing/thawing, a micro-CT image of the renal stent, in which the vasculature is filled with microfil. (**a**) is the control group, (**b**) is the direct freezing group, (**c**–**e**) is the micro-CT image of the stent after adding the cryoprotectant. (**c**) is a group loaded with 10% dimethyl sulfoxide (Me2SO) and 6% hydroxyethyl starch (HES). (**d**) is the group loaded with 20% Me2SO + 6% HES. (**e**) is the group loaded with VS83 [136]

closer to humans. Nakayama and others processed the kidney sections of cynomolgus monkeys of different ages by using a mixed reagent of SDS and TritonX-100 [137]. They found that kidney decellularized sections of macaque monkeys of any age can provide a natural kidney decellularized scaffold with appropriate structure and certain biochemical properties. In addition, these scaffolds also retain a variety of growth factors and antibacterial proteins. These advantages are likely to provide strong support for the cultivation of recellular organs in vitro in the future [138]. At the same time, they also found that the age of the donor greatly affected the efficiency of recellularization [139].

Although the macaque kidney has the advantage of being closer to human characteristics, and is considered to be a suitable material for studying kidney acellular scaffolds, its size and quantity supply cannot meet the future needs of the clinic. So the researchers gradually turned their vision to pigs. Currently the most common donor animal for clinical xenotransplantation is pigs. They are considered to be suitable, clinically relevant animal models, and can provide a certain reference for whether the decellularization technology can be used in human organs. Sullivan and colleagues evaluated the decellularization effect through a high-throughput device and found that 0.5% SDS reagent was very effective in decellularizing the kidney of adult pigs [15]. After decellularizing adult pig kidneys, Sullivan and colleagues successfully retained the inherent structure of the kidney decellularization scaffold, including the components of the renal vasculature and ECM. In order to increase the understanding of the characteristics of pig kidney acellular scaffolds, in 2018, Hussein et al. explored the biocompatibility and blood compatibility of pig acellular scaffolds [140]. Porcine kidney decellularized scaffolds were obtained by perfusion, and then SEM and DAPI staining were used to confirm the complete removal of cells. At the same time, through angiography, it was proved that the complete renal blood vessels were successfully retained. The stent showed good compatibility after being exposed to pig blood or planted cells. Finally, it was shown that the developed rapid decellularization technology can produce a renal decellularization scaffold with good compatibility.

However, the current decellularized detergents always cause certain damage to the microstructure of kidney tissue during the process of generating decellularized scaffolds, so some researchers try to minimize such adverse effects. For example, Poornejad and others combine physical and chemical steps to prevent the detergent from damaging the inherent structure of the decellularized scaffold [141]. The final result proves that this method can effectively remove natural cells and reduce the damage to ECM.

Although the kidney decellularization scaffold has achieved certain results in animal experiments, whether it can provide a donor for human kidney transplantation remains to be verified. After all, in the final analysis, human-derived kidneys are the most ideal source of clinical human kidney decellularization scaffolds, and also have better compatibility during kidney transplantation. However, almost all of the available human kidneys are discarded kidneys, so some researchers have begun to try to use discarded human kidneys as a material for tissue engineering and kidney transplantation research [142]. In 2013, Orlando and others reported for the first time the experiment of applying decellularization strategies to abandoned human kidneys [143]. They selected the SDS reagent to decellularize the kidney, and the obtained renal decellularized scaffold successfully retained the inherent structure and biochemical characteristics of ECM. It has also been found that the three-dimensional structure of natural glomeruli and the compliance of the innate vascular network were successfully retained in the human kidney decellularization scaffold, and the growth factors in ECM were also preserved [144]. Unfortunately, abandoned kidneys will be permanently damaged due to the disease or age of the donor, which will negatively affect the structure of the ECM. Therefore, the damaged kidney organs cannot be used as the source of organs for clinical kidney transplantation.

At present, the methods of decellularization of kidney mainly include whole-organ perfusion and tissue section immersion method, but it is recognized that the most commonly used and most effective decellularization method is still to deliver decellularized fluid through blood vessels. The key to making a kidney decellularization scaffold using whole-organ perfusion is the need for a complete vasculature and suitable decellularization reagents [145]. The whole-organ perfusion method prepares the decellularized scaffold by the method that the blood vessel delivers the decellularized reagent to every corner of the kidney, which can minimize the damage of the ECM structure. The purpose of the tissue section immersion method is to increase the contact area between the kidney and the decellularization reagent. Tissue section immersion method first slices kidney tissue, and then decellularizes the kidney tissue. At the same time, it should be noted that the stirring speed in the decellularization process will affect the removal of DNA and the loss of ECM to a certain extent.

Currently, due to the limitations of decellularization technology, we cannot remove all cellular components. Therefore, during the process of cell replanting, the remaining cell components may cause cell compatibility problems. At the same time, the remaining cellular components may also lead to immune rejection problems [146]. These are issues that we cannot afford to ignore. In order to avoid the problems of cell compatibility and immune rejection, and to estimate the risk of adverse reactions, after completing the decellularization treatment, the kidney decellularized scaffold needs to be evaluated for its quality from the aspects of scaffold structural integrity, cell residue, DNA content, and protein content.

The integrity of the macrostructure and microstructure of the renal decellularization scaffold is the key to the success of subsequent experiments. The integrity of the macrostructure can be judged by the appearance of the stent, the casting of blood vessels, and the imaging of blood vessel networks [134]. The microstructure can be judged by HE staining and DAPI staining. The mechanical strength of the renal decellularization scaffold will seriously decrease with the extension of the decellularization reagent's action time, so it is extremely important to explore suitable decellularization reagents and decellularization time under different experimental conditions [147]. The amount of DNA directly affects the immune rejection of the renal decellularized scaffold, so it is particularly important to detect the residual DNA content in the renal decellularized scaffold. Usually quantitative DNA measurement is achieved using mature commercial kits. It is generally believed that an ideal decellularized scaffold should meet the following three conditions. First, there is no nucleus in HE or DAPI staining. Second, the DNA content of the scaffold per milligram of dry weight is <50 ng. Third, and the length of the existing DNA fragment is <200 bp. In order to understand the types and amounts of proteins in the EMC that constitute the renal decellularization scaffold, and to achieve the purpose of adjusting the specific components of the scaffold for different cell growth, it is necessary to perform quantitative protein analysis. Gel electrophoresis and mass spectrometry are commonly used tools for screening proteins and peptides [148]. Through the combination of different protein components, a scaffold suitable for different cell growth and immune rejection can be constructed, and at the same time, it can help the scaffold to function better.

So far, the research on renal organ regeneration engineering is still in the early stages. In 2019, Kajbafzadeh et al. explored the changes in the vasculature of kidney decellularized scaffolds after transplantation into sheep [149]. Based on the perfusion method, they treated sheep kidneys with two different decellularizing agents (Reagent 1: 1% TritonX-100 + 0.5% SDS; Reagent 2: 1% SDS) and implanted them in sheep. The results show that both reagents can achieve complete decellularization. In addition, the ECM in the renal decellularized scaffold treated with reagent 1 is better preserved and the vasculature is intact. The sample treated with reagent 2 exhibited liquid leakage. The results are shown in Fig. 5.12. Although some laboratories have produced recellularized artificial kidneys that can be used for animal transplantation, there are still many difficulties and challenges that we need to continue to solve. Only by thoroughly understanding the structure of the kidney and the biology of kidney cells, can we realize the creation of acellular decellularized scaffold that can be applied to the human body at an early date.

The development of regenerative medicine provides new solutions for the treatment of kidney diseases. Although the existing decellularization scheme has been able to remove more than 99% of the cellular components, due to the side effects of the decellularization reagents, the components and mechanical strength of ECM in the renal decellularization scaffold will always be damaged to a certain extent. Therefore, achieving decellularization without damaging the structure and composition of the scaffold is a key direction for future research.

5.3.5 Pancreas

With the improvement of living standards in modern society, unfortunately, diabetes has obviously become a serious disease that endangers human health, which has aroused widespread attention in countries around the world. According to data from the World Health Organization, it can be predicted that the prevalence of diabetes between the ages of 20 and 79 will increase to 7.7% by 2030, and the proportion will be higher and higher. Diabetic patients are characterized by elevated circulating blood glucose levels. As the condition worsens, blindness, renal failure, or nerve

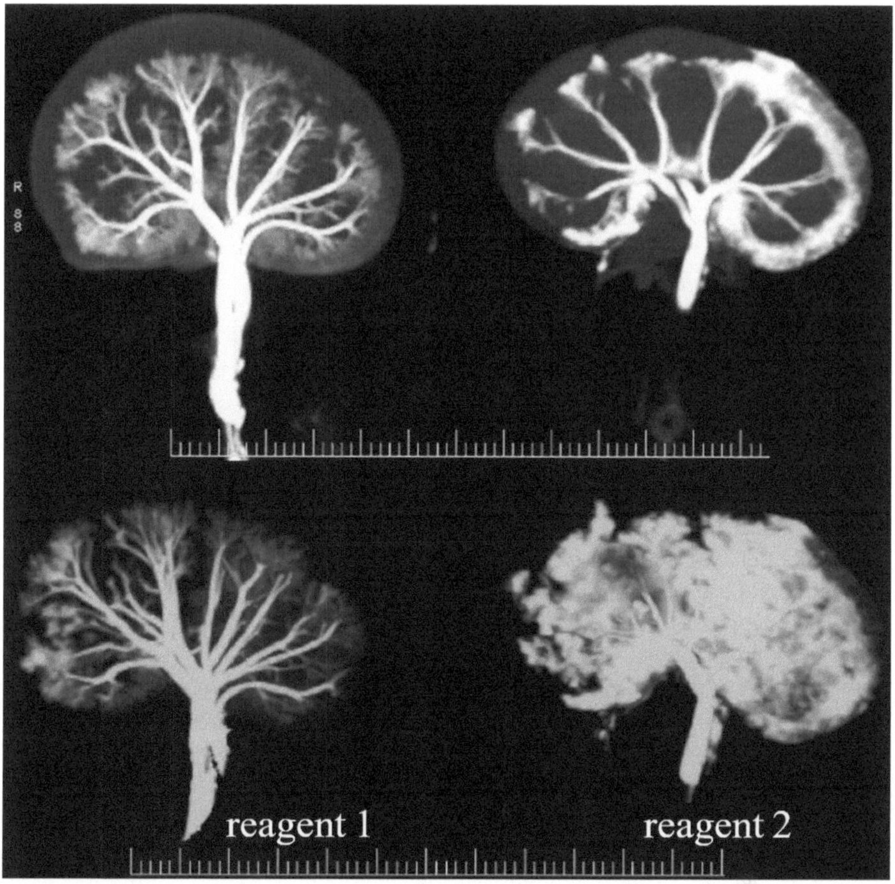

Fig. 5.12 CT angiography of goat kidney decellularized scaffold. After treatment with Reagent 1, the decellularized scaffold retains the complete vasculature. After the reagent 2 was processed, a leak of the contrast medium was detected, and the integrity of the blood vessel was destroyed [149]

damage may occur. At the same time, diabetes has brought a huge economic burden to society. Therefore, it is imminent to find a good treatment for diabetes.

Current treatment options for diabetes include insulin therapy, drug therapy, alpha-glucosidase inhibitor therapy, and pancreas transplantation and islet transplantation. However, oral hypoglycemic agent therapy or insulin injection therapy is still the most effective way to treat diabetes in clinical practice. Some insulin release devices have also been successfully designed to improve patient comfort and control and regulate insulin release. But the advent of insulin therapy has not helped tremendous progress in the field of diabetes treatment. Long-term use of insulin cannot avoid the occurrence of diabetes-related complications, and it will bring a huge burden to the patient's body, which in turn causes diabetic retinopathy, ketoacidosis, and weight gain. Therefore, the development of alternative insulin

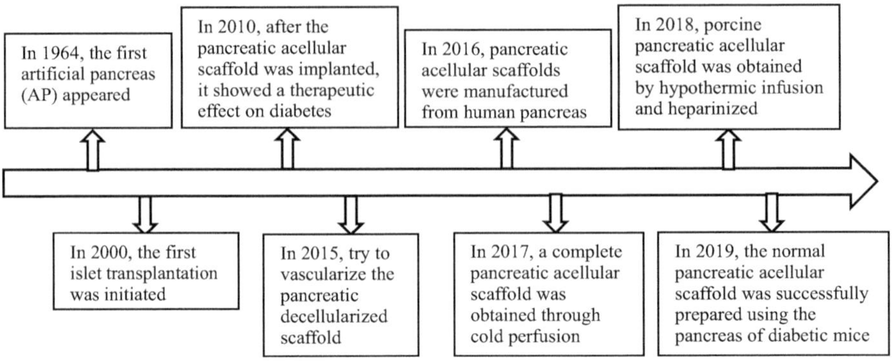

Fig. 5.13 Diagram of the progress of acellular pancreas research

therapy has become an important research direction. Tissue engineering shows good prospects in the treatment of diabetes, and offers researchers the possibility to propose better treatment options. Its core purpose is to establish a viable living tissue to treat tissue and organ diseases, and has broad application prospects [150].

The development of pancreatic acellular scaffolds is shown in Fig. 5.13. The development of artificial pancreas (AP) can be traced back to 1964, 56 years ago. Kadish produced the world's first artificial pancreas by combining portable and implantable insulin pumps, continuous blood glucose monitoring systems, and closed-loop control algorithms [151]. The development of tissue engineering has prompted people to try pancreas or islet transplantation. Since Edmoton launched islet transplantation in 2000 [152], the success of multiple trials has successfully demonstrated that cell therapy can be used as a viable method for treating diabetes. Despite the success of the Edmoton islet transplantation program in the short term, the long-term efficacy of islet transplantation has not been found to be satisfactory. In the late period of transplantation, complications such as portal vein thrombosis and excessive portal pressure appeared. But this is a great attempt of tissue engineering in the field of diabetes treatment. The rapid development of regenerative medicine provides a new strategy for diabetes treatment. Goh and others succeeded in proving that pancreas transplantation is also a potentially feasible treatment for diabetes by replanting insulin-secreting cells in pancreatic tissue [153].

Organ transplantation provides an unprecedented treatment plan for the treatment of diabetes, and is an effective method for the complete treatment of organ failure and repair of tissue damage. However, the contradiction between the increasing demand for organ transplantation and the insufficient number of donors has become the main bottleneck restricting this treatment plan. In addition, long-term immunosuppressive therapy is needed to prevent rejection [154]. Or by encapsulation, a barrier is established between the donated islets and the recipient's immune system [155]. However, the materials used for encapsulation have limited biocompatibility, and there are problems of hypoxia and delayed insulin secretion. In addition, long-term use of immunosuppressive drugs will cause the loss of grafts to a certain extent.

The development of regenerative medicine and biological tissue engineering technology has laid a foundation for the proposal of pancreatic scaffolds, which also provides new ideas for solving the problem of appeal. Currently, pancreatic stent materials include natural ingredients, chemically synthesized materials, and nanomaterials, but because of the inherent defects of these materials, they are not suitable for clinical use. The natural extracellular matrix (ECM) produced by decellularization technology avoids these problems. The decellularization technology not only retains ECM scaffolds, but also retains relevant growth factors and cytokines while removing cells and cellular material [156]. At the same time, the formation of amyloid and the loss of extracellular matrix are non-immune factors for islet death. But the pancreatic acellular scaffold reduces the formation of amyloid and improves the vitality and function of human islets, increasing the possibility of islet transplantation [157]. The pancreatic acellular scaffold can be used for cell culture and artificial tissue reconstruction, and is a new type of 3D biological scaffold suitable for cell growth. In the process of decellularization, it is theoretically believed to be able to remove antigens associated with cell membranes, which can ensure that almost no immune response is caused during the recellularization process. Research results have shown that pancreatic stents can induce capillary regeneration and regulate insulin release.

In the past period of time, considerable progress has been made in the field of decellularization. However, the current research on decellularization mainly focuses on the large organs rich in blood vessels and suitable for perfusion, such as kidney, liver, heart, and lung, and the research on the whole pancreas acellular matrix scaffold (APB) is still lagging. This may be because the pancreas is both endocrine and exocrine tissue, and the cells are complicated. However, some studies have successfully prepared pancreatic acellular matrix scaffolds with intact ECM structures by using the perfusion of the natural duct structure of pancreatic tissue. In 2013, Goh et al. studied the feasibility of pancreatic perfusion decellularization [153]. The results are shown in Fig. 5.14a–g. As a result, cells and nuclear materials were successfully removed, while the complex three-dimensional structure of the perfusionable vascular system and ECM was retained. This result promoted the development of pancreatic decellularized scaffolds. By decellularizing the pancreas by cold perfusion, a pancreatic decellularized scaffold with complete ECM structure and well-preserved vascular structure can be obtained [17]. In addition to inhibiting enzymes at low temperatures, the addition of chemical serine protease inhibitors can also improve the decellularization quality of the scaffold [158]. However, it should be noted that since the pancreas is rich in fat tissue, a fat solvent should be added to the decellularization reagent. Since the pancreatic acellular scaffold is derived from natural organs, it is very suitable as a platform for the entire organ bioengineering. The whole pancreatic acellular matrix scaffold finally obtained can be effectively used for cell seeding.

With the completion of the pancreatic decellularization scaffold, more and more researchers have set their sights on its application. De Carlo and others successfully planted the pancreatic decellularized scaffold into a tubular capsule structure and implanted it under the skin of rats [159]. The results proved that the scaffold

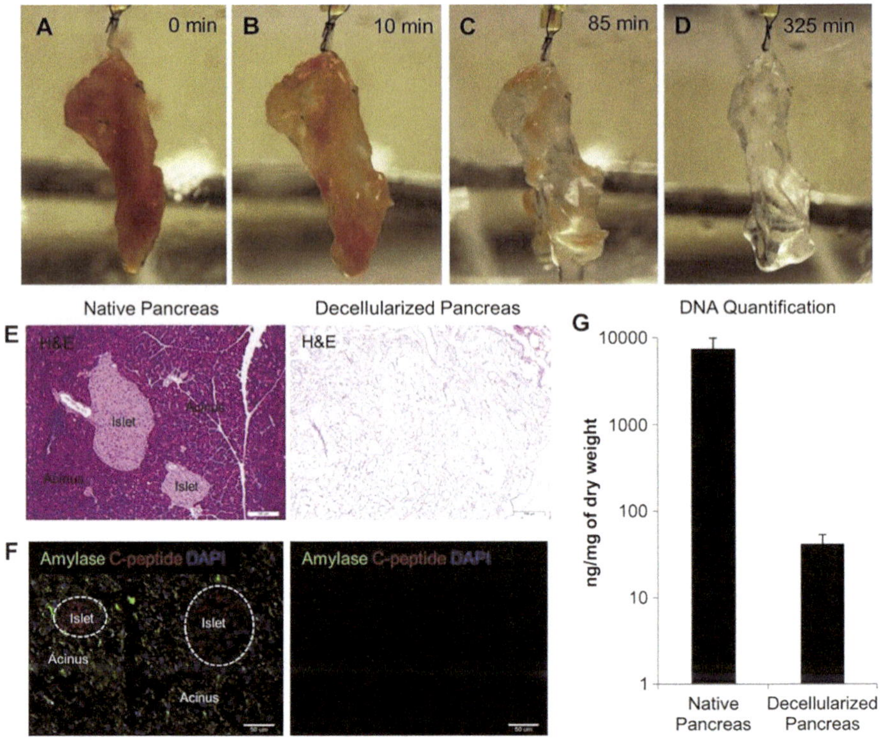

Fig. 5.14 Decellularization of mouse pancreas by perfusion decellularization. The original pancreas of the mouse (**a**), and after infusion of 0.5% SDS decellularization reagent, the color of the mouse pancreas gradually changed (**b–d**). After 325 min of perfusion, (**d**) the decellularized pancreas becomes translucent. (**e**) H&E staining of natural pancreas and acellular pancreas. (**f**) IHC assessment of natural pancreas and acellular pancreas. (**g**) Quantitative analysis of DNA content in decellularized pancreas. The data shows that the cells in the decellularized pancreas are removed without the main components of the pancreas. In addition, the DNA present in the decellularized pancreas is less than 50 ng/mg [153]

successfully differentiated to form a pancreatic islet-like structure, prolonged the action time of insulin, and finally successfully reduced the blood glucose of rats, verifying its therapeutic effect on diabetes.

As the technology matures, how to extract pancreatic acellular scaffolds from large animals and humans has attracted wide attention from researchers. Pigs are animals that are closer to humans. Porcine pancreatic acellular scaffolds is not only easy to obtain but also simple to prepare. Furthermore, the porcine pancreatic basement membrane has a loose structure and is easy to decellularize. In addition, it promotes cell adhesion during recellularization, so it has become the first choice for research. In order to explore a suitable decellularization method, Kuna and others uses the aorta as the inlet and the portal vein as the outlet, and decellularizes the pig pancreas by perfusion at 4 °C [160]. At the same time, in order to prevent the formation of thrombus, heparinization treatment was carried out, and they finally successfully obtained the porcine pancreatic acellular scaffold.

With the deepening of research, researchers gradually realized that it is necessary to strengthen the research on the preparation of non-human primate pancreatic decellularized scaffolds, and eventually deepened into the study of human own pancreatic decellularized scaffolds, in order to explore a more suitable human pancreatic decellularized scaffold. Research models have gradually evolved from pigs to human models. Peloso and others tried to produce pancreatic decellularized scaffolds from human pancreas. The results prove that the pancreatic acellular scaffold can be produced from the human body, while maintaining the spatial structure, stiffness, and important growth factors of the pancreas itself. The human pancreatic acellular scaffold inhibits the expansion of CD4 T cells by inducing apoptosis and promoting their conversion to regulatory T cells [161].

Although the research on normal pancreas has achieved gratifying results, the number of normal human pancreas donors is limited after all, so pancreatic tissue that can be used as a source of pancreatic decellularization scaffolds may often come from individuals with diabetes or pre-diabetes symptoms. Therefore, whether the pancreas of an individual with diabetes can be used as a donor for a decellularized scaffold, and whether there is a significant difference between a healthy pancreatic scaffold and a diabetic pancreatic scaffold becomes a question worth exploring. Huang et al. prepared the decellularized pancreatic scaffolds of normal mice, mice with type 1 diabetes, and mice with type 2 diabetes by using the cryo-perfusion method, and evaluated them [162]. It was found that individual mice with diabetes can successfully prepare the same pancreatic acellular scaffold as normal mice, and retain the ECM composition, ultrastructure, and various cytokines. The results of this study initially proved that the diabetic pancreatic acellular scaffold has the same effect as the normal scaffold, which provides a basis for further research on the diabetic pancreatic scaffold.

The ideal pancreatic acellular scaffold should contain a continuous and complete vascular structure. For tissue-engineered organs, a complete functional vascular network is an important channel for providing nutrients and oxygen for cell growth. The vascular network structure of pancreatic tissue is complex. Due to the lack of good blood supply, decellularized pancreatic scaffolds after transplantation are often only effective in the short term [163]. Therefore, this reflects that the vascularization of the pancreatic acellular scaffold is a special obstacle that hinders its clinical application. Many feasible vascularization strategies have been proposed in the current study, mainly using artificial scaffolds, using pre-vascularization networks, using artificial blood vessel structures, using natural vascular structures, and building new blood vessels based on cellular components [164], but the experimental results are not ideal. At the same time, researchers have formed a fixed method for detecting vascularization. They use techniques such as vascular casting, fluorescent dyes and contrast to evaluate the integrity of the vascularized structure in the pancreatic decellularized stent.

In addition, due to the dual challenges in the process of pancreatic tissue engineering research, the current research on pancreatic acellular scaffolds is still relatively lacking. Because the pancreas is composed of two different tissue types, exocrine tissue (Acinar and ductal cells) and endocrine tissue (α-cells, β-cells, ε-cells, δ-cells, and PP cells) [165]. Therefore, in the decellularization process, the requirements for the components that constitute the decellularization reagent are high, and it is necessary to pay attention to the applicability to different structures, and at the same time, pay attention to the addition of the fat-dissolving agent. In addition, because the pancreas has the function of assisting the duodenum to digest [166], the pancreatic juice secreted by the pancreas contains a variety of different digestive enzymes and bicarbonate secreted by the ductal cells. A large amount of protease and a small amount of different lipases, nucleases, and amylases constitute the main components of digestive enzymes, but these enzymes are usually produced and secreted in the form of inactive enzymes, only under the action of active trypsin the cascade reaction is activated. However, during the decellularization process, the intracellular enzymes of the lysed cells are exposed to the surrounding environment, and some enzymes may be activated under the interaction with the surrounding environment, which in turn causes other enzymes in the chain reaction to be activated. Unfortunately, the activated protease may damage the pancreatic decellularization scaffold, so during the decellularization process, these enzymes have to be inhibited by appropriate methods to ensure the smooth progress of the pancreatic decellularization process. There has been a method to inhibit the activity of these enzymes, namely to adjust the ambient temperature. Experiments to study the decellularization process of rat pancreas showed that the hydrolysis efficiency decreased to about 20% at 5 °C, and the hydrolysis efficiency decreased three times from room temperature to 5 °C compared with the optimal temperature [167].

Finally, it should be noted that when perfusion is performed on the pancreas, it is particularly important to control the perfusion pressure. However, since the substantial parts of these large organs can easily hinder the diffusion of decellularized reagents, high perfusion pressure methods are usually required when treating these organs to ensure the effect. Paradoxically, prolonged high-pressure infusion of potent and long-lasting decellularization reagents will dissolve proteins in ECM and disrupt the three-dimensional structure of the original tissue. Therefore, exploring appropriate perfusion pressure and perfusion time is particularly important to enhance the effect of decellularization and recellularization.

The ultimate goal of the pancreatic acellular scaffold is to create an organ that can be transplanted. Therefore, the first issue to consider is the biocompatibility of the stent. Currently, there are internationally accepted decellularization standards to ensure biocompatibility. In addition, what needs to be evaluated is the immunogenicity of pancreatic decellularized scaffolds, but there are few researches at present, and most of the studies are based on the morphological evaluation of implanted scaffolds in vivo [168]. But the body's remodeling process of the pancreatic decellularized scaffold implanted in the body is a complex and long process [169], which involves many different inflammatory cells and cytokines, and the most important part of the chronic inflammatory response is macrophage cell [170]. Through more in-depth studies, it has been proved that the macrophage polarization model is closely related to the remodeling process of pancreatic acellular scaffolds in vivo, and has been used by researchers as a reliable indicator to predict the results of scaffold remodeling [171].

The study of the pancreas decellularization scaffold is relatively lagging behind the study of other large organs, which is caused by the complex structure of the pancreas and its complex cellular composition. But since the study of decellularization of rat pancreas, the field has gradually developed to decellularization of pig, human, and human diabetic pancreas. The special nature of the pancreas is both endocrine and exocrine tissues, so that the decellularization reagents used in the decellularization process need to be properly adjusted. Firstly, because the pancreas is rich in adipose tissue, it is necessary to add a fat-dissolving agent to the reagent. Secondly, heparinization therapy is needed to prevent the formation of thrombus. In addition, since the pancreatic juice contains a large amount of digestive enzymes and bicarbonate, which will adversely affect the stent itself, it is necessary to perform decellularization in a low-temperature environment. All in all, the pancreatic acellular scaffold provides a promising method for the treatment of diabetes, but how to obtain a scaffold that can replace the diseased pancreas for a long time is the focus of subsequent research in this field. In addition, how to use the damaged human acellular pancreas to treat diabetic patients is still worthy of study.

5.3.6 *Intestine*

Intestinal failure (IF) is an intestinal disease characterized by the inability of the intestine to absorb nutrients. At present, the treatment method of small bowel transplantation is proposed, but the clinical application of this method is limited due to the shortage of organs. The proposal of acellular small intestinal stent provides hope for this therapy.

The development process of acellular small intestine is shown in Fig. 5.15. In 1966, Matsumoto and others replaced the damaged tissue by transplanting the obtained autologous SIS of the dog as a vascular substitute to its vena cava segment [172]. But shortly after implantation, occlusion of the graft was observed, but no dysfunction was found in the early stage. The allograft was successfully replaced and maintained its function for up to 6 months, but unfortunately eventually failed due to the occurrence of an aneurysm. In 2012, Totonelli et al. proposed to use detergent enzymatic treatment (DET) method to optimize the construction of rat intestinal decellularization scaffold in order to provide better treatment for patients with intestinal failure [173]. The results showed that after 1 cycle of treatment, the cellular components were removed, and the natural structure and connective tissue of the intestinal decellularized scaffold were retained. The results are shown in Fig. 5.16.

In 2013, in order to retain the complex structure of natural organs during decellularization, Maghsoudlou and others proposed an effective technique for decellularizing the small intestine of rats [174]. They used a mixed reagent of chemical and enzyme to decellularize the small intestine, and successfully retained

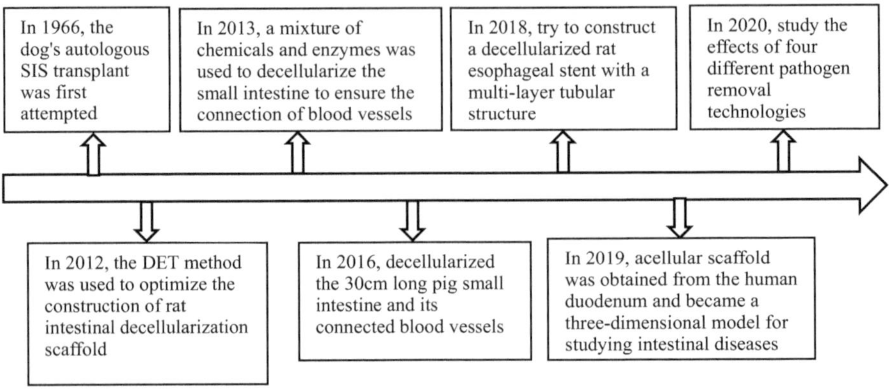

Fig. 5.15 Progress map of acellular intestinal research

the villi-crypt axis in the small intestine cavity, and at the same time ensured the blood vessel connection by isolating the small intestine. In the same year, in order to produce a tissue-engineered small intestine (TESI) that could be implanted into the human body, Patil and others decellularized the small intestine tissue and obtained a small intestine (SI) piece with a length of about 6 cm [175]. As a result, it was found that structural proteins and villi were successfully retained, and cell components were completely removed. This proves that TESI can avoid immunosuppression. In 2016, in order to study the behavior and immune response type of the porcine small intestine decellularization scaffold in the body, Nowocin and others decellularized a 30 cm long porcine small intestine and the vasculature connected to it [176]. After successful decellularization, the stent was implanted into the animal and angiogenesis was successfully achieved without causing fibrosis or foreign body reaction. After a spectrum analysis of the macrophages, it was found that in the fourth to eighth week, the immune response began to change to promote remodeling. After implantation of acellular scaffolds, tissue loss is often caused by failure to form proper vascular access. Therefore, in order to find an in vitro platform for studying angiogenesis, in 2016, Dew et al. obtained a natural vascular stent by decellularizing the intestine of rats [177]. The stent retains part of the angiogenic factors, and at the same time guarantees the integrity of the vascular channel. In the case of external stimulation, it has the ability to generate new blood vessels.

In 2018, Urbani and others tried to build a multi-layer tubular structure decellularized rat esophageal stent in order to overcome the difficulties of esophagus replacement in tissue engineering [178]. In the same year, Piccoli and others proposed a new tissue engineering method to apply the generated 3D acellular scaffold to colorectal cancer (CRC) examination [179]. They studied the difference in biological activity between normal decellularized scaffolds and CRC scaffolds, providing a powerful solution for clinical research and disease screening. The results showed that the healthy mucosal acellular scaffold and the CRC scaffold are very different in structure and secreted protein, and compared with the normal scaffold, the potential for vascularization in the acellular scaffold is reduced. Still in this year,

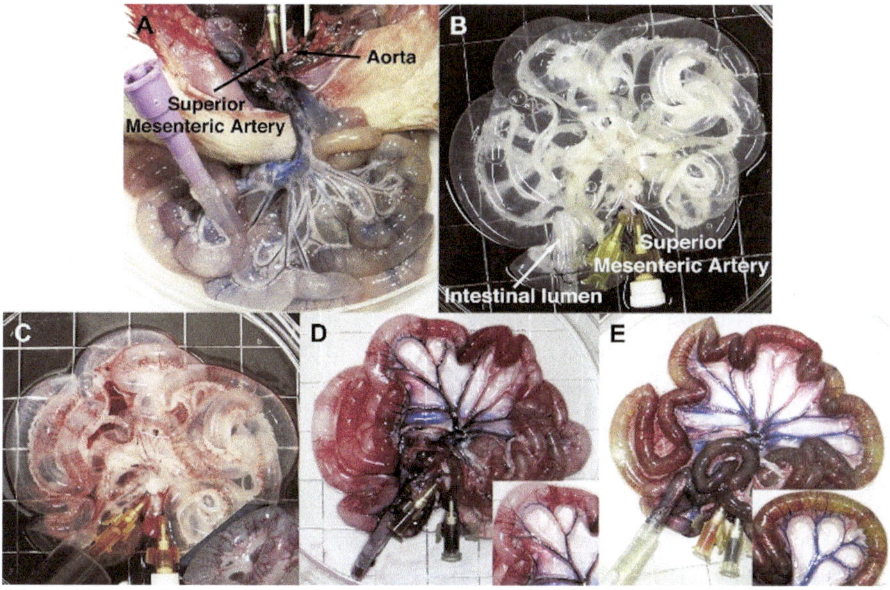

Fig. 5.16 The rat small intestine (**a**) before decellularization is shown in the figure, and a translucent rat small intestine (**b**) is obtained after one cycle of treatment. (**c**) Perfusion of rat small intestine with Rosso Ponceau, (**d**) Perfusion of rat small intestine with trypan blue, (**e**) Perfusion of rat small intestine with rhodamine green. The results show that the patency and distribution of the capillary beds of the arteries and veins are no problem, and there is no leakage [173]

Kajbafzadeh et al. obtained a human small intestine decellularized scaffold by immersing the human small intestine in a hypotonic solution [180]. Later, in order to verify its function, it was used to repair the rabbit's bladder. After verification, the scaffold decellularization effect is good. Observed 6 months after implantation, the results showed that the bladder after implantation was similar to the normal bladder, forming new muscles and blood vessels. This provides a basis for the application of small intestine decellularization scaffolds in bladder augmentation and cystoplasty.

With the development of decellularization technology, the human intestinal decellularized scaffold obtained from the human intestine may become a new available 3D model for studying intestinal diseases and promote the discovery of therapeutic targets. In 2019, Giuffrida et al. designed and verified the method of obtaining a decellularized scaffold from the human duodenum, and successfully obtained the decellularized scaffold of the human duodenum [181]. The primary human intestinal myofibroblasts cultured on them differ greatly from standard 2D culture dishes in the expression of key mRNAs involved in ECM conversion and the expression of pro-fibrotic growth factors. This proves the feasibility of the human intestinal acellular scaffold as a platform for studying diseases and drugs. In the same year, Chen and others proposed an effective decellularization method for human small intestinal stents to transform the performance of human colon implants

[182]. This provides a new method for the repair of human intestines and the identification of potential cancer driver genes.

In 2020, in order to study the effects of different pathogen removal technologies, Gosztyla et al. chose four different protocols that have appeared to sterilize intestinal tissue [183]. Quantitative and qualitative techniques were used to test the processed intestinal tissue. The results showed that all methods were able to achieve sterilization and all had similar ability to support cell growth. But all methods also have an adverse effect on the microarchitecture of intestinal tissue. Therefore, the best sterilization method still needs to be explored.

The great advantages of the acellular small intestinal scaffold make it a hot spot in the field of tissue repair. It has the advantages of strong biological activity, good absorbability, and almost no immune response, but it also has the limitations of difficult to control the degradation rate and unstable mechanical properties [184]. But its huge potential has enabled it to be used in various fields. There are still many problems to be solved. First of all, a standardized production process and quality evaluation standards should be established for the preparation of acellular small intestinal stents. In addition, the mechanism of interaction between the scaffold and stem cells should be explored. Then, the mechanism by which biologically active molecules in the scaffold influence tissue regeneration should be studied. Finally, the long-term safety of the stent should be continuously observed.

5.3.7 *Uterus*

Absolute uterine factor infertility (AUFI) is a permanent infertility disease caused by congenital or acquired uterine defects [185], which plagues hundreds of thousands of infertile women worldwide. It was not until the advent of uterine transplantation that provided an effective method for the treatment of AUFI. At the same time, the introduction of immunosuppressive drugs avoids the occurrence of immune rejection and promotes the development of clinical research on uterine transplantation. However, due to the adverse side effects caused by long-term use of immunosuppressive drugs and the shortage of organ donors [186], it has become an obstacle to the further development of uterine transplantation. The proposal of acellular uterine stent provides hope for the removal of this obstacle. The research progress of acellular uterine stent is shown in Fig. 5.17.

In 2014, Santoso et al. proposed for the first time that uterine acellular ECM scaffolds can be used as scaffolds for uterine reconstruction and promote uterine reconstruction [187]. To this end, they decellularized the uterus of SD rats by infusion of SDS or using high hydrostatic pressure (HHP). The results show that both methods can remove cells and DNA. Not only is the DNA content in the scaffolds treated with HHP lower, but the main components of ECM are also better retained. In addition, the mechanical strength of the stent after SDS treatment was significantly improved. 30 days after implanting the obtained stent into the uterus of the rat, the uterine tissue was successfully regenerated and at the same time pregnant.

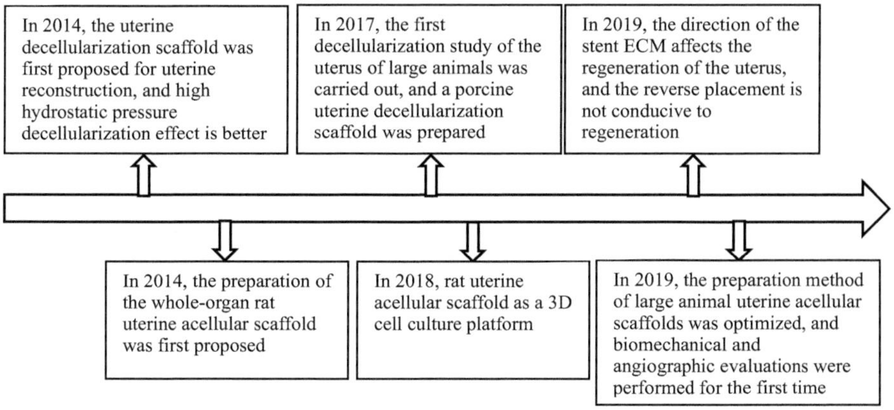

Fig. 5.17 Diagram of the progress of acellular uterus research

The results of Santoso et al. proved that uterine acellular ECM can be used as a potential scaffold for uterine reconstruction. In the same year, in order to solve the problem of insufficient uterine donors and to avoid immunosuppressive therapy, Hellström et al. proposed for the first time a preparation scheme for a whole-organ rat uterine acellular scaffold [188]. They used three different decellularization reagents (Scheme 1: 4% dimethyl sulfoxide (DMSO) + 1% TritonX-100 + PBS + 0.05% sodium azide; Scheme 2: dH2O + 0.05% sodium azide +4% DMSO +1% TritonX-100 + freeze-thaw (−80 °C); Scheme 3: dH2O + 0.05% sodium azide +2% SDC) to perfuse the rat uterus to achieve decellularization. The results are shown in Fig. 5.18a–g. The results show that Scheme 2 can produce a stent with a compact structure and weak mechanical strength, while the ECM microstructure has received some damage. Scheme 3 can prepare a porous stent closer to the natural uterus. Their research results provide a reference for the preparation of the entire uterus.

As different scientific research groups successfully obtained uterus from small animals and prepared uterine acellular scaffolds that meet the requirements, researchers gradually turned their attention to large animals. In 2017, Campo et al. tried to prepare a pig uterine acellular scaffold, which was the first decellularization study of large reproductive organs [189]. They succeeded in preparing a decellularized scaffold that can be used and that the ECM and vascular network remain intact. This research provides a reference for the clinical application of bioengineered uterus.

With the in-depth study of some disease models, researchers gradually began to try to cultivate cells in a three-dimensional (3D) environment, but this requires the construction of a complex culture environment. The ECM scaffold is rich in proteins, growth factors, and other substances, and plays an important role in cell proliferation, so it is an ideal scaffold for 3D cell culture. In 2018, in order to provide a high-quality 3D cell culture platform, Padma and others successfully prepared a rat uterine acellular scaffold with a complete vascular system through perfusion decellularization [190]. The scaffold can provide obvious value in

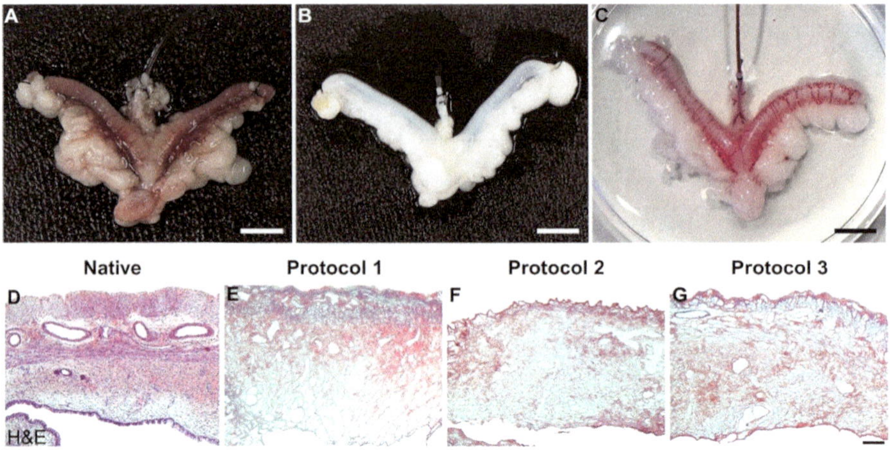

Fig. 5.18 The rat's natural uterus (**a**) turned white (**b**) after treatment with perfusion decellularization reagents, and there was no significant change in shape and size. (**c**) Infusion of dye through the vascular system to analyze the integrity of the vascular structure, because there is no obvious visual difference between different methods, so only one picture is shown. (**d**–**g**) Histological and hematoxylin and eosin (H&E) staining of the obtained scaffold and natural uterus to analyze the structural integrity of the tissue. The results show that the obtained uterine acellular scaffold retains the complete vascular structure and removes the DNA. Scale bar: 1 cm (**a**–**c**) and 100 μm (**d**–**g**) [188]

studying the invasion mechanism of endometrial cancer cells and the mechanism of embryo implantation. In addition, it has the potential to be used in uterine transplantation.

It has been shown that the rat uterine acellular scaffold can regenerate after transplantation and support pregnancy. The direction of ECM and the preservation of perfusable vascular structures are thought to be the cause of the regeneration of rat uterine acellular scaffolds. In 2019, in order to verify this guess, Miki et al. studied the regenerative capacity of a rat uterine acellular scaffold placed reversely in a rat uterine defect area [191]. The results show that, compared with the normal placed rat uterine acellular scaffold, although the reversely placed scaffold can still achieve pregnancy, it is easier to observe abnormal structures, such as the abnormal lining of the smooth muscle layer. This shows that the direction of ECM can affect the regeneration of the uterus, so transplanting the rat uterine acellular scaffold in the correct direction is beneficial to the regeneration of the uterus. In 2020, in order to optimize the method of preparing a uterine decellularization scaffold that is similar in size to the human uterus, Daryabari et al. decellularized the sheep uterus using a variety of different methods [192]. For the first time, biomechanical and angiographic evaluations of the obtained large animal uterine stents were performed. As a result, it was found that perfusion with 0.25% or 0.5% SDS solution and subsequent preservation of the obtained scaffold in 10% formalin is an effective decellularization protocol. At the same time, the obtained stent is similar to the natural uterus in mechanical properties, vascular structure, and ECM structure, and also has good biocompatibility.

Since the first introduction of the uterine acellular scaffold in 2014, research on whole-organ uterine acellular scaffolds has not stopped. This provides hope for patients who are unable to obtain a donor uterus in time, and also provides hope for the early cure of AUFI. At present, in the process of uterine decellularization, chemical reagents SDC and SDS have shown good results, and HHP has also shown good results. At present, the research on the uterus of large animals is becoming more mature, and the evaluation system is gradually improved, and it is expected to be used in clinic as soon as possible.

Organ decellularization is a transition from tissue decellularization. It is inevitable that there will be some similarities in the application of decellularization methods, but it is a qualitative leap in the field of decellularization technology from the tissue level to the organ level. In addition, because the organ has the anatomical characteristics of a complete vasculature, the perfusion method has emerged. Although there is a certain gap between the current organ decellularization technology and the tissue decellularization technology, there are still many problems, but we can still see its application prospects.

References

1. Hynes RO. The extracellular matrix: not just pretty fibrils. Science. 2009;326:1216–9.
2. Gupta SK, Mishra NC, Dhasmana A. Decellularization Methods for Scaffold Fabrication. Methods Mol Biol. 2018;1577:1–10.
3. Guler S, Aslan B, Hosseinian P, Aydin HM. Supercritical carbon dioxide-assisted Decellularization of aorta and cornea. Tissue Eng Part C Methods. 2017;23:540–7.
4. Struecker B, Butter A, Hillebrandt K, Polenz D, Reutzel-Selke A, Tang P, et al. Improved rat liver decellularization by arterial perfusion under oscillating pressure conditions. J Tissue Eng Regen Med. 2017;11:531–41.
5. Crapo PM, Gilbert TW, Badylak SF. An overview of tissue and whole organ decellularization processes. Biomaterials. 2011;32:3233–43.
6. Lü WD, Zhang L, Wu CL, Liu ZG, Lei GY, Liu J, et al. Development of an acellular tumor extracellular matrix as a three-dimensional scaffold for tumor engineering. PLoS One. 2014;9: e103672.
7. Morris AH, Chang J, Kyriakides TR. Inadequate processing of decellularized dermal matrix reduces cell viability in vitro and increases apoptosis and acute inflammation in vivo. BioRes Open Access. 2016;5:177–87.
8. Jamur MC, Oliver C. Permeabilization of cell membranes. Methods Mol Biol. 2010;588:63–6.
9. Simões IN, Vale P, Soker S, Atala A, Keller D, Noiva R, et al. Acellular urethra bioscaffold: decellularization of whole urethras for tissue engineering applications. Sci Rep. 2017;7:1–13.
10. Liem PH, Morimoto N, Ito R, Kawai K, Suzuki S. Autologous skin reconstruction by combining epidermis and acellular dermal matrix tissue derived from the skin of giant congenital melanocytic nevi. J Artif Organs. 2013;16:332–42.
11. Hellström M, El-Akouri RR, Sihlbom C, Olsson BM, Lengqvist J, Bäckdahl H, et al. Towards the development of a bioengineered uterus: comparison of different protocols for rat uterus decellularization. Acta Biomater. 2014;10:5034–42.

12. Rodi PM, Gianello MB, Corregido MC, Gennaro AM. Comparative study of the interaction of CHAPS and triton X-100 with the erythrocyte membrane. Biochim Biophys Acta. 2014;1838:859–66.
13. Brown BN, Freund JM, Han L, Rubin JP, Reing JE, Jeffries EM, et al. Comparison of three methods for the derivation of a biologic scaffold composed of adipose tissue extracellular matrix. Tissue Eng Part C Methods. 2011;17:411–21.
14. Baptista PM, Siddiqui MM, Lozier G, Rodriguez SR, Atala A, Soker S. The use of whole organ decellularization for the generation of a vascularized liver organoid. Hepatology. 2011;53:604–17.
15. Sullivan DC, Mirmalek-Sani SH, Deegan DB, Baptista PM, Aboushwareb T, Atala A, et al. Decellularization methods of porcine kidneys for whole organ engineering using a high-throughput system. Biomaterials. 2012;33:7756–64.
16. Hillebrandt K, Polenz D, Butter A, Tang P, Reutzel-Selke A, Andreou A, et al. Procedure for decellularization of rat livers in an oscillating-pressure perfusion device. J Visual Exp. 2015: e53029.
17. Elebring E, Kuna VK, Kvarnström N, Sumitran-Holgersson S. Cold-perfusion decellularization of whole-organ porcine pancreas supports human fetal pancreatic cell attachment and expression of endocrine and exocrine markers. J Tissue Eng. 2017;8:2041731417738145.
18. Park SM, Yang S, Rye S-M, Choi SW. Effect of pulsatile flow perfusion on decellularization. BioMed Central. 2018;17:1–10.
19. Say S, Dugos N, Roces S, Mondragon JM. Effect of sonication power on perfusion decellularization of cadaveric porcine kidney. MATEC Web of Conferences. 2019;268:01009.
20. Manalastas TM, Dugos N, Ramos G, Mondragon JM. Effect of decellularization parameters on the efficient production of kidney bioscaffolds. Appl Biochem Biotechnol. 2020. https://doi.org/10.1007/s12010-020-03338-2.
21. Schilling BK, Lamberti KK, Snowden MJ, Baker JS, Byrd K, Komatsu C, et al. Design and fabrication of an automatable, 3D printed perfusion device for tissue infusion and perfusion engineering. Tissue Eng Part A. 2020;26:253–64.
22. Willemse J, Verstegen MMA, Vermeulen A, Schurink IJ, Roest HP, Laan LJW, et al. Fast, robust and effective decellularization of whole human livers using mild detergents and pressure controlled perfusion. Mat Sci Eng C. 2020;108:110200.
23. Badylak SF, Valentin JE, Ravindra AK, McCabe GP, Stewart-Akers AM. Macrophage phenotype as a determinant of biologic scaffold remodeling. Tissue Eng Part A. 2008;14:1835–42.
24. Gilbert TW, Sellaro TL, Badylak SF. Decellularization of tissues and organs. Biomaterials. 2006;27:3675–83.
25. Mirmalek-Sani SH, Sullivan DC, Zimmerman C, Shupe TD, Petersen BE. Immunogenicity of decellularized porcine liver for bioengineered hepatic tissue. Am J Pathol. 2013;183:558–65.
26. Ji R, Zhang N, You N, Li Q, Liu W, Jiang N, et al. The differentiation of MSCs into functional hepatocyte-like cells in a liver biomatrix scaffold and their transplantation into liver-fibrotic mice. Biomaterials. 2012;33:8995–9008.
27. Sabetkish S, Kajbafzadeh AM, Sabetkish N, Khorramirouz R, Akbarzadeh A, Seyedian SL, et al. Whole-organ tissue engineering: Decellularization and recellularization of three-dimensional matrix liver scaffolds. J Biomed Mater Res A. 2015;103:1498–508.
28. Rosso F, Giordano A, Barbarisi M, Barbarisi A. From cell–ECM interactions to tissue engineering. J Cell Physiol. 2004;199:174–80.
29. Li Q, Uygun BE, Geerts S, Ozer S, Scalf M, Gilpin SE, et al. Proteomic analysis of naturally-sourced biological scaffolds. Biomaterials. 2016;75:37–46.
30. Mazza G, Rombouts K, Rennie Hall A, Urbani L, Vinh Luong T, Al-Akkad W, et al. Decellularized human liver as a natural 3D-scaffold for liver bioengineering and transplantation. Sci Rep. 2015;5:13079.

31. Ott HC, Matthiesen TS, Goh SK, Black LD, Kren SM, Netoff TI, et al. Perfusion-decellularized matrix: using nature's platform to engineer a bioartificial heart. Nat Med. 2008;14:213–21.
32. Mattei G, Di Patria V, Tirella A, Alaimo A, Elia G, Corti A, et al. Mechanostructure and composition of highly reproducible decellularized liver matrices. Acta Biomater. 2014;10:875–82.
33. Everwien H, Ariza de Schellenberger A, Haep N, Tzschätzsch H, Pratschke J, Sauer IM, et al. Magnetic resonance elastography quantification of the solid-to-fluid transition of liver tissue due to decellularization. J Mech Behav Biomed Mater. 2020;104:103640.
34. Mozaffarian D, Benjamin EJ, Go AS, Arnett DK, Blaha MJ, Cushman M, et al. Heart disease and stroke statistics--2015 update: a report from the American Heart Association. Circulation. 2015;131:e29–322.
35. Rossano JW, Cabrera AG, Jefferies JL, Naim MP, Humlicek T. Pediatric cardiac Intensive Care Society 2014 consensus statement: pharmacotherapies in cardiac critical care chronic heart failure. Pediat Crit Care Med. 2016;17:S20–34.
36. Tonsho M, Michel S, Ahmed Z, Alessandrini A, Madsen JC. Heart transplantation: challenges facing the field. Cold Spring Harb Perspect Med. 2014;4:a015636.
37. Harken DE, Soroff HS, Taylor WJ, Lefemine AA, Gupta SK, Lunzer S. Partial and complete prostheses in aortic insufficiency. J Thorac Cardiovasc Surg. 1960;40:744–62.
38. Bonhoeffer P, Boudjemline Y, Saliba Z, Merckx J, Aggoun Y, Bonnet D, et al. Percutaneous replacement of pulmonary valve in a right-ventricle to pulmonary-artery prosthetic conduit with valve dysfunction. Lancet. 2000;356:1403–5.
39. Ross DN. Replacement of aortic and mitral valves with A pulmonary autograft. Lancet. 1967;2:956–8.
40. Wainwright JM, Czajka CA, Patel UB, Freytes DO, Tobita K, Gilbert TW, et al. Preparation of cardiac extracellular matrix from an intact porcine heart. Tissue Eng Part C Methods. 2010;16:525–32.
41. Cebotari S, Lichtenberg A, Tudorache I, Hilfiker A, Mertsching H, Leyh R,et al. Clinical application of tissue engineered human heart valves using autologous progenitor cells. Circulation. 2006;114:I132–7.
42. Cebotari S, Tudorache I, Ciubotaru A, Boethig D, Sarikouch S, Goerler A, et al. Use of fresh decellularized allografts for pulmonary valve replacement may reduce the reoperation rate in children and young adults: early report. Circulation. 2011;124:S115–23.
43. Siddiqui RF, Abraham JR, Butany J. Bioprosthetic heart valves: modes of failure. Histopathology. 2009;55:135–44.
44. Wong ML, Griffiths LG. Immunogenicity in xenogeneic scaffold generation: antigen removal vs. decellularization. Acta Biomater. 2014;10:1806–16.
45. Liao J, Joyce EM, Sacks MS. Effects of decellularization on the mechanical and structural properties of the porcine aortic valve leaflet. Biomaterials. 2008;29:1065–74.
46. Zhou J, Fritze O, Schleicher M, Wendel HP, Schenke-Layland K, Harasztosi C, et al. Impact of heart valve decellularization on 3-D ultrastructure, immunogenicity and thrombogenicity. Biomaterials. 2010;31:2549–54.
47. Weymann A, Loganathan S, Takahashi H, Schies C, Claus B, Hirschberg K, et al. Development and evaluation of a perfusion decellularization porcine heart model–generation of 3-dimensional myocardial neoscaffolds. Circul J. 2011;4:852–60.
48. Akhyari P, Aubin H, Gwanmesia P, Barth M, Hoffmann S, Huelsmann J, et al. The quest for an optimized protocol for whole-heart decellularization. A comparison of three popular and a novel Decellularization technique and their diverse effects on crucial extracellular matrix qualities. Tissue Eng Part C Methods. 2011;17:915–26.
49. Remlinger NT, Wearden PD, Gilbert TW. Procedure for decellularization of porcine heart by retrograde coronary perfusion. J Visual Exp. 2012;70:e50059-e.
50. Weymann A, Patil NP, Sabashnikov A, Jungebluth P, Korkmaz S, Li S, et al. Bioartificial heart: a human-sized porcine model--the way ahead. PLoS One. 2014;9:e111591.

51. Oberwallner B, Brodarac A, Choi YH, Saric T, Anić P, Morawietz L, et al. Preparation of cardiac extracellular matrix scaffolds by decellularization of human myocardium. J Biomed Mater Res A. 2014;102:3263–72.
52. Crapo PM, Gilbert TW, Badylak SF. An overview of tissue and whole organ decellularization processes. Biomaterials. 2011;32:3233–43.
53. Dong J, Li Y, Mo X. The study of a new detergent (octyl-glucopyranoside) for decellularizing porcine pericardium as tissue engineering scaffold. J Surg Res. 2013;183:56–67.
54. Cicha I, Rüffer A, Cesnjevar R, Glöckler M, Agaimy A, Daniel WG, et al. Early obstruction of decellularized xenogenic valves in pediatric patients: involvement of inflammatory and fibroproliferative processes. Cardiovasc Pathol. 1900;20:222–31.
55. Rice JJ, Martino MM, De Laporte L, Tortelli F, Briquez PS, Hubbell JA. Engineering the regenerative microenvironment with biomaterials. Adv Healthc Mater. 2013;2:57–71.
56. Wong ML, Leach JK, Athanasiou KA, Griffiths LG. The role of protein solubilization in antigen removal from xenogeneic tissue for heart valve tissue engineering. Biomaterials. 2011;32:8129–38.
57. Kawasaki T, Kirita Y, Kami D, Kitani T, Ozaki C, Itakura Y, et al. Novel detergent for whole organ tissue engineering. J Biomed Mater Res A. 2015;103:3364–73.
58. Hülsmann J, Aubin H, Bandesha ST, Kranz A, Stoldt VR, Lichtenberg A, et al. Rheology of perfusates and fluid dynamical effects during whole organ decellularization: a perspective to individualize decellularization protocols for single organs. Biofabrication. 2015;7:035008.
59. Lee PF, Chau E, Cabello R, Yeh AT, Sampaio LC, Gobin AS, et al. Inverted orientation improves decellularization of whole porcine hearts. Acta Biomater. 2017;49:181–91.
60. Garcia O, Carraro G, Navarro S, Bertoncello I, McQualter J, Driscoll B, et al. Cell-based therapies for lung disease. Br Med Bull. 2012;101:147–61.
61. Guler S, Aydin HM, Lü LX, Yang Y. Improvement of decellularization efficiency of porcine aorta using dimethyl sulfoxide as a penetration enhancer. Artif Organs. 2018;42:219–30.
62. Peralta M, Steed E, Harlepp S, González-Rosa JM, Monduc F, Ariza-Cosano A, et al. Heartbeat-driven pericardiac fluid forces contribute to epicardium morphogenesis. Curr Biol. 2013;23:1726–35.
63. Garanich JS, Mathura RA, Shi Z-D, Tarbell JM. Effects of fluid shear stress on adventitial fibroblast migration: implications for flow-mediated mechanisms of arterialization and intimal hyperplasia. Am J Phys Heart Circ Phys. 2007;292:H3128–35.
64. Crapo PM, Gilbert TW, Badylak SF. An overview of tissue and whole organ decellularization processes. Biomaterials. 2011;32(12):3233-43.
65. Pereira RHA, Prado AR, Caro LFCD, Zanardo TÉC, Alencar AP, Nogueira BV. A non-linear mathematical model using optical sensor to predict heart decellularization efficacy. Sci Rep. 2019;9:12211.
66. Reddy N, Reddy R, Jiang Q. Crosslinking biopolymers for biomedical applications. Trends Biotechnol. 2015;33:362–9.
67. Assmann A, Delfs C, Munakata H, Schiffer F, Horstkötter K, Huynh K, et al. Acceleration of autologous in vivo recellularization of decellularized aortic conduits by fibronectin surface coating. Biomaterials. 2013;34:6015–26.
68. Lal MK, Manktelow BN, Draper ES, Field DJ. Chronic lung disease of prematurity and intrauterine growth retardation: a population-based study. Pediatrics. 2003;111:483–7.
69. Prakash YS, Tschumperlin DJ, Stenmark KR. Coming to terms with tissue engineering and regenerative medicine in the lung. Am J Physiol Lung Cell Mol Physiol. 2015;309:L625–38.
70. Hardy JD, Webb WR, Dalton ML Jr, Walker GR Jr. Lung homotransplantation in man: report of the initial case. JAMA. 1963;186:1065–74.
71. Pinsker KL, Koerner SK, Kamholz SL, Hagstrom JW, Veith FJ. Effect of donor bronchial length on healing: a canine model to evaluate bronchial anastomotic problems in lung transplantation. J Thorac Cardiovasc Surg. 1979;77:669–73.

72. Hayama M, Date H, Oto T, Aoe M, Andou A, Shimizu N. Improved lung function by means of retrograde flush in canine lung transplantation with non-heart-beating donors. J Thorac Cardiovasc Surg. 2003;125:901–6.
73. Veith FJ, Kamholz SL, Mollenkopf FP, Montefusco CM. Lung transplantation 1983. Transplantation. 1983;35:271–8.
74. Yusen RD, Edwards LB, Kucheryavaya AY, Benden C, Dipchand AI, Goldfarb SB, et al. The registry of the International Society for Heart and Lung Transplantation: thirty-second official adult lung and heart-lung transplantation report-2015; focus theme: early graft failure. J Heart Lung Transplant. 2015;34:1264–77.
75. Mallory GB, Spray TL. Paediatric lung transplantation. Eur Respir J. 2004;24:839–45.
76. Khan MS, Heinle JS, Adachi I, Shirkey B, Schecter MG, Mallory GB, et al. Is lung transplantation survival better in infants? Analysis of over 80 infants. J Heart Lung Transplant. 2013;32:44–9.
77. Shigemura N, Hattler B, Bermudez C, McCurry KR, Toyoda Y. Impact of graft volume reduction for oversized grafts on outcome after lung transplantation in recipients with end-stage restrictive pulmonary diseases. J Heart Lung Transplant. 2009;28:130–4.
78. Shigemura N, D'Cunha J, Bhama JK, Shiose A, Abou El Ela A, Hackmann A, et al. Lobar lung transplantation: a relevant surgical option in the current era of lung allocation score. Ann Thorac Surg. 2013;96:451–6.
79. Ott HC, Clippinger B, Conrad C, Schuetz C, Pomerantseva I, Ikonomou L, et al. Regeneration and orthotopic transplantation of a bioartificial lung. Nat Med. 2010;16:927–33.
80. Petersen TH, Calle EA, Zhao L, Lee EJ, Gui L, Raredon MB, Gavrilov K, et al. Tissue-engineered lungs for in vivo implantation. Science. 2010;329:538–41.
81. Bonvillain RW, Danchuk S, Sullivan DE, Betancourt AM, Semon JA, Eagle ME, et al. A nonhuman primate model of lung regeneration: detergent-mediated decellularization and initial in vitro recellularization with mesenchymal stem cells. Tissue Eng Part A. 2012;18:2437–52.
82. Nichols JE, Niles J, Riddle M, Vargas G, Schilagard T, Ma L, et al. Production and assessment of decellularized pig and human lung scaffolds. Tissue Eng Part A. 2013;19:2045–62.
83. O'Neill JD, Anfang R, Anandappa A, Costa J, Javidfar J, Wobma HM, et al. Decellularization of human and porcine lung tissues for pulmonary tissue engineering. Ann Thorac Surg. 2013;96:1046–56.
84. Vorotnikova E, McIntosh D, Dewilde A, Zhang J, Reing JE, Zhang L, et al. Extracellular matrix-derived products modulate endothelial and progenitor cell migration and proliferation in vitro and stimulate regenerative healing in vivo. Matrix Biol. 2010;29:690–700.
85. Murray JC, Liotta L, Rennard SI, Martin GR. Adhesion characteristics of murine metastatic and nonmetastatic tumor cells in vitro. Cancer Res. 1980;40:347–51.
86. Clark RA, Nielsen LD, Welch MP, McPherson JM. Collagen matrices attenuate the collagen-synthetic response of cultured fibroblasts to TGF-beta. J Cell Sci. 1995;108:1251.
87. Balestrini JL, Gard AL, Gerhold KA, Wilcox EC, Liu A, Schwan J, et al. Comparative biology of decellularized lung matrix: implications of species mismatch in regenerative medicine. Biomaterials. 2016;102:220–30.
88. Bonenfant NR, Sokocevic D, Wagner DE, Borg ZD, Lathrop MJ, Lam YW, et al. The effects of storage and sterilization on de-cellularized and re-cellularized whole lung. Biomaterials. 2013;34:3231–45.
89. Wagner DE, Bonenfant NR, Parsons CS, Sokocevic D, Brooks EM, Borg ZD, et al. Comparative decellularization and recellularization of normal versus emphysematous human lungs. Biomaterials. 2014;35:3281–97.
90. Scarritt ME, Bonvillain RW, Burkett BJ, Wang G, Glotser EY, Zhang Q, et al. Hypertensive rat lungs retain hallmarks of vascular disease upon decellularization but support the growth of mesenchymal stem cells. Tissue Eng Part A. 2014;20:1426–43.

91. Nichols JE, La Francesca S, Vega SP, Niles JA, Argueta LB, Riddle M, et al. Giving new life to old lungs: methods to produce and assess whole human paediatric bioengineered lungs. J Tissue Eng Regen Med. 2017;11:2136–52.
92. Stahl EC, Bonvillain RW, Skillen CD, Burger BL, Hara H, Lee W, et al. Evaluation of the host immune response to decellularized lung scaffolds derived from α-gal knockout pigs in a non-human primate model. Biomaterials. 2018;187:93–104.
93. Reing JE, Brown BN, Daly KA, Freund JM, Gilbert TW, Hsiong SX, et al. The effects of processing methods upon mechanical and biologic properties of porcine dermal extracellular matrix scaffolds. Biomaterials. 2010;31:8626–33.
94. Maghsoudlou P, Georgiades F, Tyraskis A, Totonelli G, Loukogeorgakis SP, Orlando G, et al. Preservation of micro-architecture and angiogenic potential in a pulmonary acellular matrix obtained using intermittent intra-tracheal flow of detergent enzymatic treatment. Biomaterials. 2013;34:6638–48.
95. Wang Z, Wang Z, Yu Q, Xi H, Weng J, Du X, et al. Comparative study of two perfusion routes with different flow in decellularization to harvest an optimal pulmonary scaffold for recellularization. J Biomed Mater Res A. 2016;104:2567–75.
96. Uriarte JJ, Nonaka PN, Campillo N, Palma RK, Melo E, de Oliveira LVF, et al. Mechanical properties of acellular mouse lungs after sterilization by gamma irradiation. J Mech Behav Biomed Mater. 2014;40:168–77.
97. Nonaka PN, Campillo N, Uriarte JJ, Garreta E, Melo E, de Oliveira LVF, et al. Effects of freezing/thawing on the mechanical properties of decellularized lungs. J Biomed Mater Res A. 2014;102:413–9.
98. Calle EA, Hill RC, Leiby KL, Le AV, Gard AL, Madri JA, et al. Targeted proteomics effectively quantifies differences between native lung and detergent-decellularized lung extracellular matrices. Acta Biomater. 2016;46:91–100.
99. Linhardt RJ, Toida T. Role of glycosaminoglycans in cellular communication. Acc Chem Res. 2004;37:431–8.
100. Schultz V, Suflita M, Liu X, Zhang X, Yu Y, Li L, et al. Heparan sulfate domains required for fibroblast growth factor 1 and 2 Signaling through fibroblast growth factor receptor 1c. J Biol Chem. 2017;292:2495–509.
101. Gilpin SE, Wagner DE. Acellular human lung scaffolds to model lung disease and tissue regeneration. Eur Respir Rev. 2018;27:180021.
102. da Palma RK, Fratini P, Schiavo Matias GS, Cereta AD, Guimarães LL, Anunciação ARDA, et al. Equine lung decellularization: a potential approach for in vitro modeling the role of the extracellular matrix in asthma. J Tiss Eng. 2018;9:2041731418810164.
103. Nichols WG, Steel HM, Bonny T, Adkison K, Curtis L, Millard J, et al. Hepatotoxicity observed in clinical trials of aplaviroc (GW873140). Antimicrob Agents Chemother. 2008;52:858–65.
104. GBD 2013 Mortality and Causes of Death Collaborators. Global, regional, and national age-sex specific all-cause and cause-specific mortality for 240 causes of death, 1990-2013: a systematic analysis for the global burden of disease study 2013. Lancet. 2015;385:117–71.
105. Moore FD, Smith LL, Burnap TK, Dallenbach FD, Dammin GJ, Gruber UF, et al. One-stage homotransplantation of the liver following total hepatectomy in dogs. Transplant Bull. 1959;6:103–7.
106. Starzl TE, Marchioro TL, Vonkaulla KN, Hermann G, Brittain RS, Waddell WR. Homotransplantation of the liver in humans. Surg Gynecol Obstet. 1963;117:659–76.
107. Hannoun Z, Steichen C, Dianat N, Weber A, Dubart-Kupperschmitt A. The potential of induced pluripotent stem cell derived hepatocytes. J Hepatol. 2016;65:182–99.
108. Marinho A. A study on organ transplantation waiting lines in Brazil's unified national health system. Cad Saude Publica. 2006;22:2229–39.
109. Chen AA, Thomas DK, Ong LL, Schwartz RE, Golub TR, Bhatia SN. Humanized mice with ectopic artificial liver tissues. Proc Natl Acad Sci U S A. 2011;108:11842–7.

110. Dhawan A, Puppi J, Hughes RD, Mitry RR. Human hepatocyte transplantation: current experience and future challenges. Nat Rev Gastroenterol Hepatol. 2010;7:288–98.
111. Lee CA, Dhawan A, Smith RA, Mitry RR, Fitzpatrick E. Instant blood-mediated inflammatory reaction in hepatocyte transplantation: current status and future perspectives. Cell Transplant. 2016;25:1227–36.
112. Zhang J, Zhao X, Liang L, Li J, Demirci U, Wang S. A decade of progress in liver regenerative medicine. Biomaterials. 2018;157:161–76.
113. Bhatia SN, Ingber DE. Microfluidic organs-on-chips. Nat Biotechnol. 2014;32(8):760–72.
114. Uygun BE, Soto-Gutierrez A, Yagi H, Izamis M-L, Guzzardi MA, Shulman C, et al. Organ reengineering through development of a transplantable recellularized liver graft using decellularized liver matrix. Nat Med. 2010;16:814–20.
115. Shupe T, Williams M, Brown A, Willenberg B, Petersen BE. Method for the decellularization of intact rat liver. Organogenesis. 2010;6:134–6.
116. Järveläinen H, Sainio A, Koulu M, Wight TN, Penttinen R. Extracellular matrix molecules: potential targets in pharmacotherapy. Pharmacol Rev. 2009;61:198–223.
117. De Kock J, Ceelen L, De Spiegelaere W, Casteleyn C, Claes P, Vanhaecke T, et al. Simple and quick method for whole-liver decellularization: a novel in vitro three-dimensional bioengineering tool? Arch Toxicol. 2011;85:607–12.
118. Ren H, Shi X, Tao L, Xiao J, Han B, Zhang Y, et al. Evaluation of two decellularization methods in the development of a whole-organ decellularized rat liver scaffold. Liver Int. 2013;33:448–58.
119. Jiang WC, Cheng YH, Yen MH, Chang Y, Yang VW, Lee OK. Cryo-chemical decellularization of the whole liver for mesenchymal stem cells-based functional hepatic tissue engineering. Biomaterials. 2014;35:3607–17.
120. Mattei G, Di Patria V, Tirella A, Alaimo A, Elia G, Corti A, et al. Mechanostructure and composition of highly reproducible decellularized liver matrices. Acta Biomater. 2014;10:875–82.
121. Gómez MP, Pérez B, Manyalich M. International registry in organ donation and transplantation—2013. Transplant Proc. 2014;46:1044–8.
122. Schmidt CE, Baier JM. Acellular vascular tissues: natural biomaterials for tissue repair and tissue engineering. Biomaterials. 2000;21:2215–31.
123. Wang Y, Bao J, Wu X, Wu Q, Li Y, Zhou Y, et al. Genipin crosslinking reduced the immunogenicity of xenogeneic decellularized porcine whole-liver matrices through regulation of immune cell proliferation and polarization. Sci Rep. 2016;6:24779.
124. Struecker B, Butter A, Hillebrandt K, Polenz D, Reutzel-Selke A, Tang P, et al. Improved rat liver decellularization by arterial perfusion under oscillating pressure conditions. J Tissue Eng Regen Med. 2017;11:531–41.
125. Willemse J, Verstegen MMA, Vermeulen A, Schurink IJ, Roest HP, van der Laan LJW, et al. Fast, robust and effective decellularization of whole human livers using mild detergents and pressure controlled perfusion. Mater Sci Eng C Mater Biol Appl. 2020;108:110200.
126. Mani NS, Manoj MG, Malik A. Concurrent renal tuberculosis and renal cell carcinoma: A coincidental finding. Indian J Nephrol. 2013;23:237.
127. Schieppati A, Remuzzi G. Chronic renal diseases as a public health problem: epidemiology, social, and economic implications. Kidney Int Suppl. 2005;68:S7–10.
128. Mann BS, Manns BJ, Barnieh L, Oliver MJ, Devoe D, Lorenzetti D, et al. Peritoneal dialysis: a scoping review of strategies to maximize PD utilization. Perit Dial Int. 2017;37:159–64.
129. Chacko MP, Mathan A, Daniel D, Basu G, Varughese S. Significance of pre-transplant anti-HLA antibodies detected on an ELISA mixed antigen tray platform. Indian J Nephrol. 2013;23:351–3.
130. Anil Kumar BN, Mattoo SK. Organ transplant & the psychiatrist: an overview. Indian J Med Res. 2015;141:408–16.
131. Song JJ, Guyette JP, Gilpin SE, Gonzalez G, Vacanti JP, Ott HC. Regeneration and experimental orthotopic transplantation of a bioengineered kidney. Nat Med. 2013;19:646–51.

132. Hussein KH, Park KM, Lee YS, Woo JS, Kang BJ, Choi KY, et al. New insights into the pros and cons of cross-linking decellularized bioartificial organs. Int J Artif Organs. 2017;40:136–41.
133. Ross EA, Williams MJ, Hamazaki T, Terada N, Clapp WL, Adin C, et al. Embryonic stem cells proliferate and differentiate when seeded into kidney scaffolds. J Am Soc Nephrol. 2009;20:2338–47.
134. Caralt M, Uzarski JS, Iacob S, Obergfell KP, Berg N, Bijonowski BM, et al. Optimization and critical evaluation of decellularization strategies to develop renal extracellular matrix scaffolds as biological templates for organ engineering and transplantation. Am J Transplant. 2015;15:64–75.
135. Chani B, Puri V, Sobti RC, Jha V, Puri S. Decellularized scaffold of cryopreserved rat kidney retains its recellularization potential. PLoS One. 2017;12:e0173040.
136. Feng H, Xu Y, Luo S, Dang H, Liu K, Sun WQ. Evaluation and preservation of vascular architectures in decellularized whole rat kidneys. Cryobiology. 2020;95:72–9.
137. Nakayama KH, Batchelder CA, Lee CI, Tarantal AF. Decellularized rhesus monkey kidney as a three-dimensional scaffold for renal tissue engineering. Tissue Eng Part A. 2010;16:2207–16.
138. Nakayama KH, Lee CC, Batchelder CA, Tarantal AF. Tissue specificity of decellularized rhesus monkey kidney and lung scaffolds. PLoS One. 2013;8:e64134.
139. Nakayama KH, Batchelder CA, Lee CI, Tarantal AF. Renal tissue engineering with decellularized rhesus monkey kidneys: age-related differences. Tissue Eng Part A. 2011;17:2891–901.
140. Hussein KH, Saleh T, Ahmed E, Kwak HH, Park KM, Yang SR, et al. Biocompatibility and hemocompatibility of efficiently decellularized whole porcine kidney for tissue engineering. J Biomed Mater Res A. 2018;106:2034–47.
141. Poornejad N, Momtahan N, Salehi AS, Scott DR, Fronk CA, Roeder BL, et al. Efficient decellularization of whole porcine kidneys improves reseeded cell behavior. Biomed Mater. 2016;11(2):025003.
142. Gifford S, Zambon JP, Orlando G. Recycling organs-growing tailor-made replacement kidneys. Regen Med. 2015;10:913–5.
143. Orlando G, Booth C, Wang Z, Totonelli G, Ross CL, Moran E, et al. Discarded human kidneys as a source of ECM scaffold for kidney regeneration technologies. Biomaterials. 2013;34:5915–25.
144. Peloso A, Petrosyan A, Da Sacco S, Booth C, Zambon JP, O'Brien T, et al. Renal extracellular matrix scaffolds from discarded kidneys maintain glomerular morphometry and vascular resilience and retains critical growth factors. Transplantation. 2015;99:1807–16.
145. Baptista PM, Orlando G, Mirmalek-Sani SH, Siddiqui M, Atala A, Soker S. Whole organ decellularization - a tool for bioscaffold fabrication and organ bioengineering. Annu Int Conf IEEE Eng Med Biol Soc. 2009;2009:6526-9.
146. Zhang Q, Raoof M, Chen Y, Sumi Y, Sursal T, Junger W, et al. Circulating mitochondrial DAMPs cause inflammatory responses to injury. Nature. 2010;464:104–7.
147. Destefani AC, Sirtoli GM, Nogueira BV. Advances in the Knowledge about Kidney Decellularization and Repopulation. Front Bioeng Biotechnol. 2017;5:34.
148. Mayorca-Guiliani AE, Madsen CD, Cox TR, Horton ER, Venning FA, Erler JT. ISDoT: in situ decellularization of tissues for high-resolution imaging and proteomic analysis of native extracellular matrix. Nat Med. 2017;23:890–8.
149. Kajbafzadeh AM, Khorramirouz R, Nabavizadeh B, Ladi Seyedian SS, Akbarzadeh A, Heidari R, et al. Whole organ sheep kidney tissue engineering and in vivo transplantation: effects of perfusion-based decellularization on vascular integrity. Mater Sci Eng C Mater Biol Appl. 2019;98:392–400.
150. Parmaksiz M, Dogan A, Odabas S, Elçin AE, Elçin YM. Clinical applications of decellularized extracellular matrices for tissue engineering and regenerative medicine. Biomed Mater. 2016;11:022003.

151. Kadish AH. Automation control of blood sugar. I. A servomechanism for glucose monitoring and control. Am J Med Electron. 1964;3:82–6.
152. Shapiro AM, Lakey JR, Ryan EA, Korbutt GS, Toth E, Warnock GL, et al. Islet transplantation in seven patients with type 1 diabetes mellitus using a glucocorticoid-free immunosuppressive regimen. N Engl J Med. 2000;343:230–8.
153. Goh SK, Bertera S, Olsen P, Candiello JE, Halfter W, Uechi G, et al. Perfusion-decellularized pancreas as a natural 3D scaffold for pancreatic tissue and whole organ engineering. Biomaterials. 2013;34:6760–72.
154. Rogers J, Farney AC, Al-Geizawi S, Iskandar SS, Doares W, Gautreaux MD, et al. Pancreas transplantation: lessons learned from a decade of experience at Wake Forest Baptist Medical Center. Rev Diab Stud. 2011;8:17–27.
155. Yang HK, Yoon KH. Current status of encapsulated islet transplantation. J Diabetes Complicat. 2015;29:737–43.
156. Scarritt ME, Pashos NC, Bunnell BA. A review of cellularization strategies for tissue engineering of whole organs. Front Bioeng Biotechnol. 2015;3:43.
157. Zhang Y, Jalili RB, Warnock GL, Ao Z, Marzban L, Ghahary A. Three-dimensional scaffolds reduce islet amyloid formation and enhance survival and function of cultured human islets. Am J Pathol. 2012;181:1296–305.
158. Gilbert TW, Sellaro TL, Badylak SF. Decellularization of tissues and organs. Biomaterials. 2006;27:3675–83.
159. De Carlo E, Baiguera S, Conconi MT, Vigolo S, Grandi C, Lora S, et al. Pancreatic acellular matrix supports islet survival and function ina synthetic tubular device: in vitro and in vivo studies. Int J Mol Med. 2010;25:195–202.
160. Kuna VK, Kvarnström N, Elebring E, Holgersson SS. Isolation and decellularization of a whole porcine pancreas. J Vis Exp. 2018:e58302.
161. Peloso A, Urbani L, Cravedi P, Katari R, Maghsoudlou P, Fallas ME, et al. The human pancreas as a source of protolerogenic extracellular matrix scaffold for a new-generation bioartificial endocrine pancreas. Ann Surg. 2016;264:169–79.
162. Huang YB, Mei J, Yu Y, Ding Y, Xia W, Yue T, et al. Comparative decellularization and recellularization of normal versus streptozotocin-induced diabetes mellitus rat pancreas. Artif Organs. 2019;43:399–412.
163. Wu D, Wan J, Huang Y, Guo Y, Xu T, Zhu M, et al. 3D culture of MIN-6 cells on decellularized pancreatic scaffold: in vitro and in vivo study. Biomed Res Int. 2015;2015:432645.
164. Dew L, MacNeil S, Chong CK. Vascularization strategies for tissue engineers. Regen Med. 2015;10:211–24.
165. Gittes GK. Developmental biology of the pancreas: a comprehensive review. Dev Biol. 2009;326:4–35.
166. Hegyi P, Petersen OH. The exocrine pancreas: the acinar-ductal tango in physiology and pathophysiology. Rev Physiol Biochem Pharmacol. 2013;165:1–30.
167. Mikhailova AG, Rumsh LD, Dalgalarrondo M, Chobert JM, Haertle T. Activities of anionic and cationic trypsins in the temperature range from 5 to 37 degrees C. Mutant anionic trypsins as a model of cold-adapted (psychrophilic) enzymes. Biochemistry (Mosc). 2003;68:926–33.
168. Nichols SP, Koh A, Brown NL, Rose MB, Sun B, Slomberg DL, et al. The effect of nitric oxide surface flux on the foreign body response to subcutaneous implants. Biomaterials. 2012;33:6305–12.
169. Mantovani A, Sica A, Locati M. Macrophage polarization comes of age. Immunity. 2005;23:344–6.
170. Valentin JE, Stewart-Akers AM, Gilbert TW, Badylak SF. Macrophage participation in the degradation and remodeling of extracellular matrix scaffolds. Tissue Eng Part A. 2009;15:1687–94.
171. Brown BN, Valentin JE, Stewart-Akers AM, McCabe GP, Badylak SF. Macrophage phenotype and remodeling outcomes in response to biologic scaffolds with and without a cellular component. Biomaterials. 2009;30:1482–91.

172. Matsumoto T, Holmes RH, Burdick CO, Metzger JF, Heisterkamp CA 3rd, O'Connell TJ Jr. A study of inverted intestinal graft in the major veins. Angiology. 1966;17:842–50.
173. Totonelli G, Maghsoudlou P, Garriboli M, Riegler J, Orlando G, Burns AJ, et al. A rat decellularized small bowel scaffold that preserves villus-crypt architecture for intestinal regeneration. Biomaterials. 2012;33:3401–10.
174. Maghsoudlou P, Totonelli G, Loukogeorgakis SP, Eaton S, De Coppi P. A decellularization methodology for the production of a natural acellular intestinal matrix. J Vis Exp. 2013: e50658.
175. Patil PB, Chougule PB, Kumar VK, Almström S, Bäckdahl H, Banerjee D, et al. Recellularization of acellular human small intestine using bone marrow stem cells. Stem Cells Transl Med. 2013;2:307–15.
176. Nowocin AK, Southgate A, Gabe SM, Ansari T. Biocompatibility and potential of decellularized porcine small intestine to support cellular attachment and growth. J Tissue Eng Regen Med. 2016;10:E23–33.
177. Dew L, English WR, Chong CK, MacNeil S. Investigating neovascularization in rat decellularized intestine: an in vitro platform for studying angiogenesis. Tissue Eng Part A. 2016;22:1317–26.
178. Urbani L, Camilli C, Phylactopoulos DE, Crowley C, Natarajan D, Scottoni F, et al. Multi-stage bioengineering of a layered oesophagus with in vitro expanded muscle and epithelial adult progenitors. Nat Commun. 2018;9:4286.
179. Piccoli M, D'Angelo E, Crotti S, Sensi F, Urbani L, Maghin E, et al. Decellularized colorectal cancer matrix as bioactive microenvironment for in vitro 3D cancer research. J Cell Physiol. 2018;233:5937–48.
180. Kajbafzadeh AM, Khorramirouz R, Masoumi A, Keihani S, Nabavizadeh B. Decellularized human fetal intestine as a bioscaffold for regeneration of the rabbit bladder submucosa. J Pediatr Surg. 2018;53:1781–8.
181. Giuffrida P, Curti M, Al-Akkad W, Biel C, Crowley C, Frenguelli L, et al. Decellularized human gut as a natural 3D platform for research in intestinal fibrosis. Inflamm Bowel Dis. 2019;25:1740–50.
182. Chen HJ, Shuler ML. Engineering a bioartificial human colon model through decellularization and recellularization. Methods Mol Biol. 2019;1907:91–102.
183. Gosztyla C, Ladd MR, Werts A, Fulton W, Johnson B, Sodhi C, et al. A comparison of sterilization techniques for production of decellularized intestine in mice. Tissue Eng Part C Methods. 2020;26:67–79.
184. Cao G, Huang Y, Li K, Fan Y, Xie H, Li X. Small intestinal submucosa: superiority, limitations and solutions, and its potential to address bottlenecks in tissue repair. J Mater Chem B. 2019;7:5038–55.
185. Stanton AL, Lobel M, Sears S, DeLuca RS. Psychosocial aspects of selected issues in women's reproductive health: current status and future directions. J Consult Clin Psychol. 2002;70:751–70.
186. Hellström M, Bandstein S, Brännström M. Uterine tissue engineering and the future of uterus transplantation. Ann Biomed Eng. 2017;45:1718–30.
187. Santoso EG, Yoshida K, Hirota Y, Aizawa M, Yoshino O, Kishida A, et al. Application of detergents or high hydrostatic pressure as decellularization processes in uterine tissues and their subsequent effects on in vivo uterine regeneration in murine models. PLoS One. 2014;9: e103201.
188. Hellström M, El-Akouri RR, Sihlbom C, Olsson BM, Lengqvist J, Bäckdahl H, et al. Towards the development of a bioengineered uterus: comparison of different protocols for rat uterus decellularization. Acta Biomater. 2014;10:5034–42.
189. Campo H, Baptista PM, López-Pérez N, Faus A, Cervelló I, Simón C. De- and recellularization of the pig uterus: a bioengineering pilot study. Biol Reprod. 2017;96:34–45.
190. Padma AM, Tiemann TT, Alshaikh AB, Akouri R, Song MJ, Hellström M. Protocols for rat uterus isolation and decellularization: applications for uterus yissue engineering and 3D cell culturing. Methods Mol Biol. 2018;1577:161–75.

191. Miki F, Maruyama T, Miyazaki K, Takao T, Yoshimasa Y, Katakura S, et al. The orientation of a decellularized uterine scaffold determines the tissue topology and architecture of the regenerated uterus in rats. Biol Reprod. 2019;100:1215–27.
192. Daryabari SS, Kajbafzadeh AM, Fendereski K, Ghorbani F, Dehnavi M, Rostami M, et al. Correction to: development of an efficient perfusion-based protocol for whole-organ decellularization of the ovine uterus as a human-sized model and in vivo application of the bioscaffolds. J Assist Reprod Genet. 2020;37:491.

Chapter 6
Recellularization of Decellularized Whole Organ Scaffolds: Elements, Progresses, and Challenges

Jungen Hu, Yizhou Huang, Jie Tan, Lincui Da, and Huiqi Xie

Abstract Organ transplantation is considered as the gold standard treatment for end-stage organ failure. Because of the severe shortage of donor organs, tremendous efforts, including whole organ engineering, have been made to construct biological substitutes to restore the function of diseased or injured organs. In the field of whole organ engineering, seed cells, acellular whole organ scaffolds, and recellularization strategies make up the essential elements. Both differentiated and stem cells, either autologous or allogenic, have been utilized in the proof-of-concept exploration of whole organ engineering. Based on the encouraging achievements made in the clinical application of decellularized extracellular matrix materials for various diseases, decellularized whole organ scaffolds are proposed as the ideal candidate for whole organ engineering, featuring a beneficial organ-specific microenvironment for seed cells, as well as an intact vasculature for subsequent supply of oxygen and nutrients. These merits of acellular whole organ scaffolds are too sophisticated to mimic artificially. After proper delivery of essential parenchymal and non-parenchymal cells, and following the necessary in vitro bioreactor-based culture to allow the seed cells to proliferate, differentiate, and mature within the scaffolds, bioengineered organs are promising for eventual transplantation. This chapter aims to provide an overview of the recellularization of decellularized whole organ scaffolds, with a focus on the basic elements, current research progresses, and future challenges.

Keywords Whole organ engineering · Decellularization · Recellularization · Bioreactor · Regenerative medicine

J. Hu · Y. Huang (✉) · J. Tan · H. Xie (✉)
Laboratory of Stem Cell and Tissue Engineering, Orthopedic Research Institute, State Key Laboratory of Biotherapy and Cancer Center, West China Hospital, Sichuan University and Collaborative Innovation Center of Biotherapy, Chengdu, China
e-mail: huangyizhouscu@sina.com; xiehuiqi@scu.edu.cn

L. Da
Reproductive Center, Fujian Provincial Maternity and Children's Hospital, Affiliated Hospital of Fujian Medical University, Fuzhou, China

X. Li, H. Xie (eds.), *Decellularized Materials*,
https://doi.org/10.1007/978-981-33-6962-7_6

6.1 Introduction

According to the World Health Organization, cardiovascular diseases, diabetes, and chronic respiratory diseases are the most commonly observed chronic diseases in western countries [1]. It is anticipated that the incidence of these diseases will increase continuously, since the prevalence of risk factors remains at high levels, and the life expectancy is growing worldwide. In the majority of patients with these diseases, end-stage organ failure is the inevitable clinical destination [2], which greatly affects the life quality of patients and poses a big burden to the society. In the USA alone, there are approximately six million adults suffering from heart failure, and approximately 670,000 new cases were diagnosed every year, along with more than 56,000 deaths reported annually [3]. Furthermore, nearly 27,000 deaths occur annually in the USA because of end-stage liver disease, about 120,000 deaths happen annually as a result of chronic lung diseases, and approximately 112,000 deaths occur annually due to end-stage renal diseases [4].

Organ transplantation has been widely considered as the gold standard treatment for patients with end-stage organ failure. As many progresses have been achieved such as donor/recipient patient selection, immunosuppression methods, and surgical techniques, patients considering organ transplantation for the treatment of irreversible end-stage organ failure are increasing [5]. Indeed, organ transplantation has obviously improved the life quality of the patients and their survival, representing the only durable long-term therapy in many cases. However, due to a critical shortage of donor organs, only a small number of patients are fortunate enough to benefit from this treatment. For instance, in January 2018, there were more than 14,000 people waiting for an organ in the Eurotransplant region, whereas just about 6700 organs were transplanted in 2017, and more than 1300 patients died on the waiting list in the same year [6].

The gap between patients waiting for organ transplantation and available donor organs continues to widen. For patients who fortunate enough to receive a donor organ, they are confronted with adverse side effects related to a lifelong regimen of immunosuppression and the risk of chronic rejection of the grafted organ. It is noted that, after 5 years of transplantation, some organ transplants have been reported to result in a failure rate of 25–40% [7]. To address these issues, alternative treatments are urgently required to increase the supply pool of organs, to reduce the patient wait-list time, and to improve the long-term outcomes after treatment. These attempts have intensified research on alternatives of organ procurement. Since bioengineered organs can be used as an abundant source of artificial organs for transplantation, engineering whole organs in the laboratory using tissue engineering approaches has attracted considerable interest in recent years; particularly, engineered organs from autologous stem cells may make lifelong immune suppressive therapy redundant in the recipients [8].

By integrating the principles of engineering with life sciences, tissue engineering aims primarily at creating functional tissues or organs in vitro, which can be used to replace the damaged or diseased parts of human body [9]. The key elements of tissue engineering include seed cells, scaffolds, and appropriate environments to facilitate the formation and maturation of the engineered tissues in vitro [10]. Significant progress has been made in creating tissue substitutes with relatively simple structures, especially those with flat two-dimensional structures or hollow tubular structures. Successful preclinical outcomes have been reported in several tissues, such as skin [11, 12], cartilage [13, 14], bladder [15, 16], and trachea [17, 18]. Notably, because of the relatively simple structures, these tissues do not require a preformed intact vascular network at the time of implantation. In contrast, for solid organs with complicated structures, such as the heart, lung, liver, and kidney, organ engineering remains an immense challenge.

It is generally believed that when tissues become more complex, their vascular networks become more important, since they are essential for the supplying of nutrients and oxygen to maintain the viability and growth of neo-organs. Based on this, it is well known that engineering solid organs requires an intact vascular system to allow for immediate vascular supply at the time of implantation [4]. However, due to the complexity of the vasculature of solid organs, and especially because of the sorely lacking of suitable scaffolds that can closely mimic the structure and distribution of vasculature in solid organs, the field of whole organ engineering has been impeded for a long period of time.

In recent years, the application of decellularization methods to obtain an acellular organ scaffold has facilitated tissue engineering research progress toward whole organ engineering. In the field of tissue engineering, different kinds of decellularization techniques, such as physical treatments, enzymatic methods, chemical protocols, or a combination of them [6, 19, 20], have been applied to remove cells from native tissues, remaining a three-dimensional (3D) scaffold that is composed of extracellular matrix (ECM). Decellularization of simple tissues can date back to the late 1980s [21], but it takes almost three decades to adapt this strategy to generate acellular whole organ scaffolds [22]. Unlike methods used to obtain ECM scaffolds from simple tissues, decellularization of whole organs needs more sophisticated processing because of their much thicker and more complex structure. Actually, before the introduction of whole organ decellularization methods, many efforts had been devoted to synthetic materials to fabricate scaffolds mimicking the native architecture of solid organs [23]; however, these attempts failed to replicate the complex vascular network. As decellularization strategies progress, intact hearts [24, 25], livers [26, 27], lungs [28, 29], kidneys [30, 31], and pancreases [32, 33] have been successfully decellularized from different species.

Similar to the concept of tissue engineering, three basic elements are involved in whole organ engineering, namely cell sources for recellularization, decellularized whole organ scaffolds, and recellularization techniques [4]. In general, the process of whole organ engineering contains two stages before transplantation in vivo: firstly, the decellularization of a homologous organ to create a 3D scaffold with similar architecture and components to that of the native organ; secondly, the recellularization of acellular scaffolds with appropriate seed cells to form a structural and functional unit resembling that of target organ (Fig. 6.1) [6]. Since

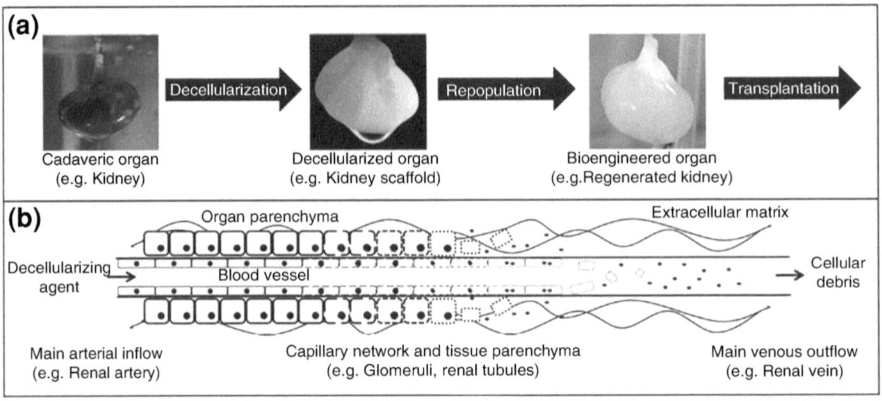

Fig. 6.1 A schematic illustration of the general process of whole organ engineering based on the recellularization of acellular organ scaffolds. Acellular organ scaffolds, which contain the ECM of original organs and preserve their native macro- and micro-architectures (**a**), are yielded by the application of decellularization methods (e.g., the perfusion-decellularization process) (**b**) to native donor organs. After decellularization, the scaffolds are seeded with regenerative cells (e.g., adult stem cells) and cultured in a bioreactor with biomimetic conditions for functional maturation, and then the engineered organs can be used to partially or completely replace the biological function of an injured/diseased organ in the recipients. (cited from reference [34])

the emergence of whole organ engineering, these essential elements and the fabrication techniques have been the focus of research.

As concerns the source of seed cells, both stem cells and differentiated cells have been considered for the recellularization of whole organ scaffolds. Owing to their robust proliferation ability, multilineage differentiation potential, and paracrine functions, stem/progenitor cells derived from fetal or adult tissues have been proposed as ideal cell sources for organogenesis in vitro. In addition to stem cells derived from adult tissues, embryonic stem cells (ESCs), and more recently the induced pluripotent stem cells (iPSCs) have shown good potential for whole organ engineering [4]. Apparently, the choice of suitable seed cells for recellularization depends on the type of organ to be fabricated. Apart from organ-specific cells, other somatic cells such as fibroblasts and endothelial cells are indispensable, because they are fundamental for the organization of organ structure.

The fabrication of an engineered organ requires not only proper sources of seed cells but also a suitable microenvironment for cell proliferation and differentiation. Decellularized organ scaffolds consist exclusively of ECM components. Both the functional molecules and the architecture are essential components of the microenvironment where seed cells reside in. The microenvironment plays a fundamental role in the regulation of seed cells. For instance, growth factors and other inductive signals preserved in the decellularized scaffolds influence the attachment, proliferation, and differentiation of seed cells [35]. Besides molecular cues, decellularized scaffolds also possess organ-specific biomechanical properties that exert great impact on the biology of seed cells such as their phenotype and migration [36]. It is, therefore, possible that decellularized scaffolds derived from homologous organs can be used as a platform to induce site-specific differentiation of stem cells.

Along with acellular organ scaffolds and seed cells, whole organ engineering also requires optimal cell seeding methods as well as organ culture systems that are practical and physiologically relevant. Inspired by cell therapy, the most practical cell seeding method seems to be intraarterial infusion [36]. Following cell seeding, it usually takes several weeks to culture the cell-scaffold constructs, which is usually accomplished in a bioreactor designed specifically for organ engineering [37]. For instance, with the help of a continuous infusion system in the bioreactor, the vascular system in the decellularized scaffolds allows for the supply of nutrients and oxygen to seed cells. To improve the function of engineered organs, bioreactors should be optimized according to the specific requirements of target organs, such as chemical induction, mechanical stimulation, and electrical stimuli [4].

To date, successful regeneration of a functional whole organ has not been realized, yet the field of whole organ engineering progresses rapidly. Several exciting achievements have been reported in the fabrication of essential solid organs such as heart, lung, liver, kidney and pancreas [24–33]. Impressively, using decellularized whole organs as scaffolds, the generation of neo-organs with rudimentary functions has been observed in small animal models [8, 38–40]. These researches provide valuable experience to address future challenges. Particularly, emphasis should be placed on the finding of clinically relevant cell sources, the optimization of bioreactor-based techniques for recellularization and cultivation, scaling up the techniques for human-sized organs, and adequate reconstruction of the vasculature and parenchyma for long-term function in vivo.

In this chapter, we attempt to provide an overview of the essential elements of whole organ engineering, with a special focus on the seed cells and current progress in recellularization strategies. The function of engineered solid organs was critically reviewed, and the emerging challenges and future directions were discussed.

6.2 Cell Sources for Whole Organ Engineering

6.2.1 General Considerations

To truly create a solid organ in vitro, it is necessary to replicate all of its structural properties and biological components, including the parenchyma, vasculature, and other supporting structures, all of which differ evidently in the cell numbers and types [41]. For instance, to engineer a heart from a decellularized heart scaffold, various kinds of cells are desired to repopulate the scaffold, including but not limited to cardiac progenitor cells, parasympathetic neurons, fibroblasts, myofibroblasts, and epicardial cells [4]. Based on current technology in the laboratory, it seems unrealistic to reproduce all of the embryogenesis process from single-cell stage to organogenesis [41]. Consequently, it remains unreasonable to ask a single population of cells to generate all kinds of cells needed to fabricate an organ in vitro.

Theoretically, the ideal cell source for whole organ engineering should meet the following criteria: (1) they can be harvested without or with minimal invasive procedure; (2) they are easy to expand in vitro and can proliferate as needed; (3) they are fully characterized and are able to differentiate into the cell types required; (4) they are nonimmunogenic after transplantation in the recipient, and are fully functional in the microenvironment [4]. Distinct cell types, including differentiated cells, adult stem cells, iPSCs, and ESCs, have been evaluated for potential applications for recellularization of acellular organ scaffolds (Fig. 6.2) [37], but none of them meets all of the aforementioned criteria. Thus far, the ideal clinically viable cell source for whole organ engineering has not been identified.

6.2.1.1 Differentiated Versus Stem Cells

Usually, differentiated cells employed in whole organ engineering are parenchymal or supportive cells (e.g., fibroblasts and endothelial cells) that can be isolated from easily accessible tissues (e.g., peripheral blood and placenta tissues) or can be derived directly from organ of interest through biopsy or surgical harvest [4]. Due to the inability of differentiated cells to proliferate extensively in vitro, it is very difficult to obtain adequate number of cells for whole organ engineering; thus, it is commonly believed that they are actually not effective in the recellularization of acellular organs. Instead, the most likely candidate source that is possible to fulfill the demands are stem or progenitor cells.

Stem cells employed in whole organ engineering can be broadly categorized as ESCs, adult stem cells, and iPSCs [4, 37]. The unlimited ability of self-renewal, along with the pluripotency, makes ESCs an extremely attractive source of seed cells for regenerative medicine. However, the clinical value of ESCs is limited by the ethical and supply issues as well as the immunogenicity after allogeneic transplantation [42].

Adult stem cells residing in human tissues represent another cell source for whole organ engineering. For most adult stem cells, current understanding about their native identity is very limited [43]. In addition, how to preserve their stem/progenitor properties after extensive expansion remains a big challenge. Our understanding of the biology of tissue resident stem/progenitor cells is primarily come from studies using mouse as the models. This makes it unsuitable to apply the knowledge directly to human, since big variations exist in the cell biology among different species [44]. Unlike ESCs and adult stem cells, iPSCs provide another good choice to obtain autologous cells with pluripotent potential, but the concern regarding genetic alteration should be resolved before clinical applications [45, 46].

6.2.1.2 Autologous Versus Allogenic Cells

Autologous cells are self-derived and are unlikely to invoke an adverse immune response after implantation, thus eliminating the need of immunosuppressive drug

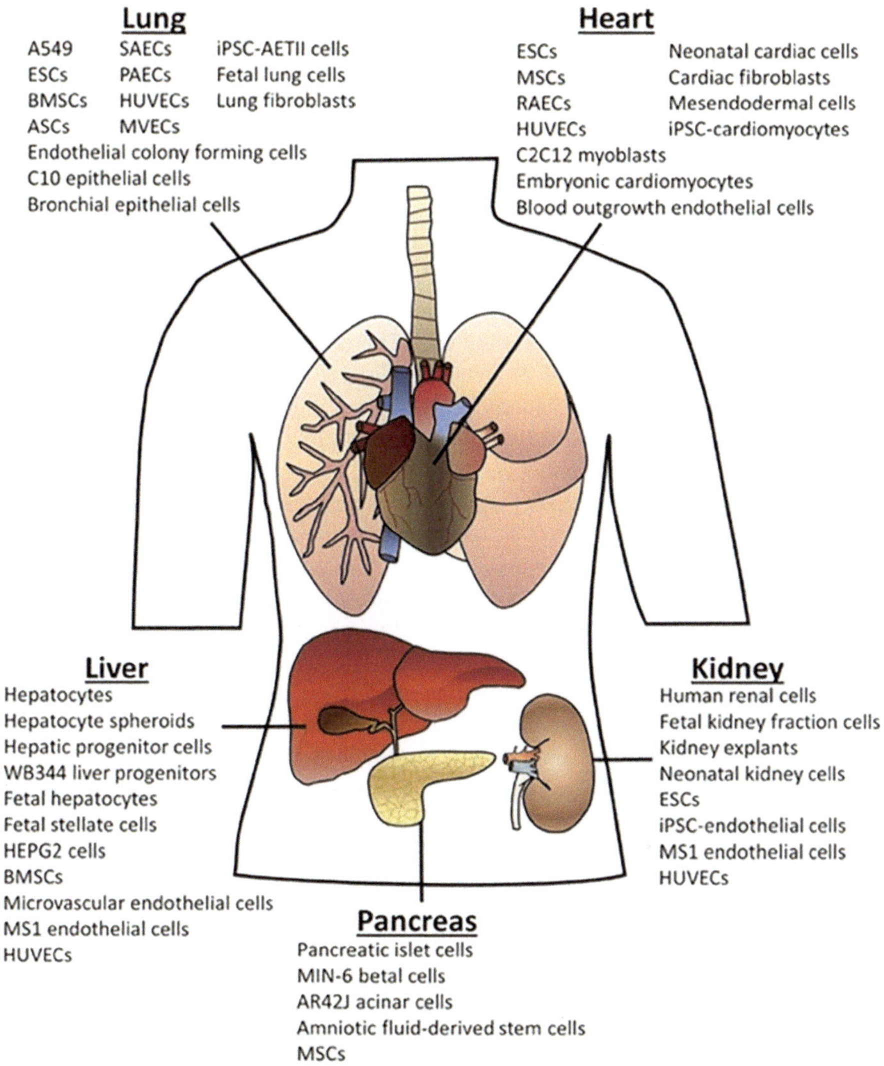

Fig. 6.2 Different type of cells, including differentiated cells (e.g., hepatocytes), stem cells (e.g., ESCs, iPSCs, and MSCs), and cell lines (e.g., MIN-6 beta cells and HUVECs), have been used for the recellularization of acellular whole organ scaffolds. Abbreviations: AETII: alveolar epithelial type II cells; ASCs: adipose tissue-derived stem cells; ESCs: embryonic stem cells; BMSCs: bone marrow-derived stem cells; HUVECs: human umbilical vein endothelial cells; MSCs: mesenchymal stem cells; iPSCs: induced pluripotent stem cells; MVECs: microvascular endothelial cells; PAECs: pulmonary alveolar epithelial cells; SAECs: small airway epithelial cells. (cited from reference [37])

regimens. This advantage, in turn, reduces the regulatory burden on first-in-human trials [4]. Furthermore, autologous cells are quite appealing to patients, because they can avoid the severe side effects of long-term immunosuppression, especially an increased risk of cancer and infection. However, it is important to keep in mind that not all autologous cells can proliferate to large numbers in vitro. In most cases of solid organs, such as lungs, hearts, and pancreases, autologous cells are difficult to harvest due to the difficulty of tissue sampling; in addition, the cell numbers after expansion are usually insufficient to create a neo-organ. Therefore, allogeneic cells are often considered for organ engineering.

Allogeneic cells are usually harvested from young and healthy donors and can be prepared in advance for an off-the-shelf use. The decision on the choice of autologous or allogeneic cells depends on many conditions, including, but not limited to, the following: (1) the immunogenicity of the cells; (2) the number of cells required; (3) the ease of cell harvest and culture; (4) the timeframe needed for cell expansion; (5) the availability of methods that can effectively induce the differentiation of cells to gain organ-specific phenotypes [4].

Actually, from a perspective of business and engineering, allogenic cells possess several advantages [4]. First, they are able to proliferate in large numbers that enable the generation of an off-the-shelf product. Second, when comparing with autologous products, allogeneic cells are with less lot-to-lot variability. Third, comparing with patient-specific products, allogeneic cells take less time to produce and are less expensive. In spite of these merits, allogeneic cells have some disadvantages, such as potential adverse immune reactions and an increased risk of the transmission of adventitial agents. There are some good examples about the application of allogeneic cells for regenerative medicine. For instance, bone marrow-derived MSCs have been used for the treatment of graft-versus-host disease [47]. Allogeneic bone marrow transplantation has been performed for nearly 60 years to treat blood disorders [48].

6.2.1.3 Other Considerations

The seed cells and scaffolds used to engineer an organ have an inherently close relationship, which is critical to the ultimate functions and clinical outcomes of the engineered organs. It is well known that organ derived ECM scaffolds provide a friendly microenvironment for seed cells and regulate their biological performance and tissue assembly. Seed cells interact directly with the architecture of ECM scaffolds, and indirectly with their cytokines and growth factors. Of note, the response of seed cells to ECM scaffolds depends on the type of cells, such as their tissue origin [49, 50]. It is therefore of great importance to choose the proper source of seed cells to facilitate the inductivity of scaffolds for organ engineering.

Besides the tissue origin of seed cells, other factors influencing the choice of seed cells should be considered, especially the cost, the regulatory issues, and the intellectual property constraints. Currently, the clinical application of allogeneic cells is regulated by the U.S. Food and Drug Administration (FDA). In the case of autologous human cells, they are subject to the regulation by the U.S. FDA, if the cells are maintained outside the body and are considered "manipulated" or if they are harvested and applied to nonhomologous locations [51]. Since many types of human stem cells or their isolation methods have been patented in the past years, the use of these cells for commercial applications should pay close attention to the intellectual property constraints.

6.2.2 Cell Lines

Cell lines have been widely used in the field of organ engineering, especially to evaluate the cytocompatibility of decellularized ECM scaffolds [52–54]. Cell lines can be derived from animals or human being, from either a patient or a healthy individual. They are immortalized to acquire the capability of indefinite proliferation in culture. As they are under the pressure of constant growing, their phenotype may change with increasing number of passages.

Based on the degree of maturation, cell lines have restricted potential to differentiate into other cell types. For instance, a human embryonic stem cell line possesses the ability to differentiate into all human cell lineages, yet a cell line derived from HUVECs has only restricted differentiation potential. To construct a functional organ containing various kinds of necessary cellular constituents, multiple cell lines are needed in organ engineering.

In the literature, different kinds of cell lines, such as ESCs, hepatic stellate cells, and cardiomyocytes, have been used to repopulate decellularized ECM scaffolds [55–58]. However, it is worth noting that the use of cell lines to repopulate acellular scaffolds has a high risk of immune rejection after implantation in the recipients, since most cell lines do not match the patients (e.g., HLA mismatching). Hence, despite their usefulness in proof-of-concept animal studies, the utilization of cell lines to construct an engineered organ for transplantation seems not applicable to human.

6.2.3 Endothelial Progenitor Cells and Endothelial Cells

6.2.3.1 Endothelial Progenitor Cells (EPCs)

First identified in the peripheral blood in 1997 by Asahara et al. [59], EPCs are a population of hematopoietic cells in the bone marrow that possess the ability to differentiate into endothelial cells. Phenotypically, EPCs are characterized by the positive expression of several cell surface antigens, including CD133 and CD34, which are hematopoietic cell markers, and vascular endothelial growth factor receptor-2, which is an endothelial marker [60]. Under the condition of diseased or tissue injury, EPCs in the bone marrow can migrate into the circulation, home to the sites of ischemia, and augment the vasculogenesis via direct differentiation or paracrine signaling [61].

The angiogenetic potential of EPCs has been investigated in many studies [62–64]. For instance, using a hyaluronic acid hydrogel as scaffold, Yee et al. have determined the capability of EPCs to form capillary-like structures in vitro. They found that the angiogenetic potential of EPCs depended greatly on the physical properties of the hydrogel. When cultured on hydrogels with decreased matrix viscoelasticity, the vascular tube length and vascular area formed by EPCs increased significantly [64].

Besides in vitro studies, the regeneration potential of EPCs has been observed in many animal models, such as an acute myocardial infarction animal model [65]. It has been found that when transplanted into ischemic hearts, EPCs improved the cardiac functions, which was mainly due to enhanced angiogenesis and improved perfusion [66]. Interestingly, it has been observed that the number of grafted cells retaining at the tissue where they are delivered remains very low [66–68]. Therefore, it is generally thought that the paracrine action of EPCs plays an important role in the therapeutic effects.

Furthermore, it is well known that EPCs, such as endothelial colony forming cells, can be harvested from a patient's blood and used for the reendothelialization of decellularized whole organ scaffolds. After cell seeding, it has been observed that EPCs could adhere and disperse throughout the vascular compartment of decellularized whole organ scaffolds, such as decellularized livers [69] and decellularized pancreas [70].

6.2.3.2 Endothelial Cells (ECs)

As the primary cell type utilized for the revascularization of biomaterials, ECs have been successfully cultured from many tissues, such as umbilical vein [71, 72], pulmonary artery [73, 74], liver [75, 76], kidney [77, 78], and heart [79, 80]. Since they require little manipulation before transplantation, ECs have been widely used as seed cells for the vascularization of tissue engineered constructs [81–84]. Also, they have been utilized in many clinical studies for endogenous cell therapy, and have been generally considered as an ideal cell source for the rebuilding of vasculature of nascent organs. After transplantation of an engineered organ in vivo, it is particularly desirable to minimize or even avoid any direct contact between the recipient's blood and foreign proteins. Consequently, relining vessels in the engineered organs by using EC populations has gained much interest in the field of whole organ engineering.

Actually, many studies have employed ECs to reendothelialize the decellularized whole organ scaffolds. In the first report of whole organ engineering, Ott et al. used rat aortic ECs for the reendothelialization of a decellularized rat heart. According to their results, it is observed that the ECs lined the coronary vessels of the scaffolds, resulting in decreased thrombogenicity upon implantation [85].

In another study, Ott et al. used the method of gravity perfusion to seed HUVECs into the pulmonary artery and the pulmonary vein of a decellularized rat lung scaffold for reendothelialization. It is observed that, after culture in a perfusion bioreactor for a period of 5 days, the cells engrafted from the pulmonary artery to the capillaries. Furthermore, when implanted orthotopically into rats, the scaffolds seeded with rat fetal lung cells and HUVECs did not show any bleeding to the airway for up to 14 days [86]. For reendothelialization of rat kidney scaffolds, HUVECs were also employed by seeding the cells through the portal vein, and the scaffolds seeded with HUVECs and rat kidney cells did not show any thrombus formation or hemorrhaging after orthotopic implantation [39].

6.2.4 Organ-Derived Cells for Whole Organ Engineering

6.2.4.1 Differentiated Cells Isolated from the Organ of Interest

The fabrication of a whole organ requires both parenchymal and non-parenchymal cells. To create an engineered organ that is capable to renew and respond to injury, an integration of stromal, immune, mural, and interstitial cells is important to complete recellularization [37]. It is well known that fibroblasts play an important role in the matrix remodeling of tissues [87, 88]. Immune cells such as alveolar macrophages are able to limit tissue damage from infections [89]. Therefore, in order to better repopulate a decellularized organ, various primary cells from the recipient would be needed.

Primary cells have been utilized for the recellularization of acellular organ scaffolds. For instance, rat neonatal cardiomyocytes and aortic endothelial cells have been used to recellularize decellularized rat hearts to produce a functional tissue [85]. Primary rat endothelial cells and hepatocytes have been employed to repopulate the decellularized rat liver scaffolds to produce a metabolically functional tissue [90]. Endothelial and fetal rat lung cells have been used to recellularize rat lung scaffolds, which leaded to the formation of lung constructs capable of gas exchange after orthotopic transplantation [86].

Fetal cells isolated from the organ of interest are commonly used as seed cells in whole organ engineering. When fetal cells are seeded on the scaffolds, they often remain their phenotypes and display functionality. For instance, fetal hepatic cells have been used for the recellularization of liver scaffolds, which showed promising outcomes [91–93]. When seeded on rat decellularized kidney scaffolds, neonatal renal cells showed similar success in the function, with the ability to produce urine in vivo [39]. However, fetal cells just provide a proof-of-concept cell source for researches but not a viable source of seed cells for clinically relevant organ engineering.

Although primary cells can be cultured from biopsy or a donor organ, they have limited proliferation ability and are difficult to expand for sufficient numbers to engineering a human-sized organ scaffold. Another source of cells that maybe of interest for whole organ engineering, especially the recellularization of vasculature, is stromal vascular fraction (SVF) derived from adipose tissue. As a mixture of heterogeneous cell populations, SVF contains endothelial, mural, and immune cells [94, 95]. Thus, SVF can provide a good environment for engineering a competent vasculature, which is very attractive for organ engineering.

6.2.4.2 Organ-Derived Stem/Progenitor Cells

Organ-specific stem cells provide another potential cell source for recellularization of decellularized scaffolds, although they have not been used in organ engineering yet. Theoretically, they should be expanded to provide a sufficient number of cells for whole organ engineering, and thus are not considered as primary cells. As a population of primary cells with great proliferation ability, organ-specific stem cells can differentiate into different cell lineages of the tissue where they reside in. In this section, the major stem/progenitor cells relevant to our topic are those derived from heart, liver, lung, kidney, and pancreas.

The human heart is composed of a wide variety of cell types originally derived from multipotent cardiac progenitors and mesodermal precursors at early embryogenesis [96, 97]. Multipotent cardiac progenitors are defined by their spatiotemporal presence and the ability to differentiate into specific cardiac lineages. The earliest cardiac progenitors are found before commitment to either of the first or second heart field [98]. In adult mammalian hearts, several populations of cardiac progenitor cells have been identified. However, it seems like that these "adult" cardiac progenitors contribute rarely to direct generation of new cardiac myocytes in the adults [99].

After injury, the liver possesses a robust regenerative capability through the proliferation of remaining healthy hepatocytes. However, when the replacement of lost hepatic mass by the remaining hepatocytes is precluded, hepatic progenitor cells (HPCs) are induced to proliferate and differentiate into hepatocytes. Thus, HPCs are regarded as a "second line of defense" in liver regeneration [100]. In healthy liver, HPCs are scarce; however, upon stimulation, it has been observed that HPCs residing in the Canals of Hering can proliferate and infiltrate the liver parenchyma [101]. As a heterogeneous population of cells, HPCs express different combinations of cell phenotypic markers, such as α-fetoprotein, transferrin, E-cadherin, and cytokeratin 7 [102]. They can differentiate into the hepatic lineages (i.e., hepatocytes and cholangiocytes) as well as non-hepatic lineages (e.g., pancreatic cell types) [103].

In the lung, the role of organ-specific stem cells in tissue repair has been showed in animal studies [36]. Resident, multipotent pulmonary stem cells with self-renewal ability and clonogenicity have been identified [104]. Also, lung-resident MSCs in mice and humans have been characterized, which are able to differentiate into club cells as well as alveolar type I or type II cells [105]. It is reported that lung-resident MSCs, although not play a direct part in epithelial renewal, can establish communication with the epithelium and serve as a local protector [104].

The kidney has been believed as an organ without true regenerative ability. This opinion is challenged by the fact that renal injuries are able to at least partially heal. Particularly, the functionality and integrity of the injured part of nephron can be restored. Previous studies about the biology of kidneys have revealed the existence of renal progenitors that are able to give rise to glomerular and tubular epithelial cells [106]. First identified in adult human kidneys, renal progenitors were positive of CD133 and CD24 [107]. They are located in the Bowman's capsule and the cortical nephron [108, 109]. In mature kidneys, renal progenitors account for <2% of total renal cells [109]. After isolation, they can differentiate into glomerular and tubular epithelial cells [107–111].

Much evidence indicates the existence of pancreatic progenitor cells in the adult pancreatic duct. For instance, in some previous studies, insulin-producing cells have been observed in the duct of adult pancreas [112]. In addition, both exocrine duct and acinar cells can differentiate into pancreatic β-cells [113]. Histological observations of β-cell regeneration by the transdifferentiation of duct cells have reported for many years [112]. In vitro, pancreatic progenitor cells can be identified by the positive expression of hepatocyte growth factor and c-Met receptor, as well as the negative expression of CD45, TER119, c-Kit, and Flk-1 [114, 115]. In vitro, they are able to form colonies, to largely expand, and to differentiate into multiple pancreatic cells; furthermore, following transplantation in vivo, they can differentiate into pancreatic endocrine and acinar cells [115]. All these properties make pancreatic progenitor cells a potential cell source for organ engineering, but further investigations are required to explore the possibility.

6.2.5 Stem Cells for Whole Organ Engineering

Because of the ease of isolation and expansion, the readily available in large quantities, the multiple-lineage differentiation potential, and the high regenerative ability, stem cells are an attractive source of seed cells for whole organ engineering. Numerous studies have showed that stem cells can adapt to a great variety of microenvironments, including the decellularized organ scaffolds which in turn control the maturation and function of stem cells.

In some studies of whole organ engineering, both ESCs and iPSCs have shown promising results [116–119], but the issues concerning their safety and efficacy remain unsolved. By contrast, adult stem cells do not form teratoma after implantation in vivo, and are more restricted in the differentiation potential. Importantly, much evidence has confirmed that endogenous adult stem cells, such as MSCs, can be mobilized into the circulation and then recruited into the injured or diseased tissues at distal sites to participate in the process of tissue repair and regeneration [120]. Apart from the differentiation into the cell types of the tissue where they reside in, adult stem cells possess the capability to cross lineage barriers to adopt a new phenotype of those cells unique to other tissues, an ability termed as the "plasticity" of adult stem cells. Currently, various kinds of adult stem cells such as adipose tissue-derived stem cells are under intensive investigation for the recellularization of biomaterials in tissue engineering researches [121].

6.2.5.1 ESCs for Whole Organ Engineering

ESCs are pluripotent stem cells, which are typically derived from the inner cell mass of blastocysts (Fig. 6.3a) [122]. They possess a high level of clonality, have unlimited self-renewal ability, and are capable of differentiating into all derivatives of the three germ layers. They express several transcription factors, including Oct-4, Nanog, and Sox2, to maintain the pluripotent state of cells (Fig. 6.3b) [123]. One of the primary risks associated with the application of human ESCs is the potential to lead to teratomas formation if grafted in an undifferentiated state (Fig. 6.3c) [124]. Other major factors that complicate possible clinical applications include the ethical debates concerning cell harvest and applications, as well as their allogeneic origin that can lead to immunological rejection after implantation.

Indeed, before human ESCs can serve as a suitable cell source for organ engineering, it is necessary to understand the signaling cascades that govern the fate decision of ESCs, which in turn can be used to prevent the tumorigenicity of cells. In addition, because they are derived early after fertilization, the genome of human ESCs is not yet modified by the environmental cues, namely without epigenetic modifications that are usually observed in other stem cells [125]. Thus, when comparing with stem or progenitor cells with more differentiated states, human ESCs are presumably more responsive to environmental cues. Based on this advantage, the first clinical study of human ESCs was carried out in the USA in 2010 [126].

ESCs have been commonly considered as a potentially unlimited source of seed cells for whole organ engineering of hearts. Upon the establishment of the cells, human ESCs have been demonstrated to be capable of differentiating into cardiomyocytes definitively. Cardiomyocytes derived from ESCs have a high proliferation ability, which is very attractive for whole heart organ engineering since a large number of cardiomyocytes are needed. When transplanted into the hearts, human ESC-derived cardiomyocytes can electromechanically integrate with the host tissue and regenerate new myocardium [127]. These properties lay the foundation of future applications of human ESCs for cardiac regeneration. However, it should be noted that a rigorous method to efficiently induce the differentiation of human ESCs into cardiomyocytes has not yet been established.

The effect of decellularized organ scaffolds on the differentiation of ESCs has been explored in a number of studies [128–132]. Without any exogenous inductive signals, mouse ESCs seeded on decellularized rat kidneys expressed both of renal and endothelial markers, whereas the expression of pluripotency markers decreased [128, 131, 132]. When comparing with other scaffolds like Gelfoam, Matrigel, and collagen hydrogel, decellularized lung scaffolds have shown improved inductivity to promote the differentiation of mouse ESCs into the epithelial and endothelial lineages [129]. Interestingly, it is also observed that human ESCs seeded on

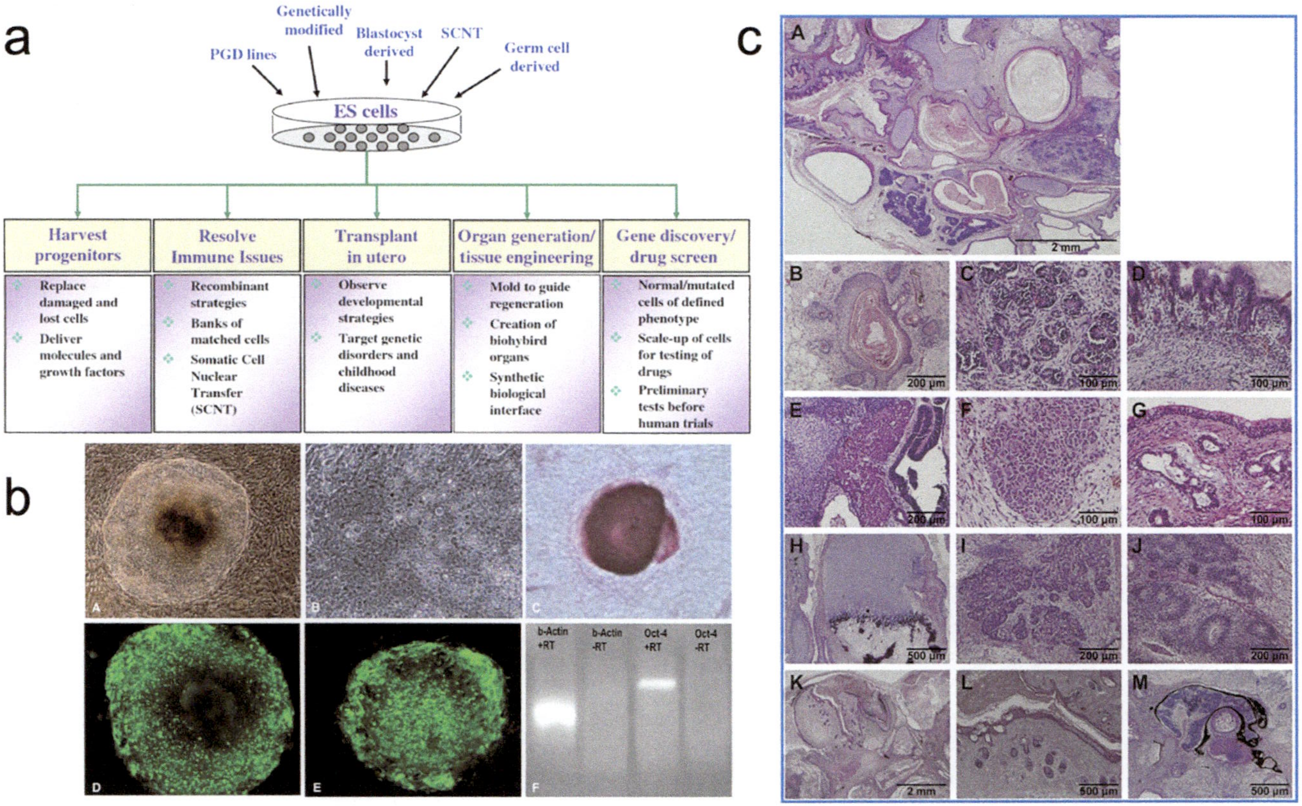

Fig. 6.3 An overview of the basic biology and potential applications of ESCs. ESCs possess promising potential for regenerative medicine, such as the application to replace damaged and lost cells (**a**) [122]. Undifferentiated human ESCs growing on a feeder layer of cells show colony morphology and are positive for alkaline phosphatase, Tra-1-60, Tra-1-81, and Oct-4 (**b**) [123]. When subcutaneous transplanted into immunodeficient mice, they result in teratoma formation (**c**) [124]. *SCNT* somatic cell nuclear transfer, *PGD* preimplantation genetic diagnosis. (adapted from references [122–124])

decellularized rhesus macaque lung and kidney scaffolds expressed both of kidney and lung markers. This study did not show any organ-specific effect of the decellularized scaffolds on the differentiation of ESCs [130], thereby indicating that decellularized organ scaffolds are likely limited in the capability to offer all of the inductive signals needed for the differentiation of ESCs. Despite this, it is still reasonable to assume that decellularized organs are likely more suitable for promoting the differentiation of ESCs when used in combination with other conditions such as appropriate differentiation cocktails.

Besides applications for the recellularization of the parenchyma of decellularized solid organs, ESCs have also been considered for the recellularization of vasculature within the scaffolds [128, 131, 132]. When seeded on the artery of decellularized rat kidney scaffolds and cultured without any supplement of exogenous growth factors, mouse ESCs showed an endothelial-like morphology in the vascular structures [131], and were positive for the endothelial-specific lectin [132]. Similarly, in another study, murine ESCs were seeded in the renal artery of decellularized rat kidneys. It is found that the cells repopulated the glomerular and peritubular capillaries and positively expressed markers of endothelial cells (Tie-2 and CD31) 24 h after seeding [128].

6.2.5.2 iPSCs for Whole Organ Engineering

iPSCs, created by reprogramming of somatic cells (e.g., skin fibroblasts) with retroviral transduction with specific pluripotent genes (e.g., Oct4, Sox2, c-Myc, and Klf4), provide a promising source of seed cells for organ recellularization. iPSCs share some of the properties of ESCs. For instance, both of them express embryonic markers (e.g., Oct3/4, Sox2, Nanog, etc.) and can differentiate into all of the three germ layers (Fig. 6.4). However, iPSCs avoid some of the limitations of ESCs, such as legal and ethical controversy and the need for allogeneic sources. Importantly, iPSCs can be expanded to large numbers to enable the production of patient-specific organs.

By definition, the generation of iPSCs requires a cell transformation process which changes the cells from a terminally differentiated state into a more primitive phenotype. When comparing with ESCs, iPSCs have a major disadvantage, namely the epigenetic modification of DNA, which can happen during the transformation of the cells or may have been present in the starting cells [134]. Indeed, epigenetic modification of the cells may reduce the safety and their responsivity. Therefore, safety issues about the tumorigenicity and transformed nature of iPSCs should be considered, since they limit the potential clinical applications.

If iPSCs were transplanted in an undifferentiated state, they can give rise to the formation of teratomas in vivo. Several factors that can lead to the tumorigenicity of iPSCs have been identified: (1) cell reprogramming methods that involve the genomic integration of DNA or the insertion of mutagenic molecules; (2) inadequate differentiation of iPSCs after induction; (3) inefficient method for the purification of differentiated cells; (4) genomic aberrations in the starting cells [8]. In recent years,

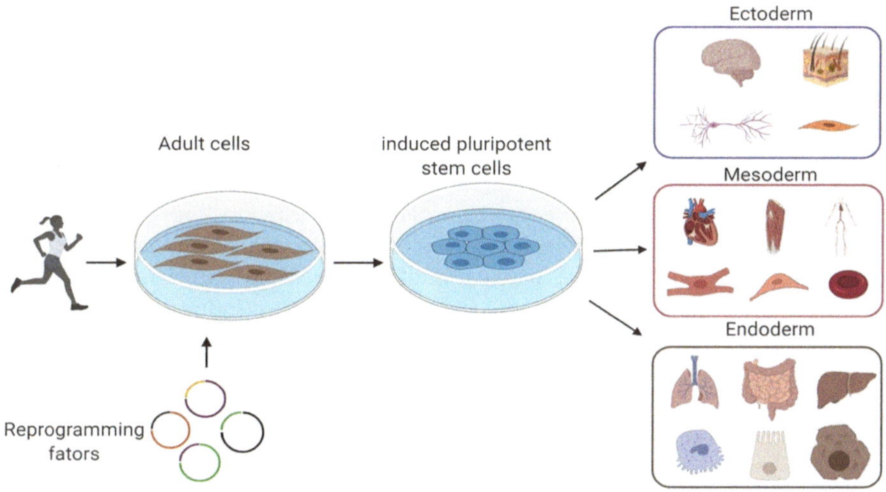

Fig. 6.4 An overview of the generation and differentiation potential of iPSCs. They are somatic cells that have been reprogrammed into a pluripotent state by using reprogramming factors such as small molecules and microRNAs. They can differentiate into all cell types of the human body. (cited from reference [133])

intense efforts have been devoted to avoiding the mutagenic risk of the cell reprogramming methods. For instance, several genomic integration-free methods have been developed to generate iPSCs, such as direct delivery of proteins, RNA-based viruses, or small molecule compounds [135, 136]. In addition, methods that are able to selectively eliminate the undifferentiated iPSCs, such as the use of targeting small compounds, also provide a feasible strategy to avoid tumor formation in vivo. Furthermore, when clinical application is considered, genetic screening should be included in the organ recellularization protocols.

The fate of stem cells can be induced in vitro using growth factors or chemical agents to regulate the cell signaling pathways and the downstream gene expressions. Using cell differentiation protocols, organ-specific cell types such as the cardiomyocytes have been induced from iPSCs with a high efficacy. Before entering clinical arena, the differentiation protocols for iPSCs should be safe and reproducible. However, after induction, immature and heterogeneous phenotypes have been observed in the human iPSC-derived cell populations. For instance, it has been reported that hepatocyte-like cells derived from human iPSCs abnormally expressed α-fetoprotein and had impaired function of detoxification enzymes [137]. In another case, a mixture of atrial, ventricular, and nodal morphologies was noted in the cardiac myocytes induced from human iPSCs, and inconsistencies in the electrophysiological and genetic analyses were recorded among cell populations [138]. Since the heterogeneity of iPSC-derived cardiac myocytes can lead to arrhythmia and inefficient force generation in vivo, it is necessary to develop reliable and reproducible cell induction protocols to generate pure ventricular cardiomyocyte lineages.

iPSCs have been used to repopulate decellularized organ scaffolds. For example, human iPSC-derived cardiovascular progenitors have been used to repopulate decellularized mouse hearts to engineer functional heart constructs [139]. After seeding on the scaffolds, the cells began to repopulate the matrix and obtained the phenotypes of cardiomyocytes, endothelial cells, and smooth muscle cells to reconstruct the heart scaffolds. After 20 days of culture, the construct began spontaneous contraction and was responsive to drugs. However, echocardiogram showed an irregular wave morphology, which indicated that there was a lack of an organized conduction system. Therefore, the use of iPSCs to engineer personalized heart constructs requires additional investigation. In addition to engineer hearts, human iPSCs have been employed for lung recellularization. Interestingly, when seeded into lung matrices, iPSC-derived alveolar epithelial type II cells were able to proliferate and maintained the expression of alveolar markers, suggesting a promising application potential for future applications [140].

6.2.5.3 Adult Stem Cells for Whole Organ Engineering

Stem cells isolated from any tissue after birth are termed as adult stem cells, differing from ESCs and iPSCs in that they are more restricted in the proliferation ability as well as the differentiation potential and do not spontaneously generate complex tissues in culture. Adult stem cells can be cultured from the biopsy material from many tissues (e.g., bone marrow, skin, synovial fluid, and peripheral blood), among which bone marrow is the most common source of tissue.

According to the Clinical Trials Data Bank at the National Institutes of Health (www.clinicaltrials.gov), adult stem cells have been widely utilized in the registered clinical trials to evaluate the therapeutic potential for a variety of difficult-to-treat diseases ranging from myocardial infarction to graft-versus-host disease and other chronic illnesses such as osteoarthritis and diabetic wounds. Similarly, in the field of organ engineering, adult stem cells have been considered as the most promising source of seed cells.

Perinatal Stem Cells

The term “perinatal” means around the time of birth, and by definition the term “perinatal period” always refers to the period about 5 months before birth and the first month of the neonatal period [141]. After birth, different types of perinatal tissues can be obtained (Fig. 6.5) [142], such as umbilical cord, cord blood, and placenta. These perinatal tissues have long been discarded as medical waste, but in recent years they have been proved to be an abundant source of adult stem cells.

Various types of perinatal stem cells, most of which belong to the category of MSCs, have been successfully isolated and characterized from different perinatal tissues that can be easily obtained after birth, such as the amnion [143, 144], umbilical cord [145, 146], cord blood [147], and placenta [148, 149]. Importantly,

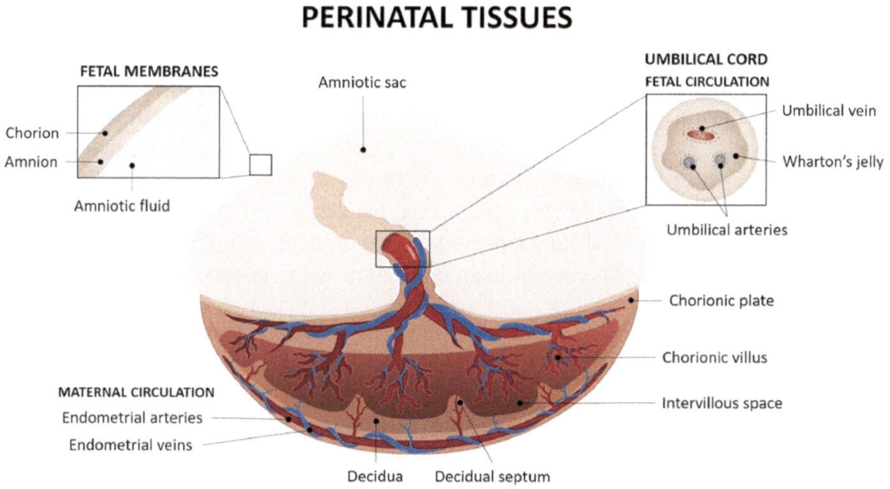

Fig. 6.5 Schematic drawing of perinatal tissues, which include umbilical cord, placenta, and fetal membranes. All of them contain abundant source of perinatal stem cells that have good potential for future applications in regenerative medicine. (cited from reference [142])

perinatal stem cells can be obtained easily, non-invasively, in large quantities, and without ethical controversies. They possess many attractive properties for future applications, especially the extensive proliferation ability, low immunogenicity, immunomodulatory properties, and multipotency.

Comparing with MSCs derived from other adult tissues, MSCs isolated from perinatal tissues have shown several advantages. For instance, MSCs derived from umbilical cord [150, 151], cord blood [152, 153], amnion [154, 155], chorion [156, 157], chorionic villi [158, 159], and decidua basalis [160, 161] appear to be more primitive, mainly based on the relatively higher expression of some of the markers typical for ESCs (e.g., SOX2, SSEA4, SSEA3, etc.). In addition, they possess a higher proliferation ability than other types of MSCs, such as bone marrow- and adipose tissue-derived MSCs [162, 163]. Hence, intensive studies have been dedicated to explore their potential clinical applications for tissue repair and organ engineering.

Besides MSCs, another type of perinatal stem cells has been successfully identified from amniotic membrane, namely amniotic epithelial cells, which form a single layer adjacent to amniotic fluid in vivo. When cultured in vitro, human amniotic epithelial cells show a cobblestone-like morphology and can proliferate for several passages [164, 165]. They express a wide range of markers commonly found in human ESCs, such as SOX2, OCT4, SSEA3, and SSEA4. Furthermore, they express low levels of Class I MHC complexes, do not express Class II MHC antigens, and lack the risk of teratoma formation. Previous studies also confirmed their ability to differentiate into all of the three germ layers.

In the field of whole organ engineering, several studies have used perinatal stem cells as seed cells. For example, using a medium specifically designed for the maintenance of the fully differentiated state of respiratory epithelial cells, Berger et al. showed that umbilical cord blood-derived progenitor cells can differentiate into respiratory epithelial cells [166], exhibiting the potential for further applications for whole lung engineering. In another study, Mirmalek-Sani et al. seeded GFP-labeled human amniotic fluid-derived stem cells on a decellularized porcine pancreas scaffold, and found that the scaffold supported the attachment, migration, and expansion of cells, suggesting that this scaffold can be used as a suitable platform for pancreas organ engineering [167].

In 2014, Oberwallner et al. used human umbilical cord blood MSCs to assess the ability of human decellularized myocardial ECM scaffold for cell seeding. They reported that the cells could attach to the scaffold, infiltrate into the porous matrix, display higher viability and metabolism than those cells cultured without the scaffold [168]. In sum, considering the many advantages of perinatal stem cells, they may be of great value for organ recellularization; further studies are needed to explore their potential applications.

Mesenchymal stem cells (MSCs)

Due to the broad clinical applicability, MSCs hold immense promise for organ engineering. MSCs can be isolated from bone marrow aspirate or other mesenchymal tissues (Fig. 6.6) [169], and have a high expansion capacity allowing the obtainment of physiologically relevant cell numbers [170]. The isolation of MSCs usually takes advantage of their plastic adherent property or by the method of density-gradient fractionation. Therefore, MSC cultures are generally a mixture of heterogeneous cell populations.

Besides self-renewal ability, MSCs possess the potential to differentiate along all mesenchymal lineages, and in some conditions the endothelial cells [171]. Despite years of research, no MSC-specific marker has been identified so far. Instead, the expression of a set of cell surface makers, including the positive expressions of CD73, CD90, and CD105, and the negative expressions of CD31, CD34, and CD45, has been used to reliably characterize MSCs. However, differences among MSCs derived from different tissues have been observed. Since a uniform characterization of MSCs is critical for data exchange among different investigators, the International Society for Cellular Therapies has proposed a set of minimal criteria to define human MSCs [172].

Because of the low expression of Class I major histocompatibility (MHC) complexes and the lack of Class II MHC molecules and co-stimulatory molecules [173, 174], MSCs cause only a mild or no immune response after implantation and are generally believed to be immune privileged. More importantly, based on their paracrine functions, MSCs can serve as potent modulators of the immune response in vivo [175]. When induced by inflammatory cytokines (e.g., interferon-γ and interleukin-1), MSCs exhibit an immunosuppressive effect, inhibiting the activation

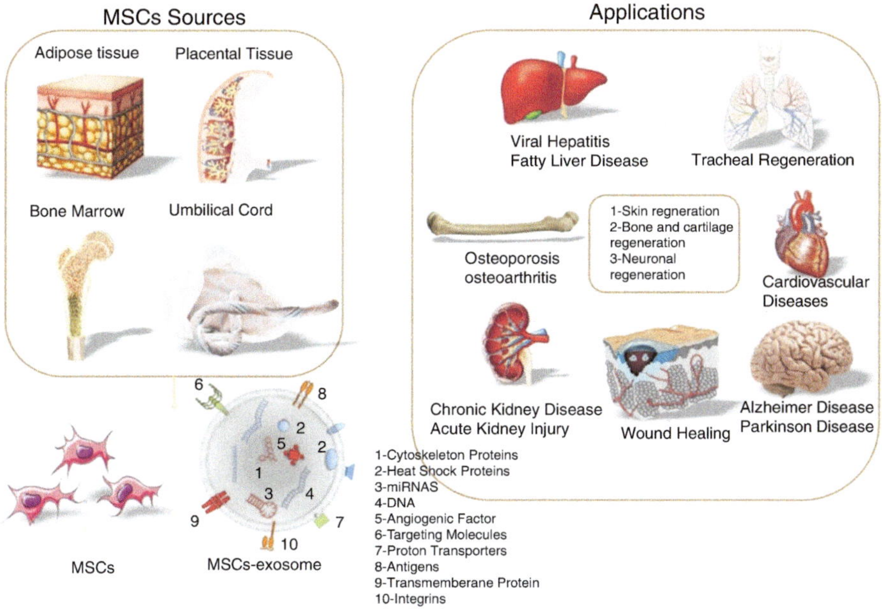

Fig. 6.6 The sources of MSCs and their applications for tissue regeneration. MSCs can be isolated from a variety of adult tissues (e.g., bone marrow, adipose tissue, etc.) as well as perinatal tissues such as placenta and umbilical cord. They have been widely used as regenerative cells for tissue repair and regeneration. Particularly, they can secrete a great number of bioactive factors such as exosomes that contain different kinds of biomacromolecules. (cited from reference [169])

and proliferation of B cells and T cells [176]. Interestingly, it is noteworthy that the expression of MHC I/II molecules can be elevated after differentiation of MSCs [177], which might lead to the possibility of an immune rejection of the allogeneic transplants.

MSCs are multipotent and can be inducted into different cell types. Interestingly, there is increasing evidence that decellularized organ scaffolds modulate the fate decision of MSCs [178–180]. For instance, when cultured on mouse decellularized liver scaffolds with hepatic differentiation media, Jiang et al. found that the 3D decellularized liver scaffolds significantly enhanced the hepatocytic differentiation of mouse bone marrow-derived MSCs when compared with those cultured on the traditional 2D cell culture flasks. Furthermore, it is reported that when the engineered constructs were transplanted into a mouse fulminant hepatic failure model, a rescued liver function was recorded in the animals [179].

In another study, Ji et al. investigated the repair potential of hepatocyte-like cells derived from MSCs that pre-differentiated on a decellularized liver scaffold. When implanted into CCl_4-injured mice, the cells showed increased survival, improved liver function, enhanced engraftment, and decreased collagen deposition to restore the architecture of liver. Furthermore, they found that the paracrine effects of grafted cells on the hepatic stellate cells and the resident hepatocytes played a critical role in the result of this treatment [178]. Likewise, other studies also reported that decellularized heart scaffolds promoted the differentiation of MSCs into cardiomyocytes [180].

The reason why decellularized organ scaffolds promote the differentiation of MSCs remains largely unknown, but it is clear that the organ-specific cell-ECM communications play a vital role in the fate decision of MSCs. It has been reported that when seeded on decellularized rat lungs, human MSCs can differentiate toward pulmonary epithelial lineages [181]. Furthermore, the organized, organ-specific stroma of liver or lung scaffolds was reported to be required for the differentiation of hepatocytes and alveolar cells [182].

In addition to differentiation into the function cells of the injured tissues, MSCs provide stromal support through their trophic functions [183]. It is well known that MSCs can secrete a range of bioactive molecules to help tissue repair and regeneration (Fig. 6.6). The secreted factors, such as growth factors, cytokines, and exosomes, are able to exert protective effects on damaged tissues, such as inhibiting scar formation, enhancing angiogenesis, suppressing the apoptosis of resident cells, and stimulating the proliferation of local stem cells in the injured area [184].

A number of animal studies have demonstrated the beneficial roles of MSCs for cardiac therapy [185–187], lung diseases [188, 189], and diabetes [190, 191]. Take heart regeneration as an example, it has been showed that transplantation of MSCs into infarcted myocardium enhanced the cardiac function of animals. In addition, the grafted MSCs have been found to be capable of differentiating into cardiomyocytes and endothelial cells in vivo [187]. However, it should be noted that only a small amount of grafted cells (approximately 2 ~ 3%) was positive for cardiomyocyte or endothelial phenotypes [185, 186]. Since the number of MSCs that differentiated into cardiac cells is too small in vivo, the transdifferentiation mechanism of MSCs is hard to explain the beneficial effects of MSC-based therapy. Therefore, the paracrine effects of MSCs have been proposed as the primary repair mechanism. According to the literature, this hypothesis has been verified in a great many preclinical studies, which demonstrated that the administration of MSC conditioned medium can recapitulate the benefits of MSC transplantation [192, 193].

To elicit better tissue regeneration, a growing number of researches use different approaches, such as hypoxic culture or cell aggregate methods, to enhance the paracrine function of MSCs. It was showed that hypoxia preconditioning of MSCs not only enhanced the secretion of paracrine factors, but also improves the survival of cells after transplantation in damaged tissues [194, 195]. These observations about the paracrine effects of MSCs open up a promising route for developing new therapeutic strategies.

6.3 Decellularization of Whole Organ Scaffolds

In whole organ engineering, the preparation of scaffolds is the first step of many to allow cells to grow, develop, and mature towards the construction of live and functional organs. In the context of scaffolds, the framework between the cells, i.e. the 3D structure of the organ and the ECM, is intricate and difficult to mimic artificially. Decellularized whole organ scaffolds do not contain any immune potential cells, but maintain the ECM components and structures, especially the microenvironmental cues for cell seeding, migration, engraftment, survival, and function, as well as the vascular structures for subsequent supply of essential oxygen and nutrients for seeded cells, thus making them ideal scaffolds for whole organ engineering. ECM has various structural and functional macromolecules with distinct biological activities. Common ECM components include collagens, elastin, laminin, fibronectin, and other proteoglycans, etc. These ECM components are evolutionarily conserved across species, which is a basis for xenogeneic transplantation. Generally, the immune responses of these ECM components are considered ignorable.

Collagen is a main component of ECM, providing strength to the ECM structures [196]. The type and content of collagens vary in different organs. Previous studies have shown that collagen I, III, and IV are presented within the decellularized scaffolds [40, 85]. Collagen I is generally most abundant, while collagen III is identified in elastic connective tissues/organs, such as the vessels, the skin, and the lung. Collagen IV is presented at the basement membrane which supports the adhesion and function of endothelial cells in the vasculature [197]. Collagens have the RGD motif, which is crucial for cell adhesion [198]. They also act as a pool to retain a variety of growth factors [199]. Elastin is another critical ECM component providing mechanical strength as well as elasticity that is a requisite to the elastic movement of tissues/organs, for example, the lung [200]. In decellularized scaffolds, elastin structures are generally retained because of its good stability and insolubility in most decellularization agents [201]. Laminin is usually present in the basement membranes [202], which defines the architecture and integrity of decellularized scaffolds [203]. Laminin plays a significant role in the adhesion, migration, and differentiation of reseeded cells. Fibronectin is a typical glycoprotein in the ECM, especially in the basement membranes, and is also associated with the adhesion, proliferation, migration, and differentiation of reseeded cells [204]. Like elastin, fibronectin is generally insoluble when exposed to most decellularization agents, thus it is often abundant in decellularized scaffolds. Cells expressing β1 integrin can tightly attach to collagens, laminins, and fibronectin [198]. In addition to fibronectin, other proteoglycans (e.g., decorin and biglycan) and glycosaminoglycans (e.g., heparan sulfate and hyaluronic acid) are also important ECM components that are capable of directing cell differentiation and protecting growth factors from proteolytic degradations [205], which eventually affecting the biology of seeded cells within the scaffold.

Though evolutionarily conserved, some concerns have been proposed about the immunogenic potentials of ECM components. For instance, there are several evidences demonstrating that collagens and elastin can induce immune responses, mainly by activation of T-cells, induction of proinflammatory cytokines, and formation of antibodies [206, 207]. Laminin, on the contrary, showed a beneficial limitation of immune responses [208]. Likewise, some proteoglycans have shown their ability to reduce immune and inflammatory responses [209]. In short, every ECM component can lead to positive and/or negative immune responses; therefore, the overall immunogenicity of decellularized whole organ scaffolds needs careful investigations. However, it should be noted that decellularized whole organ scaffolds are still the best choice for whole organ engineering at current stage, mainly due to their outstanding merits such as the unique property to closely mimic the architecture of native organs.

To fabricate scaffolds for subsequent whole organ engineering, whole organ decellularization is generally performed to remove cellular components while preserving the ECM components. In this section, we will discuss the methods used to prepare and evaluate decellularized whole organ scaffolds, and provide examples of pivotal decellularized whole organ scaffolds, including heart, liver, lung, kidney, and pancreas.

6.3.1 Decellularization Methods

Decellularization is a process that interrupts cellular interactions with the ECM or with each other, and thus removes all of the cellular components of a tissue or organ, while maintaining the surface topography as well as the resident ligands of native ECM components [20, 210]. Figure 6.7 shows a stepwise removal of cellular and extracellular antigens [211], representing the decellularization process. In fact, it is quite challenging for current decellularization methods to retain all of the important molecules of native ECM, such as glycosaminoglycans, growth factors, and surface ligands. To create a decellularized scaffold suitable for whole organ engineering, the methods used for organ decellularization should further emphasize on their ability to preserve the 3D architecture as well as the vascular tree of native organs.

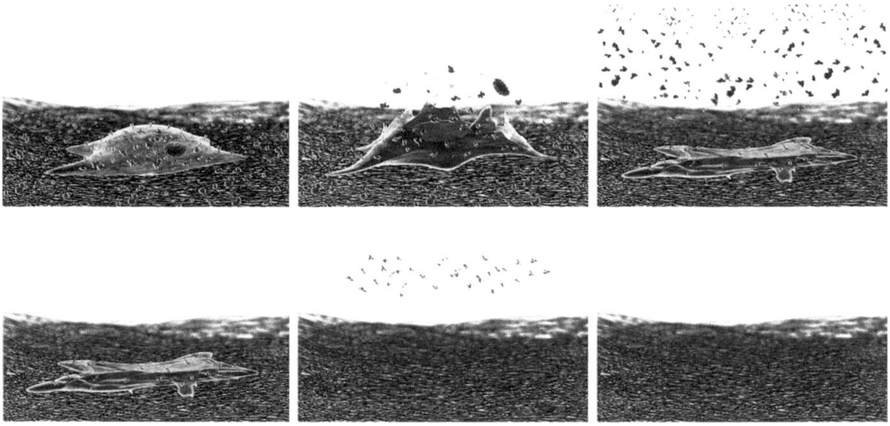

Fig. 6.7 A schematic drawing showing the stepwise removal of antigens. Native tissues contain cell, nucleus (purple), hydrophilic (red) or lipophilic (blue) antigens, and ECM (black). After cell lysis, hydrophilic antigens and nuclei were removed, then the lipophilic antigens were removed, resulting in acellular ECM. This figure also demonstrates the concept of decellularization. (adapted from reference [211])

Typical decellularization protocols are composed of cell lysis by physical and/or chemical treatments, dissociation of the cytoplasmic and nuclear components from the ECM by chemical and/or enzymatic agents, and final removal of the debris from the tissue or organ [4, 210]. Nevertheless, for different types of tissues or organs, the decellularization protocols can differ greatly from each other. The methods for decellularization can be broadly divided into the following categories: the physical methods, the chemical methods, and the enzymatic methods. For most protocols, combination of different methods and agents are often adopted for higher efficiency in cell removal and ECM retention.

For long-performed decellularization of two-dimensional tissues, such as intestine, pericardium, and amnion, the popular decellularization methods include freezing/thawing, osmotic shock, mechanical removal of redundant tissue layers (for example, the muscle or the submucosa), and treatments with detergents and specific solvents followed by adequate rinsing [212–214]. However, the ultimate goal of organ decellularization for subsequent whole organ engineering applications is the removal of all cellular materials and the preservation of the ECM's composition, architecture, and biological activity [4]. Particularly, minimal damage to vasculature is a must for following reendothelialization, anastomosis, and implantation of the scaffolds. Hence, a "gentle" decellularization process is required for whole organ scaffold fabrication.

To this end, perfusion, whether by antegrade or retrograde, has been developed to decellularize whole organs, aiming to retain the 3-D architecture of the organ as well as the ECM composition and topography [20]. The vasculature within the organs can efficiently deliver oxygen and nutrients to cells, which can also be adapted to deliver decellularization agents efficiently into the organs and transport cellular materials from the tissue [20]. The perfusion process for decellularization initially delivers decellularization reagents into the organs to lyse the cells, then transport the cellular debris out, and finally thoroughly rinse with saline or water. The rinsing step also contributes to removing the remaining detergents or enzymes used in the decellularization. Generally, this method is expected to gain the organ scaffold in a semitransparent appearance with near intact macro-architecture and micro-architecture of the whole organ. Perfusion-based decellularization has been performed on hearts, livers, lungs, kidneys, and pancreas from various species ranging from rodents to primates to obtain acellular whole organ scaffolds [38, 40, 85, 86, 91, 129, 131, 167, 215–229]. At current stage, pressure- or flow-controlled perfusion is considered as the most appropriate method for whole organ decellularization.

In the perfusion decellularization process, all the extrinsic factors have both advantages and disadvantages. Especially, the disadvantages should be considered thoroughly when setting up a decellularization protocol. For example, freeze–thawing can influence the ultrastructure of the scaffolds, pressure may influence the mechanics of the scaffolds, alkaline and acidic solutions can partially degrade ECM components, ionic detergents can disrupt protein–protein interactions and can be difficult to be cleared from the matrix, and alcohol can increase the stiffness of ECM by crosslinking collagen [230, 231]. It should be understood that all decellularization agents and methods will change ECM composition and lead to somewhat ultrastructure disruption. In decellularization, the objective is minimization of these undesirable effects rather than complete avoidance [20]. An optimized combination of the available methods and agents is beneficial.

Effective decellularization depends on a lot of factors, including morphology, thickness, density, cellularity, and lipid content [23, 232]. It is important to realize that although different protocols for decellularizing different organs have been reported, no single protocol is suitable for every tissue/organ, the agents and methods must be based on specific organ types/origins to optimize the decellularization outcomes. Typical perfusion setup, commonly used agents and their effects on decellularization, and common sterilization methods are discussed below.

6.3.1.1 Set up a Perfusion System

A perfusion system is firstly set up to perform perfusion decellularization. As exemplified in Fig. 6.8, the system is generally consisted of reservoir, pump, organ container, and tubes. Organs are kept in containers with a suitable environment, enabling perfusion of the organ with solutions pumped from the reservoir through its vasculature. It should be noted that bubbles may adversely affect the decellularization outcome, hence air trap is usually necessary in the perfusion system. In addition, the perfusion solutions may be discarded directly, or may be circulated, based on the organ type and decellularization stages. As exemplified by a porcine liver decellularization protocol in Fig. 6.9, the whole decellularization process involves stepwise procedures.

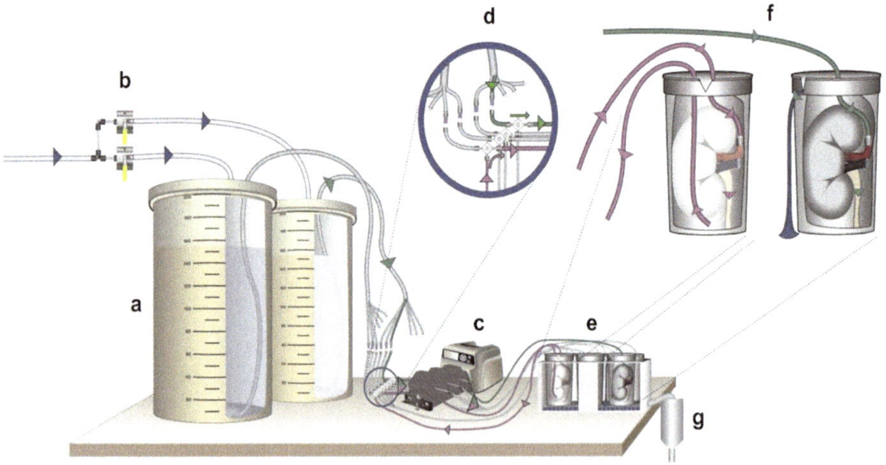

Fig. 6.8 Layout of a typical perfusion system for decellularization. Basically, a perfusion system is consisted of reservoir, pump, organ container, and tubes. (**a–b**), reservoir containing decellularization or washing solutions. (**c**), a peristaltic pump. (**d**), interchangeable connectors and a 3-way valve. (**e**), organ container. (**f**), solutions exit the organ and overflow into the general collecting bin or into (**g**) a drain for recirculation. (adapted from reference [227])

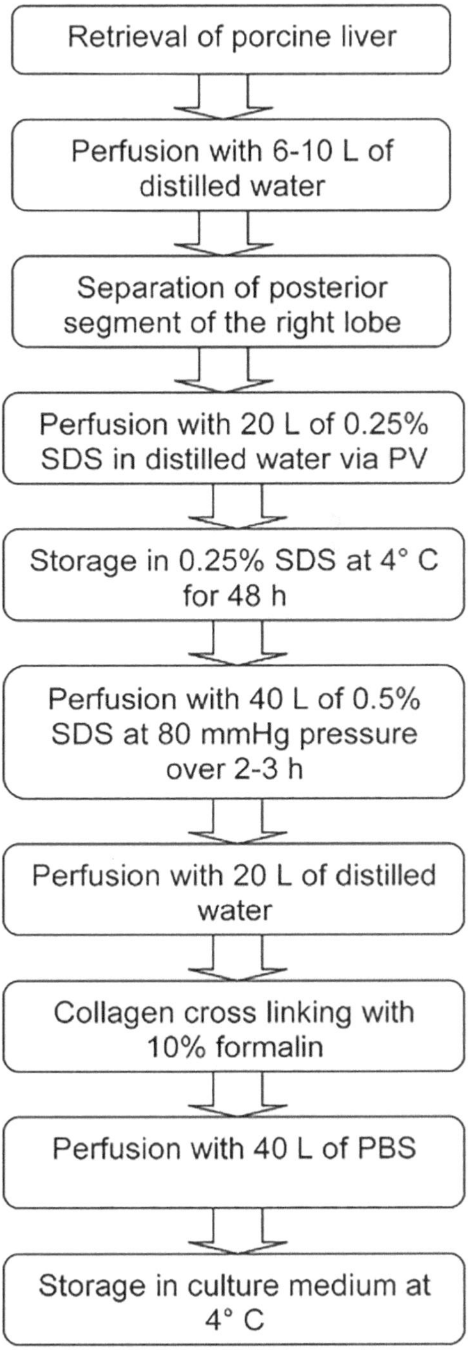

Fig. 6.9 A stepwise protocol in whole liver decellularization. The decellularization process often involves multiple stepwise procedures including pre-treatment (step 1 ~ 3), core decellularization (step 4 ~ 6), washing (step 7), and post-treatment (step 8 ~ 10). (adapted from reference [215])

6.3.1.2 Factors Used in Whole Organ Decellularization

The decellularization factors can be roughly categorized into physical, chemical, and enzymatic factors. Each factor will contain various agents with weak to strong efficiency as well as adverse effects due to different mechanism of action.

Physical Factors

In perfusion decellularization, physical factors include temperature, hydrostatic and hydrodynamic force, and osmotic pressure. Decellularization is usually performed at room temperature, but sometimes the scaffolds can be cooled to strengthen ECM preservation, or be warmed while applying enzymatic treatments [40, 91, 229, 233]. Except for constant temperature, repetitive freeze–thaw cycles can be utilized throughout decellularization and do not cause significantly loss of ECM components [129]. Applying straight pressure to tissue can also lyse cells, but the effectiveness is restricted to tissues or organs without dense ECM such as lung and liver [41]. The perfusion-associated pressure could lead to punctures within the vascular networks [23], hence the flow rate when performing perfusion needs to be considered. Tapias and Ott recommended to perform perfusion-decellularization at constant low physiological pressures with satisfactory delivery of decellularizing agents, since higher pressure leads to excessive mechanical forces which damages ECM [34]. Constant pressure, but not constant flow, can help to maintain the vascular integrity during decellularization [37]. Hypertonic and hypotonic solutions can induce osmotic shock and thus facilitate cell lysis.

Chemical Factors

Chemical agents are widely used in decellularization, but they involve different mechanisms of action. Water is the universally applicable solvent for decellularization. For fatty amorphous tissues and organs, such as adipose tissue, small intestine, brain, and pancreas, other solvents like alcohols and acetone are often used to remove lipids [20]. However, such solvents can fix and damage the ECM ultrastructure [20]. Acid (such as acetic acid or peracetic acid) and basic (such as ammonium hydroxide) agents can also effectively disrupt cell membranes and intracellular organelles, but these treatments also disrupt key molecules like glycosaminoglycans [20]. Particularly, acid agents can subtract the collagen from the matrix, which makes the main component of ECM.

Detergents are often the core decellularization agents. All detergents enhance cell lysis and cellular component solubilization. According to their charge, the detergents are categorized as non-ionic, ionic, or zwitterionic. Non-ionic detergents are considered as non-denaturing, which helps to maintain the biologically active of proteins [234]. A commonly used non-ionic detergent is Triton X-100. It has a hydrophilic head group with zero charge, and it disrupts lipid's interactions with other molecules but does not affect protein–protein interactions [234]. Ionic detergents can be represented by sodium deoxycholate and SDS. They have a negatively charged hydrophilic head group, and they generally show greater decellularization efficiencies (solubilize cytoplasmic and nuclear membranes) but also greater damage to the ECM components and structure (denature ECM proteins, disrupt native architecture) for subsequent in vitro recellularization and in vivo remodeling [234]. For example, SDS-decellularized scaffolds have weakened cellular attachment and proliferation [235]. Besides, SDS can be difficult to be removed after decellularization, residual SDS will certainly induce toxicity to reseeded cells. Even after intense risings, SDS-treated scaffolds showed less cell attachment and penetration, which is likely due to the change of ECM compositions caused by SDS. Despite the deleterious effects, decellularization with SDS can still result in a recellularized functional organ eventually. The possible explanations are that the protocols have different concentration of detergents, duration of treatment, pressure within the perfusion system, and the amount of cells to be reseeded which can compete with the adverse impacts of SDS on the scaffold [235]. Triton-X100 and SDS represent the most commonly used detergents, but other detergents have also been reported. A representative of zwitterionic detergents is 3-[(3-Cholamidopropyl) dimethylammonio]-1-propanesulfonate (CHAPS), which shows both non-ionic and ionic properties with a net charge of zero [38, 2366].

Chelating agents, such as ethylene glycol tetraacetic acid (EGTA) and ethylenediaminetetraacetic acid (EDTA) [234, 237], can also accelerate cell detachment. They can separate metal ions, especially the calcium, to promote the dissociation of cells from each other and from the ECM proteins. They are frequently utilized with enzymatic agents, like trypsin as discussed below, to break the cell-matrix linkages.

Enzymatic Factors

Protease agents can disrupt the protein–protein interactions, promote the detachment of cells from the ECM scaffolds, but also damage some protein components within the ECM. Among the various potential enzymes, trypsin is the most frequently used and effective enzyme for cell detachment and lysis [234]. Trypsin can cleave peptide chains at lysine or arginine carboxyl side, except when the amino acid is followed by proline. In fact, trypsin has limited cleavage of collagens [238], meaning the major ECM components can tolerate this enzyme in decellularization. Since DNA residual can lead to immune responses when transplanting the scaffold, and DNA content is one standard to evaluate the decellularization outcomes which is discussed below, DNase is also frequently used in the stepwise decellularization process to break down the DNA components [235]. After enzymatic treatments, the organs must be adequately rinsed to clear any remaining enzymes.

6.3.1.3 Sterilization of the Whole Organ Scaffolds

Danger of infection is unavoidably associated with the transplantation. For subsequent recellularization and further application, the decellularized whole organ scaffolds must be pathogen free, including both viruses, bacteria and fungi. Virus transmission is a nonnegligible problem in whole organ transplantation [235]. For example, potential porcine viruses include hepatitis E virus, porcine endogenous retrovirus, etc. [37]. Though after decellularization, viruses are less of a concern, remaining viral components can still potentially trigger immune responses or even lead to infection. Hence, proper source control, such as standardized breeding and rigid viral tests, is preferred to eliminate the threats of zoonotic viruses.

In decellularization, despite following a rigorous sterile whole process which is difficult to address, the whole organ scaffolds usually require a sterilization treatment to remove bacterial and fungal pathogens. One widely used approach is to treat the scaffolds with broad-spectrum antibiotics and antifungal drugs [37]. Ionizing radiation can also be alternatives, but these methods can alter ECM properties to a certain extent, thus affecting subsequent biomedical applications [37]. A crucial step as it is, sterilization should be efficient, but at the same time should have only a lowest biological impact on the scaffold.

6.3.2 Evaluation of Decellularization

The decellularization protocols should be adapted for each organ to remove cellular components that will trigger an immune response, to retain the favorable properties of the scaffold for following recellularization, and to rinse the scaffold adequately so that the scaffold will be biocompatible. After decellularization, it is necessary to comprehensively evaluate the resulting whole organ scaffolds to determine the integrity of the vascular structures, the acceptable amount of residual decellularization agents, the effectiveness of decellularization, and the maintenance of ECM integrity with both qualitative and quantitative assays. As every decellularization agent can damage the ECM to a certain extent when removing the cellular materials, optimization of the decellularization protocol for specific organ is of great significance.

6.3.2.1 Integrity of Vascular Structures

The vascular networks are critical in the decellularized whole organ scaffold for subsequent recellularization and functional restoration. Different decellularization protocols, such as detergent and delivery method used, can result in various preservation of the vascular structures. In order to evaluate the integrity of the vascular structures within the 3D decellularized whole organ scaffold, multiple techniques have been applied such as micro-CT, perfusion of dyes (colorful, fluorescent, or contrast), angiography, and corrosion casting [37, 38]. In addition, SEM was also

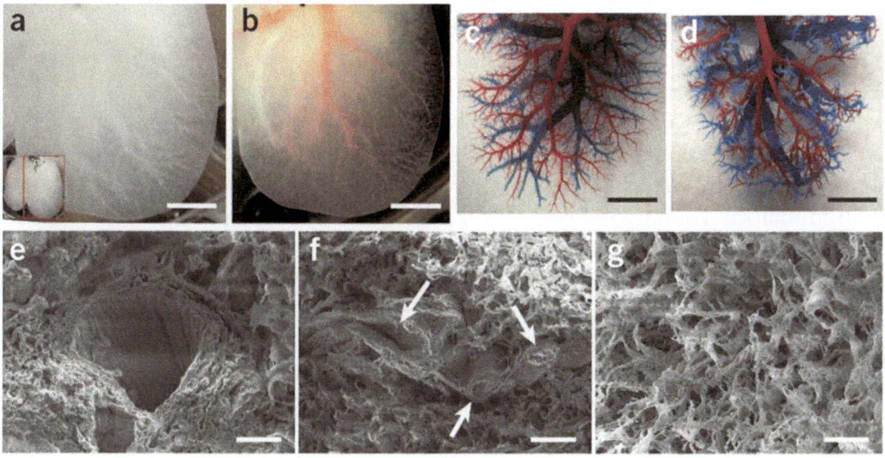

Fig. 6.10 Preserved vascular structures in decellularized livers. Integrated vascular structure is an important advantage of the decellularized whole organ scaffolds, which can be evaluated by several methods. (**a**), general view of a decellularized liver lobe with distinguishable vascular structure. (**b**), the vascular structure perfused with red dye. (**c**), corrosion cast model of a normal liver lobe. (**d**), corrosion cast model of a decellularized liver. The red part shows the portal vasculature, while the blue part shows the venous vasculature. (**e**), a vessel by SEM. (f), bile duct-like structures (arrows). (**g**), parenchymal ECM without hepatocytes. Scale bars = 10 mm in (**a**, **b**); = 5 mm in (**c**, **d**); = 20μm in (**e**–**g**). (adapted from reference [40])

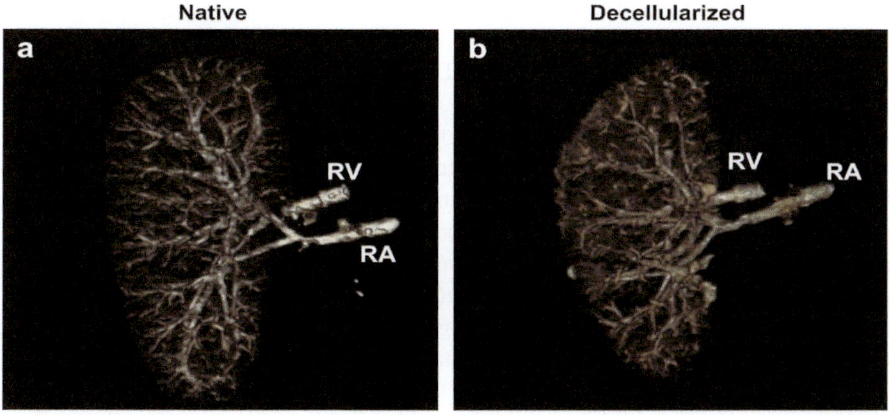

Fig. 6.11 The integrity of vasculature of decellularized whole organs can also be evaluated by non-invasive methods. In this figure, the vasculature in decellularized kidneys was imaged by CT. (**a**) native kidney. (**b**) decellularized kidney. *RA* = renal artery, *RV* = renal vein. (adapted from reference [227])

used to demonstrate the microstructures of decellularized scaffolds [131]. Figure 6.10 showed the preservation of vascular networks in a decellularized liver by colorful dye and corrosion casting, as well as the microstructures by SEM. Figure 6.11 showed the preserved vasculature in a decellularized kidney by CT imaging.

6.3.2.2 Detergent Residual

As mentioned earlier, residual detergent can damage the repopulating cells. Hence, adequately rinsing of the decellularized whole organ scaffold is an important step in the stepwise decellularization protocols. Indeed, detergent like SDS is somewhat challenging to be detached because of its ionic nature [239]. A previous study showed that the volume of solutions for valid cell removal and following clear the debris and residual agents for porcine kidneys can reach 86 L [221]. It is proposed that the solution volume required for effective decellularization and rinsing of the whole organ scaffold is approximately 100 times the volume of the organ [5]. Detection of residual detergent has utilized multiple analytic methods, such as measuring pH and conductivity, applying high performance liquid chromatography (HPLC), etc. [240]

6.3.2.3 Cellular Components

After decellularization, remaining cellular components within the scaffolds could cause immune responses. DNA residual is a hallmark for evaluating the residual cellular components. In fact, residual DNA can induce immune responses. DNA sensors, like pattern recognition receptors (PRRs), are found on endosomes, in cytoplasm and nucleus, capable of recognizing DNA and activating a variety of signaling pathways resulting in expression of inflammatory genes, and even cell death [235].

Hence, measurement of the amount of DNA residual within the scaffold is widely accepted to reflect the efficiency of cell removal. However, despite including treatment of DNase with different exposure times and concentrations in the decellularization process, it is not likely to achieve complete nucleic acid removal. Crapo et al. have proposed a standard addressing the DNA residual in decellularized scaffold, i.e. the amount of DNA should be less than 50 ng per mg dry ECM weight, the DNA fragments should be shorter than 200 bp, and no visible nuclear components [20]. Specifically, the DNA amount can be quantitatively measured by fluorometric assays, the length of DNA fragments can be measured by agarose gel electrophoresis, and the nuclei can be visualized via DAPI or H&E staining.

To evaluate the degree of decellularization, some other cellular components should also be assessed. Oligosaccharide and cellular proteins within the scaffold can also lead to immune responses, especially in dense tissues where cellular components are difficult to be removed from the surrounding ECM [235]. Theα-Gal is an oligosaccharide presented on cytomembrane in most mammals, but not in primates [235]. However, humans have antibodies against the Gal epitope, which can activate complement cascade and lead to hyperacute rejection upon recognition of the Gal epitope [235]. Hence, Gal epitope recognition has been a barrier for xenogeneic transplantation, including decellularized xenogeneic scaffolds potentially carrying the Gal epitope. There have been efforts to generate Gal knockout animals promising for reduced inflammation [235]. It has been reported that the residual gal-epitope in the scaffolds is not sufficient to activate complement or bind antibody, possibly due to their universally scattered distribution and relatively low quantity [4]. The major histocompatibility complex (MHC), also known as the HLA system, a protein complex that helps to recognize self and non-self and thus directs immune responses, exemplifies the residual cellular proteins [235]. This complex has three subtypes as HLA I (in most nucleated cells), HLA II, and HLA III (in specific cell types). In whole organ transplantation, HLA mismatching is the major reason for rejection [235]. Though its presence in decellularized scaffolds is rarely evaluated because a residual DNA-qualified scaffold is considered few residual cell components are removed, it was reported that more than 300 proteins which are normally found inside the cell were identified in the ECM after decellularization [241]. In animal experiment, antibody against the residual proteins were found [241]. Nevertheless, the presence of Gal epitope and MHC should be noted which can be detected by immunoassays such as antibody-based microscopic imaging.

It should be noted that host-transplant immune responses are not always unwanted events. The effects of xenogeneic ECM on the macrophages may induce a necessary M2 phenotype to enhance the integration and remodeling of the ECM scaffold [242].

6.3.2.4 ECM Components

Generally, one of the major objectives of whole organ decellularization is to preserve the ECM components, which can significantly influence cell behaviors such as attachment, proliferation, differentiation, and constructive regeneration [243]. All aspects of ECM, including the organ-specific 3D ultrastructure, topology, biomechanics, and composition, contribute to the biological properties of ECM scaffolds [20]. Akhyari et al. concluded that there are no ideal decellularization protocols, since preservation of ECM proteins and removal of cellular components are inevitably contradictory [244]. Cell removal and ECM damage increase simultaneously with the strength of the decellularization agents, the flow rate of perfusion, and the duration of treatment. Anyway, qualitative and quantitative assessment of the ECM components are essential in preparation of decellularized whole organ scaffold.

The ECM components can be visualized by various histology staining methods [245]. Both scanning and transmission electron microscope are employed to identify the ultrastructure within the scaffolds [245]. Immunohistochemistry and immunofluorescence are also widely applied to qualitatively evaluate structures and components of the ECM [245]. Quantitatively, enzyme-linked immunosorbent assay (ELISA) has long been used to analyze retained growth factors [5]. Figure 6.12 shows characterization of decellularized pancreas ECM by immunofluorescence staining, Alcian blue staining, and quantification assays. Furthermore, high throughput proteomics can also be utilized to analyze the ECM components [246, 247]. In fact, a quantitative proteomic analysis revealed that decellularized liver may contain many kinds of non-matrix proteins (Fig. 6.13) [247].

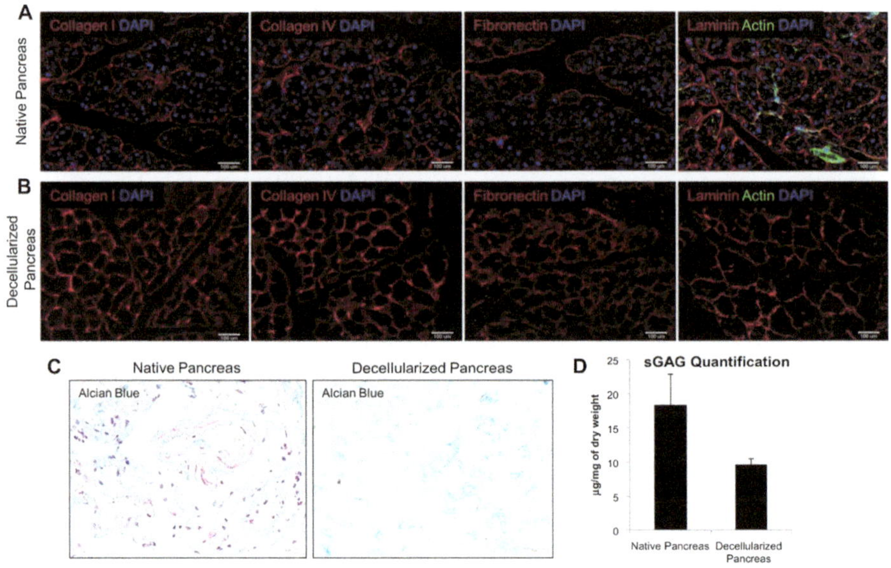

Fig. 6.12 Characterization of ECM components can be easily done by histological staining and conventional biochemical methods. In this figure, the components of decellularized pancreas ECM were identified by histological staining and quantified by photometric assay. (**a**), immunofluorescence staining of native pancreas. (**b**), immunofluorescence staining of decellularized pancreas. ECM components (collagens, fibronectin, and laminin) were well preserved, while cellular components (cytoskeleton and nuclei) were efficiently removed. (**c**), Alcian blue staining demonstrated the retention of GAGs after decellularization. (**d**), quantitative measurement of GAGs in native and decellularized pancreas by photometric assay. (adapted from reference [217])

6.3.2.5 Mechanics of the Decellularized Scaffolds

The stiffness of decellularized whole organ scaffolds may increase or decrease due to the loss of cellular components and damage of ECM components [217, 248]. Since that scaffold stiffness plays a role in cell fate decision, particularly for mechanotransduction related stem cell differentiation and phenotype maintenance [37], and that organ compliance and elastance can greatly affect the final functions of recellularized organ [86], special attentions are needed to the mechanics of the scaffold, reflecting the retention of ECM and usability of the scaffold.

Many traditional engineering techniques are applied to assess the biophysical features of the decellularized whole organ scaffolds. For example, atomic force microscopy and nano-indention are used for microscale assessment, while uni--orbi-axial mechanical testing is used for macroscale assessment [37, 249]. Besides, multiphoton microscopy combined with image correlation spectroscopy shows potentials as a non-invasive method for prediction of the decellularized scaffolds' mechanical properties [37].

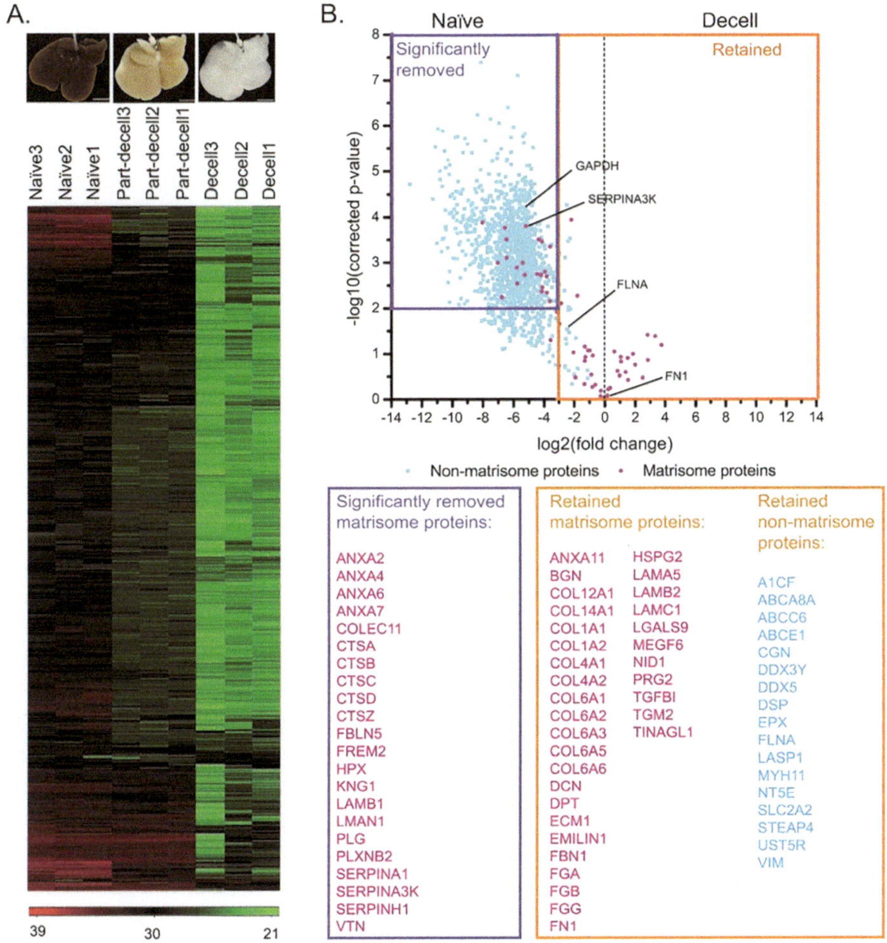

Fig. 6.13 Quantitative proteomics analysis can provide high-through data to characterize ECM proteins. An example of quantitative proteomics analysis of decellularized livers is shown. (**a**), Images of native, partly decellularized, and decellularized livers (scale bar: 10 mm), and a corresponding heatmap showing the changes in protein abundances at each stage. (**b**), Volcano plotting of quantitative comparison of the native and decellularized livers. Some non-matrix proteins were identified in decellularized livers. (adapted from reference [247])

6.3.3 Decellularized Whole Organ Scaffolds

An adequate source of donor organ for preparation of whole organ scaffolds remains challenging. Preparation of decellularized whole organ scaffolds usually starts with small animals such as rat, and then step by step scales up to the source animals, like goat, monkey, and pig. Among the source animals, organs from pig can have similar size and physiology as human organs, less ethical issues than primate organs

[5, 37]. Meanwhile, regarding the commercial production and clinical application, modern breeding industry allows for controlled and standard conditions to ensure the health and availability of donor organs.

Decellularized whole organ scaffolds have been obtained for hearts (rat, pig, and human) [85, 180, 229], liver (rat, goat, and pig) [40, 215, 226, 228], lungs (rat, pig, monkey, and human) [86, 216, 220], kidney (rat, monkey, pig) [131, 218, 221, 222, 227, 248], and pancreas (rat, human) [167, 217, 250]. Major examples for different decellularized organs are discussed in detail below. It should be noted that despite the various protocols and results reported for each organ, whether the decellularization protocols work needs to be evaluated not just by the scaffolds themselves, but more conclusively by subsequent recellularization procedures and final physiological functions of the engineered whole organs.

6.3.3.1 Decellularized Heart Scaffolds

The first report on perfusion-decellularized cardiac ECM scaffolds to bioengineer a whole heart was published in 2008 [85]. Ott et al. decellularized rat hearts by perfusion to generate a 3D scaffold with intact native organ architecture (Fig. 6.14). The aorta was cannulated for following retrograde coronary perfusion with heparinized PBS with 10μM adenosine for 15 min at a pressure of 77.4 mm Hg. Then, the hearts were perfused with 1% SDS (or 1% polyethylene glycol, or 1% Triton X-100 as comparations) for 12 h, with water for 15 min, and with 1% Triton X-100 for 30 min in sequence. Final rinsing was perfusion with antibiotic-containing PBS for 124 h. The results showed that perfusion with SDS yielded a translucent white heart scaffold. Histological staining revealed no remaining nuclei or cellular components. More than 96% DNA content was removed, while the vascular network was intact. Collagen I/III, laminin, and fibronectin were retained after decellularization. The myocardial ECM fiber was well preserved. After decellularization, the longitudinal stress but not circumferential stress increased, the tangential modulus increased while the membrane stiffness had no significant changes.

In 2010, decellularizing larger hearts from a porcine origin was reported by Wainwright et al. [229]. After primary cleaning and washing, the porcine heart was stored at −80 °C for at least 16 h to aid in cell lysis, then thawed in water at room temperature. The heart was placed in a container and submerged in water. The aorta was cannulated to perform perfusion. Initially recirculated perfusion was performed with the water for 15 min at a rate of 1 L/min. Then the heart was perfused with 2× PBS for 15 min at the same rate. The core decellularization was performed with 0.02% trypsin/0.05% EDTA/0.05% NaN_3 solution at 37 °C for 2 h at the same rate followed perfusion with 3% Triton X-100/0.05% EDTA/0.05% NaN3 solution and 4% deoxycholic acid solution for 2 h at a rate of 1.3 L/min at room temperature. Between each perfusion, non-recirculating water and recirculating 2× PBS was used to wash the heart. Sterilization was performed with 0.1% peracetic acid/4% ethanol for 1 h at a rate of 1.7 L/min. The scaffold was neutralized with PBS

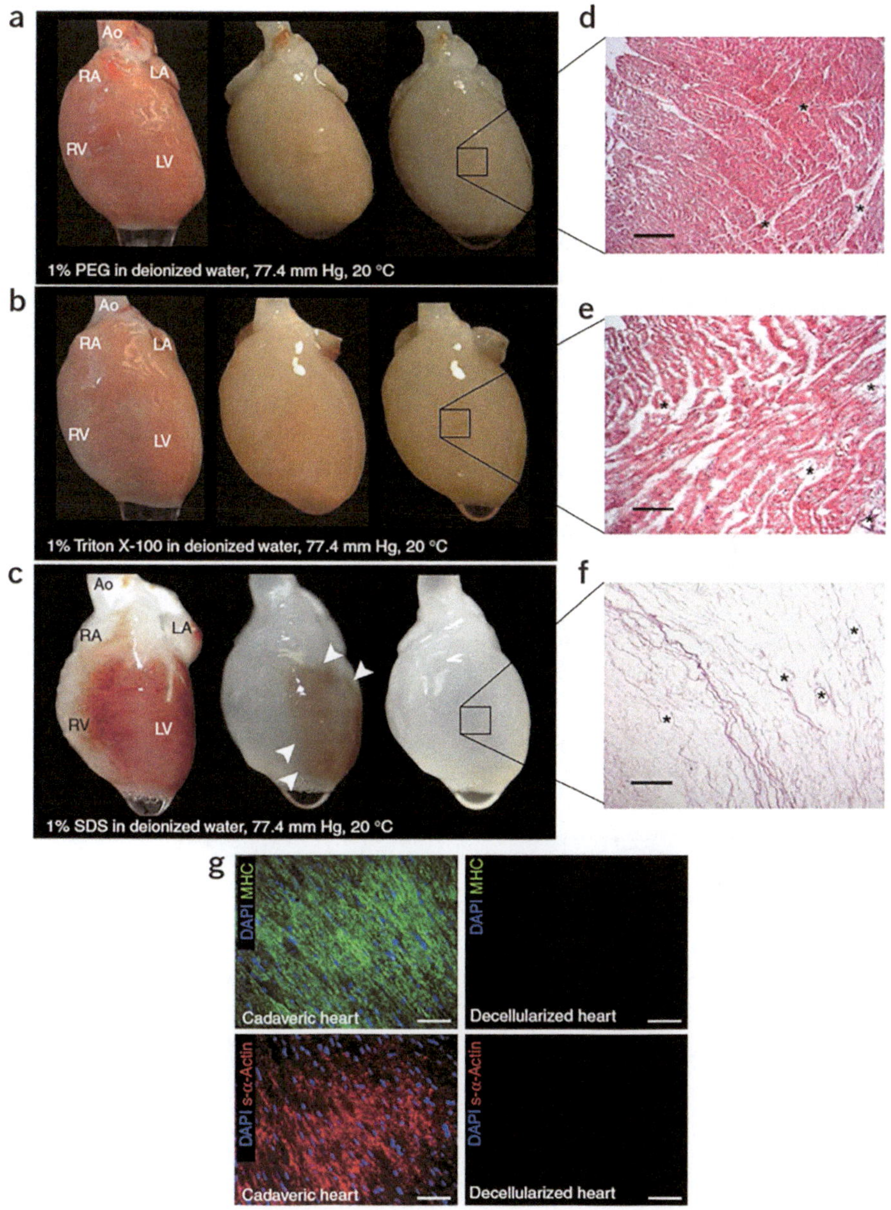

Fig. 6.14 The first reported decellularized whole organ scaffold for subsequent whole organ bioengineering. Whole rat hearts were decellularized by perfusion-based method. Macroscopic views and histological staining images were presented. (**a–c**), Images of the hearts during decellularization process with different protocols. (**d–f**), H&E staining of sections from LV showed different decellularization efficiencies. Asterisks indicated the vasculature conduits. Scale bars = 200μm. (**g**), Immunofluorescence staining of native and SDS-decellularized heart sections. Cellular components (nuclei, cardiac α-myosin heavy chain, and sarcomeric α-actin) could not be detected after decellularization. Scale bars = 50μm. (adapted from reference [85])

twice and rinsed with water three times. The authors pointed out that though the flow rate was increased during decellularization, the perfusion pressure dropped. H&E and DAPI staining demonstrated no residual nuclei or dsDNA. Movat's pentachrome staining showed the preservation of collagen I, collagen III, and elastin after decellularization. The basement membrane structures were found at endocardium, myocardium, and coronary arteries by collagen IV staining. Quantitative analysis showed that 92% of the DNA components were removed, while the GAGs and elastin were well preserved in the decellularized ECM. The decellularized heart showed no significant changes in mechanical behaviors compared to the native myocardium.

Many other efforts have been made to improve and optimize the protocols for whole heart decellularization. Akhyari et al. compared four decellularization protocols with rat hearts [244]. All protocols preserved the macro-architecture and showed biocompatibility, differences in the micro-composition (elastin, collagen IV, GAGs, DNA residual) were noticed. Other protocols also reported different cannulating routes, different concentrations and combinations of detergent agents, different duration of perfusion, etc. [219, 224].

6.3.3.2 Decellularized Liver Scaffolds

In 2010, two groups both reported preparation of decellularized rat liver scaffolds. Uygun et al. performed perfusion through the portal vein (Fig. 6.15) [40]. The frozen livers were firstly thawed at 4 °C, perfused with PBS overnight, then with 0.01%

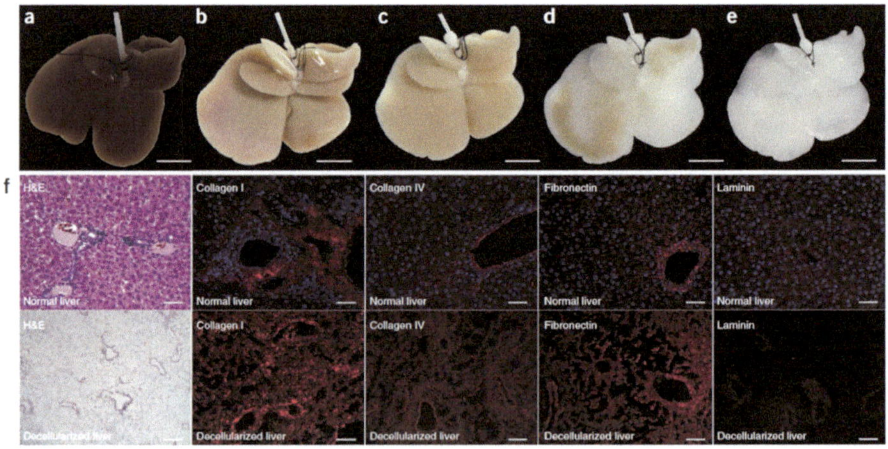

Fig. 6.15 Uygun et al. showed gradual macroscopic records of decellularized whole rat liver, and final histological staining images of the decellularized whole rat livers. (**a–e**), Images of decellularizing rat livers at different time points (0 h, 18 h, 48 h, 52 h, and 72 h, respectively). Scale bar = 10 mm. (**f**), H&E staining and immunofluorescence staining of native (top row) and decellularized (bottom row) livers. Scale bar = 100μm. (adapted from reference [40])

SDS, 0.1% SDS, and 1% SDS for 24 h each. Following procedures included perfusion with water for 15 min, with 1% Triton X-100 for 30 min, and with PBS for 1 h. After decellularization, histological and fluorescent observation demonstrated no nuclei or cytoplasmic components, but preservation of ECM components (collagen I and IV, fibronectin, and laminin). Quantitative analysis showed retention of all fibrillar collagen and approximately half GAGs, while more than 97% of DNA contents were removed. The microvasculature remained intact after decellularization. In the other study, Shupe et al. performed perfusion through the inferior vena cava before excision of the liver [226]. The liver was firstly perfused with 100 ml PBS, then 300 ml 1% Triton X-100 in PBS, 300 ml 2% Triton X-100 in PBS, and 3% Triton X-100 in PBS at a rate of 5 mL/min, followed by 300 ml 0.1% SDS in PBS, another 300 ml PBS, and final 10 ml fetal bovine serum. After decellularization, no nucleic contents were observed, but collagen IV and laminin were preserved. The vessel was strong enough to maintain cannulation, and the vascular network was able to traffic cells.

Baptista et al. decellularized livers from different animals including pig [91]. Barakat et al. in detail reported the procedures and outcomes of porcine liver decellularization (Fig. 6.16) [215]. The livers were harvested from pigs weighing 25–35 kg, and perfusion was performed via the portal vein. Firstly, 0.25% SDS solution was infused into *the liver* until the effluent became clear. Then the liver was stored in 0.25% SDS solution for 48 h at 4 °C, perfused with 40 L 0.5% SDS at 80 mmHg pressure, and finally washed with 20 L water. After decellularization, the liver had maintained the gross appearance and size, but became white and translucent with intact vascular network while weighing only 70% to 80% the original liver. Histological examination revealed that ECM components (collagen I and IV, fibronectin, and laminin) were maintained, but the nuclei were not completely removed.

As for optimization of liver decellularization protocols, Kajbafzadeh et al. compared diffusion and perfusion techniques and five protocols with ovine livers [251]. They concluded that perfusion technique can achieve better outcomes, and that perfusion with NH_4OH solution followed by Triton X-100 is the most accurate and appropriate protocol for liver decellularization. Other protocols have also been continuously reported with satisfactory results [26, 252].

6.3.3.3 Decellularized Lung Scaffolds

In 2010, four groups firstly reported decellularization of lung from rodents [38, 86, 129, 223]. Ott et al. performed retrograde perfusion via pulmonary artery (Fig. 6.17) [86]. Rat lungs were perfused with heparinized PBS, then with 0.1% SDS for 120 min, with water for 15 min, with 1% Triton-X-100 for 10 min, and finally with antibiotics in PBS for 72 h. After decellularization, histological analysis showed no remaining nuclei or intracellular elements, while the ECM proteins (collagens, proteoglycans, elastic fibers) were preserved. The matrix architecture of airways, alveoli, and vasculature was intact. Only scattered nuclear residuals were observed in the trachea rings, and quantitatively, the DNA content was 26.9 ± 13.2 ng/mg in the decellularized lungs. The authors reported that high

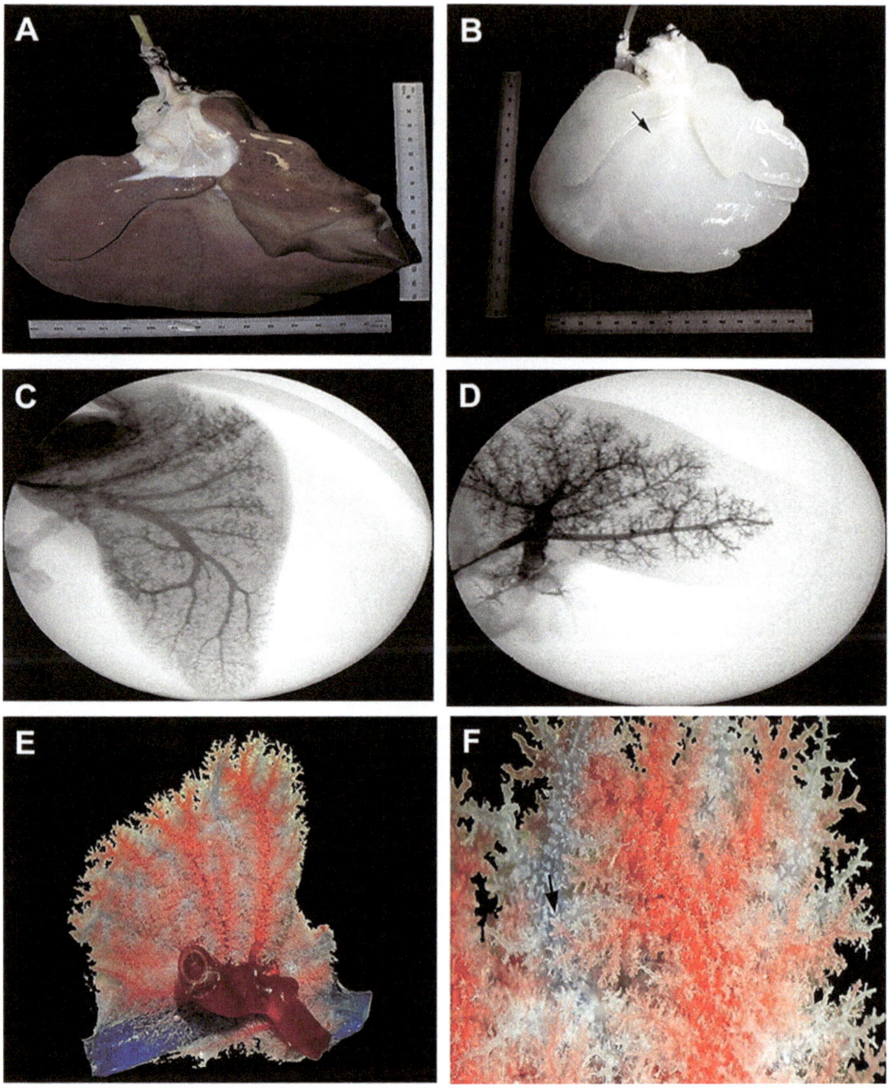

Fig. 6.16 Barakat et al. showed macroscopic images of porcine liver before and after decellularization, and they evaluated the integrity of the vasculature after decellularization. (**a**–**b**), After decellularization, the liver became white and semitranslucent in appearance, maintained the general shape and size. (**c**–**d**), Intact vascular structures imaged by fluoroscopic examination of the hepatic and portal venous system, respectively. (**e**–**f**), Corrosion casting also showed the preservation of the large vessels as well as extensive microvascular branches. (adapted from reference [215])

concentration of SDS could yield complete decellularization, but would also damage the integrity of the membrane architecture. No secretory proteins or vessel markers were identified by Western blotting in the decellularized lungs. After decellularization, a decreased fractional volume of alveolar septa, an increased

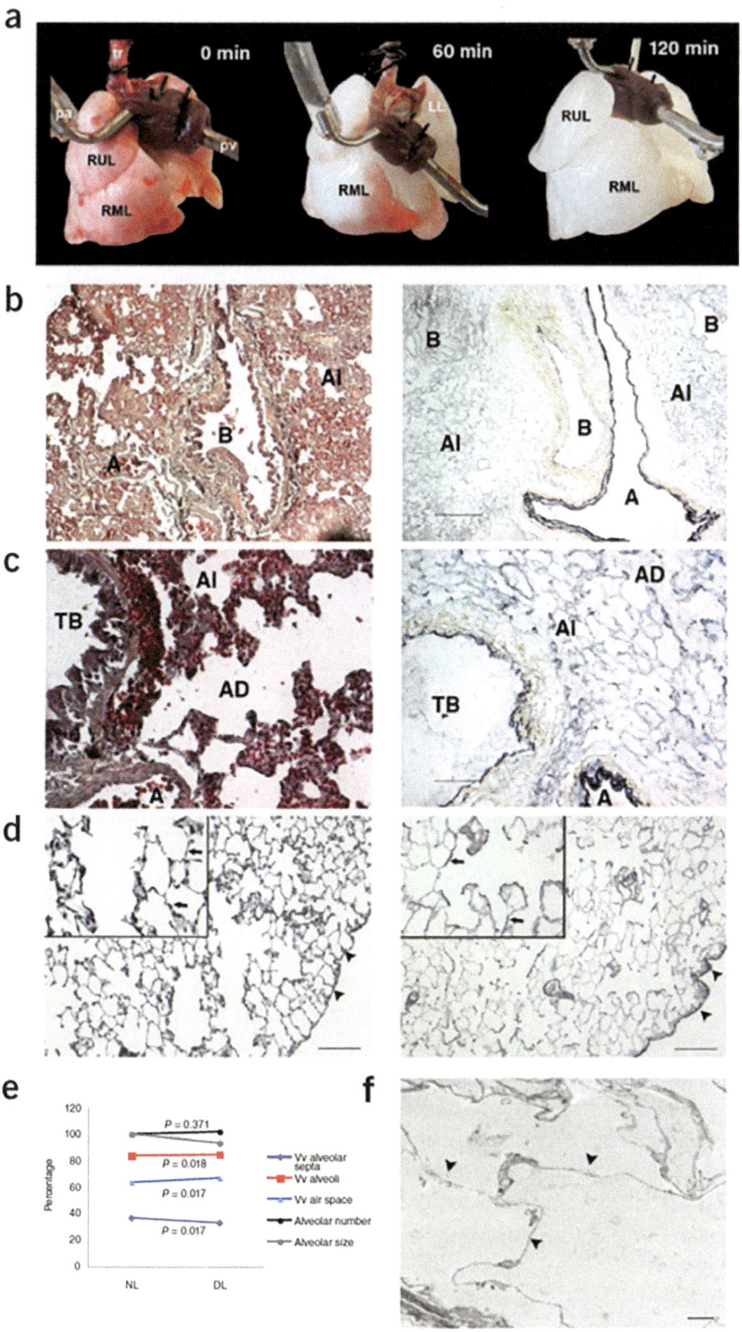

Fig. 6.17 Ott et al. showed the gradual process of whole rat lung decellularization, and histological imaging and scoring before and after decellularization. (**a**), Images of decellularizing lungs at different time points (0 min, 60 min, and 120 min). (**b–c**), Movat pentachrome staining of native (left column) and decellularized (right column) lung parenchyma, showing preservation of collagen, proteoglycans, and elastic fibers after decellularization. Scale bar = 200μm in (**b**), or 50μm in (**c**).

fractional volume of airspace and alveolar space were noted. The number or the size of alveoli showed no significant differences. The lung compliance decreased after decellularization.

Peterson et al. also decellularized rat lungs but with mild agents and a different protocol [38]. They treated the vascular and airway compartments with a zwitterionic detergent CHAPS. The perfusion pressure was kept below 20 mmHg, and the duration lasted for 2–3 h. After decellularization, cellular and nuclear contents were removed, leaving intact lung architecture, arterial tree, alveolar septal and microvasculature. Quantitative analysis demonstrated removal of approximately 99% DNA. MHC I and II were depleted from lung matrices. ECM components such as collagen, elastin, and laminin were partially preserved with different degrees in decellularized lungs. Most GAG components (>90%) were removed. Micro-CT analysis showed that acellular lung matrices were obtained.

Studies by Price et al. and Cortiella et al. both used mouse lungs for decellularization [129, 223]. Price et al. performed perfusion via the trachea and right ventricle with decellularization agents (Triton X-100, sodium deoxycholate, DNase), and water or PBS [223]. Cortiella et al. used different protocol which lasted for weeks, where freeze–thaw cycles were first applied, SDS was infused via the airway compartment with constant rotational agitation, and DNase/RNase treatment and antibiotics treatment were also adopted [129]. Both studies reported removal of nucleic materials, preservation of ECM components and lung macro−/micro- architecture.

Soon after, decellularized monkey, pig, and human lungs were reported [216, 220, 236]. Bonvillain et al. performed perfusion decellularization of monkey lungs via pulmonary artery [216]. The lungs were firstly washed with PBS containing antibiotics and EDTA, then with PBS containing heparin and sodium nitroprusside by gravity-driven flow. Core decellularization included were washed with water containing antibiotics, with Triton X-100 solution, deoxycholate solution, and hypertonic saline, and DNase in sequence. Rinsing with water was done between each step, while final rinsing was performed with PBS. After decellularization, the ECM components were well preserved. Proteomics analysis revealed enrichment of the ECM proteins because the cellular proteins were largely removed. DNA content was effectively removed (approximately 92%), with only a few disorganized and small fragments.

Nichols et al. decellularized human and pig lungs via the pulmonary artery and trachea [220]. They applied a relatively simple decellularization protocol including freeze/thaw treatment and perfusion with SDS solutions. Final rinsing was performed with water and PBS containing antibiotics. After decellularization, the lung scaffolds showed removal of cellular materials, and retention of some ECM components. Lung compliance and mechanics were also maintained.

Fig. 6.17 (continued) (**d**), Verhoeff's elastic-tissue staining of lung parenchyma), showing preservation of elastic fibers after decellularization. Scale bar = 200μm. (**e**), Morphometric and stereologic assessment of native and decellularized lungs. (**f**), TEM images identified basal membranes of alveolar septa after decellularization. Scale bar = 2μm. (adapted from reference [86])

In another study by O'Neill et al., the human and pig lungs were decellularized not by perfusion of whole organ but by soaking and agitation of organ slices [236]. Briefly, washed lungs were frozen, partially thawed, and sectioned into sheets. After washing with 2× PBS, decellularization was performed with SDS, or with CHAPS, or with Tween-20, sodium deoxycholate and peracetic acid. In final rinsing, water and PBS were used. After decellularization, CHAPS showed the best maintenance of ECM components in both human and porcine lungs, with similar retention of ECM proteins except for elastin. Detailed results are not discussed here as these protocols did not produce scaffolds suitable for whole lung engineering.

6.3.3.4 Decellularized Kidney Scaffolds

In 2009, Liu et al. first reported rat whole kidney decellularized scaffolds prepared by perfusion via the renal artery [218]. The decellularization protocol involves perfusion with heparinized PBS, 1% SDS, water, 1% Triton X-100, and PBS containing antibiotics in sequence. Final sterilization was performed with Co-60 radiation. After decellularization, no cellular residue was found, and the reticular structures consisting of basilar membrane and collagen.

Ross et al. also reported rat whole kidney decellularization scaffolds with more detailed evaluation [131]. In their study, rat kidneys were perfused following different protocols via the renal artery at a constant pressure of 100 mm Hg. The first one used non-ionic detergent, with sequential solutions which are Triton X-100 with increasing concentrations from 0.5% to 10%, a $CaCl_2$ and $MgSO_4$ solution, a NaCl solution, water, and DNase in buffer at pH 5–6. All the solution had 0.05% NaN_3. The second protocol used ionic detergent (sodium deoxycholate or SDS). The authors optimized the protocol, with sequential perfusion of Triton X-100, DNase, and SDS, and final water rinsing. As a result, both protocols produced transparent scaffolds with web-like basement membrane architecture. SEM demonstrated retention of microstructures within the scaffolds. The authors concluded that SDS was superior to sodium deoxycholate, as with more complete disruption and removal of cells while preserving the laminin and collagen IV.

In 2010, Nakayama et al. decellularized kidney sections from monkey [248]. In 2012, perfusion decellularization of whole porcine kidneys via renal artery was reported [221, 222, 227]. As reported by Park et al., the kidneys were rinsed with heparinized PBS, then perfused with circulated 1% SDS for 12 h at approximately 100 mL/min, and finally rinsed with PBS for 24 h [222]. The kidney scaffolds showed a white appearance with well-preserved morphology after decellularization. The renal architecture as well as vascular branches was maintained, while cellular materials, especially 91.9% DNA contents, were removed.

Similar results were reported by Orlando et al. [221]. The porcine kidneys were perfused with water for 12 h, then SDS for 48 h, and finally PBS for 5 days. After decellularization the scaffolds had preserved renal architecture and vessels, collagen components, but no cellular materials. Particularly, the authors pointed out that Triton X-100 alone was not sufficient to decellularize porcine kidneys. The kidney

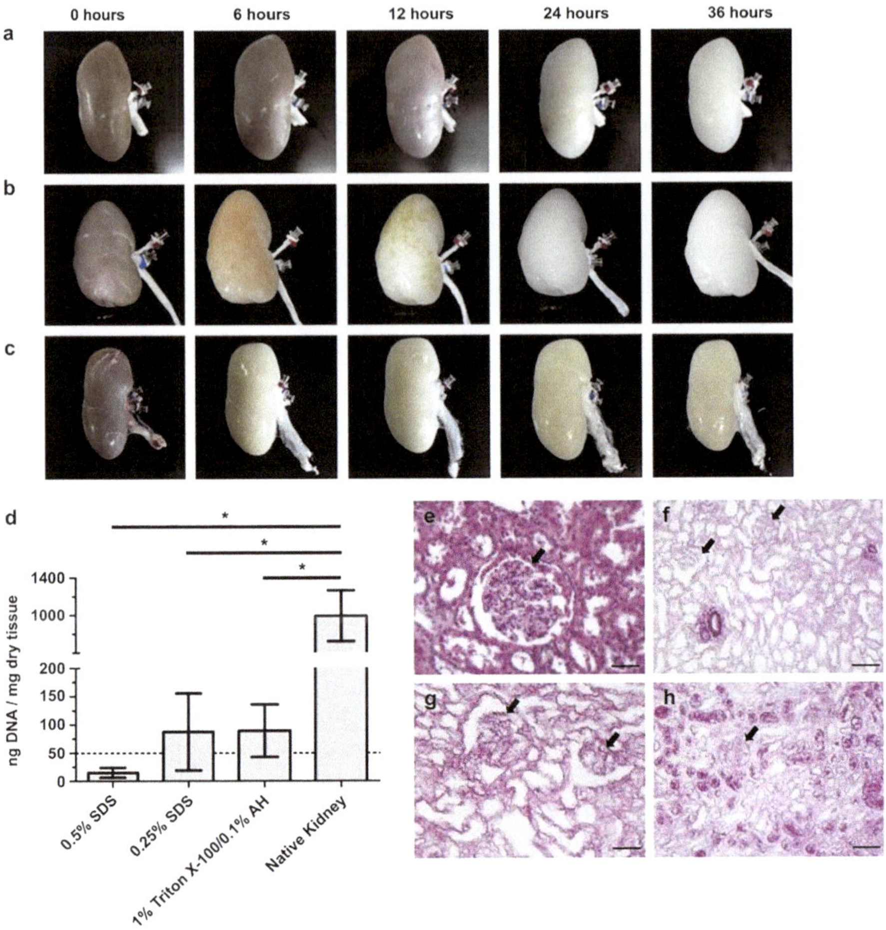

Fig. 6.18 Sullivan et al. showed gradual decellularization of porcine kidneys by different protocols, and they compared the efficiencies of the applied protocols on DNA removal. Histological staining images were also showed. (**a–c**), Images of the kidneys during the decellularization with different protocols at various time points. (**d**), DNA quantitation assessments of the decellularized kidneys. (**e–h**), H&E staining of the decellularized kidneys by different protocols, all showing removal of cellular materials. Scale bars = 50μm. (adapted from reference [227])

scaffolds retained the critical architecture and intact vasculature with uncompromised biocompatibility.

In the other study reported by Sullivan et al. [227], the porcine kidneys were perfused with solutions, initially with sodium heparin, then with 0.5% or 0.25% SDS, or 1% Triton X-100/0.1% NH_4OH. After detergent perfusion, the scaffolds were perfused with PBS, and with recirculating DNase solution (containing 0.0025% DNase and 10 mM $MgCl_2$ in PBS at neutral pH) overnight. Final rinsing was performed with PBS for 1 h. The kidney scaffolds were sterilized by 10.0 kGy gamma irradiation. As shown in Fig. 6.18, after decellularization the scaffolds showed no cellular materials. Treatment with higher SDS concentration yielded a more translucent appearance at the end of detergent perfusion, while Triton X-100/NH_4OH caused faster visual changes. The decellularized scaffolds had unchanged volume, maintained shape, intact arteriovenous renal system, and important intact nephron architecture. ECM components such as GAGs and collagens were well preserved.

6.3.3.5 Decellularized Pancreas Scaffolds

In 2010, De Carlo et al. reported decellularization of rat pancreas slices [253]. In 2011, two studies reported decellularization of whole porcine pancreas [167, 217]. Interestingly, the first report on whole pancreas decellularization used organs from pigs but not smaller animals [167]. Actually, the second report used mouse pancreas to do whole organ decellularization [217].

In the first study [167], Mirmalek-Sani et al. perfused whole pancreas via the pancreatic duct, firstly with sodium heparin in a buffer solution, then with Triton X-100/NH_4OH for 24 h, and finally with PBS for 5 d. After decellularization (Fig. 6.19), the vasculature remained patent, the cellular materials were removed, while the collagen IV, elastin, and laminin were preserved.

In the second study [217], Goh et al. performed retrograde perfusion through the hepatic portal vein. The perfusate entered the splenic vein, and then the vasculature of the pancreas (Fig. 6.20). The authors also attempted to do perfusion via the pancreatic duct, but the process was time-consuming and less effective. The perfusion applied 0.5% SDS, water, and 1% Triton X-100 in sequence at constant flow rate. Thereafter, benzonase and 10% FBS with antibiotics were perfused as final washing. As a result, no remnant cells were identified in the decellularized scaffolds, while the DNA content decreased dramatically. ECM structure and components (collagen I and IV, fibronectin, laminin, and GAGs) were preserved. Proteomics analysis identified 114 unique proteins which were mostly ECM proteins. After decellularization, the scaffolds had adequately retained pancreatic ultrastructure and architecture, but with increased Young's modulus. The decellularized whole pancreas scaffolds showed good biocompatibility.

6.4 Recellularization Methods for Whole Organ Engineering

The definition of recellularization is repopulation of acellular scaffolds with proper cell types, aiming to reconstitute the microanatomy as well as the function of specific tissue or organ [6]. Previous studies have shown that recellularization is essential for long-term survival of implanted decellularized ECM scaffolds. Except for the anatomical and functional reconstitution, recellularization can also possibly degrade or hide the various antigen presented on, particularly, the decellularized ECM

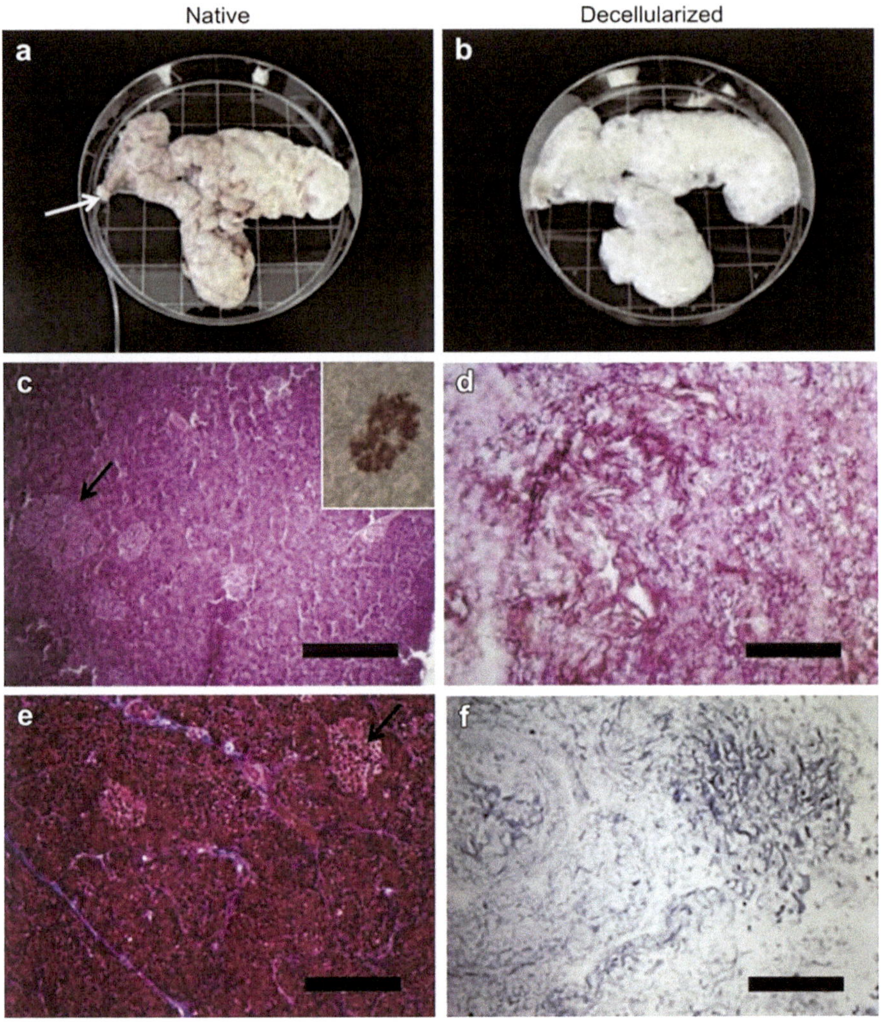

Fig. 6.19 Mirmalek-Sani et al. showed macroscopic and microscopic histological staining images of porcine pancreas before and after decellularization. (**a**–**b**), general view of the native and decellularized pancreas. (**c**–**d**), H&E staining of the native and decellularized pancreas. (**e**–**f**), Masson's trichrome staining of the native and decellularized pancreas. Scale bar = 200μm. (adapted from reference [167])

scaffolds, thus attenuate the immune responses [235]. Without proper recellularization in vitro or in vivo, the risk of ECM scaffold implantation ranges from failure of transplantation to death of recipient. For example, Tan et al. showed that decellularized small intestine submucosa seeded with bone marrow mesenchymal stem cells could promote re-epithelialization in vivo as an esophagus patch [254]. Macchiarini et al. used patient's own epithelial and progenitor cells to

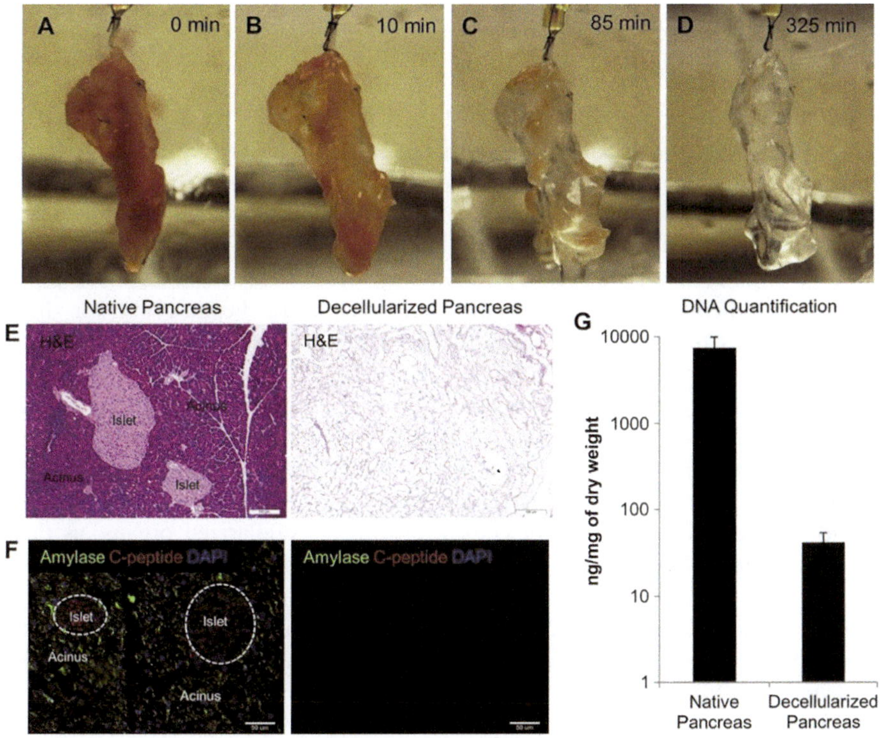

Fig. 6.20 Goh et al. showed gradual decellularization of murine pancreas. They also performed histological staining and DNA quantification to evaluate the decellularized pancreas ECM. (**a–b**), Images of the pancreas during the decellularization at various time points. (**e**), H&E staining of the native and decellularized pancreas. (**f**), immunofluorescence staining of the native and decellularized pancreas. (**g**), Quantitative assessment of the DNA contents before and after decellularization. (adapted from reference [217])

recellularize tissue-engineered airway, which was transplanted in the patient suffering with end-stage tuberculosis-induced bronchial stenosis. This construct allowed for normal functioning without rejection over at least 1 year [255]. On the contrary, it has been reported that the implanted tissue-engineered heart valve needed replacement due to stenosis [256]. Lethal events were even reported due to rupture of the implanted xenogeneic decellularized heart valves [257].

To engineer whole organs for eventual transplantation, survival, and functioning, the decellularized whole organ scaffolds as discussed in previous section need to be recellularized properly ex vivo with adequate cells which are discussed in Sect. 6.2, because immediate blood circulation as well as organ functioning is required once they are implanted. The methods applied in recellularization of whole organ scaffolds generally range from cell culture to organ perfusion culture [4]. Despite the success in repopulating decellularized tissues, repopulation of whole organs has been challenging as each organ differs from the others. Considering that each organ type has unique architecture and that numerous cell types derived from different lineages interact within one organ, recellularization of whole organ scaffolds is even more difficult [235]. Nevertheless, recellularizations of different organs share some common issues, e.g., the cells, the ECM scaffolds, and the recellularization techniques. The former two issues have been generally discussed in previous sections, but in the context of recellularization for whole organ engineering some points should be particularly noted.

Regarding the cells, they can be divided into parenchymal and non-parenchymal cells. To engineer a whole organ, parenchymal recellularization is of most importance as it defines organ's specific functions, e.g., hepatocytes for liver, cardiomyocytes for heart, epithelium for lung, etc. [4] As discussed in Sect. 6.2, the parenchymal cells can be differentiated already or stem cell-derived, can be autologous or allogenic. The key issue is that specific parenchymal cell types are needed depending on the specific organ. Non-parenchymal recellularization features reendothelialization of the vascular networks. The vascular networks of whole organ ECM scaffolds retained in the decellularization process are naturally utilized to provide nutrients and oxygen for cells within the recellularized organs, as these vital components have limited diffusion ranges [37]. It is a major advantage of decellularized whole organ scaffolds that intact vascular networks are preserved, but once the engineered organ is transplanted, the vasculature is highly thrombogenic even with anticoagulation if it is not properly reendothelialized [91, 221, 225], which can lead to graft failure. The repopulation of the vascular networks with endothelial cells is a prerequisite to block the thrombogenicity caused by exposure to ECM components [6]. Another important cell type for non-parenchymal recellularization is the fibroblast, which can secrete and remodel the ECM, and can interact with parenchymal cells improving their functions [4]. In vitro, coculture of hepatocytes with fibroblasts can enhance hepatic functions significantly. In heart, fibroblasts are coupled with cardiomyocytes and capable of propagating the electrical stimuli over longer distances. However, fibroblasts are often organ-specific [258, 259], making it difficult to perform recellularization of fibroblast in vitro. Similar to endothelial cells, the fibroblasts are recruited as necessary in vivo. Parenchymal and non-parenchymal cells can be seeded simultaneously as a mixture, or separately as single-cell type suspensions [5].

A groundbreaking example of parenchymal recellularization of decellularized whole organ scaffold was reported by Ott et al. [85]. In that study, they seeded the decellularized rat heart scaffolds with neonatal cardiomyocytes, and after a period of culture the engineered heart was capable of responding to single paces as electric and contractile behaviors, and was capable of spontaneous rhythmic depolarizations and contractions for several minutes. Similar pioneering progress in liver engineering was also reported. Uygun et al. seeded primary rat hepatocytes into the decellularized rat liver scaffolds, and the liver constructs showed positive but immature hepatic metabolic activity such as albumin production and urea synthesis [40].

An impressive example of non-parenchymal recellularization was carried out by Kaushal et al. [260]. In this study, the decellularized blood vessels were seeded with endothelial progenitor cells ex vivo, and the construct showed long-term patency and survival. Indeed, mature and well-aligned endothelial cells form a non-thrombotic barrier in the vasculature, so that the perfusate ex vivo or blood flow in vivo can be restricted to the vascular spaces and do not cause hydrodynamic shear stress to the parenchymal cells [4]. For example, Baptista et al. reported [9] a bioengineered functional humanized rat liver [91], in which the liver vascular structures were coated with endothelial cells, and the reseeded endothelial cell expressed nitric oxide synthase. When the reendothelialized constructs were perfused with blood, significant less platelet adhesion and aggregation was observed compared with the unseeded scaffold. To date, no reendothelialization studies have demonstrated full coverage of the vascular structures [86, 261, 262]. Anyhow, reendothelialization can continuously occur in vivo after transplantation of engineered constructs, so it is plausible that ex vivo reendothelialization need to be effective only long enough to ensure the initial survival until the in vivo reendothelialization occurs.

Regarding the scaffolds, whole organ scaffolds derived from small animals are feasible for proof of concept. However, the ultimate goal of whole organ engineering is clinical application, decellularized organ scaffolds derived from human scale animal (e.g., pig) are much more authentic and application-oriented in whole organ engineering from the bench to bedside.

This section will focus on the recellularization techniques for whole organ engineering. It is impossible to standardize a universal protocol of recellularization, different strategies should be employed for different organs [263]. But generally, the recellularization process can be divided into three major steps. Firstly, the scaffolds are seeded with cells with a redistribution similar to their native spatial configuration. Secondly, the scaffolds are perfused with culture media to prepare the cells for further potential integration, proliferation, differentiation, maturation, and function. Lastly, the bioengineered whole organ constructs are evaluated before transplantation.

6.4.1 Recellularization Methods

6.4.1.1 Cell Delivery

In cell transplantation therapies, two strategies have been employed to deliver the cells, such as injection directly into the organ or injection into the circulation for following cell homing. Similar strategies have been used for recellularization in whole organ engineering [4]. Accordingly, two major concerns will logically be proposed when considering the recellularization of whole organ scaffolds, including an appropriate number of cells and distribution of the cells within the scaffold which simulate the native organ as much as possible. As the two concerns vary significantly with the organ, the numbers of the reseeding cells and the routes of delivery should be determined.

Cell Number

In whole organ engineering, it is often not expected to rebuild whole organ of native complexity and cellularity, meaning a minimum of cell number is required to repopulate the scaffolds [4]. However, as the scaffolds are human scale to provide clinically relevant function, the required cell number is still quite large. For example, the bioengineered hearts require strong biomechanical functions at the time of implantation, thus require a high number of cells as well as appropriate cell types to be immediately present. In heart, cell density is as high as 10^8 cardiomyocytes per cubic centimeter [264]. To bioengineer whole heart of functions, the decellularized heart scaffold needs to be almost completely recellularized when performing implantation [4]. On the contrary, organs such as the liver mainly exert metabolic functions, hence less cell mass is acceptable when transplanted. An adult liver is estimated to have approximately 4×10^9 hepatocytes per kg of body weight [265], but in hepatocyte transplantation studies clinical improvement has been confirmed using less than one tenth of the recipient's liver mass [4, 266]. Based on these results, it can be assumed that approximately a minimum of 10^{10} cells are needed to support a patient suffering from acute liver failure [4].

Delivery Routes

Two different techniques are typically utilized in cell delivery, depending on the cell type and the goal. The first one involves recellularization via the vasculature or other hollow structures (e.g., airway and ureter, in lung and kidney, respectively) easily via perfusion [39, 90, 267, 268]. The second one implements recellularization by injection of the seeding cells directly into the parenchymal compartment. In fact, it has been reported to recellularize whole organ scaffolds by combinations of both techniques [225].

For example, in heart recellularization, cells can be perfused via the aorta, the brachiocephalic artery, or inferior vena cava [139, 225], and be directly injected into the ventricular wall [85, 225]. In liver recellularization, cells can be delivered by perfusion through the portal vein, the inferior vena cava or the hepatic artery [40, 91, 269], and by injection into the parenchyma or common bile duct [270]. In lung recellularization, cells can be delivered by perfusion via the pulmonary artery, the pulmonary vein, or the airway [38, 86, 129, 223, 271]. In kidney recellularization, cells can be delivered by perfusion or injection through the renal artery, the renal vein, or the ureter [128, 131, 272]. In pancreas recellularization, cells can be delivered by perfusion via the hepatic portal vein or the pancreatic duct [217, 273]. In general, the cells can be delivered via a single route or via multiple routes with a higher efficiency.

Cell Concentration

Cell concentration can affect the delivery efficiency. For example, high cell concentration could result in extensive cell death or occlusion of the vasculature [40]. Previous studies indicated that direct cell injections may cause cell aggregation, and thus poor engraftment and insufficient distribution [37], which is possibly due to the high cell concentration.

6.4.1.2 Perfusion Culture

After initial delivery into the decellularized whole organ scaffolds, the cells require long-term culture (for weeks) for engraftment, migration, proliferation, and maturation processes to complete [4]. For example, in vitro endothelialization of acellular vessels can take 2 weeks [274], which is therefore a reasonable initial estimate for the duration of culture [4]. In addition, more time is needed when the recellularization starts with stem/progenitor cells. Static culture is sometimes utilized for a short period of time to allow the cells to engraft, and following perfusion culture can circulate the culture medium providing nutrients and oxygen as well as carrying away the waste.

6.4.1.3 Endpoints

As a bioengineered whole organ for transplantation, endpoints of the constructs should contain both the resistance to thrombosis of the vascular system and the functionality of the organ. Properly covered blood vessels with endothelia cells would prevent thrombosis. Blood perfusion is a viable approach to assessing the resistance to thrombosis as well as the patency of the vascular system [37]. Another valuable approach is determining the activities of thrombosis-related proteins such as thrombomodulin and thrombin-mediated protein C, as reported by Robertson et al. [225].

With the foremost goal that bioengineered whole organs are used for transplantation, the constructs must show physiological functions whatever it is enhancement or replacement. Naturally different organs have different endpoints. Specifically, heart should have a minimal function of contractility, while advanced features include pumping against afterload, drug responsivity, etc. Liver should have a minimal function of albumin and urea production, while advanced features include drug metabolism and metabolic regulation. Lung should have a minimal function of oxygen exchange, while advanced feature is respiration. Kidney should have a minimal function of urine production, while advanced features include renin angiotensin system and blood pressure maintenance. Pancreas should have a minimal function of insulin/glucagon secretion, while advanced features include pancreatic digestive enzymes and endocrine hormone production [4].

6.4.2 Bioreactor

The human body itself theoretically is the ideal natural bioreactor [5], providing an authentic environment that promotes the maturation of implanted bioengineered organs, but some steps need to be performed prior to transplantation. It can be fatal if these steps are performed in vivo, such as reendothelialization of the vascular system as previously discussed. Bioreactors are designed with the concept that the universal culture requirements are met, and the special conditions for particular organ types are supplied to mimic the in vivo environment.

6.4.2.1 The Basic Framework of a Bioreactor

The universal culture requirements involve delivery of nutrients and oxygen into the entire bioengineered organ, as well as removal of the harmful products from it, meaning that a constant exchange of substances is needed. To this end, perfusion-based bioreactors are feasible for all highly vascular organs, which mimic the blood flow within the body. In the view of engineering, the bioreactors are based on organ perfusion devices, which have been used to preserve the whole organ prior to cold storage techniques [4]. Stirred tank bioreactors may also be suitable for some tissue or organ types not discussed herein.

As shown in Fig. 6.21 [5], the basic framework of a perfusion bioreactor is consisted of a pump, an organ chamber (the bioreactor in narrow sense) with at least one inlet and one outlet [92, 179, 275]. Based on this, sensors are added to monitor parameters of the system such as the flow rate and pressure, the temperature, the oxygen, the pH, etc. Furthermore, auto-control system can adjust the parameters to maintain a more reliable perfusion culture process [85, 86, 215, 276–279]. The basic framework provides the fundamental culture conditions for organ survival and maturation. For example, the perfusion system can exert fluid flow-induced shear stresses to the endothelial cells within the vascular networks. This hydrodynamic force can promote the alignment of the endothelial cells in the flow direction as its in vivo arrangement [280].

6.4.2.2 Organ-Specific Bioreactors

To fulfill the requirements of different organs, additional physiologically relevant stimuli were added in the bioreactor systems to mimic in vivo conditions and to facilitate the functional maturation of the bioengineered organs. The stimuli include mechanical, electrical, and chemical cues, or a combination of these, depending on the specific organ type.

For heart, myocardium functions to propagate as well as respond to electrical impulses, resulting in synchronized contractions and therefore pumping blood [4]. In whole heart engineering, electrical and biomechanical (preload, afterload, and

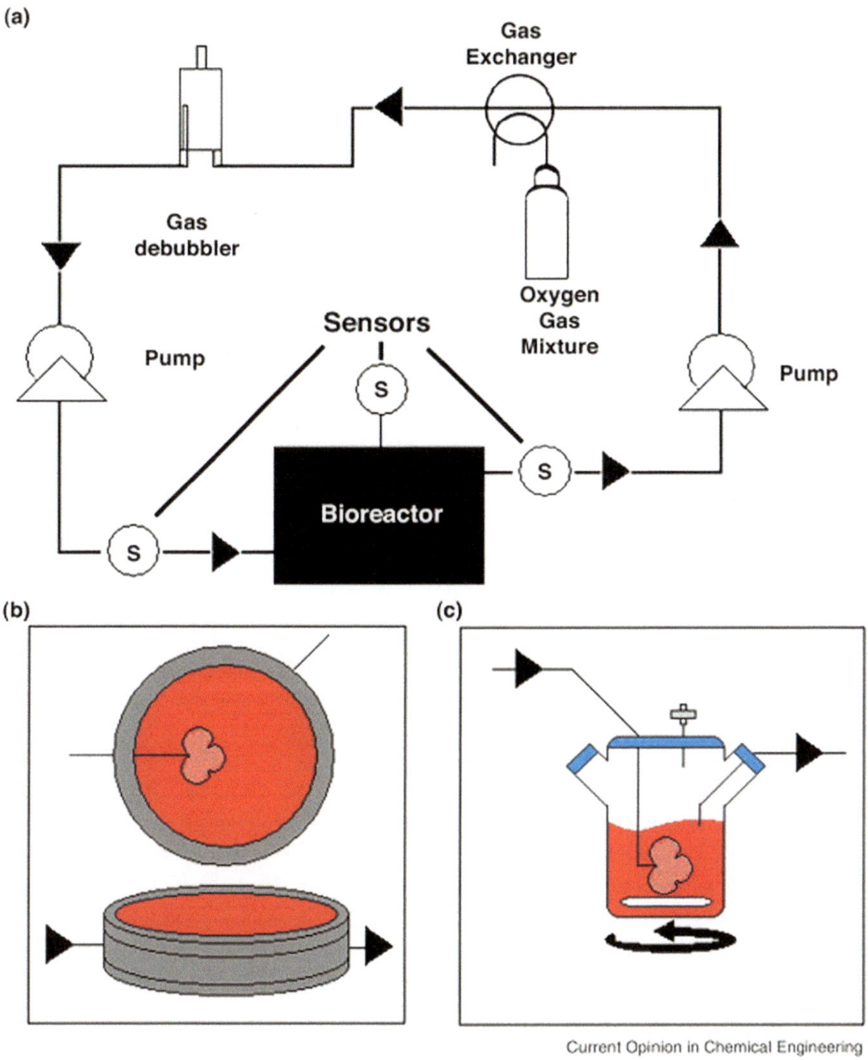

Fig. 6.21 A typical layout of a perfusion bioreactor and two bioreactor configurations with different hydrodynamic features. (**a**), For a perfusion bioreactor, a pump and an organ chamber are the essential devices. Additional devices are added to enhance the stability of the system and the reproducibility of recellularization protocol, such as sensors, oxygenator, gas debubbler. (**b**), a cylindrical configuration bioreactor can limit dead zones and promotes mixing by placing the outlet 120° from the inlet. (**c**), a spinner with a stir bar configuration promotes recirculating of cells in the bulk media. (adapted from reference [5])

intraventricular pressure) stimulations are therefore required [85, 277, 281, 282]. Mechanically stretching can enhance cell alignment and expression of cardiac markers [283]. Electrical stimulation can induce the contractile forces which is essential for myocardium development and maturation [284]. For example, Ott

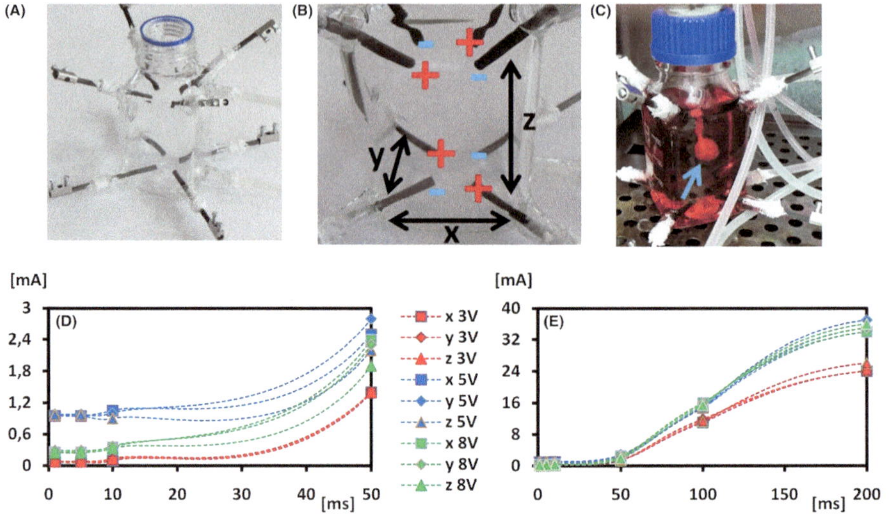

Fig. 6.22 An example of electric stimulation in whole heart engineering. (**a**), A square formed 250 mL flask with inlet and outlet on a planar side is used as the organ chamber, additional openings are added at the edges of the flask to insert electrodes. (**b**), The setting of the electric field. (**c**), Perfusion and stimulation at work. The blue arrow points at the bioengineered heart. (**d**), Measured current under alternating voltage with pulse durations of 1–50 ms stimulation in x, y, and z direction as shown in B. (**e**), Measured current with longer pulse durations (up to 200 ms). (adapted from reference [282])

et al. used pulsatile antegrade left heart perfusion to simulate the systolic and diastolic flow, while electrical pacing was provided through epicardial leads [85]. Figure 6.22 shows a layout of electric stimulation in whole heart engineering [282]. For liver, mechanical stimuli are not needed. However, due to the high metabolic activities, a dialyzer is beneficial to clear toxic by-products and thus to maintain availability of the constructs for organ transplantation [285]. For lung, breathing-induced mechanical stretch contributes to the differentiation of the pulmonary epithelium. In whole lung engineering, proper intermittent ventilation also promotes the survival proliferation and differentiation of lung epithelium [38]. Hence, the mechanical compliance must be retained appropriately so that the inflating and deflating movements and accompanying air exchanges can take place. This unique feature makes the mimic bioreactor somewhat sophisticated. The common method used for lung bioreactor is to keep the construct in a container of medium and to connect the airway to a ventilator that simulates the respiratory volume and rate [223, 271, 286]. Figure 6.23 shows a schematic bioreactor installation for whole lung engineering [286].

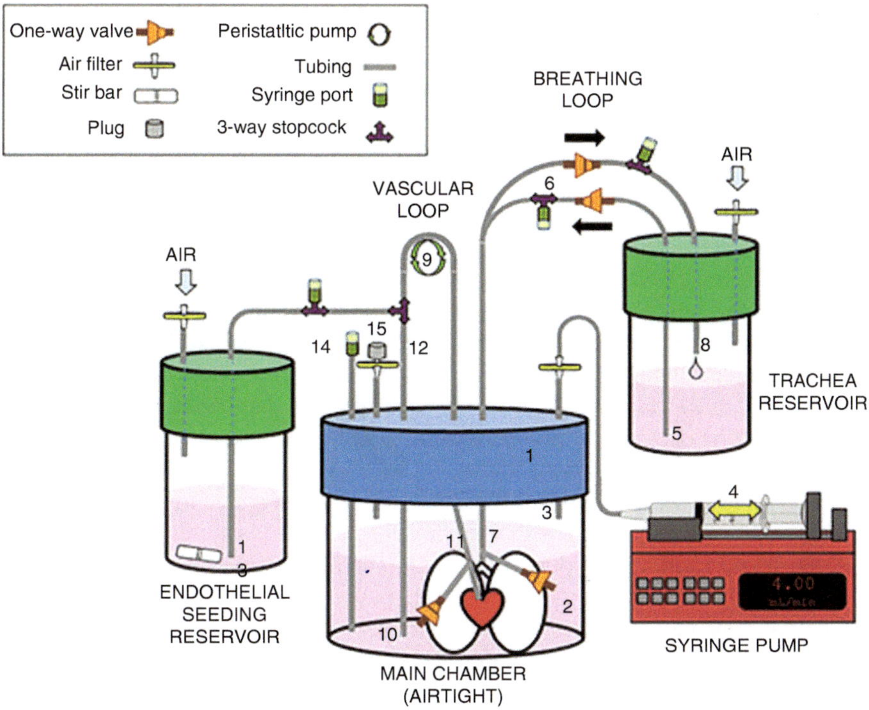

Fig. 6.23 A schematic bioreactor installation for whole lung engineering featuring simulated breathing stimulation. (1), Organ chamber lid. (2), Organ chamber. (3), Air inlet/outlet for ventilation. (4), Syringe pump for ventilation cycles. (5), Culture media for flowing into the lung airway. (6), One-way breathing loop with valves. (7), Culture media enters lung via the trachea (simulating inflating). (8), Culture media returns to the trachea reservoir (simulating deflating). (9), A pump recirculates culture media. (10), Culture media recirculating outlet. (11), The pulmonary artery as culture media recirculating inlet. (12), A 3-way stopcock for seeding endothelial cells. (13), Endothelial seeding reservoir. (14), Injection port for changing culture media. (15), air filter. (adapted from reference [286])

6.4.2.3 Monitor and Control the Recellularization Process

In whole organ engineering, endpoints evaluation is a quality control step for subsequent transplantation. Indeed, the recellularization experiments are applying the black box approach, that different outcome parameters are evaluated afterwards [6]. However, real-time monitoring and controlling the recellularization process can effectively and timely improve the feasibility and the reproducibility of the bioengineering workflow. It also helps to summarize and optimize the recellularization protocol. The environmental parameters should be monitored and thus controlled accordingly, and the recellularized whole organs should be monitored to estimate the status of the constructs.

Monitoring and Controlling the Environmental Parameters

In the bioreactor system, the environmental parameters include perfusion pressure and flow rates, temperature, pO2, pCO_2, pH, electrolyte levels, glucose or lactate concentrations, etc. These parameters need to be precisely monitored and controlled so that the recellularization process can be stable in a long-term culture period [37]. Basically, all the environmental parameters can be measured by sensor or probes [287].

The perfusion pressure and rates can affect the recellularization, especially the reendothelialization of the vascular networks [288]. Commonly and logically, near physiological pressure and flow rate are the first guess [38, 40]. However, some parenchymal cells, like hepatocytes, are susceptible to shear stress when seeded into scaffolds not well reendothelialized. Hence, keeping the flow pressure and rate below physiological values may be beneficial at the early stage of recellularization [4]. These parameters can be adjusted by controlling the pump and the tubular system. The ideal temperature for recellularization and organ culture is likely to be within the physiological ranges so that the cells can have best viability. It has been reported to perform whole organ culture in thermostatic incubators or in water-jacketed medium reservoirs to achieve temperature control [40, 85, 86, 223].

The vascular networks help to deliver oxygen, but the capillaries may be less efficient even after reendothelialization. For bioengineered organs with high metabolic activities at maturation stage such as heart and liver, oxygen shall be supplied at higher levels to support the survival of the cells [4]. Carbon dioxide is also an important component in the perfusion system as it maintains the pH [4]. An oxygen/carbon dioxide mixture can be gassed into the perfusate by an oxygenator.

Physiological concentrations of electrolytes such as Na^+, K^+ Ca^{2+}, and Mg^{2+} are of significance in maintaining the cellular phenotypes and functions. For example, the Na^+–K^+-ATPase (sodium pump) can maintain the normal electrical activity and ion gradients, generate an action potential, and control cell volume for cardiac cells [289]. Calcium and magnesium also influence cardiovascular function [290, 291]. Glucose provides energy for cellular activities, while lactate is an intermediate metabolite of anaerobic glycolysis [292]. By monitoring these indicators, the cellular metabolism can be evaluated. Furthermore, with the computer-aided methods or even artificial intelligence in the future, bioreactor systems can adjust the perfusion parameters. Some progresses have been reported in this field [85, 86, 215, 277]. Figure 6.24 shows a monitoring/controlling-integrated bioreactor system [293].

Monitoring the Recellularized Organs

Various methods have been utilized to monitor the growth and function of the reseeded cells. Performing a biopsy for histological assessment can provide authentic information. Though this operation can be done in a minimally invasive manner, disturbance and damage are unavoidable. Non-invasive methods are thus preferred to monitor the recellularized organs.

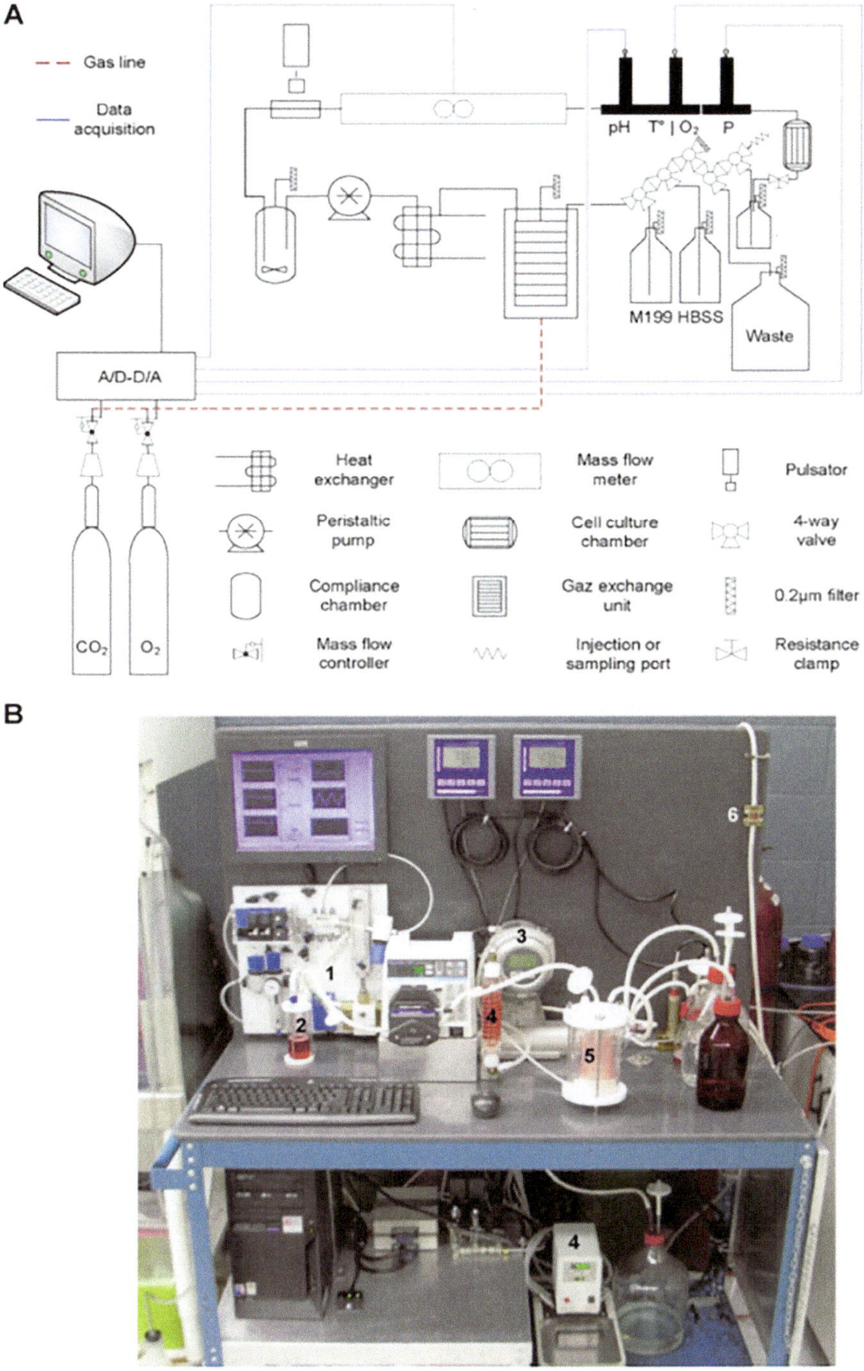

Fig. 6.24 Monitoring/controlling the environmental parameters of bioreactor system can achieve more accurate and stable whole organ engineering workflow. In this figure, a monitoring/controlling-integrated bioreactor system is shown. (**a**), Schematic illustration of the system. This system

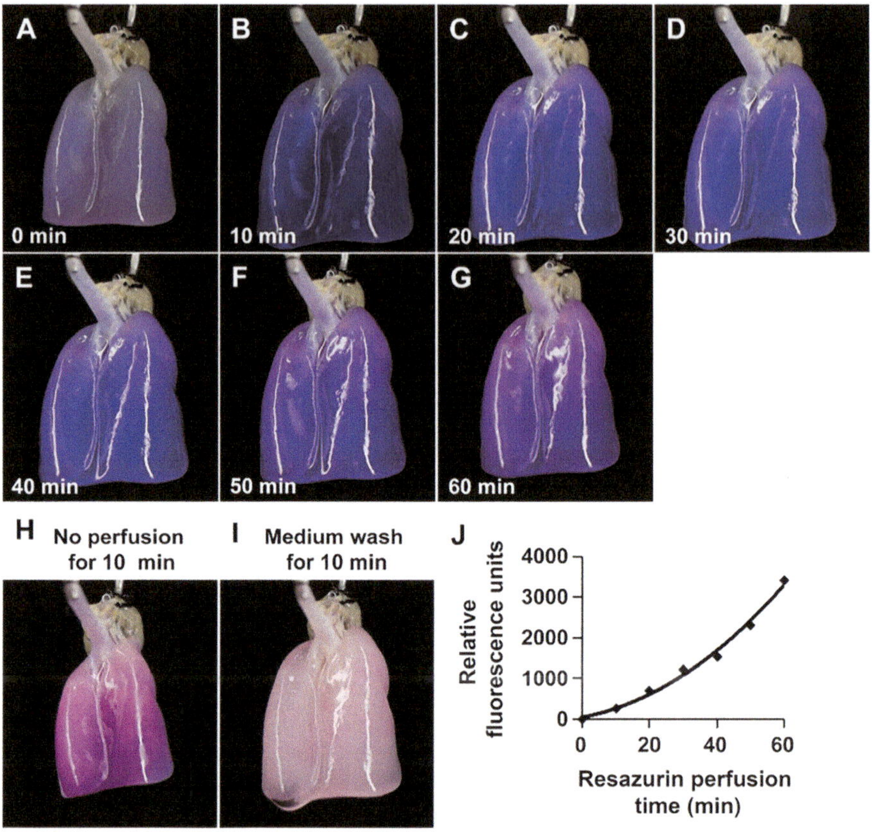

Fig. 6.25 Monitoring the growth and function of the reseeded cells can help determining the status of the recellularized organs. In this figure, a non-invasive assay is presented to monitor reendothelialized lungs. Resazurin can be metabolized by viable cells, while the product is fluorescent and cell lysis free. Thus, the agent is perfused into the recellularized lungs. (**a–g**), Time-lapse (with an interval of 10 min after the start of perfusion) photography of the reendothelialized lung perfused with resazurin. (**h**), A photo of the reendothelialized lung perfused with resazurin but after a 10 min pause. (**i**), A final photo of the lung after replacing resazurin-containing medium and an additional 10 min rinse. (**j**), Relative fluorescence intensities during the resazurin perfusion as sampled every 10 min. (adapted from reference [261])

By sampling the perfusate, the metabolites can be analyzed to evaluate the function of the constructs. For example, the levels of albumin, coagulation factors, and urea can indicate the hepatic functions. Ren et al. developed a resazurin-based method to monitor cell viability and proliferation in bioengineered organs, featuring rapid non-invasive estimation [261]. Figure 6.25 shows resazurin perfusion of

---

Fig. 6.24 (continued) contains monitoring probes (for flow pressure, temperature, O_2 and pH), and actuators to control these parameters. (**b**), Picture of the system. (adapted from reference [293])

reendothelialized lung for non-invasive monitoring the viable cell numbers. Other non-invasive methods have also been proposed to monitor bioengineered constructs, such as near-infrared imaging and micro-CT. The former allows for real-time deep tissue imaging without hazardous ionizing radiation [294], while the latter can achieve higher resolution with contrast agents even for soft tissues [295].

6.4.3 Optimization of the Recellularization Process

The recellularization process has complex parameters, which all affect the final outcomes. Currently, the recellularization protocols can be continuously optimized to achieve the ultimate goal, i.e. bioengineering whole organs for transplantation.

In terms of cell seeding, issues such as the cell concentrations, delivery routes, perfusion pressure and rates, and scaffold coating can influence the efficiency of recellularization. High cell concentrations may also lead to vessel occlusion, but low cell concentrations often need longer seeding period. Due to the fact that the cell types are mostly adhesion-dependent in the context of recellularization, the cells cannot tolerate extended periods of recirculating in the media without attachment, which may lead to anoikis [37]. Thus, it is necessary to determine the optimal cell concentration considering the cell type as well as the organ type.

Baptista et al. compared delivery routes in recellularization of liver ECM scaffolds [91]. The cells were seeded via the portal vein or via the vena cava, resulting in different spatial depositions of the cells. For liver recellularization, cell delivery via multiple routes should provide better cell distribution within the scaffolds. Robertson et al. also compared different routes for reendothelialization of heart scaffolds [225]. The cells were delivered by arteriosus or venous infusions, or a combination of both. The combination method resulted in the highest cellularity. In addition, many studies have shown that multiple infusions rather than single infusion can achieve sufficient cells seeding in liver engineering [40, 296].

The recellularization of vasculature largely depends on the perfusion process. Continuous perfusion at relatively low flow rates has been used to seed endothelial cells [37]. But it should be noted that endothelial cells prefer pulsatile flow [4], and the direction of the flow also influences the alignment of endothelia cells [280]. The perfusion pressure and rate also change with the course of recellularization. As scaffolds become recellularized, the porosity of the scaffold will decrease and the flow pressure will increase [297]. Real-time monitoring of the perfusion and sequent feedback and adjustment will be critical to optimize the organ culture protocols, controlling hydrodynamic shear stress-induced damage to seeded cells.

Scaffold coating has been proposed to strengthen cell adherence and retention, such as using laminin and fibronectin [118]. As mentioned in previous section, cell–matrix interaction can be improved by priming the scaffold with specific proteins, and thus promote the efficiency of cell seeding. However, complete seeding may not be necessary if the recellularization starts with progenitor/stem cells rather than mature cells [4].

The perfusate used for culturing bioengineered whole organ is based on the organ type. Current formulas can maintain the culture for weeks, but they are unlikely the optimal. For example, classical hepatocyte culture media can lead to cell dedifferentiation and function loss after a few weeks [4]. They also contain non-physiological levels of supplements with potential risks [4]. On the other hand, growth factors necessary for cellular development and maturation may not be retained or sufficient in the ECM scaffolds, thus additions of these components are favorable for whole organ engineering. The concentrations of the supplements need to be optimized as a proper balance is important. For example, VEGF promotes the growth and maturation of endothelial cells, while high level of VEGF can lead to giant cells and aggregates [37].

The recellularization process can also be optimized based on the recellularization progress. For example, seeded progenitor/stem cells often require hypoxic conditions to promote expansion and to minimize differentiation. Then the oxygen concentration can be increased to enhance differentiation after adequate cell expansion. Except for oxygenator devices within the bioreactor system, oxygen carriers can be supplemented to increase the oxygen transport capacity, e.g. whole blood, red blood cells or hemoglobin [4, 298]. Since whole blood introduces immune components with risks of initiating systemic inflammatory responses when activated, red blood cells also contain immunogenic factors, cell-free hemoglobin with physiological oxygen-carrying ability is plausible to be a better choice [298].

6.5 Regenerative Potential of Organ Scaffolds After Recellularization

In the past two decades, most tissue engineering researches focused on the fabrication of tissues with relatively simple structures, such as bone [299, 300], cartilage [301, 302], tendon [303, 304], skin [305, 306], urethra [307, 308], esophagus [309, 310], and bladder [311, 312]. In spite of many progresses achieved, the construction of solid organs with thick tissues and complex architectures (e.g., heart, liver and kidney) remains a significant challenge, mainly due to the difficulty to create suitable scaffolds closely mimicking the complex architecture of native organs as well as the challenge of sufficient supply of nutrients and oxygen to the cells residing within the engineered constructs.

The application of decellularization techniques to obtain an acellular organ scaffold has helped tissue engineering progress toward whole organ engineering. As for now, based on the development and utilization of a decellularized organ scaffold, many attempts have been made in engineering a solid organ. Particularly, several preclinical studies have demonstrated impressive results, such as the formation of neo-organs with rudimentary functions in different animal models. However, much work is still needed to improve the function of the engineered organs in vivo, especially the long-term physiological functions after implantation in the recipients.

In this section, we will provide a comprehensive review of current research progress in the engineering of whole organs, including the heart, liver, lung, kidney, and pancreas.

6.5.1 Heart

Currently, more than 22 million people worldwide are suffering with heart failure [85]. In the USA, heart disease is still the leading cause of death [313]. In 1967, Dr. Christiaan Barnard successfully performed the first heart transplantation [314], which provides a new hope for those patients with heart failure or other severe cardiac diseases. However, there are a great number of patients die on the waiting list for heart transplantation, mainly due to the unavailability of enough donor hearts. In view of this, alternative strategies are urgently required to generate new healthy hearts for transplantation. Whole heart organ engineering, which is based on the use of a decellularized intact heart scaffold as well as proper sources of seed cells (e.g., iPSCs), could be an alternative choice to overcome the acute shortage of donor heart (Fig. 6.26).

According to previous studies, decellularization and recellularization approach provides a promising technology for creating a personalized heart with normal cardiac functions. Basically, there are three elements that need critical considerations in the process of engineering a functional heart graft: (1) the 3D architecture of the scaffolds, (2) the appropriate composition and source of seed cells, and (3) the maturation of engineered constructs to develop cardiac functions [263]. After perfusion decellularization of native hearts, the obtained scaffolds can preserve the 3D anatomical structures, including a perfusable vascular bed, the ventricular geometry and heart valves, altogether provide a blueprint for entire organ engineering. The ideal seed cells for whole heart engineering should have robust proliferative capability and can give rise to all cardiac cell types including cardiomyocytes, smooth muscle cells, myofibroblasts, and endothelial cells. After recellularization, the engineered cell-scaffold constructs will be cultured in vitro (usually in a bioreactor) to facilitate the development of cardiac functions.

In 2008, Ott et al. developed the first whole organ decellularization technique through coronary perfusion with detergents to produce a decellularized heart scaffold, which preserved the ECM, the vascular architecture, the acellular valves, and the chamber geometry [85]. Next, they seeded the acellular heart scaffolds with intramural injection of neonatal cardiomyocytes and the perfusion of endothelial cells. After culture in a bioreactor that mimics the physiological flow properties for a period of 8 days, the recellularized constructs showed contractile function and were drug responsive [85]. This work paved the way for engineering a recellularized, functional organ scaffold. After the publication of this research, many decellularization and recellularization studies have been conducted in the field of organ engineering.

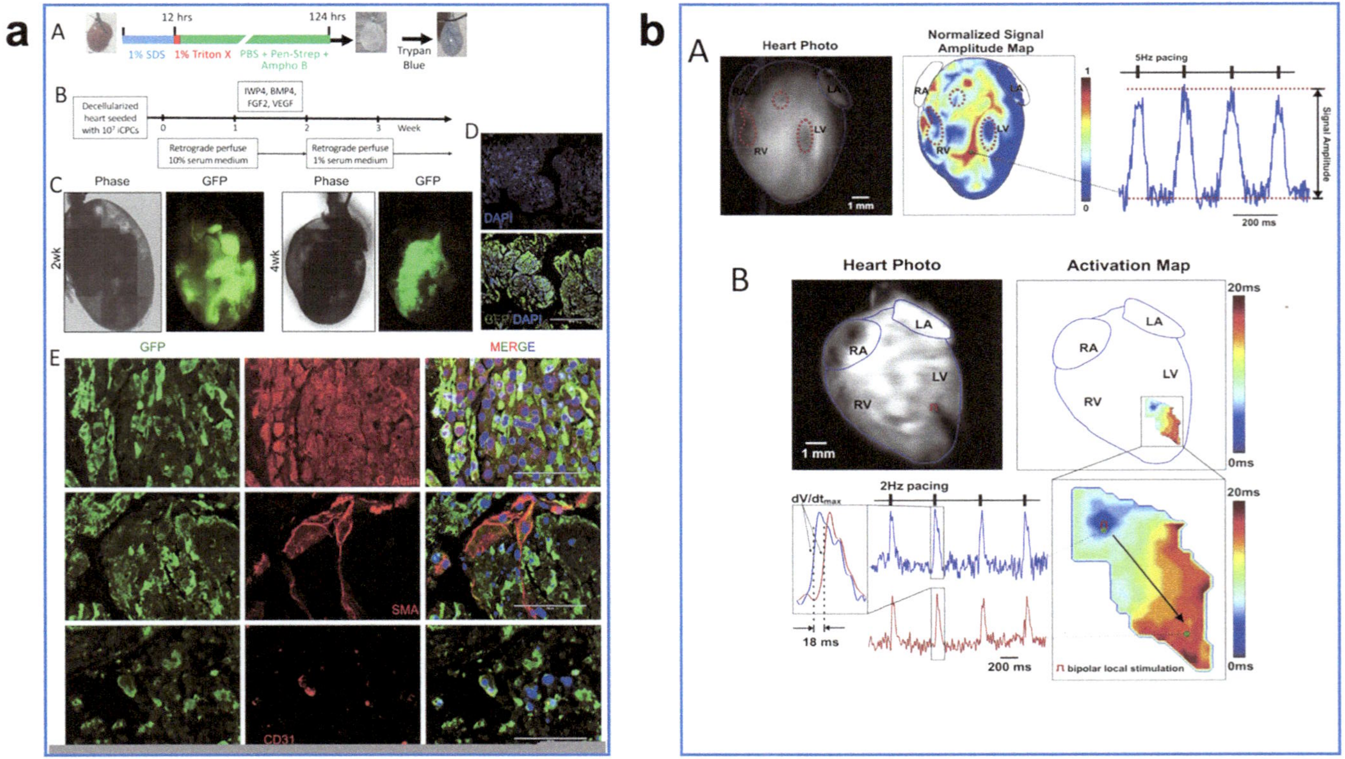

Fig. 6.26 An example of the engineering of whole heart based on the use of induced cardiac progenitor cells and decellularized mouse hearts. Induced cardiac progenitor cells were seeded on decellularized heart scaffolds and then induced to differentiate into cardiac lineages (**a**). After induction, induced cardiac progenitor cell-derived cardiomyocytes in the recellularized scaffolds showed functional maturation, such as Ca^{2+} transients and cell–cell coupling (**b**). (adapted from reference [24])

After the first study of Ott et al., Ng et al. [315] utilized mouse decellularized hearts as scaffolds and seeded them with human ESCs and mesendodermal cells (MECs) derived from human ESCs, respectively. The cardiac lineage differentiation of these cell types was compared. The results demonstrated that both cells upregulated the expression of cardiac markers, but they failed to generate beating function when subcutaneously implanted in SCID mice. Despite this deficiency, it is clearly shown that decellularized hearts can direct the differentiation of stem cells into the cardiac lineages. Following this work, Rajabi et al. [316] used human ESCs-derived cardiovascular progenitor cells (CPCs) as the seed cells to generate heart constructs. Through repopulating CPCs on bFGF-tethered decellularized rat hearts, they found that the seed cells enhanced their differentiation into the cardiomyocytes, endothelial cells, and smooth muscle cells. Notably, after 12 days of perfusion, spontaneous and synchronous contractions of the engineered hearts were observed. These researches provide valuable platform for engineering functional hearts.

Considering that personalized heart grafts hold tremendous promise for future heart disease treatment, iPSCs are attracting increasing interest for heart organ engineering, since they are reprogrammed directly from patient's adult cells and avoid potential immune rejection after implantation. Several attempts have been made to the fabrication of an engineered heart construct, based on the use of iPSCs or progenitor cells derived from iPSCs. By repopulating acellular mouse heart scaffolds with human iPSC-derived cardiovascular progenitors, Lu et al. [139] demonstrated that the seeded cells can migrate, proliferate, and differentiate into cardiac-related cells in the scaffolds such as cardiomyocytes, endothelial cells, and smooth muscle cells. Importantly, after 20 days of perfusion with growth factors, the engineered constructs showed spontaneous contractions, production of mechanical forces and responses to drug stimulation. Furthermore, it is observed that heart ECM enhanced cardiomyocyte propagation, differentiation and myofilament development from repopulated progenitor cells. Surely, this study provides a novel approach to engineer a personalized heart construct, which may find applications in preclinical testing or even for clinical practices after developments.

In another study, Guyette et al. [317] translated the perfusion decellularization method to human scale and successfully obtained human decellularized cardiac scaffolds. After this, they seeded the cardiac scaffolds with human iPSC-derived cardiomyocytes and generated tissues of increasing 3D complexity in vitro. They cultured the cardiac constructs for 120 days to show the definitive sarcomeric structure, the contractile force, the cell and matrix deformation, and the electrical conduction. To answer the question of whether human scale functional myocardial tissues could be created by this platform, they partially recellularized human decellularized hearts with iPSC-derived cardiomyocytes. After cultured under biomimetic conditions, the recellularized constructs developed into force-generating myocardial tissues and exhibited left ventricular pressure development, electrical conductivity, and metabolic functions. It is therefore seeming feasible to engineer a functional human myocardial-like tissue by using decellularized cardiac scaffolds in association with iPSC-derived cardiomyocytes.

Besides stem cells like ESCs and iPSCs as described above, cardiac-related cells (e.g., cardiomyocytes, CPCs, and ECs) and adult stem cells (e.g., MSCs) have also been used as seed cells to engineer heart grafts [225, 318–320]. For instance, Nguyen et al. have delivered HEK293 cells and primary human cardiac fibroblasts into decellularized hearts, respectively, to show the spatial distribution of cells and their integration with the scaffolds under perfusion in vitro [318]. Likewise, Robertson et al. have used rat aortic endothelial cells to reendothelialized the vascular conduits of decellularized rat hearts to reduce their thrombogenicity and to improve the contractility of the recellularized left ventricle [225].

Additionally, Yasui et al. have seeded acellular hearts with neonatal cardiac cells to study the conduction properties of the recellularized scaffolds [320]. Using high resolution cameras, they examined the excitation-propagation of spontaneous beatings of the engineered hearts. It is reported that the cardiac cells were able to migrate into the interstitial spaces via extravasation from coronary arteries. Generally, the bioengineered hearts showed well-organized stable excitation-propagation, but they also demonstrated arrhythmogenic propensity. Randomly mixed alignment of cardiomyocytes, smooth muscle cells, and endothelial cells was recorded, together with a decreased expression of connexin 43 and disarray of cardiomyocytes.

To determine the effect of decellularized cardiac scaffolds derived from fetal and adult animals on the biology of cardiac cells, Silva et al. have seeded heart-derived Sca-1$^+$ progenitors and neonatal cardiomyocytes on the scaffolds and further compared their biological performances, respectively [319]. It is observed that both scaffolds supported the 3D culture of cells, which migrated into the center of scaffolds, showed an elongated morphology and displayed excellent cell viability. However, when compared with decellularized cardiac scaffolds derived from adults, the fetal counterparts resulted in more Sca-1$^+$ cells and cardiomyocytes at the end of culture. The enhanced proliferation of Sca-1$^+$ cells on fetal decellularized cardiac scaffolds was found to rely on the beta1-integrin independent mitogenic signals. Furthermore, according to the results of gene expression profiles, Sca-1$^+$ cells cultured on the fetal hearts were anticipated to construct a permissive microenvironment for cardiomyogenesis, revealing a superior potential of fetal decellularized heart scaffolds for the proliferation and function of cardiac cells.

In another study, Sanchez et al. [180] used different cell types including human cardiac progenitor cells (CPCs), bone marrow-derived MSCs, HUVECs, H9c1 rat cardiomyocytes, and HL-1 cardiomyocytes to assess the cytocompatibility of decellularized human hearts. They cultured the cells on the slices of decellularized human left ventricle, and analyzed their survival, organization, gene expression, and electrical coupling. It is reported that neither human CPCs nor human MSCs adopted a cardiocyte morphology or organization, but both cells expressed cardiocyte genes; HUVECs lined the endocardium and vasculature of the scaffolds; cardiomyocytes formed nascent muscle bundles and showed mature calcium dynamics and electrical coupling within the scaffolds. Therefore, it is clear that decellularized human heart scaffolds possess the potential to promote the expression of cardiocyte genes in stem cells and also to organize cardiomyocytes to nascent muscles with electrical coupling ability.

As described above, many studies have explored the biological responses of seed cells cultured on acellular heart tissues. Exciting results have been reported in the literature such as good cytocompatibility of the scaffolds and especially a favorable inductive potential of decellularized hearts on the fate decision of stem/progenitor cells. Despite this, the regenerative ability of engineered hearts after in vivo implantation remains largely unknown, since only a few studies have determined the in vivo performance of recellularized heart scaffolds. In these preliminary studies, recellularized heart scaffolds have been heterotopically implanted in different animal models (e.g., rat, porcine and bovine) for up to 2 months. However, none of these researches have demonstrated long-term perfusion and proper cardiac functions of the engineered grafts in vivo [85, 225, 315, 321], although it has been reported that heterotopic transplantation of reendothelialized hearts displayed less blood clotting than acellular scaffolds [225].

In summary, current research progress has showed some success in the recellularization of decellularized whole hearts. However, to engineer an artificial heart functionally comparable to the native one, many challenges remain to be solved, such as issues regarding the selection of seed cells capable of generating all cellular components of cardiac tissues, the establishment of efficient cell seeding methods to guarantee good dispersion of seed cells within scaffolds, the optimization of organ culture methods to facilitate the development of cardiac functions of the constructs, practical approaches for sufficient reendothelialization of the vasculatures to avoid blood clotting in vivo, and especially the development of strategies to achieve long-term cardiac functions of the engineered hearts after implantation in vivo. Considering that there are many challenges ahead, clinical application of engineered whole hearts remains a remote prospect.

6.5.2 Liver

The liver is one of the most important solid organs with multiple physiological functions, mainly including albumin production, detoxification, and bile secretion. Liver dysfunction can lead to serious discomforts or even death in the patients. It is well known that a broad range of diseases are associated with liver dysfunction, such as hepatitis B virus infection, long-time alcohol consumption, and genetic disorders. All of them lead to cirrhosis, and eventually result in liver failure. Thus far, no definite treatment has been developed to cure liver failure. Although liver transplantation has achieved much progress, a great number of patients are still on the list waiting for liver transplantation because of severe deficiency of donor livers. With increased understanding of liver biology and rapid development in biotechnology, engineering an artificial liver graft has been considered as a potential alternative strategy to manage liver failure (Fig. 6.27).

In 2010, Uygun et al. [40] reported for the first time a novel method—decellularization of native liver in association with recellularization with hepatocytes—to generate a transplantable liver graft. In this study, they demonstrated that

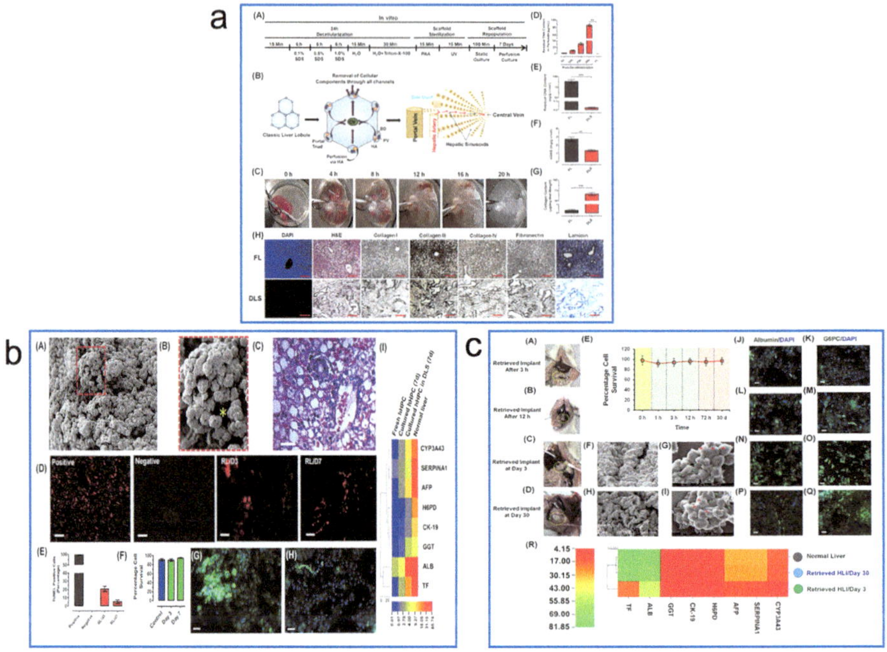

Fig. 6.27 Engineering a liver graft through the strategy of decellularization and recellularization. Decellularization of native livers can be performed by retrograde perfusion of reagents through the hepatic artery. Decellularization resulted in much lower content of residual DNA in the decellularized livers, which contained more than 50% of the GAGs of fresh liver and abundant collagen (**a**). Following decellularization, recellularization of the scaffolds was performed. Recellularized liver showed good organization and engraftment of the seed cells in the liver parenchyma and sinusoidal spaces, and the seed cells expressed liver cell markers in the scaffolds (**b**). When grafted into rat peritoneal cavity, recellularized livers showed good cell survival percentage post-transplantation. Furthermore, functional assessment of the retrieved implants showed comparable expression of liver markers in the retrieved implants at day 30 and normal liver (**c**). (adapted from reference [322])

decellularization of livers by portal perfusion with SDS could effectively remove the cellular components of native liver tissue while preserve the structure of microvascular and duct network. After recellularization with hepatocytes, the engineered liver scaffolds were capable of supporting liver-specific functions in vivo, such as urea synthesis, albumin secretion, and the expression of cytochrome P450, all of which exhibited a comparable level to normal liver counterparts. Thus, this study provided a proof-of-principle to generate a transplantable engineered liver graft for the management of liver diseases.

Since then, researches about liver decellularization and recellularization have been exploded. In 2013, Mirmalek-Sani et al. [323] decellularized porcine liver with the perfusion of PBS via hepatic artery, followed by successive SDS solution and Triton X-100 in saline buffer. After decellularization, they seeded the scaffolds with hepatoblastoma cells and implanted the cell-scaffold complexes into rodents to determine the immunogenicity of the scaffolds. According to the results, the seeded cells were found to survive for over 21 days without obviously increased immune responses, either in xenogeneic (pig-into-rat) or allogeneic (rat-into-rat) transplantation settings, as compared to the control group. This study suggests that the obtained decellularized liver scaffolds were nonimmunogenic and were suitable for hepatic bioengineering.

In order to determine whether there are some differences between homograft and xenograft of liver scaffolds, Sabetkish et al. [92] compared sheep decellularized liver matrix with rat decellularized counterpart. The results showed that homograft resulted in more organized structures than those in xenografts, indicating that transplantation to the same species would be more suitable for liver regeneration as compared to xenograft origins. To better preserve the matrix of liver, Jiang et al. [179] developed a cryo-chemical strategy to decellularize the liver and reseeded the acellular scaffolds with MSCs. They observed that MSCs cultured on the decellularized liver scaffolds were more efficient in hepatic differentiation than those cultured in two-dimensional condition, as evidenced by more expressions of hepatic-associated genes and proteins, increased albumin secretion, more urea production, and increased glycogen storage. To improve the migration and differentiation of seed cells, Nakamura et al. [324] had immobilized hepatocyte growth factor (HGF) to decellularized liver matrix (about 75% efficiency), and found that the HGF immobilized scaffolds could facilitate the migration of hepatocytes and stimulate their expressions of genes and proteins related to liver-specific functions.

Recent studies have shown that decellularized liver scaffolds provide a suitable microenvironment for the maturation of stem cell. In 2012, Ji et al. [178] explored the effect of decellularized liver scaffolds on the hepatocytic differentiation of MSCs. After cultured on the decellularized liver matrix with the presence or absence of growth factors, MSCs showed elevated expression of genes and proteins associated with liver function, revealing that decellularized liver ECM allows defined differentiation of adult stem cells into specific cell types. Additionally, when transplanted into CCl_4-injured mice, MSCs pretreated with a combination of dynamic culture on decellularized liver matrix and growth factors exhibited increased survival, engraftment, and further hepatic differentiation.

Besides MSCs, iPSCs represent another potential source of seed cell for liver organ engineering [119, 325, 326]. In the study by Park et al. [119], iPSCs-derived hepatocytes were seeded on decellularized liver scaffolds, and the interactions between cells and scaffolds were investigated. It was observed that decellularized liver scaffolds improved the hepatic maturation efficiency of iPSCs-derived hepatocytes, as evidenced by stimulated expression of albumin. Likewise, Jaramillo et al. [325] demonstrated that decellularized liver scaffold upregulated the expression of hepatic functional markers in human iPSCs as compared with standard differentiation conditions. Notably, the expression levels of hepatic transcription and nuclear factors were similar with primary human adult hepatocytes. Interestingly, analysis of the progression of differentiation of cells seeded on the decellularized liver scaffolds revealed that their expression of hepatic developmental markers was consistent with hepatic development. Additionally, the cells displayed a progressive loss of sensitivity to the substrate composition during the differentiation process. Overall, this work provided a protocol allowing for the induction of human iPSCs to differentiate into cells with phenotypically closer to mature hepatocytes.

The application potential of primary cells derived from liver tissues for engineering artificial liver organ has been studied in several reports [269, 322, 327, 328]. Hassanein et al. [327] conducted a comparison of the recellularization of decellularized liver scaffolds with either neonatal or adult liver cells. Cell slurries derived from neonatal or adult rat livers were seeded on the scaffolds via the bile duct. The cell slurries contained hepatocytes, cholangiocytes, and endothelial cells. Perfused through the portal vein with cell growth medium, the recellularized grafts showed increasing albumin concentration, and scaffolds recellularized with a neonatal liver cell slurry showed more robust albumin production than those recellularized with an adult liver cell slurry. Thus, this study suggested that neonatal slurry was superior to adult one. In another study, Butter et al. [269] had repopulated decellularized liver scaffolds with primary rat hepatocytes. To determine whether the function of the recellularized scaffolds was changed during in vitro culture, they detected the expression of hepatocyte-related genes and the secretion of albumin for consecutive 7 days. Unfortunately, a constant decrease of the function was recorded, particularly after 6 and 7 days of perfusion. Considering a constant loss of function, recellularized scaffolds should be implanted as soon as graft re-organization is completed.

Thrombosis and deficient vascularization of engineered livers are major obstacles needed to be overcome before clinical translation. Several studies have attempted to address these issues through scaffold modification or seeding with endothelial cells [90, 91, 275]. In 2011, Baptista et al. [91] successfully fabricated a liver-like tissue in vitro, providing a useful tool to create fully functional bioengineered livers for transplantation. They repopulated the decellularized liver scaffolds with endothelial cells and human fetal liver cells. The seed cells were found to locate at their putative native locations within the scaffolds and expressed typical hepatic, endothelial, and biliary epithelial markers. In 2018, Kojima et al. [90] have developed a cell seeding method to optimize the distribution of parenchymal and endothelial cells, mainly based on the structure of decellularized liver scaffolds. They seeded primary rat hepatocytes as well as liver sinusoidal endothelial cells into decellularized liver scaffolds via their biliary duct and portal vein, respectively. Appropriate distribution of the hepatocytes in the parenchymal space of the decellularized scaffolds was observed after seeding cells through biliary duct. In addition, after seeding through the portal vein, liver sinusoidal endothelial cells simultaneously lined the portal lumen. These results indicated that the recellularization method could be used to generate practical liver grafts with blood perfusion.

In addition to seeding with endothelial cells, scaffold modification with anticoagulation agents provides a practical strategy to avoid possible thrombosis of engineered livers after implantation in vivo. For instance, Bao et al. [275] have developed a 3D decellularized liver scaffold by using the method of layer-by-layer deposition of heparin. They repopulated the heparin-modified scaffolds with hepatocytes and implanted the recellularized scaffolds into the portal system of extended hepatectomized rats. The recellularized scaffolds were perfused by portal vein blood, and treatment of the extended hepatectomized rats with recellularized scaffolds resulted in prolonged survival of the animals, with the mean lifespan extending from 16 to 72 h. Also, the recellularized scaffolds improved the liver functions, with the observation of functional and viable hepatocytes at 72 h postoperation.

Some biomaterials and antibodies have showed the potential to improve the reendothelialization of decellularized liver scaffolds [329, 330]. By conjugating anti-CD31 antibody to the vasculature of acellular porcine liver scaffolds, Ko et al. [329] reestablished, for the first time, the vascular network within a scaffold of clinically relevant size in vitro, showing uniform coverage of the vessel walls with ECs. Upon blood perfusion of the scaffolds in vitro, the re-reendothelialized scaffolds greatly reduced platelet adhesion. Importantly, when grafted into recipient pigs, the re-reendothelialized scaffolds were capable of withstanding physiological blood flow and maintaining for up to 24 h, showing for the first time that re-reendothelialized liver scaffolds can be grafted and maintained in vivo.

In another study, Meng et al. [330] used gelatin hydrogel to increase the endothelial cell distribution in acellular liver scaffolds. They perfused endothelial cells in 5% gelatin hydrogel solutions via the portal vein for reendothelialization of the scaffolds. Comparing with the method of cell culture medium-based perfusion of cells, gelatin hydrogels-based perfusion of endothelial cells greatly increased the number of cells retaining in the scaffolds and the ratio of cells covering the vascular lumen. The blood retention ability of the vasculatures within engineered liver scaffolds was obviously improved after reendothelialization. Notably, 8 days post heterotopic transplantation of the reendothelialized liver scaffolds, active blood flows within the liver scaffolds were detected by Doppler ultrasound analysis.

Overall, recent progress in whole organ engineering has provided the possibility of fabricating an engineered liver with some of the basic physiological functions. However, many challenges are still needed to be overcome before clinical translation, especially the appropriate source of cells, the ideal protocols for recellularization, the design of bioreactors to optimize the engineering process of artificial livers, the establishment of a functional vasculature in the scaffolds, as well as the strategies to achieve a long-term stable biological function of the engineered livers after implantation in vivo.

6.5.3 Lung

The regeneration capacity of adult lungs is limited. Chronic lung diseases (e.g., cystic fibrosis, idiopathic pulmonary fibrosis, etc.) remain major causes of morbidity and mortality. For patients with end-stage lung disorders, lung transplantation is the only definitive treatment. However, similar to other organs, the availability of donor lungs is limited. Furthermore, graft rejections and long-term immune suppression

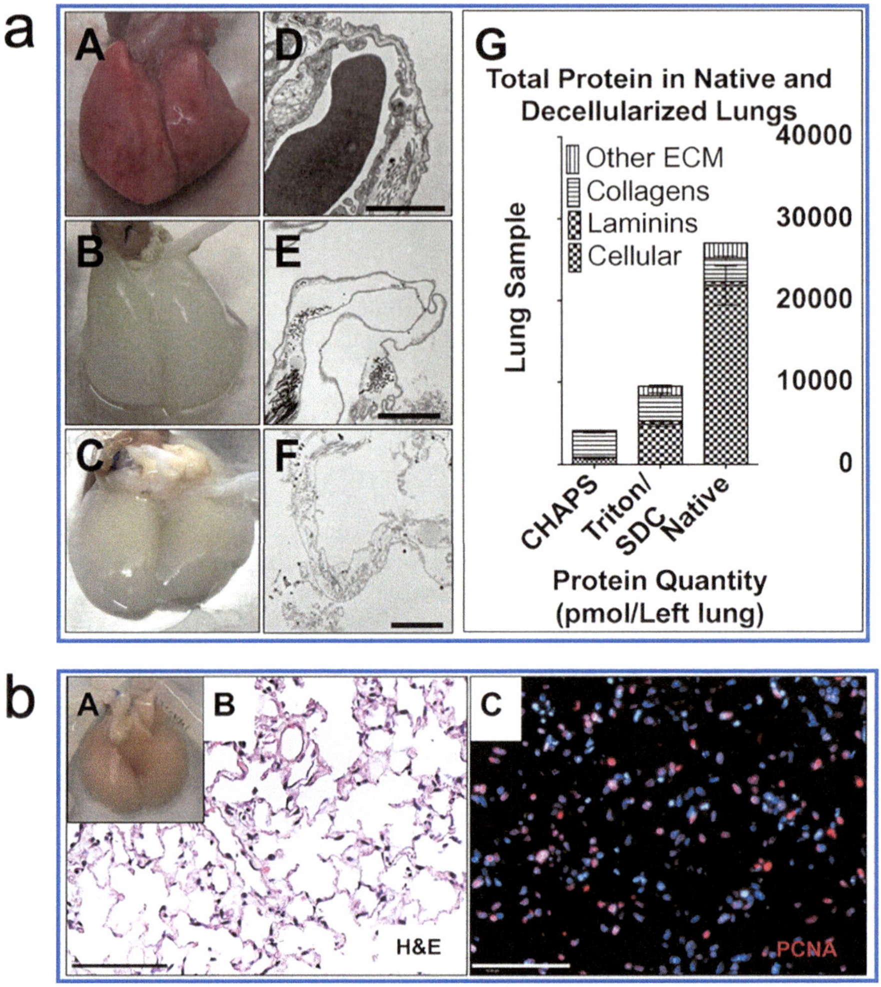

Fig. 6.28 An example of the generation of a bioengineered lung through decellularization and recellularization strategy. After decellularization, acellular lung scaffolds showed similar structure and ECM composition to that of native rat lungs (**a**). Recellularization of the acellular lungs revealed good distribution and proliferation of seed cells, as indicated in the results of HE staining and immunofluorescence staining of PCNA (**b**). (adapted from reference [331])

impede the application and effectiveness of this treatment. With the development of organ decellularization technology and stem cell biology, engineering an artificial lung graft offers an emerging solution to overcome the shortage of donor lungs. Based on the strategy of decellularization and recellularization, several studies have reported the fabrication of a bioengineered lung in vitro (Fig. 6.28), and especially the repair potential has been evaluated in some researches [38, 86, 271, 332].

The first experiments of de- and recellularization of whole lungs were reported in 2010 by Ott et al. and Petersen et al. [38, 86]. Based on the perfusion-mediated organ decellularization methods, they first generated decellularized whole rat lungs, and used them as a platform to create a bioengineered lung through the recellularization of pneumocytes and endothelial cells. The cell-seeded constructs were cultured under biomimetic conditions by using a bioreactor that perfused and ventilated the scaffolds to provide physiological environment of developing lung. The recellularized constructs were found to be able to exchange oxygen and carbon dioxide, with the gas exchange ability comparable to isolated native lungs. Importantly, the recellularized constructs retained the hierarchical structure of the airway and vasculature, which can be used to connect with the host pulmonary vein, artery, and bronchus. When orthotopically transplanted into rats, the recellularized constructs were perfused by the circulation of the recipient and showed gas exchange capacity in vivo. However, a few hours post-transplantation, the bioengineered lungs showed impaired function because of pulmonary edema [86] or bleeding into the airways [38].

In a following study conducted by Song et al. [271], bioengineered lungs showed in vivo function for as long as 14 days. The authors decellularized cadaveric rat lungs by detergent perfusion and seeded the acellular scaffolds with endothelial and fetal lung cells. After biomimetic organ culture for 7 days, the engineered constructs were orthotopically implanted in rats with left pneumonectomy. The animals could be fully recovered, and the engineered lungs provided oxygenation at levels comparable to that of rats received cadaveric lung transplantation for 7 days. Nevertheless, because of the progressive consolidation of the grafts and inflammation, the compliance and oxygenation of the grafts declined gradually, being similar with pneumonectomized rats at day 14 after transplantation.

Although several milestones have been achieved in rodent models, scale-up of whole lung engineering is necessary for future application. Up to date, only a few studies have successfully created bioengineered lungs with clinically relevant size and determined their repair potential in large animal models. Recently, Zhou et al. have performed a large animal study of orthotopic transplantation of bioengineered lungs [332]. They seeded the acellular porcine lungs with HUVECs and human airway epithelial progenitor cells and cultured the recellularized constructs for 6 days for maturation, which was critical for sufficient anti-thrombotic function of the pulmonary vasculature and the gas exchange of the grafts. After transplantation in porcine recipients, these constructs could withstand the blood flow from the pulmonary circulation of the recipients. During the 1-h in vivo observation of the grafts, it was noted that the grafts could exchange gases upon ventilation. However, when comparing with native lungs, gas exchange and compliance of the engineered lungs were significantly lower.

Indeed, recent years have witnessed the successful application of perfusion decellularization approach to obtain acellular organs, such as the fabrication of acellular lungs from porcine [220, 332], rhesus macaques [216], and human [140, 220, 262, 333]. Notably, decellularization of human lungs that are not suitable for transplantation may provide an "ideal" source for whole organ engineering.

However, the availability of such lungs is very limited and whether diseased organs could be used in whole organ engineering remains unclear. Sokocevic et al. [334] have reported that acellular lungs obtained from mice with induced emphysema or fibrosis significantly affected the growth and differentiation of seed cells, which indicates that the choice of proper donors is important.

Besides the research of human acellular lungs, other research groups have determined whether acellular porcine lungs could serve as a proper substitute for decellularized human lungs [236, 335]. Encouraging results have been reported: For instance, O'Neil et al. demonstrated that decellularized lungs from pigs possessed similar properties to those of human origin [236]. Furthermore, both scaffolds have showed good biocompatibility. For example, Gilpin et al. have reported that decellularized human lungs supported the proliferation and the lineage commitment of lung progenitor cells derived from human iPSCs [335]. These studies indicate the possibility of porcine acellular lungs as a platform for engineering lungs for transplantation. However, it should be noted that successful engineering of an artificial lung graft with the size of human lungs has not been reported so far. Taken together, despite substantial progress that has been achieved, we are only just beginning to grasp the complexity of this field.

6.5.4 Kidney

The number of patients with end-stage kidney disease (ESKD) is increasing. Currently, kidney transplantation is regarded as the only definitive treatment for ESKD patients, especially concerning the morbidity of patients, their survival, and health economics. However, a severe shortage of kidney donors has limited the application of this treatment, resulting in an extremely long list of patients awaiting transplant and an obvious increase of mortality in the wait list. For instance, approximately 100,000 Americans are waiting for a donor kidney. However, only 18,000 transplants are performed annually [336]. Thus, alternative treatments are urgently needed.

Delivery of stem cells into the sites of renal failure may alleviate some of the burden of chronic renal failure. However, according to previous evidence, it seems unlikely that transplantation of stem cells alone could successfully treat chronic renal failure [337]. Therefore, several attempts have been made to create a bioartificial device to partially substitute the structure and function of native kidneys, such as current hemodialysis techniques [338]. However, although these techniques are capable of clearing solute metabolic wastes in the blood, they fail to provide other physiological functions of kidneys, such as the metabolic, endocrine, and transport functions.

To address the shortcomings mentioned above, researches aiming to engineer a viable tubular device to supplement the hemofiltration have been carried out. For instance, Humes et al. have developed a renal assist device through the combination of traditional hemofiltration system and the endocrine and metabolic functions of

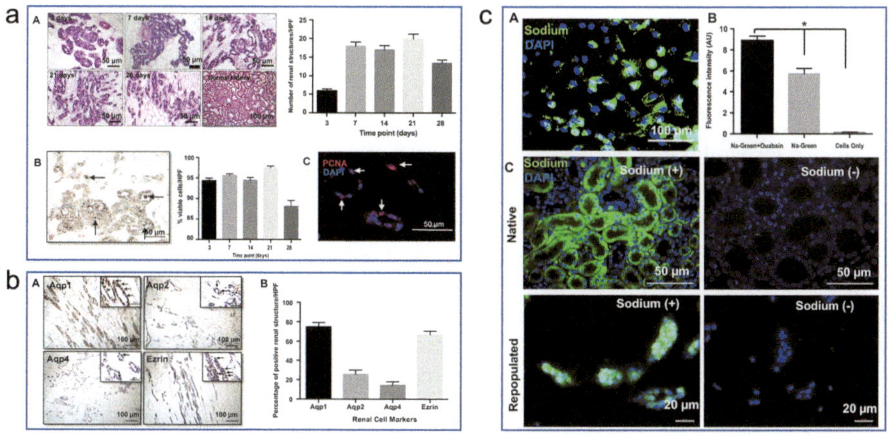

Fig. 6.29 An example of the recellularization of acellular kidney scaffolds to generate an engineered kidney graft. After cell seeding and cultured in a bioreactor, renal cells formed tubular structures within the acellular kidney scaffolds (**a**), maintained renal specific phenotypes (**b**), and demonstrated the ability of re-adsorption of electrolyte (**c**). (adapted from reference [340])

renal tubular cells [339]. These renal assist devices hold great promise to improve the life quality of patients with ESKD. They provided some of the physiological function of kidneys, but still fail to provide lifelong protection and cannot serve as an alternative to a natural normal kidney.

The idea of fabricating an engineered kidney has the potential to address the challenges mentioned above by providing engineered kidney transplants that fully replicate renal functions. Therefore, a whole organ approach to permanently replace kidney functions has gained considerable interest, especially the potential clinical application as an alternative to donor kidney transplantation for patients with ESKD.

Similar to the other organs discussed above, the implementation of decellularization-recellularization strategy has shown promise for engineering a bioartificial kidney (Fig. 6.29). After the development of detergent-based organ decellularization techniques, acellular kidneys have been shown to maintain the characteristics of renal ECM [341]. Several groups have determined the performance of acellular kidney scaffolds after implantation in animals without previous recellularization in vitro [31, 221, 342–345]. For instance, to assess whether acellular porcine kidneys would sustain normal blood pressure after implantation in vivo, Orlando et al. [221] have implanted the scaffolds in pigs without previous recellularization and measured the mechanical characteristics of the vascular tree. The results showed that the scaffolds maintained an intact vascular tree and the vasculature would sustain normal blood pressure.

Similarly, Kajbafzadeh et al. [31] have compared the effects of different decellularization protocols on the preservation of ECM and the vasculature integrity of acellular kidney scaffolds. After implantation in the sheep, the efficacy of well-preserved acellular scaffolds and their vasculature networks in post renal transplant outcomes have been recorded, as demonstrated that scaffolds with an intact vasculature avoided fluid extravasation or leakage after implantation in vivo.

Likewise, in another research, Zambon et al. [345] have addressed a common concern regarding the decellularization of kidneys, namely the disruption of glomerular microarchitectures after perfusion of detergents through renal artery. In this study, they first demonstrated the damage of glomeruli microarchitectures and peritubular capillaries after decellularization of the kidneys. To better preserve the intact and functional microvascular architectures, they developed an optimized decellularization protocol through the modification of decellularization parameters, such as detergent concentration, flow rate, and decellularization time. Following the improved protocol, they recellularized the obtained scaffolds with ECs and determined the in vitro blood perfusion of the reendothelialized constructs, which demonstrated successful perfusion with whole porcine blood for 24 h without thrombosis. This research may contribute to long-term vascular patency of acellular kidneys after implantation.

Using the methods of resin casting at electronic microscopy and pulse-wave measurement, Peloso et al. [342] have evaluated the morphometry and resilience of vasculature in the acellular scaffolds derived from discarded human kidneys. Their results demonstrated that after decellularization, the scaffolds preserved the morphometric characteristics of glomerulus as well as the physiological parameters of the vasculature bed (pressures and resistances). Because of the well-preserved structures and the good function of vasculature, this work implied that discarded human kidneys could be used as a valuable source of organs for producing acellular renal scaffolds.

Several reports have indicated that decellularized kidney scaffolds would be useful in promoting renal recovery in the treatment of kidney diseases [343, 344]. For instance, Yu et al. [343] have showed the potential of acellular kidney scaffolds in renal function recovery. They first grafted decellularized kidney scaffolds into partially nephrectomized rat kidneys, and then they found an increase of renal size as well as significant recovery of renal functions. In the repair area where grafted scaffold was observed, regenerated renal parenchyma cells were observed. Furthermore, renal progenitor cells in the scaffold-grafted kidneys were more abundant than that in the control group. In the following study, they verified that the improved renal regeneration mediated by acellular kidney scaffolds could be attributed to the regenerated glomeruli [344]. Further in-depth mechanism analysis revealed that hypoxia-inducible factor (HIF)-1alpha would be the critical factor involved in the induced renal regeneration, providing new insights into the treatment of kidney diseases.

As described above, successful decellularization of whole kidneys derived from different species like rat, sheep, and human have been reported in the literature. Despite many successes in the decellularization protocols, recellularization of acellular kidney scaffolds is much more challenging. In 2009, Ross et al. [131] reported for the first time the de- and recellularization of acellular kidneys. It is well known that there are more than 26 types of cells in the kidney. Therefore, identification of a suitable cell source is critical for whole kidney engineering. Since ESCs can generate all renal cell types, Ross et al. used murine ESCs as the seed cell to recellularize rat acellular kidneys through the renal artery and ureter, and further determined the proliferation and differentiation of the seed cells. It was observed that the cells successfully repopulated the glomerular, vascular, and tubular structures of the scaffolds. Interestingly, the cells lost embryonic phenotypes and expressed markers for differentiation, which indicates that acellular kidney scaffolds may direct kidney regeneration.

In 2017, Remuzzi et al. [346] have determined the possibility of in vitro or in vivo recellularization strategies for acellular kidney scaffolds. According to their results, the decellularized kidneys implanted in rats failed to be repopulated with host cells. Therefore, in vitro recellularization is necessary. Nevertheless, due to the permeability of renal ECM, uniform cell delivery was not achieved in those cell infusion protocols they employed, which just resulted in very limited and inconsistent cell seeding. This study highlighted the physical barriers that hamper recellularization of decellularized kidney scaffolds.

The first successful orthotopic transplantation of a bioengineering kidney which resulted in partial restoration of renal function was reported by Song et al. in 2013 [39]. In this work, they decellularized cadaveric rat kidneys by detergent perfusion and repopulated the acellular scaffolds by ECs through the renal artery to prevent thrombosis. Meanwhile, rat neonatal kidney cells were seeded into the scaffolds through the ureter. To provide whole organ culture conditions, the scaffolds were maintained in a perfusion bioreactor with controlled conditions. For reendothelialization, the scaffolds were cultured in the bioreactor for 3–5 days. Following this, they were cultured for 12 days to mature the function of neonatal kidney cells. After bioreactor-based culture, the recellularized constructs were found to be able to produce rudimentary urine in vitro. When transplanted orthotopically in rats, the recellularized constructs were perfused by the circulation of the recipient and showed urine production and clearance of metabolites. When compared with controls, the regenerated kidneys produced a lesser amount of filtrate, which contains higher levels of glucose and albumin, and lower levels of urea and creatinine. Despite functional immaturity of bioengineered kidneys, this landmark research offered an innovative technology and paved the road for further research.

Because of the lack of efficient cell seeding methods, the fabrication of engineered kidneys with sufficient renal function is still extremely challenging. To overcome this obstacle, Abolbashiri et al. [340] have reported the establishment of an effective recellularization approach by using pig kidney cells to repopulate acellular porcine kidney scaffolds. To achieve robust colonization of the acellular kidney scaffolds, they first developed cell culture methods to efficiently expand primary porcine renal cells and maintain their normal renal phenotype. Next, an effective cell seeding method was employed, following with bioreactor-based culture of the constructs. It is demonstrated that the seeded cells formed tubule-like structures and expressed normal renal tubule markers. Importantly, cells within the engineered constructs showed normal renal functions such as absorption of electrolyte and production of erythropoietin. These outcomes highlight the possible applications of engineered kidneys.

According to the studies presented above, many progresses have been made in the field of kidney engineering in recent years. However, a great number of obstacles remain to be overcome before successful reconstruction of an engineered kidney that possesses a complete renal function. None of the currently engineered kidneys can perform the vital renal functions in the recipient animals for longer than a few hours. Many challenges need to be addressed before engineered kidneys can be used in clinic.

6.5.5 Pancreas

The pancreas, constituted predominantly by exocrine cells and pancreatic matrix, is an organ with the function of secreting digestive enzymes and buffer solution into the duodenum. Approximately 1% of the pancreas is dominated by islets of Langerhans that are widely dispersed throughout the whole pancreas. Type one diabetes (T1D), an intractable autoimmune disorder, is characterized by the dysfunction of pancreas islets due to the specific destruction of insulin-producing β-cells. T1D patients can suffer from sustained hyperglycemia or toxic acids accumulated in the urine and blood. They frequently predispose to many complications, such as infections, cardiovascular disorders, chronic kidney diseases, and osteoporosis.

As reported in the literature, three million individuals in the USA have T1D [347], and the estimated population of patients will exceed five million by 2050. Effective strategies to reduce the medical expenditure as well as improve the life quality of the patients are urgently needed. In the clinic, healthy dieting, regular exercise, and exogenous insulin supplement are beneficial to the management of the levels of blood glucose, yet there are some drawbacks to be overcome. For example, long-time insulin treatment has many side effects, such as renal dysfunction and hepatic impairment. Thus, to address these issues, new treatments for T1D patients are warranted.

The fabrication of decellularized pancreas scaffolds from mouse, rat, pig, and human has been reported [32, 70, 167, 217, 233, 268, 348–353]. Following these achievements, many studies have demonstrated successful applications of decellularized pancreas scaffolds for cell seeding, proliferation, and the secretion of insulin [217]. An appropriate combination of decellularized pancreas scaffolds, stem cells, and bioactive molecules would be of great therapeutic potential for diabetic patients.

In 2013, Goh et al. first investigated the potential of decellularized pancreas scaffolds for whole pancreas engineering in vitro [217] (Fig. 6.30). In this study, a specific perfusion-decellularization method was utilized to remove the nuclear as well as the cellular components of native pancreas, while concurrently did not destruct the intricate vascular architecture, the duct network, and the innate ECM of native organ. After decellularization, the decellularized mouse pancreas scaffolds were recellularized with acinar and beta cell lines, and then cultured in vitro for

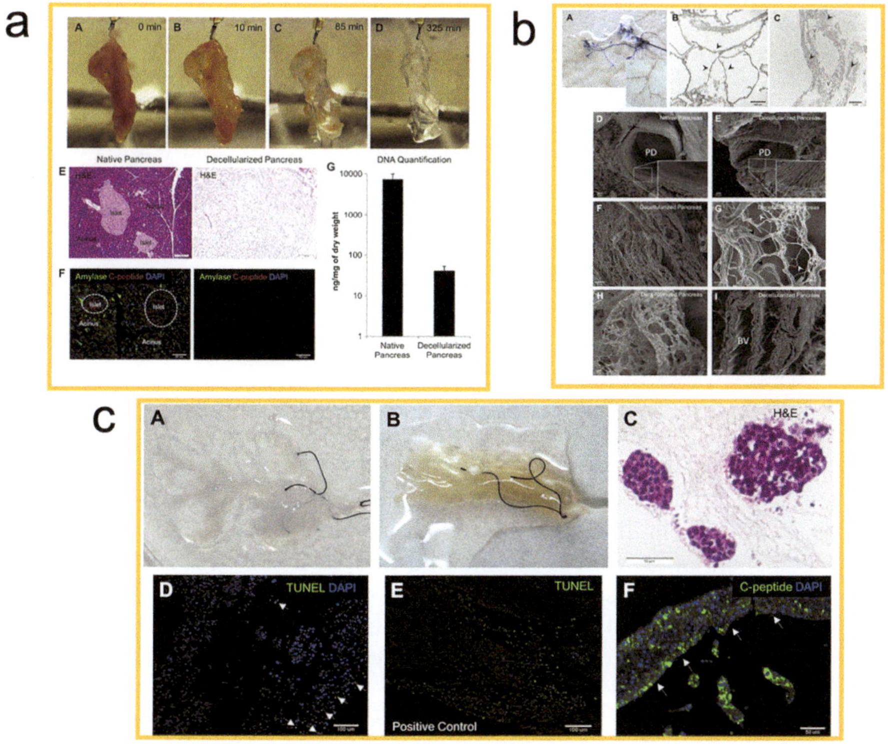

Fig. 6.30 An example of the recellularization of decellularized pancreas scaffolds. Decellularized mouse pancreas was fabricated by a perfusion-decellularization method, which resulted in the removal of resident cells and major pancreatic components (**a**). The ultrastructure of mouse pancreas was well-preserved, as revealed in an intact vasculature and basement membrane of the scaffolds, and the preservation of microstructure of the pancreatic duct (**b**). Recellularization of the decellularized pancreas via multi-step cell infusion showed a good cytocompatibility of the scaffolds as well as the engraftment of seeded cells at the surrounding parenchymal region (**c**). (adapted from reference [217])

5 days. It is observed that the pancreatic cells seeded on the scaffolds upregulated their gene expression of insulin. This work indicates that decellularized pancreas scaffolds can provide a promising material to enhance the function of pancreatic cells.

Pancreatic islets, alone or in combination with stem cells, have been used to repopulate decellularized pancreas. For instance, after successful fabrication of porcine decellularized pancreas, Mirmalek-Sani et al. have determined whether it is a suitable platform for pancreas engineering with the use of stem cells and pancreatic islets [167]. The scaffolds supported cell adhesion and maintained cell functions. When compared with those islets plated in Petri dishes, islets seeded on decellularized pancreas significantly increased insulin secretion, suggesting a good potential of the scaffolds for whole organ engineering.

In 2016, Katsuki et al. used a multi-step infusion technique to seed decellularized porcine pancreas with isolated primary porcine islets [350]. They found that the decellularized pancreas maintained the cellular integrity of the islets, which repopulated the parenchymal space of the scaffold. After repopulation, the islets survived for 4 days; in the glucose stimulation-insulin secretion (GSIS) test, the concentration of secreted insulin was comparable to those islets cultivated in normal medium. Likewise, in another study, Napierala et al. [268] repopulated decellularized rat pancreas with islets of Langerhans. After infusion of the islets, they were viable and functional, as revealed in TUNEL staining and GSIS test. Interestingly, the infused islets could leak from pancreatic duct into peri-ductular space. This work provides first proof-of-concept for repopulation of decellularized rat pancreas with functional islets.

Considering the generation of patient-specific pancreas, several groups have used iPSCs or iPSC-derived cells as the seed cells that were in association with decellularized pancreas scaffolds [32, 348, 351]. For example, Wan et al. [351] have evaluated whether decellularized pancreas scaffolds could improve the production of β-like cells from iPSCs. The results demonstrated that decellularized pancreas support the survival of mouse iPSC-derived ß-like cells; and importantly, the gene expression of insulin was accelerated when compared to plate-based cell culture. This work provides preliminary evidence for the generating of bioengineered pancreas from decellularized scaffolds that were in combination with iPSCs-derived cells.

More recently, Berger et al. [32] have investigated the regulatory effects of decellularized matrix on the multilineage differentiation of human iPSCs. They demonstrated that organ-specific ECM cues, including the physical structure, composition, and biophysical properties, have a significant impact on the fate decision of human iPSCs. Furthermore, after investigating the proteomics of decellularized rat pancreatic ECM hydrogel, Bi et al. [348] demonstrated that type V collagen is a critical regulator in promoting the generation of islet organoids from human iPSCs. These studies highlighted the valuable role of particular biomaterial as an important regulator in pancreatic bioengineering.

Since the components of ECM scaffolds play critical roles in guiding the responses of stem cells, several studies have tried to optimize the fabrication of decellularized pancreas, with the aim to preserve a more physiologically relevant microenvironment for seed cells [233, 349]. For instance, to retain the ECM components to the greatest extent and to improve the efficiency of decellularization, Elebring et al. have used cold-perfusion technology to produce decellularized pancreas. After recellularized with human fetal pancreatic stem cells and cultured for 14 days, the cell-scaffold constructs showed positive expression of exocrine and endocrine markers [233]. In another study, normal and streptozotocin-induced T1DM or T2DM rat pancreas were decellularized and compared [349]. The results showed that both scaffolds possessed excellent cytocompatibility, supporting the attachment and proliferation of MIN-6 beta cells and HUVECs. According to this pilot study, it is possible that decellularized pancreas derived from diabetic patients may be useful for pancreas bioengineering.

Revascularization remains a big challenge for engineering a functional pancreas suitable for transplantation in vivo. An adequate reendothelialization of the vascular network of decellularized organs is a prerequisite for successful revascularization of the engineered organs in vivo. EPCs are widely used as seed cells for reendothelialization of decellularized pancreas [70, 352, 353]. In 2018, Guo et al. seeded EPCs on rat decellularized pancreas scaffolds to obtain a pre-vascularized scaffold, and then implanted the reendothelialized scaffolds in vivo to detect their ability to regenerate new blood vessels [70]. The results showed that, after reendothelialization in vitro, the vessel structures of the decellularized pancreas were covered with EPCs. After implantation in vivo, the anastomosis with host vascular network via new blood vessel formation was earlier in the pre-vascularized pancreas scaffolds than the control group.

Surface modification of the vasculature of decellularized pancreas scaffolds has been utilized in some studies to improve the efficiency of endothelialization of the scaffolds. Using heparin modification under mild conditions, Xu et al. [353] demonstrated that heparin-pancreatic decellularized scaffolds promoted the proliferation of HUVECs in vitro. When implanted in vivo, the modified scaffolds showed improved angiogenetic ability. Similarly, Wan et al. [352] reported an efficient approach to enhance the reendothelialization of rat decellularized pancreas scaffolds via covalent conjugation of a functional peptide, namely GRGDSPC peptide, to the vasculature of scaffolds. The results showed that GRGDSPC-conjugated scaffolds promoted the proliferation of HUVECs in vitro, and the angiogenetic ability of the modified scaffolds was significantly improved.

Despite many in vitro experiments as described above, few studies have determined the performance of engineered pancreases in vivo. It is generally acknowledged that decellularized pancreas can provide a favorable microenvironment to induce the differentiation of stem cells toward pancreatic cells. However, the question regarding which one is the key factor in guiding the fate decision of stem cells remains elusive. Further investigations are needed to facilitate the revascularization of engineered pancreas for future applications.

6.6 Conclusion and Outlook

As a general conclusion, whole organ engineering is a breakthrough in the field of tissue engineering and regeneration medicine, as it integrates the state-of-the-art progresses of biology, medicine, and engineering. There is no doubt that encouraging achievements have been made in this field in the past decade. With the basis of seeding cells, decellularized whole organ scaffolds and recellularization techniques, bioengineered whole organ is promising for future clinic application as alternative transplants addressing the shortage of donor organ for chronic end-stage organ failure.

However, bioengineered whole organs are still generally immature at current stage and not yet available for clinical applications. Optimization concerning the seeding cells, decellularization protocol, recellularization strategies, and bioreactor design have been discussed in previous sections. Moreover, additional considerations are needed for expected translation of whole organ engineering from bench to bedside.

6.6.1 Innervation

Innervation describes the distribution of nerves and the regulation of target body part. Innervation of organs is very important for relevant physiological functions, especially for heart, liver, etc. For example, sympathetic and parasympathetic nerves can modulate the cardiac physiology activities including heart rate, conduction velocity, and contraction force [354]. After heart transplantation, denervation can lead to exhaustion of circulating catecholamines, elevation of resting heart rate, and dysregulation of nocturnal blood pressure, and reduction of exercise capacity [355]. Reinnervation may happen in vivo, though usually inconsistently and incompletely. It can improve cardiac contraction, blood circulation, and life quality for heart transplantation recipients [356]. Besides, adequate reinnervation after heart transplantation enables angina to warn the individual as a lifesaving symptom in case of myocardial infarction [357]. For bioengineered whole heart, electric stimuli can be supplied to enhance the maturation of cardiomyocytes. However, it is likely very challenging to achieve proper innervation of bioengineered whole heart and other organs. With some progresses having been at tissue level or mini organ level such as organ on chip [358], scaling up to whole organ level is somewhat possible. A future direction for whole organ engineering is adequate innervation before or after in vivo transplantation.

6.6.2 Good Manufacturing Practice

Doubtlessly, fabrication of clinical-grade bioengineered whole organs should follow good manufacturing practice (GMP) guidelines in the view of translational medicine to ensure their consistency and quality for eventual transplantation, which involves all the procedures in whole organ engineering, including the harvesting, preparation, and evaluation of the seeding cells and decellularized whole organ scaffolds, as well as the design and utilization of recellularization strategies and bioreactors devices, and whole agents, equipment, and environment, etc.

But due to the fact that bioengineered organs have living cells that are almost impossible to be standardized in the same way as conventional pharmaceuticals, the establishment of GMP for whole organ engineering is truly challenging. As a reference, some parameters for cryotherapy's GMP may be adapted for the emerging demands. Such parameters comprise absence of infectious pathogens or process-related impurities, demonstration of bioengineered construct's contents and functionality, documentation of every result (www.eurostemcell.org).

6.6.3 Long-Term Investigation

As introduced in previous sections, bioengineered organs showed somewhat practicability in pioneering proof-of-concept researches. For example, Ott et al. showed that bioengineered lungs can survive for a few hours after transplantation into rats, with inspiring O_2 and CO_2 exchange functions for the first time, but subsequently pulmonary edema happened and reintubation was required [86]. Indeed, bioengineered whole organs are still at early stage, safety and effectiveness need to be fully demonstrated with the advancement of strategies and techniques in future studies.

Furthermore, most studies only focused on short time evaluations, thus data of long-term investigation are insufficient. As transplanted organs are expected to serve for a prolonged period of time, whether bioengineered whole organs are competent remains to be clarified in future researches.

6.6.4 Regulations

Any medical researches should comply with generally accepted medical ethics, and be regulated properly by local agencies. Whole organ engineering involves multiple elements and steps, making its regulation complicated. Previously, ECM scaffold materials have been usually considered as biological products or medical devices regarding the regulation by authorities. In the USA, stem cells are supervised by FDA as biological drugs since 2017 [359]. In China, current regulation for clinical applications of stem cells employs a dual-track approach since 2015, i.e. either as medical techniques or as biological drugs, co-supervised by former National Health and Family Planning Commission of the PRC and former China Food & Drug Administration (now by National Health Commission of the PRC and National Medical Products Administration). Current regulation rules for stem cells provide a reference for future policy setting.

In whole organ engineering, the decellularized whole organ scaffolds become recellularized to produce a combinatorial product for the goal of transplantation, accordingly regulation of this field or this industry in the future may need to comply with more restrict rules. Based on the primary mechanism of function of whole organs, the category can be as a "biologic" or as a "medical device," with arguments made for either [37], but we hold the view that biologic products should be more suitable for a more rigorous regulation. Nevertheless, it is definitely necessary to regulate bioengineered whole organs for supervision as well as promotion of the field.

References

1. World Health Organization. World health statistics. Geneva: World Health Organization; 2013.
2. Song JJ, Ott HC. Organ engineering based on decellularized matrix scaffolds. Trends Mol Med. 2011;17(8):424–32.
3. Go AS, Mozaffarian D, Roger VL, Benjamin EJ, Berry JD, Borden WB, Bravata DM, Dai S, Ford ES, Fox CS, Franco S, Fullerton HJ, Gillespie C, Hailpern SM, Heit JA, Howard VJ, Huffman MD, Kissela BM, Kittner SJ, Lackland DT, Lichtman JH, Lisabeth LD, Magid D, Marcus GM, Marelli A, Matchar DB, McGuire DK, Mohler ER, Moy CS, Mussolino ME, Nichol G, Paynter NP, Schreiner PJ, Sorlie PD, Stein J, Turan TN, Virani SS, Wong ND, Woo D, Turner MB, C. American Heart Association Statistics, S. Stroke Statistics. Heart disease and stroke statistics–2013 update: a report from the American Heart Association. Circulation. 2013;127(1):e6–e245.
4. Badylak SF, Taylor D, Uygun K. Whole-organ tissue engineering: decellularization and recellularization of three-dimensional matrix scaffolds. Annu Rev Biomed Eng. 2011;13 (1):27–53.
5. Bijonowski BM, Miller WM, Wertheim JA. Bioreactor design for perfusion-based, highly-vascularized organ regeneration. Curr Opin Chem Eng. 2013;2(1):32–40.
6. Hillebrandt KH, Everwien H, Haep N, Keshi E, Pratschke J, Sauer IM. Strategies based on organ decellularization and recellularization. Transpl Int. 2019;32(6):571–85.
7. Pollara J, Edwards RW, Lin L, Bendersky VA, Brennan TV. Circulating mitochondria in deceased organ donors are associated with immune activation and early allograft dysfunction. JCI Insight. 2018;3(15):e121622.
8. Moser PT, Ott HC. Recellularization of organs: what is the future for solid organ transplantation? Curr Opin Organ Transplant. 2014;19(6):603–9.
9. Langer R, Vacanti JP. Tissue engineering. Science (New York, NY). 1993;260(5110):920–6.
10. Griffith LG, Naughton G. Tissue engineering–current challenges and expanding opportunities. Science (New York, NY). 2002;295(5557):1009–14.
11. Huang YZ, Gou M, Da LC, Zhang WQ, Xie HQ. Mesenchymal stem cells for chronic wound healing: current status of preclinical and clinical studies. Tissue Eng Part B Rev. 2020;
12. Zhang XR, Huang YZ, Gao HW, Jiang YL, Hu JG, Pi JK, Chen AJ, Zhang Y, Zhou L, Xie HQ. Hypoxic preconditioning of human urine-derived stem cell-laden small intestinal submucosa enhances wound healing potential. Stem Cell Res Ther. 2020;11(1):150.
13. Campos Y, Almirall A, Fuentes G, Bloem HL, Kaijzel EL, Cruz LJ. Tissue engineering: an alternative to repair cartilage. Tissue Eng Part B Rev. 2019;25(4):357–73.
14. Huang YZ, Xie HQ, Silini A, Parolini O, Zhang Y, Deng L, Huang YC. Mesenchymal stem/progenitor cells derived from articular cartilage, synovial membrane and synovial fluid for cartilage regeneration: current status and future perspectives. Stem Cell Rev Rep. 2017;13 (5):575–86.
15. Adamowicz J, Pokrywczynska M, Van Breda SV, Kloskowski T, Drewa T. Concise review: tissue engineering of urinary bladder; we still have a long way to Go? Stem Cells Transl Med. 2017;6(11):2033–43.
16. Lin HK, Madihally SV, Palmer B, Frimberger D, Fung KM, Kropp BP. Biomatrices for bladder reconstruction. Adv Drug Deliv Rev. 2015;82-83:47–63.
17. Dhasmana A, Singh A, Rawal S. Biomedical grafts for tracheal tissue repairing and regeneration "tracheal tissue engineering: an overview". J Tissue Eng Regen Med. 2020;14(5):653–72.
18. Gao B, Jing H, Gao M, Wang S, Fu W, Zhang X, He X, Zheng J. Long-segmental tracheal reconstruction in rabbits with pedicled tissue-engineered trachea based on a 3D-printed scaffold. Acta Biomater. 2019;97:177–86.
19. Choudhury D, Yee M, Sheng ZLJ, Amirul A, Naing MW. Decellularization systems and devices: state-of-the-art. Acta Biomater. 2020;115:51–9.

20. Crapo PM, Gilbert TW, Badylak SF. An overview of tissue and whole organ decellularization processes. Biomaterials. 2011;32(12):3233–43.
21. Rojkind M, Gatmaitan Z, Mackensen S, Giambrone MA, Ponce P, Reid LM. Connective tissue biomatrix: its isolation and utilization for long-term cultures of normal rat hepatocytes. J Cell Biol. 1980;87(1):255–63.
22. Baptista PM, Orlando G, Mirmalek-Sani SH, Siddiqui M, Atala A, Soker S. Whole organ decellularization - a tool for bioscaffold fabrication and organ bioengineering. In: Annual international conference of the IEEE engineering in medicine and biology society. IEEE engineering in medicine and biology society. Annual international conference 2009. Piscataway: IEEE; 2009. p. 6526–9.
23. Peloso A, Dhal A, Zambon JP, Li P, Orlando G, Atala A, Soker S. Current achievements and future perspectives in whole-organ bioengineering. Stem Cell Res Ther. 2015;6(1):107.
24. Alexanian RA, Mahapatra K, Lang D, Vaidyanathan R, Markandeya YS, Gill RK, Zhai AJ, Dhillon A, Lea MR, Abozeid S, Schmuck EG, Raval AN, Eckhardt LL, Glukhov AV, Lalit PA, Kamp TJ. Induced cardiac progenitor cells repopulate decellularized mouse heart scaffolds and differentiate to generate cardiac tissue. Biochim Biophys Acta, Mol Cell Res. 2020;1867(3):118559.
25. Ozlu B, Ergin M, Budak S, Tunali S, Yildirim N, Erisken C. A bioartificial rat heart tissue: perfusion decellularization and characterization. Int J Artif Organs. 2019;42(12):757–64.
26. Ansari T, Southgate A, Obiri-Yeboa I, Jones LG, Greco K, Olayanju A, Mbundi L, Somasundaram M, Davidson B, Sibbons PD. Development and characterization of a porcine liver scaffold. Stem Cells Dev. 2020;29(5):314–26.
27. Willemse J, Verstegen MMA, Vermeulen A, Schurink IJ, Roest HP, van der Laan LJW, de Jonge J. Fast, robust and effective decellularization of whole human livers using mild detergents and pressure controlled perfusion. Mater Sci Eng C Mater Biol Appl. 2020;108:110200.
28. Obata T, Tsuchiya T, Akita S, Kawahara T, Matsumoto K, Miyazaki T, Masumoto H, Kobayashi E, Niklason LE, Nagayasu T. Utilization of natural detergent potassium laurate for Decellularization in lung bioengineering. Tissue Eng Part C Methods. 2019;25(8):459–71.
29. Stahl EC, Bonvillain RW, Skillen CD, Burger BL, Hara H, Lee W, Trygg CB, Didier PJ, Grasperge BF, Pashos NC, Bunnell BA, Bianchi J, Ayares DL, Guthrie KI, Brown BN, Petersen TH. Evaluation of the host immune response to decellularized lung scaffolds derived from alpha-gal knockout pigs in a non-human primate model. Biomaterials. 2018;187:93–104.
30. Feng H, Xu Y, Luo S, Dang H, Liu K, Sun WQ. Evaluation and preservation of vascular architectures in decellularized whole rat kidneys. Cryobiology. 2020;95:72–9.
31. Kajbafzadeh AM, Khorramirouz R, Nabavizadeh B, Ladi Seyedian SS, Akbarzadeh A, Heidari R, Masoumi A, Azizi B, Beigi RSH. Whole organ sheep kidney tissue engineering and in vivo transplantation: effects of perfusion-based decellularization on vascular integrity. Mater Sci Eng C Mater Biol Appl. 2019;98:392–400.
32. Berger C, Bjorlykke Y, Hahn L, Muhlemann M, Kress S, Walles H, Luxenhofer R, Raeder H, Metzger M, Zdzieblo D. Matrix decoded - a pancreatic extracellular matrix with organ specific cues guiding human iPSC differentiation. Biomaterials. 2020;244:119766.
33. Berman A, Klak M, Adamiok A, Kaczynski L, Tymicki G, Gomolka M, Kowalska P, Kosowska K, Cywoniuk P, Turowski P, Bryniarski T, Wyrwicki M, Wszola M. The influence of the flow of detergent and donor characteristics on the extracellular matrix composition after human pancreas decellularization. Transplant Proc. 2020;52(7):2043–9.
34. Tapias LF, Ott HC. Decellularized scaffolds as a platform for bioengineered organs. Curr Opin Organ Transplant. 2014;19(2):145–52.
35. Figliuzzi M, Bonandrini B, Remuzzi A. Decellularized kidney matrix as functional material for whole organ tissue engineering. J Appl Biomaterials Funct Materials. 2017;15(4):e326–33.
36. Stoltz JF, Zhang L, Ye JS, De Isla N. Organ reconstruction: dream or reality for the future. Biomed Mater Eng. 2017;28(s1):S121–7.

37. Scarritt ME, Pashos NC, Bunnell BA. A review of cellularization strategies for tissue engineering of whole organs. Front Bioeng Biotechnol. 2015;3:43.
38. Petersen TH, Calle EA, Zhao L, Lee EJ, Gui L, Raredon MB, Gavrilov K, Yi T, Zhuang ZW, Breuer C, Herzog E, Niklason LE. Tissue-engineered lungs for in vivo implantation. Science (New York, NY). 2010;329(5991):538–41.
39. Song JJ, Guyette JP, Gilpin SE, Gonzalez G, Vacanti JP, Ott HC. Regeneration and experimental orthotopic transplantation of a bioengineered kidney. Nat Med. 2013;19(5):646–51.
40. Uygun BE, Soto-Gutierrez A, Yagi H, Izamis ML, Guzzardi MA, Shulman C, Milwid J, Kobayashi N, Tilles A, Berthiaume F, Hertl M, Nahmias Y, Yarmush ML, Uygun K. Organ reengineering through development of a transplantable recellularized liver graft using decellularized liver matrix. Nat Med. 2010;16(7):814–20.
41. Fu RH, Wang YC, Liu SP, Shih TR, Lin HL, Chen YM, Sung JH, Lu CH, Wei JR, Wang ZW, Huang SJ, Tsai CH, Shyu WC, Lin SZ. Decellularization and recellularization technologies in tissue engineering. Cell Transplant. 2014;23(4–5):621–30.
42. Swijnenburg RJ, Tanaka M, Vogel H, Baker J, Kofidis T, Gunawan F, Lebl DR, Caffarelli AD, de Bruin JL, Fedoseyeva EV, Robbins RC. Embryonic stem cell immunogenicity increases upon differentiation after transplantation into ischemic myocardium. Circulation. 2005;112(9 Suppl):I166–72.
43. Murray IR, Peault B. Q&A: mesenchymal stem cells – where do they come from and is it important? BMC Biol. 2015;13:99.
44. Huang YZ, Cai JQ, Lv FJ, Xie HL, Yang ZM, Huang YC, Deng L. Species variation in the spontaneous calcification of bone marrow-derived mesenchymal stem cells. Cytotherapy. 2013;15(3):323–9.
45. Yoshihara M, Hayashizaki Y, Murakawa Y. Genomic instability of iPSCs: challenges towards their clinical applications. Stem Cell Rev Rep. 2017;13(1):7–16.
46. Yoshihara M, Oguchi A, Murakawa Y. Genomic instability of iPSCs and challenges in their clinical applications. Adv Exp Med Biol. 2019;1201:23–47.
47. Muroi K, Miyamura K, Okada M, Yamashita T, Murata M, Ishikawa T, Uike N, Hidaka M, Kobayashi R, Imamura M, Tanaka J, Ohashi K, Taniguchi S, Ikeda T, Eto T, Mori M, Yamaoka M, Ozawa K. Bone marrow-derived mesenchymal stem cells (JR-031) for steroid-refractory grade III or IV acute graft-versus-host disease: a phase II/III study. Int J Hematol. 2016;103(2):243–50.
48. Yee GC, McGuire TR. Allogeneic bone marrow transplantation in the treatment of hematologic diseases. Clin Pharm. 1985;4(2):149–60.
49. Diekman BO, Rowland CR, Lennon DP, Caplan AI, Guilak F. Chondrogenesis of adult stem cells from adipose tissue and bone marrow: induction by growth factors and cartilage-derived matrix. Tissue Eng Part A. 2010;16(2):523–33.
50. Hodonsky C, Mundada L, Wang S, Witt R, Raff G, Kaushal S, Si MS. Effects of scaffold material used in cardiovascular surgery on mesenchymal stem cells and cardiac progenitor cells. Ann Thorac Surg. 2015;99(2):605–11.
51. Lysaght T, Campbell AV. Regulating autologous adult stem cells: the FDA steps up. Cell Stem Cell. 2011;9(5):393–6.
52. Backer H, Polgar L, Soos P, Lajko E, Lang O, Merkely B, Szabo G, Dohmen PM, Weymann A, Kohidai L. Impedimetric analysis of the effect of Decellularized porcine heart scaffold on human Fibrosarcoma, endothelial, and cardiomyocyte cell lines. Med Sci Monitor International Med J Exp Clin Res. 2017;23:2232–40.
53. Guruswamy Damodaran R, Vermette P. Decellularized pancreas as a native extracellular matrix scaffold for pancreatic islet seeding and culture. J Tissue Eng Regen Med. 2018;12(5):1230–7.
54. Hassanein W, Uluer MC, Langford J, Woodall JD, Cimeno A, Dhru U, Werdesheim A, Harrison J, Rivera-Pratt C, Klepfer S, Khalifeh A, Buckingham B, Brazio PS, Parsell D, Klassen C, Drachenberg C, Barth RN, LaMattina JC. Recellularization via the bile duct

supports functional allogenic and xenogenic cell growth on a decellularized rat liver scaffold. Organogenesis. 2017;13(1):16–27.
55. Mazza G, Rombouts K, Rennie Hall A, Urbani L, Vinh Luong T, Al-Akkad W, Longato L, Brown D, Maghsoudlou P, Dhillon AP, Fuller B, Davidson B, Moore K, Dhar D, De Coppi P, Malago M, Pinzani M. Decellularized human liver as a natural 3D-scaffold for liver bioengineering and transplantation. Sci Rep. 2015;5:13079.
56. Pan J, Li H, Fang Y, Shen YB, Zhou XY, Zhu F, Zhu LX, Du YH, Yu XF, Wang Y, Zhou XH, Wang YY, Wu YJ. Regeneration of a bioengineered thyroid using decellularized thyroid matrix. Thyroid. 2019;29(1):142–52.
57. Wu D, Wan J, Huang Y, Guo Y, Xu T, Zhu M, Fan X, Zhu S, Ling C, Li X, Lu J, Zhu H, Zhou P, Lu Y, Wang Z. 3D culture of MIN-6 cells on decellularized pancreatic scaffold: in vitro and in vivo study. Biomed Res Int. 2015;2015:432645.
58. Zhou Q, Ye X, Ran Q, Kitahara A, Matsumoto Y, Moriyama M, Ajioka Y, Saijo Y. Trachea engineering using a centrifugation method and mouse-induced pluripotent stem cells. Tissue Eng Part C Methods. 2018;24(9):524–33.
59. Asahara T, Murohara T, Sullivan A, Silver M, van der Zee R, Li T, Witzenbichler B, Schatteman G, Isner JM. Isolation of putative progenitor endothelial cells for angiogenesis. Science (New York, NY). 1997;275(5302):964–7.
60. Hristov M, Weber C. Endothelial progenitor cells in vascular repair and remodeling. Pharmacol Res. 2008;58(2):148–51.
61. Haider KH, Aziz S, Al-Reshidi MA. Endothelial progenitor cells for cellular angiogenesis and repair: lessons learned from experimental animal models. Regen Med. 2017;12(8):969–82.
62. Deng C, Zhang P, Vulesevic B, Kuraitis D, Li F, Yang AF, Griffith M, Ruel M, Suuronen EJ. A collagen-chitosan hydrogel for endothelial differentiation and angiogenesis. Tissue Eng Part A. 2010;16(10):3099–109.
63. Peters EB, Christoforou N, Leong KW, Truskey GA, West JL. Poly(ethylene glycol) hydrogel scaffolds containing cell-adhesive and protease-sensitive peptides support microvessel formation by endothelial progenitor cells. Cell Mol Bioeng. 2016;9(1):38–54.
64. Yee D, Hanjaya-Putra D, Bose V, Luong E, Gerecht S. Hyaluronic acid hydrogels support cord-like structures from endothelial colony-forming cells. Tissue Eng Part A. 2011;17(9–10):1351–61.
65. Demetz G, Oostendorp RAJ, Boxberg AM, Sitz W, Farrell E, Steppich B, Steinsiek AL, Rudelius M, Ott I. Overexpression of insulin-like growth Factor-2 in expanded endothelial progenitor cells improves left ventricular function in experimental myocardial infarction. J Vasc Res. 2017;54(6):321–8.
66. Gaffey AC, Chen MH, Venkataraman CM, Trubelja A, Rodell CB, Dinh PV, Hung G, MacArthur JW, Soopan RV, Burdick JA, Atluri P. Injectable shear-thinning hydrogels used to deliver endothelial progenitor cells, enhance cell engraftment, and improve ischemic myocardium. J Thorac Cardiovasc Surg. 2015;150(5):1268–76.
67. Aday S, Zoldan J, Besnier M, Carreto L, Saif J, Fernandes R, Santos T, Bernardino L, Langer R, Emanueli C, Ferreira L. Synthetic microparticles conjugated with VEGF165 improve the survival of endothelial progenitor cells via microRNA-17 inhibition. Nat Commun. 2017;8(1):747.
68. Kim KL, Han DK, Park K, Song SH, Kim JY, Kim JM, Ki HY, Yie SW, Roh CR, Jeon ES, Kim DK, Suh W. Enhanced dermal wound neovascularization by targeted delivery of endothelial progenitor cells using an RGD-g-PLLA scaffold. Biomaterials. 2009;30(22):3742–8.
69. Zhou P, Huang Y, Guo Y, Wang L, Ling C, Guo Q, Wang Y, Zhu S, Fan X, Zhu M, Huang H, Lu Y, Wang Z. Decellularization and recellularization of rat livers with hepatocytes and endothelial progenitor cells. Artif Organs. 2016;40(3):E25–38.
70. Guo Y, Wu C, Xu L, Xu Y, Xiaohong L, Hui Z, Jingjing L, Lu Y, Wang Z. Vascularization of pancreatic decellularized scaffold with endothelial progenitor cells. J Artif Organs. 2018;21(2):230–7.

71. Jaffe EA, Nachman RL, Becker CG, Minick CR. Culture of human endothelial cells derived from umbilical veins. Identification by morphologic and immunologic criteria. J Clin Invest. 1973;52(11):2745–56.
72. Lei J, Peng S, Samuel SB, Zhang S, Wu Y, Wang P, Li YF, Liu H. A simple and biosafe method for isolation of human umbilical vein endothelial cells. Anal Biochem. 2016;508:15–8.
73. Ryan US, Clements E, Habliston D, Ryan JW. Isolation and culture of pulmonary artery endothelial cells. Tissue Cell. 1978;10(3):535–54.
74. Visner GA, Staples ED, Chesrown SE, Block ER, Zander DS, Nick HS. Isolation and maintenance of human pulmonary artery endothelial cells in culture isolated from transplant donors. Am J Phys. 1994;267(4 Pt 1):L406–13.
75. Dingle AM, Yap KK, Gerrand YW, Taylor CJ, Keramidaris E, Lokmic Z, Kong AM, Peters HL, Morrison WA, Mitchell GM. Characterization of isolated liver sinusoidal endothelial cells for liver bioengineering. Angiogenesis. 2018;21(3):581–97.
76. Meyer J, Lacotte S, Morel P, Gonelle-Gispert C, Buhler L. An optimized method for mouse liver sinusoidal endothelial cell isolation. Exp Cell Res. 2016;349(2):291–301.
77. Akis N, Madaio MP. Isolation, culture, and characterization of endothelial cells from mouse glomeruli. Kidney Int. 2004;65(6):2223–7.
78. Muczynski KA, Leca N, Anderson SK. Proliferative potential of human kidney endothelial cells: bone marrow-derived cells may not be required for high proliferation. Nephrol Dialysis Transplant. 2010;25(9):2953–60.
79. Grafe M, Auch-Schwelk W, Graf K, Terbeek D, Hertel H, Unkelbach M, Hildebrandt A, Fleck E. Isolation and characterization of macrovascular and microvascular endothelial cells from human hearts. Am J Phys. 1994;267(6 Pt 2):H2138–48.
80. Pratumvinit B, Reesukumal K, Janebodin K, Ieronimakis N, Reyes M. Isolation, characterization, and transplantation of cardiac endothelial cells. Biomed Res Int. 2013;2013:359412.
81. Freiman A, Shandalov Y, Rosenfeld D, Shor E, Ben-David D, Meretzki S, Levenberg S, Egozi D. Engineering vascularized flaps using adipose-derived microvascular endothelial cells and mesenchymal stem cells. J Tissue Eng Regen Med. 2018;12(1):e130–41.
82. Geenen IL, Molin DG, van den Akker NM, Jeukens F, Spronk HM, Schurink GW, Post MJ. Endothelial cells (ECs) for vascular tissue engineering: venous ECs are less thrombogenic than arterial ECs. J Tissue Eng Regen Med. 2015;9(5):564–76.
83. Peters EB. Endothelial progenitor cells for the vascularization of engineered tissues. Tissue Eng Part B Rev. 2018;24(1):1–24.
84. Wang K, Lin RZ, Melero-Martin JM. Bioengineering human vascular networks: trends and directions in endothelial and perivascular cell sources. Cell Mol Life Sci. 2019;76(3):421–39.
85. Ott HC, Matthiesen TS, Goh SK, Black LD, Kren SM, Netoff TI, Taylor DA. Perfusion-decellularized matrix: using nature's platform to engineer a bioartificial heart. Nat Med. 2008;14(2):213–21.
86. Ott HC, Clippinger B, Conrad C, Schuetz C, Pomerantseva I, Ikonomou L, Kotton D, Vacanti JP. Regeneration and orthotopic transplantation of a bioartificial lung. Nat Med. 2010;16(8):927–33.
87. Costa Fernandes CJD, Zambuzzi WF. Fibroblast-secreted trophic factors contribute with ECM remodeling stimulus and upmodulate osteocyte gene markers in osteoblasts. Biochimie. 2020;168:92–9.
88. Erdogan B, Webb DJ. Cancer-associated fibroblasts modulate growth factor signaling and extracellular matrix remodeling to regulate tumor metastasis. Biochem Soc Trans. 2017;45(1):229–36.
89. Allard B, Panariti A, Martin JG. Alveolar macrophages in the resolution of inflammation, tissue repair, and tolerance to infection. Front Immunol. 2018;9:1777.
90. Kojima H, Yasuchika K, Fukumitsu K, Ishii T, Ogiso S, Miyauchi Y, Yamaoka R, Kawai T, Katayama H, Yoshitoshi-Uebayashi EY, Kita S, Yasuda K, Sasaki N, Komori J, Uemoto S. Establishment of practical recellularized liver graft for blood perfusion using primary rat

hepatocytes and liver sinusoidal endothelial cells. Am J Transplant Off J Am Soc Transplant Am Soc Transplant Surg. 2018;18(6):1351–9.

91. Baptista PM, Siddiqui MM, Lozier G, Rodriguez SR, Atala A, Soker S. The use of whole organ decellularization for the generation of a vascularized liver organoid. Hepatology (Baltimore, Md.). 2011;53(2):604–17.
92. Sabetkish S, Kajbafzadeh AM, Sabetkish N, Khorramirouz R, Akbarzadeh A, Seyedian SL, Pasalar P, Orangian S, Beigi RS, Aryan Z, Akbari H, Tavangar SM. Whole-organ tissue engineering: decellularization and recellularization of three-dimensional matrix liver scaffolds. J Biomed Mater Res A. 2015;103(4):1498–508.
93. Wang X, Cui J, Zhang BQ, Zhang H, Bi Y, Kang Q, Wang N, Bie P, Yang Z, Wang H, Liu X, Haydon RC, Luu HH, Tang N, Dong J, He TC. Decellularized liver scaffolds effectively support the proliferation and differentiation of mouse fetal hepatic progenitors. J Biomed Mater Res A. 2014;102(4):1017–25.
94. Bora P, Majumdar AS. Adipose tissue-derived stromal vascular fraction in regenerative medicine: a brief review on biology and translation. Stem Cell Res Ther. 2017;8(1):145.
95. Han S, Sun HM, Hwang KC, Kim SW. Adipose-derived stromal vascular fraction cells: update on clinical utility and efficacy. Crit Rev Eukaryot Gene Expr. 2015;25(2):145–52.
96. Buckingham M, Meilhac S, Zaffran S. Building the mammalian heart from two sources of myocardial cells. Nat Rev Genet. 2005;6(11):826–35.
97. Musunuru K, Domian IJ, Chien KR. Stem cell models of cardiac development and disease. Annu Rev Cell Dev Biol. 2010;26:667–87.
98. Witman N, Zhou C, Grote Beverborg N, Sahara M, Chien KR. Cardiac progenitors and paracrine mediators in cardiogenesis and heart regeneration. Semin Cell Dev Biol. 2020;100:29–51.
99. Aguilar-Sanchez C, Michael M, Pennings S. Cardiac stem cells in the postnatal heart: lessons from development. Stem Cells Int. 2018;2018:1247857.
100. Sangan CB, Tosh D. Hepatic progenitor cells. Cell Tissue Res. 2010;342(2):131–7.
101. Knight B, Matthews VB, Olynyk JK, Yeoh GC. Jekyll and Hyde: evolving perspectives on the function and potential of the adult liver progenitor (oval) cell. BioEssays. 2005;27(11):1192–202.
102. Chen J, Chen L, Zern MA, Theise ND, Diehl AM, Liu P, Duan Y. The diversity and plasticity of adult hepatic progenitor cells and their niche. Liver Int. 2017;37(9):1260–71.
103. Yang L, Li S, Hatch H, Ahrens K, Cornelius JG, Petersen BE, Peck AB. In vitro trans-differentiation of adult hepatic stem cells into pancreatic endocrine hormone-producing cells. Proc Natl Acad Sci U S A. 2002;99(12):8078–83.
104. Chistiakov DA. Endogenous and exogenous stem cells: a role in lung repair and use in airway tissue engineering and transplantation. J Biomed Sci. 2010;17(1):92.
105. Wagner DE, Bonvillain RW, Jensen T, Girard ED, Bunnell BA, Finck CM, Hoffman AM, Weiss DJ. Can stem cells be used to generate new lungs? Ex vivo lung bioengineering with decellularized whole lung scaffolds. Respirology (Carlton, Vic). 2013;18(6):895–911.
106. Romagnani P, Remuzzi G. Renal progenitors in non-diabetic and diabetic nephropathies. Trends Endocrinol Metab. 2013;24(1):13–20.
107. Sagrinati C, Netti GS, Mazzinghi B, Lazzeri E, Liotta F, Frosali F, Ronconi E, Meini C, Gacci M, Squecco R, Carini M, Gesualdo L, Francini F, Maggi E, Annunziato F, Lasagni L, Serio M, Romagnani S, Romagnani P. Isolation and characterization of multipotent progenitor cells from the Bowman's capsule of adult human kidneys. J Am Soc Nephrol. 2006;17(9):2443–56.
108. Angelotti ML, Ronconi E, Ballerini L, Peired A, Mazzinghi B, Sagrinati C, Parente E, Gacci M, Carini M, Rotondi M, Fogo AB, Lazzeri E, Lasagni L, Romagnani P. Characterization of renal progenitors committed toward tubular lineage and their regenerative potential in renal tubular injury. Stem Cells. 2012;30(8):1714–25.
109. Lazzeri E, Crescioli C, Ronconi E, Mazzinghi B, Sagrinati C, Netti GS, Angelotti ML, Parente E, Ballerini L, Cosmi L, Maggi L, Gesualdo L, Rotondi M, Annunziato F, Maggi E, Lasagni L,

Serio M, Romagnani S, Vannelli GB, Romagnani P. Regenerative potential of embryonic renal multipotent progenitors in acute renal failure. J Am Soc Nephrol. 2007;18(12):3128–38.
110. Romagnani P. Toward the identification of a "renopoietic system"? Stem Cells. 2009;27(9):2247–53.
111. Sagrinati C, Ronconi E, Lazzeri E, Lasagni L, Romagnani P. Stem-cell approaches for kidney repair: choosing the right cells. Trends Mol Med. 2008;14(7):277–85.
112. Kim HS, Lee MK. Beta-cell regeneration through the transdifferentiation of pancreatic cells: pancreatic progenitor cells in the pancreas. J Diabetes Investig. 2016;7(3):286–96.
113. Bouwens L. Transdifferentiation versus stem cell hypothesis for the regeneration of islet beta-cells in the pancreas. Microsc Res Tech. 1998;43(4):332–6.
114. Ramiya VK, Maraist M, Arfors KE, Schatz DA, Peck AB, Cornelius JG. Reversal of insulin-dependent diabetes using islets generated in vitro from pancreatic stem cells. Nat Med. 2000;6(3):278–82.
115. Suzuki A, Nakauchi H, Taniguchi H. Prospective isolation of multipotent pancreatic progenitors using flow-cytometric cell sorting. Diabetes. 2004;53(8):2143–52.
116. Batchelder CA, Martinez ML, Tarantal AF. Natural scaffolds for renal differentiation of human embryonic stem cells for kidney tissue engineering. PLoS One. 2015;10(12): e0143849.
117. Carvalho JL, de Carvalho PH, Gomes DA, Goes AM. Characterization of decellularized heart matrices as biomaterials for regular and whole organ tissue engineering and initial in-vitro recellularization with Ips cells. J Tissue Sci Eng. 2012;11:002.
118. Lecht S, Stabler CT, Rylander AL, Chiaverelli R, Schulman ES, Marcinkiewicz C, Lelkes PI. Enhanced reseeding of decellularized rodent lungs with mouse embryonic stem cells. Biomaterials. 2014;35(10):3252–62.
119. Park KM, Hussein KH, Hong SH, Ahn C, Yang SR, Park SM, Kweon OK, Kim BM, Woo HM. Decellularized liver extracellular matrix as promising tools for transplantable bioengineered liver promotes hepatic lineage commitments of induced pluripotent stem cells. Tissue Eng Part A. 2016;22(5–6):449–60.
120. Chen FM, Wu LA, Zhang M, Zhang R, Sun HH. Homing of endogenous stem/progenitor cells for in situ tissue regeneration: promises, strategies, and translational perspectives. Biomaterials. 2011;32(12):3189–209.
121. Bacakova L, Zarubova J, Travnickova M, Musilkova J, Pajorova J, Slepicka P, Kasalkova NS, Svorcik V, Kolska Z, Motarjemi H, Molitor M. Stem cells: their source, potency and use in regenerative therapies with focus on adipose-derived stem cells - a review. Biotechnol Adv. 2018;36(4):1111–26.
122. Choong C, Rao MS. Human embryonic stem cells. Neurosurg Clin N Am. 2007;18(1):1–14, vii.
123. Baharvand H, Ashtiani SK, Valojerdi MR, Shahverdi A, Taee A, Sabour D. Establishment and in vitro differentiation of a new embryonic stem cell line from human blastocyst. Differentiation. 2004;72(5):224–9.
124. Akutsu H, Nasu M, Morinaga S, Motoyama T, Homma N, Machida M, Yamazaki-Inoue M, Okamura K, Nakabayashi K, Takada S, Nakamura N, Kanzaki S, Hata K, Umezawa A. In vivo maturation of human embryonic stem cell-derived teratoma over time. Regen Ther. 2016;5:31–9.
125. Fagnocchi L, Mazzoleni S, Zippo A. Integration of signaling pathways with the epigenetic machinery in the maintenance of stem cells. Stem Cells Int. 2016;2016:8652748.
126. Ilic D, Ogilvie C. Concise review: human embryonic stem cells-what have we done? What are we doing? Where are we going? Stem Cells. 2017;35(1):17–25.
127. Jameel MN, Zhang J. Stem cell therapy for ischemic heart disease. Antioxid Redox Signal. 2010;13(12):1879–97.
128. Bonandrini B, Figliuzzi M, Papadimou E, Morigi M, Perico N, Casiraghi F, Dipl C, Sangalli F, Conti S, Benigni A, Remuzzi A, Remuzzi G. Recellularization of well-preserved acellular kidney scaffold using embryonic stem cells. Tissue Eng Part A. 2014;20(9–10):1486–98.

129. Cortiella J, Niles J, Cantu A, Brettler A, Pham A, Vargas G, Winston S, Wang J, Walls S, Nichols JE. Influence of acellular natural lung matrix on murine embryonic stem cell differentiation and tissue formation. Tissue Eng Part A. 2010;16(8):2565–80.
130. Nakayama KH, Lee CC, Batchelder CA, Tarantal AF. Tissue specificity of decellularized rhesus monkey kidney and lung scaffolds. PLoS One. 2013;8(5):e64134.
131. Ross EA, Williams MJ, Hamazaki T, Terada N, Clapp WL, Adin C, Ellison GW, Jorgensen M, Batich CD. Embryonic stem cells proliferate and differentiate when seeded into kidney scaffolds. J Am Soc Nephrol. 2009;20(11):2338–47.
132. Ross EA, Abrahamson DR, John PS, Clapp WL, Williams MJ, Terada N, Hamazaki T, Ellison GW, Batich CD. Mouse stem cells seeded into decellularized rat kidney scaffolds endothelialize and remodel basement membranes. Organogenesis. 2012;8(2):49–55.
133. Delsing L, Herland A, Falk A, Hicks R, Synnergren J, Zetterberg H. Models of the blood-brain barrier using iPSC-derived cells. Mol Cell Neurosci. 2020;107:103533.
134. Brix J, Zhou Y, Luo Y. The epigenetic reprogramming roadmap in generation of iPSCs from somatic cells. J Genet Genom. 2015;42(12):661–70.
135. Borgohain MP, Haridhasapavalan KK, Dey C, Adhikari P, Thummer RP. An insight into DNA-free reprogramming approaches to generate integration-free induced pluripotent stem cells for prospective biomedical applications. Stem Cell Rev Rep. 2019;15(2):286–313.
136. Nordin F, Ahmad RNR, Farzaneh F. Transactivator protein: an alternative for delivery of recombinant proteins for safer reprogramming of induced pluripotent stem cell. Virus Res. 2017;235:106–14.
137. Schwartz RE, Fleming HE, Khetani SR, Bhatia SN. Pluripotent stem cell-derived hepatocyte-like cells. Biotechnol Adv. 2014;32(2):504–13.
138. Ivashchenko CY, Pipes GC, Lozinskaya IM, Lin Z, Xiaoping X, Needle S, Grygielko ET, Hu E, Toomey JR, Lepore JJ, Willette RN. Human-induced pluripotent stem cell-derived cardiomyocytes exhibit temporal changes in phenotype. Am J Physiol Heart Circ Physiol. 2013;305(6):H913–22.
139. Lu TY, Lin B, Kim J, Sullivan M, Tobita K, Salama G, Yang L. Repopulation of decellularized mouse heart with human induced pluripotent stem cell-derived cardiovascular progenitor cells. Nat Commun. 2013;4:2307.
140. Ghaedi M, Calle EA, Mendez JJ, Gard AL, Balestrini J, Booth A, Bove PF, Gui L, White ES, Niklason LE. Human iPS cell-derived alveolar epithelium repopulates lung extracellular matrix. J Clin Investig. 2013;123(11):4950–62.
141. Perinatal period. In: Kirch W, editor. Encyclopedia of public health. Dordrecht: Springer; 2008. p. 1096.
142. Deus IA, Mano JF, Custodio CA. Perinatal tissues and cells in tissue engineering and regenerative medicine. Acta Biomater. 2020;110:1–14.
143. Insausti CL, Blanquer M, Bleda P, Iniesta P, Majado MJ, Castellanos G, Moraleda JM. The amniotic membrane as a source of stem cells. Histol Histopathol. 2010;25(1):91–8.
144. Umezawa A, Hasegawa A, Inoue M, Tanuma-Takahashi A, Kajiwara K, Makino H, Chikazawa E, Okamoto A. Amnion-derived cells as a reliable resource for next-generation regenerative medicine. Placenta. 2019;84:50–6.
145. Chandravanshi B, Bhonde RR. Human umbilical cord-derived stem cells: isolation, characterization, differentiation, and application in treating diabetes. Crit Rev Biomed Eng. 2018;46 (5):399–412.
146. Ding DC, Chang YH, Shyu WC, Lin SZ. Human umbilical cord mesenchymal stem cells: a new era for stem cell therapy. Cell Transplant. 2015;24(3):339–47.
147. Chakraborty SK, Banu LA, Rahman MF, Paul S. Cord blood stem cells - a dream for future medicine. Mymensingh Med J. 2014;23(3):614–20.
148. Antoniadou E, David AL. Placental stem cells, best practice & research. Clin Obstet Gynaecol. 2016;31:13–29.
149. Matikainen T, Laine J. Placenta–an alternative source of stem cells. Toxicol Appl Pharmacol. 2005;207(2 Suppl):544–9.

150. Higuchi O, Okabe M, Yoshida T, Fathy M, Saito S, Miyawaki T, Nikaido T. Stemness of human Wharton's jelly mesenchymal cells is maintained by floating cultivation. Cell Reprogram. 2012;14(5):448–55.
151. Li X, Bai J, Ji X, Li R, Xuan Y, Wang Y. Comprehensive characterization of four different populations of human mesenchymal stem cells as regards their immune properties, proliferation and differentiation. Int J Mol Med. 2014;34(3):695–704.
152. Bieback K, Kern S, Kluter H, Eichler H. Critical parameters for the isolation of mesenchymal stem cells from umbilical cord blood. Stem Cells. 2004;22(4):625–34.
153. Kern S, Eichler H, Stoeve J, Kluter H, Bieback K. Comparative analysis of mesenchymal stem cells from bone marrow, umbilical cord blood, or adipose tissue. Stem Cells. 2006;24 (5):1294–301.
154. Chang YJ, Hwang SM, Tseng CP, Cheng FC, Huang SH, Hsu LF, Hsu LW, Tsai MS. Isolation of mesenchymal stem cells with neurogenic potential from the mesoderm of the amniotic membrane. Cells Tissues Organs. 2010;192(2):93–105.
155. Nogami M, Tsuno H, Koike C, Okabe M, Yoshida T, Seki S, Matsui Y, Kimura T, Nikaido T. Isolation and characterization of human amniotic mesenchymal stem cells and their chondrogenic differentiation. Transplantation. 2012;93(12):1221–8.
156. Kim MJ, Shin KS, Jeon JH, Lee DR, Shim SH, Kim JK, Cha DH, Yoon TK, Kim GJ. Human chorionic-plate-derived mesenchymal stem cells and Wharton's jelly-derived mesenchymal stem cells: a comparative analysis of their potential as placenta-derived stem cells. Cell Tissue Res. 2011;346(1):53–64.
157. Koo BK, Park IY, Kim J, Kim JH, Kwon A, Kim M, Kim Y, Shin JC, Kim JH. Isolation and characterization of chorionic mesenchymal stromal cells from human full term placenta. J Korean Med Sci. 2012;27(8):857–63.
158. Abumaree MH, Al Jumah MA, Kalionis B, Jawdat D, Al Khaldi A, AlTalabani AA, Knawy BA. Phenotypic and functional characterization of mesenchymal stem cells from chorionic villi of human term placenta. Stem Cell Rev Rep. 2013;9(1):16–31.
159. Castrechini NM, Murthi P, Gude NM, Erwich JJ, Gronthos S, Zannettino A, Brennecke SP, Kalionis B. Mesenchymal stem cells in human placental chorionic villi reside in a vascular niche. Placenta. 2010;31(3):203–12.
160. Abomaray FM, Al Jumah MA, Alsaad KO, Jawdat D, Al Khaldi A, AlAskar AS, Al Harthy S, Al Subayyil AM, Khatlani T, Alawad AO, Alkushi A, Kalionis B, Abumaree MH. Phenotypic and functional characterization of mesenchymal stem/multipotent stromal cells from decidua basalis of human term placenta. Stem Cells Int. 2016;2016:5184601.
161. In't Anker PS, Scherjon SA, der Keur C K-v, de Groot-Swings GM, Claas FH, Fibbe WE, Kanhai HH. Isolation of mesenchymal stem cells of fetal or maternal origin from human placenta. Stem Cells. 2004;22(7):1338–45.
162. Barlow S, Brooke G, Chatterjee K, Price G, Pelekanos R, Rossetti T, Doody M, Venter D, Pain S, Gilshenan K, Atkinson K. Comparison of human placenta- and bone marrow-derived multipotent mesenchymal stem cells. Stem Cells Dev. 2008;17(6):1095–107.
163. Sangeetha KN, Vennila R, Secunda R, Sakthivel S, Pathak S, Jeswanth S, Surendran R. Functional variations between mesenchymal stem cells of different tissue origins: a comparative gene expression profiling. Biotechnol Lett. 2020;42(7):1287–304.
164. Evron A, Goldman S, Shalev E. Human amniotic epithelial cells cultured in substitute serum medium maintain their stem cell characteristics for up to four passages. Int J Stem Cells. 2011;4(2):123–32.
165. Miki T. Stem cell characteristics and the therapeutic potential of amniotic epithelial cells. Am J Reproduct Immunol. 2018;80(4):e13003.
166. Berger MJ, Adams SD, Tigges BM, Sprague SL, Wang XJ, Collins DP, McKenna DH. Differentiation of umbilical cord blood-derived multilineage progenitor cells into respiratory epithelial cells. Cytotherapy. 2006;8(5):480–7.

167. Mirmalek-Sani SH, Orlando G, McQuilling JP, Pareta R, Mack DL, Salvatori M, Farney AC, Stratta RJ, Atala A, Opara EC, Soker S. Porcine pancreas extracellular matrix as a platform for endocrine pancreas bioengineering. Biomaterials. 2013;34(22):5488–95.
168. Oberwallner B, Brodarac A, Choi YH, Saric T, Anic P, Morawietz L, Stamm C. Preparation of cardiac extracellular matrix scaffolds by decellularization of human myocardium. J Biomed Mater Res A. 2014;102(9):3263–72.
169. Maqsood M, Kang M, Wu X, Chen J, Teng L, Qiu L. Adult mesenchymal stem cells and their exosomes: sources, characteristics, and application in regenerative medicine. Life Sci. 2020;256:118002.
170. Naji A, Eitoku M, Favier B, Deschaseaux F, Rouas-Freiss N, Suganuma N. Biological functions of mesenchymal stem cells and clinical implications. Cell Mol Life Sci. 2019;76 (17):3323–48.
171. Oswald J, Boxberger S, Jorgensen B, Feldmann S, Ehninger G, Bornhauser M, Werner C. Mesenchymal stem cells can be differentiated into endothelial cells in vitro. Stem Cells. 2004;22(3):377–84.
172. Dominici M, Le Blanc K, Mueller I, Slaper-Cortenbach I, Marini F, Krause D, Deans R, Keating A, Prockop D, Horwitz E. Minimal criteria for defining multipotent mesenchymal stromal cells. Int Soc Cell Therapy Position Statement Cytotherapy. 2006;8(4):315–7.
173. Liu S, Yuan M, Hou K, Zhang L, Zheng X, Zhao B, Sui X, Xu W, Lu S, Guo Q. Immune characterization of mesenchymal stem cells in human umbilical cord Wharton's jelly and derived cartilage cells. Cell Immunol. 2012;278(1–2):35–44.
174. Weiss ML, Anderson C, Medicetty S, Seshareddy KB, Weiss RJ, VanderWerff I, Troyer D, McIntosh KR. Immune properties of human umbilical cord Wharton's jelly-derived cells. Stem Cells. 2008;26(11):2865–74.
175. Bernardo ME, Fibbe WE. Mesenchymal stromal cells: sensors and switchers of inflammation. Cell Stem Cell. 2013;13(4):392–402.
176. De Miguel MP, Fuentes-Julian S, Blazquez-Martinez A, Pascual CY, Aller MA, Arias J, Arnalich-Montiel F. Immunosuppressive properties of mesenchymal stem cells: advances and applications. Curr Mol Med. 2012;12(5):574–91.
177. Le Blanc K, Tammik C, Rosendahl K, Zetterberg E, Ringden O. HLA expression and immunologic properties of differentiated and undifferentiated mesenchymal stem cells. Exp Hematol. 2003;31(10):890–6.
178. Ji R, Zhang N, You N, Li Q, Liu W, Jiang N, Liu J, Zhang H, Wang D, Tao K, Dou K. The differentiation of MSCs into functional hepatocyte-like cells in a liver biomatrix scaffold and their transplantation into liver-fibrotic mice. Biomaterials. 2012;33(35):8995–9008.
179. Jiang WC, Cheng YH, Yen MH, Chang Y, Yang VW, Lee OK. Cryo-chemical decellularization of the whole liver for mesenchymal stem cells-based functional hepatic tissue engineering. Biomaterials. 2014;35(11):3607–17.
180. Sanchez PL, Fernandez-Santos ME, Costanza S, Climent AM, Moscoso I, Gonzalez-Nicolas MA, Sanz-Ruiz R, Rodriguez H, Kren SM, Garrido G, Escalante JL, Bermejo J, Elizaga J, Menarguez J, Yotti R, Perez del Villar C, Espinosa MA, Guillem MS, Willerson JT, Bernad A, Matesanz R, Taylor DA, Fernandez-Aviles F. Acellular human heart matrix: a critical step toward whole heart grafts. Biomaterials. 2015;61:279–89.
181. Mendez JJ, Ghaedi M, Steinbacher D, Niklason LE. Epithelial cell differentiation of human mesenchymal stromal cells in decellularized lung scaffolds. Tissue Eng Part A. 2014;20(11–12):1735–46.
182. Shamis Y, Hasson E, Soroker A, Bassat E, Shimoni Y, Ziv T, Sionov RV, Mitrani E. Organ-specific scaffolds for in vitro expansion, differentiation, and organization of primary lung cells. Tissue Eng Part C Methods. 2011;17(8):861–70.
183. Samsonraj RM, Raghunath M, Nurcombe V, Hui JH, van Wijnen AJ, Cool SM. Concise review: multifaceted characterization of human mesenchymal stem cells for use in regenerative medicine. Stem Cells Transl Med. 2017;6(12):2173–85.

184. Caplan AI, Dennis JE. Mesenchymal stem cells as trophic mediators. J Cell Biochem. 2006;98(5):1076–84.
185. Fang CH, Jin J, Joe JH, Song YS, So BI, Lim SM, Cheon GJ, Woo SK, Ra JC, Lee YY, Kim KS. In vivo differentiation of human amniotic epithelial cells into cardiomyocyte-like cells and cell transplantation effect on myocardial infarction in rats: comparison with cord blood and adipose tissue-derived mesenchymal stem cells. Cell Transplant. 2012;21(8):1687–96.
186. Guo Y, Yu Y, Hu S, Chen Y, Shen Z. The therapeutic potential of mesenchymal stem cells for cardiovascular diseases. Cell Death Dis. 2020;11(5):349.
187. Tan MY, Zhi W, Wei RQ, Huang YC, Zhou KP, Tan B, Deng L, Luo JC, Li XQ, Xie HQ, Yang ZM. Repair of infarcted myocardium using mesenchymal stem cell seeded small intestinal submucosa in rabbits. Biomaterials. 2009;30(19):3234–40.
188. Behnke J, Kremer S, Shahzad T, Chao CM, Bottcher-Friebertshauser E, Morty RE, Bellusci S, Ehrhardt H. MSC based therapies-new perspectives for the injured lung. J Clin Med. 2020;9(3):682.
189. Cruz FF, Rocco PRM. The potential of mesenchymal stem cell therapy for chronic lung disease. Expert Rev Respir Med. 2020;14(1):31–9.
190. Bohacova P, Holan V. Mesenchymal stem cells and type 1 diabetes treatment. Vnitrni Lekarstvi. 2018;64(7–8):725–8.
191. Path G, Perakakis N, Mantzoros CS, Seufert J. Stem cells in the treatment of diabetes mellitus - focus on mesenchymal stem cells. Metab Clin Exp. 2019;90:1–15.
192. Bogatcheva NV, Coleman ME. Conditioned medium of mesenchymal stromal cells: a new class of therapeutics. Biochem Biokhimiia. 2019;84(11):1375–89.
193. Sagaradze G, Grigorieva O, Nimiritsky P, Basalova N, Kalinina N, Akopyan Z, Efimenko A. Conditioned medium from human mesenchymal stromal cells: towards the clinical translation. Int J Mol Sci. 2019;20(7):1656.
194. Hu C, Li L. Preconditioning influences mesenchymal stem cell properties in vitro and in vivo. J Cell Mol Med. 2018;22(3):1428–42.
195. Zhao L, Hu C, Zhang P, Jiang H, Chen J. Preconditioning strategies for improving the survival rate and paracrine ability of mesenchymal stem cells in acute kidney injury. J Cell Mol Med. 2019;23(2):720–30.
196. Kular JK, Basu S, Sharma RI. The extracellular matrix: structure, composition, age-related differences, tools for analysis and applications for tissue engineering. J Tissue Eng. 2014;5:2041731414557112.
197. Fratzl P. Collagen structure and mechanics. In: Collagen. Boston: Springer; 2008. p. 1–13.
198. Barczyk M, Carracedo S, Gullberg D. Integrins. Cell Tissue Res. 2010;339(1):269–80.
199. Hynes RO. The extracellular matrix: not just pretty fibrils. Science (New York, NY). 2009;326(5957):1216–9.
200. Mecham RP. Elastin in lung development and disease pathogenesis. Matrix Biol J Int Soc Matrix Biol. 2018;73:6–20.
201. Vindin H, Mithieux SM, Weiss AS. Elastin architecture. Matrix biol J Int Soc Matrix Biol. 2019;84:4–16.
202. Mak KM, Mei R. Basement membrane type IV collagen and laminin: an overview of their biology and value as fibrosis biomarkers of liver disease. Anat Rec (Hoboken). 2017;300(8):1371–90.
203. Jandl K, Marsh LM, Hoffmann J, Mutgan AC, Baum O, Bloch W, Thekkekara-Puthenparampil H, Kolb D, Sinn K, Klepetko W, Heinemann A, Olschewski A, Olschewski H, Kwapiszewska G. Basement membrane remodeling controls endothelial function in idiopathic pulmonary arterial hypertension. Am J Respir Cell Mol Biol. 2020;63(1):104–17.
204. Singh P, Carraher C, Schwarzbauer JE. Assembly of fibronectin extracellular matrix. Annu Rev Cell Dev Biol. 2010;26:397–419.
205. Rnjak-Kovacina J, Tang F, Whitelock JM, Lord MS. Glycosaminoglycan and proteoglycan-based biomaterials: current trends and future perspectives. Adv Healthc Mater. 2018;7(6):e1701042.

206. Bayrak A, Tyralla M, Ladhoff J, Schleicher M, Stock UA, Volk HD, Seifert M. Human immune responses to porcine xenogeneic matrices and their extracellular matrix constituents in vitro. Biomaterials. 2010;31(14):3793–803.
207. Lee SH, Goswami S, Grudo A, Song LZ, Bandi V, Goodnight-White S, Green L, Hacken-Bitar J, Huh J, Bakaeen F, Coxson HO, Cogswell S, Storness-Bliss C, Corry DB, Kheradmand F. Antielastin autoimmunity in tobacco smoking-induced emphysema. Nat Med. 2007;13 (5):567–9.
208. Warren KJ, Iwami D, Harris DG, Bromberg JS, Burrell BE. Laminins affect T cell trafficking and allograft fate. J Clin Investig. 2014;124(5):2204–18.
209. Groeneveld TW, Oroszlan M, Owens RT, Faber-Krol MC, Bakker AC, Arlaud GJ, McQuillan DJ, Kishore U, Daha MR, Roos A. Interactions of the extracellular matrix proteoglycans decorin and biglycan with C1q and collectins. J Immunol. 2005;175(7):4715–23.
210. Gilbert TW, Sellaro TL, Badylak SF. Decellularization of tissues and organs. Biomaterials. 2006;27(19):3675–83.
211. Wong ML, Griffiths LG. Immunogenicity in xenogeneic scaffold generation: antigen removal vs. decellularization. Acta Biomater. 2014;10(5):1806–16.
212. Luo JC, Chen W, Chen XH, Qin TW, Huang YC, Xie HQ, Li XQ, Qian ZY, Yang ZM. A multi-step method for preparation of porcine small intestinal submucosa (SIS). Biomaterials. 2011;32(3):706–13.
213. Mendoza-Novelo B, Avila EE, Cauich-Rodriguez JV, Jorge-Herrero E, Rojo FJ, Guinea GV, Mata-Mata JL. Decellularization of pericardial tissue and its impact on tensile viscoelasticity and glycosaminoglycan content. Acta Biomater. 2011;7(3):1241–8.
214. Wilshaw SP, Kearney JN, Fisher J, Ingham E. Production of an acellular amniotic membrane matrix for use in tissue engineering. Tissue Eng. 2006;12(8):2117–29.
215. Barakat O, Abbasi S, Rodriguez G, Rios J, Wood RP, Ozaki C, Holley LS, Gauthier PK. Use of decellularized porcine liver for engineering humanized liver organ. J Surg Res. 2012;173 (1):e11–25.
216. Bonvillain RW, Danchuk S, Sullivan DE, Betancourt AM, Semon JA, Eagle ME, Mayeux JP, Gregory AN, Wang G, Townley IK, Borg ZD, Weiss DJ, Bunnell BA. A nonhuman primate model of lung regeneration: detergent-mediated decellularization and initial in vitro recellularization with mesenchymal stem cells. Tissue Eng Part A. 2012;18(23–24):2437–52.
217. Goh SK, Bertera S, Olsen P, Candiello JE, Halfter W, Uechi G, Balasubramani M, Johnson SA, Sicari BM, Kollar E, Badylak SF, Banerjee I. Perfusion-decellularized pancreas as a natural 3D scaffold for pancreatic tissue and whole organ engineering. Biomaterials. 2013;34 (28):6760–72.
218. Liu CX, Liu SR, Xu AB, Kang YZ, Zheng SB, Li HL. Preparation of whole-kidney acellular matrix in rats by perfusion. Nan fang yi ke da xue xue bao = Journal of Southern Medical University. 2009;29(5):979–82.
219. Methe K, Backdahl H, Johansson BR, Nayakawde N, Dellgren G, Sumitran-Holgersson S. An alternative approach to decellularize whole porcine heart. BioRes Open Access. 2014;3 (6):327–38.
220. Nichols JE, Niles J, Riddle M, Vargas G, Schilagard T, Ma L, Edward K, La Francesca S, Sakamoto J, Vega S, Ogadegbe M, Mlcak R, Deyo D, Woodson L, McQuitty C, Lick S, Beckles D, Melo E, Cortiella J. Production and assessment of decellularized pig and human lung scaffolds. Tissue Eng Part A. 2013;19(17–18):2045–62.
221. Orlando G, Farney AC, Iskandar SS, Mirmalek-Sani SH, Sullivan DC, Moran E, AbouShwareb T, De Coppi P, Wood KJ, Stratta RJ, Atala A, Yoo JJ, Soker S. Production and implantation of renal extracellular matrix scaffolds from porcine kidneys as a platform for renal bioengineering investigations. Ann Surg. 2012;256(2):363–70.
222. Park KM, Woo HM. Porcine bioengineered scaffolds as new frontiers in regenerative medicine. Transplant Proc. 2012;44(4):1146–50.

223. Price AP, England KA, Matson AM, Blazar BR, Panoskaltsis-Mortari A. Development of a decellularized lung bioreactor system for bioengineering the lung: the matrix reloaded. Tissue Eng Part A. 2010;16(8):2581–91.
224. Remlinger NT, Wearden PD, Gilbert TW. Procedure for decellularization of porcine heart by retrograde coronary perfusion. J Visual Exp. 2012;(70):e50059.
225. Robertson MJ, Dries-Devlin JL, Kren SM, Burchfield JS, Taylor DA. Optimizing recellularization of whole decellularized heart extracellular matrix. PLoS One. 2014;9(2): e90406.
226. Shupe T, Williams M, Brown A, Willenberg B, Petersen BE. Method for the decellularization of intact rat liver. Organogenesis. 2010;6(2):134–6.
227. Sullivan DC, Mirmalek-Sani SH, Deegan DB, Baptista PM, Aboushwareb T, Atala A, Yoo JJ. Decellularization methods of porcine kidneys for whole organ engineering using a high-throughput system. Biomaterials. 2012;33(31):7756–64.
228. Verstegen MMA, Willemse J, van den Hoek S, Kremers GJ, Luider TM, van Huizen NA, Willemssen F, Metselaar HJ, JNM IJ, van der Laan LJW, de Jonge J. Decellularization of whole human liver grafts using controlled perfusion for transplantable organ bioscaffolds. Stem Cells Dev. 2017;26(18):1304–15.
229. Wainwright JM, Czajka CA, Patel UB, Freytes DO, Tobita K, Gilbert TW, Badylak SF. Preparation of cardiac extracellular matrix from an intact porcine heart. Tissue Eng Part C Methods. 2010;16(3):525–32.
230. Agmon G, Christman KL. Controlling stem cell behavior with decellularized extracellular matrix scaffolds. Curr Opin Solid State Mater Sci. 2016;20(4):193–201.
231. Pulver A, Shevtsov B, Leybovich I, Artyuhov Y, Maleev A. Peregudov, production of organ extracellular matrix using a freeze-thaw cycle employing extracellular cryoprotectants. Cryo Lett. 2014;35(5):400–6.
232. Sackett SD, Tremmel DM, Ma F, Feeney AK, Maguire RM, Brown ME, Zhou Y, Li X, O'Brien C, Li L, Burlingham WJ, Odorico JS. Extracellular matrix scaffold and hydrogel derived from decellularized and delipidized human pancreas. Sci Rep. 2018;8(1):10452.
233. Elebring E, Kuna VK, Kvarnstrom N, Sumitran-Holgersson S. Cold-perfusion decellularization of whole-organ porcine pancreas supports human fetal pancreatic cell attachment and expression of endocrine and exocrine markers. J Tissue Eng. 2017;8:2041731417738145.
234. Faulk DM, Johnson SA, Zhang L, Badylak SF. Role of the extracellular matrix in whole organ engineering. J Cell Physiol. 2014;229(8):984–9.
235. Bilodeau C, Goltsis O, Rogers IM, Post M. Limitations of recellularized biological scaffolds for human transplantation. J Tissue Eng Regen Med. 2020;14(3):521–38.
236. O'Neill JD, Anfang R, Anandappa A, Costa J, Javidfar J, Wobma HM, Singh G, Freytes DO, Bacchetta MD, Sonett JR, Vunjak-Novakovic G. Decellularization of human and porcine lung tissues for pulmonary tissue engineering. Ann Thorac Surg. 2013;96(3):1046–55; discussion 1055-6.
237. Yagi H, Soto-Gutierrez A, Kitagawa Y. Whole-organ re-engineering: a regenerative medicine approach to digestive organ replacement. Surg Today. 2013;43(6):587–94.
238. Bornstein P, Kang AH, Piez KA. The limited cleavage of native collagen with chymotrypsin, trypsin, and cyanogen bromide*. Biochemistry. 2002;5(12):3803–12.
239. Gilpin A, Yang Y. Decellularization strategies for regenerative medicine: from processing techniques to applications. Biomed Res Int. 2017;2017:9831534.
240. London AS, Japutra C, Planck K, Lihon M, Nguyen AA. A novel method to determine residual detergent in biological samples post endotoxin reduction treatment and evaluation of strategies for subsequent detergent removal. Int Immunopharmacol. 2016;37:16–22.
241. Boer U, Lohrenz A, Klingenberg M, Pich A, Haverich A, Wilhelmi M. The effect of detergent-based decellularization procedures on cellular proteins and immunogenicity in equine carotid artery grafts. Biomaterials. 2011;32(36):9730–7.

242. Badylak SF, Valentin JE, Ravindra AK, McCabe GP, Stewart-Akers AM. Macrophage phenotype as a determinant of biologic scaffold remodeling. Tissue Eng Part A. 2008;14 (11):1835–42.
243. Hussey GS, Dziki JL, Badylak SF. Extracellular matrix-based materials for regenerative medicine. Nat Rev Materials. 2018;3(7):159–73.
244. Akhyari P, Aubin H, Gwanmesia P, Barth M, Hoffmann S, Huelsmann J, Preuss K, Lichtenberg A. The quest for an optimized protocol for whole-heart decellularization: a comparison of three popular and a novel decellularization technique and their diverse effects on crucial extracellular matrix qualities. Tissue Eng Part C Methods. 2011;17(9):915–26.
245. Arenas-Herrera JE, Ko IK, Atala A, Yoo JJ. Decellularization for whole organ bioengineering. Biomed Mater. 2013;8(1):014106.
246. Gilpin SE, Guyette JP, Gonzalez G, Ren X, Asara JM, Mathisen DJ, Vacanti JP, Ott HC. Perfusion decellularization of human and porcine lungs: bringing the matrix to clinical scale. J Heart Lung Transplantation Off Publication Int Soc Heart Transplant. 2014;33(3):298–308.
247. Li Q, Uygun BE, Geerts S, Ozer S, Scalf M, Gilpin SE, Ott HC, Yarmush ML, Smith LM, Welham NV, Frey BL. Proteomic analysis of naturally-sourced biological scaffolds. Biomaterials. 2016;75:37–46.
248. Nakayama KH, Batchelder CA, Lee CI, Tarantal AF. Decellularized rhesus monkey kidney as a three-dimensional scaffold for renal tissue engineering. Tissue Eng Part A. 2010;16 (7):2207–16.
249. Freudenberg U, Liang Y, Kiick KL, Werner C. Glycosaminoglycan-based biohybrid hydrogels: a sweet and smart choice for multifunctional biomaterials. Adv Mater. 2016;28 (40):8861–91.
250. Peloso A, Urbani L, Cravedi P, Katari R, Maghsoudlou P, Fallas ME, Sordi V, Citro A, Purroy C, Niu G, McQuilling JP, Sittadjody S, Farney AC, Iskandar SS, Zambon JP, Rogers J, Stratta RJ, Opara EC, Piemonti L, Furdui CM, Soker S, De Coppi P, Orlando G. The human pancreas as a source of protolerogenic extracellular matrix scaffold for a new-generation bioartificial endocrine pancreas. Ann Surg. 2016;264(1):169–79.
251. Kajbafzadeh AM, Javan-Farazmand N, Monajemzadeh M, Baghayee A. Determining the optimal decellularization and sterilization protocol for preparing a tissue scaffold of a human-sized liver tissue. Tissue Eng Part C Methods. 2013;19(8):642–51.
252. Struecker B, Hillebrandt KH, Voitl R, Butter A, Schmuck RB, Reutzel-Selke A, Geisel D, Joehrens K, Pickerodt PA, Raschzok N, Puhl G, Neuhaus P, Pratschke J, Sauer IM. Porcine liver decellularization under oscillating pressure conditions: a technical refinement to improve the homogeneity of the decellularization process. Tissue Eng Part C Methods. 2015;21 (3):303–13.
253. De Carlo E, Baiguera S, Conconi MT, Vigolo S, Grandi C, Lora S, Martini C, Maffei P, Tamagno G, Vettor R, Sicolo N, Parnigotto PP. Pancreatic acellular matrix supports islet survival and function in a synthetic tubular device: in vitro and in vivo studies. Int J Mol Med. 2010;25(2):195–202.
254. Tan B, Wei RQ, Tan MY, Luo JC, Deng L, Chen XH, Hou JL, Li XQ, Yang ZM, Xie HQ. Tissue engineered esophagus by mesenchymal stem cell seeding for esophageal repair in a canine model. J Surg Res. 2013;182(1):40–8.
255. Macchiarini P, Jungebluth P, Go T, Asnaghi MA, Rees LE, Cogan TA, Dodson A, Martorell J, Bellini S, Parnigotto PP, Dickinson SC, Hollander AP, Mantero S, Conconi MT, Birchall MA. Clinical transplantation of a tissue-engineered airway. Lancet. 2008;372(9655):2023–30.
256. Schreiber C, Eicken A, Seidl S, Lange R. Unexpected early failure of a decellularized right ventricle to pulmonary artery graft. Ann Thorac Surg. 2008;86(6):2026–7; author reply 2027-8.
257. Simon P, Kasimir MT, Seebacher G, Weigel G, Ullrich R, Salzer-Muhar U, Rieder E, Wolner E. Early failure of the tissue engineered porcine heart valve SYNERGRAFT in pediatric patients. Eur J Cardio-Thoracic Surg. 2003;23(6):1002–6; discussion 1006.

258. Higuchi Y, Kojima M, Ishii G, Aoyagi K, Sasaki H, Ochiai A. Gastrointestinal fibroblasts have specialized, diverse transcriptional phenotypes: a comprehensive gene expression analysis of human fibroblasts. PLoS One. 2015;10(6):e0129241.
259. Muhl L, Genove G, Leptidis S, Liu J, He L, Mocci G, Sun Y, Gustafsson S, Buyandelger B, Chivukula IV, Segerstolpe A, Raschperger E, Hansson EM, Bjorkegren JLM, Peng XR, Vanlandewijck M, Lendahl U, Betsholtz C. Single-cell analysis uncovers fibroblast heterogeneity and criteria for fibroblast and mural cell identification and discrimination. Nat Commun. 2020;11(1):3953.
260. Kaushal S, Amiel GE, Guleserian KJ, Shapira OM, Perry T, Sutherland FW, Rabkin E, Moran AM, Schoen FJ, Atala A, Soker S, Bischoff J, Mayer JE Jr. Functional small-diameter neovessels created using endothelial progenitor cells expanded ex vivo. Nat Med. 2001;7(9):1035–40.
261. Ren X, Tapias LF, Jank BJ, Mathisen DJ, Lanuti M, Ott HC. Ex vivo non-invasive assessment of cell viability and proliferation in bio-engineered whole organ constructs. Biomaterials. 2015;52:103–12.
262. Ren X, Moser PT, Gilpin SE, Okamoto T, Wu T, Tapias LF, Mercier FE, Xiong L, Ghawi R, Scadden DT, Mathisen DJ, Ott HC. Engineering pulmonary vasculature in decellularized rat and human lungs. Nat Biotechnol. 2015;33(10):1097–102.
263. Khan AA, Vishwakarma SK, Bardia A, Venkateshwarulu J. Repopulation of decellularized whole organ scaffold using stem cells: an emerging technology for the development of neo-organ. J Artif Organs. 2014;17(4):291–300.
264. Vunjak-Novakovic G, Tandon N, Godier A, Maidhof R, Marsano A, Martens TP, Radisic M. Challenges in cardiac tissue engineering. Tissue Eng Part B Rev. 2010;16(2):169–87.
265. Sussman NL, Kelly JH. Artificial liver: a forthcoming attraction. Hepatology. 1993;17(6):1163–4.
266. Bilir BM, Guinette D, Karrer F, Kumpe DA, Krysl J, Stephens J, McGavran L, Ostrowska A, Durham J. Hepatocyte transplantation in acute liver failure. Liver Transplantation. 2000;6(1):32–40.
267. Kitano K, Schwartz DM, Zhou H, Gilpin SE, Wojtkiewicz GR, Ren X, Sommer CA, Capilla AV, Mathisen DJ, Goldstein AM, Mostoslavsky G, Ott HC. Bioengineering of functional human induced pluripotent stem cell-derived intestinal grafts. Nat Commun. 2017;8(1):765.
268. Napierala H, Hillebrandt KH, Haep N, Tang P, Tintemann M, Gassner J, Noesser M, Everwien H, Seiffert N, Kluge M, Teegen E, Polenz D, Lippert S, Geisel D, Reutzel Selke A, Raschzok N, Andreou A, Pratschke J, Sauer IM, Struecker B. Engineering an endocrine neo-pancreas by repopulation of a decellularized rat pancreas with islets of Langerhans. Sci Rep. 2017;7:41777.
269. Butter A, Aliyev K, Hillebrandt KH, Raschzok N, Kluge M, Seiffert N, Tang P, Napierala H, Muhamma AI, Reutzel-Selke A, Andreou A, Pratschke J, Sauer IM, Struecker B. Evolution of graft morphology and function after recellularization of decellularized rat livers. J Tissue Eng Regen Med. 2018;12(2):e807–16.
270. Chen Y, Devalliere J, Bulutoglu B, Yarmush ML, Uygun BE. Repopulation of intrahepatic bile ducts in engineered rat liver grafts. Technology. 2019;7(1–2):46–55.
271. Song JJ, Kim SS, Liu Z, Madsen JC, Mathisen DJ, Vacanti JP, Ott HC. Enhanced in vivo function of bioartificial lungs in rats. Ann Thorac Surg. 2011;92(3):998–1005; discussion 1005-6.
272. Ciampi O, Bonandrini B, Derosas M, Conti S, Rizzo P, Benedetti V, Figliuzzi M, Remuzzi A, Benigni A, Remuzzi G, Tomasoni S. Engineering the vasculature of decellularized rat kidney scaffolds using human induced pluripotent stem cell-derived endothelial cells. Sci Rep. 2019;9(1):8001.
273. Goh SK, Bertera S, Vaidya V, Dumpe S, Barner S, Mathew S, Banerjee I. Development of perfusion bioreactor for whole organ engineering - a culture system that enhances cellular engraftment, survival and phenotype of repopulated pancreas. Technology. 2018;6(3–4):118–34.

274. Deutsch M, Meinhart J, Vesely M, Fischlein T, Groscurth P, von Oppell U, Zilla P. In vitro endothelialization of expanded polytetrafluoroethylene grafts: a clinical case report after 41 months of implantation. J Vasc Surg. 1997;25(4):757–63.
275. Bao J, Shi Y, Sun H, Yin X, Yang R, Li L, Chen X, Bu H. Construction of a portal implantable functional tissue-engineered liver using perfusion-decellularized matrix and hepatocytes in rats. Cell Transplant. 2011;20(5):753–66.
276. Gullon L, Gutierrez-Gutierrez J, Sanchez Del Arco RT, Rivera A, Fernandez I, Del Canizo JF. Development of an automated liver perfusion system: the benefit of a hemofilter. Int J Artif Organs. 2018;41(11):723–9.
277. Hulsmann J, Aubin H, Kranz A, Godehardt E, Munakata H, Kamiya H, Barth M, Lichtenberg A, Akhyari P. A novel customizable modular bioreactor system for whole-heart cultivation under controlled 3D biomechanical stimulation. J Artif Organs. 2013;16(3):294–304.
278. Schmid J, Schwarz S, Meier-Staude R, Sudhop S, Clausen-Schaumann H, Schieker M, Huber R. A perfusion bioreactor system for cell seeding and oxygen-controlled cultivation of three-dimensional cell cultures. Tissue Eng Part C Methods. 2018;24(10):585–95.
279. Uzarski JS, Bijonowski BM, Wang B, Ward HH, Wandinger-Ness A, Miller WM, Wertheim JA. Correction to: dual-purpose bioreactors to monitor noninvasive physical and biochemical markers of kidney and liver scaffold recellularization. Tissue Eng Part C Methods. 2015;21 (10):1032–43. https://doi.org/10.1089/ten.tec.2014.0665, by Uzarski JS, Bijonowski BM, Wang B, Ward HH, Wandinger-Ness A, Miller WM, and Wertheim JA, Tissue engineering. Part C, Methods 24(7) (2018) 441–442.
280. Lovett M, Lee K, Edwards A, Kaplan DL. Vascularization strategies for tissue engineering. Tissue Eng Part B Rev. 2009;15(3):353–70.
281. Hulsmann J, Aubin H, Wehrmann A, Lichtenberg A, Akhyari P. The impact of left ventricular stretching in model cultivations with neonatal cardiomyocytes in a whole-heart bioreactor. Biotechnol Bioeng. 2017;114(5):1107–17.
282. Hulsmann J, Aubin H, Sugimura Y, Lichtenberg A, Akhyari P. Electrophysiological stimulation of whole heart constructs in an 8-pole electrical field. Artif Organs. 2018;42(12):E391–405.
283. Maidhof R, Tandon N, Lee EJ, Luo J, Duan Y, Yeager K, Konofagou E, Vunjak-Novakovic G. Biomimetic perfusion and electrical stimulation applied in concert improved the assembly of engineered cardiac tissue. J Tissue Eng Regen Med. 2012;6(10):e12–23.
284. Hernandez D, Millard R, Sivakumaran P, Wong RC, Crombie DE, Hewitt AW, Liang H, Hung SS, Pebay A, Shepherd RK, Dusting GJ, Lim SY. Electrical stimulation promotes cardiac differentiation of human induced pluripotent stem cells. Stem Cells Int. 2016;2016:1718041.
285. Tolboom H, Pouw R, Uygun K, Tanimura Y, Izamis ML, Berthiaume F, Yarmush ML. A model for normothermic preservation of the rat liver. Tissue Eng. 2007;13(8):2143–51.
286. Bonvillain RW, Scarritt ME, Pashos NC, Mayeux JP, Meshberger CL, Betancourt AM, Sullivan DE, Bunnell BA. Nonhuman primate lung decellularization and recellularization using a specialized large-organ bioreactor. J Visual Exp. 2013;(82):e50825.
287. O'Mara P, Farrell A, Bones J, Twomey K. Staying alive! Sensors used for monitoring cell health in bioreactors. Talanta. 2018;176:130–9.
288. Lichtenberg A, Cebotari S, Tudorache I, Sturz G, Winterhalter M, Hilfiker A, Haverich A. Flow-dependent re-endothelialization of tissue-engineered heart valves. J Heart Valve Dis. 2006;15(2):287–93; discussion 293-4.
289. Elzwiei F, Bassien-Capsa V, St-Louis J, Chorvatova A. Regulation of the sodium pump during cardiomyocyte adaptation to pregnancy. Exp Physiol. 2013;98(1):183–92.
290. Karbassi E, Fenix A, Marchiano S, Muraoka N, Nakamura K, Yang X, Murry CE. Cardiomyocyte maturation: advances in knowledge and implications for regenerative medicine. Nat Rev Cardiol. 2020;17(6):341–59.
291. Schwinger RH, Erdmann E. Heart failure and electrolyte disturbances. Methods Find Exp Clin Pharmacol. 1992;14(4):315–25.

292. Tsuruoka N, Ishii K, Matsunaga T, Nagatomi R, Haga Y. Lactate and glucose measurement in subepidermal tissue using minimally invasive microperfusion needle. Biomed Microdevices. 2016;18(1):19.
293. Chouinard JA, Gagnon S, Couture MG, Levesque A, Vermette P. Design and validation of a pulsatile perfusion bioreactor for 3D high cell density cultures. Biotechnol Bioeng. 2009;104 (6):1215–23.
294. Hong GS, Antaris AL, Dai HJ. Near-infrared fluorophores for biomedical imaging. Nat Biomed Eng. 2017;1(1):0010.
295. Lusic H, Grinstaff MW. X-ray-computed tomography contrast agents. Chem Rev. 2013;113 (3):1641–66.
296. Soto-Gutierrez A, Zhang L, Medberry C, Fukumitsu K, Faulk D, Jiang H, Reing J, Gramignoli R, Komori J, Ross M, Nagaya M, Lagasse E, Stolz D, Strom SC, Fox IJ, Badylak SF. A whole-organ regenerative medicine approach for liver replacement. Tissue Eng Part C Methods. 2011;17(6):677–86.
297. Lawrence BJ, Devarapalli M, Madihally SV. Flow dynamics in bioreactors containing tissue engineering scaffolds. Biotechnol Bioeng. 2009;102(3):935–47.
298. Shonaka T, Matsuno N, Obara H, Yoshikawa R, Nishikawa Y, Gouchi M, Otani M, Takahashi H, Azuma H, Sakai H, Furukawa H. Application of perfusate with human-derived oxygen carrier solution under Subnormothermic machine perfusion for donation after cardiac death liver grafts in pigs. Transplant Proc. 2018;50(9):2821–5.
299. Huang Y-Z, Xie H-Q, Li X. Scaffolds in bone tissue engineering: research Progress and current applications. In: Zaidi M, editor. Encyclopedia of bone biology. Oxford: Academic Press; 2020. p. 204–15.
300. Peric Kacarevic Z, Rider P, Alkildani S, Retnasingh S, Pejakic M, Schnettler R, Gosau M, Smeets R, Jung O, Barbeck M. An introduction to bone tissue engineering. Int J Artif Organs. 2020;43(2):69–86.
301. Rahmani Del Bakhshayesh A, Babaie S, Tayefi Nasrabadi H, Asadi N, Akbarzadeh A, Abedelahi A. An overview of various treatment strategies, especially tissue engineering for damaged articular cartilage. Artif Cells Nanomed Biotechnol. 2020;48(1):1089–104.
302. Zhao Z, Fan C, Chen F, Sun Y, Xia Y, Ji A, Wang DA. Progress in articular cartilage tissue engineering: a review on therapeutic cells and macromolecular scaffolds. Macromol Biosci. 2020;20(2):e1900278.
303. Jaiswal D, Yousman L, Neary M, Fernschild E, Zolnoski B, Katebifar S, Rudraiah S, Mazzocca AD, Kumbar SG. Tendon tissue engineering: biomechanical considerations. Biomed Materials. 2020;15(5):052001.
304. Youngstrom DW, Barrett JG. Engineering tendon: scaffolds, bioreactors, and models of regeneration. Stem Cells Int. 2016;2016:3919030.
305. Nourian Dehkordi A, Mirahmadi Babaheydari F, Chehelgerdi M, Raeisi Dehkordi S. Skin tissue engineering: wound healing based on stem-cell-based therapeutic strategies. Stem Cell Res Ther. 2019;10(1):111.
306. Yu JR, Navarro J, Coburn JC, Mahadik B, Molnar J, Holmes JHT, Nam AJ, Fisher JP. Current and future perspectives on skin tissue engineering: key features of biomedical research, translational assessment, and clinical application. Adv Healthc Mater. 2019;8(5):e1801471.
307. Mangir N, Wilson KJ, Osman NI, Chapple CR. Current state of urethral tissue engineering. Curr Opin Urol. 2019;29(4):385–93.
308. Versteegden LRM, de Jonge P, IntHout J, van Kuppevelt TH, Oosterwijk E, Feitz WFJ, de Vries RBM, Daamen WF. Tissue engineering of the urethra: a systematic review and meta-analysis of preclinical and clinical studies. Eur Urol. 2017;72(4):594–606.
309. Dorati R, De Trizio A, Marconi S, Ferrara A, Auricchio F, Genta I, Modena T, Benazzo M, Benazzo A, Volpato G, Conti B. Design of a bioabsorbable multilayered patch for esophagus tissue engineering. Macromol Biosci. 2017;17(6):1600426.
310. Saxena AK. Esophagus tissue engineering: designing and crafting the components for the "hybrid construct" approach. Eur J Pediatr Surg. 2014;24(3):246–62.

311. Gill BC, Damaser MS, Chermansky CJ. Future perspectives in bladder tissue engineering. Current Bladder Dysfunction Rep. 2014;10(4):443–8.
312. Lam Van Ba O, Aharony S, Loutochin O, Corcos J. Bladder tissue engineering: a literature review. Adv Drug Deliv Rev. 2015;82–83:31–7.
313. Hoyert DL, Xu J. Deaths: preliminary data for 2011. National Vital Statistics Rep. 2012;61 (6):1–51.
314. Barnard CN. Human cardiac transplantation. An evaluation of the first two operations performed at the Groote Schuur hospital, Cape Town. Am J Cardiol. 1968;22(4):584–96.
315. Ng SL, Narayanan K, Gao S, Wan AC. Lineage restricted progenitors for the repopulation of decellularized heart. Biomaterials. 2011;32(30):7571–80.
316. Rajabi S, Pahlavan S, Ashtiani MK, Ansari H, Abbasalizadeh S, Sayahpour FA, Varzideh F, Kostin S, Aghdami N, Braun T, Baharvand H. Human embryonic stem cell-derived cardiovascular progenitor cells efficiently colonize in bFGF-tethered natural matrix to construct contracting humanized rat hearts. Biomaterials. 2018;154:99–112.
317. Guyette JP, Charest JM, Mills RW, Jank BJ, Moser PT, Gilpin SE, Gershlak JR, Okamoto T, Gonzalez G, Milan DJ, Gaudette GR, Ott HC. Bioengineering human myocardium on native extracellular matrix. Circ Res. 2016;118(1):56–72.
318. Nguyen DT, O'Hara M, Graneli C, Hicks R, Miliotis T, Nystrom AC, Hansson S, Davidsson P, Gan LM, Magnone MC, Althage M, Heydarkhan-Hagvall S. Humanizing miniature hearts through 4-flow cannulation perfusion decellularization and recellularization. Sci Rep. 2018;8 (1):7458.
319. Silva AC, Rodrigues SC, Caldeira J, Nunes AM, Sampaio-Pinto V, Resende TP, Oliveira MJ, Barbosa MA, Thorsteinsdottir S, Nascimento DS, Pinto-do OP. Three-dimensional scaffolds of fetal decellularized hearts exhibit enhanced potential to support cardiac cells in comparison to the adult. Biomaterials. 2016;104:52–64.
320. Yasui H, Lee JK, Yoshida A, Yokoyama T, Nakanishi H, Miwa K, Naito AT, Oka T, Akazawa H, Nakai J, Miyagawa S, Sawa Y, Sakata Y, Komuro I. Excitation propagation in three-dimensional engineered hearts using decellularized extracellular matrix. Biomaterials. 2014;35 (27):7839–50.
321. Kitahara H, Yagi H, Tajima K, Okamoto K, Yoshitake A, Aeba R, Kudo M, Kashima I, Kawaguchi S, Hirano A, Kasai M, Akamatsu Y, Oka H, Kitagawa Y, Shimizu H. Heterotopic transplantation of a decellularized and recellularized whole porcine heart. Interact Cardiovasc Thorac Surg. 2016;22(5):571–9.
322. Vishwakarma SK, Bardia A, Lakkireddy C, Raju N, Paspala SAB, Habeeb MA, Khan AA. Intraperitoneal transplantation of bioengineered humanized liver grafts supports failing liver in acute condition. Mater Sci Eng C Mater Biol Appl. 2019;98:861–73.
323. Mirmalek-Sani SH, Sullivan DC, Zimmerman C, Shupe TD, Petersen BE. Immunogenicity of decellularized porcine liver for bioengineered hepatic tissue. Am J Pathol. 2013;183(2):558–65.
324. Nakamura S, Ijima H. Solubilized matrix derived from decellularized liver as a growth factor-immobilizable scaffold for hepatocyte culture. J Biosci Bioeng. 2013;116(6):746–53.
325. Jaramillo M, Yeh H, Yarmush ML, Uygun BE. Decellularized human liver extracellular matrix (hDLM)-mediated hepatic differentiation of human induced pluripotent stem cells (hIPSCs). J Tissue Eng Regen Med. 2018;12(4):e1962–73.
326. Minami T, Ishii T, Yasuchika K, Fukumitsu K, Ogiso S, Miyauchi Y, Kojima H, Kawai T, Yamaoka R, Oshima Y, Kawamoto H, Kotaka M, Yasuda K, Osafune K, Uemoto S. Novel hybrid three-dimensional artificial liver using human induced pluripotent stem cells and a rat decellularized liver scaffold. Regen Ther. 2019;10:127–33.
327. Hassanein W, Cimeno A, Werdesheim A, Buckingham B, Harrison J, Uluer MC, Khalifeh A, Rivera-Pratt C, Klepfer S, Woodall JD, Dhru U, Bromberg E, Parsell D, Drachenberg C, Barth RN, LaMattina JC. Liver scaffolds support survival and metabolic function of multilineage neonatal allogenic cells. Tissue Eng Part A. 2018;24(9–10):786–93.

328. Zhou P, Lessa N, Estrada DC, Severson EB, Lingala S, Zern MA, Nolta JA, Wu J. Decellularized liver matrix as a carrier for the transplantation of human fetal and primary hepatocytes in mice. Liver Transplant. 2011;17(4):418–27.
329. Ko IK, Peng L, Peloso A, Smith CJ, Dhal A, Deegan DB, Zimmerman C, Clouse C, Zhao W, Shupe TD, Soker S, Yoo JJ, Atala A. Bioengineered transplantable porcine livers with re-endothelialized vasculature. Biomaterials. 2015;40:72–9.
330. Meng F, Almohanna F, Altuhami A, Assiri AM, Broering D. Vasculature reconstruction of decellularized liver scaffolds via gelatin-based re-endothelialization. J Biomed Mater Res A. 2019;107(2):392–402.
331. Calle EA, Hill RC, Leiby KL, Le AV, Gard AL, Madri JA, Hansen KC, Niklason LE. Targeted proteomics effectively quantifies differences between native lung and detergent-decellularized lung extracellular matrices. Acta Biomater. 2016;46:91–100.
332. Zhou H, Kitano K, Ren X, Rajab TK, Wu M, Gilpin SE, Wu T, Baugh L, Black LD, Mathisen DJ, Ott HC. Bioengineering human lung grafts on porcine matrix. Ann Surg. 2018;267 (3):590–8.
333. Ghaedi M, Le AV, Hatachi G, Beloiartsev A, Rocco K, Sivarapatna A, Mendez JJ, Baevova P, Dyal RN, Leiby KL, White ES, Niklason LE. Bioengineered lungs generated from human iPSCs-derived epithelial cells on native extracellular matrix. J Tissue Eng Regen Med. 2018;12(3):e1623–35.
334. Sokocevic D, Bonenfant NR, Wagner DE, Borg ZD, Lathrop MJ, Lam YW, Deng B, Desarno MJ, Ashikaga T, Loi R, Hoffman AM, Weiss DJ. The effect of age and emphysematous and fibrotic injury on the re-cellularization of de-cellularized lungs. Biomaterials. 2013;34 (13):3256–69.
335. Gilpin SE, Ren X, Okamoto T, Guyette JP, Mou H, Rajagopal J, Mathisen DJ, Vacanti JP, Ott HC. Enhanced lung epithelial specification of human induced pluripotent stem cells on decellularized lung matrix. Ann Thorac Surg. 2014;98(5):1721–9; discussion 1729.
336. Moran EC, Dhal A, Vyas D, Lanas A, Soker S, Baptista PM. Whole-organ bioengineering: current tales of modern alchemy. Translational Res. 2014;163(4):259–67.
337. Choi S, Park M, Kim J, Hwang S, Park S, Lee Y. The role of mesenchymal stem cells in the functional improvement of chronic renal failure. Stem Cells Dev. 2009;18(3):521–9.
338. Liew A. Perspectives in renal replacement therapy: haemodialysis. Nephrology (Carlton). 2018;23(Suppl 4):95–9.
339. Saito A. Development of bioartificial kidneys. Nephrology (Carlton). 2003;8(Suppl):S10–5.
340. Abolbashari M, Agcaoili SM, Lee MK, Ko IK, Aboushwareb T, Jackson JD, Yoo JJ, Atala A. Repopulation of porcine kidney scaffold using porcine primary renal cells. Acta Biomater. 2016;29:52–61.
341. Poornejad N, Momtahan N, Salehi AS, Scott DR, Fronk CA, Roeder BL, Reynolds PR, Bundy BC, Cook AD. Efficient decellularization of whole porcine kidneys improves reseeded cell behavior. Biomed Materials. 2016;11(2):025003.
342. Peloso A, Petrosyan A, Da Sacco S, Booth C, Zambon JP, O'Brien T, Aardema C, Robertson J, De Filippo RE, Soker S, Stratta RJ, Perin L, Orlando G. Renal extracellular matrix scaffolds from discarded kidneys maintain glomerular morphometry and vascular resilience and retains critical growth factors. Transplantation. 2015;99(9):1807–16.
343. Yu YL, Shao YK, Ding YQ, Lin KZ, Chen B, Zhang HZ, Zhao LN, Wang ZB, Zhang JS, Tang ML, Mei J. Decellularized kidney scaffold-mediated renal regeneration. Biomaterials. 2014;35 (25):6822–8.
344. Yu Y, Cui H, Chen C, Wen G, Xu J, Zheng B, Zhang J, Wang C, Chai Y, Mei J. Hypoxia-inducible factor-1alpha directs renal regeneration induced by decellularized scaffolds. Biomaterials. 2018;165:48–55.
345. Zambon JP, Ko IK, Abolbashari M, Huling J, Clouse C, Kim TH, Smith C, Atala A, Yoo JJ. Comparative analysis of two porcine kidney decellularization methods for maintenance of functional vascular architectures. Acta Biomater. 2018;75:226–34.

346. Remuzzi A, Figliuzzi M, Bonandrini B, Silvani S, Azzollini N, Nossa R, Benigni A, Remuzzi G. Experimental evaluation of kidney regeneration by organ scaffold Recellularization. Sci Rep. 2017;7:43502.
347. Ni Q, Pham NB, Meng WS, Zhu G, Chen X. Advances in immunotherapy of type I diabetes. Adv Drug Deliv Rev. 2019;139:83–91.
348. Bi H, Ye K, Jin S. Proteomic analysis of decellularized pancreatic matrix identifies collagen V as a critical regulator for islet organogenesis from human pluripotent stem cells. Biomaterials. 2020;233:119673.
349. Huang YB, Mei J, Yu Y, Ding Y, Xia W, Yue T, Chen W, Zhou MT, Yang YJ. Comparative decellularization and recellularization of normal versus Streptozotocin-induced diabetes mellitus rat pancreas. Artif Organs. 2019;43(4):399–412.
350. Katsuki Y, Yagi H, Okitsu T, Kitago M, Tajima K, Kadota Y, Hibi T, Abe Y, Shinoda M, Itano O, Takeuchi S, Kitagawa Y. Endocrine pancreas engineered using porcine islets and partial pancreatic scaffolds. Pancreatology. 2016;16(5):922–30.
351. Wan J, Huang Y, Zhou P, Guo Y, Wu C, Zhu S, Wang Y, Wang L, Lu Y, Wang Z. Culture of iPSCs derived pancreatic beta-like cells in vitro using Decellularized pancreatic scaffolds: a preliminary trial. Biomed Res Int. 2017;2017:4276928.
352. Wan J, Wang L, Huang Y, Fan H, Chen C, Yuan X, Guo Y, Yin L. Using GRGDSPC peptides to improve re-endothelialization of decellularized pancreatic scaffolds. Artif Organs. 2020;44 (4):E172–80.
353. Xu L, Guo Y, Huang Y, Xiong Y, Xu Y, Li X, Lu J, Wang L, Wang Y, Lu Y, Wang Z. Constructing heparin-modified pancreatic decellularized scaffold to improve its re-endothelialization. J Biomater Appl. 2018;32(8):1063–70.
354. Vegh AMD, Duim SN, Smits AM, Poelmann RE, Ten Harkel ADJ, DeRuiter MC, Goumans MJ, Jongbloed MRM. Part and parcel of the cardiac autonomic nerve system: unravelling its cellular building blocks during development. J Cardiovasc Dev Dis. 2016;3(3):28.
355. Awad M, Czer LS, Hou M, Golshani SS, Goltche M, De Robertis M, Kittleson M, Patel J, Azarbal B, Kransdorf E, Esmailian F, Trento A, Kobashigawa JA. Early denervation and later reinnervation of the heart following cardiac transplantation: a review. J Am Heart Assoc. 2016;5(11):e004070.
356. Bengel FM, Ueberfuhr P, Schiepel N, Nekolla SG, Reichart B, Schwaiger M. Effect of sympathetic reinnervation on cardiac performance after heart transplantation. N Engl J Med. 2001;345(10):731–8.
357. Bildirici U, Celikyurt U, Ural E, Sahin T, Vural A, Koz C, Ural D. Successful percutaneous intervention to acute myocardial infarction presenting with typical chest pain in transplanted heart. Circulation J. 2009;73(11):2166–8.
358. Park D, Lee J, Chung JJ, Jung Y, Kim SH. Integrating organs-on-chips: multiplexing, scaling, vascularization, and innervation. Trends Biotechnol. 2020;38(1):99–112.
359. Charo RA, Sipp D. Rejuvenating regenerative medicine regulation. N Engl J Med. 2018;378 (6):504–5.

Chapter 7
The Applications of the Recellularization Organs in Organ Replacement at the Stage of Animal Research

Xili Ding, Yuqi He, and Xiaoming Li

Abstract Organ transplantation is the best treatment option for patients with end-stage organ failure. Unfortunately, the availability of donor organs and immune rejection severely limited organ transplantation, which results in a large number of patient deaths. Organ bioengineering could provide new ideas for the global shortage of critical organs. The use of decellularization-recellularization technology of natural tissues and organs has promising results in heart, lung, liver, pancreas, intestine, and kidney engineering. Decellularized whole organ with preserved structures and vasculature can provide a decellularized tissue platform for organ regeneration. Successful decellularization has provided suitable scaffolds to repopulate cells to grow fully transplantable organs ex vivo. Recellularized organs can perform organ-specific functions in a short time, which indicates that engineered organs have potential in future clinical applications. Selecting the best cell type, obtaining the required numbers of cells, delivering the cells to the scaffold, and maturing these cells into functional organs are all steps. Although the decellularization and recellularization of various organs including the heart, liver, lung, kidney, bladder, etc. have yielded exciting results, it is not possible to obtain donor organs for this method from humans. In this review, we focus on the progress of recellularization of whole organs in vivo and highlight future challenges in this field.

Keywords Decellularization · Recellularization · Whole organ · Tissue engineering · Animal research

---

X. Ding (✉) · Y. He · X. Li
Key Laboratory for Biomechanics and Mechanobiology of Ministry of Education, Beijing Advanced Innovation Center for Biomedical Engineering, School of Biological Science and Medical Engineering, Beihang University, Beijing, China
e-mail: xiliding@buaa.edu.cn; lixm@buaa.edu.cn; x.m.li@hotmail.com

X. Li, H. Xie (eds.), *Decellularized Materials*,
https://doi.org/10.1007/978-981-33-6962-7_7

7.1 Introduction

Orthotopic transplantation is the ideal treatment for end-stage organ failure. However, the shortage of donor organs has led to an increase in morbidity and mortality among thousands of patients each year. In the USA alone, there are nearly 27,000 deaths in patients with end-stage liver disease each year, of which 425,000 are due to coronary heart disease [1], 112,000 are due to end-stage kidney failure, and 120,000 patients as a result of chronic lung disease [2]. As of May 2015, in the USA alone, more than 123,000 people waiting for organ transplantation. Due to a severe organ shortage, in 2014, 6885 patients in the USA died while waiting for organ transplantation [3]. Some patients are lucky to get organs, but they still bear the burden of chronic rejection and the morbidity because of a lifelong immunosuppressant therapy. Although the development of engineered tissues such as blood vessels [1, 4], urinary bladder [5], and trachea [6] made great progress, these tissues can be connected to the host circulation after implantation and do not need an immediate vascular supply.

In recent years, a promising method of functional organ transplantation has attracted our attention. Acellular scaffold material can be regenerated by decellularization of allogeneic or xenogeneic donor organs such as lung [7–9], liver [10], and heart [11], which can then be repopulated with selected cell types or stem cells. Functional tissue can be formed by a combination of these seeded cells and three-dimensional matrix in short-term preclinical animal models [10, 12, 13]. This method provides the opportunity to connect directly to the patient's vasculature in an ectopic or in situ location. Many challenges of this three-dimensional organ-engineering method still exist, including the identification of candidate species from which the donor organ can be obtained, optimized donor organ decellularization methods, and optimized recellularization techniques. Moreover, the proper use of ex vivo bioreactors, endothelialization of the donor matrix vasculature, and the most suitable cell populations are also needed to investigate. Recellularization mainly refers to the repopulating the cells into organ matrix and then the cells can attach, migrate, proliferate, and have specialized functions to regain organ functions as in natural tissue [14].

Once a satisfactory organ scaffold is produced, it needs to be recellularized to obtain functional organ products for implantation. The process of recellularization involves obtaining a source of renewable cells, seeding the cells on a scaffold, and organizing a specific culture [15]. In order to establish effective and reproducible protocols to generate functional organ products, much work has been done.

Tissue engineering could offer different solutions for reducing the waiting list by creating functional in vitro bioengineered constructs as an alternative source of transplantable tissue, which could be readily available [16]. In addition, these constructs can be regenerated with autologous (stem) cells to reduce the complexity of rejection and immunosuppression. Figure 7.1 schematically depicts this process, which shows the traditional tissue engineering paradigm. A functional tissue construct can be produced by combining specific cells with signaling molecules (such as

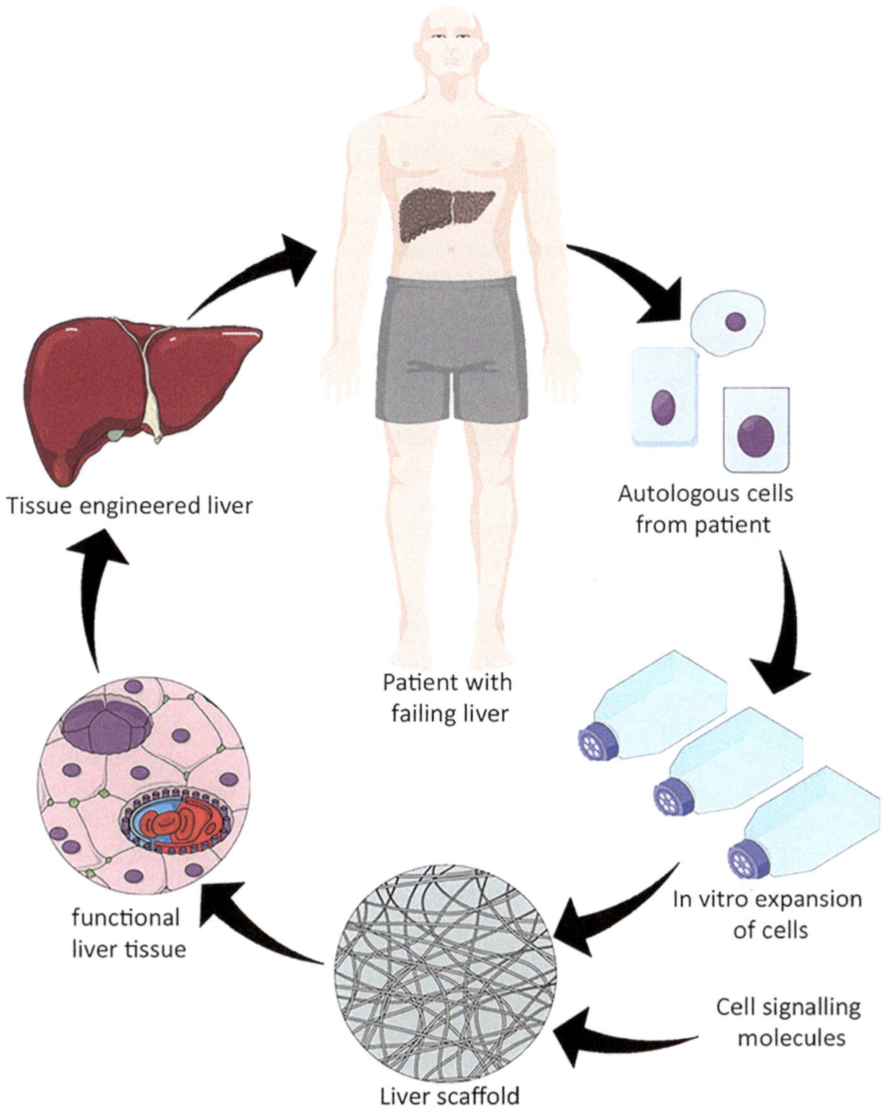

Fig. 7.1 Schematic depiction of the liver tissue engineering paradigm. Different types of patient cells are obtained and then expanded in vitro. These cells are then cultured on the liver scaffold with tissue-specific cell signaling molecules. Together, these components form small pieces of functional tissue that can be scaled up into transplantable grafts [17]. (Copyright permission from Elsevier Science Ltd.)

tissue-specific growth factors) and materials that mimic the tissues' extracellular matrix (ECM). This can be cultured in vitro to form a functional tissue construct [17, 18].

The cell seeding technique is organ-specific to a large extent. It is common for all organs to use perfusion seeding through the vasculature but non-vascular routes (such as ureter and airway, in kidney or lung, respectively) have also shown good results [15]. Cell concentration and diffusion over the scaffold have been improved by the use of mechanical environment and rate of cell perfusion, optimization of scaffold coating, and multiple perfusion routes. Successful seeding of the scaffold vascular system is critical to the survival of bioengineered organs [15]. To date, most recellularization protocols have used endothelial cells but recent studies have shown the potential advantages of induced pluripotent stem cells (iPSCs) [11, 19]. Interestingly, it has been shown that perfusion flow rates could determine the success of vasculature seeding techniques [20]. After inoculating enough cells into the organ scaffold, cell culture is required. Delivery of nutrients to the whole 3-dimensional organ needs a perfusion bioreactor to provide a microenvironment, which can encourage cell growth in an organ-specific manner [15].

For the whole organ tissue engineering to fully realize its clinical potential, engineered tissues and organs must be structurally and functionally similar to healthy organs. The acellular matrix was cultured in bioreactors and reseeded with cells in some earlier experiments [11, 21]. However, the bioreactors cannot fully simulate the physiological environment in vivo. Animal models are valuable tools for understanding the physiology and developing recellularized organs for diseases. Indeed, animal research provides invaluable data to alleviate human suffering and has made a significant contribution to human healthcare [22]. Compared with the information obtained from in vitro, in vivo research will obtain more updated information [23]. Thus, implanting the recellularized organs into animals is necessary to investigate their effectiveness. This not only fully reflects the physiological environment in vivo but also provides support for the further application of recellularized organs. Performing experimenting under controlled situations and mimicking biological conditions of human and animal diseases strengthen the creation of animal biological models and the development of scientific methods [24].

For the following reasons, animals are used for recellularized organs research [23].

(i) Feasibility. It is relatively easy to manage animal models since the compounding influence of dietary intake and environmental factors (such as temperature and lighting) can be controlled. Compared with human studies, there is relatively less environmental variation. Detailed experimental and biomolecular investigations can be performed with the isolation of their organs. Animals typically have a shorter life span than humans. Therefore, they can be studied over their whole life cycle or even across several generations, which makes them good models [25, 26].

(ii) Similarities to Human. Due to their anatomical similarity and physiological functions with humans, many animals are suitable. For instance, mice and chimpanzees share about 98% and 99% of DNA with humans, respectively [27, 28]. As a result, animals tend to be influenced by many health troubles afflicting humans. Thus, animal models are suitable for the study of human diseases.

7.2 Applications of Decellularized Heart After Recellularization

7.2.1 The Need for Tissue-Engineered Hearts

Cardiovascular diseases are the primary cause of death all the world, claiming more lives each year than all forms of cancer and respiratory diseases combined. Coronary heart disease is the most common form of cardiovascular disease in the USA, followed by stroke and heart failure [29]. Heart disease at the late stage will lead to end-stage heart failure. Although medical and surgical treatments improved, the incidence of end-stage heart failure continues to increase and presents a challenge to healthcare systems and clinicians [30]. End-stage heart failure is a major public health problem, which affects up to 15 million patients in Europe and more than five million Americans [31, 32].

The first heart transplantation was performed in 1967 and this therapy makes a new era for patients with end-stage heart failure, but is limited by donor organ shortage and waiting list mortality [21, 33, 34].

Complications resulting from immunosuppressive therapy and acute rejection of the transplanted organ are the primary limitations of heart transplantation [35–37]. Moreover, it is difficult to control and monitor organ rejection after transplantation. The limited number of donor organs and the lack of satisfactory immunosuppression have led researchers to find alternatives for transplantation. Furthermore, the donor organs are not enough for the people who need heart transplantation even if the number of heart transplantations has increased [38]. Tissue-engineered whole organs have emerged as a promising therapeutic option, which uses synthetic or natural biomaterials as a scaffold for cell proliferation, differentiation to construct tissues and organs [39].

As a natural biologic cardiac scaffold, a decellularized whole heart is composed of ECM, sugars, and proteins [11]. The structure and components of the ECM should be retained with minimal damage to maintain the intrinsic mechanical and geometric properties of the organ [40, 41]. The mechanics and geometry of the scaffolds can directly influence the processes of chemotaxis, mitosis, and differentiation of cells within the scaffolds and these processes are essential for successful recellularization [42]. Tissue-engineered whole organs maybe can alleviate the shortage of donor organs since they can reduce rejection after transplantation with recipient's cells.

The heart valves are complex soft tissue structures that are responsible for unidirectional blood flood in the heart. Valvular heart disease (VHD) can be broadly characterized by stenosis—the stiffening of the leaflets such that the valves do not fully open or close—and/or regurgitation—the process of blood flowing back through the valve [43].

VHD is still the cause of morbidity and death around the world [44]. In the USA, about 2.5% of the population suffers from aortic heart valve disease, and for the population over 75 years old, this number soars to 13.2% [45, 46]. The reasons for heart valve disease involve infective endocarditis, rheumatic fever, valve calcification, and congenital defects. The multiple options for repair or replacement each have limitations after non-surgical options are exhausted. It is estimated that the number of patients requiring heart valve surgery in the world is 290,000 each year. The number is expected to triple to over 850,000 in the next 5 years as the world's population continues to grow and age [47]. Although the ideal heart valve prosthesis has been pursued for decades, such a valve still does not exist. Mechanical valves require lifelong anticoagulation therapy since they have poor hemocompatibility. Lack durability is a problem with biological valves (such as homografts) and xenogeneic biological repair valves. Due to calcification and/or degeneration, it usually takes about 10 years to replace it [48, 49]. Children have restricted the appropriate choice of heart valve replacement surgery. The clinical need for the ideal pediatric valve is greater although pediatric patients only represent a part of the total heart valve replacement population [50]. Each type of valve replacement available for clinical use has inherent complications and is amplified in children. The cryopreserved allograft is the current gold-standard valve replacement for children. But due to patient-valve size mismatch and the lack of available grafts, this is not without problems. Most importantly, there is currently no alternative to allow the growth of somatic cells after implantation, which requires multiple reoperations on children and affects surgical strategies such as serial valve selections, timing, and sequence of surgeries [50]. Therefore, for successful heart valve replacement surgery, especially for pediatric patients, there is an urgent need to meet clinical needs [51].

Tissue engineering is a promising method and its novel constructs may overcome the limitations of current valve prosthetics. With the combination of porous scaffolds, signaling factors, and cell populations, the tissue-engineered heart valve (TEHV) has the potential to provide (1) growth characteristics, preventing the need for reoperation; (2) active tissue remodeling, preventing degeneration; and (3) excellent hemodynamics, eliminating the need for anticoagulation therapy [51].

Decellularized heart valves contain biological materials that can promote cell differentiation and works as building blocks during tissue remodeling [52, 53]. Moreover, decellularized valve generally keeps the mechanical anisotropy of the native valves since it does not require complete biodegradation [54–58]. To date, decellularized heart valves are more clinically relevant than polymeric valves since they have been implanted in animals and humans as independent valve replacements and TEHVs although the results have been mixed [59–63].

7.2.2 *The Native Human Heart: Structure and Function*

The heart is a vital organ that pumps blood throughout the body via the circulatory system, which lacks capacity to regenerate after injury [64]. To mimic the native structure of the heart, a fully functional heart should have its natural cells, and micro/macrostructure, as well as its function.

The heart has four chambers: two upper chambers (the right atrium and left atrium), two lower ones (right ventricle and left ventricle). The atria are separated from the ventricles by valves that prevent blood backflow and the right side is separated from the left by a complete septum (Fig. 7.2). Compared to the thinner walls of the atria, the ventricles have thicker walls to pump the blood to the rest of the body or the lungs [65].

The walls of the heart are layered in a unique, complex spatial organization [66]. The innermost layer is the endocardium, the middle layer is the myocardium, and the outer membranous layer is the epicardium. A variety of cell types exist within the heart, including smooth muscle cells, cardiac fibroblasts, endothelial cells, neurons, cardiomyocytes, and pacemaker cells. Although cardiomyocytes represent only a third of the total cell number in the adult heart, they account for the majority (75%) of the heart's volume [67]. Endothelial cells (40% of heart cells) and smooth muscle cells are responsible for the unique degree of cardiac perfusion that is required for heart function. Cardiac fibroblasts contribute to ECM synthesis and remodeling [67]. Finally, nerve cells are responsible for maintaining heart rate in response to certain drugs and other stimuli.

In the myocardium, ECM has two parts: the interstitium and the basement membrane. The interstitium provides mechanical properties. The basement membrane directly contacts with cells that provides signaling molecules. The interstitium is composed of elastin and collagen I. The interstitial collagen network provides the stiffness and strength required for contraction [68]. Elastin exists in the interstitium

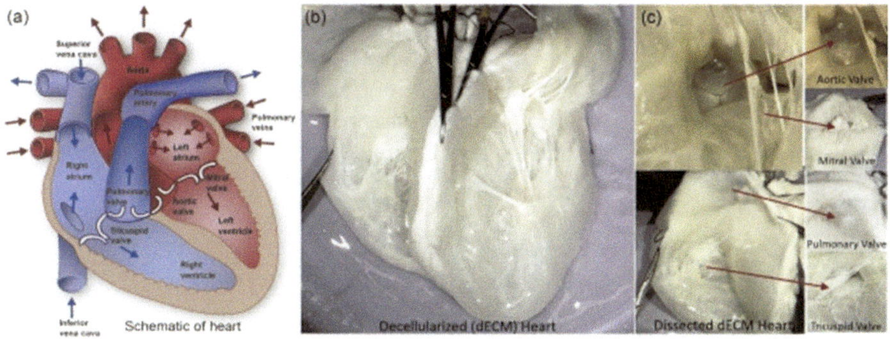

Fig. 7.2 Heart chambers and valves of mammals. (**a**) Schematic of a normal heart: the right atrium, left atrium, right ventricle, and left ventricle. (**b**) A decellularized porcine heart. (**c**) The four heart valves in a dissected, decellularized porcine heart [65]. (Copyright permission from Elsevier Science Ltd.)

and the walls of the blood vessels, but its role is unknown [69]. Laminin and collagen IV are the major components of the basement membrane. Basement membrane components play a key role in promoting cardiomyocyte–ECM interaction in the myocardium.

The aortic valve is composed of three compliant leaflets attached to the wall of the fibrous annulus. These two leaflets are mechanically anisotropic and structurally heterogeneous [43]. The leaflets have functional structures that act like unidirectional valves in the heart. The internal heterogeneous structure of the leaflets can be divided into three layers, each layer contained different cells and molecules [70]. The ventricle faces towards the left ventricle and is composed of a laminated material of collagen arranged in multiple directions and elastin arranged radially. The fibrosa is located on the other side of the leaflet and is densely packed with longitudinally directed collagen fibers, facing the aorta. The fibrosa is the major load-bearing layer during diastole, but the collagen and elastin in the ventricles open and close in a coordinated and rapid manner [71]. Finally, during the movement of the lobules, the sponges rich in glycosaminoglycans and proteoglycans act as a shear slip buffer between the ventricles and fibrosa [72].

The aortic valve consists of three semilunar cusps (also called "leaflets") attached to a fibrous annulus connected to the distal end of the left ventricular outflow tract (Fig. 7.3, [73]. Part of the ring is attached to the cardiac muscle, while the other half is connected to the fibrous leaflet of the mitral valve. However, the functional unit of the valve includes the cusps and its respective aortic sinus complexes, the latter collectively referred to as the aortic root. The histological cross-section of the aortic root is similar to that of large arteries, with intima, media, and adventitia layers. The valve cusps are very thin and flexible structures to come together to seal the valve orifice during diastole [71]. The aortic valve leaflets cooperate with the aortic root to maintain efficient periodic opening and closing.

Cardiac valves maintain unidirectional blood flow during the cardiac cycle. When the heart contracts and relaxes, the valve passively opens and closes due to the pressure gradient across the valve, resulting in alternate blood flow from the atria to the ventricles and from the ventricles to the great vessels. The outlet or semilunar (SL) valves prevent reverse flow from the arteries to the ventricles when they relax. The inlet or atrioventricular (AV) valves prevent backflow from the ventricles to the atria when the ventricles contract. Valve failure occurs when the valve cannot be opened properly (stenosis) or cannot completely close (regurgitation) [72].

7.2.3 *The In Vivo Evaluation of Decellularized Heart after Recellularization*

Decellularization is a promising and new technique that has been successfully used in a variety of tissue engineering applications, including cardiac tissue engineering [30].

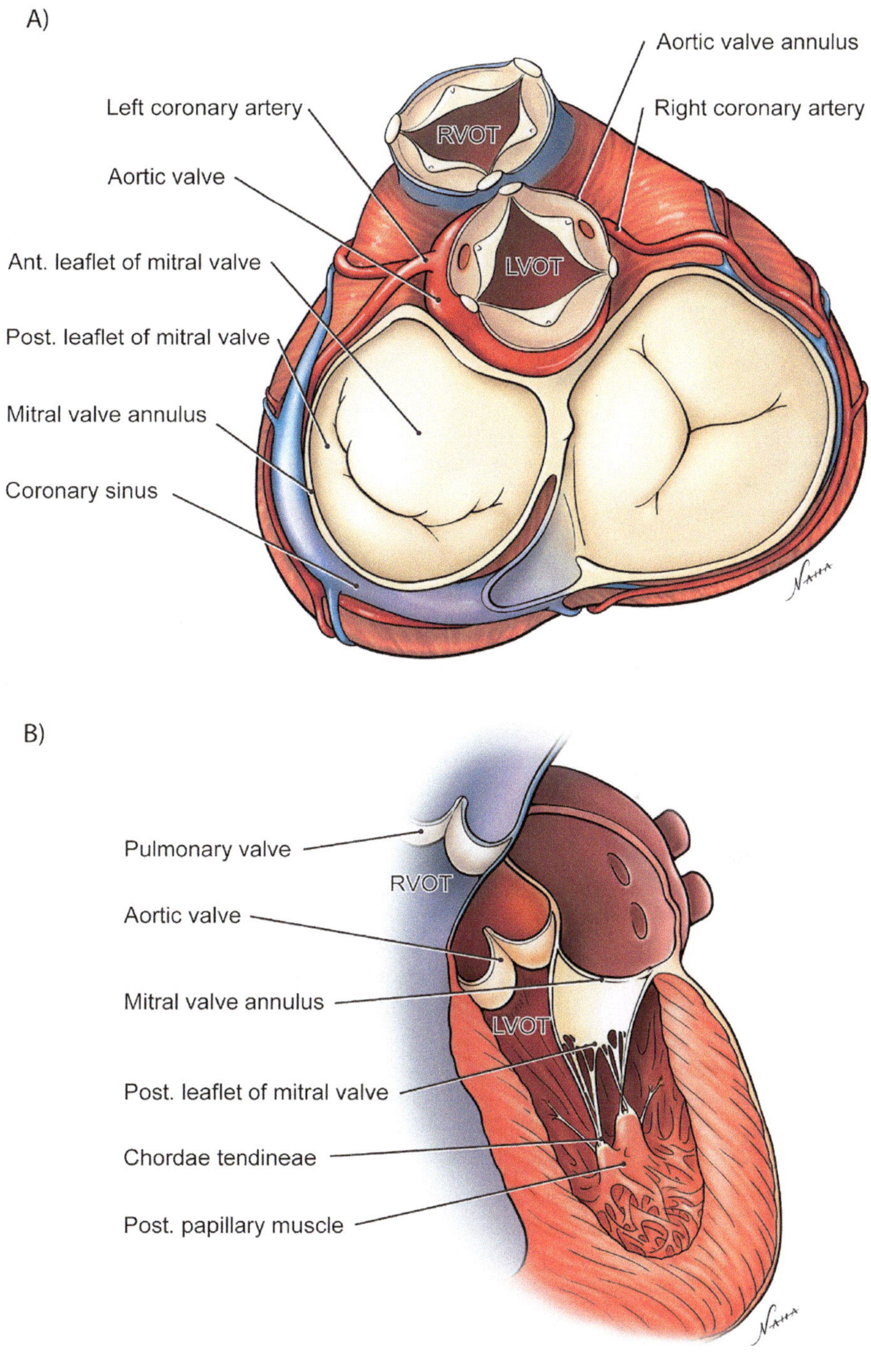

Fig. 7.3 Superior (**a**) and anterior (**b**) views of the heart showing the left ventricular outflow tract from the left ventricle bounded by the aortic valve, and posterior leaflet of the mitral valve

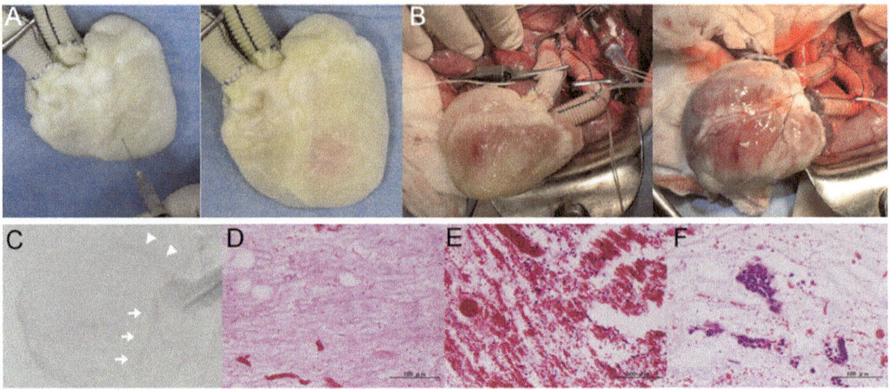

Fig. 7.4 Heterotopic transplantation of the recellularized porcine heart with mesenchymal stem cells. (**a**) Muscular injection of pig mesenchymal stem cells in the decellularized heart. (**b**) Before (left) and after (right) de-clamping of the scaffold. (**c**) Digital subtraction angiography of the coronary artery. (**d–f**) Hematoxylin and eosin staining of the recellularized transplanted heart. Scale bar: 100μm [74]. (Copyright permission from Oxford University Press)

Kitahara et al. first reported heterotopic transplantation of the decellularized whole porcine heart with mesenchymal stem cells in a pig model [74]. In this work, they transplanted a decellularized heart and a recellularized heart into another pig with anastomosis to the abdominal aorta and inferior vena cava (shown in Fig. 7.4). They modified the decellularization as the method used in a rat model [11]. Then, they confirmed the integrity of the decellularized heart to endure blood pressure and surgical procedures. A recellularized heart with cardiomyocytes was reported to have electrical activity in a rat model [11].

However, in this work, they did not observe the electrical activity since they applied mesenchymal stem cells (MSCs) but not cardiomyocytes. Moreover, they use appropriate doses of immunosuppressive drugs and steroids to show a patent for coronary arteries in a decellularized heart. They claimed that during the systolic phase, the blood flow reached the distal end of the artery. However, due to thrombosis, blood flow was short-lived. Muscular injection of pMSCs or coronary perfusion did not help prevent thrombosis.

Decellularized whole hearts may act as niches for repopulated cells and vessels for supplying the nutrients to the engineered myocardial constructs [75]. It was indicated that ECM components and preserved mechanical properties of the decellularized heart had directed differentiation of the stem/progenitor cells into the cardiac lineage. Complete recellularization is needed to use decellularized-ECM (dECM) to construct functional cardiac tissue. To mimic the natural heart functions

---

Fig. 7.3 (continued) [73]. (Illustration by David B. Nahabedian ©2019, provided under CC-BY-NC-ND, Copyright permission from Elsevier Science Ltd.)

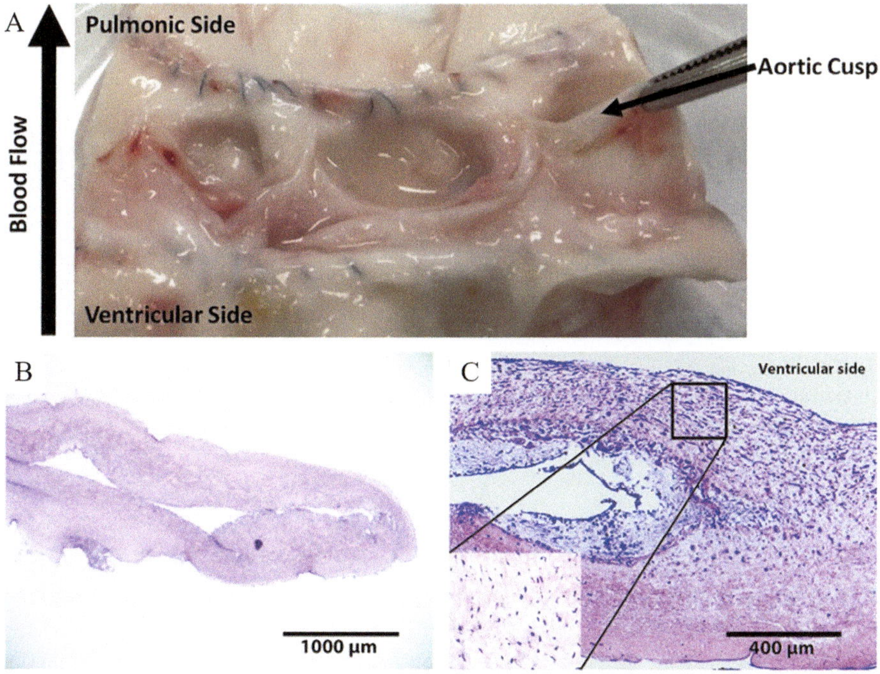

Fig. 7.5 (**a**) Gross examination. Explanted aortic valves showing aortic cusp without evidence of thrombosis, increased thickness, or calcification. (**b**) H&E staining of decellularized porcine cusp before implantation. (**c**) H&E staining of recellularized Aortic cusp of sheep [77]. (Copyright permission from Plos One Ltd)

including drug response, electrical conduction, and contractility, the cardiac dECM scaffold must be recellularized by reseeding specific cell types [15, 76].

Current valve heart repair research focuses on TEHV because it has a growth potential similar to that of natural heart valves. Decellularized xenografts are a promising solution, but host recellularization remains challenging. In the study of Hennessy et al. [77], decellularized porcine aortic valves were implanted into the right ventricular outflow tract (RVOT) of sheep to study its recellularization potential. The non-fixed decellularized porcine aortic valve sterilized with supercritical carbon dioxide ($scCO_2$) and implanted into the RVOT of sheep to verify whether it could be recellularized in an in vivo environment. The porcine aortic valves were decellularized with sodium dodecyl sulfate (SDS), were sterilized by $scCO_2$, and then implanted into the RVOT of five juvenile polypay sheep for 5 months. This study proved the feasibility of implanting non-fixed acellularized porcine aortic valves into sheep's RVOT across species (Fig. 7.5). The implantation leads to recellularization of the valve, which has sufficient hemodynamic functions for a 5-month study. Therefore, this study provides a potential role of TEHV in the treatment of human valve diseases.

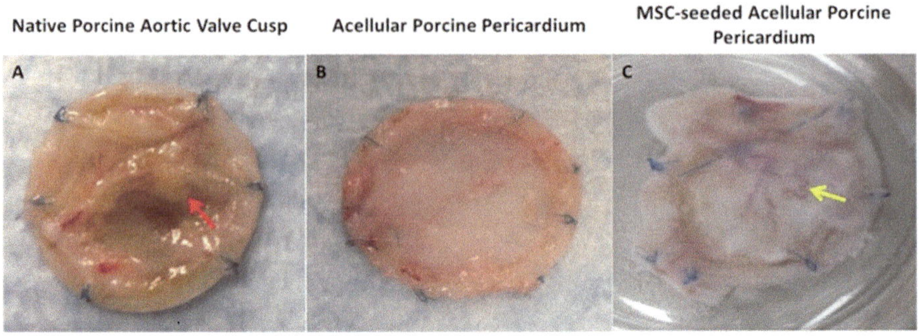

Fig. 7.6 (**a**) Severe inflammation in native porcine aortic valve cusps (red arrow). (**b**) No signs of inflammation in acellular porcine pericardium. **(C)** An area of neovascularization appeared in MSC-seeded acellular porcine pericardium after 8 weeks (yellow arrow) [78]. (Copyright permission from © 2020 Springer Nature Limited)

The potential of using MSCs to introduce in vitro seeding has been studied to help them recellularize and remodel. The current study of Khorramirouz evaluated the in vivo behavior of native porcine aortic valve cusps, MSC-seeded porcine pericardium, and acellular porcine pericardium in an in vivo model. To evaluate biomechanical properties, recellularization, and immune reaction, subcutaneous implantation in a rat model was used [78].

Tissue biomechanics, extracellular matrix evaluation, and immunohistochemistry were studied for the transplanted tissue. Vimentin, CD31, and CD163 were expressed on acellular valve scaffolds at different time points, which indicated that the host cells were recellularized. However, greater recellularization was observed on MSC-seeded tissue. Native porcine pericardial tissue expressed CD3 (inflammatory cells) throughout the study. There were no inflammation observed in MSC-seeded or acellular scaffolds. After 1 week implantation, the mechanical properties of all groups decreased. Compared with natural porcine aortic valve cusps, acellular porcine and MSC-seeded and pericardium reduced inflammatory response and promote host cell recellularization.

At all three time points, implanted native porcine aortic valve cusps expressed severe inflammation (Fig. 7.6). After 1 week, the acellular porcine pericardium was adhered to the adjoining muscle. However, after 8 weeks, tissue was easily separated by the host tissue. An area of neovascularization was observed in MSC-seeded acellular porcine pericardium after 8 weeks (Fig. 7.6). At all time points, the collagen architecture of MSC-seeded acellular porcine pericardium and acellular porcine pericardium remained intact. After 4 weeks, MSC-seeded porcine pericardium showed faster angiogenesis.

Acellular and MSC-seeded porcine pericardial scaffolds have been demonstrated to have good host cell recellularization. This study demonstrated that acellular and MSC-seeded porcine pericardium showed improved biomechanics, better host cell recellularization, and minimal inflammation, which indicate that acellular and

MSC-seeded porcine pericardium hold promise for tissue-engineered TAVR applications [78].

One method in heart valve tissue engineering is to use decellularized biological matrix and autologous cells for reseeding [79]. Despite the growing interest in bioengineered cell-seeded heart valve stents, this is a technically difficult and time-consuming process. Jordan et al. reported self-seeding heart valves that mature quickly in vivo and have a shorter preparation time [80]. They use decellularized porcine valves with detergent methods. Left untreated, conjugated with CD133 antibodies, or reseeded with autologous endothelial progenitor cell-derived endothelial cells were used. Their results showed that 1 month after implantation, endothelial cells lined the ventricular and arterial surfaces of conjugated valves. The content of interstitial cell and structural protein of conjugated valves increased from 1 month to 3 months. However, at any time point, there were few endothelial or interstitial cells on the cell-seeded or unconjugated valves.

The first clinical case of a tissue-engineered heart valve to reconstruct the RVOT was performed in 2000, as published by Dohmen et al. [61]. In this study, a viable heart valve was created from decellularized cryopreserved pulmonary allograft that was seeded with viable autologous vascular endothelial cells. A 43-year-old patient suffering from aortic valve stenosis underwent a Ross operation. Transthoracic echocardiography and magnetic resonance imaging showed the excellent hemodynamic function of the tissue-engineered heart valve and the neoaortic valve as well. Valvular calcification was not observed by multislice computed tomography. After 1 year of follow-up, the patient is in good condition without any restrictions, and the aortic and pulmonary valve function was normal. Ten-year clinical results of these tissue-engineered heart valves are promising; however, only a limited number of patients were included [62]. The advantage of this strategy is the use of end-differentiation cell types in which all cell functions are preserved. However, the growth potential is limited [81]. They also pointed out that the optimal matrix for tissue-engineered heart valves may be porcine valves since they are relatively easy to obtain and cheap.

The Ross procedure is attractive as a means for providing a living valve replacement for the aortic position. From May 2000 until June 2002, 11 consecutive patients, mean age 39.6 ± 10.3 years, received a tissue-engineered heart valve. Two to 4 weeks prior to the Ross operation a piece of forearm vein or saphenous vein was harvested to isolate, characterize, and expand endothelial cells. A cryopreserved pulmonary allograft was decellularized, coated, and seeded with autologous vascular endothelial cells, using a specially developed bioreactor. All patients survived the operation. There was no obvious source of fever after surgery. All patients are Class I patients of the New York Heart Association. Transthoracic echocardiography assessment of the tissue-engineered heart valve showed an average pressure gradient of 5.4 ± 2.0 mm Hg at 10 years. No calcification was found by multislice computed tomography for up to 10 years. During long-term follow-up, tissue-engineered heart valves exhibited excellent hemodynamic performance and could prevent degeneration. Decellularization of heart valves and seeding with autologous vascular endothelial cells may prevent tissue degeneration and improve valve durability

[62]. There are still limitations although the results are good. First, harvesting, isolating, and proliferating endothelial cells in vitro are an extremely delicate process. There is a risk of interstitial cells contamination, which overgrows the endothelial cells. Another weakness was the small number of patient, but 100% of follow-up were obtained over 10 years.

7.3 Applications of Decellularized Lung After Recellularization

7.3.1 The Need for Tissue-Engineered Lung

Lung disease is the third highest cause of death in the USA after cardiovascular disease and cancer; however, unlike the latter two, death rates from pulmonary disease continue to increase [82]. In the USA, nearly 30 million patients are currently affected by end-stage lung disease, and 12.1 million adults suffer from chronic obstructive pulmonary disease, which is the fourth leading cause of death [13, 83]. Lung transplantation is the only definitive treatment of terminal lung disorders such as chronic obstructive pulmonary disorder [76, 84]. There are more than 2000 lung transplants each year in the USA [85]. Definitely, lung transplantation could improve survival and provide an improved quality of life for patients [86]. However, it is still a challenge to perform lung transplantation.

The shortage of transplantable organs is the biggest problem. In 2011, 2280 patients were waiting for lung transplantation in the USA [87]. The average waiting time is about 2 years and 9 months in Japan.9 The key issue for lung transplantation is that donor lungs often suffer edema, atelectasis, or pneumonia caused by preoperative systemic control or donor systemic problems [88]. In addition, lung transplantation is further complicated by high rates of both acute and chronic rejection. The recipients need lifelong administration of immunosuppression drugs, which increases the risks of infections and cancer. Therefore, lung transplantation is needed for lung disease but currently there is not enough lung organ for transplantation [14]. As a result of these challenges, the current 5-year survival rate is 50%. Despite significant efforts to improve pre- and postoperative care, the average survival rates have remained relatively unchanged over the past two decades [44]. New approaches are urgently needed for these patients.

Tissue-engineered lung should involve the reconstruction of the lung and have the physiologic functions of lung. Functionally, lung is composed of branching epithelial cell lined tubes. These tubes support the conduction of air to the numerous alveoli, where circulating blood is oxygenated. The alveoli form the air/blood or endothelial cell/epithelial cell interface and used for gas exchange. Cell types within the lung are very different and the development of tissue-engineered lung depends on the number of different cell types that make up this organ. There has been some progress in the tissue-engineered lung. Many developments have been made in the

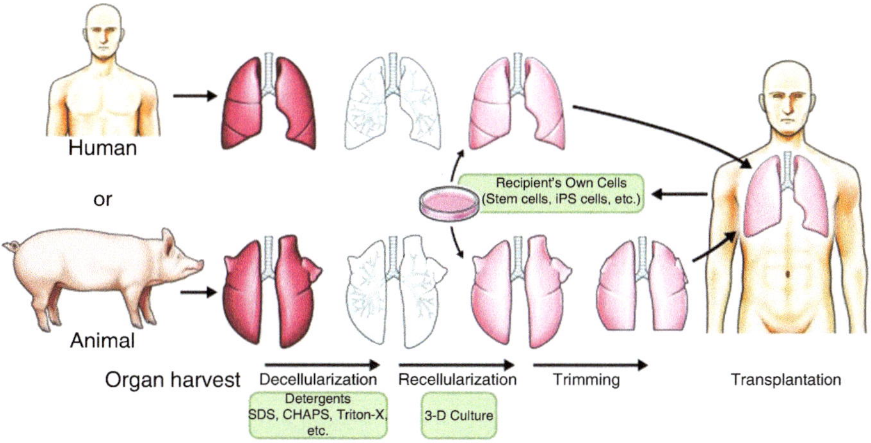

Fig. 7.7 Schematic of decellularization and recellularization of whole lung regeneration for transplantation [14]. (Copyright permission from Elsevier Science Ltd.)

identification of cell populations with the potential to be used as a source for development of engineered tissues. A recent interesting study is the recellularization of acellular cadaveric trachea with mesenchymal stem cells to engineer the upper respiratory [89, 90]. This is a major step forward for the development of tissue-engineered lung, which could provide the recipient with a functional conducting airway [91].

The decellularization process involves removing all cellular contents from the scaffold while remaining the macro-nano mechanical properties of the source tissue, the molecular and protein composition as well as the complex natural three-dimensional structure. The decellularized scaffolds will then ideally support the colonization of patient-derived stem/progenitor cells and produce transplantable bioengineered lungs, thereby significantly reducing the risk of rejection and immunosuppression requirements [92, 93].

Numerous studies have shown that acellular scaffolds can be derived from a variety of species, including rodents, non-human primates, pigs, and humans [94–97]. At the same time, techniques have been developed to produce decellularized lung scaffolds, which are derived from animals and patients with lung diseases [95, 98, 99]. The structure and composition of the diseased organs are preserved by these acellular scaffolds. When cells are reseeded onto these scaffolds, they will adopt a phenotype similar to those observed in vivo. Thus, they have become a new three-dimensional model system for studying human diseases ex vivo, especially cell–matrix interactions [100].

The bioengineering of the whole organ, especially the bioengineering of the lungs is a development area in biomedical engineering [101, 102]. Figure 7.7 showed flow diagram of whole lung bioengineering [14].

7.3.2 The Native Human Lung: Structure and Function

The lungs are located in the chest cavities and other mediastinal structures on both sides of the heart. The left lung is smaller and has two lobes while the right lung has three lobes (Fig. 7.8-a). The right lung consists of ten segments and the left lung consists of eight parts (Fig. 7.8-b). These segmental bronchi can be identified by a bronchoscopist. The large fibrous fissures separate the upper and lower lobes of the right and left lungs. The other small fissures separate the right middle lobe. These cracks may be complete or incomplete anatomically. This may be important because if fissures are incomplete, air can be transferred from one section or one leaf to another through incidental ventilation [103]. The lower segment contains the tracheobronchial tree, which is composed of the trachea and the bronchi, bronchioles, and alveoli of the paired lungs [104].

The vascular compartments, interstitial and epithelial of the lung are formed by more than 40 different cell types. The epithelium consists of secretory cells (such as club, serous, mucous, and type II), which are scattered in the airways (kulchitsky, ciliated, neuroendocrine, basal, and brush) and airspaces (alveolar type I and type II), each with unique and different functions. Interstitial cells (fibroblasts, chondrocytes, myofibroblasts, and smooth muscle cells) form the necessary scaffolds to shape the airways and parenchyma as well as nerve and lymphatic blood vessels and cells (basophils, eosinophils, mast cells, lymphocytes, macrophages, monocytes, and neutrophils) and substances that travel between the blood and airspaces. The vascular cavity of the lung allows blood flow in which blood exchange occurs between the heart and pulmonary capillaries. Gas exchange across the thin air barrier to blood tissue composed of endothelial cells and pericytes, as well as the alveolar epithelium and interstitial matrix (basement membrane) that form part of the alveolar septa [104].

The lung parenchyma consists of alveoli, alveolar ducts, and alveolar sacs. This gas exchange part of the lung accounts for 80–90% of the total lung volume. Alveoli (300 mm in diameter) are the smallest basic anatomical unit related to gas exchange. It is composed of the airspace enclosed (defined) by the alveolar wall and its opening into the alveolar duct [104].

The total surface area of the alveoli is about 143 m^2. The alveoli are arranged in ducts and sacs or as simple patches on or protruding from the walls of respiratory bronchiolar. There are alveolar cells and the basal layer of capillaries, migratory alveolar macrophages, elastic, and type III collagen fibers in the alveolar compartment. The septa are perforated by interalveolar pores. The lining cells of the alveoli (pneumocytes) are of two types:

- Type I cells. These simple squamous epithelial cells account for about 90% of the alveolar surface area and are the cells that perform gas change.
- Type II cells. These smaller rounded cells account for approximately 10% of the alveolar surface area although they may exceed the Type I cells in number. These cells produce and recycle surfactants [105].

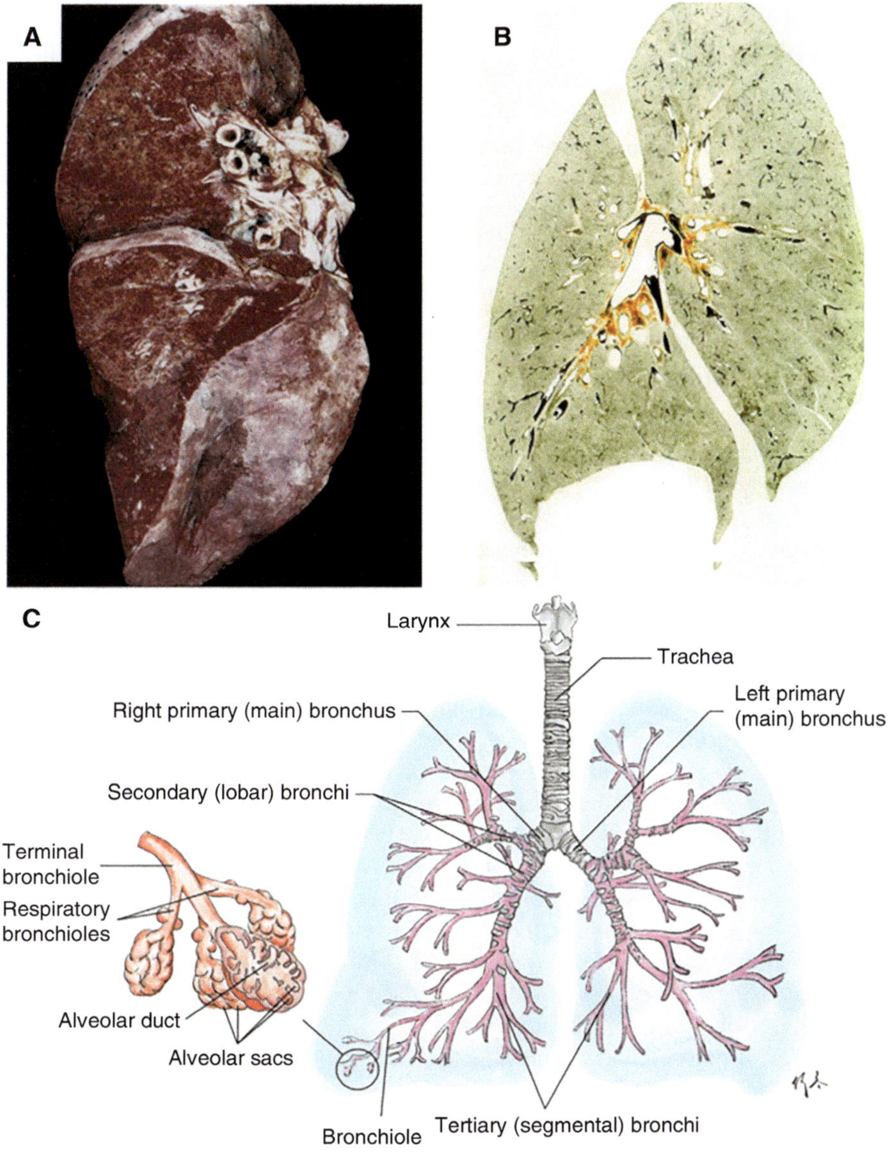

Fig. 7.8 (**a**) The adult right lung has three lobes, separated by a large fissure and a small fissure. (**b**) A thin Gough section of the adult left lung [103]. (**c**) Structure of lower respiratory tract [104]. (Copyright permission from Elsevier Science Ltd.)

The function of lungs is to extract oxygen from the air for the body and to remove the gaseous waste from the body. To maintain baseline energy levels needed for survival, humans require an enormous quantity of oxygen [106, 107]. In order to maximize the surface area available for gas exchange, lungs have evolved to form a complex network of tube-like epithelial branches. Their branches are known as the conducting airway, which includes the trachea, bronchi, and bronchioles. The tubes in this branched network get progressively smaller until they terminate with alveoli. Alveoli are closely associated with the capillary network to allow diffusion of oxygen into the bloodstream and removal of carbon dioxide. When lung function is compromised because of developmental defects or disease, the consequences can be severe [108].

7.3.3 The In Vivo Evaluation of Decellularized Lung After Recellularization

Lung transplantation is the only potentially curative option for severely damaged lungs since adult lung tissue has limited regeneration capacity. Although whole-organ regeneration is in its very early stage [83], some efforts have successfully recapitulated different steps of lung development, which provide useful milestones for tissue-engineered lung.

Because the extracellular matrix template should have three-dimensional (3D) architecture and regional-specific cues for cellular adhesion, acellular matrices can provide attractive scaffolds for repopulation with lung-specific cells for lung engineering [9]. To generate functional lung in vivo, the composite should (i) display the branching geometry of the airways and contain a perfusing microvasculature, (ii) provide barrier function to separate blood from air, (iii) have mechanical properties that allow ventilation at physiological pressures, and (iv) contain lung-specific cells [9].

Unlike simple organs (such as skin, heart valves, or bones), the lung has a complex 3D structure, so the decellularized lung itself cannot be used clinically. However, decellularized lungs have been developed as scaffolds to clarify the differentiation and behavior of stem cells. Most recellularization studies are conducted using mouse lungs, and the recellularization approach is limited to trachea or directly insolated on lung slices [14].

Ott et al. [8]. decellularized rat lungs by detergent perfusion and obtained scaffolds with acellular vasculature, airways, and alveoli. They seeded scaffolds with epithelial and endothelial cells to regenerate gas exchange tissue. They perfused and ventilated cell-seeded constructs in a bioreactor simulating the physiologic environment of developing lung to establish function. By day 5, constructs could be perfused with blood and ventilated using physiologic pressures. Regenerated lungs were transplanted into the orthotopic position of a rat to show in vivo function. They used the recipient's circulation and airway and respiratory muscles to perfuse and ventilate the constructs by. The results showed that they provided gas exchange in vivo for up to 6 h after extubating.

Further study is needed to be performed to develop their approach. It is also needed to improve the differentiation and maturation of cells in the reseeded graft.

Differentiation of the endothelial and epithelial cells was observed to infuse into their scaffolds. However, they also found that the squamous epithelium lined the large conducting airways, rather than ciliated or secretory cells. Moreover, some alveolar regions did not regenerated since the presence of parenchymal regions lacking surfactant or T1α staining. In future work, it will be important and meaningful to evaluate whether these regions include clonal, multipotent, and proliferative responses. Before successful long-term transplantation, it is necessary to study how much in vitro maturation is required. In addition, the degree of maturity and incorporation of host cells in the body will complete the organ regeneration process.

In initial experiments, graft function in vivo was limited to a few hours owing to pulmonary edema and secretions from bioartificial lung grafts [8]. Song et al. [83] regenerate lungs based on perfusion decellularized cadaveric rat lungs and examined graft function in vivo for14 days after implantation (Fig. 7.9). Their data showed that bioartificial lung grafts based on perfusion decellularized-ECM scaffolds could be implanted in an orthotopic position. The regenerated lung could provide gas exchange in vivo over a prolonged period without negatively affecting the contralateral native lung. After using improved graft preservation and postoperative weaning protocols, animals could be fully recovered, and bioartificial lung constructs provided oxygenation as long as 7 days at levels comparable to cadaveric lung transplants. Gas exchange, compliance, and radiographic appearance gradually declined over the subsequent 7 days because of inflammation and progressive graft consolidation. However, graft longevity and homeostasis are still limited. Moreover, more complete recellularization of airways with appropriate epithelial phenotypes along the proximodistal spectrum and regeneration of a functional mucociliary apparatus need to be achieved to maintain aeration and reduce inflammatory changes in the graft over a prolonged period.

Zhou et al. [109] seeded porcine decellularized lung scaffolds with human airway epithelial progenitor cells and banked human umbilical vein endothelial cells. Whole organ culture was performed to enable tissue formation. Pulmonary vasculature with mature endothelial lining was regenerated by reseeded human endothelial cells. The epithelial surface was repopulated with human epithelial progenitor cells to generate a functioning gas exchange graft. The bioengineered lung grafts were transplanted in a large animal model in vivo. Results suggested that the grafts could withstand physiological blood flow and could exchange gases during ventilation within the 1-hour observation. This study demonstrated the feasibility of engineering and transplantation of decellularized porcine lung scaffolds inoculated with epithelial and endothelial cells. To achieve higher levels of functions, such as ventilation-perfusion matching and mucociliary clearance graft maturation will be required.

Whole bioengineered lungs (BELs) based on acellular lung scaffolds have been transplanted in small animal models, but lungs failed due to intravascular coagulation and defects in endothelial barrier function leading to pulmonary edema [9, 11, 83, 109].

The inability to generate perfusable microvasculature networks that can support tissue survival and withstand physiological pressures without leakage is a basic problem facing the field of tissue engineering. The microvasculature is essential

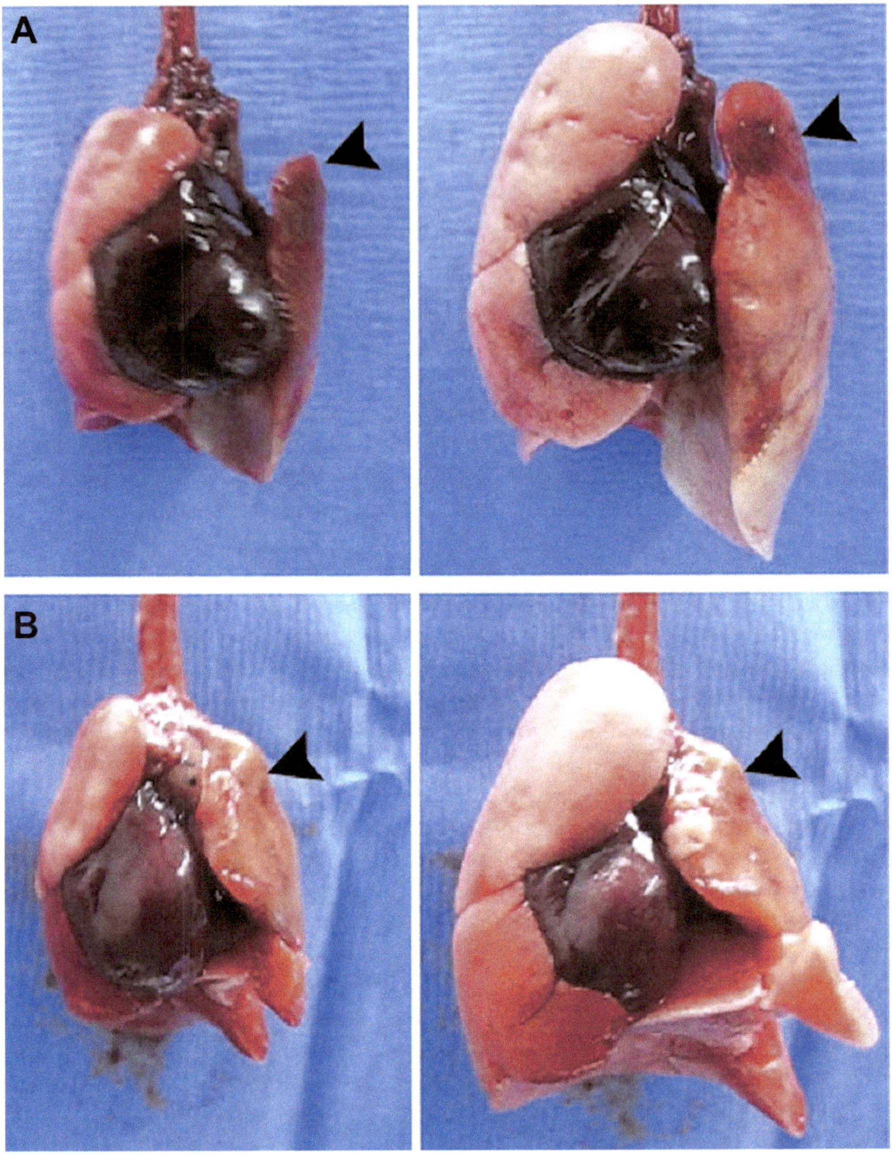

Fig. 7.9 (**a**) The isolated heart and lungs of bioartificial lung transplant recipients were removed as a whole, ventilated through the trachea, and inflated and deflated 24 h after transplantation. (**b**) 14 days after transplantation, inflated (left), and deflated (right), black arrowhead marking the transplanted bioartificial left lung [83]. (Copyright permission from Elsevier Science Ltd.)

for the production of BELs. It requires systemic circulation to support the survival of tissues and to coordinate the circulatory and respiratory systems to ensure proper gas exchange.

Nichols et al. [110] focused on the development of the bronchial systemic circulation in non-immunosuppressed pigs to support BEL growth and survival after transplantation. They performed a pilot study to establish the feasibility of BEL transplantation, with an airway anastomosis but without a vascular (pulmonary) anastomosis. A single pneumonectomy was performed 1 month before BEL implantation, which provides the source of autologous cells used to bioengineer the organ on an acellular lung scaffold. Systemic vessel was developed using growth factor-loaded microparticles after 30 days of bioreactor culture. They evaluated recipient survival, autograft (BEL) vascular and parenchymal tissue development, graft rejection, and microbiome reestablishment in autografted animals 10 h, 2 weeks, 1 month, and 2 months after transplant. 2 weeks after the transplant, BEL became well vascularized. Alveolar tissue was observed in all animals ($n = 4$). Transplant rejection was not found.

7.4 Applications of Decellularized Liver After Recellularization

7.4.1 *The Need for Tissue-Engineered Liver*

About 30 million people have a liver disorder in the USA and about 27,000 people died from liver disease [111]. The main therapeutic methods for liver cancer are hepatectomy and liver transplantation, which also result in some complications [112]. Acute liver failure (ALF) encompasses a devastating condition resulting in severe hepatocyte injury which leads to rapid deterioration of liver functions and deleterious clinical manifestations such as encephalopathy, coagulopathy, and progressive multi-organ failure. With an incidence of fewer than 10 cases per million persons per year in the developed world, acute liver failure is seen most commonly in previously healthy adults in their 30s with a very high mortality rate due to multiple organ failure and presents unique challenges in clinical management [113].

Transplantation is the only lifesaving therapy for severe hepatic failure. Dr. Starzl [114] performed the first liver transplant. However, short-term success was achieved with 1-year survival after transplantation until 1967 [114]. Owing to improved donor-recipient pairing, surgical techniques, and immunosuppressive regimens, the survival rates of liver transplantation have increased [115]. However, today, the number of transplants performed is not enough for the patients waiting for a liver transplant.

According to a report published by the Organ Procurement and Transplantation Network in 2014, 14,600 people were waiting to receive a donor's liver and only 6729 liver transplants were performed [116]. The deficit of livers was about 4000 per year because of the critical shortage of organs [117]. Many patients die of liver failure since they cannot find a suitable donor.

Tissue-engineered decellularized liver organ scaffolds provide promise in addressing the issue of donor organ shortage [118]. Three-dimensional bioprinting and recellularization of natural scaffolds have the potential to generate an unlimited source of grafts for liver transplantation through endogenous regeneration [119].

Although liver transplantation is the main treatment for liver cancer [112], it also brings many complications, which include small for size syndrome (SFSS) [120]. The problem with SFSS is liver dysfunction and postoperative coagulopathy because of portal hyperperfusion and insufficient functional liver mass. The tumor mass could be removed by in vivo decellularization and then the liver lobe was repopulated with non-parenchymal cells and normal hepatocytes. Therefore, it acts like a supportive liver lobe, and the complication of SFSS could be prevented [121].

Different from renal and cardiac failure, it is impossible to use only mechanical devices to provide temporary assistance due to the complexity of liver function. Live hepatocytes are needed for liver support devices to provide various functions. In addition, whole organ transplantation causes more mortality and morbidity than primary hepatocyte transplantation, and primary hepatocyte transplantation could provide enough cells to correct inherited metabolic deficiencies [122].

However, hepatocyte transplantation is still not a conventional method for treating liver diseases. For instance, hepatocytes cannot survive long after transplantation, and the survival and proliferation rates of primary or fetal hepatocytes in the animal experiments are usually very low [123]. Moreover, the widely used approaches of the portal vein or spleen injection can only transplant a limited number of liver progenitor cells or hepatocytes.

After several days' culture, primary hepatocytes may be under epithelial-mesenchymal transition or dedifferentiation and then lose their typical morphology and function [124]. Thus, it is very important to create a liver microenvironment to maintain hepatocyte function. The ECM can provide a scaffold where cells can grow. It can also regulate the adhesion, proliferation, and differentiation of cells as well as the interactions between different cell types [125]. With the organ and tissue decellularization development, specific ECM tissues can be obtained from whole organs by the perfusion of the organs with different kinds of detergents [11]. The entire vascular network beds could be maintained in the whole organ decellularized matrix. These vascular network beds could provide a microenvironment for the infusion of desired cell types. Moreover, compared with thin layers of decellularized matrix in the 2-dimensional environment, 3-dimensional environment could be provided for the infused cells [126].

In the UK, more than 40% of LT transplant recommendations are rejected due to the judgment criteria or comorbidities exceeding the marginal criteria. This provides an important opportunity to explore alternative uses of human livers that are not suitable for LT after organ procurement. In particular, although cellular viability is easily compromised, ECM can be better maintained in discarded livers. It can be used as a scaffold for normal human hepatocytes and reconstruct functional human liver tissue in vitro [119].

7.4.2 *The Native Human Liver: Structure and Function*

The liver is the largest gland and the second largest organ in the body (second only to skin). Its weight can vary between 1.2 and 1.8 kg in humans (1.4–1.8 kg in adult men and 1.2–1.4 kg in adult women) [127]. Anatomically the liver has four lobes: right, left, caudate, and quadrate. The quadrate lobe is located on the inferior surface of the right lobe. The caudate lobe is located between the left and right lobes in an anterior and superior location (Fig. 7.10, [128].

Many types of cells make up the liver. These cells included Kupffer cells, hepatocytes, sinusoidal endothelial cells, stellate cells, and biliary epithelial cells (cholangiocytes). These have their own functions and regulate the liver functions in different ways. Hepatocytes perform many of the functions ascribed to the liver, which are the primary epithelial cell population of the liver and make up the majority of the liver volume [129].

The functional structural unit of the liver is the lobule. The lobule is composed of chords of hepatocytes, which are organized in a typically hexagonal shape around the central vein.

The liver, the largest internal gland in the body, is a very important organ in metabolic homeostasis since it is a critical hub for many essential metabolic exocrine and endocrine functions. The liver is responsible for various physiological processes. These include lipid and cholesterol homeostasis, endocrine control of growth signaling pathways, blood volume regulation, macronutrient metabolism, and immune system support. In addition, the liver's function is to store glucose in the form of glycogen and assemble glucose via the gluconeogenic pathway. The liver oxidizes lipids, but can also package excess lipid for secretion to and storage in other tissues, such as adipose. Finally, the liver is responsible for the majority of proteins secreted in the blood, disposal of nitrogenous waste from protein degradation in the form of urea metabolism, and the processing of amino acids for energy [129]. Moreover, the liver possesses a unique regenerative capacity since it can regenerate most of its function after losing up to three-quarters of its mass as a result of partial hepatectomy or toxic injury [130].

7.4.3 *The In Vivo Evaluation of Decellularized Liver After Recellularization*

Decellularization is a relatively hot scaffold preparation technology in tissue regeneration because it can retain the original architecture of the tissue especially the native microvasculature [10, 131]. Great progress has been made in the field of liver tissue engineering using scaffolds obtained by decellularizing donor organs over the last decade. Several million primary rat hepatocytes were used in the first report of a transplantable liver biomatrix [10].

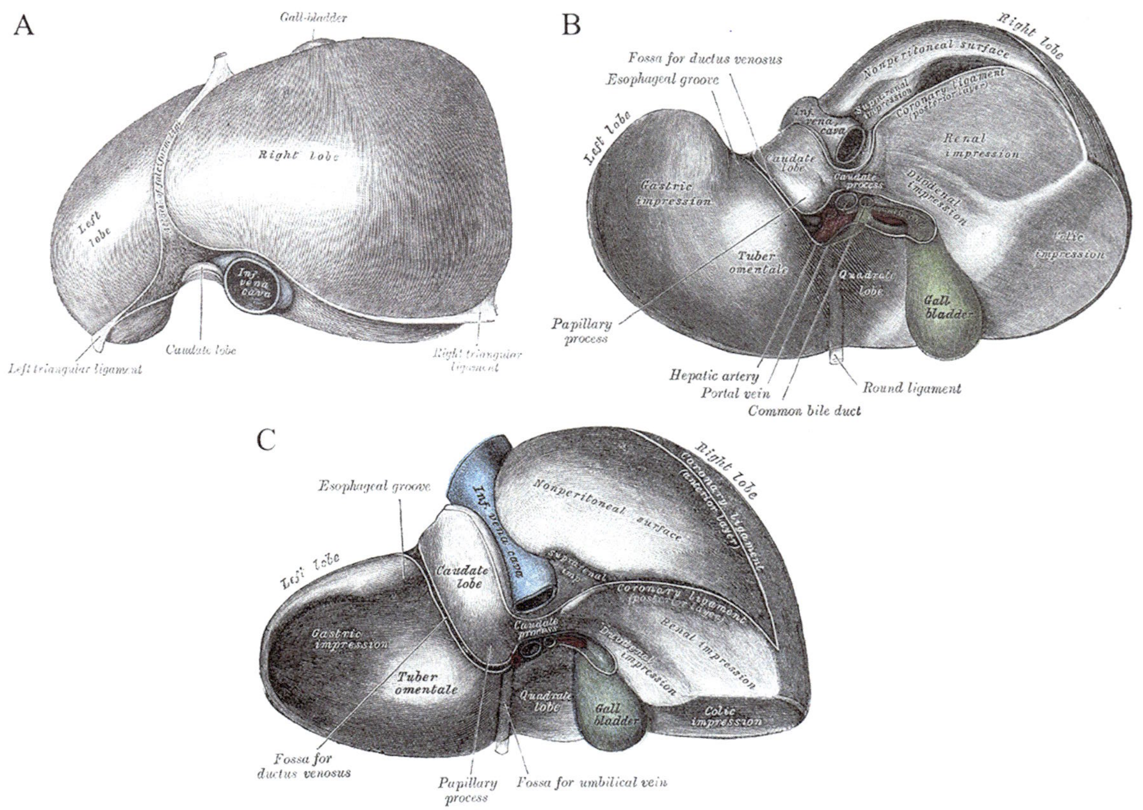

Fig. 7.10 The Liver (**a**) The superior surface of the liver. (**b**) Inferior surface of the liver (**c**) Posterior and inferior surfaces of the liver. Contributed by Gray's Anatomy Plates [128]. (Copyright permission from © 2020, StatPearls Publishing LLC)

However, the engineering of complex solid organs (such as kidney and liver) constitutes other challenges that must be used clinically. When creating engineered functional solid organs, a major obstacle is to reconstruct organ's vasculature, which can transfer blood to the entire organ, prevent thrombosis, and withstand physiological blood flow conditions. This challenge was successfully solved in one study and the results showed that re-endothelialized liver scaffolds can withstand physiologic blood pressure and maintain blood flow in the bioengineered livers for 24 h [132]. This study showed for the first time to overcome a major obstacle in the process of transplantable liver engineering and the establishment of a functional vasculature. Moreover, these findings were also of significant importance for the future development of fully functional bioengineered organ transplants.

Yang et al. remove livers from inbred male C57BL/6 mice [133]. They injected primary hepatocytes and neuroglia antigen-2 (NG2)-expressing hepatic progenitor cells (NG2 + HPCs) derived from adult mouse liver into the decellularized liver scaffold (DLS) via the portal vein and inferior vena cava, respectively. A DLS scaffold was placed into the region from which a kidney had been removed and connected to the recipient by auxiliary heterotopic liver transplantation. Their study was also feasible in larger animals (BaMa pigs). In conclusion, this decellularized liver bioscaffold supported primary hepatocyte survival, promoted blood vessel patency, and provided growth factors for hepatocyte proliferation and survival, angiogenesis, endothelial cell differentiation, and organogenesis (Fig. 7.11).

In a sense, the liver ECM is the best scaffold for liver tissue engineering. Different methods and different targets (liver organoid proliferation, 3D hydrogel, or 2D coating) can be used to refill decellularized liver scaffolds with functional cells [10]. The latter could be used to assist local LT [17]. The development of human livers in pigs may ameliorate the severe shortage of grafts for human liver transplantation. It also could make liver transplantation possible without the need for dead or living donors. In addition, human hepatocytes could be added to a porcine ECM, leading to the development of a chimeric liver (CL).

In 2013, Hata et al. [134] evaluated the feasibility of this process and a rodent model of CL was developed by repopulating rat hepatocytes in mice and auxiliary CL was successfully transplanted into rat recipients through vessel reconstruction. It is not clear whether such engineered CLs can be transplanted into recipients. They developed a rodent chimeric liver transplantation model to study its feasibility (Fig. 7.12). After transplanting the chimeric liver into a rat recipient with vessel remodeling, they evaluated the functions and long-term viability of the transplanted CL grafts. Engineered CLs have been successfully transplanted and can maintain their structure and function for a long time.

In another study, whole human liver and lobes were completely decellularized to form extracellular matrix scaffolds with preserved architectures [135]. Using human cell lines hepatic stellate cells, hepatoblastoma, and hepatocellular carcinoma, decellularized human liver scaffolds were refilled for up to 21 days and it has excellent viability, motility, and extracellular matrix proliferation. Biocompatibility was investigated by transplanting the liver stent cube omentum or subcutaneous

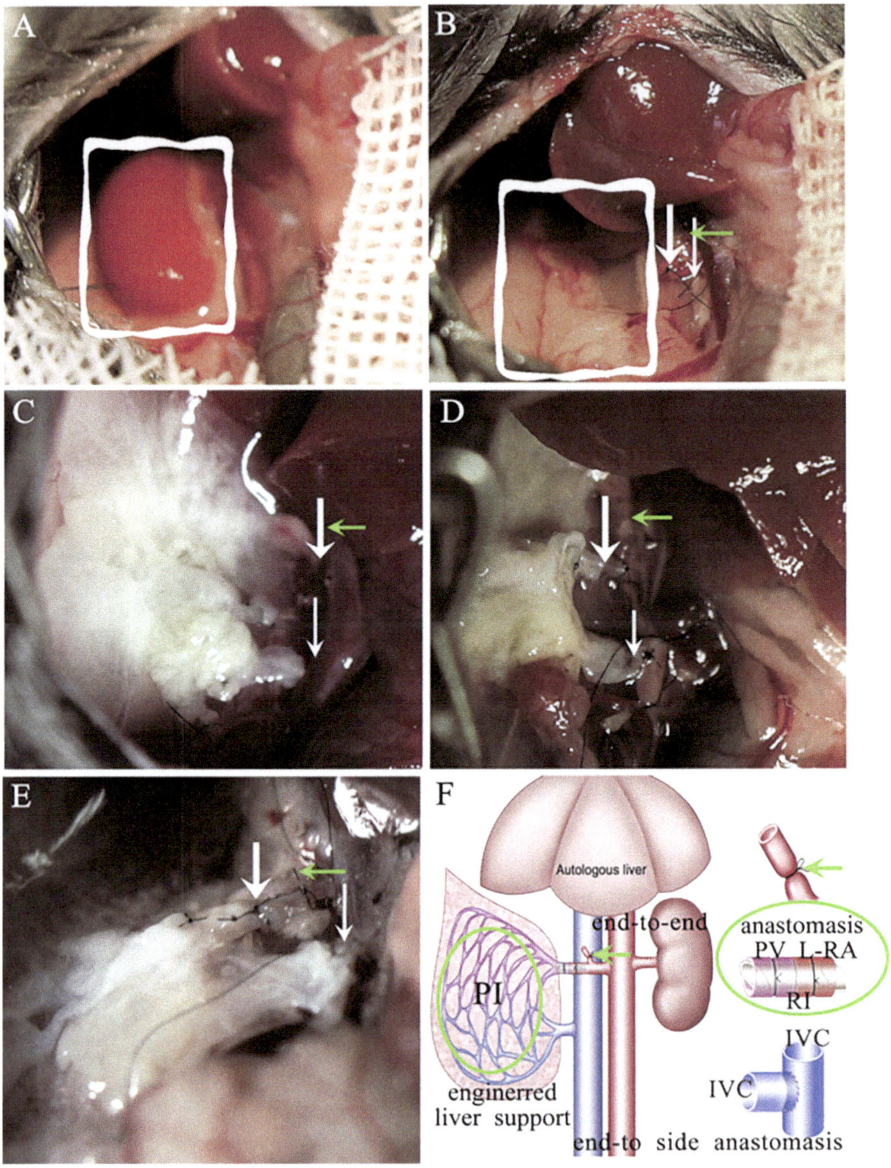

Fig. 7.11 A surgery of portal-renal arterialized auxiliary heterotopic liver transplantation was taken on the mice. This image shows the surgery progress. (**a**) Expose the right-side kidney (the square indicates the kidney). (**b**) Ectomy of the right-side kidney (the square indicates the kidney removed). (**c**) The DLS was seeded with cells and placed in the area of removed-kidney. (**d**) The DLS without cells was sutured to the recipient by the same procedure above, where the kidney had been removed (**e**). (**f**) Diagram of the surgery steps. The left green circle implied the DLS [133]. (Copyright permission from Elsevier Science Ltd.)

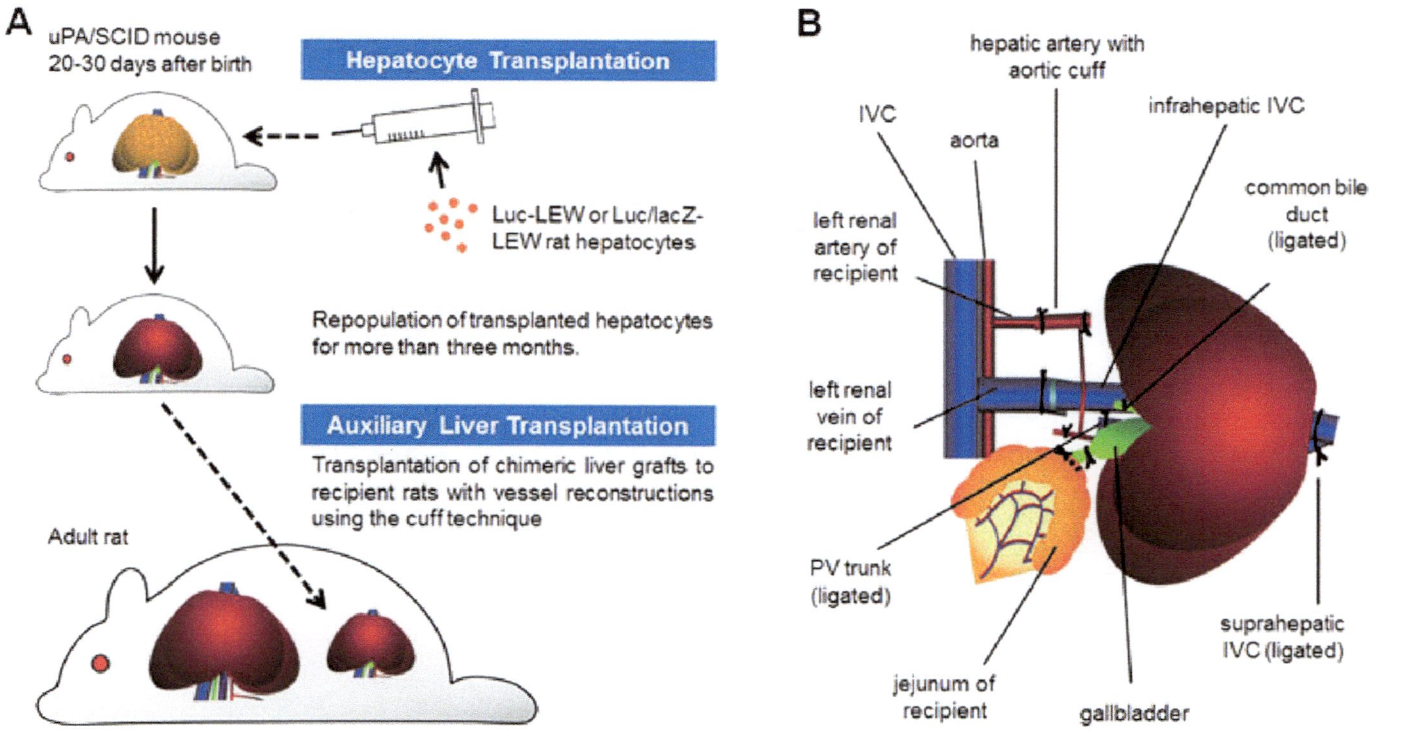

Fig. 7.12 (**a**) Experimental protocol for Chimeric liver transplantation. (**b**) Schematic diagram of engraftment of CL into recipient rat [134]. (Copyright permission from © 2020 Ovid Technologies, Inc., and its partners and affiliates)

xenotransplantation into immune-competent mice resulting in a lack of foreign body response. They demonstrated the decellularization of the human liver and repopulation with derived human liver cells, which is a key process in the development of a bioartificial liver. The purpose of this study is first to prove the feasibility of an innovative solution for single leaf or whole human liver decellularization, and to comprehensively evaluate the quality and the in vivo biocompatibility of the obtained liver ECM scaffold. The results of this study provide clear conceptual data that can support the development of bioartificial liver tissue by using decellularized human ECM liver scaffolds. Therefore, the usage of donor human livers offers more new possibilities and is suitable for transplantation [135].

In one study, human umbilical vein endothelial cells were used to revascularize decellularized whole porcine liver and implanted heterotopically into immunosuppressed pigs whose spleens had been removed [136]. The results demonstrated that the regenerated liver could be perfused for up to 15 days. Without anticoagulation therapy, implantation of rBELs into pigs ectopically resulted in continuous perfusion for 3 days. The immune response to human endothelial cells was obvious. Ten days steroid-based immunosuppression regimen and splenectomy at the time of rBEL implantation reduced the immune responses and resulted in continuous perfusion of the rBELs for over 2 weeks. The revascularized liver scaffolds that can maintain blood perfusion at physiological pressures may eventually bring hope to human liver transplantation.

The decellularized liver scaffold provides a microvascular network for the transport of nutrients and oxygen and the excretion of metabolites [131]. In addition, the antigenicity of decellularized organ scaffolds is low [137]. With the latest developments in liver tissue regeneration, people have been tried to develop artificial liver scaffolds by decellularization and repopulating with hepatocytes, followed by cell perfusion [132, 138].

Shimoda et al. [139] believed that the decellularized liver scaffold itself could induce the proliferation and differentiation of liver progenitor cells since it contains sufficient extracellular matrix. This is the first attempt to use decellularized liver scaffolds after hepatectomy and a large model is used to evaluate whether the acellular scaffolds contribute to liver regeneration. In terms of liver regeneration after hepatectomy, this study showed that the scaffolds could help repair injured or damaged tissue. However, long-term regeneration of hepatocytes did not studied in this model. In addition, this study only examined the microscopic aspects of liver regeneration and did not evaluate clinical recovery after liver injury. A large liver resection model is needed to carry similar experiments and investigations to show the regeneration of clinical liver volume (Fig. 7.13).

The application of whole organ decellularization technology is limited to the use of donor decellularized organs for the reconstruction and transplantation of recipients [140]. Since detergents have the ability to remove cells or active viruses, decellularization of autologous organs to preserve the extracellular matrix and its architecture may be another way to rescue organs and chemically remove cancer. Once the bioengineered liver is constructed and implanted, vascular patency is

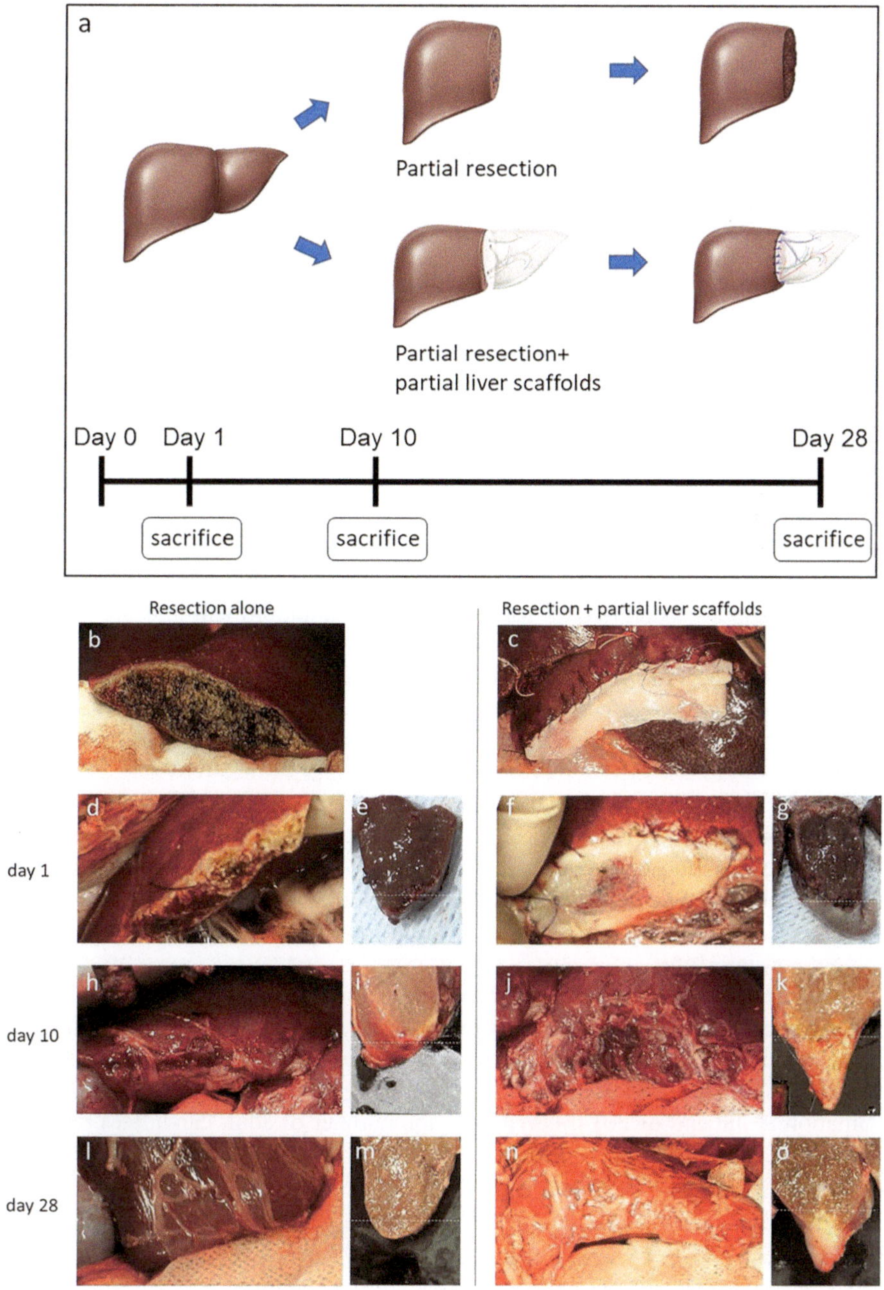

Fig. 7.13 This image illustrated that the engineered scaffolds were sutured from the liver of partially hepatectomized porcine model to examine their potential for liver regeneration. (**a**) Exhibiting the grafting DC liver scaffolds procedures. (**b, c**) Partial hepatectomy was performed on the control group without suture of any scaffolds, while for the treatment group, the liver scaffolds were sutured to the surface of the same size, partially resected pig liver, and completely covered the incision area. Other images show gross view on 1, 10, and 28 days after surgery.

essential for the steady-state survival of the artificial organ and repair and remodeling in the event of injury [141].

7.5 Applications of Decellularized Kidney after Recellularization

7.5.1 The Need for Tissue-Engineered Kidney

The number of patients with kidney insufficiency increased since the aging of the population combined with the rise in the prevalence of diabetes and hypertension [142]. In addition, the rapid increase of chronic diseases (such as diabetes and hypertension) leads to the rise of end-stage renal disease (ESRD) [143]. Although renal replacement therapy (RRT) by dialysis or renal transplantation is lifesaving, people with ESRD cannot afford this expensive treatment [143]. The global prevalence of maintenance dialysis has increased 1.7 times from 165 pmp (Per Million People) patients in 1990 to 284 pmp in 2010 [143]. It is estimated that the expected number of people receiving RRT (dialysis or transplantation) will increase from 2.6 billion in 2010 to 5,4 billion in 2030 worldwide. It is noticeable that many patients died because they did not have access to treatment [144]. Moreover, developing countries such as Asia, Africa, and Latin America, RRT is still unaffordable for the majority of patients and causes severe financial hardship for those who have access to it [145].

Chronic kidney disease (CKD) is a key determinant of the poor health outcomes of major non-communicable diseases. CKD is related with an 8- to 10-fold increase in cardiovascular mortality and is a risk multiplier in patients with diabetes and hypertension. 5–7% of the world population is affected by milder CKD (often because of diabetes and hypertension) and milder CKD is more common in developing countries [145].

Compared with dialysis, kidney transplant has significant advantages in terms of cost, survival, and quality-of-life. However, but the number of candidates for kidney transplantation on the waiting list is still increasing every year. Almost 1500 pediatric candidates and 100,000 adult candidates were waiting for kidney transplantation at the end of 2014. Obviously, the number of adult candidates removed

---

Fig. 7.13 (continued) (**h**, **j**, **l**, **n**) On the tenth and 28th days after the operation, slight adhesion to the peritoneum and omentum was observed around the surgical sites in both groups. (**f, g**) The scaffolds retained their premier structural characteristics. (**j, k**) The line of demarcation becomes vague, and it can be seen that in the sutured decellularized scaffold, the liver tissue has undergone macroscopic regeneration. (**n, o**) From a macro-perspective, the volume of the sutured stent appears to be small, and the actual stent space has been mildly collected aggregated [139]. (Copyright permission from © 2020 Springer Nature Limited)

from the list increased because of deteriorating medical condition or death and waiting times also increased [146].

In the foreseeable future, given that the number of patients in need of treatment is expected to increase, even in the richest countries, the cost of dialysis will be a heavy financial burden because current dialysis technology costs tens of thousands of dollars per patient per year [147], which is similar in Singapore [148]. In terms of quality of life, cost, and survival, renal transplantation is considered to be the best available treatment for ESRD [146, 147]. However, the shortage of donor organ transplants has limited the global scope of this treatment [146]. Among the various regenerative medicine technologies currently under development, cells implanted throughout supporting scaffolds are the most advantageous candidate in practical clinical applications [149].

To overcome the limits of current RRT, numerous investigators have suggested that tissue engineering may be a promising approach for regenerating damaged kidneys in vitro [150]. Tissue engineering would help to address this shortcoming by regeneration of damaged kidney using naturally occurring scaffolds seeded with precursor renal cells [151]. Theoretically, producing a decellularized kidney scaffold and then seeding it with stem cells to repopulate the kidney's ECM [152, 153] could be a promising technique. However, after obtaining promising results with kidney decellularization, scientists have started to realize that effectively advancing the recellularization of natural scaffolds is more complex than was previously hoped and that it necessitates the development of novel techniques to obtain uniform cell seeding and growth within acellular tissues [154, 155] and this is particularly challenging for the peculiar 3D structure of the renal vasculature and tubules.

7.5.2 The Native Human Kidney: Structure and Function

The kidneys are paired organs that lie retroperitonially on the posterior abdominal wall. Each kidney is of a characteristic shape, having a superior and inferior pole, a convex border placed laterally, and a concave medial border. The medial border has a marked depression, the hilum, which contains the renal vessels and the renal pelvis [44].

The two kidneys are located outside the peritoneal cavity and very close to the posterior abdominal wall, one on each side of the vertebral column. All the kidneys have bean-shaped structure [156]. Figure 7.14 shows the structure of kidney [157].

The kidney is an organ that plays a key role in homeostasis. They reabsorb precious nutrients and maintain endocrine balance, as well as excrete inorganic and organic wastes [158]. Each kidney contains about one million functional units, nephrons, which include many special cell types from different embryological lineages. Each nephron contains a filter body, the glomerulus, and a long tubule composed of twelve differentiated segments. The collecting ducts were formed by the final parts of these tubules, which lead to the renal pelvis [158]. The endocrine, metabolic, and excretory functions of the kidney mediate the necessary interactions

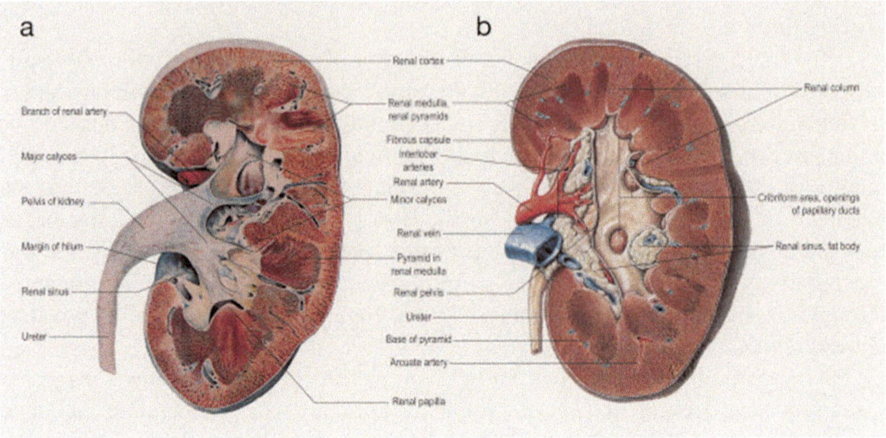

Fig. 7.14 Coronal sections through the left kidney showing pelvicalyceal system and renal sinus [157]. (Copyright permission from © 2019 Elsevier Ltd. All rights reserved)

with several organs to maintain a range of important functions. These functions include regulation of water and thirst, erythropoiesis, potassium balance, ventilation, drug metabolism, calcium and phosphate metabolism, blood pressure, and acid-base homoeostasis [158].

Regulating the excretion of substances is the main function of kidneys and its rate can accurately balance their input to the body, thereby maintaining the overall body balance. The second main function of kidneys is to regulate blood osmolarity, blood volume, and total body sodium content in a manner that determines average blood pressure [156].

7.5.3 *The In Vivo Evaluation of Decellularized Kidney After Recellularization*

Constructing transplantable grafts to restore kidney functions will resolve the shortage of donor organs and the incidence of immunosuppression. This bioengineered graft should have the architecture and function of kidney and allow urine to be perfused, filtered, secreted, absorbed, and drained.

Detergent perfusion was used to prepare decellularized kidneys, which have intact and perfusable vascular, glomerular, and tubular compartments [159]. The rat kidney scaffolds were seeded with endothelial and epithelial cells and then perfused in a bioreactor to regenerate functional tissue. When perfused through their inherent vascular bed, the regenerated grafts produced rudimentary urine in vitro. We performed experimental left nephrectomies and transplanted regenerated left kidneys in an orthotopic position. When transplanted in orthotopic position in rat, the grafts were perfused through the circulation of the recipient, and produced urine via the ureteral conduit in vivo. Throughout the test period, regenerated kidney grafts appeared well perfused without any evidence of bleeding from vasculature, collecting system, or parenchyma. In vivo data showed that the regenerated kidneys group could produce a urine-like solution although their levels were lower than those of native kidneys. In addition, creatinine, urea, and albuminuria in the regenerated grafts group were lower than that of native kidneys but higher than that of decellularized kidneys.

Consistent with previous studies [160, 161], Song's study confirmed that cells and cellular debris were removed from whole human, porcine, and rat kidneys by detergent decellularization and tubular, glomerular, and vascular ultrastructure were not damaged. The viable endothelium and epithelium can regulate an acid-base equilibrium, secretion and reabsorption of metabolites and electrolytes, and the retention of macromolecule, so they reseeded endothelial and immature epithelial cells within acellular scaffolds [162]. Compared with cadaveric kidneys, the glomerular filtration rate in regenerated kidneys is lower-partly because increased vascular resistance leads to reduced graft perfusion. The creatinine clearance rate improved with increasing renal artery perfusion pressure. Partial reabsorption of electrolytes has not reached the level of cadaveric kidneys, which may be related to incomplete vaccination and immature stage of neonatal epithelial cells inoculated [163].

Despite the successful development of solutions for obtaining organ scaffolds, the problem of scaffold recellularization has not been resolved. Remuzzi et al. evaluated the potential of in vivo and in vitro kidney scaffold recellularization procedures [164]. Results suggested that acellular scaffolds implanted in rats could not be repopulated with host cells, and the acellular scaffolds needed recellularization. According to the previous protocols, they evaluated the strategies for in vitro cell seeding using different experimental protocols [153, 159], to obtain the cellularization of the different structural regions of the kidney scaffold. However, when using different infusion protocols, we obtained very limited and inconsistent cell seeding regardless of the injection site. Moreover, the experimental and theoretical data suggested that it was not possible to use these infusion protocols to deliver uniform cells to renal scaffolds, due to the permeability of the extracellular matrix of the scaffold. This study highlighted the main physical barriers that limited the recellularization of cell-free renal scaffolds in vitro, and the barriers that must be studied to effectively advance this regenerative medicine strategy.

In view of all the processes in the decellularization and refilling of organs observed in kidney bioengineering, it has broad prospects for confirming the recovery of kidney function after transplantation [165]. Researchers' expectation of for the next 10 years is that research related to decellularized and recellularized kidneys is in the preclinical phase. Compared with other methods of tissue engineering to repair the kidneys (such as 3D bioprinting renal or organoids), recellularized kidneys have very good potential [149].

The strategy to recellularized ECM scaffolds is still in the early stages, and the results are too preliminary. In addition, it seems difficult to extend to clinically relevant models. Although current technology allows the production of natural ECM scaffolds from almost all animal species, there is still a lack of comprehensive characterization of the renal ECM, and the remodeling of the cell compartment after decellularization is too far from the feasible range [149].

7.6 Applications of Decellularized Pancreas After Recellularization

7.6.1 The Need for Tissue-Engineered Pancreas

The incidence of pancreatic diseases is high and can be life-threatening. Since 1988, about 300,000 people in the USA every year are currently affected by acute pancreatitis, and the incidence has been increasing [166]. Although the incidence of pancreatic cancer and chronic pancreatitis is lower than that of other organs, compared with other types of cancer, it is usually accompanied by cancer with exocrine and/or endocrine insufficiency combined with higher mortality. At present, organ transplantation is the best treatment for pancreatic failure. However, the limited number of donors and organ rejection limit the application of pancreatic transplantation [167]. Thus, there is an urgent need for alternative treatment options for pancreatic diseases [168].

One of the most common gastrointestinal causes of hospitalization in the USA is acute pancreatitis. Although the incidence of chronic pancreatitis is low, it will significantly affect the patient's quality of life. Pancreatic cancer has the characteristics of high mortality and is one of the top five causes of death from cancer [166].

The treatment of diabetes is still insufficient. Although exogenous insulin therapy can effectively prevent acute metabolic decompensation in type 1 diabetes, less than 40% of patients achieve and maintain a therapeutic goal [169]. Consequently, an important cause of morbidity and mortality among diabetic people is hyperglycemia-related organ damage. Oral hypoglycemics, exogenous insulin, physical activity, and dietary modification can achieve glycemic control. It can reduce but not eliminate the microvascular and macrovascular complications of diabetes [170].

Lifelong lifestyle and drug intervention are the current best practice guidelines regarding diabetes management. Although these treatments reduce the incidence of diabetes complications and emergencies, they cannot cure or relieve diabetes. The only treatment for establishing long-term and stable normal blood glucose in type 1 diabetic patients is β-cell replacement by pancreas or islet cell transplantation [170].

Through the development of bioengineered microenvironments and microencapsulation technology, regenerative medicine is expected to promote the development of islet transplantation. As a means of immunoisolation, encapsulation can "disguise" the foreign antigens from the allogeneic or xeno-transplanted islets of the host's immune surveillance [171]. Encapsulation scheme involves packaging the islets in a bio-inert and semi-permeable membrane, which selectively allows the oxygen, waste products, nutrients, insulin, and glucose to pass through and prevent immune cell penetration [171, 172]. In theory, successful encapsulation eliminates the need for aggressive and lifelong immunosuppression, thereby improving the viability of β cell viability and the incidence of the host. However, despite the encouraging results obtained in early animal studies, the clinical value of islet encapsulation is limited by the following barriers [172]: (1) hypoxia secondary to failure of revascularization; (2) insufficient immune isolation because of the penetration of small immune mediators, like nitric oxide, cytokines, and chemokines; (3) poor biocompatibility of capsule materials.

Cutting-edge technology in regenerative medicine has recently enabled investigators to take the advantages of preserving inherent ECM for organ bioengineering research [173, 174]. Regenerative medicine is exploring the possibility of using ECM from animal or human whole organs for bioengineering organs. Detergent-based decellularization can generate ECM scaffolds from whole animal organs [2, 174]. Current decellularization protocols can remove cell surface antigens, cellular material, and DNA from the ECM scaffold while retaining vascular channels, structural integrity, and attachment sites [40].

The main goal of regeneration medicine is the regeneration of insulin-producing β-cells. The limited portability of human islets and serious complications of immunosuppression therapy have promoted investigators to seek a cell-based method that can produce many transplantable β-cells or islets to meet the needs of the millions of diabetic patients [174].

Although islet transplantation can effectively treat type I diabetes [175], islet transplantation often fails after 3–5 years. Factors that lead to graft failure include cell cytotoxicity [176, 177], immediate blood-mediated inflammatory reaction (IBMIR) [178], as well as insufficient vascularization leading to ischemia and subsequent cell death [179, 180]. In addition, the process of isolating islets from pancreatic tissue also involves the use of severe proteolytic enzymes, trypsin, clostripain, neutral protease, and containing collagenase [181]. Although these enzymes can effectively break down the ECM connections between exocrine and endocrine tissue, they also severely damage islet vasculature and ECM, which reduce the quality of the islets [176, 180]. The compounded effect of all these factors leads to the failure of islet transplantation and relapse of diabetes in patients [182].

7.6.2 The Native Human Pancreas: Structure and Function

The pancreas is a large, retroperitoneal organ located behind the posterior wall of the lesser sac, on the floor of the supracolic compartment of the abdominal cavity. The pancreas is a lobulated, flat, soft, elongated, and yellowish gland located on the posterior abdominal wall, more-or-less laterally. It is a retroperitoneal structure and has a thin capsule (Fig. 7.15) [183]. In addition to being a key accessory exocrine digestive gland, the pancreas has a vital endocrine component, which is composed of

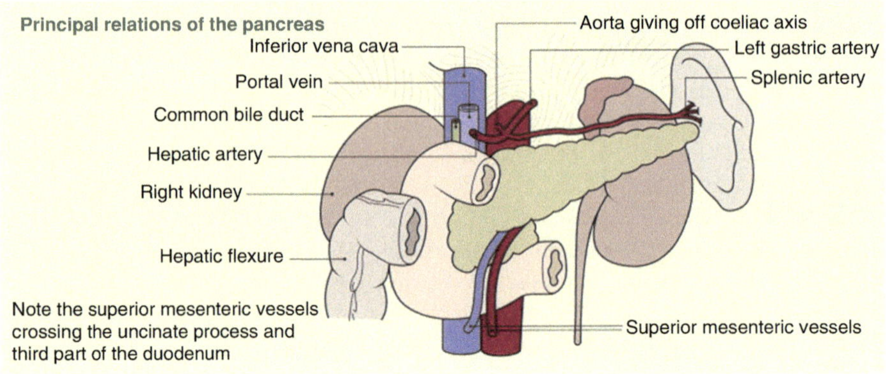

Fig. 7.15 Principal relations of the pancreas [183]. (Copyright permission from Elsevier Science Ltd.)

approximately one million Langerhans islets distributed throughout the substance of the pancreas [184].

Pancreas is a highly vascular organ, which has both endocrine and exocrine functions [185]. Structurally, the human pancreas is composed of two organs: an endocrine gland and an exocrine gland. Duct cells and pancreatic acinar cells formed the exocrine gland, which can produce sodium bicarbonate and digestive enzymes, respectively. Secreting the digestive enzymes is the main function of the exocrine pancreas, which can digest and absorb daily foodstuffs, and finally, absorb nutrients into our bodies. The endocrine glands are composed of five types of secretory islet cells, which secrete peptide hormones to maintain glucose homeostasis. The secretory function of the pancreatic is well regulated by nerve secretion, endocrine, paracrine, and intracrine mechanisms [186].

Although the pancreas is mainly an exocrine gland, it also has vital endocrine functions. Digestive enzymes are secreted by the exocrine pancreas, which is produced by the pancreatic acini and released into a complex catheter system that eventually leads to the second part of the duodenum. Its depth and close terrain relationship with several important structures make pancreatic surgery both dangerous and challenging. A correct understanding of the topography of the pancreas, the anatomy of blood vessels, and catheters is essential for the successful management of pancreatic cancer [184].

The main duct of the gland, the Wirsung's duct, extends along the length of the gland and is usually connected to the end of the common bile duct to form the ampulla of Vater. The accessory pancreatic duct passes from the upper part of the head and opens into the duodenal mucosa at the minor papilla [184].

7.6.3 *The In Vivo Evaluation of Decellularized Pancreas after Recellularization*

The ECM scaffolds were developed by the perfusion decellularization of cadaveric rat and human pancreases to preserve the natural islet cell niche to improve the function of transplanted islets. Functional endocrine was generated by seeding with human islets in vitro, which reversed the diabetic state after transplantation in a rat model [187]. The feasibility of in vivo transplantation has been demonstrated in a rat model.

Regeneration of decellularized pancreas of the whole organ will need recellularization of the parenchyma, as well as the vasculature and supporting cell components. Actually, the clinical success of tissue-engineered scaffolds largely depends on their success in vascular inosculation and subsequent integration with host vasculature. However, the vascular system of decellularized organ is prone to thrombosis [188]. Thus, a complete endothelium needs to be formed and preformed microvasculature successful transplant [182]. Prevascularization has become a new and promising concept in tissue engineering, specifically referring to the pre-formation of microvasculature in tissue structures before implantation in the body.

A pre-vascularized pancreatic decellularized scaffold was developed by endothelializing with endothelial progenitor cells (EPCs) in a bioreactor [189]. Afterwards, in order to observe the anastomosis with the host vascular network, the scaffold was transplanted into the rat in vivo. The data showed that there were EPCs around the blood vessel wall. Compared with the non-endothelialized control group, the re-endothelialized scaffold connected with the host earlier by new blood vessel formation. All these data suggested that the pre-reseeded EPCs could accelerate the formation of blood vessels within the scaffold. These findings confirmed that it is possible to engineering a whole pancreatic vasculature, which will allow integration into the circulatory system and to support the development of bioengineered pancreas in the future.

In order to build a microenvironment similar to the natural pancreas, the microenvironment is not only suitable for cell growth, but also can enhance the cell function. Thus, decellularized mouse pancreases were developed as natural 3D scaffolds [190]. MIN-6 β cells were used to repopulate the decellularized mouse pancreas. Compared with traditional plate-cultured cells, the qRT-PCR data confirmed that the insulin gene expression of recellularized pancreas was upregulated. A diabetic mice model was used to further study the function of the recellularized scaffolds. First, they observed that bioscaffold planted with cells in the body has the potential to control blood glucose. However, a gradual increase in blood sugar was observed, which indicated that the seeded cells may die due to a lack of continuous supply of nutrition. Finally, they confirmed that the cause for cell death was the lack of nutrient supply, not the bioscaffold itself. In summary, decellularized pancreas can be used as an ideal platform for studying pancreatic bioengineering. The

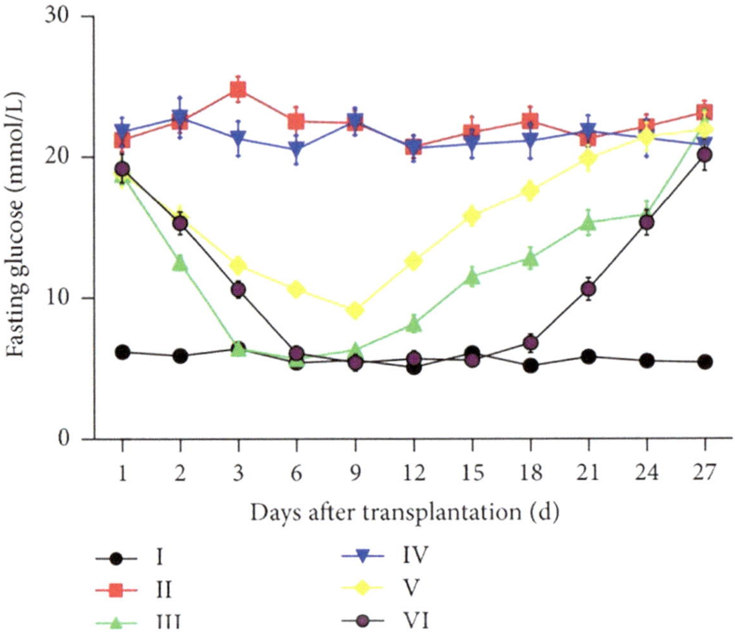

Fig. 7.16 Fasting blood glucose monitoring after in vivo implantation. I: healthy mice; II: diabetic mice without interference; III: diabetic mice with subcapsular injection of the liver of MIN-6 β; IV: diabetic mice with scaffold implantation group; V: diabetic mice with recellularized scaffold implantation group; VI: diabetic mice with recellularized scaffold cubes implantation group [190]. (Copyright © 2015 Di Wu et al.)

platform has a 3D structure and is essential for the survival and normal function of seeded cells (Fig. 7.16).

7.7 Applications of Decellularized Intestine After Recellularization

7.7.1 The Need for Tissue-Engineered Intestine

Under pathological conditions, when major intestine damage occurs, the intestinal mucosa regenerative capacity will be greatly reduced. Extensive intestinal resections can treat many diseases, including mesenteric vascular disease, congenital atresia, inflammatory bowel disease, neonatal necrotizing enterocolitis, volvulus, and trauma [191].

Small bowel functional defects caused by diseases such as short bowel syndrome (SBS) [192], severe inflammatory bowel disease (IBD) [193], and radioactive enteritis [194] greatly reduce the quality of life and increase the risk of other diseases

[192]. Function loss may be due to surface mucosa rupture (e.g. IBD) or reduced SI surface area (e.g. SBS). The main reason for SBS is massive SI resection (about 50–75% of small intestinal length [195]) due to many reasons, such as trauma, congenital abnormalities, or newborn surgical emergencies (like inflammation of the intestine wall associated with premature birth and necrotizing enterocolitis, a disease related to bacterial invasion) [196, 197]. SBS patients suffer from diarrhea, malnutrition, metabolic disorders, prolonged hospitalization, and SBS-related liver disease [198]. The 5-year mortality of SBS is about 30% [196]. The common treatment for SBS is long-term total parenteral nutrition [199], but it is expensive [200] and has many side effects, including cholestasis, inflammation, hepatic fibrosis, infection, and electrolyte abnormalities [196]. IBD is a chronic immune-mediated disease, including ulcerative colitis and Crohn's disease (CD). Transmural inflammation is characteristic of CD, which leads to anatomic damage that develops from aphthous ulcers to deep ulcers, strictures, and/or fistulae. As a chronic disease, IBD often needs frequent hospitalizations and long-term medication, which brings a high cost and a huge psychosocial burden [201, 202]. Although many attempts have been made to restore the function of the intestinal epithelium [203], standardized procedures have not yet been developed [204]. 70–80% of CD patients will eventually need to surgically remove damaged parts of the SI and many people need many surgeries to remove the other parts of the small intestine since the damage caused by CD is irreversible. Therefore, CD patients not only have a huge SBS risk due to these surgeries, but also suffer impaired SI function because of inflammation.

About 60–70 million people are affected by intestinal diseases. In the USA alone, the annual direct medical cost more than $100 billion [205]. Since the intestine plays a key role in digesting food and absorbing nutrients, intestinal diseases significantly affect the quality of life of the patient. These diseases may result from autoimmune, vascular diseases, inflammatory immune responses, metabolic, bacterial or viral infections, or cancerous growths. Symptom management and surgery are the current treatment options for these diseases, resulting in a poor quality of life for the patient after treatment. Long-term use of systemic anti-inflammatory drugs may be involved in symptom management, which may hurt the kidneys and liver over time. It maybe also involves immunosuppressants, which reduce the body's overall immune response and then suppress an autoimmune response. Surgical treatments may require removal of diseased sections of the intestine, which may involve long-term measures such as total parenteral nutrition and will reduce the quality of life after treatment. Patients lucky enough to undergo bowel transplant may also require lifetime immunosuppression to reduce the higher risk of intestinal graft transplantation rejection [206].

In order to provide a better quality of life, treatment strategies aimed at tissue regeneration over disease management are needed. Tissue regeneration is to provide treatment by combing biological and medical knowledge with engineering concepts and using biological materials to achieve localized treatment of more systemic drugs, thereby promoting tissue regeneration. The center of the tissue engineering field is a scaffold, which is a biomaterial structure that can provide a growth environment for cells, or provide a support structure for cell infiltration and growth. Various natural and synthetic biological materials have been studied to utilize the body's own regenerative capabilities to restore damaged tissue and provide structural assistance to cells to restore lost functions. In order to develop any therapeutic strategy for tissue regeneration, understanding the physiology of the tissue and the physical and chemical properties of the biomaterial are required [206].

7.7.2 *The Native Human Intestine: Structure and Function*

The small intestine extends from the pylorus to the ileocaecal valve, which is divided into the duodenum, jejunum, and ileum [207]. The average length of the small intestine is about 6–7 m [208]. It originates from the original intestine formed by endoderm. The duodenum comes from the foregut, and the jejunum and ileum come from the midgut. From the mucosa to mesenteric nodes, then to nodes near SMA and SMV, there is also a large number of lymphatic drainage, which then flows into the cisterna chyli and the thoracic ducts [209].

The small intestine also houses the mesentery, which is recently considered as an organ [5]. The double-layered visceral peritoneum forms a mesentery with a continuous structure. It anchors the small intestine to the posterior abdominal wall. Thus, when the body maintains an upright posture, this can prevent it from collapsing into the pelvis. Moreover, the mesentery contains nerves, blood, and lymphatic vessels [210].

The wall of the small intestine can be divided into four layers, namely the mucosa, muscularis propria, submucosa, and adventitia. Mucosa the folded portion of the small intestine faces the lumen. The surface area is large because of the 600-fold amplification from plicae circulares, villi, microvilli, or muscular flaps [211, 212]. Embryologically, small intestine is arranged by cubic cells until the ninth week of gestation when the villi begin to form. Each villus is about 1 mm long and contains a network of veins, arteries, lymphatic vessels, and capillary and is covered with microvilli, forming a "brush-like border."

Intestinal epithelium is an efficient physical barrier covering the surface of the GI tract, which can selectively absorb fluids, electrolytes, and nutrients while protecting the host from environmental pathogens [213]. This epithelial cell layer includes absorptive (enterocytes) and secretory cells (goblet cells and Paneth cells) [213]. Enterocytes are the most abundant epithelial cells in the intestine with dense, ordered brush-like borders on their top surface composed of organized microvilli (Fig. 7.17) [214].

Digestion and Absorption:

Small intestine is the main place to absorb water, nutrients, and electrolytes. About 9 L of water per day (7 L of secretions and 2 L in the diet) enter the small intestine [215]. Jejunum is the main place to absorb water and electrolyte. The small intestine secretes 2 L of sodium-containing fluids, which comes mainly from the mucus of goblet cells and intestinal crypt cells [216].

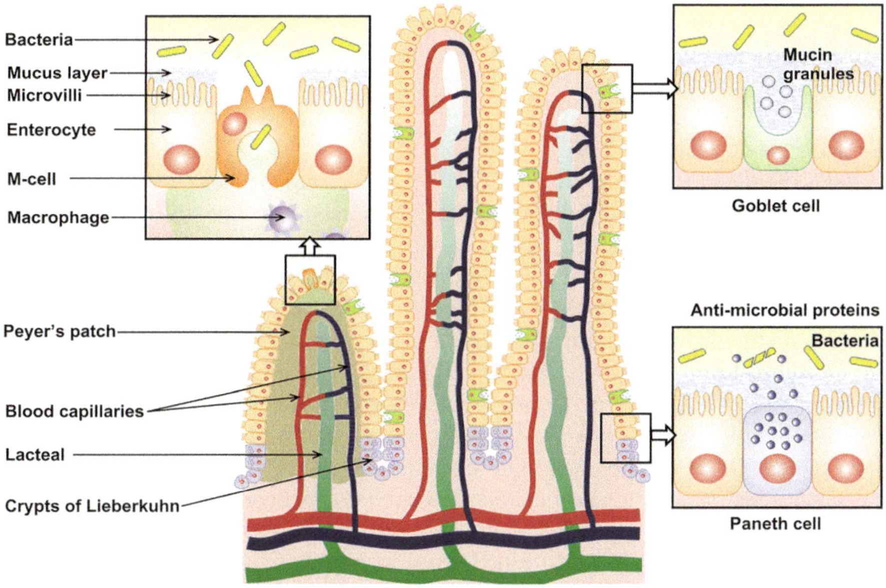

Fig. 7.17 Schematic diagram of the structure of the intestinal epithelium and its barrier function against exogenous pathogens; the cell types including enterocytes, goblet cells, paneth cells, and the M cells in Peyer's patches [214]. (Copyright permission from Elsevier Science Ltd.)

Motility of the Small Intestine:
The key function of the small intestine is exercise, which can mix and transfer digested contents and secretions, and can also remove toxic or excess non-absorbed contents from the system [209].

Immune and Barrier Function:
Small intestine has complex folds and epithelium, which can serve a barrier to a large number of microbiotas in the intestinal ecosystem. It can be considered a transition zone between the sparse microbiota zone of stomach and the abundant microbiota zone of large intestine [209].

The first line of defense is mucus in the small intestine. Mucus is a physical barrier that prevents microorganisms or toxins from reaching or damaging the surface of the epithelium and the accumulation of lymph. Mucin, especially high molecular weight glycoprotein is the main component of mucus gel [217]. Absorbing water is the main digestive function of the large intestine. It converts the 1–2 L of iso-osmotic chyme entering it into 200–250 mL of semi-solid feces [212].

The colon is a continuous structure. I t starts from the ileocecal valve and extends to the anus. The first part of the colon is the cecum, which is located behind the right iliac fossa, and is oval. This cavity is larger than the rest of the colon. The appendix (outside pocket at the blind end) starts from the cecum and is usually visible during colonoscopy [218].

The length of the adult large intestine is about 1.5 m. The large intestine is divided into the caecum and appendix, colon, rectum and anal canal in turn. The colon is further subdivided into ascending, transverse, descending and sigmoid segments. The sigmoid colon enters the pelvic cavity through the edge of the pelvis, where it is connected to the rectum [219].

The colon and caecum are easily distinguishable from the small intestine because the presence of three longitudinal muscle bands called taeniae coli on their outer surface (Fig. 7.18-a). Each of these bands is approximately 1.5 cm wide, and represents a condensation of the longitudinal muscle fibers in the outer layer of the muscularis propria of the bowel wall. There is no boletus in the rectum and appendix. Another distinguishing feature of the colon is the presence of many small fat tags on the serosal surface. These are called appendices epiploic (Fig. 7.18-b). They are most abundant on the surface of the sigmoid colon and distal descending colon. Its function is unknown. By the way, there is no appendix on the rectum, appendix, and caecum. Other features that represent the large intestine and distinguish it from the small intestine are its larger caliber and curved appearance [219].

The last part of the gastrointestinal tract extending from the cecum to the anal canal is the colon. The main two functions of human colon are absorbing water and pushing waste and storing it in the distal colon until defecation. As a metabolically active organ, the colon mainly reabsorbs water/electrolytes and unabsorbed nutrients, and plays an important role in maintaining the overall health of the human body. The colon also plays a role in the formation and storage of feces and its elimination through defecation [220]. In addition, the colon is a tubular cavity with immunological activity, which plays an important role in the host's immune response and defense against pathogens. Important in various disease and health states is the interaction with the colonic microbiome, including inflammatory bowel disease, diarrhea, and obesity. The colon can also directly or in directly promotes the metabolism of energy and nitrogen through the microbiome. Chronic conditions including chronic inflammatory disease and protein losing enteropathy, which have an impact on energy growth and expenditure [218].

7.7.3 The In Vivo Evaluation of Decellularized Intestine after Recellularization

The small intestinal submucosa (SIS) is a tissue source used for extensive studies of decellularization. Chen and Badylak [221] used 0.1% peracetic acid decellularized porcine SIS slices as patches to repair partially resected bowel in dogs. The SIS patch was remodeled by migratory cells in 2–3 months and histological staining indicated that all layers of the intestinal wall were present within 6 months. However, similarly configured tubular patches were unsuccessful and were prone to leaks and stenosis.

The best treatment option for patients with irreversible bowel failure that has been recognized worldwide is small bowel transplantation [222]. The subject-specific

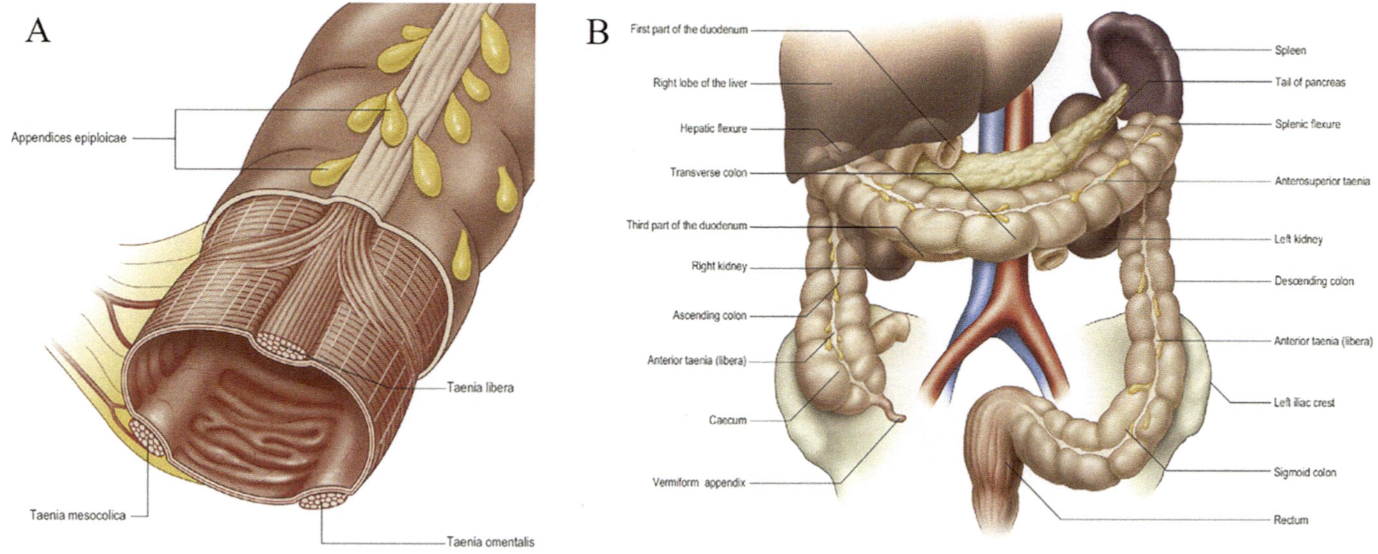

Fig. 7.18 (**a**) External features and subdivisions of large intestine; (**b**) Colonic wall showing taeniae coli and appendices epiploicae [219]. (Copyright permission from Elsevier Science Ltd.)

organ generated by a patient's own primary cells or patient-derived iPSCs has great potential to overcome these intestinal transplantation obstacles. Studies have shown that the immunogenicity of terminally differentiated cells derived from autologous iPSCs is negligible [223]. In addition, iPSCs can directly differentiate into intestinal progenitor cells and then differentiate into mature epithelium [224].

Using patient-derived cells to generate an implantable bioengineered intestinal graft would provide an alternative treatment strategy for patients. Kitnao et al. [225] evaluated the humanized intestinal grafts generated by repopulated decellularized rat intestinal matrix with human induced pluripotent stem cell-derived intestinal epithelial cells and human endothelial cells. After 28 days of in vitro culture, human iPSC-derived progenitor cells differentiate into polarized intestinal epithelial monolayers. Human endothelial cells seeded through native vasculature can restore perfusion. The isolated perfusion test confirmed that glucose and medium-chain fatty acids were transferred from the lumen to venous effluent. Four weeks after transplantation to RNU rats, grafts showed regenerative survival and maturity. Whole-body vein sampling and positron emission tomography confirmed glucose and fatty acids uptake in vivo. The bioengineered intestine on a vascularized natural scaffold could bridge the gap between cell/tissue-scale regeneration and the entire organ-scale technology needed to treat patients with intestinal failure (Fig. 7.19).

7.8 The Applications of Other Recellularization Tissue In Vivo

7.8.1 Skin

The skin, the largest organ of the body, has function as a primary barrier to the chemical, physical, and biological aggressions of the environment. Skin as a mechanical barrier between the inner and outer environment of our body protects us against infection and electrolyte loss. This organ consists of 3 layers: the epidermis, dermis, and hypodermis. Skin lesions are normal to all species, regardless of gender or age. In animals, these lesions may be due to fights and/or predations, also as in humans, there is a very common cause of dermal lesions that are caused by burns and carcinomas [226, 227]. Skin damages are one of the most challenging issues in regenerative medicine. The wounds are generally considered as injuries in the external or internal surface of the body, caused by burning, cutting, pressure, or trauma. As a category, chronic wounds are divided into four main types: diabetic ulcers, venous ulcers, pressure ulcers, and postsurgical wounds.

Acellular materials, such as amniotic membrane, porcine dermal xenografts, and cadaveric skin allografts, are used in reconstructive surgeries, the latter being safer and more economically viable [228].

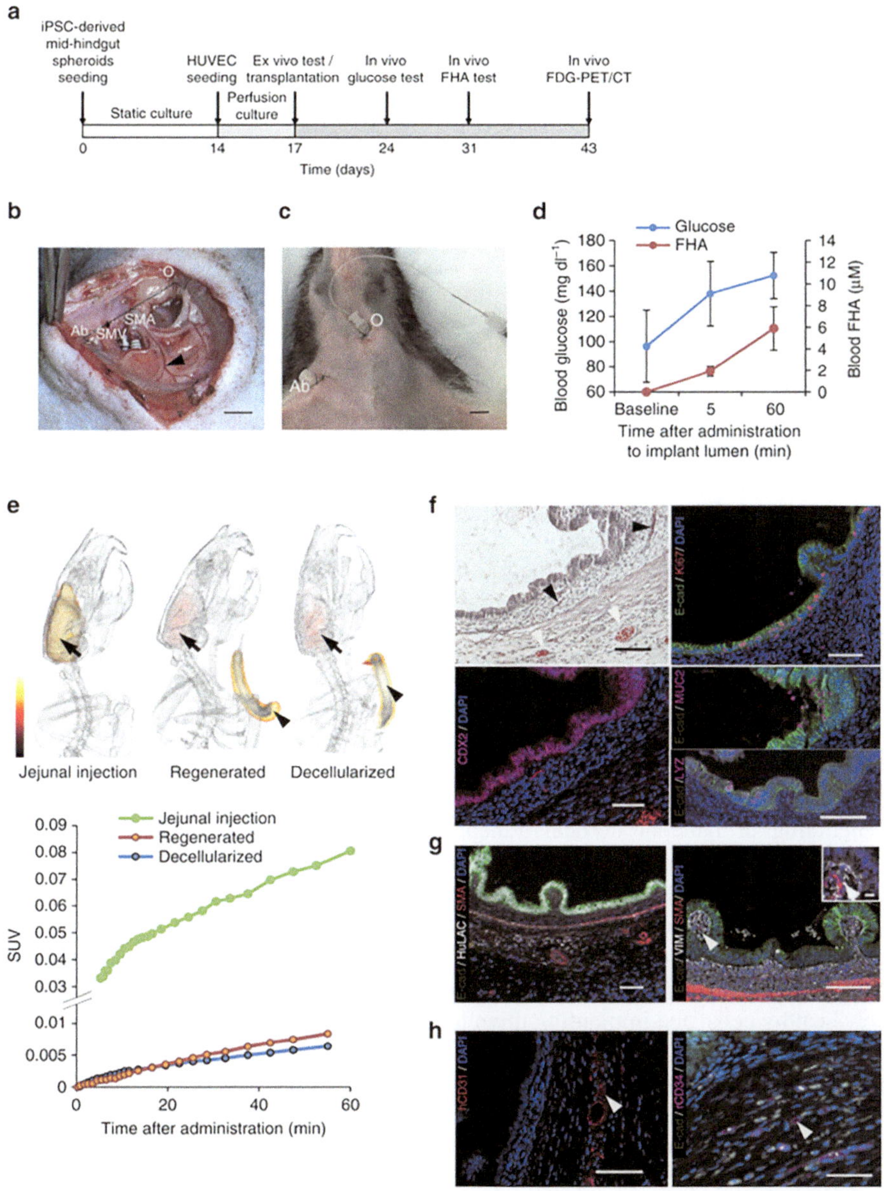

Fig. 7.19 In order to prove the survival and function of the bioengineered intestine in the body, heterotopic transplantation and in vivo functional tests were performed. (**a**) In vivo test sequence after the initial inoculation of human iPSC-derived spheroids. (**b**) Photograph of gross observation of ectopic transplanted regenerated intestine construct into the subcutaneous pocket of rat right neck. (**c**) The glucose solution is administered through the stomata. (**d**) After in vivo administration into the regenerated intestinal lumen, an increase in the concentration of systemic glucose and FHA was detected in whole blood. (**e**) 60 minutes after administration of 18F-fluorodeoxyglucose, recipient rats were subjected to positron emission tomography and 3D reconstruction based on CT images. (**f**) Histological evaluation of the transplanted regenerated intestinal graft 4 weeks after transplantation confirmed that the transplanted columnar epithelial cells derived from iPSC

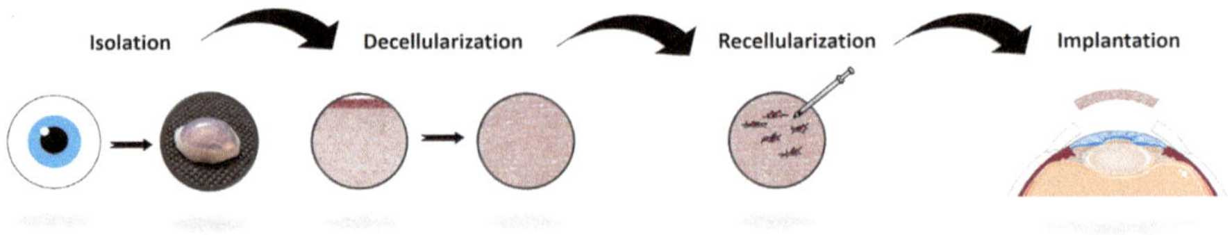

Fig. 7.20 General workflow for the use of decellularized corneas as donor alternatives [234]. (Copyright permission from Elsevier Science Ltd.)

7.8.2 Cornea

The cornea is the outmost structure of the eye. It is comprised of three main cellular layers: the endothelium, the stroma, and the epithelium. The corneal epithelium is composed of layered epithelial cells whose main function is to act as a barrier to protect the inner corneal layers [229].

About ten million people globally in the world suffer from bilateral corneal blindness [230]. Corneal disease is the fourth leading cause of blindness worldwide [231]. Severe cases can only be treated by corneal transplantation or keratoplasty. Although the cornea is an immune-privileged organ due to its avascular nature, it also has disadvantages, such as the risk of immune-mediated rejection and the lack of high-quality donor corneas [232, 233]. For these reasons, people are very interested in corneal bioengineering as a method to produce transplantable tissues [234].

Corneal ectasias, such as keratoconus, are characterized by a gradual thinning, bulging, and deformation of the cornea, which in turn leads to loss of vision due to highly irregular astigmatism. Visual rehabilitation of late corneal ectasias needs penetrating or lamellar corneal transplantation techniques. This technology has several disadvantages, due to the high level of postoperative astigmatism associated with sutures, which leads to graft rejection, failure, and slow vision recovery [235].

The use of decellularized corneas as a scaffold for corneal engineering is attractive because it can accurately mimic the structure of fibril and the composition of the corneal. The process includes separation of donor tissue, removal of cells, adding new healthy cells, and implanting them in the patient (Fig. 7.20) [234]. Before being used with patients, the corneas need to be tested in animal models in vivo to assess the host's response to the implant.

Barrio et al. investigated the in vivo biocompatibility of decellularized human corneal stroma with or without the recellularization of human adipose-derived adult stem cells (h-ADASC) [236]. The human corneal stromal sheet with a thickness of 90μm was decellularized, and the cytotoxicity was assayed. The recellularization

---

Fig. 7.19 (continued) continued to transfer into the lumen of the intestinal stent. (**g**) The relationship between epithelial and mesenchymal cells derived from human iPSC and tissues derived from the host. (**h**) The presence of human endothelial cells in small blood vessels [225]. (Copyright permission from © 2020 Springer Nature Limited)

was obtained by injecting 2×10^5 labeled h-ADASC in the graft and culture for 5 days. The grafts were implanted in vivo for 3 months, the bio-integration of the graft, the viability of the stem cells, and the expression of keratocan were evaluated (Fig. 7.21). During the whole follow-up process, completely stable graft transparency was observed without any clinical rejection. The postmortem analysis demonstrated the survival of the transplanted human stem cells inside the graft and their differentiation into functional keratocytes, as assessed by the expression of human keratocan.

Further, in phase 1 human pilot study, the safety and effectiveness of acellular human corneal stroma transplantation were evaluated in patients with advanced keratoconus with or without autologous adipose-derived adult stem cell recellularization. Twenty-four hours before implantation, the lamina of the patient who received the recellularized tissue was placed in tissue culture wells and autologous ADASCs were used for recellularization (Fig. 7.22 **Middle**). After the recellularization process is completed, the laminas were transferred to the theater in PBS at room temperature for their implantation (Fig. 7.22 **Right**) [237]. Recellularized laminas did not show postoperative haze because they were colonized initially by ADASCs rather than adult keratocytes. During the 6 months of follow-up, they did not show the potential advantage of adding mesenchymal stem cells to the decellularized matrix layer. Using corneal OCT, the recellularized graft did not show faster visual recovery, enhanced outcomes, or new collagen production areas. However, we cannot discard the possible advantages of the transplanted ADASCs in the long-term maintenance of the collagen lenticule.

7.9 Difficulties and Possible Strategies for the Application of Decellularized Organs

These exciting studies on the aforementioned complex organs such as the heart, liver, and lungs provide insights into the power of this method. This technology has many advantages. First, bioartificial organs could theoretically solve the problem of lifelong immunosuppression after organ transplantation. Second, the three-dimensional scaffold and extracellular matrix could provide microenvironments for engraftment of repopulated cells, the survival of transplanted cells, and the function of new organs. Third, the vascular bed in the decellularized scaffolds can quickly transport oxygen and nutrients after recellularization and reconnected to the circulatory system [238]. The studies of the recellularization organs in organ replacement at the stage of animal research are showed in Table 7.1.

The cell type and cell sources used to fill the matrix of acellular organs are the most important issues in this technology. Different cell types are required to rebuild the parenchyma, vasculature, and supporting structures below. The ideal strategy is that stem cells or progenitor cells can differentiate into specific cells to form functional organs. Currently, cell sources are widely classified. Non-stem or

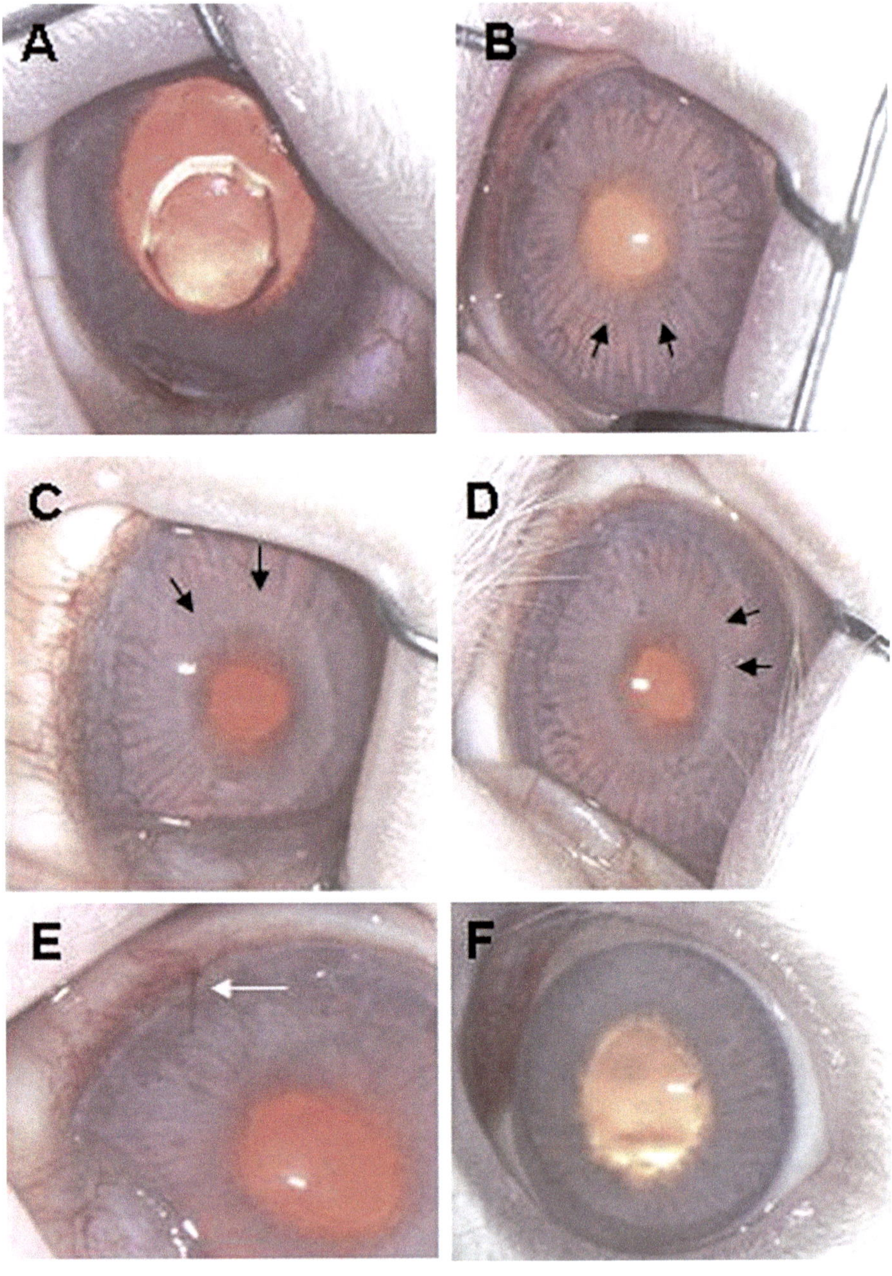

Fig. 7.21 Example of the transparency evolution of the sheet (rabbit 8) before (**a**) and after implantation and at 1 week (**b**), 1 month (**c**), and 3 months (**d**). None of the control eyes showed a clinical abnormality during the follow-up except a mild superior peripheral neovascularization (white arrow) at 1 week (**e**) that regressed at 3 months (**f**) [236]. (Copyright permission from Elsevier Science Ltd.)

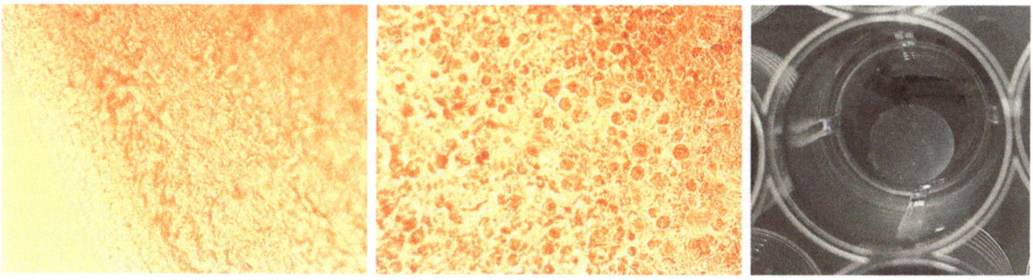

Fig. 7.22 **(Left)** Phase-contrast image of decellularization lamina. **(Middle)** Phase-contrast image of a lamina 11 hours after seeding ADASCs. **(Right)** Macroscopic photograph of the lamina in a dish containing phosphate buffer saline [237]. (Copyright permission from Elsevier Science Ltd.)

progenitor cells used for repopulation are usually vascular cells (e.g. endothelial cells), parenchymal cells (e.g. pneumonocytes, hepatocytes, cardiomyocytes), and supportive cells (e.g. fibroblasts). Stem or progenitor cells used for recellularization include adult tissue-derived iPSCs, embryonic stem cells (ESCs), adult-derived stem or progenitor cells, and fetal cells. Whether to use autologous or allogeneic cells will depend on the number of cells needed and whether they need to be used immediately [238].

Although the decellularization and recellularization of various organs such as the heart, liver, lung, kidney, bladder, and other organs have achieved exciting results, it is not feasible to obtain donor organs for this method from the human beings [238].

The precise positioning of specific cell types in their corresponding specific organ chambers is one of the main problems in the field of organ bioengineering. The method of seeding cells into whole organ-derived dECM scaffolds largely depends on the organ itself and usually requires the use of bioreactors. Complex organs including kidney, heart, lung, and liver require the development of cell seeding and culture methods to accelerate tissue formation and maturation. The diffusion of oxygen in engineered tissues is limited to a maximum tissue thickness of 200 micrometers [239], which means that higher tissue thicknesses must require functional vasculature to provide oxygen and nutrients to the cells and promote the removal of metabolic waste. In this regard, bioreactor technology for the entire organ engineering still needs to overcome many problems: from the definition of the best cell seeding method to the possibility of non-invasively monitoring biochemical and biophysical markers that indicate organ maturity and function. Other issues are related to the application of physiological relevant stimuli, which can enhance tissue formation and function [240].

Bioengineering has various problems and ambiguities. The most critical challenge is to select the most suitable organ source, standardize the scaffold, and implant it in cells, to achieve proper organ revascularization, and its storage. In addition, there are many ethical and legal issues [155]. In terms of bioengineering, it is essential to understand the developmental processes of the organs and understand the mechanisms behind many organ pathologies and their treatment methods [155].

With the advancement of bioengineering methods and the improvement of transplant functions, extensive preclinical testing in large animal models will be an

Table 7.1 The studies of the recellularization organs in organ replacement at the stage of animal research

Study	Organ	Cell types	Animals	Highlights of the study
Kitahara [74]	Heart	Porcine mesenchymal stem cells	Pig	They successfully performed heterotopic transplantation Of a bioengineered porcine heart using decellularization And recellularization of the whole-organ scaffold.
Jordan [80]	Heart valve	Endothelial progenitor cell-derived endothelial Cells	Sheep	Conjugated valves have several other distinct advantages over autologous pre-cell-seeded valves.
Hennessy [77]	Heart valve	No	Sheep	The implantation resulted in recellularization of the valve with sufficient hemodynamic function for the 5-month study
Khorramirouz [78]	Heart valve	Mesenchymal stem cell	Rat	MSC-seeded porcine pericardium showed faster onset of blood vessel formation after 4 weeks.
Dohmen [61]	Heart valve	Autologous vascular endothelial cells	Human	The first clinical case with implantation of a tissue-engineered heart valve. After 1 year of follow-up the patient is in excellent condition without limitation and exhibits normal aortic and pulmonary valve function.
Dohmen [62]	Heart valve	Autologous vascular endothelial cells	Human	Tissue-engineered heart valves showed Excellent hemodynamic performance and may prevent Degeneration during long-term follow-up.
Ott [8]	Lung	Epithelial; endothelial cells	Rat	They transplanted regenerated lungs into an orthotopic position. After transplantation, they provided gas exchange in vivo for up to 6 h after extubation.
Song [83]	Lung	Human umbilical cord endothelial cells	Rat	Regenerate lungs based on perfusion decellularized cadaveric rat lungs could

(continued)

Table 7.1 (continued)

Study	Organ	Cell types	Animals	Highlights of the study
				provide gas exchange as long as 7 days at levels comparable to cadaveric lung transplants.
Zhou [109]	Lung	Human umbilical vein endothelial cells; human airway epithelial Progenitor cells	Pig	Grafts were able to withstand physiological blood flow from the recipient's pulmonary circulation and exchanged gases upon ventilation during the 1-hour observation.
Nichols [110]	Lung	Mesenchymal stem cells; M2 macrophages	Pig	This study focused on the development of the bronchial systemic circulation in non-immunosuppressed pigs to support BEL growth and survival after transplantation.
Yang [132]	Liver	Hepatocytes and NG2 + HPCs	Mice and BaMa pigs	This decellularized liver bioscaffold supported primary hepatocyte survival, promoted blood vessel patency.
Hata [133]	Liver	Hepatocytes	Rats	Engineered CLs were successfully transplanted and showed maintenance of function and structure for a long period.
Shaheen [135]	Liver	Human umbilical vein endothelial cells	Pig	Revascularized liver scaffolds that can maintain blood perfusion at physiological pressures.
Shimoda [138]	Liver	No	Pig	This is the first attempt to use decellularized liver scaffolds after hepatectomy and to determine whether the acellular scaffolds facilitate liver regeneration in partially hepatectomized animals, using a large model.
Ko [131]	Liver	Murine endothelial cells	Pig	a strategy for overcoming a major hurdle in the engineering of transplantable liver, the establishment of a functional vasculature.

(continued)

Table 7.1 (continued)

Study	Organ	Cell types	Animals	Highlights of the study
Song [158]	Kidney	Endothelial and epithelial cells	Rat	The regenerated scaffolds could produce urine via the ureteral conduit in vivo.
Remuzzi [163]	Kidney	Murine embryonic stem	Rat	Acellular scaffolds implanted in rats could not be repopulated with host cells.
Guo [188]	Pancreas	Endothelial progenitor cells	Rat	A pre-vascularized pancreatic decellularized scaffold was developed by endothelializing with EPCs in a bioreactor
Wu [189]	Pancreas	MIN-6 β cells	Mice	Decellularized mouse pancreas were reseeded with MIN-6 β cells and show potential of controlling the blood glucose.
Kitano [224]	Intestine	Human induced pluripotent stem cell-derived intestinal epithelial cells and human endothelial cells	Rat	Four weeks after transplantation to RNU rats, systemic venous sampling and positron emission tomography confirm uptake of glucose and fatty acids in vivo.

important requirement for human transplantation. Due to the possible large inter-patient heterogeneity and complexity of translational parameters, a sufficient number of studies are required to thoroughly validate any clinical-grade procedure and effectively translate this technology to a therapy [100].

7.9.1 Heart

Decellularized heart valves have some limitations. Decellularized valves need human or animal tissue to be manufactured, which is limited in supply and requires cryopreservation for storage. The freeze-drying of biologic heart valves has been explored to promote long-term preservation. However, freeze-drying can destroy the structure of ECM and biomolecules [241]. Research with lyoprotectants may overcome this limitation in the future. In addition, the decellularization process and the potential immune response after implantation will determine the success of decellularized heart valves [51].

Preclinical experiments in sheep using in vitro engineered heart valves have found two major problems, thickening and retraction of the leaflets, resulting in unacceptable valve regurgitation [242–244]. However, the limitation of donors may affect the widespread use of acellular homografts, especially for young patients [72]. Therefore, due to the regulatory history, long clinical experience with homografts, and in-depth research by many research groups, acellular valve scaffolds have the greatest potential in the development of a TEHV [43, 50, 81]. One major challenge must be overcome before the TEHV can be achieved: the development of a physiologically suitable cell population within the leaflet tissue [51].

For young patients who need growth, host integration, and controlled biomechanical remodeling, valve replacement is essential. The latest development in manufacturing TEHVs all has the potential for development and risk management. Regardless of the strategy adopted, rapid cellularization (whether in vitro seeding or in vivo recruitment) is required to sense and respond to local biochemical and mechanical cues. Many TEHVs can withstand the hemodynamic load of the lungs tested in vivo and in vitro, but persistent stenosis has dimmed the long-term prospects. The natural aortic valve exhibits a highly complex multi-scale structure, which may be essential for long-term function due to the limitation of manufacturing strategies, this feature has not been realized in macros-TEHV. Advances in fabrication strategies and technologies (such as 3D bioprinting, self-assembly, etc.) are crucial to overcome the barriers in incorporating heterogeneity and anatomical fidelity into TEHVs [43].

7.9.2 Liver

It is vital to refill the ECM scaffold with different types of hepatocytes and confirm that the 3D structure can effectively homing and target the cells and locate them in the correct position. This is an essential step to characterize the potential use of ECM scaffolds for regenerative medicine [135].

The vascular system needs to be completely filled with a layer of endothelium to separate the internal environment from direct exposure to blood. Severe inflammatory reactions could occur after implantation if this barrier is not continuous because of incomplete recellularization [10, 245]. In addition, to ensure the functionality of the bile ducts and the normal secretion of bile by the liver, the biliary tree must be refilled with cholangiocytes [17]. To ensure that cells are ultimately in the correct location is a further challenge. For example, hepatocytes can be injected into the empty ECM in the form of blood vessels or as cholangiocytes through the bile ducts. In the first case, hepatocytes vascular obstruction and thrombosis may occur after hepatocyte transplantation. The vascular system and biliary tree also need to be filled with a layer of endothelium and cholangiocytes, respectively. The biliary ducts and the integrity of the vessels barrier are essential to distinguish the structure and to ensure the function of the recellularized scaffold. In addition, different types of cells in specific ratios are required. For example, hepatocytes are needed in large quantities. Cells from organoids can be differentiated into hepatocytes or cholangiocytes in vitro before being injected into the liver graft. However, cells can be injected as undifferentiated organoids or as a mixture of a certain population of differentiated and undifferentiated cells [119].

Since the human liver is estimated to contain approximately 300 billion cells, to obtain a sufficient number of liver cells from healthy donors is the first challenge [246]. In addition, hepatocytes can quickly dedifferentiate and lose their function when they are cultured in vitro on relatively hard plastic [247]. Adult stem cells and pluripotent cells have the potential to differentiate toward hepatocytes [248]. Induced pluripotent stem cells have the potential to differentiate into cells with hepatocyte-like morphology and function. However, these cells cannot fully mimic all the characteristics of hepatocytes [249]. MSCs are another source of adult stem cells. MSCs can be used as an alternative transplantation method and they can reduce or prevent ischemia/reperfusion injury in donor livers [247]. MSCs can differentiate into hepatocyte-like cells [250]. However, at present, they cannot be differentiated into fully functional hepatocytes [119, 251].

7.9.3 Lung

In recent years, researches on lung tissue regeneration have made some progress. However, the lung is a very complex organ. Lung is composed of chemical ECM components, multiple cell types, and intrinsic mechanical properties [252], which offers many challenges. For example, the perfusion of chemical and biological agents is the most commonly used method for lung decellularization. This method is often combined with a physical method, which may change or destroy important ECM components that affect cell adhesion and proliferation [253, 254]. New advanced bioreactor systems are in urgent need of research and development to provide mechanical ventilation, oxygen, and nutrition to bioengineered lung as well as real-time measurements of glucose, pH, electrolytes, and lactate [50, 255]. Standardized sterilization processes that do not affect the cells or scaffolds also need to be optimized [256–258].

The development of biocompatible lung scaffold initially focused on the production of a highly elastic matrix to form the complex lung structure needed to support gas exchange [91]. Secondly, bioreactor systems will be needed to update for prolonged organ culture ex vivo [259]. Finally, an adequate but non-destructive sterilization plan needs to be developed [259].

The donor lungs for decellularization and ex vivo bioengineering may come from elderly donors, and the donors may have structural lung diseases such as emphysema or pulmonary fibrosis [99]. Studies have shown that emphysematous lungs support initial engraftment and proliferation of many human cell lines. But researchers do not know why they do not support prolonged viability. In addition, other acute pulmonary pathologies when collecting lungs from autopsy may also affect the decellularization and recellularization of the acellular scaffolds [98]. The characteristics of normal and diseased lungs need to be further investigated to assess the potential use of these lungs as scaffolds for recellularization and final clinical applications. Further exploration of the potential uses of aged and diseased lungs will enable people to continue to understand the potential clinical utility of these lungs. This also could reveal the mechanisms by which the residual ECM scaffold can affect the growth and differentiation of seeded cells [99].

7.9.4 Kidney

The recellularization of the scaffolds is the most complicated phase, especially for the kidney, which consists of more than 30 types of cells. An adequate kidney scaffold requires various cell types (supporting cells, parenchymal cells, and vascular cells) for a recellularization [260].

Due to the various differentiated cells that present in the kidney, the recellularization of the scaffolds is the most complex and challenging stage. Proper reassembly of the kidney scaffold requires a large number of cells and multiple cell types. No studies have provided sufficient results in terms of proper cell types and numbers in the repopulation of renal scaffolds, and/or with proper cell differentiation. Although seeding methods have been implemented, different cell types, several cell injections, and several seeding techniques have achieved the most desirable results. Based on the results obtained so far, we conclude that renal tissue engineering is still in its early stages and the production of functional kidneys in the laboratory will not be ready soon. However, these studies have shown that functional acellular renal ECM can be used in the laboratory for cell seeding under dynamic flow conditions. This could help to understand kidney cell biology, cell differentiation, and remodeling. Moreover, these experimental settings may help to obtain new evidence about the pathophysiological mechanism of the kidney [261].

The results of cell seeding procedures showed that the extent and location of recellularization are limited [164]. Cells were often confined in the focal area and not uniformly distributed in the cortex and the medullary volume of the kidney scaffolds.

A technique for operating a cell-free kidney scaffolds is needed to develop to reduce or completely block hydraulic permeability of vascular and tubular membranes, to prevent fluid leakage, and to allow the entire kidney structure (especially the venous circulation) cell perfusion. Theoretically, it is possible to increase its resistance to fluid filtration by adhering compounds on the ECM. In addition, the use of endothelial cell layers or programmed stem cells to recellularized the vascular lumen can also improve the results of cell seeding. This may ultimately also facilitate the differentiation of cells that are seeded and delivered to the renal tubules [164].

The number of cells that can be injected is another key limiting factor for recellularization of the acellular scaffolds. The number of cells that must be injected should be equal to the native kidneys since there is evidence that the proliferation of seeded cells is quite limited. Some studies have shown that injecting such a large number of cells may cause cell entrapment and then block the circulation of arteries and capillaries, rather than uniform cell distribution of cells through the renal parenchyma. It is more difficult for porcine kidney recellularization because the number of cells must be increased to hundreds of billions of cells. It is very difficult to produce so many cells in culture, new and more effective methods must be developed to expand the expansion of stem cells [164].

In terms of malleable, iPSCs proved to be a good candidate for the cell source. In addition, autologous cells can be used together with fibroblasts and endothelial cells for reseeding the scaffold without underestimating the difficulty of finding the best cell mixture [142]. In the case of the kidney, if the cell arrangement is very different, the coordination of each required somatic cell type is laborious at best. Thus, the use of pluripotent precursor cells is an appealing alternative, which can be derived from normal adult embryonic tissues or repair pathways [152]. In addition to supporting the kidney scaffold to generate the culture environment necessary for kidney capable of replicating natural kidney functions, the understanding of the differentiation and proliferation of each cell type needs further study [262].

There are two major compartments that must be refilled for transplant, which are the endothelium to allow the necessary implantation and perfusion and the parenchymal compartment for organ-specific function. Definitely, the most complex phase is the recellularization of the ECM scaffolds. Until now, no study has provided enough results of renal scaffold repopulation and differentiation with adequate cells, which will realize organ transplantation. Thus, many efforts are focused on the study of the optimal cell source and the best growth environment of reseeded cells [155].

By evaluating the morphology and function of the orthotopic organs during prolonged periods of time, the long-term viability and functionality of the bioengineered tissue and the possibility of post-transplant follow-up should be further investigated. A few alternatives are also under development, among the most interesting being kidney organoids [263]. Exploring factors affecting decellularization and recellularization is critical for potential clinical application; these factors are summarized in Fig. 7.23 [262].

7.9.5 Pancreas

The production of the scaffold and the best plan for cell infusion have not yet to be determined since the pancreas is a fragile tissue and the islet itself is too fragile to be efficiently injected into the parenchymal space [264]. In addition, it is important to determine whether islets can be refilled with the scaffold and retain their spheroidal form in the parenchymal space rather than in the vessel. This could prevent shear stress and instant IBMIR, which are reported to be critical issues of naked islet transplantation [265, 266]. This is essential for maintaining islet function. Before successful clinical translation, the prolonged islet survival and implantation in vivo still need to be studied.

Scaffolds described in the previous study did not adequately reflect the size and structure of the adult human pancreas [267]. The porcine pancreas is a more suitable platform, because of its physiological and structural capabilities, which can support the number of β-cells needed to meet human insulin requirements [170]. Future

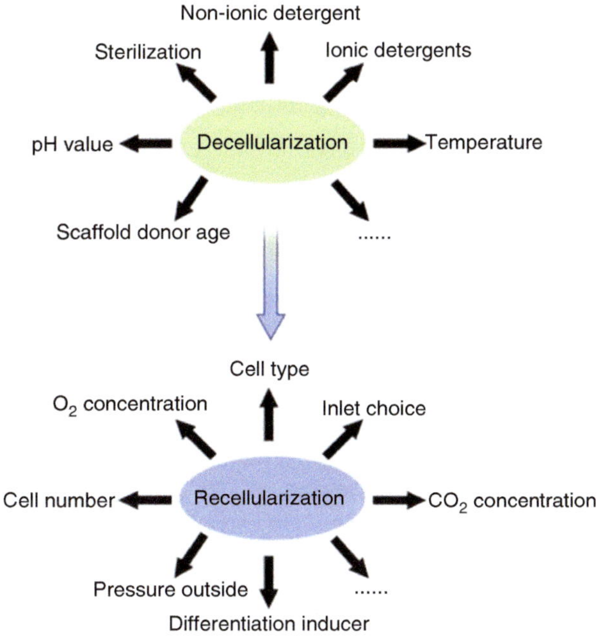

Fig. 7.23 Summarized factors based on recent studies that associated kidney decellularization and recellularization [262]. (Copyright permission from Elsevier Science Ltd.)

research will evaluate whether the complement of ECM proteins expressed by the adult porcine pancreas is sufficient to enhance β-cell differentiation and function. Further study is needed to evaluate correct cell types, cell seeding strategies, cell sources, and bioreactor culture to generate an intact and functional pancreas [268].

The spare organ is liver, because of its rich vascular network structure and endodermal origin, which may promote better implantation and oxygen perfusion of the injected islets [269, 270]. Similarly, some people try to reuse kidney scaffold for pancreas engineering to take advantage of the abundant vasculature of the kidney [271]. However, a potential limitation to this approach is that the extracellular milieu of the acellular organ may not be optimal for viability and function of the pancreatic islets [272]. Thus, this strategy may require more careful analysis and possibly further supplementation with relevant growth factors [273]. This aspect may become more critical when repopulating the scaffold using stem cell-derived islet-like cells or undifferentiated progenitor populations.

Special attention to the recellularization route is needed for the recellularization of the acellular pancreas scaffold with intact islets. Compared with isolated beta cells or beta cell lines, the intact islets are 3D aggregates in the range of 100–200μm. The inside of the islets is also prone to hypoxia, so the oxygen demand is higher. For the acellular scaffold to reproduce sufficiently and provide oxygen and nutrients to the intact islets, the islets will need to escape and enter the parenchymal space. Compared with isolated single cells, the use of 3D spheroidal islets is more challenging. In addition to the problem of oxygenation, human islet cell transplantation also failed due to inflammation and rejection of cadaveric donor tissue. Many researchers are actively studying the use of synthetic or semi-synthetic substances as immune barriers to wrap islet cells [274, 275].

7.9.6 *Intestine*

Although this technology has successful clinical applications in skin or vessels, many structural and functional problems need to be solved before it can be used in the digestive organs. Until now, this method has only achieved short-term functionality in vivo and then only in rodent models. Nevertheless, previous decades of work in the field of tissue engineering and current research in stem cells are producing encouraging results, which 1 day might solve the problem of organ assembly on demand [276].

It has been a challenge to imitate the natural 3D structure of human digestive organs to reproduce the functions and metabolism necessary for clinical use (such as the peristaltic movement of mucous and acid secretion). This requires the creation of micro-structures and large vessels or ducts for external secretions, including bile or pancreatic exocrine factors [276].

7.10 Conclusion

This chapter on the applications of the recellularization organs presented a comprehensive overview of challenges and opportunities that currently exist in this area. The bioengineering of whole organs is promising for organ transplantation. There have been considerable advances that have led to the development of scale up recellularization to large animal models and human tissue. However, most of the current research is still in their infancy and clinical translation cannot be foreseen based on the current knowledge. One of the most significant challenges in whole-organ bioengineering is to find a suitable source of cells to refill acellular scaffolds and obtain complete revascularization of the decellularized scaffolds. Moreover, essential work is needed to ensure that the decellularized scaffolds cannot be recognized by the host immune system. Hence, a series of hurdles must be overcome before clinical translation can be carried out. Altogether, the application of the recellularization organs discussed in this chapter may have an impact on bioengineering of whole organs-based therapy, which is an alternative and effective therapy for organ transplantation.

References

1. Shin'oka T, Imai Y, Ikada Y. Transplantation of a tissue-engineered pulmonary artery. N Engl J Med. 2001;344(7):532–3.
2. Badylak SF, Taylor D, Uygun K. Whole-organ tissue engineering: decellularization and recellularization of three-dimensional matrix scaffolds. In: Yarmush ML, Duncan JS, Gray ML, editors. Annual review of biomedical engineering, vol. 13. Palo Alto: Annual Reviews; 2011. p. 27–53.
3. Welman T, Michel S, Segaren N, Shanmugarajah K. Bioengineering for organ transplantation: progress and challenges. Bioengineered. 2015;6(5):257–61.
4. L'Heureux N, McAllister TN, de la Fuente LM. Tissue-engineered blood vessel for adult arterial revascularization. N Engl J Med. 2007;357(14):1451–3.
5. Atala A, Bauer SB, Soker S, Yoo JJ, Retik AB. Tissue-engineered autologous bladders for patients needing cystoplasty. Lancet. 2006;367(9518):1241–6.
6. Macchiarini P, Walles T, Biancosino C, Mertsching H. First human transplantation of a bioengineered airway tissue. J Thorac Cardiovasc Surg. 2004;128(4):638–41.
7. Price AP, England KA, Matson AM, Blazar BR, Panoskaltsis-Mortari A. Development of a decellularized lung bioreactor system for bioengineering the lung: the matrix reloaded. Tissue Eng Part A. 2010;16(8):2581–91.
8. Ott HC, Clippinger B, Conrad C, Schuetz C, Pomerantseva I, Ikonomou L, Kotton D, Vacanti JP. Regeneration and orthotopic transplantation of a bioartificial lung. Nat Med. 2010;16 (8):927–33.
9. Petersen TH, Calle EA, Zhao L, Lee EJ, Gui L, Raredon MB, Gavrilov K, Yi T, Zhuang ZW, Breuer C, Herzog E, Niklason LE. Tissue-engineered lungs for in vivo implantation. Science. 2010;329(5991):538–41.
10. Uygun BE, Soto-Gutierrez A, Yagi H, Izamis ML, Guzzardi MA, Shulman C, Milwid J, Kobayashi N, Tilles A, Berthiaume F, Hertl M, Nahmias Y, Yarmush ML, Uygun K. Organ reengineering through development of a transplantable recellularized liver graft using decellularized liver matrix. Nat Med. 2010;16(7):814–20.
11. Ott HC, Matthiesen TS, Goh SK, Black LD, Kren SM, Netoff TI, Taylor DA. Perfusion-decellularized matrix: using nature's platform to engineer a bioartificial heart. Nat Med. 2008;14(2):213–21.
12. Niklason L, Gao J, Abbott W, Hirschi K, Houser S, Marini R, Langer R. Functional arteries grown in vitro. Science. 1999;284(5413):489–93.
13. Cortiella J, Niles J, Cantu A, Brettler A, Pham A, Vargas G, Winston S, Wang J, Walls S, Nichols JE. Influence of acellular natural lung matrix on murine embryonic stem cell differentiation and tissue formation. Tissue Eng Part A. 2010;16(8):2565–80.
14. Tsuchiya T, Sivarapatna A, Rocco K, Nanashima A, Nagayasu T, Niklason LE. Future prospects for tissue engineered lung transplantation: decellularization and recellularization-based whole lung regeneration. Organogenesis. 2014;10(2):196–207.
15. Scarritt ME, Pashos NC, Bunnell BA. A review of cellularization strategies for tissue engineering of whole organs. Front Bioeng Biotechnol. 2015;3:43.
16. Williams DF. To engineer is to create: the link between engineering and regeneration. Trends Biotechnol. 2006;24(1):4–8.
17. Willemse J, Lieshout R, van der Laan LJW, Verstegen MMA. From organoids to organs: bioengineering liver grafts from hepatic stem cells and matrix. Best Pract Res Clin Gastroenterol. 2017;31(2):151–9.
18. Griffith LG, Naughton G. Tissue engineering--current challenges and expanding opportunities. Science. 2002;295(5557):1009–14.
19. Lu T-Y, Lin B, Kim J, Sullivan M, Tobita K, Salama G, Yang L. Repopulation of decellularized mouse heart with human induced pluripotent stem cell-derived cardiovascular progenitor cells. Nat Commun. 2013;4(1):2307.

20. Lichtenberg A, Cebotari S, Tudorache I, Sturz G, Winterhalter M, Hilfiker A, Haverich A. Flow-dependent re-endothelialization of tissue-engineered heart valves. J Heart Valve Dis. 2006;15(2):287–93. discussion 293-4
21. Weymann A, Patil NP, Sabashnikov A, Jungebluth P, Korkmaz S, Li S, Veres G, Soos P, Ishtok R, Chaimow N, Patzold I, Czerny N, Schies C, Schmack B, Popov AF, Simon AR, Karck M, Szabo G. Bioartificial heart: a human-sized porcine model--the way ahead. PLoS One. 2014;9(11):e111591.
22. Robinson NB, Krieger K, Khan FM, Huffman W, Chang M, Naik A, Yongle R, Hameed I, Krieger K, Girardi LN, Gaudino M. The current state of animal models in research: a review. Int J Surg. 2019;72:9–13.
23. Leong XF, Ng CY, Jaarin K. Animal models in cardiovascular research: hypertension and atherosclerosis. Biomed Res Int. 2015;2015:528757.
24. Andersen ML, Winter LMF. Animal models in biological and biomedical research - experimental and ethical concerns. An Acad Bras Cienc. 2019;91(suppl 1):e20170238.
25. Franklin TB, Russig H, Weiss IC, Gräff J, Linder N, Michalon A, Vizi S, Mansuy IM. Epigenetic transmission of the impact of early stress across generations. Biol Psychiatry. 2010;68(5):408–15.
26. Erickson ZT, Falkenberg EA, Metz GA. Lifespan psychomotor behaviour profiles of multigenerational prenatal stress and artificial food dye effects in rats. PLoS One. 2014;9(6): e92132.
27. Waterson RH, Lander ES, Wilson RK, The Chimpanzee S, Analysis C. Initial sequence of the chimpanzee genome and comparison with the human genome. Nature. 2005;437 (7055):69–87.
28. Mural RJ, Adams MD, Myers EW, Smith HO, Miklos GL, Wides R, Halpern A, Li PW, Sutton GG, Nadeau J, Salzberg SL, Holt RA, Kodira CD, Lu F, Chen L, Deng Z, Evangelista CC, Gan W, Heiman TJ, Li J, Li Z, Merkulov GV, Milshina NV, Naik AK, Qi R, Shue BC, Wang A, Wang J, Wang X, Yan, Yane JY, Yooseph S, Zhao Q, Zheng L, Zhu SC, Biddick K, Bolanos R, Delcher AL, Dew IM, Fasulo D, Flanigan MJ, Huson DH, Kravitz SA, Miller JR, Mobarry CM, Reinert K, Remington KA, Zhang Q, Zheng XH, Nusskern DR, Lai Z, Lei Y, Zhong W, Yao A, Guan P, Ji RR, Gu Z, Wang ZY, Zhong F, Xiao C, Chiang CC, Yandell M, Wortman JR, Amanatides PG, Hladun SL, Pratts EC, Johnson JE, Dodson KL, Woodford KJ, Evans CA, Gropman B, Rusch DB, Venter E, Wang M, Smith TJ, Houck JT, Tompkins DE, Haynes C, Jacob D, Chin SH, Allen DR, Dahlke CE, Sanders R, Li K, Liu X, Levitsky AA, Majoros WH, Chen Q, Xia AC, Lopez JR, Donnelly MT, Newman MH, Glodek A, Kraft CL, Nodell M, Ali F, An HJ, Baldwin-Pitts D, Beeson KY, Cai S, Carnes M, Carver A, Caulk PM, Center A, Chen YH, Cheng ML, Coyne MD, Crowder M, Danaher S, Davenport LB, Desilets R, Dietz SM, Doup L, Dullaghan P, Ferriera S, Fosler CR, Gire HC, Gluecksmann A, Gocayne JD, Gray J, Hart B, Haynes J, Hoover J, Howland T, Ibegwam C, Jalali M, Johns D, Kline L, Ma DS, MacCawley S, Magoon A, Mann F, May D, McIntosh TC, Mehta S, Moy L, Moy MC, Murphy BJ, Murphy SD, Nelson KA, Nuri Z, Parker KA, Prudhomme AC, Puri VN, Qureshi H, Raley JC, Reardon MS, Regier MA, Rogers YH, Romblad DL, Schutz J, Scott JL, Scott R, Sitter CD, Smallwood M, Sprague AC, Stewart E, Strong RV, Suh E, Sylvester K, Thomas R, Tint NN, Tsonis C, Wang G, Wang G, Williams MS, Williams SM, Windsor, Wolfe K, Wu MM, Zaveri J, Chaturvedi K, Gabrielian AE, Ke Z, Sun J, Subramanian G, Venter JC, Pfannkoch CM, Barnstead M, Stephenson LD. A comparison of whole-genome shotgun-derived mouse chromosome 16 and the human genome. Science. 2002;296(5573):1661–71.
29. Sanchis-Gomar F, Perez-Quilis C, Leischik R, Lucia A. Epidemiology of coronary heart disease and acute coronary syndrome. Ann Transl Med. 2016;4(13)
30. Zia S, Mozafari M, Natasha G, Tan A, Cui Z, Seifalian AM. Hearts beating through decellularized scaffolds: whole-organ engineering for cardiac regeneration and transplantation. Crit Rev Biotechnol. 2016;36(4):705–15.

31. W.G.t.R.N. Evidence, U.t.G.f.t.M.o.P.w.C.H.F.W.o.B.o.t.H.F.W. Committee, Jessup M, Abraham WT, Casey DE, Feldman AM, Francis GS, Ganiats TG, Konstam MA, Mancini DM, Rahko PS. 2009 focused update: ACCF/AHA guidelines for the diagnosis and management of heart failure in adults: a report of the American College of Cardiology Foundation/American Heart Association Task Force on Practice Guidelines: developed in collaboration with the International Society for Heart and Lung Transplantation. Circulation. 2009;119(14):1977–2016.
32. Smit FE, Dohmen PM. Bio-artificial heart as ultimate treatment of end-stage heart failure. Med Sci Monit Basic Res. 2014;20:161–3.
33. Lund LH, Edwards LB, Kucheryavaya AY, Dipchand AI, Benden C, Christie JD, Dobbels F, Kirk R, Rahmel AO, Yusen RD, Stehlik J, H. International Society for, Lung T. The registry of the international society for heart and lung transplantation: thirtieth official adult heart transplant report--2013; focus theme: age. J Heart Lung Transplant. 2013;32(10):951–64.
34. Taylor DA, Parikh RB, Sampaio LC. Bioengineering hearts: simple yet complex. Curr Stem Cell Rep. 2017;3(1):35–44.
35. Jessup M. Mechanical cardiac-support devices - dreams and devilish details. N Engl J Med. 2001;345(20):1490–3.
36. Kedzierska K, Sindrewicz K, Sporniak-Tutak K, Bober J, Stanczyk-Dunaj M, Dolegowska B, Kaliszczak R, Sienko J, Kabat-Koperska J, Golembiewska E, Ciechanowski K. Effect of immunosuppressive therapy on Proteinogram in rats. Med Sci Monit. 2016;22:1987–98.
37. Levitsky J, Salomon DR, Abecassis M, Langfelder P, Horvath S, Friedewald J, Wang E, Kurian SM, Mondala T, Gil S, McDade R, Ballard K, Gallon L. Clinical and plasma proteomic markers correlating with chronic kidney disease after liver transplantation. Am J Transplant. 2011;11(9):1972–8.
38. Davies RR, Farr M, Silvestry S, Callahan LR, Edwards L, Meyer DM, Uccellini K, Chan KM, Rogers JG. The new United States heart allocation policy: progress through collaborative revision. J Heart Lung Transplant. 2017;36(6):595–6.
39. Soto-Gutierrez A, Yagi H, Uygun BE, Navarro-Alvarez N, Uygun K, Kobayashi N, Yang YG, Yarmush ML. Cell delivery: from cell transplantation to organ engineering. Cell Transplant. 2010;19(6):655–65.
40. Gilbert TW, Sellaro TL, Badylak SF. Decellularization of tissues and organs. Biomaterials. 2006;27(19):3675–83.
41. Witzenburg C, Raghupathy R, Kren SM, Taylor DA, Barocas VH. Mechanical changes in the rat right ventricle with decellularization. J Biomech. 2012;45(5):842–9.
42. Weymann A, Loganathan S, Takahashi H, Schies C, Claus B, Hirschberg K, Soos P, Korkmaz S, Schmack B, Karck M, Szabo G. Development and evaluation of a perfusion decellularization porcine heart model--generation of 3-dimensional myocardial neoscaffolds. Circ J. 2011;75(4):852–60.
43. Cheung DY, Duan B, Butcher JT. Current progress in tissue engineering of heart valves: multiscale problems, multiscale solutions. Expert Opin Biol Ther. 2015;15(8):1155–72.
44. Valarmathi MT, Goodwin RL, Fuseler JW, Davis JM, Yost MJ, Potts JD. A 3-D cardiac muscle construct for exploring adult marrow stem cell based myocardial regeneration. Biomaterials. 2010;31(12):3185–200.
45. Go AS, Mozaffarian D, Roger VL, Benjamin EJ, Berry JD, Blaha MJ, Dai S, Ford ES, Fox CS, Franco S, Fullerton HJ, Gillespie C, Hailpern SM, Heit JA, Howard VJ, Huffman MD, Judd SE, Kissela BM, Kittner SJ, Lackland DT, Lichtman JH, Lisabeth LD, Mackey RH, Magid DJ, Marcus GM, Marelli A, Matchar DB, McGuire DK, Mohler ER 3rd, Moy CS, Mussolino ME, Neumar RW, Nichol G, Pandey DK, Paynter NP, Reeves MJ, Sorlie PD, Stein J, Towfighi A, Turan TN, Virani SS, Wong ND, Woo D, Turner MB. Heart disease and stroke statistics--2014 update: a report from the American Heart Association. Circulation. 2014;129(3):e28–e292.
46. Nkomo VT, Gardin JM, Skelton TN, Gottdiener JS, Scott CG, Enriquez-Sarano M. Burden of valvular heart diseases: a population-based study. Lancet. 2006;368(9540):1005–11.

47. Yacoub MH, Takkenberg JJM. Will heart valve tissue engineering change the world? Nat Clin Pract Cardiovasc Med. 2005;2(2):60–1.
48. Bourguignon T, Bergöend E, Mirza A, Ayegnon G, Neville P, Aupart MR, Marchand M. Risk factors for valve-related complications after mechanical heart valve replacement in 505 patients with long-term follow up. J Heart Valve Dis. 2011;20(6):673–80.
49. Siddiqui RF, Abraham JR, Butany J. Bioprosthetic heart valves: modes of failure. Histopathology. 2009;55(2):135–44.
50. Hopkins R. From cadaver harvested homograft valves to tissue-engineered valve conduits. Prog Pediatr Cardiol. 2006;21(2):137–52.
51. VeDepo MC, Detamore MS, Hopkins RA, Converse GL. Recellularization of decellularized heart valves: progress toward the tissue-engineered heart valve. J Tissue Eng. 2017;8:2041731417726327.
52. Li F, Li W, Johnson S, Ingram D, Yoder M, Badylak S. Low-molecular-weight peptides derived from extracellular matrix as chemoattractants for primary endothelial cells. Endothelium. 2004;11(3–4):199–206.
53. Iop L, Renier V, Naso F, Piccoli M, Bonetti A, Gandaglia A, Pozzobon M, Paolin A, Ortolani F, Marchini M, Spina M, De Coppi P, Sartore S, Gerosa G. The influence of heart valve leaflet matrix characteristics on the interaction between human mesenchymal stem cells and decellularized scaffolds. Biomaterials. 2009;30(25):4104–16.
54. Liao J, Joyce EM, Sacks MS. Effects of decellularization on the mechanical and structural properties of the porcine aortic valve leaflet, Biomaterials 2008;29(8):1065–74.
55. Korossis SA, Booth C, Wilcox HE, Watterson KG, Kearney JN, Fisher J, Ingham E. Tissue engineering of cardiac valve prostheses II: biomechanical characterization of decellularized porcine aortic heart valves. J Heart Valve Dis. 2002;11(4):463–71.
56. Vedepo MC, Buse EE, Quinn RW, Williams TD, Detamore MS, Hopkins RA, Converse GL. Species-specific effects of aortic valve decellularization. Acta Biomater. 2017;50:249–58.
57. Argento G, Simonet M, Oomens CWJ, Baaijens FPT. Multi-scale mechanical characterization of scaffolds for heart valve tissue engineering. J Biomech. 2012;45(16):2893–8.
58. Converse GL, Armstrong M, Quinn RW, Buse EE, Cromwell ML, Moriarty SJ, Lofland GK, Hilbert SL, Hopkins RA. Effects of cryopreservation, decellularization and novel extracellular matrix conditioning on the quasi-static and time-dependent properties of the pulmonary valve leaflet. Acta Biomater. 2012;8(7):2722–9.
59. Simon PK, Kasimir MT, Seebacher G, Weigel G, Wolner E. Early failure of the tissue engineered porcine heart valve SYNERGRAFT (TM) in pediatric patients. Eur J Cardiothorac Surg. 2003;23(6):1002–6. discussion 1006
60. Dohmen PM, Lembcke A, Holinski S, Kivelitz D, Braun JP, Pruss A, Konertz W. Mid-term clinical results using a tissue-engineered pulmonary valve to reconstruct the right ventricular outflow tract during the Ross procedure. Ann Thorac Surg. 2007;84(3):729–36.
61. Dohmen PM, Lembcke A, Hotz H, Kivelitz D, Konertz WF. Ross operation with a tissue-engineered heart valve. Ann Thorac Surg. 2002;74(5):1438–42.
62. Dohmen PM, Lembcke A, Holinski S, Pruss A, Konertz W. Ten years of clinical results with a tissue-engineered pulmonary valve. Ann Thorac Surg. 2011;92(4):1308–14.
63. Quinn RW, Hilbert SL, Bert AA, Drake BW, Bustamante JA, Fenton JE, Moriarty SJ, Neighbors SL, Lofland GK, Hopkins RA. Performance and morphology of Decellularized pulmonary valves implanted in juvenile sheep. Ann Thorac Surg. 2011;92(1):131–7.
64. Vunjak-Novakovic G, Lui KO, Tandon N, Chien KR. Bioengineering heart muscle: a paradigm for regenerative medicine. Annu Rev Biomed Eng. 2011;13:245–67.
65. Taylor DA, Hochman-Mendez C, Elgalad A, Sampaio LC. Whole-heart scaffolds—how to build a heart. In: Mozafari M, Sefat F, Atala A, editors. Handbook of tissue engineering scaffolds, vol. 1: Woodhead Publishing; 2019. p. 617–42.
66. Torrent-Guasp F, Kocica MJ, Corno AF, Komeda M, Carreras-Costa F, Flotats A, Cosin-Aguillar J, Wen H. Towards new understanding of the heart structure and function. Eur J Cardiothorac Surg. 2005;27(2):191–201.

67. Parker KK, Ingber DE. Extracellular matrix, mechanotransduction and structural hierarchies in heart tissue engineering. Philos Trans R Soc Lond Ser B Biol Sci. 2007;362(1484):1267–79.
68. Weber KT. Cardiac interstitium in health and disease: the fibrillar collagen network. J Am Coll Cardiol. 1989;13(7):1637–52.
69. Fomovsky GM, Thomopoulos S, Holmes JW. Contribution of extracellular matrix to the mechanical properties of the heart. J Mol Cell Cardiol. 2010;48(3):490–6.
70. Latif N, Sarathchandra P, Taylor PM, Antoniw J, Yacoub MH. Localization and pattern of expression of extracellular matrix components in human heart valves. J Heart Valve Dis. 2005;14(2):218–27.
71. Butcher JT, Mahler GJ, Hockaday LA. Aortic valve disease and treatment: The need for naturally engineered solutions. Adv Drug Deliv Rev. 2011;63(4–5):242–68.
72. MacGrogan D, Luxán G, Driessen-Mol A, Bouten C, Baaijens F, De La Pompa JL. How to make a heart valve: from embryonic development to bioengineering of living valve substitutes. Cold Spring Harb Perspect Med. 2014;4(11):a013912.
73. Whiteman S, Alimi Y, Carrasco M, Gielecki J, Zurada A, Loukas M. Anatomy of the cardiac chambers: a review of the left ventricle. Transl Res Anat. 2020;100095
74. Kitahara H, Yagi H, Tajima K, Okamoto K, Yoshitake A, Aeba R, Kudo M, Kashima I, Kawaguchi S, Hirano A, Kasai M, Akamatsu Y, Oka H, Kitagawa Y, Shimizu H. Heterotopic transplantation of a decellularized and recellularized whole porcine heart. Interact Cardiovasc Thorac Surg. 2016;22(5):571–9.
75. Ng SLJ, Narayanan K, Gao S, Wan ACA. Lineage restricted progenitors for the repopulation of decellularized heart. Biomaterials. 2011;32(30):7571–80.
76. Pellegata AF, Tedeschi AM, De Coppi P. Whole organ tissue vascularization: engineering the tree to develop the fruits. Front Bioeng Biotechnol. 2018;6(1):56.
77. Hennessy RS, Go JL, Hennessy RR, Tefft BJ, Jana S, Stoyles NJ, Al-Hijji MA, Thaden JJ, Pislaru SV, Simari RD, Stulak JM, Young MD, Lerman A. Recellularization of a novel off-the-shelf valve following xenogenic implantation into the right ventricular outflow tract. PLoS One. 2017;12(8):–e0181614.
78. Khorramirouz R, Go JL, Noble C, Morse D, Lerman A, Young MD. In vivo response of acellular porcine pericardial for tissue engineered Transcatheter aortic valves. Sci Rep. 2019;9 (1):1094.
79. Schmidt D, Stock UA, Hoerstrup SP. Tissue engineering of heart valves using decellularized xenogeneic or polymeric starter matrices. Philosophical Transact R Soc B Biol Sci. 2007;362 (1484):1505–12.
80. Jordan JE, Williams JK, Lee S-J, Raghavan D, Atala A, Yoo JJ. Bioengineered self-seeding heart valves. J Thorac Cardiovasc Surg. 2012;143(1):201–8.
81. Dohmen PM. Clinical results of implanted tissue engineered heart valves. HSR Proc Intensive Care Cardiovasc Anesth. 2012;4(4):225–31.
82. Firth AL, Dargitz CT, Qualls SJ, Menon T, Wright R, Singer O, Gage FH, Khanna A, Verma IM. Generation of multiciliated cells in functional airway epithelia from human induced pluripotent stem cells. Proc Natl Acad Sci U S A. 2014;111(17):E1723–30.
83. Song JJ, Kim SS, Liu Z, Madsen JC, Mathisen DJ, Vacanti JP, Ott HC. Enhanced in vivo function of bioartificial lungs in rats. Ann Thorac Surg. 2011;92(3):998–1005. discussion 1005–1006
84. Murphy SL, Xu J, Kochanek KD. Deaths: final data for 2010. Natl Vital Stat Rep. 2013;61 (4):1–117.
85. Orens JB, Garrity ER Jr. General overview of lung transplantation and review of organ allocation. Proc Am Thorac Soc. 2009;6(1):13–9.
86. Kotsimbos T, Williams TJ, Anderson GP. Update on lung transplantation: programmes, patients and prospects. Eur Respiratory Soc 2012;21(126):271–305.
87. Valapour M, Skeans M, Heubner B, Smith J, Hertz MI, Edwards L, Cherikh W, Callahan E, Snyder J, Israni A. OPTN/SRTR 2013 annual data report: lung. Am J Transplant. 2015;15 (S2):1–28.

88. O'Grady NP, Murray PR, Ames N. Preventing ventilator-associated pneumonia: does the evidence support the practice? JAMA. 2012;307(23):2534–9.
89. Macchiarini P, Jungebluth P, Go T, Asnaghi MA, Rees LE, Cogan TA, Dodson A, Martorell J, Bellini S, Parnigotto PP. Clinical transplantation of a tissue-engineered airway. Lancet. 2008;372(9655):2023–30.
90. Sato T, Nakamura T. Tissue-engineered airway replacement. Lancet. 2008;9655(372):2003–4.
91. Nichols JE, Niles JA, Cortiella J. Design and development of tissue engineered lung: Progress and challenges. Organogenesis. 2009;5(2):57–61.
92. Wagner DE, Bonvillain RW, Jensen T, Girard ED, Bunnell BA, Finck CM, Hoffman AM, Weiss DJ. Can stem cells be used to generate new lungs? Ex vivo Lung bioengineering with decellularized whole Lung scaffolds. Respirology. 2013;18(6):895–911.
93. Ren X, Moser PT, Gilpin SE, Okamoto T, Wu T, Tapias LF, Mercier FE, Xiong L, Ghawi R, Scadden DT, Mathisen DJ, Ott HC. Engineering pulmonary vasculature in decellularized rat and human lungs. Nat Biotechnol. 2015;33(10):1097–102.
94. Gilpin SE, Guyette JP, Gonzalez G, Ren X, Asara JM, Mathisen DJ, Vacanti JP, Ott HC. Perfusion decellularization of human and porcine lungs: bringing the matrix to clinical scale. J Heart Lung Transplant. 2014;33(3):298–308.
95. Booth AJ, Hadley R, Cornett AM, Dreffs AA, Matthes SA, Tsui JL, Weiss K, Horowitz JC, Fiore VF, Barker TH, Moore BB, Martinez FJ, Niklason LE, White ES. Acellular normal and fibrotic human lung matrices as a culture system for in vitro investigation. Am J Respir Crit Care Med. 2012;186(9):866–76.
96. Wagner DE, Bonenfant NR, Sokocevic D, DeSarno MJ, Borg ZD, Parsons CS, Brooks EM, Platz JJ, Khalpey ZI, Hoganson DM, Deng B, Lam YW, Oldinski RA, Ashikaga T, Weiss DJ. Three-dimensional scaffolds of acellular human and porcine lungs for high throughput studies of lung disease and regeneration. Biomaterials. 2014;35(9):2664–79.
97. Balestrini JL, Gard AL, Gerhold KA, Wilcox EC, Liu A, Schwan J, Le AV, Baevova P, Dimitrievska S, Zhao L, Sundaram S, Sun H, Rittié L, Dyal R, Broekelmann TJ, Mecham RP, Schwartz MA, Niklason LE, White ES. Comparative biology of decellularized lung matrix: implications of species mismatch in regenerative medicine. Biomaterials. 2016;102:220–30.
98. Wagner DE, Bonenfant NR, Parsons CS, Sokocevic D, Brooks EM, Borg ZD, Lathrop MJ, Wallis JD, Daly AB, Lam YW, Deng B, DeSarno MJ, Ashikaga T, Loi R, Weiss DJ. Comparative decellularization and recellularization of normal versus emphysematous human lungs. Biomaterials. 2014;35(10):3281–97.
99. Sokocevic D, Bonenfant NR, Wagner DE, Borg ZD, Lathrop MJ, Lam YW, Deng B, DeSarno MJ, Ashikaga T, Loi R, Hoffman AM, Weiss DJ. The effect of age and emphysematous and fibrotic injury on the re-cellularization of de-cellularized lungs. Biomaterials. 2013;34 (13):3256–69.
100. Gilpin SE, Wagner DE. Acellular human lung scaffolds to model lung disease and tissue regeneration. Eur Respir Rev. 2018;27(148):180021.
101. Uriarte JJ, Uhl FE, Rolandsson Enes SE, Pouliot RA, Weiss DJ. Lung bioengineering: advances and challenges in lung decellularization and recellularization. Curr Opin Organ Transplant. 2018;23(6):673–8.
102. Gilpin SE, Charest JM, Ren X, Tapias LF, Wu T, Evangelista-Leite D, Mathisen DJ, Ott HC. Regenerative potential of human airway stem cells in lung epithelial engineering. Biomaterials. 2016;108:111–9.
103. Kradin RL. Chapter 2 - anatomy of the Lung. In: Kradin RL, editor. Understanding pulmonary pathology. Boston: Academic Press; 2017. p. 5–19.
104. Mack SM, Wu C-W, Zhang C, Richard H, Pinkerton KE. Lung anatomy and aging, reference module in biomedical sciences: Elsevier; 2020.
105. Newell RLM. Anatomy of the post-laryngeal airways, lungs and diaphragm. Surgery (Oxford). 2011;29(5):199–203.
106. Alvarado A, Arce I. Metabolic functions of the lung, disorders and associated pathologies. J Clin Med Res. 2016;8(10):689.

107. Bernhard W. Lung surfactant: function and composition in the context of development and respiratory physiology. Ann Anatomy-Anatomischer Anzeiger. 2016;208:146–50.
108. Miller AJ, Spence JR. In vitro models to study human Lung development, disease and homeostasis. Physiology (Bethesda). 2017;32(3):246–60.
109. Zhou H, Kitano K, Ren X, Rajab TK, Wu M, Gilpin SE, Wu T, Baugh L, Black LD, Mathisen DJ, Ott HC. Bioengineering human Lung grafts on porcine matrix. Ann Surg. 2018;267 (3):590–8.
110. Nichols JE, La Francesca S, Niles JA, Vega SP, Argueta LB, Frank L, Christiani DC, Pyles RB, Himes BE, Zhang R, Li S, Sakamoto J, Rhudy J, Hendricks G, Begarani F, Liu X, Patrikeev I, Pal R, Usheva E, Vargas G, Miller A, Woodson L, Wacher A, Grimaldo M, Weaver D, Mlcak R, Cortiella J. Production and transplantation of bioengineered lung into a large-animal model. Sci Transl Med. 2018;10(452)
111. Heron M, Hoyert DL, Murphy SL, Xu J, Kochanek KD, Tejada-Vera B. Deaths: final data for 2006, national vital statistics reports: from the Centers for Disease Control and Prevention, National Center for Health Statistics. Natl Vital Statist Syst. 2009;57(14):1–134.
112. Mikhail S, Cosgrove D, Zeidan A. Hepatocellular carcinoma: systemic therapies and future perspectives. Expert Rev Anticancer Ther. 2014;14(10):1205–18.
113. Bernal W, Wendon J. Acute liver failure. N Engl J Med. 2013;369(26):2525–34.
114. Starzl TE, Marchioro TL, Vonkaulla KN, Hermann G, Waddell WR. Homotransplantation of the liver in humans. Surg Gynecol Obstet. 1963;117:659–76.
115. Wertheim JA, Petrowsky H, Saab S, Kupiec-Weglinski JW, Busuttil RW. Major challenges limiting liver transplantation in the United States. Am J Transplant. 2011;11(9):1773–84.
116. Kim WR, Lake JR, Smith JM, Skeans MA, Schladt DP, Edwards EB, Harper AM, Wainright JL, Snyder JJ, Israni AK, Kasiske BL. Liver. Am J Transplant. 2016;16(S2):69–98.
117. Punch JD, Hayes DH, LaPorte FB, McBride V, Seely MS. Organ donation and utilization in the United States, 1996-2005. Am J Transplant. 2007;7(5 Pt 2):1327–38.
118. Damania A, Kumar A, Teotia AK, Kimura H, Kamihira M, Ijima H, Sarin SK, Kumar A. Decellularized liver matrix-modified Cryogel scaffolds as potential hepatocyte carriers in bioartificial liver support systems and implantable liver constructs. ACS Appl Mater Interfaces. 2018;10(1):114–26.
119. Cesaretti M, Zarzavajian Le Bian A, Moccia S, Iannelli A, Schiavo L, Diaspro A. From deceased to bioengineered graft: new frontiers in liver transplantation. Transplant Rev. 2019;33(2):72–6.
120. Tucker ON, Heaton N. The ‘small for size’ liver syndrome. Curr Opin Crit Care. 2005;11 (2):150–5.
121. Pan J, Yan S, Gao J-j, Wang Y-y, Lu Z-j, Cui C-w, Zhang Y-h, Wang Y, Meng X-q, Zhou L, Xie H-y, Zheng J, Zheng MH, Zheng S-s. In-vivo organ engineering: perfusion of hepatocytes in a single liver lobe scaffold of living rats. Int J Biochem Cell Biol. 2016;80:124–31.
122. Fox IJ, Chowdhury JR, Kaufman SS, Goertzen TC, Chowdhury NR, Warkentin PI, Dorko K, Sauter BV, Strom SC. Treatment of the Crigler–Najjar syndrome type I with hepatocyte transplantation. N Engl J Med. 1998;338(20):1422–7.
123. Choi MS, Catana AM, Wu J, Kim YS, Yoon SJ, Borowsky AD, Gambhir SS, Gupta S, Zern MA. Use of bioluminescent imaging to assay the transplantation of immortalized human fetal hepatocytes into mice. Cell Transplant. 2008;17(8):899–909.
124. Yamamoto N, Wu J, Zhang Y, Catana AM, Cai H, Strom S, Novikoff PM, Zern MA. An optimal culture condition maintains human hepatocyte phenotype after long-term culture. Hepatol Res. 2006;35(3):169–77.
125. Svegliati-Baroni G, De Minicis S, Marzioni M. Hepatic fibrogenesis in response to chronic liver injury: novel insights on the role of cell-to-cell interaction and transition. Liver Int. 2008;28(8):1052–64.
126. Zhou P, Lessa N, Estrada DC, Severson EB, Lingala S, Zern MA, Nolta JA, Wu J. Decellularized liver matrix as a carrier for the transplantation of human fetal and primary hepatocytes in mice. Liver Transpl. 2011;17(4):418–27.

127. Mitra V, Metcalf J. Functional anatomy and blood supply of the liver. Anaesth Intens Care Med. 2012;13(2):52–3.
128. Vernon H, Wehrle CJ, Kasi A. Anatomy, Abdomen and Pelvis, Liver. StatPearls: StatPearls Publishing. Copyright © 2020, StatPearls Publishing LLC., Treasure Island (FL), 2020
129. Trefts E, Gannon M, Wasserman DH. The liver. Curr Biol. 2017;27(21):R1147–51.
130. Sainz-Arnal P, Pla-Palacín I, Sánchez-Romero N, Almeida M, Morini S, Solanas E, Lue A, Serrano-Aulló T, Baptista PM. Chapter 62 - bioengineering of liver tissue. In: Atala A, Lanza R, Mikos AG, Nerem R, editors. Principles of regenerative medicine. 3rd ed. Boston: Academic Press; 2019. p. 1101–13.
131. Badylak SF. The extracellular matrix as a biologic scaffold material. Biomaterials. 2007;28 (25):3587–93.
132. Ko IK, Peng L, Peloso A, Smith CJ, Dhal A, Deegan DB, Zimmerman C, Clouse C, Zhao W, Shupe TD, Soker S, Yoo JJ, Atala A. Bioengineered transplantable porcine livers with re-endothelialized vasculature. Biomaterials. 2015;40(1):72–9.
133. Yang W, Chen Q, Xia R, Zhang Y, Shuai L, Lai J, You X, Jiang Y, Bie P, Zhang L, Zhang H, Bai L. A novel bioscaffold with naturally-occurring extracellular matrix promotes hepatocyte survival and vessel patency in mouse models of heterologous transplantation. Biomaterials. 2018;177:52–66.
134. Hata T, Uemoto S, Fujimoto Y, Murakami T, Tateno C, Yoshizato K, Kobayashi E. Transplantation of engineered chimeric liver with autologous hepatocytes and xenobiotic scaffold. Ann Surg. 2013;257(3):542–7.
135. Mazza G, Rombouts K, Rennie Hall A, Urbani L, Vinh Luong T, Al-Akkad W, Longato L, Brown D, Maghsoudlou P, Dhillon AP, Fuller B, Davidson B, Moore K, Dhar D, De Coppi P, Malago M, Pinzani M. Decellularized human liver as a natural 3D-scaffold for liver bioengineering and transplantation. Sci Rep. 2015;5:13079.
136. Shaheen MF, Joo DJ, Ross JJ, Anderson BD, Chen HS, Huebert RC, Li Y, Amiot B, Young A, Zlochiver V, Nelson E, Mounajjed T, Dietz AB, Michalak G, Steiner BG, Davidow DS, Paradise CR, van Wijnen AJ, Shah VH, Liu M, Nyberg SL. Sustained perfusion of revascularized bioengineered livers heterotopically transplanted into immunosuppressed pigs. Nat Biomed Eng. 2020;4(4):437–45.
137. Hawkins JA, Hillman ND, Lambert LM, Jones J, Di Russo GB, Profaizer T, Fuller TC, Minich LL, Williams RV, Shaddy RE. Immunogenicity of decellularized cryopreserved allografts in pediatric cardiac surgery: comparison with standard cryopreserved allografts. J Thorac Cardiovasc Surg. 2003;126(1):247–52. discussion 252-3
138. Jiang WC, Cheng YH, Yen MH, Chang Y, Yang VW, Lee OK. Cryo-chemical decellularization of the whole liver for mesenchymal stem cells-based functional hepatic tissue engineering. Biomaterials. 2014;35(11):3607–17.
139. Shimoda H, Yagi H, Higashi H, Tajima K, Kuroda K, Abe Y, Kitago M, Shinoda M, Kitagawa Y. Decellularized liver scaffolds promote liver regeneration after partial hepatectomy. Sci Rep. 2019;9(1):12543.
140. Minami T, Ishii T, Yasuchika K, Fukumitsu K, Ogiso S, Miyauchi Y, Kojima H, Kawai T, Yamaoka R, Oshima Y, Kawamoto H, Kotaka M, Yasuda K, Osafune K, Uemoto S. Novel hybrid three-dimensional artificial liver using human induced pluripotent stem cells and a rat decellularized liver scaffold. Regener Therapy. 2019;10:127–33.
141. Hussein KH, Park K-M, Kang K-S, Woo H-M. Heparin-gelatin mixture improves vascular reconstruction efficiency and hepatic function in bioengineered livers. Acta Biomater. 2016;38:82–93.
142. Fedecostante M, Onciu OG, Westphal KGC, Masereeuw R. Towards a bioengineered kidney: recellularization strategies for decellularized native kidney scaffolds. Int J Artif Organs. 2017;40(4):150–8.
143. Thomas B, Wulf S, Bikbov B, Perico N, Cortinovis M, Courville de Vaccaro K, Flaxman A, Peterson H, Delossantos A, Haring D, Mehrotra R, Himmelfarb J, Remuzzi G, Murray C,

Naghavi M. Maintenance dialysis throughout the world in years 1990 and 2010. J Am Soc Nephrol. 2015;26(11):2621–33.

144. Liyanage T, Ninomiya T, Jha V, Neal B, Patrice HM, Okpechi I, Zhao MH, Lv J, Garg AX, Knight J, Rodgers A, Gallagher M, Kotwal S, Cass A, Perkovic V. Worldwide access to treatment for end-stage kidney disease: a systematic review. Lancet. 2015;385 (9981):1975–82.
145. Couser WG, Remuzzi G, Mendis S, Tonelli M. The contribution of chronic kidney disease to the global burden of major noncommunicable diseases. Kidney Int. 2011;80(12):1258–70.
146. Hart A, Smith JM, Skeans MA, Gustafson SK, Stewart DE, Cherikh WS, Wainright JL, Boyle G, Snyder JJ, Kasiske BL, Israni AK. Kidney. Am J Transpl. 2016;16(Suppl 2):11–46.
147. Winkelmayer WC, Weinstein MC, Mittleman MA, Glynn RJ, Pliskin JS. Health economic evaluations: the special case of end-stage renal disease treatment. Med Decis Mak. 2002;22 (5):417–30.
148. Yang F, Lau T, Luo N. Cost-effectiveness of haemodialysis and peritoneal dialysis for patients with end-stage renal disease in Singapore. Nephrology (Carlton). 2016;21(8):669–77.
149. Peloso A, Tamburrini R, Edgar L, Wilm B, Katari R, Perin L, Murray P, Orlando G. Extracellular matrix scaffolds as a platform for kidney regeneration. Eur J Pharmacol. 2016;790:21–7.
150. Uzarski JS, Xia Y, Belmonte JC, Wertheim JA. New strategies in kidney regeneration and tissue engineering. Curr Opin Nephrol Hypertens. 2014;23(4):399–405.
151. Bonandrini B, Figliuzzi M, Papadimou E, Morigi M, Perico N, Casiraghi F, Dipl C, Sangalli F, Conti S, Benigni A, Remuzzi A, Remuzzi G. Recellularization of well-preserved acellular kidney scaffold using embryonic stem cells. Tissue Eng Part A. 2014;20(9–10):1486–98.
152. Ross EA, Williams MJ, Hamazaki T, Terada N, Clapp WL, Adin C, Ellison GW, Jorgensen M, Batich CD. Embryonic stem cells proliferate and differentiate when seeded into kidney scaffolds. J Am Soc Nephrol. 2009;20(11):2338–47.
153. Caralt M, Uzarski JS, Iacob S, Obergfell KP, Berg N, Bijonowski BM, Kiefer KM, Ward HH, Wandinger-Ness A, Miller WM, Zhang ZJ, Abecassis MM, Wertheim JA. Optimization and critical evaluation of decellularization strategies to develop renal extracellular matrix scaffolds as biological templates for organ engineering and transplantation. Am J Transpl. 2015;15 (1):64–75.
154. Burch PT, Kaza AK, Lambert LM, Holubkov R, Shaddy RE, Hawkins JA. Clinical performance of decellularized cryopreserved valved allografts compared with standard allografts in the right ventricular outflow tract. Ann Thorac Surg. 2010;90(4):1301–5. discussion 1306
155. Karczewski M, Malkiewicz T. Scaffolds from surgically removed kidneys as a potential source of organ transplantation. Biomed Res Int. 2015;2015:325029.
156. Eaton DC. Vander's renal physiology. New York: McGraw-Hill Medical; 2009.
157. Mahadevan V. Anatomy of the kidney and ureter. Surgery (Oxford). 2019;37(7):359–64.
158. Eckardt K-U, Coresh J, Devuyst O, Johnson RJ, Köttgen A, Levey AS, Levin A. Evolving importance of kidney disease: from subspecialty to global health burden. Lancet. 2013;382 (9887):158–69.
159. Song JJ, Guyette JP, Gilpin SE, Gonzalez G, Vacanti JP, Ott HC. Regeneration and experimental orthotopic transplantation of a bioengineered kidney. Nat Med. 2013;19(5):646–51.
160. Orlando G, Farney AC, Iskandar SS, Mirmalek-Sani SH, Sullivan DC, Moran E, AbouShwareb T, De Coppi P, Wood KJ, Stratta RJ, Atala A, Yoo JJ, Soker S. Production and implantation of renal extracellular matrix scaffolds from porcine kidneys as a platform for renal bioengineering investigations. Ann Surg. 2012;256(2):363–70.
161. Nakayama KH, Batchelder CA, Lee CI, Tarantal AF. Decellularized rhesus monkey kidney as a three-dimensional scaffold for renal tissue engineering. Tissue Eng Part A. 2010;16 (7):2207–16.
162. Humes HD. Acute renal failure: prevailing challenges and prospects for the future. Kidney Int Suppl. 1995;50:S26–32.
163. Falk G. Maturation of renal function in infant rats. Am J Phys. 1955;181(1):157–70.

164. Remuzzi A, Figliuzzi M, Bonandrini B, Silvani S, Azzollini N, Nossa R, Benigni A, Remuzzi G. Experimental evaluation of kidney regeneration by organ scaffold recellularization. Sci Rep. 2017;7:43502.
165. Destefani AC, Sirtoli GM, Nogueira BV. Advances in the knowledge about kidney Decellularization and repopulation. Front Bioeng Biotechnol. 2017;5:34.
166. Yadav D, Lowenfels AB. The epidemiology of pancreatitis and pancreatic cancer. Gastroenterology. 2013;144(6):1252–61.
167. Zheng C-X, Sui B-D, Hu C-H, Qiu X-Y, Zhao P, Jin Y. Reconstruction of structure and function in tissue engineering of solid organs: toward simulation of natural development based on decellularization. J Tissue Eng Regen Med. 2018;12(6):1432–47.
168. Pineda Molina C, Lee YC, Badylak SF. Chapter 38 - pancreas whole organ engineering. In: Orlando G, Piemonti L, Ricordi C, Stratta RJ, Gruessner RWG, editors. Transplantation, bioengineering, and regeneration of the endocrine pancreas: Academic Press; 2020. p. 527–36.
169. Orlando G, Stratta RJ, Light J. Pancreas transplantation for type 2 diabetes mellitus. Curr Opin Organ Transplant. 2011;16(1):110–5.
170. Mirmalek-Sani S-H, Orlando G, McQuilling JP, Pareta R, Mack DL, Salvatori M, Farney AC, Stratta RJ, Atala A, Opara EC, Soker S. Porcine pancreas extracellular matrix as a platform for endocrine pancreas bioengineering. Biomaterials. 2013;34(22):5488–95.
171. Opara EC, Mirmalek-Sani S-H, Khanna O, Moya ML, Brey EM. Design of a bioartificial pancreas+: BMJ Publishing Group Limited; 2010.
172. Vaithilingam V, Tuch BE. Islet transplantation and encapsulation: an update on recent developments, the review of diabetic studies. RDS. 2011;8(1):51.
173. Song JJ, Ott HC. Organ engineering based on decellularized matrix scaffolds. Trends Mol Med. 2011;17(8):424–32.
174. Orlando G, Baptista P, Birchall M, De Coppi P, Farney A, Guimaraes-Souza NK, Opara E, Rogers J, Seliktar D, Shapira-Schweitzer K, Stratta RJ, Atala A, Wood KJ, Soker S. Regenerative medicine as applied to solid organ transplantation: current status and future challenges. Transpl Int. 2011;24(3):223–32.
175. Lau J, Jansson L, Carlsson PO. Islets transplanted intraportally into the liver are stimulated to insulin and glucagon release exclusively through the hepatic artery. Am J Transpl. 2006;6 (5p1):967–75.
176. de Vos P, Smink AM, Paredes G, Lakey JRT, Kuipers J, Giepmans BNG, de Haan BJ, Faas MM. Enzymes for pancreatic islet isolation impact chemokine-production and polarization of insulin-producing β-cells with reduced functional survival of immunoisolated rat islet-allografts as a consequence. PLoS One. 2016;11(1)
177. Donath MY, Böni-Schnetzler M, Ellingsgaard H, Halban PA, Ehses JA. Cytokine production by islets in health and diabetes: cellular origin, regulation and function. Trends Endocrinol Metabolism. 2010;21(5):261–7.
178. Kanak MA, Takita M, Itoh T, SoRelle JA, Murali S, Kunnathodi F, Shahbazov R, Lawrence MC, Levy MF, Naziruddin B. Alleviation of instant blood-mediated inflammatory reaction in autologous conditions through treatment of human islets with NF-κB inhibitors. Transplantation. 2014;98(5):578.
179. Lammert E, Gu G, McLaughlin M, Brown D, Brekken R, Murtaugh LC, Gerber H-P, Ferrara N, Melton DA. Role of VEGF-A in vascularization of pancreatic islets. Curr Biol. 2003;13(12):1070–4.
180. Smink AM, Hertsig DT, Schwab L, van Apeldoorn AA, de Koning E, Faas MM, de Haan BJ, de Vos P. A retrievable, efficacious polymeric scaffold for subcutaneous transplantation of rat pancreatic islets. Ann Surg. 2017;266(1):149–57.
181. de Haan BJ, Faas MM, Spijker H, van Willigen J-W, De Haan A, de Vos P. Factors influencing isolation of functional pancreatic rat islets. Pancreas. 2004;29(1):e15–22.
182. Banerjee I. Chapter 17 - ECM-based scaffolds for pancreas bioengineering. In: Orlando G, Piemonti L, Ricordi C, Stratta RJ, Gruessner RWG, editors. Transplantation, bioengineering, and regeneration of the endocrine pancreas: Academic Press; 2020. p. 243–55.

183. Ellis H. Anatomy of the pancreas. Surgery (Oxford). 2007;25(2):72–3.
184. Mahadevan V. Anatomy of the pancreas and spleen. Surgery (Oxford). 2019;37(6):297–301.
185. Shelat VG, Kapoor VK. Pancreas; anatomy and development. In: Kuipers EJ, editor. Encyclopedia of gastroenterology. 2nd ed. Oxford: Academic Press; 2020. p. 7–9.
186. Leung PS. Overview of the pancreas. In: Leung PS, editor. The renin-angiotensin system: current research progress in the pancreas: the RAS in the pancreas. Dordrecht: Springer Netherlands; 2010. p. 3–12.
187. Conrad C, Schuetz C, Clippinger B, Vacanti JP, Markmann JF, Ott HC. Bio-engineered endocrine pancreas based on decellularized pancreatic matrix and mesenchymal stem cell/islet cell coculture. J Am Coll Surg. 2010;211(3):S62.
188. Baptista PM, Siddiqui MM, Lozier G, Rodriguez SR, Atala A, Soker S. The use of whole organ decellularization for the generation of a vascularized liver organoid. Hepatology. 2011;53(2):604–17.
189. Guo Y, Wu C, Xu L, Xu Y, Xiaohong L, Hui Z, Jingjing L, Lu Y, Wang Z. Vascularization of pancreatic decellularized scaffold with endothelial progenitor cells. J Artif Organs. 2018;21(2):230–7.
190. Wu D, Wan J, Huang Y, Guo Y, Xu T, Zhu M, Fan X, Zhu S, Ling C, Li X, Lu J, Zhu H, Zhou P, Lu Y, Wang Z. 3D culture of MIN-6 cells on decellularized pancreatic scaffold: in vitro and in vivo study. Biomed Res Int. 2015;2015:–432645.
191. Patil PB, Chougule PB, Kumar VK, Almström S, Bäckdahl H, Banerjee D, Herlenius G, Olausson M, Sumitran-Holgersson S. Recellularization of acellular human small intestine using bone marrow stem cells. Stem Cells Transl Med. 2013;2(4):307–15.
192. Fallon EM, Mitchell PD, Nehra D, Potemkin AK, O'Loughlin AA, Gura KM, Puder M. Neonates with short bowel syndrome: an optimistic future for parenteral nutrition independence. JAMA Surg. 2014;149(7):663–70.
193. Harrison E, Allan P, Ramu A, Vaidya A, Travis S, Lal S. Management of intestinal failure in inflammatory bowel disease: small intestinal transplantation or home parenteral nutrition? World J Gastroenterol: WJG. 2014;20(12):3153.
194. Wang R, Yuan W, Zhao Q, Song P, Yue J, Lin S-D, Zhao T-B. An experimental study of preventing and treating acute radioactive enteritis with human umbilical cord mesenchymal stem cells. Asian Pac J Trop Med. 2013;6(12):968–71.
195. Stollman NH, Neustater BR, Rogers AI. Short-bowel syndrome. Gastroenterologist. 1996;4(2):118–28.
196. Schall KA, Thornton ME, Isani M, Holoyda KA, Hou X, Lien CL, Grubbs BH, Grikscheit TC. Short bowel syndrome results in increased gene expression associated with proliferation, inflammation, bile acid synthesis and immune system activation: RNA sequencing a zebrafish SBS model. BMC Genomics. 2017;18(1):1–13.
197. Ghory MJ, Sheldon CA. Newborn surgical emergencies of the gastrointestinal tract. Surg Clin N Am. 1985;65(5):1083–98.
198. Pereira-Fantini PM, Byars SG, Pitt J, Lapthorne S, Fouhy F, Cotter PD, Bines JE. Unravelling the metabolic impact of SBS-associated microbial dysbiosis: insights from the piglet short bowel syndrome model. Sci Rep. 2017;7:43326.
199. Spurrier RG, Grikscheit TC. Tissue engineering the small intestine. Clin Gastroenterol Hepatol. 2013;11(4):354–8.
200. McMellen ME, Wakeman D, Longshore SW, McDuffie LA, Warner BW. Growth factors: possible roles for clinical management of the short bowel syndrome. Semin Pediatr Surg. 2010;19(1):35–43.
201. Brant SR, Okou DT, Simpson CL, Cutler DJ, Haritunians T, Bradfield JP, Chopra P, Prince J, Begum F, Kumar A, Huang C, Venkateswaran S, Datta LW, Wei Z, Thomas K, Herrinton LJ, Klapproth J-MA, Quiros AJ, Seminerio J, Liu Z, Alexander JS, Baldassano RN, Dudley-Brown S, Cross RK, Dassopoulos T, Denson LA, Dhere TA, Dryden GW, Hanson JS, Hou JK, Hussain SZ, Hyams JS, Isaacs KL, Kader H, Kappelman MD, Katz J, Kellermayer R, Kirschner BS, Kuemmerle JF, Kwon JH, Lazarev M, Li E, Mack D, Mannon P, Moulton DE,

Newberry RD, Osuntokun BO, Patel AS, Saeed SA, Targan SR, Valentine JF, Wang M-H, Zonca M, Rioux JD, Duerr RH, Silverberg MS, Cho JH, Hakonarson H, Zwick ME, McGovern DPB, Kugathasan S. Genome-wide association study identifies African-specific susceptibility Loci in African Americans with inflammatory bowel disease. Gastroenterology, 217.e2. 2017;152(1):206.
202. Chouliaras G, Margoni D, Dimakou K, Fessatou S, Roma-Giannikou E. Disease impact on the quality of life of children with inflammatory bowel disease. World J Gastroenterol. 2017;23 (6):1067.
203. Jeppesen PB. Pharmacologic options for intestinal rehabilitation in patients with short bowel syndrome. JPEN J Parenteral Enteral Nutr. 2014;38(1 Suppl):45S.
204. Tetsuya N, Toshiro S. Advancing intestinal organoid technology toward regenerative medicine: Keio University; 2017.
205. H. National Institutes of, U.S.D.o. Health, S. Human. Opportunities and challenges in digestive diseases research: recommendations of the national commission on digestive diseases. Bethesda, MD: National Institutes of Health; 2009.
206. Kanetkar NS, Ekenseair AK. 57 - scaffolds for intestinal tissue engineering. In: Mozafari M, Sefat F, Atala A, editors. Handbook of tissue engineering scaffolds, vol. 2: Woodhead Publishing; 2019. p. 593–632.
207. Mahadevan V. Anatomy of the small intestine. Surgery (Oxford). 2020;
208. Mowat AM, Agace WW. Regional specialization within the intestinal immune system. Nat Rev Immunol. 2014;14(10):667–85.
209. Ma ZF, Lee YY. Chapter 7 - small intestine anatomy and physiology. In: Rao SSC, Lee YY, Ghoshal UC, editors. Clinical and basic neurogastroenterology and motility: Academic Press; 2020. p. 101–11.
210. Coffey JC, O'Leary DP. The mesentery: structure, function, and role in disease. Lancet Gastroenterol Hepatol. 2016;1(3):238–47.
211. Höllwarth ME. Short bowel syndrome: pathophysiological and clinical aspects. Pathophysiology. 1999;6(1):1–19.
212. Campbell I. The mouth, stomach and intestines. Anaesth Intensive Care Med. 2012;13 (2):56–8.
213. Watson AJM, Duckworth CA, Guan Y, Montrose MH. Mechanisms of epithelial cell shedding in the mammalian intestine and maintenance of barrier function. Ann N Y Acad Sci. 2009;1165(1):135.
214. Chen M-C, Sonaje K, Chen K-J, Sung H-W. A review of the prospects for polymeric nanoparticle platforms in oral insulin delivery. Biomaterials. 2011;32(36):9826–38.
215. DeSesso JM, Jacobson CF. Anatomical and physiological parameters affecting gastrointestinal absorption in humans and rats. Food Chem Toxicol. 2001;39(3):209–28.
216. Campbell I. Digestion and absorption. Anaesth Intensive Care Med. 2012;13(2):62–3.
217. Cornick S, Tawiah A, Chadee K. Roles and regulation of the mucus barrier in the gut. Tissue Barriers. 2015;3(1–2):e982426.
218. Maqbool A. Colon: structure, function, and disorders. In: Caballero B, editor. Encyclopedia of human nutrition. 3rd ed. Waltham: Academic Press; 2013. p. 378–96.
219. Mahadevan V. Anatomy of the caecum, appendix and colon. Surgery (Oxford). 2020;38 (1):1–6.
220. Arbizu R, Nurko S. Colon: structure and function. In: Caballero B, Finglas PM, Toldrá F, editors. Encyclopedia of food and health. Oxford: Academic Press; 2016. p. 259–64.
221. Chen MK, Badylak SF. Small bowel tissue engineering using small intestinal submucosa as a scaffold. J Surg Res. 2001;99(2):352–8.
222. Fishbein TM. Intestinal transplantation. N Engl J Med. 2009;361(10):998–1008.
223. Araki R, Uda M, Hoki Y, Sunayama M, Nakamura M, Ando S, Sugiura M, Ideno H, Shimada A, Nifuji A, Abe M. Negligible immunogenicity of terminally differentiated cells derived from induced pluripotent or embryonic stem cells. Nature. 2013;494(7435):100–4.

224. Spence JR, Mayhew CN, Rankin SA, Kuhar MF, Vallance JE, Tolle K, Hoskins EE, Kalinichenko VV, Wells SI, Zorn AM, Shroyer NF, Wells JM. Directed differentiation of human pluripotent stem cells into intestinal tissue in vitro. Nature. 2011;470(7332):105–9.
225. Kitano K, Schwartz DM, Zhou H, Gilpin SE, Wojtkiewicz GR, Ren X, Sommer CA, Capilla AV, Mathisen DJ, Goldstein AM, Mostoslavsky G, Ott HC. Bioengineering of functional human induced pluripotent stem cell-derived intestinal grafts. Nat Commun. 2017;8(1):765.
226. Martins AR, Matias GSS, Batista VF, Miglino MA, Fratini P. Wistar rat dermis recellularization. Res Vet Sci. 2020;131:222–31.
227. Nunez-Villaveiran T, Feasel P, Keenan S, Aliotta R, Bosler D, Stearns D, Bergfeld W, Gurunluoglu R. Donor skin allograft survival after bone marrow transplantation: case report and systematic review of the literature☆. J Plast Reconstr Aesthet Surg. 2019;72(1):23–34.
228. Voigt CD, Williamson S, Kagan RJ, Branski LK. 14 - The skin Bank. In: Herndon DN, editor. Total burn care. 5th ed: Elsevier; 2018. p. 158–166.e2.
229. Bergmanson JPG. 3 - anatomy and physiology of the cornea and related structures. In: Phillips AJ, Speedwell L, editors. Contact lenses. 6th ed. London: Elsevier; 2019. p. 33–64.
230. Pascolini D, Mariotti SP. Global estimates of visual impairment: 2010. Br J Ophthalmol. 2012;96(5):614–8.
231. da Mata Martins TM, da Silva Cunha P, Rodrigues MA, de Carvalho JL, de Souza JE, de Carvalho Oliveira JA, Gomes DA, de Goes AM. Epithelial basement membrane of human decellularized cornea as a suitable substrate for differentiation of embryonic stem cells into corneal epithelial-like cells. Mater Sci Eng C. 2020;116:111215.
232. Tan DTH, Dart JKG, Holland EJ, Kinoshita S. Corneal transplantation. Lancet. 2012;379 (9827):1749–61.
233. Ramasamy B, Tey A, Kaye SB. Graft rejection and graft failure after anterior lamellar versus penetrating keratoplasty. Am J Ophthalmol. 2011;152(6):–0-1082.
234. Fernández-Pérez J, Ahearne M. Decellularization and recellularization of cornea: Progress towards a donor alternative. Methods. 2020;171:86–96.
235. Arnalich-Montiel F, del Barrio JLA, Alió JL. Corneal surgery in keratoconus: which type, which technique, which outcomes? Eye Vision. 2016;3(1):2.
236. Alio del Barrio JL, Chiesa M, Garagorri N, Garcia-Urquia N, Fernandez-Delgado J, Bataille L, Rodriguez A, Arnalich-Montiel F, Zarnowski T, Álvarez de Toledo JP, Alio JL, De Miguel MP. Acellular human corneal matrix sheets seeded with human adipose-derived mesenchymal stem cells integrate functionally in an experimental animal model. Exp Eye Res. 2015;132:91–100.
237. Alió del Barrio JL, El Zarif M, Azaar A, Makdissy N, Khalil C, Harb W, El Achkar I, Jawad ZA, de Miguel MP, Alió JL. Corneal stroma enhancement with decellularized stromal laminas with or without stem cell recellularization for advanced keratoconus. Am J Ophthalmol. 2018;186:47–58.
238. Liu Y, Yang R, He Z, Gao W-Q. Generation of functional organs from stem cells. Cell Regenerat. 2013;2(1):2:1.
239. Folkman J, Hochberg M. Self-regulation of growth in three dimensions. J Exp Med. 1973;138 (4):745–753
240. Garreta E, Oria R, Tarantino C, Pla-Roca M, Prado P, Fernández-Avilés F, Campistol JM, Samitier J, Montserrat N. Tissue engineering by decellularization and 3D bioprinting. Mater Today. 2017;20(4):166–78.
241. Wang S, Goecke T, Meixner C, Haverich A, Hilfiker A, Wolkers WF. Freeze-dried heart valve scaffolds, tissue engineering. Part C Methods. 2012;18(7):517–25.
242. Syedain ZH, Bradee AR, Kren S, Taylor DA, Tranquillo RT. Decellularized tissue-engineered heart valve leaflets with Recellularization potential. Tissue Eng Part A. 2013;19(5–6):759–69.
243. Gottlieb D, Kunal T, Emani S, Aikawa E, Brown DW, Powell AJ, Nedder A, Engelmayr GC, Melero-Martin JM, Sacks MS, Mayer JE. In vivo monitoring of function of autologous engineered pulmonary valve. J Thorac Cardiovasc Surg. 2010;139(3):723–31.

244. Flanagan TC, Jrg SS, Frese J, Heike S, Jockenhoevel S. In vivo remodeling and structural characterization of fibrin-based tissue-engineered heart valves in the adult sheep model. Tissue Eng Part A. 2009;15(10):2965–76.
245. Borschel GH, Huang YC, Calve S, Arruda EM, Lynch JB, Dow DE, Kuzon WM, Dennis RG, Brown DL. Tissue engineering of recellularized small-diameter vascular grafts. Tissue Eng. 2005;11(6):778–86.
246. Bianconi E, Piovesan A, Facchin F, Beraudi A, Casadei R, Frabetti F, Vitale L, Pelleri MC, Tassani S, Piva F. An estimation of the number of cells in the human body. Ann Hum Biol. 2013;40(6):463–71.
247. Rowart P, Erpicum P, Detry O, Weekers L, Grégoire C, Lechanteur C, Briquet A, Beguin Y, Krzesinski J-M, Jouret F. Mesenchymal stromal cell therapy in ischemia/reperfusion injury. J Immunol Res. 2015;2015:602597.
248. Pai M, Zacharoulis D, Milicevic MN, Helmy S, Jiao LR, Levicar NA, Tait P, Scott M, Marley SB, Jestice K. Autologous infusion of expanded mobilized adult bone marrow-derived CD34+ cells into patients with alcoholic liver cirrhosis. Am J Gastroenterol. 2008;103(8):1952–8.
249. Roy-Chowdhury N, Wang X, Guha C, Roy-Chowdhury J. Hepatocyte-like cells derived from induced pluripotent stem cells. Hepatol Int. 2017;11(1):54–69.
250. Zhou X, Cui L, Zhou X, Yang Q, Wang L, Guo G, Hou Y, Cai W, Han Z, Shi Y. Induction of hepatocyte-like cells from human umbilical cord-derived mesenchymal stem cells by defined microRNAs. J Cell Mol Med. 2017;21(5):881–93.
251. Meyer U, Wiesmann H-P. Tissue engineering: a challenge of today's medicine. Head Face Med. 2005;1(1):1–2.
252. Franks TJ, Colby TV, Travis WD, Tuder RM, Mary CW. Resident cellular components of the human Lung: current knowledge and goals for research on cell phenotyping and function. Proc Am Thorac Soc. 2008;5(7):763–6.
253. Burgess JK, Mauad T, Tjin G, Karlsson JC, Westergren-Thorsson G. The extracellular matrix - the under-recognized element in lung disease? J Pathol. 2016;240(4):397–409.
254. Bellezzia MA, Cruz FF, Martins V, de Castro LL, Lopes-Pacheco M, Vilanova EP, Mourão PA, Rocco PRM, Silva PL. Impact of different intratracheal flows during lung decellularization on extracellular matrix composition and mechanics. Regen Med. 2018;13 (05):519–30.
255. Panoskaltsis M. Angela, bioreactor development for Lung tissue engineering. Curr Transplan Rep. 2015;2(1):90–7.
256. Uriarte JJ, Nonaka PN, Campillo N, Palma RK, Melo E, De Oliveira LVF, Navajas D, Farré R. Mechanical properties of acellular mouse lungs after sterilization by gamma irradiation. J Mech Behav Biomed Mater. 2014;40:168–77.
257. Kajbafzadeh A-M, Javan-Farazmand N, Monajemzadeh M, Baghayee A. Determining the optimal decellularization and sterilization protocol for preparing a tissue scaffold of a human-sized liver tissue. Tissue Eng Part C Methods. 2013;19(8):642–51.
258. Bonenfant NR, Sokocevic D, Wagner DE, Borg ZD, Lathrop MJ, Lam YW, Deng B, DeSarno MJ, Ashikaga T, Loi R. The effects of storage and sterilization on de-cellularized and re-cellularized whole lung. Biomaterials. 2013;34(13):3231–45.
259. Gilpin SE, Ott HC. Using nature's platform to engineer bio-artificial lungs. Ann Am Thorac Soc. 2015;12(Suppl 1):S45–9.
260. Figliuzzi M, Remuzzi G, Remuzzi A. Chapter 63 - Recellularization of kidney scaffold with stem cells. In: Orlando G, Remuzzi G, Williams DF, editors. Kidney transplantation, bioengineering and regeneration: Academic Press; 2017. p. 877–86.
261. Figliuzzi M, Bonandrini B, Remuzzi A. Decellularized kidney matrix as functional material for whole organ tissue engineering. J Appl Biomater Funct Mater. 2017;15(4):e326–33.
262. Lin Y-Q, Wang L-R, Pan L-L, Wang H, Zhu G-Q, Liu W-Y, Wang J-T, Braddock M, Zheng M-H. Kidney bioengineering in regenerative medicine: an emerging therapy for kidney disease. Cytotherapy. 2016;18(2):186–97.

263. Takasato M, Er PX, Becroft M, Vanslambrouck JM, Stanley EG, Elefanty AG, Little MH. Directing human embryonic stem cell differentiation towards a renal lineage generates a self-organizing kidney. Nat Cell Biol. 2013;16(1):118–26.
264. Katsuki Y, Yagi H, Okitsu T, Kitago M, Tajima K, Kadota Y, Hibi T, Abe Y, Shinoda M, Itano O, Takeuchi S, Kitagawa Y. Endocrine pancreas engineered using porcine islets and partial pancreatic scaffolds. Pancreatology. 2016;16(5):922–30.
265. Naziruddin B, Iwahashi S, Kanak MA, Takita M, Itoh T, Levy MF. Evidence for instant blood-mediated inflammatory reaction in clinical autologous islet transplantation. Am J Transplant. 2014;14(2):428–37.
266. Ramnath RD, Maillard E, Jones K, Bateman PA, Hughes SSJ, Gralla J, Johnson PR, Gray DWR. In vitro assessment of human islet vulnerability to instant blood-mediated inflammatory reaction (IBMIR) and its use to demonstrate a beneficial effect of tissue culture. Cell Transplant. 2015;24(12):2505–12.
267. De Carlo E, Baiguera S, Conconi MT, Vigolo S, Grandi C, Lora S, Martini C, Maffei P, Tamagno G, Vettor R. Pancreatic acellular matrix supports islet survival and function in a synthetic tubular device: in vitro and in vivo studies. Int J Mol Med. 2010;25(2):195–202.
268. Kuna VK, Kvarnström N, Elebring E, Holgersson SS. Isolation and decellularization of a whole porcine pancreas. J Vis Exp. 2018;140:58302.
269. Xu T, Zhu M, Guo Y, Wu D, Huang Y, Fan X, Zhu S, Lin C, Li X, Lu J. Three-dimensional culture of mouse pancreatic islet on a liver-derived perfusion-decellularized bioscaffold for potential clinical application. J Biomater Appl. 2015;30(4):379–87.
270. Shirakigawa N, Takei T, Ijima H. Base structure consisting of an endothelialized vascular-tree network and hepatocytes for whole liver engineering. J Biosci Bioeng. 2013;116(6):740–5.
271. Willenberg BJ, Oca-Cossio J, Cai Y, Brown AR, Clapp WL, Abrahamson DR, Terada N, Ellison GW, Mathews CE, Batich CD. Repurposed biological scaffolds: kidney to pancreas. Organogenesis. 2015;11(2):47–57.
272. Orlando G, Gianello P, Salvatori M, Stratta RJ, Soker S, Ricordi C, Domínguez-Bendala J. Cell replacement strategies aimed at reconstitution of the β-cell compartment in type 1 diabetes. Diabetes. 2014;63(5):1433–44.
273. Vernon RB, Preisinger A, Gooden MD, D'Amico LA, Yue BB, Bollyky PL, Kuhr CS, Hefty TR, Nepom GT, Gebe JA. Reversal of diabetes in mice with a bioengineered islet implant incorporating a type I collagen hydrogel and sustained release of vascular endothelial growth factor. Cell Transplant. 2012;21(10):2099–110.
274. Lim F, Sun AM. Microencapsulated islets as bioartificial endocrine pancreas. Science. 1980;210(4472):908–10.
275. Scharp DW, Marchetti P. Encapsulated islets for diabetes therapy: history, current progress, and critical issues requiring solution. Adv Drug Deliv Rev. 2014;67–68:35–73.
276. Yagi H, Soto-Gutierrez A, Kitagawa Y. Whole-organ re-engineering: a regenerative medicine approach to digestive organ replacement. Surg Today. 2013;43(6):587–94.

Chapter 8
The Challenges and Development Directions of Decellularized Materials

Jie Liao, Lincui Da, Bo Xu, Huiqi Xie, and Xiaoming Li

Abstract Expanded applications of decellularized materials (DMs) in all kinds of tissue repair and organ replacement have been discussed in the former chapters. DMs have exhibited outstanding superiority in tissue engineering and might be hopeful to achieve bigger success after further studies. In this chapter, some challenges and perspectives of DMs were proposed in detail including optimizing the preparation methods of DMs, furthering researching the composition of DMs, improving the mechanical properties of DMs, controlling the degradation rate of DMs, furthering promoting the bioactivity of DMs, and promoting clinical applications.

Keywords Decellularized materials · Challenges · Perspectives

8.1 Introduction

As discussed in Chaps. 2 and 5, decellularized materials (DMs) including decellularized tissues and decellularized organs were prepared by removing cells and other antigenic components from tissues and organs through physical, chemical, biochemical, and other methods. After decellularization, antigen components and

J. Liao · B. Xu · X. Li (✉)
Key Laboratory for Biomechanics and Mechanobiology of Ministry of Education, Beijing Advanced Innovation Center for Biomedical Engineering, School of Biological Science and Medical Engineering, Beihang University, Beijing, China
e-mail: lixm@buaa.edu.cn; x.m.li@hotmail.com

L. Da
Reproductive Center, Fujian Provincial Maternity and Children's Hospital, Affiliated Hospital of Fujian Medical University, Fuzhou, China

H. Xie (✉)
Laboratory of Stem Cell and Tissue Engineering, Orthopedic Research Institute, State Key Laboratory of Biotherapy and Cancer Center, West China Hospital, Sichuan University and Collaborative Innovation Center of Biotherapy, Chengdu, China
e-mail: xiehuiqi@scu.edu.cn

X. Li, H. Xie (eds.), *Decellularized Materials*,
https://doi.org/10.1007/978-981-33-6962-7_8

cells are removed as clean as possible. At the same time, active ingredients and non-antigenic ingredients including polysaccharides, collagen, glycoproteins, fibronectin, etc., are retained to maintain the shape, the structure, and the activity of natural tissues and organs. As the native extracellular matrix (ECM), DMs showed unparalleled performances in tissue repair and organ replacement as they possess the native ingredients and structures, providing not only mechanical support for embedded cells but also mechanical stimulations to regulate various cellular behaviors, which are hardly able to be imitated by synthetic materials. Until now, DMs have been an important research hotspot in the field of tissue repair and organ replacement. And as discussed in Chaps. 3 and 4, DMs derived from different tissues and DMs processed into different traits have been applied in all kinds of tissue repair.

Compared with the traditional tissue engineering scaffold, DMs have excellent activity as a result of abundant bioactive factors and macro and micro structures similar to natural tissues and organs. In addition, it also showed excellent biodegradability. Many studies have shown that the release of certain molecules in the extracellular matrix may cause complete biodegradation. More importantly, the removal of antigens and other cell components eliminates the risk of adverse reactions, such as inflammation and immune rejection at the transplant site, thus reducing immunogenicity, which limits the application of allograft. With extensive researches on DMs, it was found that the clinical effect of this material in some application fields was close to that of autologous transplantation, however, it had great potential to overcome the disadvantages of autografts such as secondary injury and the limited tissue or organ source. Specially, superiority is reflected in decellularized organs. Due to the complex structure and function, organ repair could only be achieved by whole-organ transplantation. And, the natively derived whole-organ ECM scaffolds or acellular organs play crucial role in organ replacement as they get as close as possible to the original organs compared to artificial organs. As discussed in Chaps. 6 and 7, acellularized organs after recellularization have been tried in organ reconstruction and have shown many outstanding advantages.

However, there are still some aspects needing improvement. For example, it is still a challenging problem to achieve a coordination of precise and thorough decellularization, as well as undamaged retention of non-antigen components and active ingredients owing to the limited researches on the mechanism of decellularization and decellularization techniques. This directly determines the host response after the implantation of DMs in vivo and limits the material types of DMs. For example, many organs are difficult to maintain their complex structures after decellularization. Simultaneously, improper acellular treatment can also affect the activity and mechanical properties of the decellularized materials. Besides, the effect of organ reconstruction depends on the recellularization effectiveness. It is necessary to pay important attention to the recellularization of decellularized materials. In addition, further improving the performance of decellularized materials including mechanical property, degradation property, biocompatibility, and bioactivity also needs numerous researches.

All in all, DMs have exhibited outstanding superiority in tissue engineering applications and are hopeful to achieve bigger success after further studies. In this chapter, some challenges and perspectives of DMs will be proposed in detail, which include optimizing the decellularization methods, furthering characterizing and researching the composition of DMs, improving the mechanical properties of DMs, furthering promoting the bioactivity of DMs, controlling the degradation rate of DMs, and promoting clinical application.

8.2 Optimizing the Decellularization Methods

Since Peer proposed the idea of decellularization and removed cells through biological method in 1955, it has been more than half a century devoted to developing decellularization approaches. Even though the unnecessary damages on the composition and structure of tissues or organs have been decreased, and the antigenicity of the acellular matrix has been reduced to some extent with the development of decellularization methods, the decellularization methods and techniques still need further optimization to achieve more precise control of components. There are mainly two directions for decellularization optimization, which will be discussed as follows.

On the one hand, achieving a coordination between undamaged retention of active ingredients as well as non-antigen components, and precise and thorough decellularization is a key direction, which would limit the types of DMs and determine the host response, degraded and mechanical properties after their implantation in vivo. For example, physical methods were reported of great possibility to destroy the structure of ECM, which would affect the mechanical properties to some extent [1, 2]. Specifically, the widely used physical method—repeated freeze and thaw cycles—was reported to easily inactivate some growth factors and damage the proteins and ultrastructure contained in ECM [3]. In addition, some physical methods are only suitable for certain specific tissues. For example, electroporation method is only suitable for small and thin tissues or organs. And, chemical methods were demonstrated to frequently reduce some growth factors in tissue and organ, decreasing the biocompatibility and the bioactivity of DMs in most instances and increasing the rate of unwanted immune response because of residual reagents [4–7]. Specifically, triton X and sodium dodecyl sulfate (SDS) had been shown to easily remove GAG, among which SDS was also demonstrated to usually damage collagen and growth factors [8]. And, hypertonic and hypotonic solutions were demonstrated to be difficult in achieving effective removal of cellular residues [9]. Simultaneously, some biological decellularization methods were reported to damage the microstructure of protein, deplete collagen, laminin, fibronectin, GAG, and elastin [2, 9, 10]. In addition, it could cause immune response because of residual specific enzyme. In addition, the appropriate time of enzyme treatment is not easy to hold. For example, nucleases are used to deprotect nucleotides after cell lysis, but it has strong damage to cells, so the treatment time should be short [11]. However, cells should be given sufficient trypsin contact time, when the nuclei are removed from dense tissues [12]. Therefore, it is necessary to maintain effective components and tissue structures by optimizing decellularization methods.

On the other hand, precisely remaining and removing the component such as some bio-substances according to the requirements of target organizations would be a promising direction, which would enable DMs wider applications. First of all, bioactive cues required are tissue-specific, as a result, finding out whether the bioactive cues contained in DMs are required or not for the target tissue repair is quite important. Some bioactive cues that might put bad effect on tissue repair must be removed clearly. For example, it is necessary to remove those components promoting mineralization and calcification when repairing soft tissues. Consequently, precisely remain and removal of components would contribute to overcoming the problem that each type of DMs could be only applied in limited tissue repair. Simultaneously, the composition of DMs and the proportion of each component would directly put effect on the post-processing performance of DMs including the gelatinizable ability, the rheology as hydrogels or bio inks, the homogeneity and the agglomeration tendency as powders, the capabilities to achieve the surface modification, etc. Hence, precise control of the composition through advanced methods might contribute to an expanded range of applications. Furthermore, basing on precisely removing the components and appropriately post-processing, personalized design according to different sexes and ages could become reality. In that way, it would overcome the problem bringing by individual differentiation, an example of which is that transplanting the FDA-proved decellularized heart valve (e.g., S Matrix P [13] and YNERGRAFT [14]) to patients of young children caused risk in rupture, stenosis, calcific deposits, and immune rejection problem.

The effectiveness of decellularization depends on various of factors such as the decellularization methods and the type of tissues (the density and thickness). Achieving precisely control of composition is quite difficult because of the complexity of natural compositions and the limitation of technologies. Hence, the endeavor was mainly focused on the first aspect, that is to say, reducing unnecessary component damage while removing cells. Up to date, comparing different methods to finding out the most appropriate one, further promoting the existed methods, and developing fire-new methods were the main three kinds of solutions.

Firstly, because of the multiformity and complexity of different tissues and organs caused by the individual differences and intra-individual differences, it needs finding out the most befitting one through the comparison of different methods [15]. For instance, comparing to sodium dodecyl sulfate (SDS), sodium hydroxide (NaOH) has been demonstrated as a more appropriate reagent for decellularizing ovarian tissues because of its better performance in promoting follicular remodeling [16]. Likewise, 20 mM ammonia/Triton X-100 (AT) exhibited better decellularization effect of substance formed by the 14-days culture of Human Wharton's Jelly-derived MSCs in ascorbic acid supplemented medium compared to 1% SDS/Triton X-100 (ST) because of its better retention of matrix (about 30%) [17]. In addition, systematically researching the influence of some factors on the acellular effect or even establishing a comparatively complete database might be tremendously beneficial for DMs decellularization. For example, Xia et al.

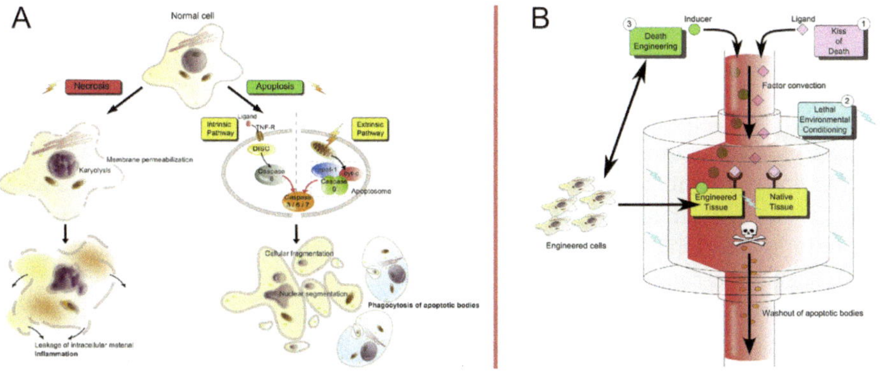

Fig. 8.1 The new method of inducing apoptosis for decellularization. (**a**) Processes of cell death. Cell death occurs through either necrosis or apoptosis. During the apoptotic process, the cellular content is degraded and strictly kept within the cell membrane through either the intrinsic or extrinsic pathway, avoiding an inflammatory reaction. Abbreviations: TNFR = tumor necrosis factor receptor; apaf-1 = apoptotic protease activating factor 1; cyt-c = cytochrome-c. (**b**) Tissue decellularization by apoptosis induction within 3D perfusion bioreactor. Three different approaches were proposed to decellularize a tissue by apoptosis induction: (1) Kiss-of-death, (2) Lethal-environmental-conditioning, (3) Death-engineering [20]

comprehensively compared the differences of cartilage decellularization ratio, mechanical property, and immunogenicity between different decellularization protocols, which would provide researchers a good reference of cartilage decellularization [18].

Secondly, many endeavors have been paid in further promoting the existed methods. For instance, the key factors of perfusate concentration and perfusion time significantly influence the eventual acellular effects and might be tightly regulated to achieve better effect. Specially, Schmitt et al. regulated the concentration of SDS and perfusion time to obtain the most appropriate protocol that could completely decellularize organs without detectable cell toxicity and obvious damage on inherent vascular network in the shortest time and gentlest way [19].

Thirdly, developing new methods is the most hopeful way to optimizing acellular methods. Currently, inducing cells apoptosis to decellularize tissues or organs is an effectively novel method. The named apoptosis is a genetically determined process of cells suicide, which could be also named as "programmed cell death." There are two kinds of pathways with different initial activation mechanisms in inducing apoptosis for decellularization, which include intrinsic and extrinsic ones (Fig. 8.1a). Among them, intrinsic pathway, also referred to as the "mitochondrial pathway," could be triggered through stress stimuli and be initiated by the permeabilization of the mitochondrial outer membrane, leading to the release of cytochrome-c (cyt-c) and subsequently activating the effectors caspase-9, and 3, 6, 7 [21, 22]. The extrinsic pathway is initiated by external signals that bind to specific cell surface death receptors of the tumor necrosis factor family. In this pathway, death inducing signaling complex (DISC) would format to result in the transduction of signal within the cytosol. And, the effectors caspase 3, 6, and 7 would be activated after caspase-8 dimerization [23]. Different from necrosis, through which the cell membrane would rapidly become permeable to form intra-cellular content leakage with big possibility of inducing serious inflammation, such immunogenic components would be wrapped by cytomembrane during apoptosis, preventing the leakage of cellular components into ECM [24, 25]. Specifically, during apoptosis process, cells could firstly change as "apoptotic bodies" containing the cellular content with decreased size (0.5-2 μm) to prevent the contact with ECM. As a result, antigenic component would not be leaked into ECM, and the space between the membrane and the ECM caused by the decreased "apoptotic bodies" provides the possibility to remove cells with few damage on ECM. Some gentle solutions (e.g., hypertonic phosphate buffered saline), not those solutions causing big damage on the tissue composition and structure, could be applied to effectively clean the dead cells. Besides, the release of Prostaglandin E_2 (PGE_2) and the activation of effector caspases during apoptosis were reported to promote tissue regeneration through stimulating proliferation of progenitor cells, which enables DMs prepared through apoptosis release key paracrine signals to prime regeneration programs [26].

Applying apoptosis for decellularization was first proposed by Bourgine et al. [20] in 2013. As he mentioned, there were mainly three strategies to induce apoptosis according to different types of pathway to be activated, including kiss-of-death, lethal-environmental-conditioning, and death-engineering (Fig. 8.1b). Among them, kiss-of-death depends on the extrinsic pathway and activates through the delivery of specific ligands binding their corresponding death receptors of the TNF superfamily. Because death receptors are expressed in hardly all tissues, kiss-of-death could be considered as an almighty way for all kinds of tissues and organs. However, it must be carefully considered to choose appropriate ligand and the type of death receptor according to the specific cellular system. Lethal-environmental-conditioning induces apoptosis through the intrinsic "mitochondrial" path way under environmental stress including pH, temperature, as well as H_2O_2, nitric oxide (NO), and CO_2/O_2 content. The minimal cost and simplicity of the setup for controlling and modulating the operating conditions are the advantages of this way. However, the efficiency of decellularization depends on an extensive multi-parametric screening of optimal conditions. Finally, death-engineering, using genetic approach, depends on the activation of any of the two apoptotic pathways. These three strategies are different in protocol, however, are compatible. After devitalization, it needs to remove the cellular fragments. And, perfusion bioreactor system could be a good choice because it establishes controlled environmental conditions (such as gas content or temperature), facilitates the convection of the apoptotic inducer, and increases the washout of apoptotic bodies. Especially, it could be a streamlined process to generate DMs and induce apoptosis in a closed, single device (Fig. 8.1b).

Currently, applying apoptosis for decellularization has been demonstrated successful. For instance, cell-derived ECM no matter as coating for 3D scaffold or as 2D slice was reported to be successfully decellularized through apoptosis and the subsequent perfusion, through which the collagen would be remained with few damage compared to those through traditional freeze/thaw protocol (with about

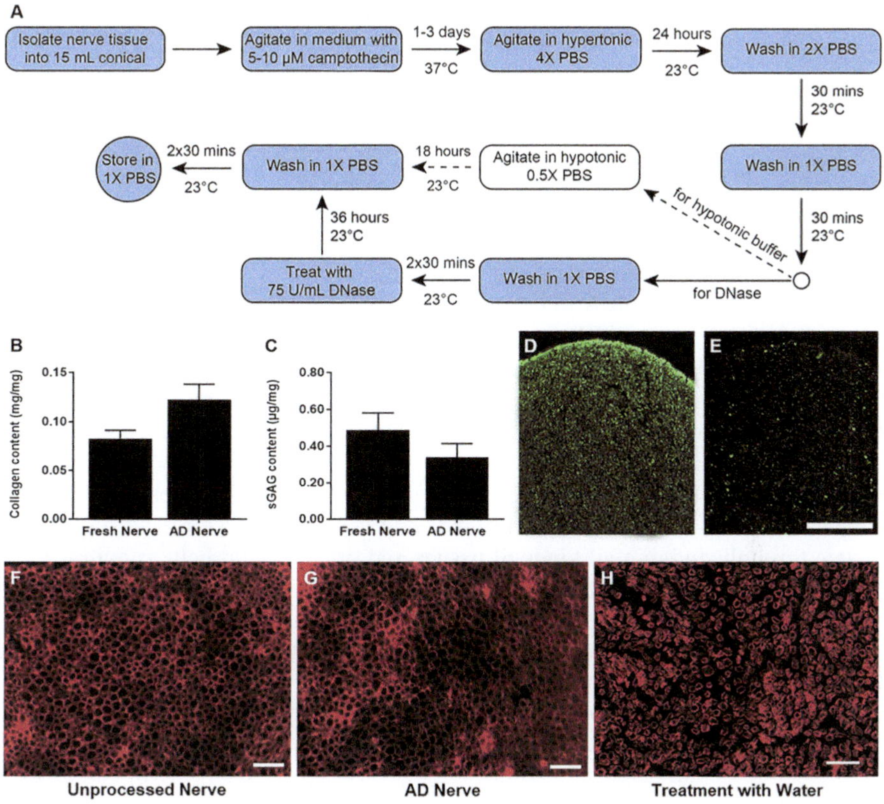

Fig. 8.2 One example of apoptosis-assisted decellularization method using camptothecin. (**a**) Flow diagram of the generalized process for apoptosis-assisted decellularization. Quantification of collagen content (**b**, $n = 4$) and sulfated GAGs (**c**, $n = 5$) before and after apoptosis-based decellularization, showing no significant difference. Immunoreactivity for myelin basic protein of unprocessed nerve (**d**) and apoptosis decellularized nerve (**e**). Fluorescence micrographs of laminin staining in fresh nerve (**f**), apoptosis decellularized nerve (**g**), and fresh nerve processed with water for cell lysis (**h**). Scale bars represent 50 μm. Apoptosis decellularization preserved desirable structure and components but not myelin proteins [27]

74% loss of collagen), showing the effectiveness of apoptosis method [20]. Besides, the research of Cornelison et al. also proved such effectiveness. They used camptothecin for apoptosis and used hypertonic phosphate buffered saline to clean dead cells. The acellular nerve tissue exhibited low immunogenicity and remained comparatively intact basal lamina micro-architecture [27] (Fig. 8.2). Even though such practice is still limited to some extent, it could be predictable that applying this strategy to decellularize tissues or organs might be hopeful.

8.3 Furthering Characterizing and Researching the Composition of DMs

As the bio-substances needed for the regeneration of unique tissues are tissue specific, it is necessary to comprehensively characterize the composition of DMs and deeply study the corresponding cellular responses caused by all kinds of components. For example, it is laminin, not collagen I, collagen II, or fibronectin contained in the decellularized bone matrix that was demonstrated as the key factor for proper epithelial barrier formation in lung tissue (with alveolar-capillary barrier) [28]. Currently, Bi et al. investigated the proteomics of decellularized rat pancreatic extracellular matrix (dpECM) hydrogel by applying advanced bioinformatics analysis, and identified that type V collagen (Col V), as a critical regulator for islet organogenesis from human pluripotent stem cells, was constantly and abundantly present in dpECM hydrogel [29]. Their conclusion could help the application of dpECM hydrogel and the repair of islet.

Therefore, advanced techniques with higher precision in characterizing the components of DMs should be further designed and developed. To establish a reference system through studying all kinds of tissues and organs could help to optimize the decellularization methods based on the characteristics of tissue thickness, lipid content, and original cell density, etc. In addition, precisely removing or remaining the key component and appropriately modifying DMs based on the requirements of expected cellular responses would contribute to establishing a series of quantitative standards and achieving bigger volume production.

8.4 Improving the Mechanical Properties of DMs

Decellularization techniques use harsh detergents to remove cellular fragments in most cases, drastically changing the ratio of remained ECM components. Most matrix proteins such as collagens, glycosaminoglycans (GAGs), laminin, elastin, and fibronectin could be preserved after the decellularization process, however, it has been reported that the loss of most of these components was up to a 50% compared to native ECM, influencing natural structures and decreasing mechanical properties to some extent [30–32]. And, changes in ECM stiffness have great effect on cell adhesion, attachment, phenotype, etc. [33, 34]. Specifically, DMs with poor mechanical properties are difficult to provide cells with good attachment sites, thereby, affecting their subsequently proliferation and differentiation. In addition, DMs with poor mechanical properties could exacerbate cell-mediated stent contractions, which will randomly change the shape of the materials and greatly reduce the mechanical property [35, 36]. Hence, these DMs would be limited in repairing those load-bearing parts, and their mechanical properties improvement is quite important. There are mainly two means to improve the mechanical properties of DMs.

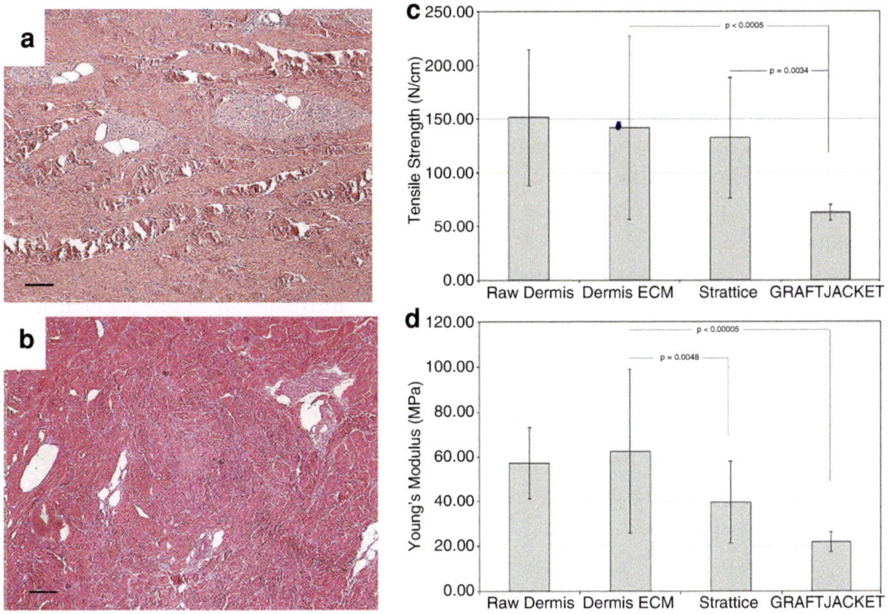

Fig. 8.3 Example of improving the mechanical properties by ameliorating decellularization methods. The comparation between raw reticular porcine dermis, the decellularized ECM of the study, and two comparative decellularized dermis materials, showing the decellularized dermis tissue prepared by optimized methods had obviously better mechanical properties. HE staining of native dermis tissue (**a**) and decellularized dermis tissue (**b**). Scale bars = 100 μm. Mechanical testing of strength of the decellularized materials. (**c**) Tensile strength and (**d**) Young's modulus comparation between raw reticular porcine dermis, the decellularized ECM of the study and two comparative decellularized dermis materials [37]

Firstly, optimizing decellularization methods have been demonstrated as a good way for enhancing the DMs mechanical properties. For instance, regulating the solution mixing time and velocity, the mechanical properties of obtained acellular porcine mesothelial could be improved. Furthermore, fabricating multilayer scaffolds through pressing could further enhance the mechanical strength of decellularized sheet materials [37] (Fig. 8.3). Likewise, the decellularized effect of different reagents would be diverse. Comparing the reagents decellularizing pig common carotid arteries of ionic agent, hypotonic and hypertonic solution, and nonionic reagent, research demonstrated that ionic agent indicated the best effect because it cleanly removed cells with the smallest damage on DMs mechanical properties [38]. Besides, it was also found that decellularized vascular grafts obtained by deoxycholic acid (DEOX) maintained a better intact ECM and exhibited excellent mechanical properties compared to decellularized vascular grafts obtained by Triton X-100 (TRITON), proving the superiority of DEOX in decellularizing vascular grafts [39].

Secondly, there are several ways to strengthen mechanical properties through further processes including pressing, compounding with other materials [40], crosslinking inside [41], coating with other materials [42], etc. On the one hand, pressing is usually used for DMs in sheet to furtherly strengthen. Gregory et al. demonstrated that the pressed multilaminate (4-layer) porcine small intestinal submucosa (MLSIS) exhibited much bigger mechanical properties in resistance to tensile, push-through, and suture pullout testing than single-layer porcine small intestinal (SIS) [43]. The strengthened multilaminate SIS have been applied in many kinds of tissue repair, such as cardiovascular, musculotendinous, abdominal wall, bone, gastrointestinal, etc., overcoming the difficulty of being applied for a long time in a serious mechanical environment [44]. On the other hand, introducing other materials to construct composites has been demonstrated to effectively improve DMs mechanical properties, which was mainly through decreasing DMs' own heterogeneity in pore size. For instance, poly (lactic-co-glycolic) acid nanoparticles (PLGA NPs) introduced into SIS significantly promoted the consistency of SIS, enhancing its mechanical properties [45].

On the aspect of crosslinking inside, the physical and chemical crosslinking methods are the two commonly used crosslinking methods. The physical methods mainly involved dehydration heat treatment, and UV irradiation [46]. And, the commonly used reagents in chemical methods include genipin, glutaraldehyde, and carbodiimide. Usually, the effect of physical crosslinking on improving mechanical properties is weaker than that of chemical crosslinking because the crosslinking degree of physical methods is limited compared to that of chemical methods. However, the chemical crosslinking might have impact on the bioactivity of DMs for the side effect of reagents [47]. Despite that, numerous studies have demonstrated the effectiveness of crosslinking inside in strengthening mechanical properties of DMs. For example, Wang et al. crosslinked SIS with genipin, and found that the intensive cyclic structure introduced by crosslinking between intermolecular and intramolecular obviously increased the ultimate tensile strength of SIS (Fig. 8.4) [48]. Amanda et al. demonstrated that decellularized cartilage crosslinked with glutaraldehyde or genipin raised the accumulation modulus by over 50% and reestablished it to the strength of native cartilage [49]. Xiang et al. crosslinked collagen in decellularized liver scaffolds through treating with riboflavin/UVA and found obvious improvement of Young's modulus average and the peak load in decellularized matrix after cross-linking [45, 50].

In fact, constructing coatings is the most used method for strengthening the mechanical properties. There were three kinds of coatings used commonly including electrospinning fiber coating, nanoparticle coating, and solution coating. Firstly, reinforcement by electrospinning coating might be attributed to the interactions of electrospinning with DMs matrix and its high mechanic properties. And, the fiber diameter, distribution, orientation, as well as the thickness of electrospinning film are the main key factors deciding the improving effects on mechanical properties. Up to date, electrospinning fibers with good mechanical properties have been successfully applied as DMs coating through mechanical compression or electrospinning [51]. For instance, it has been demonstrated that the yield stress of tubular SIS

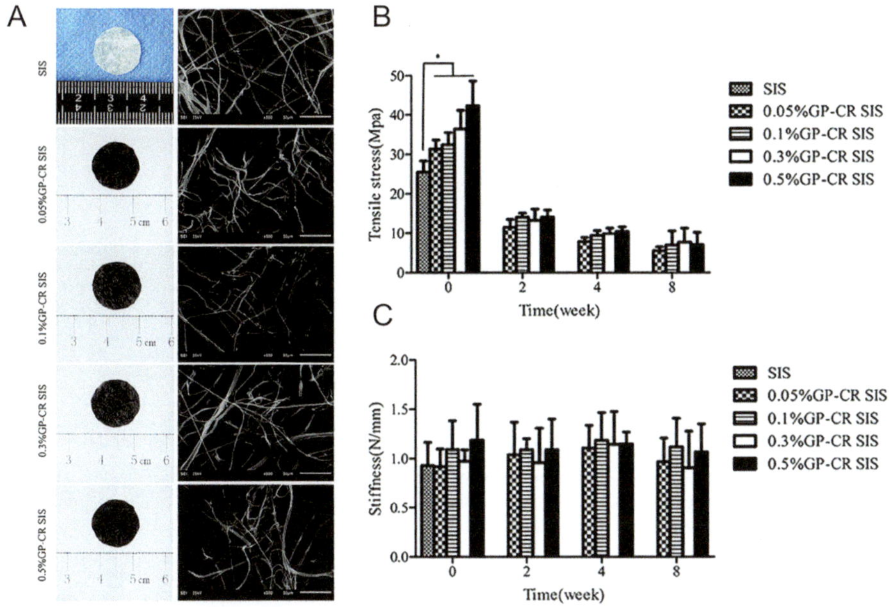

Fig. 8.4 Example of improving the mechanical properties by crosslinking inside. (**a**) Macroscopic observation of SIS and GP-CR SIS (left column); SEM photographs of SIS and GP-CR SIS (right column); SEM scale bar: 50 μm; (**b**) Tensile stress and (**c**) Stiffness of SIS and GP-CR SIS before (0 w) and after treatment with SGF for 2, 4, and 8 weeks. *: significantly different from SIS ($P < 0.05$). Data showed that genipin crosslinked SIS obviously raised the final tensile strength of SIS [48]

mechanically reinforced by PLGA electrospinning was improved by 200% and 400% along with the longitudinal and circumferential ways, respectively (Fig. 8.5a) [52]. At the same time, Yang et al. coated decellularized rat aorta (DRA) with electrospinning polycaprolactone (PCL) and PCL nanofibers combined with rapamycin (RM) (hybrid small-diameter vascular grafts (HTEV)) and RM-loaded hybrid tissue-engineered vascular graft (RM-HTEV), respectively, and found that both HTEV and RM-HTEV presented strong mechanical properties compared to DRA [53].

For nanoparticle coating, the reinforcement on mechanical properties of composites might be because of the decrease of the pore size of DMs and the heterogeneity of pore distribution, as well as the promotion of surface properties. And, the nanoparticles density and thickness are the main factors deciding the improvement effect. Some nanoparticles with good mechanical properties, such as chitosan/nanohydroxyapatite [54] and CNT [42], have been applied as coating on DMs to improve their mechanical properties. For example, Ghassemi et al. constructed single-walled carbon nanotube (CNT) coating for acellular cow articular cartilage (AC), and found the CNT coating obviously improved the mechanical properties of AC. Specifically, the Young's modulus of CNT-spiked samples was

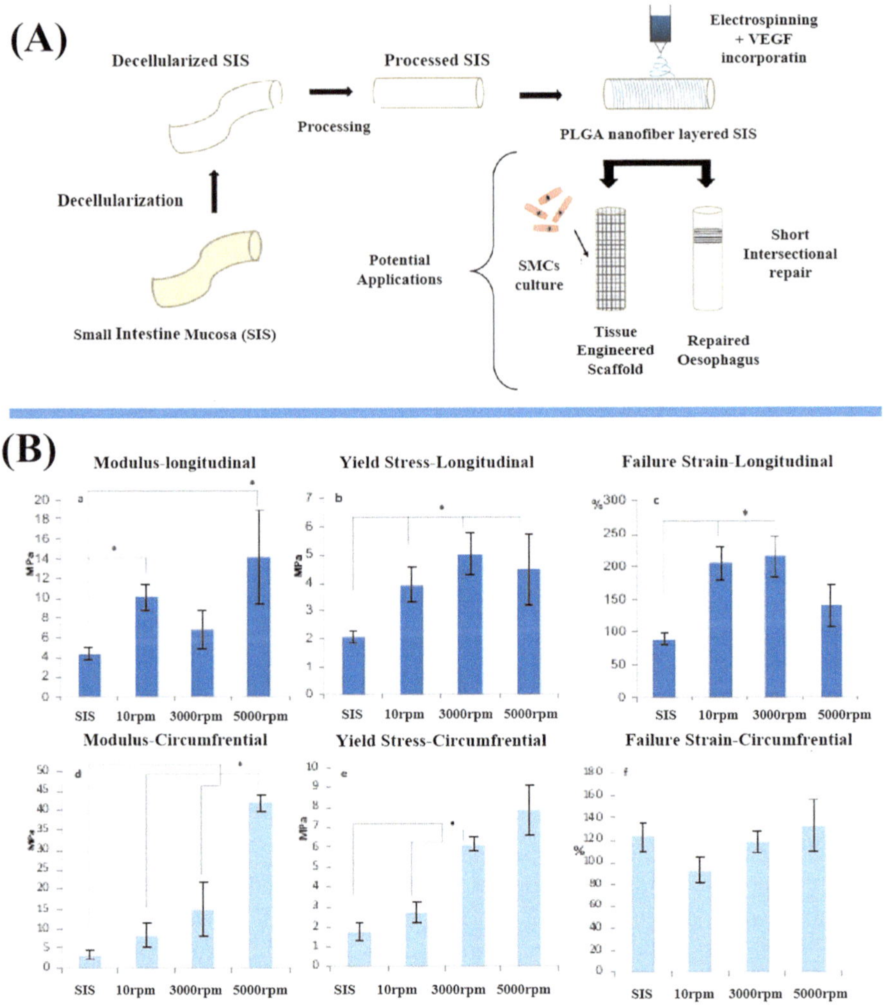

Fig. 8.5 Example of improving the mechanical properties by constructing electrospinning fiber coating. (**a**) Schematic of the process for constructing PLGA electrospinning for tubular SIS to develop hybrid PLGA-SIS scaffolds. (**b**) Mechanical testing results from the longitudinal tensile testing (*a–c*) and the circumferential tensile testing (*d–f*) of the hybrid scaffold groups. Data showed that tubular SIS could be mechanically reinforced by PLGA electrospinning [52, 53]

0.67 ± 0.09 MPa, however, the Young's modulus of pure AC samples was 0.43 ± 0.06 MPa [42]. Simultaneously, Sweta et al. constructed CS/nHAp coating for decellularized goat-lung scaffold and the compressive stiffness of the scaffold was increased [54].

In addition, there are some effective endeavors enhancing DMs through solution coating. Currently, we coated the typical extracellular matrix scaffold-small intestinal submucosa (SIS) with silk fibroin (SF) layer-by-layer construction to improve its

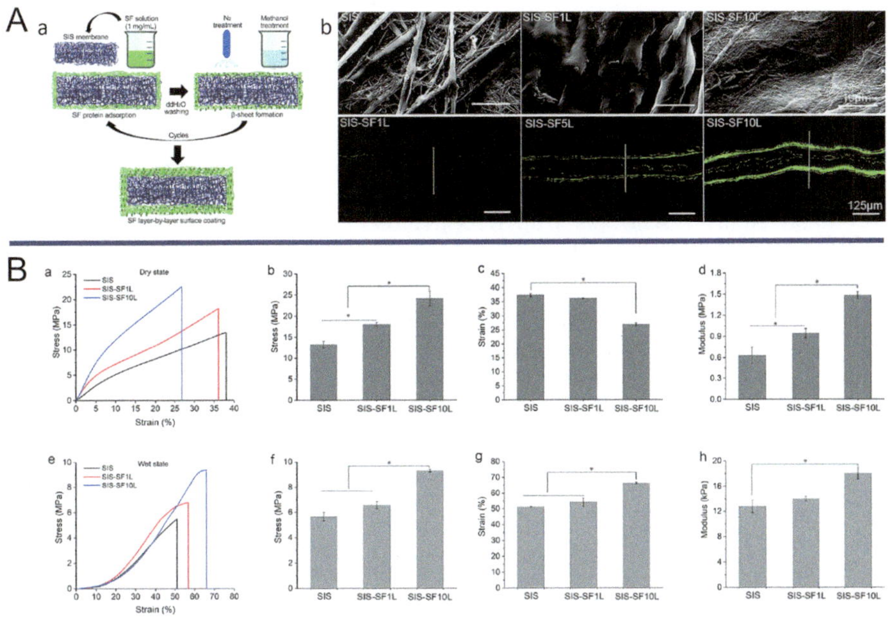

Fig. 8.6 Example of improving the mechanical properties by constructing solution coating. (**a**) (*a*) Schematic of the process of silk fibroin (SF) layer-by-layer coated on the surface of the SIS. (*b*) SEM photographs of the SIS alone and SIS coated with 1 and 10 layers of SF, as well as CLSM images of the SIS with 1, 5, and 10 layers of FITC-SF coatings, showing the changes of different layers of SF coatings. (**b**) Mechanical property analysis (stress–strain curves, stress, strain, and Young's modulus) of the SIS and SF modified SIS membranes in the dry (*a–d*) and hydrated (*e–h*) states of different membranes in the dry state. Data are expressed as the mean ± SD ($n = 5$). *$p < 0.05$. Data showed that SF layer-by-layer coating could promote the tensile strength and ductility of the SIS, especially in the wet state [55]

mechanical stability, the results of which showed SF layer-by-layer coating could increase the tensile strength and ductility of the SIS, especially in the wet state, and this effect depended on the layers [55] (Fig. 8.6).

All in all, appropriate mechanical properties are important for the applications of DMs. Even though there have been some effective measures in improving the mechanical properties, much more efforts would be needed to develop new methods.

8.5 Controlling the Degradation Rate of DMs

Ideal degradation rate of biomaterials should be matched with tissue regeneration rate. However, because the natural components are easily degraded in vivo under the effect of enzyme, bad effect on tissue repair caused by too fast degradation at the implantation site is still a serious problem faced by DMs. Specifically, Zheng et al. demonstrated that the decellularized liver scaffolds induced few parenchymal cells or vessels generation because of their fast degradation [50]. In addition, Cao et al. also demonstrated that cartilage regeneration was tremendously affected by too fast degradation of DMs [56]. There have been many endeavors to decrease the degradation rate of DMs, which could be mainly divided into two aspects, including crosslinking inside and coating other materials.

Firstly, crosslinking inside have been demonstrated as one kind of effective way to decrease the degradability of DMs [57]. For example, Xiang et al. treated decellularized liver scaffolds with riboflavin/UVA to induce the crosslinking of inside collagen and found that the in vitro degradation rate of the crosslinked scaffolds was obviously slow. Specifically, there was 0.03% degradation for the crosslinked scaffolds in 10 days, however, 0.1% degradation for non-crosslinked scaffolds in 10 days [50]. Kyle et al. crosslinked acellular tendon by glutaraldehyde and found that crosslinked acellular tendons retained both macroscopic structure and fiber orientation while non-crosslinked acellular tendons were rapidly degraded and lost fiber morphology cross (Fig. 8.7) [58]. Zhai et al. crosslinked porcine heart value matrix using procyanidins (2.5, 5, 10, 20 mg/mL), and found that crosslinked matrix could resist the enzymatic hydrolysis and decrease the degradation rate of decellularized porcine heart [59].

Secondly, it was also demonstrated that coating other materials could effectively decrease degradation rate of DMs. For example, Zheng et al. coated electrospinning PCL nanofiber mesh on decellularized human amniotic membrane (dAM) under mechanical compression. Testing the degradation rate of dAM and the PCL-dAM composite membrane in a pancreatin buffer solution (1 U), they found that the degradation rate of dAM was slowed down by PCL-dAM composite membrane by about 40% [51]. Recently, we constructed an electrospinning chitosan/elastin (CS/ES) nanofibers coating on multilayer SIS (CS/ES-SIS) (Fig. 8.8a). Using a constant deformation loading device in vitro, which was used to simulate the environment in vivo, we investigated the degradation behavior of SIS and CS/ES-SIS composites within biomechanical and enzymatic environment and found that CS/ES-SIS composites exhibited obviously slower degradation rate and could better keep the stability of their micromechanical properties and microstructures compared to SIS (Fig. 8.8b). It might be the electrostatic interactions between the positively charged CS in the nanofibers and the negatively charged collagen in the SIS that combined the components with the composites together, promoting the stability of the composites of the degradation process, especially in the environment of biomechanical [11]. Simultaneously, we fabricated a layer-by-layer SF coating for multilayer SIS and found the obvious decrease effect on SIS degradation that the SIS degraded 100% in 10-days degradation with enzymes, yet, after 2 weeks of degradation process, the resting weights of SIS-SF1L was $10.16 \pm 1.53\%$ and SIS-SF10L was $36.69 \pm 3.34\%$ (Fig. 8.9a). And, compared with the SIS group, the coating could significantly enhance the whole mechanical strength of the SIS membrane and keep higher tensile strength and modulus in the degradation within the enzymatic environment (Fig. 8.9b) [55].

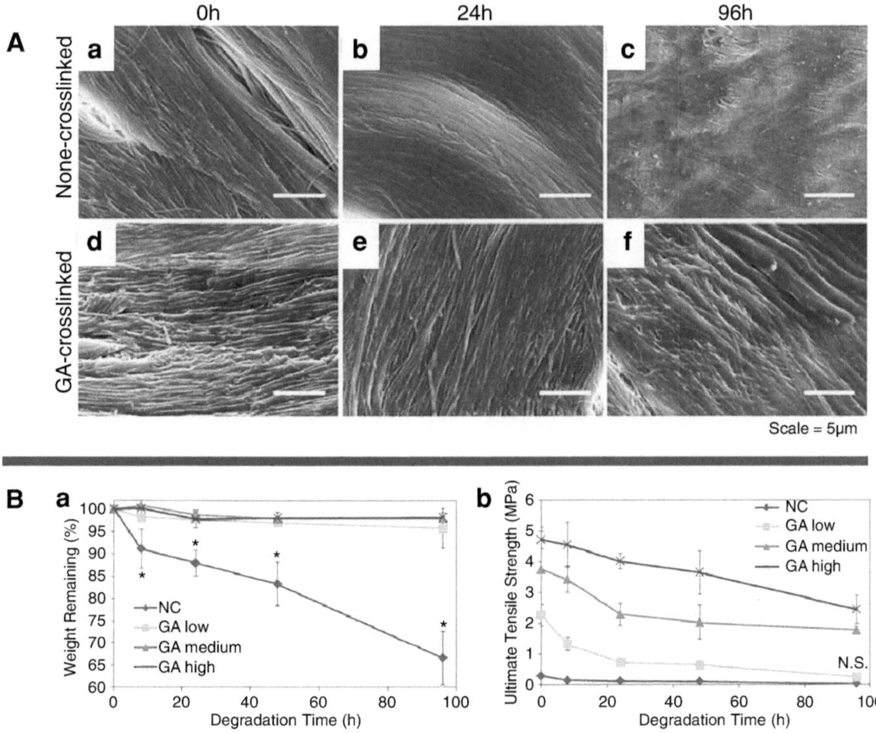

Fig. 8.7 Example of decreasing degradation rate by crosslinking inside. Crosslinking acellular tendon by glutaraldehyde. (**a**) SEM images of: (*a*) non-crosslinked samples before collagenase degradation, (*b*) after 24 h, (*c*) after 96 h and (*d*) GA-crosslinked samples before degradation, (*e*) after 24 h, (*f*) after 96 h. (**b**) (*a*) Mass resting in bioskiving scaffolds along with bacterial collagenase degradation. (*b*) final tensile strength of bioskiving scaffolds along with bacterial collagenase degradation. ($n = 4$, $* = P$ values <0.05) Data showed that crosslinking acellular tendon using GA significantly decreased the degradation of acellular tendon [58]

8.6 Furthering Promoting the Bioactivity of DMs

As described in Chap. 2, decellularization processes easily result in slight changes in the three-dimensional structure of certain bioactive substances or the damage of some active factors in DMs, decreasing the bioactivity of DMs to some extent. Hence, furthering improving the bioactivities could be beneficial for target tissue repair. And, changing properties of DMs to prevent adverse host responses affecting tissue regeneration and introducing other bioactive substances to directly promote tissue regeneration are the commonly used methods.

On the one hand, suppressing adverse host responses through decreasing immunogenicity of DMs or optimizing the chemical and physical properties of DMs to change the immunomodulatory role has been demonstrated to indirectly improve DMs bioactivities, because severe host reactions including immune and

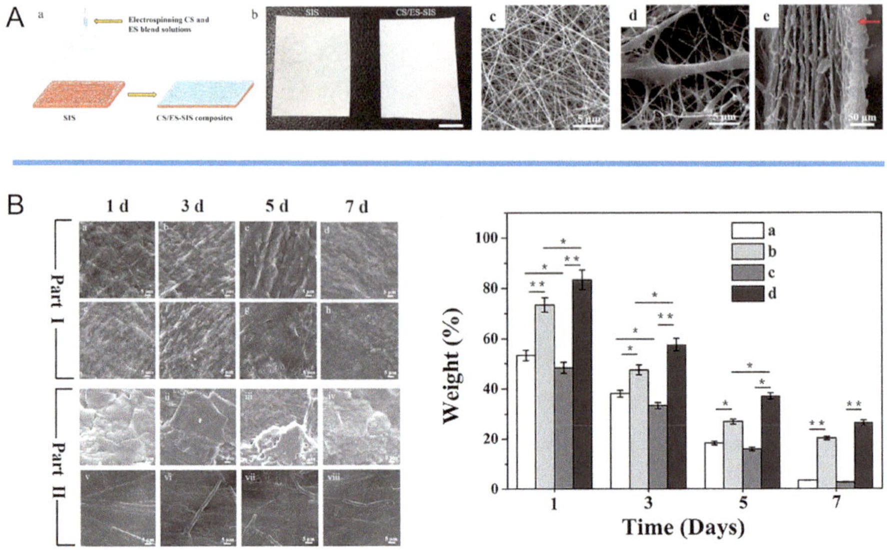

Fig. 8.8 Example of decreasing degradation rate by constructing electrospinning fiber coating. (**a**) Schematic demonstration of the preparation process of CS/ES-SIS composites (*a*); macro structure of SIS and CS/ES-SIS composites (*b*); SEM photographs of upper (*c*) and lower (*d*) appearance of CS/ES-SIS composites; cross-sectional SEM photographs of CS/ES-SIS composites (*e*). CS/ES electrospinning fibers in the composites are pointed out by red arrowheads. (**b**) SEM photographs of SIS (*a–d*) and CS/ES-SIS composites (*e–h*) at different degradation times without biomechanical load (Part I), and under biomechanical environment (Part II) (i—vii). In vitro degradation of SIS and CS/ES-SIS composites under no biomechanical load (*a* and *b*) and under biomechanical environment (*c* and *d*). Data showed that the CS/EC electrospinning fiber coating obviously decreased the degradation rate of SIS [11]

inflammatory response would do harm to tissue repair. On the aspect of promoting bioactivity by decreasing immunogenicity of DMs, there were some endeavors showing effectiveness. For example, Lu et al. used a dye-mediated photooxidation technique to crosslink acellular bovine jugular vein conduits (BJVCs) as the connection of dog right ventricles and pulmonary arteries, the results of which showed that immunogenicity of the crosslinked acellular BJVCs was obviously reduced compared to that of the acellular BJVCs, exhibiting excellent vascularization function and salient ability to resist thrombosis and calcification [12]. On the aspect of promoting bioactivities through changing immunomodulatory, efforts were mainly focused on changing the macrophages from M1 type (proinflammatory) to M2 type (anti-inflammatory and prohealing) [60]. For instance, it was demonstrated that decellularized pig bone hydrogel significantly promoted periodontal tissue regeneration compared with particles, which was mainly through improving pro-remodeling phenotypes and the changes of M1 type macrophages to M2 type [61]. In addition, we developed newly enhanced methods for traditional multilayer SIS to replace the chemical crosslinking methods and found that our approach could effectively keep the natural molecular activity of the ECM and so that regulate macrophage

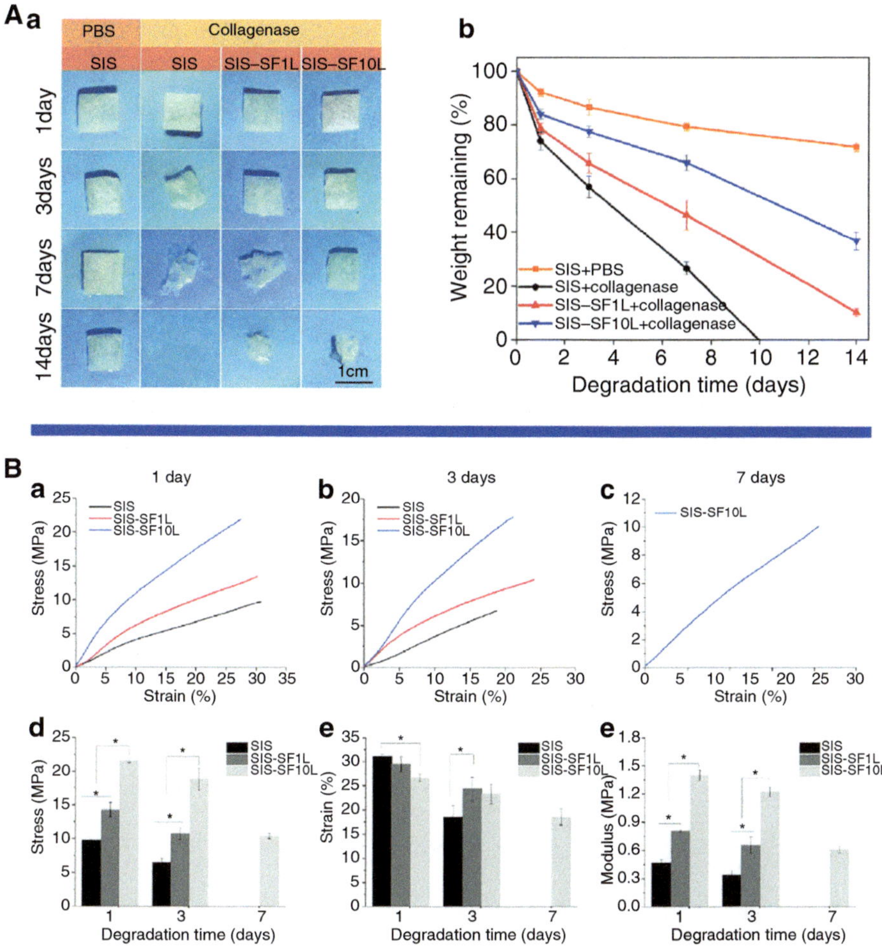

Fig. 8.9 Example of decreasing degradation rate by constructing SF layer-by-layer coating. (**a**) Analysis of membrane stability against enzyme degradation. (*a*) Macroscopic images of the SIS, SIS-SF1L, and SIS-SF10L membranes after 1, 3, 7, and 14 days of degradation. (*b*) Resting weight ratio of the membranes after treatment with PBS and collagenase solution. (**b**) Mechanical properties of the SIS and SF modified SIS membranes during enzymatic degradation. (*a–c*) Stress–strain curves and (*d–f*) Stress, strain, and Young's modulus of different kinds of membranes under enzymatic degradation at 1, 3, and 7 days. Data showed that SF modification could effectively increase the structural stability of materials, decreasing the degradation rate of SIS [55]

polarization, benefiting wound repair and tissue regeneration. Specifically, the recruited macrophages were mainly of the M2 and M1 phenotype individually for the SIS and SIS-GA groups, however, increased numbers of M2 macrophages and higher ratios of M2 phenotype macrophages were observed in the SIS-SF groups along the implantation (Fig. 8.10) [55].

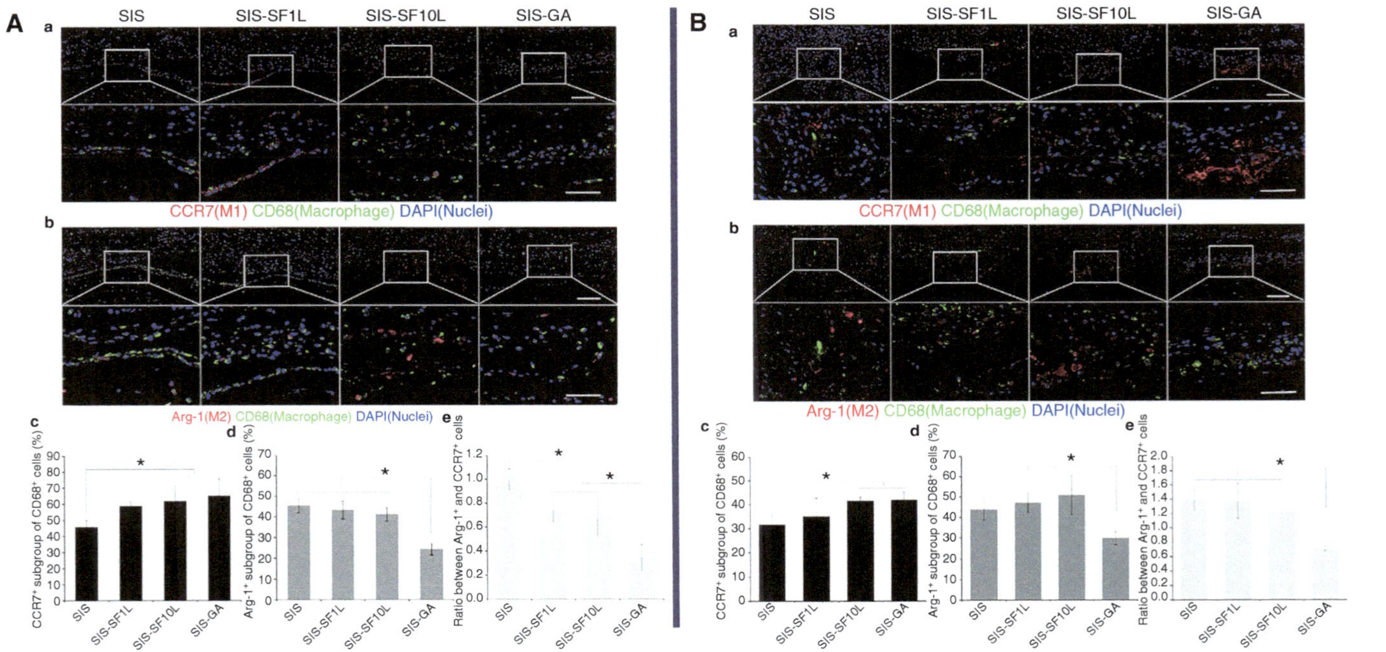

Fig. 8.10 Representative photographs of immunohistochemical staining for CD68, CCR7, and Arg-1 on sections of various kinds of membranes after implantation for 3 days (**a**) and after implantation for 14 days (**b**). (*a*) Representative photographs of immunohistochemical staining for CD68 and CCR7 on the sections. (*b*) Representative photographs of immunohistochemical staining for CD68 and Arg-1 on the sections. (*c*, *d*) Percentage of $CCR7^+$ and Arg-1^+ cells in the $CD68^+$ population. (*e*) Ratio between Arg-1^+ and $CCR7^+$ cells in the $CD68^+$ population. Data are presented as the mean ± SD ($n = 3$). $*p < 0.05$. Data showed that layer-by-layer SF coated SIS could effectively keep the natural molecular activity of the ECM and direct macrophage polarization compared to GA-crosslinked SIS, benefiting wound repair and tissue regeneration [55]

On the other hand, adding other bioactive substances, such as bioactive ECM component loss during decellularization [62], active factors (related peptides and proteins), and functional materials has achieved good effect on furthering improving the bioactivity of DMs. Firstly, Liu et al. combined human-like collagen I with acellular vascular matrix and found that the complex behaved more similarly to the native rabbit artery compared to the pure acellular vascular matrix [63]. Secondly, Lee et al. crosslinked adhesion peptides (RGD, YIGSR) onto the surface of acellular vein using dopamine and found that the composites improved endothelial progenitor cells function, and eventually expedited the endothelialization compared to unmodified acellular vein [64]. Besides, Hyeongjin et al. prepared DM/PLLA and IGF-1/DM/PLLA scaffolds by injecting the solution of DM and mixture of DM and myogenic insulin growth factor-1 (IGF-1) into PLLA, and found that IGF-1/DM/PLLA scaffolds exhibited an obviously greater myofibers at 1/2-months implantation compared to DM/PLLA groups in the repair of rabbit skeletal muscle, which indicated the significant improvement of bioactivity for promoting the muscle tissue regeneration of the added growth factors [65]. Another study also demonstrated that combination of exogenous transforming growth factor β3 (TGF-β3)/IGF-1 and dECM derived from meniscus tissue synergistically enhanced cartilaginous regeneration [66]. Thirdly, we added the functional materials—Cu into SIS patches for esophageal repair and found that the addition of Cu obviously promoted the regeneration of muscle, epithelium, and vascularization progress [67]. Moreover, we recently developed a SIS-based biocomposite (chitosan/elastin-SIS, CS/ES-SIS) in the biomimetic way for abdominal wall repair, the further study results of which demonstrated that the introduction of CS and ES significantly promoted the antibacterial rates confine *E. coli* and *S. aureus*, the viability and proliferation of fibroblast cell L929 in vitro, and promoted tissue regeneration without serious inflammatory reaction in a rat model with abdominal wall defect. Especially, the number of vascular generated in the CS/ES-SIS composites was found to be as 1.69 times as that in the SIS at 4 weeks after implantation possibly because of the synergistic interactions between ES and VEGF contained in SIS (Fig. 8.11) [11].

8.7 Promoting Clinical Application

Many of these DMs products, which have received the approval of Food and Drug Administration and have been currently widely used in the clinic, are favored by doctors and patients with their advantages of maintain the cell phenotype or promote the differentiation of stem cells into specific tissue cells which plays an important role in promoting the repair and reconstruction of functional tissues and organs [68, 69]. A variety of commercialized DMs, derived from either human sources or animal sources, are applied in treating the tissue damage and these include: porcine SIS (e.g. Oasis®, FortaFlex®, Restore™, CorMatrix ECM™, etc.), human dermis (e.g., AlloDerm®, AlloPatch HD™, FlexHD®, GraftJacket®, NeoForm™, DermACELL®, AlloSkin™ AC, etc.), porcine dermis (e.g., Strattice™, Zimmer

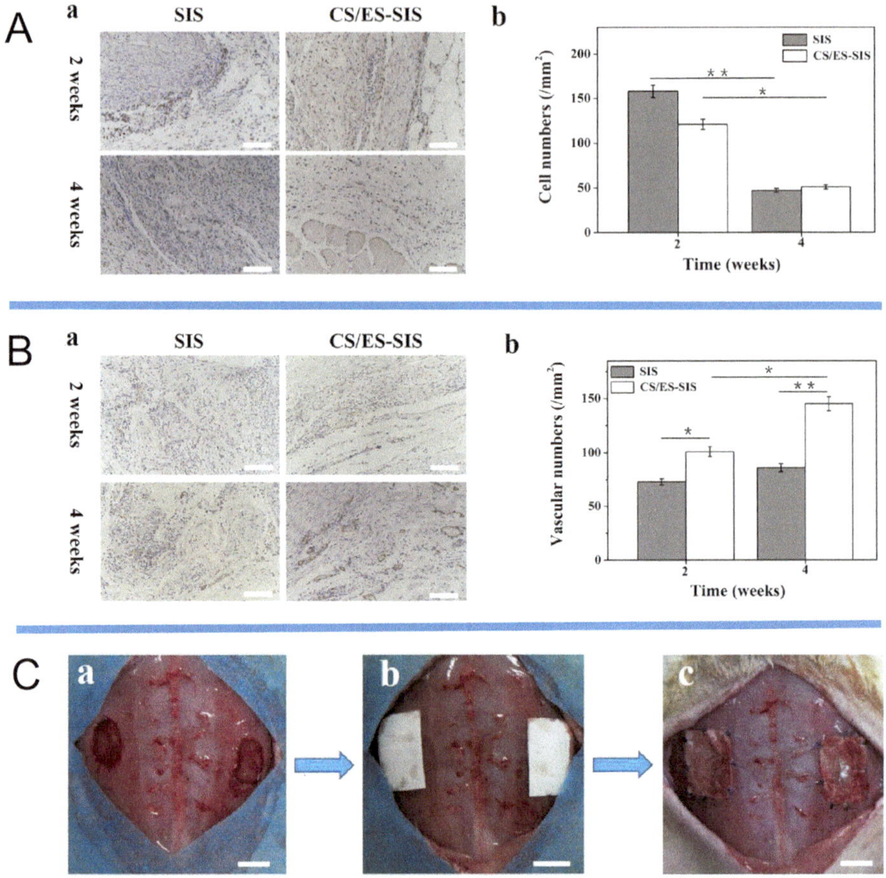

Fig. 8.11 Furthering improving the bioactivity of SIS by constructing bioactive CS/ES electrospinning fiber coating. (**a**) (*a*) Immunohistochemistry CD68 staining and (*b*) quantification of CD68 positive macrophages of SIS and CS/ES-SIS composites after implantation for 2 and 4 weeks. (**b**) (*a*) Immunohistochemistry CD31 staining and (*b*) CD31 positive vascular density of SIS and CS/ES-SIS composites after implantation for 2 and 4 weeks. Positive CD31 was indicated by black arrows. (**c**) Surgery processes of the abdominal wall defects healing (*a–c*). Macroscopic images of the implanted (*d*) SIS and (*e*) CS/ES-SIS composites at 2-weeks implantation, and the implanted (*f*) SIS and (*g*) CS/ES-SIS composites at 4-weeks implantation. Data showed that the bioactive CS/ES electrospinning fiber coating significantly improved the bioactivity of SIS [11]

Collagen Repair™, Conexa™, etc.), bovine dermis (e.g., Matriderm®, SurgiMend™, etc.), bovine pericardium (e.g., Lyoplant®, Perimount®, CopiOs®, Tutopatch®, Veritas®, PhotoFix™, SJM™ Pericardial Patch, etc.), demineralized bone matrix (e.g., Puros® DBM, Grafton®, DBX®, Progenix™ Plus, Accell Connexus®&TBM®, InterGro®, etc.), and the like [70–72].

Since the DMs were successfully applied for III degree burn wounds by Wainwright et al. in 1995, numerous examples of DMs that applied in the

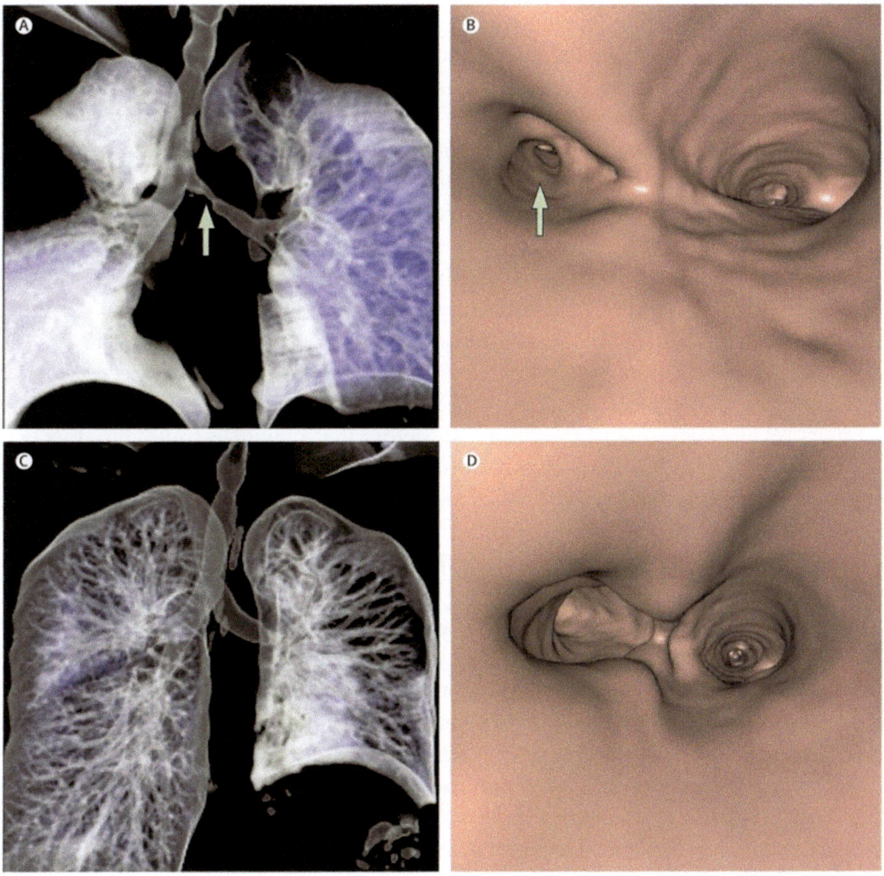

Fig. 8.12 Volume-rendering CT and virtual bronchoscopic reconstructions before (**a** and **b**) and after (**c** and **d**) decellularized trachea transplantation [82]

reconstitution of lost or damaged organs and tissues including, but not limited to, skin, cardiovascular, trachea, urinary tract, and nerve, have appeared in preclinical or clinical studies in recent years [71, 73–81]. On the one hand, substitute damaged tissue directly is currently one of the main applications of DMs with the advantage of usage simple, as well as excellent safety and efficacy. For instance, decellularized human trachea repopulated with autologous epithelial cells and mesenchymal stem-cell-derived chondrocytes with a bioreactor has been used to replace the left main bronchus of a 30-year-old woman with end-stage bronchomalacia [82]. In postoperative 1 month, the appearance of the airway improved from near-total collapse preoperative (Fig. 8.12a, b) to wide patency (Fig. 8.12c, d) [82]. And the results of 5-year follow-up show that, although there is a progressive cicatricial stenosis in the trachea anastomosis, the cell-scaffold is well vascularized and epithelialization with normal ciliary function and mucus clearance and the patient have a normal life

[83]. On the other hand, implantation or tissue filling, which is mainly used in cosmetic plastic surgery, including without limitation, facial plastic surgery, closed gum wounds, patch for breast repair surgery, is another major application [71, 84–87].

In brief, the DMs have significant clinical value in regenerative therapy. However, there are still a series of problems that need to be solved including minimizing the immunization rejection especially in the initial stage of transplantation, developing DMs products that are suitable for more complex tissue repair, and avoiding the safety risks caused by the virus and other dangerous factors carried by the animal tissues or their derivatives, and the immunotoxicity risks caused by the different animals species [78]. What is more, the decellularization method, which affects the composition, mechanical properties of the products, and the host response after implantation, is the key technique difficulty in DMs fabrication [88]. And the antigens residues are one of the principal factors that may cause immune rejection and immune toxicity which may further affect the healing of injury. Thus, different improving decellularization measures should be taken on the basis of specific tissue to obtain DMs products with maximum retention of the extracellular component while minimizing antigens residues [89]. The functionality of the reconstructed tissue-engineered organs still cannot meet the clinical needs. It is necessary to further optimize material preparation methods, seed cell types and culture methods, and implantation ways [90]. In the meantime, due to DMs are originate from the nature, the mechanical properties and degradation properties still could not meet the needs for tissue repair with special performance requirements [44]. At present, it has become a research hotspot in recent years to modify it or give it new properties by crosslinking or compounding with polymer synthetic materials and inorganic materials [91]. This strategy is expected to solve the shortcomings of DMs in terms of thermal stability, mechanical properties, and degradation properties, so as to prepare products that meet the needs of clinical applications.

8.8 Summary

Decellularized materials (DMs) have exhibited great attraction and development potential because of their unique natural structures, chemical and physical signals with spatiotemporal accuracy, comparatively high bioactivity due to their tissue-specific compositions and structures, good biodegradability, as well as low immunogenicity and outstanding immunomodulatory abilities, overcoming some limitations of synthetic materials in the traditional sense and bringing new hope for tissue regeneration and organ replacement. Even though the above challenges, there have been some efforts tried to solve and preliminary effects have been achieved. With the continuing research, more and more new methods would be developed. In addition, some complicated problems that restrict the application of decellularized materials would be put on the agenda, such as the accurate mechanism of decellularized materials to promote tissue regeneration, which involves the precise characterization of components and the further study on biological effects of single component and combination of multiple components. Overall, with the continuous explorations and improvements, the DMs will hopefully play more important role in tissue repair and organ replacement.

References

1. Sakakibara S, Ishida Y, Hashikawa K, Yamaoka T, Terashi H. Intima/medulla reconstruction and vascular contraction-relaxation recovery for acellular small diameter vessels prepared by hyperosmotic electrolyte solution treatment. J Artif Organs. 2014;17:169–77.
2. Crapo PM, Gilbert TW, Badylak SF. An overview of tissue and whole organ decellularization processes. Biomaterials. 2011;32:3233–43.
3. Wainwright JM, Czajka CA, Patel UB, Freytes DO, Tobita K, Gilbert TW, Badylak SF. Preparation of cardiac extracellular matrix from an intact porcine heart. Tissue Eng Part C Methods. 2010;16:525–32.
4. Meyer SR, Chiu B, Churchill TA, Zhu L, Lakey JR, Ross DB. Comparison of aortic valve allograft decellularization techniques in the rat. J Biomed Mater Res A. 2006;79:254–62.
5. Meezan E, Hjelle JT, Brendel K, Carlson EC. A simple, versatile, nondisruptive method for the isolation of morphologically and chemically pure basement membranes from several tissues. Life Sci. 1975;17:1721–32.
6. Lumpkins SB, Pierre N, McFetridge PS. A mechanical evaluation of three decellularization methods in the design of a xenogeneic scaffold for tissue engineering the temporomandibular joint disc. Acta Biomater. 2008;4:808–16.
7. Singh K, Gopinathan A, Sangeetha P, Kumar N, Singh KP, Raina OK. Development and clinical application of decellularized porcine SIS and cornea for the repair of corneal defects in animals. Indian J Anim Sci. 2016;86:1391–5.
8. Lopera HM, Griffiths LG. Antigen removal process preserves function of small diameter venous valved conduits, whereas SDS-decellularization results in significant valvular insufficiency. Acta Biomater. 2020;107:115–28.
9. Yang B, Zhang Y, Zhou L, Sun Z, Zheng J, Chen Y, Dai Y. Development of a porcine bladder acellular matrix with well-preserved extracellular bioactive factors for tissue engineering. Tissue Eng Part C Methods. 2010;16:1201–11.
10. Brooker JE, Camison LB, Bykowski MR, Hurley ET, Yerneni SS, Campbell PG, Weiss LE, Mooney MP, Cray J, Gilbert JR, Cooper GM, Losee JE. Reconstruction of a calvarial wound complicated by infection: comparing the effects of biopatterned bone morphogenetic protein 2 and vascular endothelial growth factor. J Craniofac Surg. 2019;30:260–4.
11. Cao G, Wang C, Fan Y, Li X. Biomimetic SIS-based biocomposites with improved biodegradability, antibacterial activity and angiogenesis for abdominal wall repair. Mater Sci Eng C Mater Biol Appl. 2020;109:110538.
12. Lu WD, Zhang M, Wu ZS, Hu TH, Xu ZJ, Liu W, Hu YR. The performance of photooxidatively crosslinked acellular bovine jugular vein conduits in the reconstruction of connections between pulmonary arteries and right ventricles. Biomaterials. 2010;31:2934–43.
13. Voges I, Brasen JH, Entenmann A, Scheid M, Scheewe J, Fischer G, Hart C, Andrade A, Pham HM, Kramer HH, Rickers C. Adverse results of a decellularized tissue-engineered pulmonary valve in humans assessed with magnetic resonance imaging. Eur J Cardiothorac Surg. 2013;44: e272–9.
14. Simon P. Early failure of the tissue engineered porcine heart valve SYNERGRAFT™ in pediatric patients. Eur J Cardiothorac Surg. 2003;23:1002–6.

15. Obata T, Tsuchiya T, Akita S, Kawahara T, Matsumoto K, Miyazaki T, Masumoto H, Kobayashi E, Niklason LE, Nagayasu T. Utilization of natural detergent potassium laurate for decellularization in lung bioengineering. Tissue Eng Part C Methods. 2019;25:459–71.
16. Eivazkhani F, Abtahi NS, Tavana S, Mirzaeian L, Abedi F, Ebrahimi B, Montazeri L, Valojerdi MR, Fathi R. Evaluating two ovarian decellularization methods in three species. Mater Sci Eng C Mater Biol Appl. 2019;102:670–82.
17. Ng WH, Ramasamy R, Yong YK, Ngalim SH, Lim V, Shaharuddin B, Tan JJ. Extracellular matrix from decellularized mesenchymal stem cells improves cardiac gene expressions and oxidative resistance in cardiac C-kit cells. Regen Ther. 2019;11:8–16.
18. Xia C, Mei S, Gu C, Zheng L, Fang C, Shi Y, Wu K, Lu T, Jin Y, Lin X, Chen P. Decellularized cartilage as a prospective scaffold for cartilage repair. Mater Sci Eng C Mater Biol Appl. 2019;101:588–95.
19. Schmitt A, Csiki R, Tron A, Saldamli B, Tubel J, Florian K, Siebenlist S, Balmayor E, Burgkart R. Optimized protocol for whole organ decellularization. Eur J Med Res. 2017;22:31.
20. Bourgine PE, Pippenger BE, Todorov A Jr, Tchang L, Martin I. Tissue decellularization by activation of programmed cell death. Biomaterials. 2013;34:6099–108.
21. Jiang XJ, Wang XD. Cytochrome C-mediated apoptosis. Annu Rev Biochem. 2004;73:87–106.
22. Lakhani SA, Masud A, Kuida K, Porter GA, Booth CJ, Mehal WZ, Inayat I, Flavell RA. Caspases 3 and 7: key mediators of mitochondrial events of apoptosis. Science. 2006;311:847–51.
23. Bhardwaj A, Aggarwal BB. Receptor-mediated choreography of life and death. J Clin Immunol. 2003;23:317–32.
24. Savill J, Fadok V. Corpse clearance defines the meaning of cell death. Nature. 2000;407:784–8.
25. Ziegler U, Groscurth P. Morphological features of cell death. News Physiol Sci. 2004;19:124–8.
26. Li F, Huang Q, Chen J, Peng YL, Roop DR, Bedford JS, Li CY. Apoptotic cells activate the "phoenix rising" pathway to promote wound healing and tissue regeneration. Sci Signal. 2010;3:ra13.
27. Cornelison RC, Wellman SM, Park JH, Porvasnik SL, Song YH, Wachs RA, Schmidt CE. Development of an apoptosis-assisted decellularization method for maximal preservation of nerve tissue structure. Acta Biomater. 2018;77:116–26.
28. Young BM, Shankar K, Tho CK, Pellegrino AR, Heise RL. Laminin-driven Epac/Rap1 regulation of epithelial barriers on decellularized matrix. Acta Biomater. 2019;100:223–34.
29. Bi H, Ye K, Jin S. Proteomic analysis of decellularized pancreatic matrix identifies collagen V as a critical regulator for islet organogenesis from human pluripotent stem cells. Biomaterials. 2020;233:119673.
30. Rosmark O, Ahrman E, Muller C, Rendin LE, Eriksson L, Malmstrom A, Hallgren O, Larsson-Callerfelt AK, Westergren-Thorsson G, Malmstrom J. Quantifying extracellular matrix turnover in human lung scaffold cultures. Sci Rep. 2018;8:5409.
31. Zhou HY, Kitano K, Ren X, Rajab TK, Wu M, Gilpin SE, Wu T, Baugh L, Black LD, Mathisen DJ, Ott HC. Bioengineering human lung grafts on porcine matrix. Ann Surg. 2018;267:590–8.
32. Rowland CR, Colucci LA, Guilak F. Fabrication of anatomically-shaped cartilage constructs using decellularized cartilage-derived matrix scaffolds. Biomaterials. 2016;91:57–72.
33. Koval M, Ward C, Findley MK, Roser-Page S, Helms MN, Roman J. Extracellular matrix influences alveolar epithelial claudin expression and barrier function. Am J Respir Cell Mol Biol. 2010;42:172–80.
34. Godin LM, Sandri BJ, Wagner DE, Meyer CM, Price AP, Akinnola I, Weiss DJ, Panoskaltsis-Mortari A. Decreased laminin expression by human lung epithelial cells and fibroblasts cultured in acellular lung scaffolds from aged mice. PLoS One. 2016;11:e0150966.
35. Lee CR, Grodzinsky AJ, Spector M. The effects of cross-linking of collagen-glycosaminoglycan scaffolds on compressive stiffness, chondrocyte-mediated contraction, proliferation and biosynthesis. Biomaterials. 2001;22:3145–54.

36. Cheng N-C, Estes BT, Awad HA, Guilak F. Chondrogenic differentiation of adipose-derived adult stem cells by a porous scaffold derived from native articular cartilage extracellular matrix. Tissue Eng Part A. 2009;15:231–41.
37. Hoganson DM, O'Doherty EM, Owens GE, Harilal DO, Goldman SM, Bowley CM, Neville CM, Kronengold RT, Vacanti JP. The retention of extracellular matrix proteins and angiogenic and mitogenic cytokines in a decellularized porcine dermis. Biomaterials. 2010;31:6730–7.
38. Dahl SLM, Koh J, Prabhakar V, Niklason LE. Decellularized native and engineered arterial scaffolds for transplantation. Cell Transplant. 2003;12:659–66.
39. Dimitrievska S, Cai C, Weyers A, Balestrini JL, Lin T, Sundaram S, Hatachi G, Spiegel DA, Kyriakides TR, Miao J, Li G, Niklason LE, Linhardt RJ. Click-coated, heparinized, decellularized vascular grafts. Acta Biomater. 2015;13:177–87.
40. Ventura RD, Padalhin AR, Kim B, Park M, Lee BT. Evaluation of bone regeneration potential of injectable extracellular matrix (ECM) from porcine dermis loaded with biphasic calcium phosphate (BCP) powder. Mater Sci Eng C Mater Biol Appl. 2020;110:110663.
41. La W-G, Jang J, Kim BS, Lee MS, Cho D-W, Yang HS. Systemically replicated organic and inorganic bony microenvironment for new bone formation generated by a 3D printing technology. RSC Adv. 2016;6:11546–53.
42. Ghassemi T, Saghatolslami N, Matin MM, Gheshlaghi R, Moradi A. CNT-decellularized cartilage hybrids for tissue engineering applications. Biomed Mater (Bristol, Engl). 2017;12:065008.
43. Arnold GA, Mathews KG, Roe S, Mente P, Seaboch T. Biomechanical comparison of four soft tissue replacement materials: an in vitro evaluation of single and multilaminate porcine small intestinal submucosa, canine fascia lata, and polypropylene mesh. Vet Surg. 2009;38:834–44.
44. Cao G, Huang Y, Li K, Fan Y, Xie H, Li X. Small intestinal submucosa: superiority, limitations and solutions, and its potential to address bottlenecks in tissue repair. J Mater Chem B. 2019;7:5038–55.
45. Mondalek FG, Lawrence BJ, Kropp BP, Grady BP, Fung KM, Madihally SV, Lin HK. The incorporation of poly(lactic-co-glycolic) acid nanoparticles into porcine small intestinal submucosa biomaterials. Biomaterials. 2008;29:1159–66.
46. Al-Nasiry S, Geusens N, Hanssens M, Luyten C, Pijnenborg R. The use of Alamar Blue assay for quantitative analysis of viability, migration and invasion of choriocarcinoma cells. Hum Reprod. 2007;22:1304–9.
47. Rowland CR, Lennon DP, Caplan AI, Guilak F. The effects of crosslinking of scaffolds engineered from cartilage ECM on the chondrogenic differentiation of MSCs. Biomaterials. 2013;34:5802–12.
48. Wang M, Li YQ, Cao J, Gong M, Zhang Y, Chen X, Tian MX, Xie HQ. Accelerating effects of genipin-crosslinked small intestinal submucosa for defected gastric mucosa repair. J Mater Chem B. 2017;5:7059–71.
49. Pinheiro A, Cooley A, Liao J, Prabhu R, Elder S. Comparison of natural crosslinking agents for the stabilization of xenogenic articular cartilage. J Orthop Res. 2016;34:1037–46.
50. Xiang JX, Zheng XL, Gao R, Wu WQ, Zhu XL, Li JH, Lv Y. Liver regeneration using decellularized splenic scaffold: a novel approach in tissue engineering. Hepatobiliary Pancreat Dis Int. 2015;14:502–8.
51. Zhou Z, Long D, Hsu CC, Liu H, Chen L, Slavin B, Lin H, Li X, Tang J, Yiu S, Tuffaha S, Mao HQ. Nanofiber-reinforced decellularized amniotic membrane improves limbal stem cell transplantation in a rabbit model of corneal epithelial defect. Acta Biomater. 2019;97:310–20.
52. Syed O, Kim JH, Keskin-Erdogan Z, Day RM, El-Fiqi A, Kim HW, Knowles JC. SIS/aligned fibre scaffold designed to meet layered oesophageal tissue complexity and properties. Acta Biomater. 2019;99:181–95.
53. Yang Y, Lei D, Zou H, Huang S, Yang Q, Li S, Qing FL, Ye X, You Z, Zhao Q. Hybrid electrospun rapamycin-loaded small-diameter decellularized vascular grafts effectively inhibit intimal hyperplasia. Acta Biomater. 2019;97:321–32.

54. Gupta SK, Dinda AK, Potdar PD, Mishra NC. Modification of decellularized goat-lung scaffold with chitosan/nanohydroxyapatite composite for bone tissue engineering applications. Biomed Res Int. 2013;2013:651945.
55. Bi XW, Li LH, Mao ZN, Liu B, Yang LB, He W, Fan YB, Li XM. The effects of silk layer-by-layer surface modification on the mechanical and structural retention of extracellular matrix scaffolds. Biomater Sci. 2020;8:4026–38.
56. Cao Z, Dou C, Dong S. Scaffolding biomaterials for cartilage regeneration. J Nanomater. 2014;2014:1–8.
57. Wassenaar JW, Braden RL, Osborn KG, Christman KL. Modulating in vivo degradation rate of injectable extracellular matrix hydrogels. J Mater Chem B. 2016;4:2794–802.
58. Alberti KA, Xu Q. Biocompatibility and degradation of tendon-derived scaffolds. Regen Biomater. 2016;3:1–11.
59. Zhai W, Chang J, Lin K, Wang J, Zhao Q, Sun X. Crosslinking of decellularized porcine heart valve matrix by procyanidins. Biomaterials. 2006;27:3684–90.
60. Zhao JY, Bass KD. Skeletal muscle regeneration by extracellular matrix biological scaffold: a case report. J Wound Care. 2018;27:S11–4.
61. Wu RX, He XT, Zhu JH, Yin Y, Li X, Liu X, Chen FM. Modulating macrophage responses to promote tissue regeneration by changing the formulation of bone extracellular matrix from filler particles to gel bioscaffolds. Mater Sci Eng C. 2019;101:330–40.
62. Bhrany AD, Lien CJ, Beckstead BL, Futran ND, Muni NH, Giachelli CM, Ratner BD. Crosslinking of an oesophagus acellular matrix tissue scaffold. J Tissue Eng Regen Med. 2008;2:365–72.
63. Liu X, Wang J, Dong F, Song P, Tian S, Li H, Hou Y. Cytocompatibility and biologic characteristics of synthetic scaffold materials of rabbit acellular vascular matrix combining with human-like collagen I. J Biomater Appl. 2017;32:463–71.
64. Lee JS, Lee K, Moon SH, Chung HM, Lee JH, Um SH, Kim DI, Cho SW. Mussel-inspired cell-adhesion peptide modification for enhanced endothelialization of decellularized blood vessels. Macromol Biosci. 2014;14:1181–9.
65. Lee H, Ju YM, Kim I, Elsangeedy E, Lee JH, Yoo JJ, Atala A, Lee SJ. A novel decellularized skeletal muscle-derived ECM scaffolding system for in situ muscle regeneration. Methods. 2020;171:77–85.
66. Liang Y, Idrees E, Szojka ARA, Andrews SHJ, Kunze M, Mulet-Sierra A, Jomha NM, Adesida AB. Chondrogenic differentiation of synovial fluid mesenchymal stem cells on human meniscus-derived decellularized matrix requires exogenous growth factors. Acta Biomater. 2018;80:131–43.
67. Tan B, Wang M, Chen X, Hou J, Chen X, Wang Y, Li-Ling J, Xie H. Tissue engineered esophagus by copper--small intestinal submucosa graft for esophageal repair in a canine model. Sci China Life Sci. 2014;57:248–55.
68. Valente P, Gomes C, Tomada N. Small intestinal submucosa grafting for peyronie disease: outcomes and patient satisfaction. Urology. 2017;100:117–24.
69. Rashidbenam Z, Jasman MH, Hafez P, Tan GH, Goh EH, Fam XI, Ho CCK, Zainuddin ZM, Rajan R, Nor FM, Shuhaili MA, Kosai NR, Imran FH, Ng MH. Overview of urethral reconstruction by tissue engineering: current strategies, clinical status and future direction. Tissue Eng Regen Med. 2019;16:365–84.
70. Heath DE. A review of decellularized extracellular matrix biomaterials for regenerative engineering applications. Regen Eng Transl Med. 2019;5:155–66.
71. Parmaksiz M, Dogan A, Odabas S, Elcin AE, Elcin YM. Clinical applications of decellularized extracellular matrices for tissue engineering and regenerative medicine. Biomed Mater (Bristol, Engl). 2016;11:022003.
72. Guruswamy Damodaran R, Vermette P. Tissue and organ decellularization in regenerative medicine. Biotechnol Prog. 2018;34:1494–505.

73. Maisel Lotan A, Ben Yehuda D, Allweis TM, Scheflan M. Comparative study of meshed and nonmeshed acellular dermal matrix in immediate breast reconstruction. Plast Reconstr Surg. 2019;144:1045–53.
74. Wang L, Li L, Lou W. Repair of a cervical skin defect using xenogeneic acellular dermal matrix in a patient with advanced laryngeal carcinoma. J Laryngol Otol. 2015;129:715–7.
75. Ali M, Kumar Pr A, Yoo JJ, Zahran F, Atala A, Lee SJ. A photo-crosslinkable kidney ECM-derived bioink accelerates renal tissue formation. Adv Healthc Mater. 2019;8:e1800992.
76. Turchan A, Rochman TF, Ibrahim A, Fauziah D, Wahyuhadi J, Parenrengi MA, Fauzi AA, Sufarnap E, Bajamal AH, Ferdiansyah, Suroto H, Purwati, Rantam FA, Paramadini AW, Lumenta CB. Duraplasty using amniotic membrane versus temporal muscle fascia: a clinical comparative study. J Clin Neurosci. 2018;50:272–6.
77. Xu Q, Chen C, Xu Z, Chen F, Yu Y, Hong X, Xu S, Chen J, Ding Q, Chen H. Ureteral reconstruction with decellularized small intestinal submucosa matrix for ureteral stricture: a preliminary report of two cases. Asian J Urol. 2020;7:51–5.
78. Yao Q, Zheng YW, Lan QH, Kou L, Xu HL, Zhao YZ. Recent development and biomedical applications of decellularized extracellular matrix biomaterials. Mater Sci Eng C Mater Biol Appl. 2019;104:109942.
79. Xing H, Lee H, Luo L, Kyriakides TR. Extracellular matrix-derived biomaterials in engineering cell function. Biotechnol Adv. 2019;42:107421.
80. Rajab TK, O'Malley TJ, Tchantchaleishvili V. Decellularized scaffolds for tissue engineering: current status and future perspective. Artif Organs. 2020;44(10):1031–43. https://doi.org/10.1111/aor.13701.
81. Mosala Nezhad Z, Poncelet A, de Kerchove L, Gianello P, Fervaille C, El Khoury G. Small intestinal submucosa extracellular matrix (CorMatrix(R)) in cardiovascular surgery: a systematic review. Interact Cardiovasc Thorac Surg. 2016;22:839–50.
82. Macchiarini P, Birchall M, Hollander A, Mantero S, Conconi MT. Clinical transplantation of a tissue-engineered airway. Lancet. 2009;373:781.
83. otti AG, Jaus MO, Daniel Barale SB, Comin C, Lavorini F, Fontana G, Sibila O, Rombola G, Jungebluth P, Macchiarini P. The first tissue-engineered airway transplantation: 5-year follow-up results. Lancet. 2014;383:238–44.
84. Chun YS, Verma K, Rosen H, Lipsitz S, Morris D, Kenney P, Eriksson E. Implant-based breast reconstruction using acellular dermal matrix and the risk of postoperative complications. Plast Reconstr Surg. 2010;125:429–36.
85. Wang F, Sbitany H. Acellular dermal matrix-assisted prosthetic breast reconstruction: efficacy and morbidity. In: Shiffman M, editor. Breast reconstruction. Cham: Springer; 2016. p. 1053–69.
86. Van Nieuwenhove I, Tytgat L, Ryx M, Blondeel P, Stillaert F, Thienpont H, Ottevaere H, Dubruel P, Van Vlierberghe S. Soft tissue fillers for adipose tissue regeneration: from hydrogel development toward clinical applications. Acta Biomater. 2017;63:37–49.
87. Mohiuddin OA, Campbell B, Poche JN, Thomas-Porch C, Hayes DA, Bunnell BA, Gimble JM. Decellularized adipose tissue: biochemical composition, in vivo analysis and potential clinical applications. Adv Exp Med Biol. 2020;1212:57–70.
88. Taylor DA, Sampaio LC, Ferdous Z, Gobin AS, Taite LJ. Decellularized matrices in regenerative medicine. Acta Biomater. 2018;74:74–89.
89. Park JY, Lee TG, Kim JY, Lee MC, Chung YK, Lee WJ. Acellular dermal matrix to treat full thickness skin defects: follow-up subjective and objective skin quality assessments. Arch Craniofac Surg. 2014;15:14.
90. Sun Y, Yan L, Chen S, Pei M. Functionality of decellularized matrix in cartilage regeneration: a comparison of tissue versus cell sources. Acta Biomater. 2018;74:56–73.
91. Da LC, Huang YZ, Xie HQ. Progress in development of bioderived materials for dermal wound healing. Regen Biomater. 2017;4:325–34.